AF353120

Biomass Processing for Biofuels, Bioenergy and Chemicals

Biomass Processing for Biofuels, Bioenergy and Chemicals

Special Issue Editors

Wei-Hsin Chen
Hwai Chyuan Ong
Thallada Bhaskar

MDPI • Basel • Beijing • Wuhan • Barcelona • Belgrade • Manchester • Tokyo • Cluj • Tianjin

Special Issue Editors

Wei-Hsin Chen
National Cheng Kung University
Taiwan

Hwai Chyuan Ong
University of Malaya
Malaysia

Thallada Bhaskar
CSIR-Indian Institute of Petroleum (IIP)
India

Editorial Office
MDPI
St. Alban-Anlage 66
4052 Basel, Switzerland

This is a reprint of articles from the Special Issue published online in the open access journal *Energies* (ISSN 1996-1073) (available at: https://www.mdpi.com/journal/energies/special_issues/biomass_processing).

For citation purposes, cite each article independently as indicated on the article page online and as indicated below:

LastName, A.A.; LastName, B.B.; LastName, C.C. Article Title. *Journal Name* **Year**, *Article Number*, Page Range.

ISBN 978-3-03928-909-7 (Pbk)
ISBN 978-3-03928-910-3 (PDF)

Contents

About the Special Issue Editors

Wei-Hsin Chen, Distinguished Professor. He received a B.S. from the Department of Chemical Engineering, Tunghai University, in 1988, and he completed his Ph.D. at the Institute of Aeronautics and Astronautics, National Cheng Kung University, in 1993. After receiving his Ph.D., Dr. Chen worked in an iron and steel corporation as a process engineer for one and a half years (1994–1995). He joined the Department of Environmental Engineering and Science, Fooyin University, in 1995 and was promoted to a full professor in 2001. In 2005, he moved to the Department of Marine Engineering, National Taiwan Ocean University. Two years later (2007), he moved to the Department of Greenergy, National University of Tainan. Currently, he is a faculty member and distinguished professor in the Department of Aeronautics and Astronautics, National Cheng Kung University. Professor Chen served as a visiting professor and invited lecturer at Princeton University, USA, from 2004 to 2005; the University of New South Wales, Australia, in 2007; the University of Edinburgh, UK, in 2009; the University of British Columbia, Canada, from 2012 to 2013; and the University of Lorraine, France, in 2017, 2019, and 2020. His teaching courses at the National Cheng Kung University include Bioenergy, Materials Engineering and Science; Energy Experiments; and Engineering Mathematics. His research topics include bioenergy, hydrogen energy, clean energy, carbon capture, and aerosol physics. He has published over 550 papers in international and domestic journals and conferences. He is an editor, associate editor, guest editor, and editorial member of a number of international journals, including *Applied Energy; Energy Conversion and Management; International Journal of Hydrogen Energy; International Journal of Energy Research; Energies; and Sustainability*. He is also the author of several books on energy science and air pollution. He has received several prestigious awards, including the 2015 and 2018 Outstanding Research Award (Ministry of Science and Technology, Taiwan), 2015 Highly Cited Paper Award (Applied Energy, Elsevier), 2017 Outstanding Engineering Professor Award (Chinese Institute of Engineers, Taiwan), 2019 Highly Cited Review Article Award (Bioresource Technology, Elsevier), as well as 2016, 2017, 2018, and 2019 Highly Cited Researcher Award (Web of Science).

Hwai Chyuan Ong obtained his B.Eng. (Hons.) in Mechanical Engineering from the Faculty of Engineering, University of Malaya, with distinction. Then, he obtained his Ph.D. in Mechanical Engineering from the same university in December 2012. His research interests are wide-ranging under the general umbrella of renewable energy. In particular, these include biofuel and bioenergy, solar thermal energy, green technology, and environmental engineering. He is currently appointed as a Senior Lecturer at the Department of Mechanical Engineering, University of Malaya. He is also a Chartered Engineer of Engineering Council (CEng) for the Institution of Mechanical Engineers (IMechE), United Kingdom. He has published more than 100 high-impact SCI journal papers with an H-index of 30 (WOS). He has received several awards, including the 2019 Highly Cited Researcher Award (Engineering) by Web of Science, Malaysia's Research Star Award (frontier researcher) in 2018 and 2017, and Malaysia's Rising Star Award (young researcher) in 2016 from the Ministry of Higher Education and Clarivate Analytics. In 2018, he also received the Outstanding Research Award and the Most Highly Cited Paper Award during the University of Malaya Excellence Awards. Currently, he is an associate editor of *Alexandria Engineering Journal, Journal of Renewable and Sustainable Energy, and Energies* and a guest editor of *Biomass Conversion and Biorefinery and Energies*.

Thallada Bhaskar (http://thalladabhaskar.weebly.com), Senior Principal Scientist, is currently heading the Material Resource Efficiency Division (MRED) at CSIR-Indian Institute of Petroleum, Dehradun, India, and the Biomass Conversion Area (BCA). He received a Ph.D. for his work at CSIR-Indian Institute of Chemical Technology (IICT), Hyderabad. He carried out postdoctoral research at Okayama University, Okayama, Japan, which he subsequently joined as Assistant Professor for 5 years. He has authored 150 publications in SCI journals of international repute with an *h-index* of 45 and more than 6050 citations; he has contributed 32 book chapters to renowned publishers (Elsevier, ACS, John Wiley, Woodhead Publishing, CRC Press, Asiatech, etc.) and produced 14 patents in his field of expertise, in addition to 300 national and international symposia presentations. He has received the Distinguished Researcher Award from AIST (2013), Japan, and the Most Progressive Researcher Award from FSRJ, Japan (2008). He was also a JSPS Visiting Scientist at Tokyo Institute of Technology, Japan, in 2009. He is also a Fellow of Royal Society of Chemistry (FRSC in 2016), UK; Fellow of Biotech Research Society of India (FBRS in 2012); Fellow of International Society of Environment, Engineering and Sustainability (FISEES in 2017) and Scientist of the Year Award (2016) from National Environmental Science Academy (NESA); Fellow of International Bioprocessing Association (FIBA in 2017); Fellow of Telangana Academy of Sciences (2017); and a member of the Board of Directors (BRSI) and General Secretary, Management Council of BRSI (2017–2019). He received the Raman Research Fellowship for the period 2013–2014. Dr. Bhaskar has received the CAS Presidential Award for Foreign Fellows in 2016 and worked as a visiting professor (Visiting Scientist at SINTEF, Norway, from 2013 Dec to 2014 Feb). He is a member of the scientific and organizing committee of several national/international symposia in India and abroad, and he has visited several countries to deliver invited/plenary lectures and technical conferences.

Article

Study of the Application of Alkaline Extrusion to the Pretreatment of Eucalyptus Biomass as First Step in a Bioethanol Production Process

Aleta Duque *, Paloma Manzanares, Alberto González and Mercedes Ballesteros

Biofuels Unit, Energy Department-CIEMAT, Avda. Complutense, 40, 28040 Madrid, Spain;
p.manzanares@ciemat.es (P.M.); alb.gonzalez@ciemat.es (A.G.); m.ballesteros@ciemat.es (M.B.)
* Correspondence: aleta.duque@ciemat.es; Tel.: +34-91-346-6737

Received: 8 October 2018; Accepted: 23 October 2018; Published: 31 October 2018

Abstract: Eucalyptus biomass was studied as a feedstock for sugars release using an alkaline extrusion plus a neutralization-based pretreatment. This approach would be a first step in a bioconversion process aimed at obtaining fuel bioethanol from eucalyptus biomass. The best operation conditions of extrusion (screw speed, temperature, liquid to solid ratio and NaOH amount) that lead to an effective destructuration of lignocellulose and enhanced sugar release were investigated. Two process configurations, with and without filtration inside the extruder, were tested. In the case without filtration, washed and not washed extrudates were compared. It was demonstrated that filtration step was convenient to remove inorganic salts resulting from neutralization and to promote the mechanical effect of extrusion, but limitations in the machine used in the work prevented testing of temperatures above 100 °C using this configuration. In the no filtration strategy, a temperature of 150 °C allowed attaining the highest glucan and xylan conversion rates by enzymatic hydrolysis of extruded biomass, almost 40% and 75%, respectively, of the maximum yield that could be attained if all carbohydrates contained in raw eucalyptus were converted to sugars. Some of the mechanisms and individual effects underlying alkaline extrusion of eucalyptus were figured out in this work, providing guidelines for a successful pretreatment design that needs to be further studied.

Keywords: lignocellulose; pretreatment; hardwood; extrusion; enzymatic digestibility; bioethanol

---

1. Introduction

Two of the biggest problems the world faces today are climate change and depletion of natural resources caused by increasing consumption of fossil fuels. In this context, renewable energies play a key role in relevant sectors as electricity, heat energy and transport by contributing to alleviate the harmful effects of such global tendency. Particularly for the transport sector, bioenergy in the form of liquid biofuels constitutes the major part (90%) of the renewables energy contribution, which has been recently estimated at 3.1% by Renewable Energy Policy Network organization [1]. According to this report, 65% of liquid biofuels input is bioethanol, 29% is biodiesel and 6% is hydrobiodiesel, mainly derived from used vegetable oils and animal fats. Bioethanol is predominantly produced at global level from feedstocks that can also be consumed as human food such as sugar and starch, but concerns about the impact of biofuels on food security have focused the attention to advanced or second generation bioethanol produced from lignocellulosic type feedstocks to avoid food competition, continue renewable transport fuel supply and move towards a more sustainable scenario for biofuels production and use.

Among different lignocellulosic biomass sources, bioenergy crops and cropping systems are well positioned to produce second generation feedstocks provided that they are grown in land unsuitable for agriculture and show positive energy and carbon balances [2]. Tree species such as eucalyptus have received much attention in the last years due to several positive features such as capability of rapid growth in a wide range of climates and poor soils that make it a valuable candidate for bioenergy production.

Lignocellulosic biomass, such as eucalyptus wood, is composed of cellulose, a linear glucose polymer; hemicellulose, a heterogeneous polymer of pentoses and hexoses with certain amounts of uronic acids and acetyl substitutions; and lignin, a polymer of phenyl propane units that wraps the ensemble [3]. The amount of each component varies depending on the type of biomass, for instance, herbaceous biomass typically has 25–40% cellulose, 35–50% hemicellulose and 15–30% lignin. On the other hand, woods are composed of 40–55% cellulose, 24–40% hemicellulose and 18–35% lignin (softwood having higher lignin content than hardwood) [4]. Specifically, xylose is the most abundant sugar in hemicelluloses of grasses and hardwood, while mannose is the predominant component of softwood hemicellulose.

Lignocellulosic biomass is designed to resist degradation by ambient conditions or biological agents, which complicates the process of obtaining fermentable sugars to be converted to ethanol. However, by using a suitable method, lignocellulose can be altered and rendered more accessible to biological catalyst such as hydrolytic enzymes. Several pretreatment technologies have been proposed and studied [5,6], each having their own strengths and weaknesses. For instance, steam explosion has been proved to be an effective method to pretreat wood residues, especially hardwood type, since softwoods are more recalcitrant and typically require higher pretreatment severity and possibly the addition of an acid catalyst [7]. Several references can be found in the literature about eucalyptus pretreatment for sugar release. For example, very high cellulose recovery and enzymatic hydrolysis yield were obtained by pretreating *Eucalyptus globulus* at 195 °C for 6 min [8]. Furthermore, dilute acid pretreatment aided by microwave cooking of eucalyptus wood chips resulted in maximum glucose hydrolysis yield of 74% of theoretical, although severe conditions that involved some xylose degradation were used [9]. After testing several methods and conditions for the pretreatment of eucalyptus bark, including hot water extraction, acid pretreatment, alkali pretreatment and a combination of both, Lima et al. [10] reported that autoclaving of this biomass with NaOH 4% at 1.05 bar for 1 h gave the best results in terms of glucose release (59–65%). This result indicates that the use of an alkaline catalyst is a possibility worth exploring.

On the other hand, extrusion has been reported as a versatile pretreatment with promising features [7] that single it out as an interesting option for treating eucalyptus to obtain enhanced enzymatic digestibility while avoiding sugar degradation. Although preferred for the treatment of herbaceous biomasses, some woody type materials have also been successfully processed by extrusion. For instance, pine wood chips were pretreated in a single-screw extruder at optimal conditions of 180 °C, 150 rpm and 25% moisture, obtaining sugar recoveries over 65% [11]. Nevertheless, Lee et al. [12] alleged a limited effect of mechanical kneading with addition of only water for the fibrillation of Douglas fir, and so they proposed an extrusion in a twin-screw extruder aided by ethylene glycol [13], or a combination of hot-compressed water followed by extrusion [14]. The latter combination was tested by the authors also for eucalyptus, resulting in a glucose release yield over 30% (of original wood weight) and xylose release around 4%. The addition of chemicals to the extrusion process has been reported in several papers. Extrusion of poplar sawdust under acidic conditions (4% wt. H_2SO_4) and high temperature of 185 °C resulted in saccharification yields of cellulose over 65% [15]. Moreover, Senturk-Ozer et al. [16] proved that a good flowability of hard-wood type biomass could be achieved inside a twin-screw extruder within a context of alkaline pretreatment. Alkaline extrusion was used to pretreat Eucalyptus Forest Residues, among other biomasses, in a configuration that included the alkaline pretreatment, followed by neutralization with H_3PO_4, filtration and addition of hydrolytic enzymes inside the extruder [17]. In that work, the amount of NaOH employed was 8% and the temperature and screw speed were set to 75 °C and 200 rpm, respectively. The saccharification yield of cellulose obtained was 14 g/100 g raw material. However, only one set of conditions were tested and thus, limited conclusions can be drawn about extrusion performance on eucalyptus biomass.

One of the difficulties of working with extrusion is the complex inter-relations among the operation variables [7]. In this exploratory work about extrusion of eucalyptus, the effect of the screw speed, the temperature, the liquid to solid (L/S) ratio and the proportion of NaOH to dry biomass (NaOH/DM) were investigated separately to better understand their individual effects on the mechanical energy requirements of the pretreatment (Specific Mechanical Energy, SME), composition of the extrudates and their enzymatic digestibility when submitted to incubation with hydrolytic enzymes. This approach was tested in an extrusion process strategy that includes a filtration step to remove part of the liquid added into the extruder, that the authors had previously tested on other biomasses [18]. Moreover, an extrusion strategy without filtration and a process scheme where the pretreated material is washed with water outside the extruder were also studied aimed at testing the effect of extrusion process strategy for sugar release by enzymatic hydrolysis of extrudates.

2. Materials and Methods

2.1. Raw Material

Eucalyptus grandis, de-barked trunk portion of diameter between 190 and 60 mm, was provided by the National Institute of Agricultural Research (INIA, Uruguay). Eucalyptus trees were harvested during October and November, 2017, from two different stands located in the Department of Rivera (Uruguay). They were cut into logs, debarked on site, and moved to the research station where they were stored for two months to be air-dried. Afterwards, logs diameter was measured with tree calipers and portions between 60 and 190 mm in diameter were selected as feedstock for this work. Logs were chipped and stored again until moist was around 20%. Finally, chips were milled to 2 mm and kept until use (11.9% moisture).

The composition of eucalyptus biomass was close to 60% carbohydrates (46.9% cellulose, 12.9% hemicellulose), 31.1% lignin and <1% ash (see Section 2.3).

2.2. Extrusion Pretreatment

Eucalyptus biomass was pretreated by alkaline extrusion in a co-rotating twin-screw extruder (Clextral Processing Platform Evolum® 25 A110, Clextral, Firminy, France) with 6 barrels and length to diameter ratio (L/D) equal to 24. The configuration used was one conceived for the alkaline extrusion with filtration inside the extruder, adapted from [18]. In this configuration, the alkaline solution (diluted NaOH) is introduced in barrel #2, the acid solution entered in barrel #4 and barrel #5 is a filter to separate the liquid and solid after neutralization. The exact screw configuration used for the present work is presented in Figure 1 and discussed in detail later in Section 3.1.

A first control sample was run with addition of water instead of chemical catalyst to assess the mechanichal-thermal and chemical effects separately. Afterwards, the influence of four operation variables: Screw speed (SS), temperature (T), L/S ratio in the reaction zone and NaOH/DM, in the pretreatment performance was tested. Table 1 shows the different experiments carried out to determine the singular effect of each of the variables. The objective was to vary one of the variables while keeping the rest of the variables constant. All deviations from this principle were directed towards the achievement of a regular flow inside the extruder and the adjustment of the parameters was made upon observation of the course of extrusion experiments.

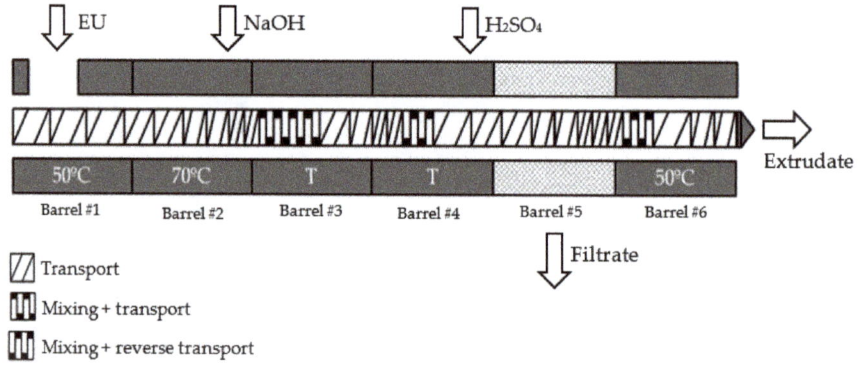

Figure 1. Barrel and screw configuration for the alkaline extrusion of eucalyptus with neutralization and filtration in a twin-screw extruder.

Table 1. List of extrusion experiments carried out, values of operation variables and information about the processing of the samples.

Experiment	Assays	R (%*w/w*)	L/S (*w/w*)	T (°C)	SS (rpm)	Filter	Washed
W	Control	-	1.2	75	200	Yes	No
SS	1	8.5	1.2	75	100	Yes	No
	2	8.5	1.2	75	200	Yes	No
	3	8.5	1.2	75	300	Yes	No
T	4	8.5	1.2	100	200	Yes	No
	5	8.5	1.2	125	200	[1]	[1]
L/S	6	8.5	0.6	75	200	Yes	No
	7	8.5	1.5	75	200	Yes	No
NaOH	8	5	0.7	100	150	Yes	No
	9	10	1.2	100	150	Yes	No
	10	20	1.2	100	250	Yes	No
NF	11	8.5	1.2	125	300	No	No
	12	8.5	1.2	150	300	No	No
NFW	13	8.5	1.2	125	300	No	Yes
	14	8.5	1.2	150	300	No	Yes

W—extrusion with water; SS—screw speed; T—temperature; L/S—liquid to solid ratio; NaOH—amount of alkali; NF—without filtration; NFW—washing of extrudates without filtration; [1] Failed run.

In another set of experiments the process configuration was changed due to the pretreatment needs, as it will be explained later in Section 3.4.1.

2.3. Materials Characterization

Untreated eucalyptus and pretreated materials (extrudates) were analyzed according to the National Renewable Energy Laboratory (NREL, Golden, CO, USA) laboratory analytical procedures (LAP) for biomass analysis [19].

2.4. Evaluation of the Enzymatic Digestibility

Enzymatic hydrolysis of untreated and alkaline extruded eucalyptus was carried out by triplicate in Erlenmeyer flasks at 5% *w/w* solids load with a total volume of 50 mL. 15 FPU/g dry matter of a commercial cellulolytic cocktail Cellic® CTec2, kindly provided by Novozymes A/S (Copenhagen,

Denmark), were added to each flask. The hydrolysis tests were done in citrate buffer 50 mM, pH 4.8, and with the addition of 1% v/v of sodium azide. The Erlenmeyer flasks were agitated in an orbital shaker at 50 °C and 150 rpm for 72 h. Samples were taken each 24 h and glucose and xylose were analyzed by HPLC as explained below.

The sugar release yield was calculated as the amount of glucose (GR) or xylose (XR) measured in the hydrolysis media divided per 100 g of dry extrudate. Alternatively, glucan and xylan conversions (GC and XC) values were obtained by dividing the corresponding sugar release by the glucose or xylose content of the extrudate and expressed as percentage.

2.5. Analytical Methods

Monomeric sugars were analyzed by high-performance liquid chromatography (HPLC) in a Waters 2695 liquid chromatograph with refractive index detector. A CARBOSep CHO-782 LEAD column (Transgenomic, Omaha, NE, USA) was used, operating at 70 °C with Milli-Q water (Millipore) as mobile-phase (0.5 mL/min).

Acetic acid was analyzed by HPLC in a Waters 2414 liquid chromatograph with refractive index detector. An ionic exclusion column Aminex HPX-87H (BioRAd Labs, Hercules, CA, USA) was operated at 65 °C with sulphuric acid 0.05 M as mobile-phase (0.6 mL/min).

Furfural and HMF in the filtrates were analyzed by HPLC (Hewlett Packard, Palo Alto, CA, USA), using an Aminex ion exclusion HPX-87H cation-exchange column (Bio-Rad Labs, Hercules, CA, USA) at 65 °C. Mobile phase was 89% 5 mM H_2SO_4 and 11% acetonitrile at a flow rate of 0.7 mL/min. Column eluent was detected with a 1040A Photodiode-Array detector (Agilent, Waldbronn, Germany).

3. Results and Discussion

3.1. Screw Configuration

One of the features that make extrusion such a versatile pretreatment is the possibility to change the screw elements that, arranged one after another, constitute the screw configuration. These elements have different shapes and effects on the biomass and the way they are placed helps creating and separating different zones along the extrusion machine. The screw configuration has not only a big influence on the severity of the pretreatment, but it also determines in first instance the flow inside the extruder, this is, the feasibility or not of the pretreatment. Thus, the first approach to the extrusion of a new biomass is to study a configuration in which it can be operated.

The screw configuration was designed to be divided into several zones to perform the alkaline extrusion, neutralization and filtration of the biomass all inside the extruder (Figure 1). The first zone was dedicated to the feeding, where conveying screws were used, and it comprised barrels #1 and #2. NaOH solution entered at the end of barrel #2 and it was mixed with the eucalyptus biomass and heated along barrels #3 and part of #4; this was the reaction zone, terminated with a reverse kneading block. The acid solution was pumped at the end of barrel #4 and the neutralization plus filtration took place in barrel #5. After the reverse kneading block placed at the beginning of barrel #6, the discharge zone started. The proposed configuration is based in a similar one reported by Duque et al. [18] for the extrusion of barley straw. In that screw profile, reverse flow screws were placed instead of reverse kneading blocks. The exact configuration was tested in a first trail with eucalyptus with no success. The different characteristics of eucalyptus with respect to barley straw (more hardness and lower water retention) caused blockage of the reverse screws, leading to full stop of the machine. To soften the screw profile, while maintaining the neutralization and filtration zone separated inside the extruder, the reverse flow screws were replaced by the above-mentioned reverse kneading blocks as depicted in Figure 1.

Many authors have established the importance of constraint elements in extrusion for the fibrillation of lignocellulosic biomass and improvement of the enzymatic digestibility [16,20–22]. Furthermore, Vandenbossche et al. [23] observed that insufficient backpressure caused unstable flow inside the extruder and prevented a good filtration of dehydrated sweet corn co-products treated inside the machine, first with NaOH and then neutralized with H_3PO_4. Following the same pretreatment concept of alkaline extrusion plus neutralization and filtration, Brault [17] had to adapt the screw profile used for the pretreatment of sweet corn co-products to the extrusion of Eucalyptus residues, by placing an additional reverse flow element in the reaction zone to increase destructuration and by removing one constraint element after the filter to ensure the formation of a stable dynamic plug.

Therefore, the configuration for the extrusion of eucalyptus in the present paper was designed to provide high shearing and mixing, while keeping a continuous flow and four separated zones with different temperatures: Feeding, reaction zone, neutralization and filtration zone, and discharge.

3.2. Extrusion with Water

A control run was carried out by pumping water instead of alkali and acid solutions to determine the effect of extrusion alone on eucalyptus biomass. Some authors have emphasized the importance of adding a rheological modifier to help the flow inside the extruder [16,24–26]. In agreement with this idea, Lamsal et al. [25] reported problems in the extrusion of soybean hulls at low moisture (<35%) in a twin-screw extruder. In the present work, flow constraints were also observed during the extrusion trial with water, but a representative sample could be obtained to be used as a control for the study. The difficulty in the flow is revealed by the SME value, presented in Figure 2, which is the highest of all the extrusion trials. In alkaline extrusion, NaOH acts as a flux modifier for the biomass suspension, increasing the viscosity and reducing the shear strength, in comparison to extrusion with only water, which requires a higher energy input.

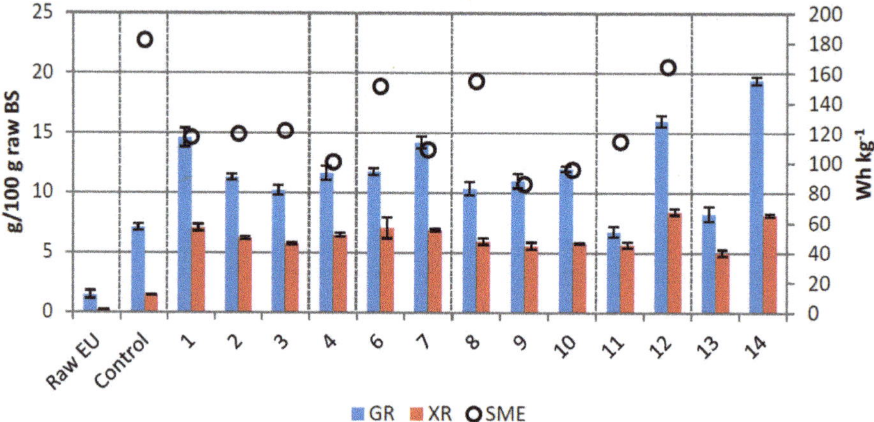

Figure 2. Glucose (GR) and xylose (XR) release (in g per 100 g) of raw biomass and eucalyptus extruded with water (control) and with NaOH at varying screw speed, temperature, liquid to solid ratio and catalyst ratio (experiments 1 to 10). Specific mechanical energy (SME) of each experiment (in Wh kg^{-1}). Experiment #1, 2 & 3—study of SS; Experiment #2 & 4—study of T; Experiment #2, 6 & 7—study of L/S; Experiment #8, 9 & 10—study of NaOH/DM; Experiment #11 & 12—without filtration; Experiment #13 & 14—washing of extrudates without filtration.

The control sample extruded with water was submitted to EH to check the effect of the mechanical-thermal effects of extrusion (excluding the chemical pretreatment) on the enzymatic digestibility of eucalyptus and the results can be seen in the third and fourth columns of Figure 2. Glucose and xylose release (GR and XR) from control extrudate are 7.1 and 1.5 g/100 g extrudate, respectively, which means 5- and 7-fold times more glucose and xylose produced than by hydrolysis of the untreated eucalyptus. Nevertheless, the glucan and xylan conversions were still under 15% of theoretical levels. These low yields were highly improved by the combination of alkaline pretreatment with extrusion in the next trials, particularly in the case of xylan, as will be discussed later.

Control extrudate composition was not significantly affected with respect to the raw material (Table 2, rows 1 and 2). In absence of a chemical catalyst, and at the mild temperature tested (75 °C), there was no sugar solubilization or degradation. This is confirmed by the composition of the filtrate reported in Table 3, where only traces (<0.1%) of glucose, xylose and acetic acid were found in control sample, indicating again the low effect of the water-based pretreatment on lignocellulose fractionation. Moreover, eucalyptus is a biomass with a very low content of ash (<1%) and extractives (2.4%, data not shown), so there is not a concentration effect from the partial solubilization of those compounds, as was the case for barley straw [27]. The efficiency of the filtration, defined as the weight of filtrate in the total output weight (filtrate and extrudate), compiled in Table 3, was only 19%, which is a result of the bad flowability of the mixture, so contributing to a low sugar extraction.

Table 2. Main components (in dwb) of raw eucalyptus and extrudates, with standard deviation.

Experiment	Cellulose (%)	Hemicellulose (%)	Lignin (%)	Ash (%)
Raw EU	46.90 ± 1.21 [a]	12.87 ± 0.35 [a,b,c,d]	31.15 ± 0.40 [a,b]	0.86 ± 0.00 [a]
Control	44.90 ± 1.86 [a,b,c]	13.71 ± 0.32 [a,c,d,e]	32.97 ± 0.86 [c,d]	0.57 ± 0.05 [a]
1	42.56 ± 3.51 [b,c]	12.13 ± 0.94 [a,b,c]	32.03 ± 1.00 [a,c]	3.78 ± 0.37 [b]
2	42.56 ± 1.19 [b,c]	11.76 ± 1.51 [a,b]	31.22 ± 0.41 [a,b]	7.35 ± 0.15 [c]
3	41.79 ± 0.88 [c]	12.16 ± 0.20 [a,b,c]	28.86 ± 0.10 [e]	9.28 ± 0.01 [d]
4	43.14 ± 1.54 [a,b]	11.47 ± 0.68 [d,e]	29.67 ± 0.65 [f]	7.01 ± 0.10 [c]
6	41.71 ± 0.56 [b,c]	12.04 ± 0.43 [c,d,e]	30.43 ± 0.67 [d]	7.37 ± 0.06 [e]
7	46.97 ± 1.21 [a]	15.59 ± 2.61 [e]	31.21 ± 0.31 [a,b]	5.65 ± 0.10 [f]
8	45.66 ± 0.10 [a,b]	14.95 ± 1.70 [d,e]	32.90 ± 0.61 [c,d]	4.16 ± 0.07 [g]
9	45.48 ± 0.82 [a,b,c]	11.36 ± 0.21 [a,b]	31.70 ± 0.18 [a]	5.22 ± 0.03 [h]
10	44.60 ± 1.77 [a,b,c]	11.02 ± 1.26 [b]	30.32 ± 0.22 [b,f]	11.17 ± 0.08 [i]
11	39.23 ± 2.31 [a,b,c]	11.78 ± 0.47 [a,b]	27.69 ± 0.45 [g]	14.91 ± 0.12 [j]
12	38.76 ± 0.36 [b,c]	11.24 ± 0.16 [b]	27.99 ± 0.22 [g]	14.54 ± 0.10 [j]
13	50.05 ± 0.38 [d]	12.72 ± 0.13 [a,b,c]	32.60 ± 0.15 [e,f]	2.04 ± 0.02 [k]
14	49.33 ± 0.71 [d]	11.55 ± 0.23 [a,b]	31.98 ± 0.20 [e]	2.62 ± 0.07 [l]

Values followed by the same letters are not significantly different at $p = 0.05$.

Table 3. Filtration parameters and sugar and acetyl groups recovery yields in the filtrates and washing liquid of the different extrusion experiments.

Experiment	Filtration Efficiency	L/S Filtration	Total Glucose	Total Xylose	Acetic Acid
	g Filtrate/100 g Total Output		% of Gluc in Raw EU	% of xyl in Raw EU	% of ac Acid in Raw EU
Control	19.0	3.9	0.0	0.3	0.8
1	69.6	3.9	0.4	8.0	27.3
2	55.1	3.9	0.1	5.9	30.8
3	58.2	3.9	0.1	5.6	27.3
4	59.2	3.9	0.2	8.0	24.0
6	48.3	3.4	0.2	5.3	36.0
7	50.8	4.2	0.3	6.9	35.5
8	43.0	2.9	0.3	2.0	36.0
9	68.8	4.8	0.4	8.7	64.5
10	64.6	5.0	0.3	11.1	58.5
13	Washing liquid		0.0	2.2	42.6
14	Washing liquid		0.1	3.2	33.4

Thus, there is a certain contribution of extrusion alone to the destructuration of eucalyptus biomass structure, resulting in a small increase of GC value, but a chemical catalyst (in the present study, NaOH) is proposed to further open the fiber and boost the enzymatic digestibility.

3.3. Alkaline Extrusion with Filtration

3.3.1. Effect of Screw Speed

The screw speed is a parameter with a complex influence on the performance of the pretreatment. On the one hand, it is inversely proportional to the duration of the pretreatment; the higher the screw speed, the shorter the residence time inside the extruder. On the other hand, higher motor speeds correspond to higher torque values [7].

In the present work, three SS were tested: 100, 200 and 300 rpm. The results depicted in Figure 2 (columns 1, 2 and 3), show that lower screw speeds favor the release of sugars by EH. Specifically, at 100 rpm, 14.6 g glucose and 7.1 g of xylose per 100 g extrudate were released after 72 h of incubation with enzymes (which is 29.6% and 55.1% of theoretical glucan and xylan, respectively). These values are in agreement with the ones reported by Brault [17], working in a similar configuration. A statistical analysis following the ANOVA methodology confirmed that the differences in the enzymatic digestibility of the extrudates between the lower and the highest SS values were significant ($p \leq 0.05$).

The screw speed had also a remarkable effect on the performance of the filtration barrel. At 100 rpm, the filtration efficiency is much higher than at 200 or 300 rpm, as can be seen in Table 3. However, the amount of sugars solubilized is very similar in all three conditions and negligible in the case of glucose. The xylose found in the filtrates is $\leq$8% of the xylose contained in the raw material, which means about 1% of the weight of the dry eucalyptus. In spite of the low amount of sugars solubilized in the filtrate, the filtration step is necessary, since there is a significant amount of liquid coming from the neutralization inside the extruder that needs to be removed before processing the material.

The fact that the EH works better with a material extruded at low screw speed shown in the present work is in contrast with the results of Karunanithy et al. [11], who obtained a better digestibility of extruded pine wood chips, as the screw speed increased from 100 to 200 rpm. The authors attribute this improvement of the EH yield to the higher energy exerted at higher SS, which would result in shortening of the fibers. However, they also pointed out that the shear forces responsible for this fiber length reduction are relatively low in the single-extrusion process that they used, compared to the intermeshing co-rotating twin-screw extrusion, as the one employed in the present work. This means that higher speeds could be needed in a single-screw extrusion to reach the level of shear displayed at 100 rpm in the twin-screw extrusion. Zhang et al. [28] extruded corn stover with water in a twin-screw extruder at speeds ranging from 40 to 140 rpm, and the SS was found to have a significant effect on the release of glucose by EH, increasing the yields as the speed increased up to 80 rpm and beyond that point having a negative effect. Similar trends were observed by Yoo et al. [24] on soybean hulls and Karunanithy et al. [29] on switchgrass, although the optimum SS were different, 350 and 100 rpm, respectively. Interestingly, the prairie cord grass extruded by Karunanithy et al. [29] showed a similar behavior to the one of eucalyptus in the present work, and the glucose yield increased as the SS decreased. The effect of SS (70 and 150 rpm) on the EH yield of alkali-extruded olive tree pruning was not significant according to the work of Negro et al. [30]. The biomass used, the type of extruder and the presence or not of a catalyst seem to have a certain influence on the effect of the screw speed, therefore, the results have to be interpreted within the conditions at which the experiments are carried out. In the case of eucalyptus, a hardwood, lower speed seems to be necessary to reinforce the destructuration of the fibers.

3.3.2. Effect of Temperature

Temperature is not constant along the barrels of the extruder; instead, a temperature profile is set to accompany the material flow through the different zones. The key elements in this work were barrels #3 and 4 (Figure 1), where the alkaline reaction took place. The temperature profile beyond those barrels was set aiming at achieving a proper flow inside the machine. The effect of temperature of the reaction zone on the extrusion performance was evaluated at 75, 100 and 125 °C. At the highest temperature backflow problems appeared, making it impossible to get a representative sample from trial #5. These problems can be attributed to the high temperature reached at the filtration barrel. Due to the limited length of the twin-screw extruder used in the present work, the temperature in barrel #5 is close to 100 °C, causing evaporation and affecting the formation of a dynamic plug necessary for the correct running of the machine [23].

Comparing the results from extrusion at 75 °C and 100 °C (trials #2 and #4) shown in Figure 2 and Table 2, there are no significant differences in terms of enzymatic digestibility or composition between both samples. Both filtrates are also comparable, as can be concluded from data in Table 3. The glucose and xylose release at these values were close to 11.5 and 6.5 g/100 g extrudate, respectively, corresponding to 23% and 43% of hydrolysis of the potential glucan and xylan of the extrudate. In addition, it was observed that the energy input required for the extrusion at 100 °C (trial #4) was lower than the one at 75 °C (trial #2), which is due to the thermal softening of the eucalyptus and subsequent decrease of the viscosity of the biomass suspension [31].

The somehow contradictory effects of temperature on extrusion depending on the type of biomass and configuration of the pretreatment have been discussed elsewhere [7]. In general, there is a positive effect of the increase of temperature on the enzymatic digestibility of extrudates [11], but in some cases a threshold is reached, beyond which the further increase of temperature is detrimental [27,29,32]. The interval of temperature tested in this work seems to be too narrow to see any effect of this variable on any of the studied categories. A woody biomass as eucalyptus may need higher temperatures to effectively disrupt the lignocellulosic matrix. In the scientific literature, pine wood chips were successfully extruded at temperatures as high as 180 °C [11]. Douglas fir and eucalyptus previously pretreated by autohydrolysis were submitted to extrusion at 170 and 180 °C, respectively [14]. Furthermore, acid extrusion of poplar sawdust was carried out at 185 °C [15], while extrusion of hardwood with the help of flux modifiers (CMC, black liquor) was possible at temperatures up to 120 °C [16]. Thus, temperature above 100 °C seems to be a constant for the extrusion (alone or with a chemical catalyst) of woody biomass. The increase of temperature could not be addressed with this first process configuration, so it was contemplated in a next set of trials (see Section 3.4).

3.3.3. Effect of L/S Ratio in the Reaction Zone

As the biomass moves forward inside the extruder, and the liquids (in this case, alkaline and acidic solutions) are pumped into it, the proportion between liquid and solid varies in the proposed working configuration (Figure 1). The L/S ratio in the reaction zone (barrels #3 and 4) is the most important for the pretreatment, since it influences the rate of shear developed during the alkaline contact time. Meanwhile, the L/S ratio in the neutralization zone affects the performance of the filtration, and consequently, the effectiveness of the separation and extraction of soluble compounds in the liquid fraction, and hence, the solids content in the extrudate.

For the purpose of this work, the L/S ratio studied is the one in the reaction zone. Since it is a variable closely related to the amount of catalyst added, to keep the NaOH/DM constant, while testing different L/S ratios, the concentration of the alkaline solution was varied accordingly. The values tested were 0.6, 1.2 and 1.5, corresponding to trials #6, 2 and 7.

Looking at the results presented in Figure 2, the increase in the L/S ratio seems to have a positive effect on the release of glucose by EH. This is confirmed by the statistical analysis of the data, which found significant differences between trials #2 and 6 and trial #7. Nevertheless, no significant effect was observed for xylose release. The maximum GR was, then, 14.2 g/100 g extrudate, equivalent to

27.5% of glucan conversion, and was obtained with the extrudate produced with 1.5 L/S (experiment #7). At this condition, the XR was close to 7 g/100 g extrudate (44% xylan conversion). The increase of GR found at the highest L/S ratio tested could be related to the better impregnation of the eucalyptus with the NaOH solution when the amount of liquid is high. Silva et al. [32] also obtained better results at high L/S ratios for the extrusion of sugarcane bagasse with an ionic liquid.

At the same time, the recorded SME values clearly decrease as the L/S ratio increases from 0.6 to 1.2 and 1.5. Low amounts of liquids have been correlated to higher friction forces, thus explaining the high energy input necessary at 0.6 L/S [16,24].

Concerning the extrudate composition shown in Table 2, the effect of the increase of the L/S ratio reflects on a certain increase of the glucose and xylose content of the extrudate #7 compared to #2 and 6. Moreover, the filtration efficiency (see Table 3) was lower when L/S was 0.6 (experiment #6) with respect to 1.2 (experiment #2). This can be due to a lower amount of liquid which would imply lower filtration flow in the case of trial #6. The amount of sugars recovered in the filtrate is in any case the same. Supporting these findings, Choi et al. [33] found that the glucan content in the extrudate increased and the lignin content decreased as the L/S ratio increased from 4 to 8 for the alkaline pretreatment of empty fruit bunches, especially at high NaOH loadings (>5% NaOH/DM).

Flow considerations aside, the results indicate that the L/S parameter should be kept as high as possible to obtain the better glucan digestibility, which would result as well in a lower energy consumption of the machine.

3.3.4. Effect of NaOH/DM Ratio

Three NaOH/DM ratios were tested: 5%, 10% and 20%. Taking into account the previous results, the evaluation of the amount of catalyst was made at 100 °C, the highest temperature that allows the operation of the machine in the proposed configuration. Some adjustments were made to the operation values with respect to the previous results, in order to favor flowability inside the extruder. Thus, the SS was 150 rpm in trials #8 and 9, and 250 rpm in trial #10. Moreover, the L/S ratio was kept at the intermediate value of 1.2, except for trial #8, where it was 0.7 to be able to reach a stable production.

As can be seen in Figure 2, in these experiments (#8, 9 and 10) the glucose release leveled up with the increase of the NaOH/DM ratio and so the GR value at 20% was statistically significant. Although it appeared to be a certain decrease of the XR as the amount of catalyst increased, the differences turned out to be non-significant. The best results were then 12 g of glucose and 5.8 g xylose/100 g extrudate, which means 24.5% of glucan conversion and 56.5% of xylan conversion. In which concerns the extrudate composition presented in Table 2, cellulose is not affected by the amount of alkali; however, there are significant differences between the hemicellulose and lignin composition at 5% and 10% or 20% NaOH/DM. This is confirmed by a higher solubilization of xylose in the filtrate (see Table 3). Deacetylation is also higher at high NaOH/DM ratios. As expected, the values shown in Table 2 for the amount of ash increase as the amount of alkali added increases, due to the greater amount of H_2SO_4 needed to neutralize the mixture and the subsequently formation of salts. The greater flow of acid added also influences the efficiency of the filtrate reported in Table 3.

Sodium hydroxide is responsible for the cleavage of the ether and ester bonds between lignin and hemicellulose and also affects the ether and carbon to carbon bonds intra-lignin, resulting in deacetylation and delignification of the biomass and solubilization of hemicelluloses [34]. The results from the present work support the deacetylation phenomena that occurred during alkaline extrusion, based on significant acetic acid recovery yields in the filtrates and washing liquid shown in Table 3. However, no furfural neither HMF were found and monomeric phenols such as vainillin or syringaldehyde were neither detected (data not shown). The presence of acetic acid in the filtrate implies that the extrudates, which in this strategy are not washed out of the extruder, also contain certain amounts of this compound that may affect enzymes performance [35], but the dilution up to the desired consistency (in this work, 5% (w/w)) makes the acetic acid concentration to decrease, so alleviating the potential inhibition effect.

Other authors have reported deacetylation and/or delignification in alkaline extrusion of different biomasses. For example, Choi et al. [33] extruded empty fruit bunches at high temperature (170 °C) and low screw speed (5 rpm) and found that, as the amount of catalyst increased from 5% to 20% NaOH/DM, the glucan content in the extrudate increases, corresponding to a decrease of the hemicellulosic sugars and lignin due to solubilization. Similar behavior was reported by Han et al. [35] and Duque et al. [27] on alkali-extruded barley straw. Xylan solubilization related to the increase of NaOH was also reported by Liu et al. [36] and Um et al. [37] working with corn stover and rape straw, respectively. Nevertheless, delignification due to alkaline extrusion did not always occurred, as noted by Liu et al. [36], Duque et al. [18] and Kang et al. [38] on different herbaceous biomasses.

To sum up the results from this part of the work with eucalyptus, extrusion with filtration was successfully carried out and some favorable operation conditions have been identified: Lowest possible SS, high L/S ratios (>1) and moderate to high NaOH/DM ratios (≥8%). The effect of increasing the temperature over 100 °C was not possible to investigate in the present configuration, therefore, a new concept was tested, aiming at achieving good flowability at high temperature.

3.4. Alkaline Extrusion without Filtration

3.4.1. Modification of the Screw Configuration

As can be seen in Figure 3, for the purpose of testing high temperatures, the problematic zone (i.e., filtration) was removed from the process configuration. Two reverse flow mixing screws were placed in barrel #3 and at the end of barrel #5 to delimit the high temperature zone by the formation of dynamic plugs. Other mixing screws along the profile helped the mixture of the biomass with the alkaline and acid solutions. A profile temperature was set with increasing values, reaching its highest point in barrels #4 and 5.

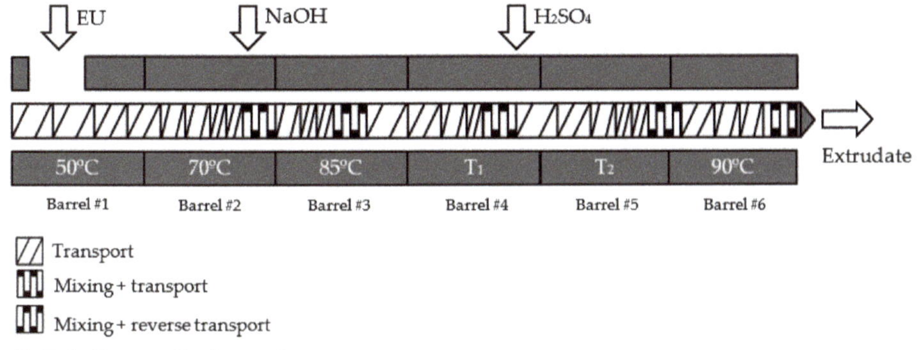

Figure 3. Barrel and screw configuration for the alkaline extrusion of eucalyptus with neutralization in a twin-screw extruder. Without filtration.

With the removal of the filter, only one material output is left. The resulting pretreated material will be called hereafter complete extrudate. With this configuration, temperatures of 125 and 150 °C could be attained (Table 1, experiments 11 and 12).

3.4.2. Performance of Complete Extrudates

No significant differences were found between the compositions of the two complete extrudates (see Table 2, rows 11 and 12), but in comparison to the previous experiments, some aspects differed. The absence of filtration had a clear consequence in the form of increase of the amount of ash up to almost 15% of the weight of extrudate. This also affected the concentration of cellulose and lignin, which decreased by 10% and 6%, compared, for example, to experiment #4.

In spite of being associated to a SME comparable to those of the experiments with filtration (114 Wh kg^{-1}), the sugar release of the complete extrudate at 125 °C (see Figure 2, column 11) was very low, particularly in the case of glucose: GR was only 6.8 g/100 g extrudate (equivalent to 15% of glucan conversion), whereas XR was 5.7 g/100 g dry extrudate, more in the line of previous trials. This indicates that there is a mechanical effect associated to the filtration step that is necessary to obtain a good pretreatment of eucalyptus. Moreover, this mechanical effect seems to be more important for cellulose, since in its absence, the glucose release drops to levels even below to the ones of the control extrudate with water and filtration. The xyloserelease, nevertheless, seems to be more affected by the chemical action of NaOH, as it reaches similar values for all the trials with chemical catalysts.

At 150 °C (experiment #12), a much higher torque than in trial #11 is developed, greater than any of the NaOH-extrusion with filtration trials (experiment 1 to 10). This resulted in a significant increase of the glucose and xylose release, which reached 16 and 8.4 g/100 g extrudate, respectively, corresponding to 37.6% and 74.6% of glucan and xylan conversion. This means that the GR of the extrusion at 150 °C (trial #12) more than doubled the one of the experiment at 125 °C (trial #11).

The enzymatic digestibility of complete extrudates was not as good as expected, especially in the case of 125 °C (trial #11). The high amount of ash in the extrudates produced in this configuration may be hindering the performance of enzymes. It has been demonstrated that the washing of the extrudates removes chemical residues and enzyme inhibitors and is beneficial for the enzymatic hydrolysis step [25,39]. Therefore, the complete extrudates were washed with ten times the weight of distillate water and the resulting cake was submitted to EH as usual.

3.4.3. Performance of Washed Extrudates

The washing of the extrudates removed most of the inorganic salts formed during the neutralization step, as can be seen by the content of ash measured in extrudates 13 and 14 shown in Table 2. This resulted in a concentration of cellulose and lignin, mainly. Comparing both washed extrudates, there seem to be a small loss of hemicellulosic sugars at 150 °C (trial #14) compared to 125 °C (trial #13). In fact, as can be seen in Table 3 (rows 13 and 14), almost all glucose remained in the extrudate after washing, however, there was some xylose solubilized in the liquid (more in trial #14), which could account for the difference in the hemicellulose content between both extrudates. The deacetylation effect of the pretreatment at 125 and 150 °C was in the line of the results from previous experiments with filtration.

Thanks to the washing, the glucose release by EH was increased 1.2-fold for both extrudates, attaining 19.4 g/100 g extrudate in trial #14 and 8.3 g/100 g extrudate in the case of experiment #13 (see Figure 2), which, in terms of GC, would be 35.8% and 15%, respectively. Nevertheless, the washing had no noticeable effect on the xylose release.

Results show that washing improved the glucose release compared to complete extrudates, and also a positive effect of increasing extrusion temperature from 125 to 150 °C. However, when talking about the glucan and xylan conversion, the yield is slightly lower in the case of the washed extrudates with respect to the complete ones and no clear effect of washing in the enzymatic digestibility of the substrates can be concluded. Thus, the increase in the sugar release found in washed substrates can be mostly attributed to the concentration of cellulose in the washing step.

4. Conclusions

Eucalyptus wood valorization through an extrusion pretreatment requires a careful planning of the configuration and setting of the operation conditions. In the present work, low screw speed, high L/S ratio inside the extruder, NaOH/DM concentrations $\geq$8% and high temperature (>100 °C) have been demonstrated to favor the glucose and xylose release by EH of the resulting substrates from the alkaline extrusion with neutralization and filtration inside the extruder.

The best results were obtained in the configuration without filtration at 150 °C and with a posterior washing of the extrudate obtained. In spite of this, the improvement in the glucose release does not lead to an equal rise of the glucan conversion. Therefore, the convenience of the washing step must be addressed in an overall assessment of the downstream processing operations on the pretreatment, since the benefits are not so clear and this adds complexity and cost to the process.

Filtration, however, has been proved to have a beneficial effect not only by the removal of inorganic salts from the extrudate, but also at a mechanical pretreatment level. Taking this into account, the ideal extrusion pretreatment process for eucalyptus biomass would comprehend a configuration with filtration and temperature around 150 °C. Such a process cannot be developed in the extrusion pilot plant used in this work, but could be investigated in a bigger facility.

Author Contributions: Conceptualization, A.D. and P.M.; methodology, A.D.; investigation, A.D.; analytical tools, A.G.; formal analysis, A.D. and A.G.; writing—original draft preparation, A.D.; writing—review and editing, P.M. and A.D.; supervision, P.M. and M.B.; project administration, P.M. and M.B.

Funding: This research was funded by the European Commission in the frame of the BABET-REAL5 project (Horizon 2020 Program, Project No. 654365).

Acknowledgments: The authors want to thank the INIA (Uruguay) for providing the eucalyptus biomass to complete this work.

Conflicts of Interest: The authors declare no conflict of interest. The funders had no role in the design of the study; in the collection, analyses, or interpretation of data; in the writing of the manuscript, or in the decision to publish the results.

References

1. REN21. *Renewables 2018 Global Status Report*; REN21 Secretariat: Paris, France, 2018.
2. Shepherd, M.; Bartke, J.; Lee, D.J.; Brawner, J.; Bush, D.; Turnbull, P.; Macdonel, P.; Brown, T.R.; Simmons, B.; Henry, R. Eucalyptus as biofuel feedstock. *Biofuels* **2011**, *2*, 639–657. [CrossRef]
3. Taherzadeh, M.J.; Karimi, K. Pretreatment of lignocellulosic wastes to improve ethanol and biogas production: A review. *Int. J. Mol. Sci.* **2008**, *9*, 1621–1651. [CrossRef] [PubMed]
4. Bajpai, P. Structure of Lignocellulose. In *Pretreatment of Lignocellulosic Biomass for Biofuel Production*; Springer: Singapore, 2016.
5. Volynets, B.; Ein-Mozaffari, F.; Dahman, Y. Biomass processing into ethanol: Pretreatment, enzymatic hydrolysis, fermentation, rheology, and mixing. *Green Process. Synth.* **2017**, *6*, 1–22. [CrossRef]
6. Alvira, P.; Tomas-Pejó, E.; Ballesteros, M.; Negro, M.J. Pretreatment technologies for an efficient bioethanol production process based on enzymatic hydrolysis: A review. *Bioresour. Technol.* **2010**, *101*, 4851–4861. [CrossRef] [PubMed]
7. Duque, A.; Manzanares, P.; Ballesteros, M. Extrusion as a pretreatment for lignocellulosic biomass: Fundamentals and applications. *Renew. Energy* **2017**, *114*, 1427–1441. [CrossRef]
8. Romaní, A.; Garrote, G.; Parajó, J.C. Bioethanol production from autohydrolyzed Eucalyptus globulus by Simultaneous Saccharification and Fermentation operating at high solids loading. *Fuel* **2012**, *94*, 305–312. [CrossRef]
9. McIntosh, S.; Vancov, T.; PAlmer, J.; Spain, M. Ethanol production from Eucalyptus plantation thinnings. *Bioresour. Technol.* **2012**, *110*, 264–272. [CrossRef] [PubMed]
10. Lima, M.A.; Lavorente, G.B.; da Silva, H.K.P. Effects of pretreatment on morphology, chemical composition and enzymatic digestibility of eucalyptus bark: A potentially valuable source of fermentable sugars for biofuel production—Part 1. *Biotechnol. Biofuels* **2013**, *6*, 75. [CrossRef] [PubMed]
11. Karunanithy, C.; Muthukumarappan, K.; Gibbons, W.R. Extrusion pretreatment of pine wood chips. *Appl. Biochem. Biotechnol.* **2012**, *167*, 81–99. [CrossRef] [PubMed]
12. Lee, S.H.; Teramoto, Y.; Endo, T. Enhancement of enzymatic accessibility by fibrillation of woody biomass using batch-type kneader with twin-screw elements. *Bioresour. Technol.* **2010**, *101*, 769–774. [CrossRef] [PubMed]
13. Lee, S.-H.; Teramoto, Y.; Endo, T. Enzymatic saccharification of woody biomass micro/nanofibrillated by continuous extrusion process I—Effect of additives with cellulose affinity. *Bioresour. Technol.* **2009**, *100*, 275–279. [CrossRef] [PubMed]

14. Lee, S.H.; Inoue, S.; Teramoto, Y.; Endo, T. Enzymatic saccharification of woody biomass micro/nanofibrillated by continuous extrusion process II: Effect of hot-compressed water treatment. *Bioresour. Technol.* **2010**, *101*, 9645–9649. [CrossRef] [PubMed]

15. Kim, T.H.; Choi, C.H.; Keun, K. Bioconversion of sawdust into ethanol using dilute sulfuric acid-assisted continuous twin screw-driven reactor pretreatment and fed-batch simultaneous saccharification and fermentation. *Bioresour. Technol.* **2013**, *130*, 306–313. [CrossRef] [PubMed]

16. Senturk, S.; Gevgilili, H.; Kaylon, D.M. Biomass pretreatment strategies via control of rheoligal behavior of biomass suspensions and reactive twin screw extrusion processing. *Bioresour. Technol.* **2011**, *102*, 9068–9075. [CrossRef] [PubMed]

17. Brault, J. Développement d'un Procédé Innovant de Dégradation Enzymatique des Parois Végétales Pour la Production de Bioéthanol Seconde Génération. Ph.D. Thesis, INPT Toulouse, Toulouse, France, 2013.

18. Duque, A.; Manzanares, P.; Ballesteros, I.; Negro, M.J.; Oliva, J.M.; Saez, F.; Ballesteros, M. Study of process configuration and catalyst concentration in integrated alkaline extrusion of barley straw for bioethanol production. *Fuel* **2014**, *134*, 448–454. [CrossRef]

19. National Renewable Energy Laboratory (NREL). *Chemical Analysis and Testing Laboratory Analytical Procedures*; NREL: Golden, CO, USA, 2007.

20. Choi, C.H.; Oh, K.K. Application of a continuous twin screw-driven process for dilute acid pretreatment of rape straw. *Bioresour. Technol.* **2012**, *110*, 349–354. [CrossRef] [PubMed]

21. Kuster Moro, M.; Sposina, R.; Sant'Ana da Silva, A.; Fujimoto, M.D.; Melo, P.A.; Secchi, A.R.; da Silva Bon, E.P. Continuous pretreatment of sugarcane biomass using a twin-screw extruder. *Ind. Crop. Prod.* **2017**, *97* (Suppl. C), 509–517. [CrossRef]

22. Zheng, J.; Choo, K.; Rehmann, L. Xylose removal from lignocellulosic biomass via a twin-screw extruder: The effects of screw configurations and operating conditions. *Biomass Bioenergy* **2016**, *88*, 10–16. [CrossRef]

23. Vandenbossche, V.; Brault, J.; Vilarem, G.; Rigal, L. Bio-catalytic action of twin-screw extruder enzymatic hydrolysis onthe deconstruction of annual plant material: Case of sweet cornco-products. *Ind. Crop. Prod.* **2015**, *67*, 239–248. [CrossRef]

24. Yoo, J.; Alavi, S.; Vadlani, P.; Behnke, K.C. Soybean hulls pretreated using thermo-mechanical extrusion—Hydrolysis efficiency, fermentation inhibitors, and ethanol yield. *Appl. Biochem. Biotechnol.* **2012**, *166*, 576–589. [CrossRef] [PubMed]

25. Lamsal, B.; Yoo, J.; Brijwani, K.; Alavi, S. Extrusion as a thermo-mechanical pre-treatment for lignocellulosic ethanol. *Biomass Bioenergy* **2010**, *34*, 1703–1710. [CrossRef]

26. Scott, C.T.; Samaniuk, J.R.; Klingenberg, D.J. Rheology and extrusion of high-solids biomass. *Tappi J.* **2011**, *10*, 47–53.

27. Duque, A.; Manzanares, P.; Ballesteros, I. Optimization of integrated alkaline-extrusion pretreatment of barley straw for sugar production by enzymatic hydrolysis. *Process Biochem.* **2013**, *48*, 775–781. [CrossRef]

28. Zhang, S.; Xu, Y.; Hanna, M. Pretreatment of Corn Stover with Twin-Screw Extrusion Followed by Enzymatic Saccharification. *Appl. Biochem. Biotechnol.* **2012**, *166*, 458–469. [CrossRef] [PubMed]

29. Karunanithy, C.; Muthukumarappan, K. Effect of Extruder Parameters and Moisture Content of Switchgrass, Prairie Cord Grass on Sugar Recovery from Enzymatic Hydrolysis. *Appl. Biochem. Biotechnol.* **2010**, *162*, 1785–1803. [CrossRef] [PubMed]

30. Negro, M.J.; Duque, A.; Manzanares, P.; Sáez, F.; Oliva, J.M.; Ballesteros, I.; Ballesteros, M. Alkaline twin-screw extrusion fractionation of olive-treepruning biomass. *Ind. Crop. Prod.* **2015**, *74*, 336–341. [CrossRef]

31. Karunanithy, C.; Muthukumarappan, K. A comparative study on torque requirement during extrusion pretreatment of different feedstocks. *Bioenergy Res.* **2012**, *5*, 263–276. [CrossRef]

32. Da Silva, A.; Sposina, R.; Endo, T.; Bon, E.P.; Lee, S.H. Continuous pretreatment of sugarcane bagasse at high loading in an ionic liquid using a twin-screw extruder. *Green Chem.* **2013**, *15*, 1991–2001. [CrossRef]

33. Choi, W.I.; Oh, K.K.; Park, J.Y.; Lee, J.S. Continuous sodium hydroxide-catalyzed pretreatment of empty fruit bunches (EFB) by continuous twin-screw-driven reactor (CTSR). *J. Chem. Technol. Biotechnol.* **2014**, *89*, 290–296. [CrossRef]

34. Kim, J.S.; Lee, Y.Y.; Kim, T.H. A review on alkaline pretreatment technology for bioconversion of lignocellulosic biomass. *Bioresour. Technol.* **2016**, *199*, 42–48. [CrossRef] [PubMed]

35. Van Walsum, G.P.; Um, B.H. Effect of pretreatment severity on accumulation of major degradation products from dilute acid pretreated corn stover and subsequent inhibition of enzymatic hydrolysis of cellulose. *Appl. Biochem. Biotechnol.* **2012**, *168*, 406–420.

36. Han, M.; Kang, K.E.; Kim, Y.; Choi, G.W. High efficiency bioethanol production from barley straw using a continuous pretreatment reactor. *Process Biochem.* **2013**, *48*, 488–495. [CrossRef]

37. Liu, C.; van der Heide, E.; Wang, H.; Li, B.; Yu, G.; Mu, X. Alkaline twin-screw extrusion pretreatment for fermentable sugar production. *Biotechnol. Biofuels* **2013**, *6*, 97. [CrossRef] [PubMed]

38. Um, B.-H.; Choi, C.H.; Oh, K.K. Chemicals effect on the enzymatic digestibility of rape straw over the thermo-mechanical pretreatment using a continuous twin screw-driven reactor (CTSR). *Bioresour. Technol.* **2013**, *130*, 38–44. [CrossRef] [PubMed]

39. Kang, K.E.; Han, M.; Moon, S.K.; Kang, H.W.; Kim, Y.; Cha, Y.L.; Choi, G.W. Optimization of alkali-extrusion pretreatment with twin-screw for bioethanol production from Mischantus. *Fuel* **2013**, *109*, 520–526. [CrossRef]

Review

An Overview of Recent Developments in Biomass Pyrolysis Technologies

M. N. Uddin [1], Kuaanan Techato [2,3], Juntakan Taweekun [4], Md Mofijur Rahman [5,6,*], M. G. Rasul [5], T. M. I. Mahlia [6] and S. M. Ashrafur [7]

[1] Sustainable Energy Management, IGES, Prince of Songkla University, Hatyai, Songkhla 90110, Thailand; 5910930005@psu.ac.th

[2] Environmental Assessment & Technology for Hazardous Waste Management Research Center, Faculty of Environmental Management, Prince of Songkla University, Hatyai, Songkhla 90110, Thailand; kuaanan.t@psu.ac.th

[3] Center of Excellence on Hazardous Substance Management (HSM), Bangkok 10330, Thailand

[4] Department of Mechanical Engineering, Faculty of Engineering, Prince of Songkla University, Hatyai, Songkhla 90112, Thailand; jantakan.t@psu.ac.th

[5] School of Engineering & Technology, Central Queensland University, Rockhampton, QLD 4701, Australia; m.rasul@cqu.edu.au

[6] Faculty of Engineering and Information Technology, University of Technology, Sydney, NSW 2007, Australia; TMIndra.Mahlia@uts.edu.au

[7] Biofuel Engine Research Facility (BERF), Queensland University of Technology, Brisbane, QLD 4000, Australia; s2.rahman@qut.edu.au

* Correspondence: m.rahman@cqu.edu.au; MdMofijur.Rahman@uts.edu.au

Received: 3 October 2018; Accepted: 8 November 2018; Published: 10 November 2018

Abstract: Biomass is a promising sustainable and renewable energy source, due to its high diversity of sources, and as it is profusely obtainable everywhere in the world. It is the third most important fuel source used to generate electricity and for thermal applications, as 50% of the global population depends on biomass. The increase in availability and technological developments of recent years allow the use of biomass as a renewable energy source with low levels of emissions and environmental impacts. Biomass energy can be in the forms of biogas, bio-liquid, and bio-solid fuels. It can be used to replace fossil fuels in the power and transportation sectors. This paper critically reviews the facts and prospects of biomass, the pyrolysis process to obtain bio-oil, the impact of different pyrolysis technology (for example, temperature and speed of pyrolysis process), and the impact of various reactors. The paper also discusses different pyrolysis products, their yields, and factors affecting biomass products, including the present status of the pyrolysis process and future challenges. This study concluded that the characteristics of pyrolysis products depend on the biomass used, and what the pyrolysis product, such as bio-oil, can contribute to the local economy. Finally, more research, along with government subsidies and technology transfer, is needed to tackle the future challenges of the development of pyrolysis technology.

Keywords: renewable energy; biofuel; environment; technology development

---

1. Introduction

Nowadays, energy usage is prodigious, and a significant key factor for the advancement of a nation, and the scarcity of energy has become an economic threat for the development of nations around the world [1,2]. It is said that "Energy is a critical component of our lives. Without energy, we can't even dream of economic growth. But despite its central role, not everyone has access to modern energy services" [3,4]. Today's energy requirement is increasing in trend, due to population growth and ongoing economic and technological advancement around the world [4]. Currently, fossil fuels are the main source of energy because of their high calorific values, good anti-knocking properties, and high heating values; meanwhile, reserves are limited. Therefore, the development of alternative energy resources can lower the depletion of fossil fuel by reducing their consumption [5–7]. On the other hand, the world's heating condition is increasing every day. The atmospheric CO_2 level has crossed the risky level that was forecast to happen in another 10 years [8]. Furthermore, the depletion of fossil fuels and extreme change of climate have driven the search for alternative energies and renewable energy sources that can meet the world's energy demand, reduce greenhouse gas emissions, curb pollution, and maintain the planet's temperature at a stable level [9–11].

Among the alternative energy sources, biomass can become a promising sustainable energy source, due to its high diversity and availability [12]. Biomass can be defined as all biodegradable organic material derived from animals, plants, or microorganisms. This definition also includes products, by-products, waste originating in agricultural activities, as well as non-fossil organic waste produced by industrial and municipal waste [13]. Biomass is the third most important source used to generate electricity and thermal applications [14,15]. The most common biomass feedstocks are banana peel, rice and coffee husks, sugarcane bagasse, palm oil processing residues, and the waste of animals [16,17]. Biomass can be considered as a blend of organic resources and minor amounts of minerals, which also contains carbon, oxygen, hydrogen, nitrogen, sulphur, and chlorine [18].

Different types of energy can be produced through the thermal conversion of biomass, such as combustion, pyrolysis, gasification, fermentation, and anaerobic decomposition. Combustion is a thermochemical process used for the production of heat, which consists of a chemical reaction in which a fuel is oxidised, and a large amount of energy is released in the form of heat (exothermic reaction). Pyrolysis is a thermal decomposition process which takes place in the absence of oxygen [19,20]. In combustion and gasification processes, the first step is pyrolysis, followed by total or partial oxidation of primary products. Gasification is the process of generating electricity by applying heat to organic material in the presence of less oxygen. In the fermentation process, organic materials are used to produce alcohol, with the help of yeast, to generate power in automobiles. Anaerobic decomposition is the process of producing biogas, and generates electricity.

Among all the conversion techniques of biomass conversion, the pyrolysis process offers a number of benefits, including less emissions and that all the by-products can be reused. In addition, during the process, pyrolysis produces solid or carbonised products, liquid products (bio-oils, tars, and water) and a gas mixture composed mainly of CO_2, CO, H_2, and CH_4 [21–23]. The oil resulting from the pyrolysis of biomass, usually referred to as bio-oil, is a renewable liquid fuel, which is the main advantage over petroleum products. It can be used for the production of various chemical substances [24]. The pyrolysis process has three stages: the dosing and feeding of the raw material, the transformation of the organic mass and, finally, the obtaining and separation of the products (coke, bio-oil, and gas). The factors that influence the distribution of the products are the heating rate, final temperature, composition of the raw material, and pressure [25].

The pyrolysis process has great market potential; in this process, biomass is used as raw material in order to produce energy. Therefore, intense research is taking place around the world to improve this method of energy production. Among the technologies, such as digestion, fermentation, and mechanical conversion, thermo-conversion for producing energy from biomass is relatively newer from a commercial perspective, and gaining more attention because of its technical and strategical advantages. In addition, the production of waste is constantly increasing, and the economic activity linked to it is becoming increasingly important. The elimination or attenuation of environmental problems and obtaining profitability in the process of managing them is a very favourable step. Therefore, pyrolysis could be an alternative means of energy recovery, obtaining different fractions that are also recoverable not only from the energy point of view.

Though the research into pyrolysis technology indicated that pyrolysis is a more promising option to the sustainable development, pyrolysis technology still needs further improvement, and several challenges need to be tackled to gain its full potential benefits. Furthermore, several types of research have been carried out recently, focusing on the use of pyrolysis technology, but only a few papers have been analysed and reviewed by the researchers. Thus, the main aims of this study are to present a brief review of the development of pyrolysis technology, including their present status and future challenges, to provide information to the researchers who are interested in pyrolysis technology. A number of studies from highly rated journals in scientific indexes are reviewed, including the most recent publications.

2. Biomass Pyrolysis

Biomass is a renewable source for the production of energy, and it is profusely obtainable everywhere in the world [26,27]. The sustainable use of biomass energy is an alternative to partially replace the use of fossil fuels and nuclear energy. Rural people in developing countries, representing about 50% of the global population, depend on biomass energy [9]. Biomass assists the world in meeting greenhouse gas reduction goals [9,28,29]. The increase in availability and technological developments of recent years allow the use of biomass as a renewable energy source with low levels of emissions and environmental impacts. Biomass energy can be in the forms of biogas, bio-liquid, and bio-solid fuels. It can be used to replace fossil fuels in power and transportation. It is considered as a renewable energy source because the energy mainly comes from the sun and, also, it needs a short time period to re-grow.

Pyrolysis process is mainly characterised by solid fuel thermal degradation, which involves the rupture of carbon–carbon bonds and the formation of carbon–oxygen bonds. Pyrolysis requires temperatures of up to 400–550 °C, although it can be done at temperatures even higher [30–33]. Figure 1 shows the percentage yield during the pyrolysis of biomass.

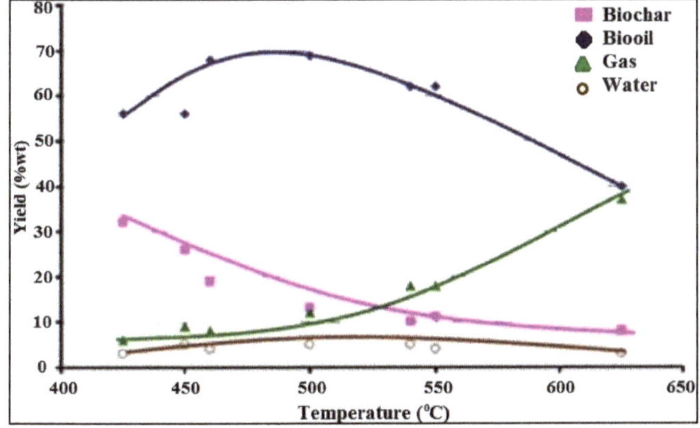

Figure 1. The % yield of the end products the pyrolysis of biomass [9].

One part of the biomass is reduced to carbon, while the remaining part is oxidised and hydrolysed to carbohydrates, phenols, aldehydes, ketones, alcohols, and carboxylic acids, which combine to form more complex molecules such as esters, polymer products, and others [34–36]. Pyrolysis can be achieved by the complete absence of the oxidising agent. The practice of using air to perform pyrolysis is achieved by feeding air in an amount below stoichiometric; combustion occurs in only a small part of the biomass and, thus, the heat released in the combustion is used to keep the temperature of the reactor constant, while processing the reactions related to pyrolysis [37].

The products formed during pyrolysis, namely, coal fines, gases, acid extract, and bio-oil, have high calorific value, and have had several applications in both the chemical and power generation industries. In ancient Egyptian times, the pyrolysis process was used to generate tar for sealing boats [16], and the ancient Egyptians performed wood decontamination by assembling tars and pyrolignous acid for use in their mummifying industry [38,39]. Pyrolysis has gained more attention as an effective and practical method in converting biomass into bio-fuel recent years [40]. Pyrolysis is not only part of the combustion and gasification processes, but it is also the first stage of both of these processes. The gas is composed of carbon monoxide, carbon dioxide, and light hydrocarbons. This dark-coloured liquid is called bio-oil and charcoal solid. The yields and quality of the products are influenced by the operating conditions. Pyrolysis receives different denominations depending on the conditions used. In slow pyrolysis or carbonisation, low temperatures and long residence times are employed, favouring the production of charcoal. High temperatures and long residence times favour the formation of gases. Whereas moderate temperatures and low residence time of the gases favour the production of liquids (bio-oil). Figure 2 shows the chemical reaction during the pyrolysis process.

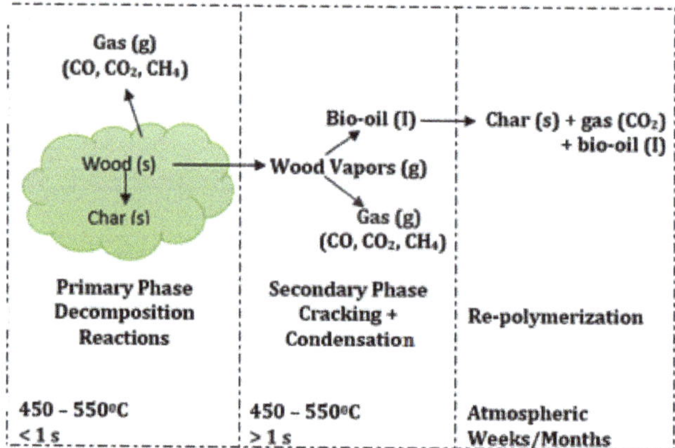

Figure 2. Representation of the reaction paths for wood pyrolysis [41].

3. Mechanism of Pyrolysis Process

The biomass pyrolysis can be divided into two categories, such as primary and secondary mechanisms [41]. Figure 3 shows the detailed mechanism of the pyrolysis process. In the primary mechanism, volatile compounds are released, while the chemical bonds within the polymers are broken during biomass heating process [42,43]. Furthermore, rearrangement reactions within the matrix of the residue take place. Some of the volatile compounds which are unstable further undergo additional reactions, which are defined as a secondary mechanism.

The primary mechanism can be described using three different approaches, namely char formation, depolymerisation, and fragmentation. In the char formation process, initially, benzene rings are formed, and these rings combine into a solid residue known as char, which is an aromatic polycyclic structure [44]. During this process, water or incondensable gas is also released [45,46]. In the depolymerisation process, the polymers are broken into monomer units, which reduce the degree of polymerisation. This process continues until the volatile molecules are produced [47]. Finally, in fragmentation, incondensable gas and small chain organic compounds are formed through the linkage of many covalent bonds of the polymer, even within the monomer units [42].

The secondary mechanism consists of cracking, recombination, and others [42,48]. In cracking, lower molecular weight molecules are formed by breaking volatile compounds [49]. By contrast, in the recombination process, volatile compounds combine into high molecular weight compounds, which may or may not be volatile [43,50]. In some cases, a secondary mechanism leads to the formation of secondary char [48,51].

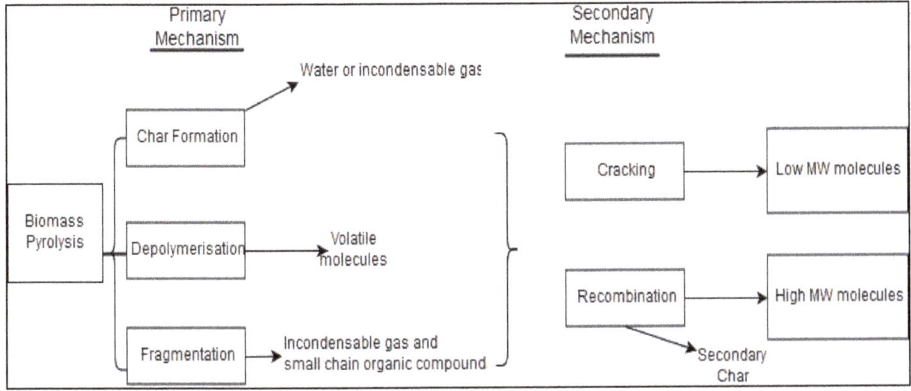

Figure 3. The detailed mechanism of the pyrolysis process.

4. Sources of Biomass and Their Properties

Biomass energy is currently recognised as the third largest global energy source. In many developing countries which have significantly large forest and agricultural land, 40–50% of energy usage is based on biomass. Green plants can directly/indirectly produce biomass using the photosynthesis process, by transforming sunlight into plant material [29,52]. The resources of biomass include various natural and derived materials, such as agricultural crops and residues, forest wood and leaf residues, municipal solid wastes (MSW), forest and mill residues, animal residues, and sewage. Agricultural crops and wastage (sugarcane, cassava, and corn) provide carbohydrate and starch. Roughly, the biomass species contain woody biomass, straw, beech wood, seedcakes, bagasse, and municipal solid waste (MSW) [53–59]. The available sources of biomass are shown in Figure 4.

Biomass is a very versatile feedstock in its morphology and physical characteristics. It can be quite wet or dry dense or fluffy, high or low ash containing, small in shape or large, homogeneous or inhomogeneous, and so on. This makes the use of biomass fuels in dedicated gasifier reactors quite difficult and, in most cases, some pre-treatment of the biomass is needed. The feedstocks used for pyrolysis and their physical and chemical properties are more important. The highest bio-char yields are achieved when feedstocks with high lignin content are pyrolysed at moderate temperatures (approx. 500 °C). Furthermore, some other indicators of pyrolysis product yields are the ratios of fixed carbon, moisture, volatile matter, and ash content. Generally, biomass containing significant volatile matter offers a large amount of syngas and bio-oil, while fixed carbon raises the production of biochar. Moisture content in biomass influences the heat transfer process, as well as significantly affects product distribution. Tables 1 and 2 show the physical and chemical properties of biomass. Biomass consists of elements such as carbon, hydrogen, oxygen, and nitrogen. Sulphur is present in smaller proportions, and some types of biomass also contain significant portions in inorganic species. The chemicals obtained from co-products and residues can improve the biomass production chains, due to the strategic participation of the chemical industry in the supply of inputs and final products to various economic sectors, for example, agribusiness, petrochemical, automotive, pharmaceutical, cosmetics, civil construction, and so on [60].

Table 1. Physical characteristics of biomass [9].

Feedstock	Density (kg/m^3)	Moisture Content (%)	Ash Content (%)	Volatile Matter (%)	Fixed Carbon (%)
Wood	380	20	0.4–1	82	17
Bituminous coal	700	11	8–11	35	45
Wheat straw	18	16	4	59	21
Barley straw	210	30	6	46	18
Pine	124	17	0.03	-	16
Polar	120	16.8	0.007	-	-
Switchgrass	108	13–15	4.5–5.8	-	-

Table 2. Chemical characteristics of biomass [9].

Feedstock	Carbon (%)	Hydrogen (%)	Oxygen (%)	Nitrogen (%)	Ash (%)
Wood	51.6	6.3	41.5	0.1	1
Bituminous coal	73.1	5.5	8.7	1.4	9
Wheat straw	48.5	5.5	3.9	0.3	4
Barley straw	45.7	6.1	38.3	0.4	6
Pine	45.7	7	47	0.1	0.03
Polar	48.1	5.30	46.10	0.14	0.007
Switchgrass	44.77	5.79	49.13	0.31	4.30

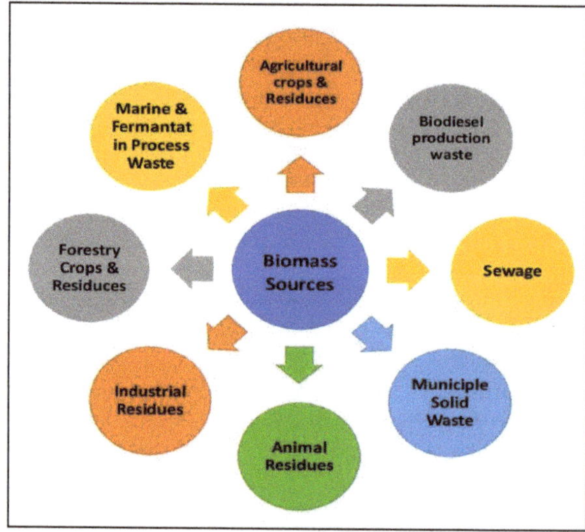

Figure 4. Available sources of biomass [61].

5. Pyrolysis Technology

Pyrolysis technology is the decomposition of heated organic matter in the absence of atmospheric oxygen, where heating is controlled by temperature ranges and provides the energy needed to break down the structures of the macromolecules present in biomass [62]. In the process of pyrolysis, biomass degradation occurs through heating, in which the formation of three products occurs: coal, oil, and pyrolytic gas, and, depending on the conditions in the reactor, one of these products can be maximised [63,64]. Currently, there are basically three pyrolysis processes in the world: slow pyrolysis, fast pyrolysis, and ultrafast pyrolysis. Biomass is first put into the reactor feed system, usually an endless screw. Then, the biomass enters the reactor and undergoes thermal degradation. Any gas that does not condense and has no energetic ends returns to the process and is used as entrainment gas in the reactor.

5.1. Slow Pyrolysis

Slow or conventional pyrolysis consists of systems known as "charcoal" or continuous systems, with slow biomass heating above 400 °C in the absence of oxygen [65]. In this process, the biomass is pyrolysed with low heating rates, around 5 to 7 °C/minimum, where the liquid and gaseous products are minimal, and the coal production is maximised [66,67]. Slow pyrolysis of wood, with a 24 h endurance, was a very common technology in industries until the early 1900s, where coal, acetic acid, methanol, and ethanol were obtained from wood [68,69]. Slow pyrolysis is characterised by small heating rates and a maximum temperature range of around 600 °C, and the biomass time in the reactor is between 5 and 30 min. The main products are bio-oil, coal, and gases [68].

5.2. Rapid Pyrolysis

Rapid pyrolysis is a promising method for conversion of biomass into a liquid product. The produced pyrolysis oil (bio-oil) is an intermediate dense energy fuel, which is possible to upgrade to hydrocarbons in diesel and gasoline [70]. In rapid pyrolysis, the biomass decomposes very quickly, generating mainly vapours and aerosols, and a small amount of coal and gas. After cooling and condensation, a homogeneous mobile dark brown liquid is formed, which has a calorific value corresponding to half of the conventional fuel oil [71]. Rapid pyrolysis technology is used globally, in large scale, for the production of liquids (bio-oils), and there is a lot of interest regarding this technology among biofuel researchers. Several reactors are used in the rapid pyrolysis process. Among them are the dragged-flow reactor, vacuum furnace reactor, vortex reactor, rotary reactor, bubbling fluidised bed reactor, and others; many researchers have contributed in the field of pyrolysis using one of these reactors [67].

5.3. Ultrafast Pyrolysis

The ultrafast pyrolysis has, as its main characteristics, very high heating rates and very low residence time of the biomass in the reactor. These characteristics favour the production of vapours, and make the process very similar to gasification. Due to the high heating rate, where biomass residence times are only a few seconds, reactors are needed to meet these heating needs [67]. These reactors have a fluidised bed and are flow-dragged. The fluidised bed reactor is used in the execution of multiphase chemical reactions, where a catalyst, usually sand, is used, working the same as with a fluid inside [72,73]. According to Laird et al. [65], ultrafast pyrolysis for coal production involves heating the biomass, under moderate to high pressure, in a reactor. In this particular case, the coal yield reaches 60%, and is volatile (bio-oil and synthesis gas) to 40%; this technology is more likely to use heat recovery equipment. Table 3, below, demonstrates some operating parameters of the three types of pyrolysis process [72,73].

Table 3. Operating parameters of different pyrolysis processes.

Process	Time (s)	Rate (K/s)	Size (mm)	Temp. (K)	Oil Yield	Char Yield	Gas Yield
Slow	450–550	0.1–1	5–50	550–950	30	35	35
Fast	0.5–10	10–200	<1	850–1250	50	20	30
Flash	<0.5	>1000	<0.2	1050–1300	75	12	13

5.4. Flash Pyrolysis

This process is also known as fast pyrolysis, due to the high speed of the process. However, in this process, not only kinetics play an important role, but heat and mass transfer processes, such as phase change phenomena, are also important. In this process, the biomass decomposes to generate mainly vapours, aerosols, and a certain amount of coke. After cooling and condensation, a dark brown liquid (bio-oil) is formed, with a calorific value that is half the value corresponding to that of diesel. Unlike traditional processes, this is an advanced process with carefully controlled parameters to obtain high liquid yields [74]. In order to carry out this process, the following must be observed: (a) subjecting the biomass particles to an optimum temperature so that they react, and (b) minimising their exposure to low intermediate temperatures that stimulate coke formation. One method to achieve these objectives is to use small particles, for example, those that are present in fluidised bed processes (a fluidised bed is a packed bed with a fine-grained solid). Another possibility is to transfer heat quickly, only to the surface of the particles that are in contact with the heat source, which is applied in ablation processes [75,76].

6. The Products of Pyrolysis Process

The pyrolysis of biomass produces three primary products, namely char, permanent gases, and vapours which condense to a viscous liquid (dark brown in colour) at ambient temperature. Biomass pyrolysis product yields can be improved as follows: (1) charcoal—less temperature and lower heating rate procedure, (2) liquid products—lower temperature but higher heating rate procedure, and (3) fuel gas—higher temperature and lower heating rate procedure. Table 4 shows the pyrolysis processes at different temperatures.

Table 4. Pyrolysis processes at different temperature.

Condition	Processes	Products
<350 °C	Free radical formation, water elimination, and depolymerisation	Formation of carbonyl and carboxyl, the evolution of CO and CO_2, and mainly a charred residue
350–450 °C	The split of glycosidic connections of polysaccharide by substitution	A combination of levoglucosan, anhydrides, and oligosaccharides as a tar segment
450–500 °C	Dehydration, rearrangement, and fission of sugar units	Formation of carbonyl compounds
>500 °C	A combination of all the above processes	A combination of all the above products
Condensation	Unsaturated products shrink and split to the char	A highly reactive char remainder comprising trapped free radicals

6.1. Bio-Oil

Bio-oil, also known as pyrolysis oil, crude bio-oil, pyrolytic tar, pyrolignous tar, pyrolignous liquor, wood liquid, wood oil, smoke condensate, and distilled from wood, is a dark brown-coloured liquid, almost black, with a characteristic odour of smoke, and an elemental composition similar to the biomass. It is a complex mixture, containing oxygenated compounds and a high volume of water, which originates from the moisture of the biomass and the reactions. It might also contain some amount of coal particles and dissolved alkali metals from the ash. The composition of the total mixture depends on the type of biomass, process conditions, equipment, and the efficiency in the separation of the coal and the condensation.

The bio-oil can be considered as a micro emulsion, in which the continuous phase is an aqueous solution of the products of cellulose and hemicellulose fragmentation which stabilises the discontinuous phase of the pyrolytic lignin macromolecules [71]. There is a very current and praising demand to convert biomass into liquid fuels in order to use in ships, trains, and aeroplanes, to substitute petrol and diesel [77,78].

Bio-oil is the main product from the pyrolysis process. Several types of research around the world, in order to maximise and improve the quantity and quality of bio-oil produced, are currently being carried out. Reactor designs are the primary target of researchers to achieve a better-quality bio-oil. As shown in Figure 5, the bio-oil product has a number of applications: it can be improved to be used as a transport fuel or used as a chemical, and it can also be used in turbines and electric power generation engines, or in boilers to generate heat. In summary, the bio-oil product has many applications and deserves large investments in research.

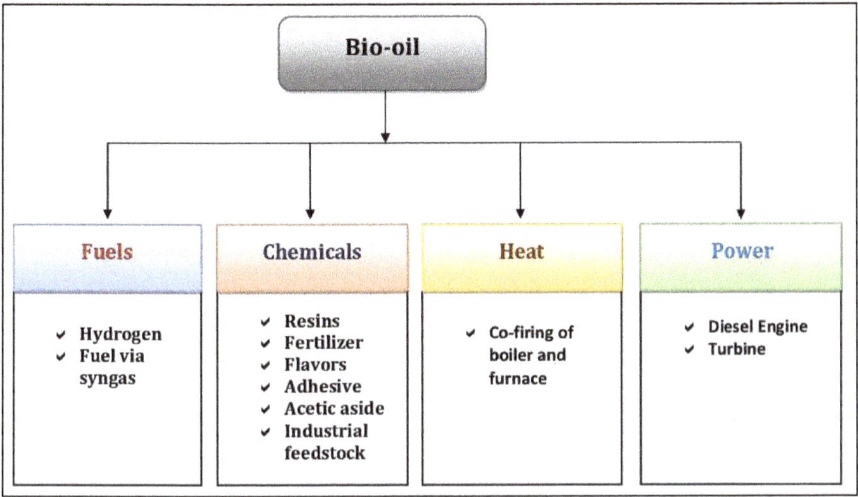

Figure 5. Various applications of pyrolysis bio-oil [77,78].

6.2. Biochar

The production of biochar is an emerging technology which can improve countries' food security and mitigate climate change [79]. In the literature, the potential benefits of applying biochar as soil enrichment have been highlighted heavily, addressing issues such as waste management, bioenergy production, increased soil fertility through alteration of soil pH, retention of nutrients through cation adsorption, reduction of emissions of nitrous oxide (N_2O), methane (CH_4), and carbon dioxide (CO_2), adsorption of organic pollutants, and improvements in productivity [80]. As a promising modifier to soil, biochar attracts the attention of policymakers in developed countries, such as the United States, Japan, Europe, and some developing countries. Sustainable biochar is one of the few technologies that is relatively cheap, widely applicable, and rapidly scalable. These benefits are confirmed by many investigations [79,81,82], including:

- Reduced nitrogen leaching in groundwater
- Possible reduced emissions of nitrous oxide
- Increased cation exchange capacity resulting in better soil fertility
- Moderation of soil acidity
- Greater water retention
- Increase in the number of beneficial soil microbes

6.3. Syngas

In slow pyrolysis processes, around 10–35% of biogas is produced which is similar to char. Syngas produced from biomass pyrolysis can be used as an alternative renewable source of fuel for industrial combustion processes, as well as for internal combustion (IC) engines. In power generation, transportation, and other sectors, gaseous fuel can be used in converted commercial petrol and diesel engines [83], which was quite common between 1901 and 1920 and, after that, due to the availability of cheap liquid fuels, the usage of gaseous fuels in IC engines. However, in recent years, as the focus has moved towards renewable fuels for engines, the use of syngas in IC engines has, once again, gained interest [84].

Syngas yield is highly influenced by the pyrolysis temperature, and it is possible to achieve a higher yield in flash pyrolysis with high temperatures. He et al. [85] investigated syngas production in a bench-scale downstream fixed-bed reactor from pyrolysis of MSW over a temperature range of 750–900 °C [86]. The researchers used calcined dolomite as a catalyst, and reported a 78.87% gas yield at 900 °C. In another study, Tang and Huang reported 76.64% syngas yield in a radio frequency plasma pyrolysis reactor [87].

Another factor that greatly influences pyrolysis processes and the resulting product distribution is the reactor temperature. With the increase of pyrolysis temperature, the inner moisture of the biomass evaporates first, followed by thermal degradation and devolatilisation of the dried particle portion. Simultaneously, tar is produced, and volatile species are slowly released from the particles' surface, which then undergoes a series of secondary reactions, such as decarboxylation, dehydrogenation, deoxygenation, and cracking, to form components of syngas. Thus, higher temperatures favour tar decomposition and the thermal cracking of tar to increase the proportion of syngas, which reduces oil and char yields [85]. Some researchers have also reported that when the reactor temperature is increased, the syngas flow rate also increases; however, this lasts for a short time, and then dramatically reduces [88].

Syngas mainly consists of hydrogen (H_2) and carbon monoxide (CO). It may also contain a small volume of nitrogen (N_2), water, carbon dioxide (CO_2), hydrocarbons such as C_2H_4, CH_4, C_2H_6, ash, tar, and so on, which depend on biomass feedstock and pyrolysis conditions [89]. These components are obtained during several endothermic reactions at high pyrolysis temperatures.

7. Reactors Employed in the Pyrolysis Process

The heart of the pyrolysis process is the reactor. This is the place where all reactions occur [90–93]. However, to perform flash pyrolysis, it is necessary to have special reactors. For this process, an oxygen-free atmosphere is required in the reactor, and a temperature range between 475 and 550 °C. When the gas flows through the bed, the solid behaves like a liquid [94,95].

The reactor is at the core of any sort of pyrolysis procedure that has been the content of invention, significant research, and advancement, to expand the indispensable physiognomies [96–98]. In the beginning, the developers of the pyrolysis reactor presumed that a minor biomass particle size and very short residence time could obtain prominent bio-oil yields, but further research has found divergent consequences. Component part size and vapour residence time have a slight impact on bio-oil yield, while the parameters significantly trace bio-oil composition [99,100]. The pyrolytic reactor is undoubtedly the most important equipment in the pyrolysis process. Currently, several types of reactors have been designed, most with the aim of maximising the main product of pyrolysis, the bio-oil. There are many pyrolytic reactors used lately, the main ones being those of fluidised bed (bubbling and circulating). Besides these, we also find the fixed bed, jet bed, rotary cylinder, cyclonic reactor, rotary cone, and others. The reactors can be classified into two general systems, either a batch system or a continuous system (continuous flow of biomass occurs, and continuous collection of the products generated). Table 5 shows the comparison of different pyrolysis reactors. The summary of previous research using different reactors and outcomes is listed in Table 6.

Table 5. Comparison of various biomass pyrolysis reactors based on overall performance and efficiency [96].

Pyrolyser	Status (units)	Bio-Oil Yield (wt %)	Operational Complexity	Particle Size	Biomass Variability	Scale-Up	The Inert Gas Flow Rate
Fixed bed	Pilot (single), lab (multiple)	75	Medium	Large	High	Hard	Low
Fluidised bed	Demo (multiple), lab (multiple)	75	Medium	Small	Low	Easy	High
Recirculating bed	Pilot (multiple), lab (multiple)	75	High	Medium	Low	Hard	High
Rotating cone	Demo (single)	70	Medium	Medium	High	Medium	Low
Ablative	Pilot (single), lab (multiple)	75	High	Large	High	Hard	Low
Screw/auger reactor	Pilot (multiple), lab (multiple)	70	Low	Medium	High	Easy	Low
Vacuum	Pilot (single), lab (few)	60	High	Large	Medium	Hard	Low

7.1. Fixed Bed Reactor

The fixed bed pyrolysis system is simple, reliable, and proven for fuels that are relatively uniform in size and have a low content of coal fines which consist of a reactor with a gas cooling and cleaning system, and it was customarily used to produce charcoal [101,102]. The fixed bed reactors generally function with high carbon preservation, low gas velocity, and low residue conveyed over a long solid residence time. A major problem of fixed bed reactors is the formation of tar, although the recent evolution in thermal and catalytic conversion of tar has given feasible opportunities for confiscating tar [103,104]. Figure 6 shows the fixed bed reactor, which is considered simple, and includes the following basic units: drying, granulation, heating, and cooling. In the fixed bed pyrolysis process, the "temperature" ensures that the variables, such as temperature program, heating rates, and residence time in the temperatures, remain within the limits established by the operator and final pyrolysis temperatures between 450 and 750 °C, with heating rates fluctuating between 5 and 100 °C min/min [105].

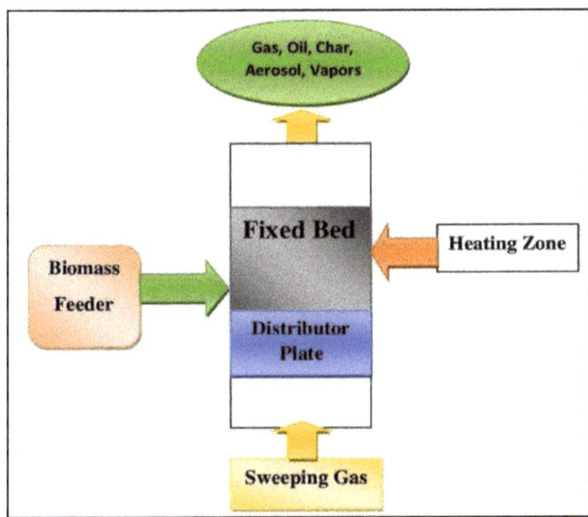

Figure 6. Fixed bed reactor.

7.2. Fluidised Bed Reactor

The fluidised bed reactors (bubbling and circulating) have a well-known technology, and they have a series of industrial applications, where they present themselves as advantageous on a commercial scale, unlike other technologies that are still in the process of improvement [106]. There are several reactors that employ the principle of the fluidised bed, among them, the vortex reactor and the abrasive reactor [105]. Fluidised bed reactors are used in many projects to maximise the liquid product (bio-oil) produced, and several projects demonstrate their real ability to produce good quality bio-oil. As biomass has a very low density, it is common in fluidised bed reactors to use an inert element, usually sand, to give fluid dynamic stability to the process and help biomass heating [107].

The fluidised bed reactor comprises a fluid–solid blend that shows similar properties to the fluid [108]. Fluidised bed reactors seem to be widespread and popular because they offer rapid reaction and heat transfer, a wide and high shallow area of contact between the fluid and solid, and high comparative velocity [108,109]. Different types of fluidised bed reactors are available include bubbling fluidised bed reactors and circulating fluidised bed reactors.

7.2.1. Bubbling Fluidised Beds

Bubbling fluidised bed gasifier is categorised as having high reaction rates, well-understood technology, simple construction and operation, virtuous temperature control, efficient heat transfer to biomass particles, and it has superior lenience to particle size range [110,111]. It is very prevalent, since it generates high quality bio-oil from a dry source. A significant feature of bubble fluidising bed reactors is they require small biomass particle sizes to attain high biomass heating rates [112].

7.2.2. Circulating Fluidised Bed (CFB) Reactors

CFB reactors are comparable with bubbling fluidised bed reactors, and this type of reactor is suitable for large quantities [113]. There are two types of CFB reactors: single circulating and double circulating. The CFB gasifier is considered by all features of the bubbling fluidised bed reactors, along with a higher charge at a lower volume. The CFB pyrolyser is notable for a decent temperature regulator in the reactor [114,115]. Figure 7 shows the circulating fluidised bed reactor.

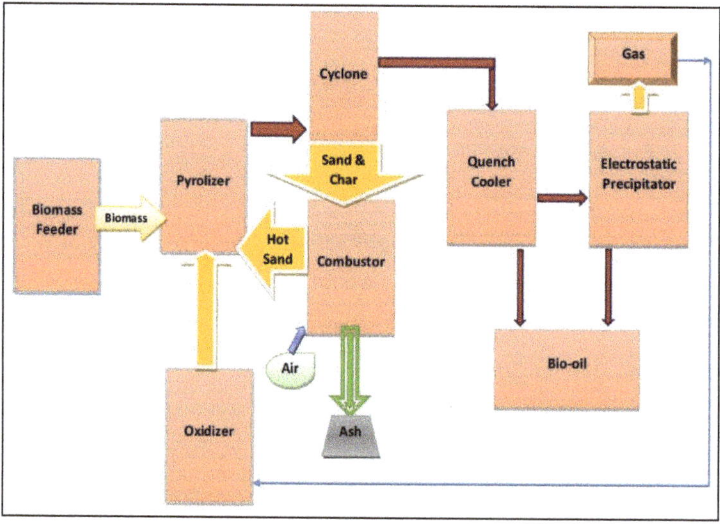

Figure 7. Circulating fluid bed reactor.

7.3. Ablative Reactor

Ablative pyrolysis is primarily dissimilar from fluid bed procedures in the absence of a fluidising gas. Material connected to the wall fundamentally melts, and the residual oil evaporates as pyrolysis vapours. The ablative pyrolysis reactors have good heat transfer with high heating rates and a relatively small contact surface. They also have high energy and cost efficiency, as no heating and cooling of fluidising gases is required, furthermore, they tolerate fixing of condensation units with a small volume in requiring less space at lower costs [116–118].

7.4. Vacuum Pyrolysis Reactor

Vacuum reactors represent a sluggish pyrolysis process with lower heat transfer rates conveyed with the fluidised bed technologies. An induction and burner heater is used with molten salts [80]. For this reactor, the vapours formed are quickly detached from the vacuum. This reactor is categorised by lengthier residence time; it is known to produce larger particles than most fast pyrolysis reactors. There is also no requirement for carrier gas, and the process is mechanically complicated; it needs high investment costs. Consistent operation of vacuum pyrolyser entails a superior feedstock input apparatus which discourages latent investors [116–118].

7.5. Rotating Cone Reactor

The rotating cone reactor is an innovative reactor for flash pyrolysis with tiny char formation. Biomass ingredients, like rice husks, wood, palm kernel, coffee husk, and so on, can be milled in the rotating cone reactor. There is no big scale of commercial implementation for a rotating cone reactor. Nonetheless, high-speed rotation provokes dynamic mixing of biomass that sequentially proceeds to fast heat transfer [114,115,119]. Figure 8 shows the rotating cone reactor.

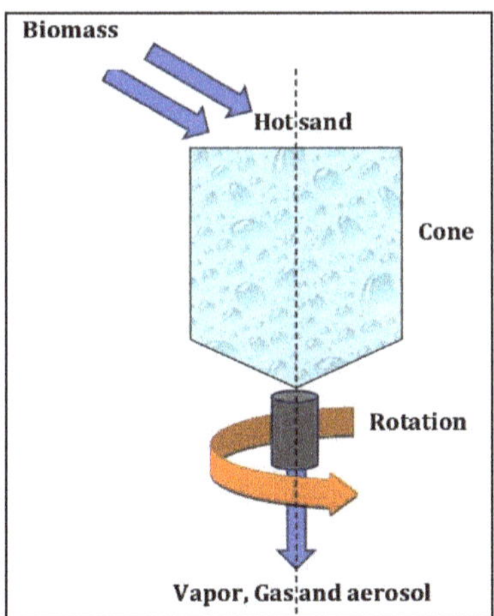

Figure 8. Rotating cone reactor.

Table 6. Summary of previous researches on biomass conversion.

Feedstock	Reactor Type	Temperature (°C)	Yields (wt %)			References
			Char	Bio-Oil	Gas	
Corn stover	Fluidised bed	450–600	28–46	35–50	11–14	[120]
Rice husk	Fluidised bed	450	29	56	15	[121]
Corn cob	Fluidised bed	500	20	62	17	[121]
Sugarcane bagasse	Fluidised bed	500	23	73	4	[122]
Switchgrass	Fluidised bed	480	13	61	11	[123]
Miscanthus	Fluidised bed	505	29	51	12	[124]
Wheat straw	Fluidised bed	550	24	54	24	[125]
Sunflower hulls	Fluidised bed	500	23	57	20	[125]
Rice husk	Fixed bed	100–500	42–48	28–35	-	[126]
Sugarcane bagasse	Vacuum	530	26	51	22	[127]
Rice straw	Vacuum	500	35	47	18	[128]
Douglas fir	Fixed bed	500	22	66	8	[129]
Pine	Vacuum	500	20	50	30	[130]
Wood	Ablative	650	6	60	34	[131]
Barley straw	Ablative	549	32	50	12	[128]
Rice straw	Auger	500	45	26	13	[120]
Hardwood	Auger	500	15	66	18	[132]
Eucalyptus	Conical spouted	500	18	75	6	[133]
Rice husk	Conical spouted	450	26	70	4	[134]
Pine chips	Fixed bed	500	31	15	18	[135]
Softwood	Auger	500	15	69	16	[132]
Olive stone	Rotary kilns	500	26	38	55	[136]

7.6. Auger Reactor

Auger reactors are used to interchange biomass feedstock over an oxygen-free cylindrical tube. In this reactor, vapour residence time could be altered by fluctuating the heated zone. Auger reactors are getting more consideration from many mid-size industries. Challenges for the auger reactor include stirring parts in the hot precinct and temperature transmission on a large scale [114,115]. Figure 9 shows the Auger pyrolysis reactor.

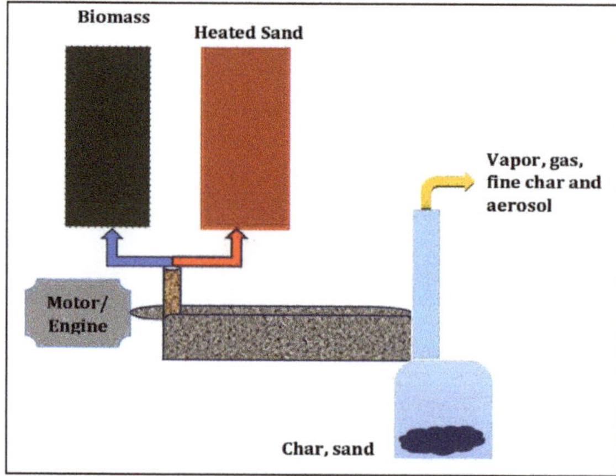

Figure 9. Auger pyrolysis reactor [96].

8. Current Status of Pyrolysis Technology

The deteriorating reserves of fossil fuels have posed a great threat and challenge to the quality of life, the world economy, and the environment [137–139]. Biomass pyrolysis possibly will help reduce CO_2 and the world's dependence on oil production [137,140,141]. These bio-oils have the potential to lower CO_2 discharges; they are derived from plants which use CO_2 for growing. An amalgamation of technologies is required to assimilate reactor design and operational procedure to recover the efficiency of biomass [142,143]. Fast pyrolysis systems process small elements to maximise bio-oil yield, whereas low pyrolysis technologies use wood to produce char chunks [138,139,141]. The recognition of the environmental matters are allied with the use of carbonisation technologies and the technical difficulties of operating fast pyrolysis reactors. Intermediate pyrolysis reactors propose prospects for the extensive balanced production of bio-oil and char [144,145]. Presently, the foremost interests in pyrolysis technology are for CO_2 mitigation, electricity generation from biomass, and energy independence. The pyrolysis technologies can be considered as slow, intermediate, fast, and flash pyrolysis [146–149] but, then again, the most frequently used systems, meanwhile, are the fast and slow pyrolysis processes. Biochar is the key product of the slow pyrolysis, and transpires with moderate temperature, longer residence time, and small heating system rate [150]. Dissimilarly, bio-oil is the key product of fast pyrolysis which formed with a fast heating rate within short residence time [151,152].

Fast pyrolysis produces a higher quality and quantity of bio-oil than the slow pyrolysis [76,153]. It is expected that environmental and economic performance will increase the effectiveness of the pyrolysis process. Various actions are needed to overcome the technical challenges, including plummeting parasitic energy losses, improving pyrolysis reactor outlines, improving feedstock logistics, and enhancing biomass heating rate [9]. Biomass feedstocks are most important to increase the pyrolysis products on a large scale [154,155]. This can be attained by producing energy-condensed products from biomass. Accumulation of metal and ash in reactor bed materials impedes pyrolysis which can reduce bio-oil yields [156,157]. Controlling pyrolysis temperature and heating rate, and using smaller particle sizes can reduce accumulation [158–160]. Recently, a study revealed that an ablative reactor can convert entire wood chips and produce more energy [161,162]. To conclude, cohesive pyrolysis systems that associate gasification or fast pyrolysis are one more important approach for making pyrolysis commercially viable and improving environmental performance [163–167]. Table 7 shows the available pyrolysis plants worldwide.

Table 7. Current pyrolysis plants worldwide [9,168–174].

Reactor Technology	Organisation/Location	Capacity (kg/h)	Desired Product
Fixed bed	Bio-alternative, USA	2000	Char
Bubbling fluidised bed	THEE	500	Gas
	Dyna Motive, Canada	400	Oil
	BEST Energy, Australia	300	Oil
	Wellman, UK	250	Oil
	Union Fenosa, Spain	200	Oil
	Zhejiang University, China	20	Oil
	RTI, Canada	20	Oil
	Waterloo University	3	Oil
	Zhejiang University, China	3	Oil
Circulating fluidised bed	Red Arrow, WI; Ensyn	1700	Chemicals
	Red Arrow, WI; Ensyn	1500	Chemicals
	Ensyn Engineering	30	Oil
	VTT, Finland, Ensyn	20	Oil
Rotating cone	BTG, Netherlands	200	Oil
	University Twente	10	Oil
Vacuum	Pyrovac, Canada	350	Oil
	Laval University	30	Oil
Ablative	PYTEC, Germany	250	Oil
	BBC, Canada	10–15	Char
	PYTEC, Germany	15	Oil
Vortex	Solar energy research Ins.	30	Oil
Another type	Fortum, Finland	350	Oil
	University Zaragoza	100	Gas
	Georgia Tech. Research Ins.	50	Oil

9. Future Challenges

To gain the full potential of biomass pyrolysis technology, such as to enable improved understanding and successful commercialisation, additional research and development are needed. In addition, a couple of issues must be overcome, including the lack of markets for pyrolysis oils and lack of biochar-derived products with well-defined performance characteristics. Also, it is recommended to speed up the improvement and deployment of bio-oil refineries. The improvement of flexible designs for pyrolysis units for producing higher yields of the product is a technical challenge. This review clearly indicates that different pyrolysis technologies have different ranges of product yields. Thus, the selection of pyrolysis technologies, feedstocks, and their operating parameters should be based on the economic trade-offs. However, in addition to the fundamental challenges, a few more important challenges for future biomass pyrolysis research are listed below:

- Understanding the proper working of pyrolysis reactors and processes
- Development of a new reactor that is cost-effective and highly efficient
- Development of catalysts for bio-oil upgrading
- Development of proper solar system reactors
- Post-pyrolysis processing to improve product bio-oil properties
- Understanding the limitations and potential for improvements of the quality of products obtained by biomass pyrolysis
- Development of both fast pyrolysis and bio-oil upgrading, ensuring these are focused on delivering useful and valuable products

For the full implementation of pyrolysis technology, more research is needed to determine designs that will remove oxygen in the gas phase from pyrolysis oil. Pyrolysis technology has the potential to be applied in a vast diversity of situations and, through this process, diversity of products can be obtained. Hence, it is quite difficult to explore a sustainable design for all prospective applications. In addition, balanced financial investments to create new knowledge, technology, and markets for the purpose of building a united vision for the utilisation of pyrolysis technologies is crucial.

10. Conclusions

Biomass is a renewable source for the production of energy which is profusely obtainable globally. The sustainable use of biomass energy can be a supplement for fossil fuels and nuclear energy. Biomass consists of elements, such as carbon, hydrogen, oxygen, and nitrogen. Sulphur is present in smaller proportions, and some types of biomass also contain significant portions of inorganic species. There are several types of pyrolysis processes, namely slow, rapid, ultrafast, and flash pyrolysis which can be used to convert the biomass, which depends upon the process used and also depends on the temperature of the pyrolysis. The main products of the pyrolysis of biomass are bio-oil, biochar, and syngas. The physical and chemical properties of these pyrolysis products depend on the quality of biomass. The bio-oil product has a number of applications; it can be improved to be used as a transport fuel or used as a chemical. Reactor designs are the primary target of researchers to achieve a better-quality bio-oil. For example, fluidised bed reactors are used in many projects to maximise the liquid product (bio-oil) produced, and several projects demonstrate their real ability to produce good quality bio-oil. Auger reactors have the potential to be used in small-scale production. Bio-char is considered as a promising addition to the soil. It can be concluded that the development of biomass pyrolysis technology offers more sustainable products compared to the other available technologies. Finally, in order to gain the full potential of biomass pyrolysis technology and address future challenges, additional research and development are needed.

Author Contributions: Original draft preparation, M.N.U.; Supervision, K.T. and J.T.; Review & Editing, M.M. and T.M.I.M.; Revision, M.G.R. and S.M.A.

Funding: This research received no external funding.

Conflicts of Interest: The authors declare no conflict of interest.

References

1. Mofijur, M.; Masjuki, H.; Kalam, M.; Atabani, A.; Shahabuddin, M.; Palash, S.; Hazrat, M. Effect of biodiesel from various feedstocks on combustion characteristics, engine durability and materials compatibility: A review. *Renew. Sustain. Energy Rev.* **2013**, *28*, 441–455. [CrossRef]
2. Kusumo, F.; Silitonga, A.S.; Ong, H.C.; Masjuki, H.H.; Mahlia, T.M.I. A comparative study of ultrasound and infrared transesterification of sterculia foetida oil for biodiesel production. *Energy Sources Part A Recovery Util. Environ. Effects* **2017**, *39*, 1339–1346. [CrossRef]
3. Islam, M. Challenges of Adopting Strategic Procurement Policies: A Case Study of Infrastructure Development Company Limited. Ph.D. Thesis, BRAC University, Dhaka, Bangladesh, 2017.
4. Bass, S.; Dalal-Clayton, B. *Sustainable Development Strategies: A Resource Book*; Routledge: Abingdon-on-Thames, UK, 2012.
5. Habibullah, M.; Masjuki, H.; Kalam, M.; Rahman, S.A.; Mofijur, M.; Mobarak, H.; Ashraful, A. Potential of biodiesel as a renewable energy source in Bangladesh. *Renew. Sustain. Energy Rev.* **2015**, *50*, 819–834. [CrossRef]
6. Rahman, M.; Rasul, M.; Hassan, N. Study on the tribological characteristics of Australian native first generation and second generation biodiesel fuel. *Energies* **2017**, *10*, 55. [CrossRef]
7. Rahman, M.; Rasul, M.; Hassan, N.; Hyde, J. Prospects of biodiesel production from macadamia oil as an alternative fuel for diesel engines. *Energies* **2016**, *9*, 403. [CrossRef]
8. Damanik, N.; Ong, H.C.; Tong, C.W.; Mahlia, T.M.I.; Silitonga, A.S. A review on the engine performance and exhaust emission characteristics of diesel engines fueled with biodiesel blends. *Environ. Sci. Pollut. Res.* **2018**, *25*, 15307–15325. [CrossRef] [PubMed]
9. Jahirul, M.I.; Rasul, M.G.; Chowdhury, A.A.; Ashwath, N. Biofuels production through biomass pyrolysis—A technological review. *Energies* **2012**, *5*, 4952–5001. [CrossRef]
10. Dharma, S.; Masjuki, H.H.; Ong, H.C.; Sebayang, A.H.; Silitonga, A.S.; Kusumo, F.; Mahlia, T.M.I. Optimization of biodiesel production process for mixed jatropha curcas-ceiba pentandra biodiesel using response surface methodology. *Energy Convers. Manag.* **2016**, *115*, 178–190. [CrossRef]
11. Kusumo, F.; Silitonga, A.S.; Masjuki, H.H.; Ong, H.C.; Siswantoro, J.; Mahlia, T.M.I. Optimization of transesterification process for ceiba pentandra oil: A comparative study between kernel-based extreme learning machine and artificial neural networks. *Energy* **2017**, *134*, 24–34. [CrossRef]
12. Gordon, I.E.; Rothman, L.S.; Hill, C.; Kochanov, R.V.; Tan, Y.; Bernath, P.F.; Birk, M.; Boudon, V.; Campargue, A.; Chance, K. The hitran 2016 molecular spectroscopic database. *J. Quant. Spectrosc. Radiat. Transf.* **2017**, *203*, 3–69. [CrossRef]
13. Basu, P. *Biomass Gasification and Pyrolysis: Practical Design and Theory*; Academic Press: Cambridge, MA, USA, 2010.
14. Basu, P. *Combustion and Gasification in Fluidized Beds*; CRC Press: Boca Raton, FL, USA, 2006.
15. Forero Núñez, C.A.; Castellanos Contreras, J.U.; Sierra Vargas, F.E. Control de una planta prototipo de gasificación de biomasa mediante redes neuronales. *Ingeniería Mecánica, Tecnología y Desarroll* **2013**, *4*, 161–168.
16. Antonio, P.; Alejandra, J.; Martinez Guerrero, P.A.; Cortés Rodriguez, M.F.; Chiviri Torres, N.; Mendoza Geney, L. Uso energético de la biomasa a través del proceso de gasificación. *Rev. Investig.* **2017**, *10*, 165–181.
17. Rincón, J.G.G.; Toscano, J.A.; Gómez, G.G. Análisis exergético de un horno de lecho fijo en la producción de panela. *Revista Colombiana De Tecnologias De Avanzada (Rcta)* **2017**, *1*. [CrossRef]
18. Szyszlak-Bargłowicz, J.; Zając, G.; Piekarski, W. Energy biomass characteristics of chosen plants. *Int. Agrophys.* **2012**, *26*, 175–179. [CrossRef]
19. Castro, A.M.; Carvalho, D.F.; Freire, D.M.G.; Castilho, L.D.R. Economic analysis of the production of amylases and other hydrolases by aspergillus awamori in solid-state fermentation of babassu cake. *Enzyme Res.* **2010**, *2010*, 576872. [CrossRef] [PubMed]
20. de Castro, A.M.; de Andréa, T.V.; dos Reis Castilho, L.; Freire, D.M.G. Use of mesophilic fungal amylases produced by solid-state fermentation in the cold hydrolysis of raw babassu cake starch. *Appl. Biochem. Biotechnol.* **2010**, *162*, 1612–1625. [CrossRef] [PubMed]

21. Saletnik, B.; Zagula, G.; Bajcar, M.; Czernicka, M.; Puchalski, C. Biochar and biomass ash as a soil ameliorant: The effect on selected soil properties and yield of giant miscanthus (miscanthus × giganteus). *Energies* **2018**, *11*, 2535. [CrossRef]

22. Chaloupková, V.; Ivanova, T.; Ekrt, O.; Kabutey, A.; Herák, D. Determination of particle size and distribution through image-based macroscopic analysis of the structure of biomass briquettes. *Energies* **2018**, *11*, 331. [CrossRef]

23. Castello, D.; Rolli, B.; Kruse, A.; Fiori, L. Supercritical water gasification of biomass in a ceramic reactor: Long-time batch experiments. *Energies* **2017**, *10*, 1734. [CrossRef]

24. Valentim, B.; Guedes, A.; Rodrigues, S.; Flores, D. Case study of igneous intrusion effects on coal nitrogen functionalities. *Int. J. Coal Geol.* **2011**, *86*, 291–294. [CrossRef]

25. Miandad, R.; Barakat, M.; Aburiazaiza, A.S.; Rehan, M.; Ismail, I.; Nizami, A. Effect of plastic waste types on pyrolysis liquid oil. *Int. Biodeterior. Biodegrad.* **2017**, *119*, 239–252. [CrossRef]

26. Ramage, J.; Scurlock, J. "Biomass". In *Renewable Energy-Power for a Sustainable Future*; Boyle, G., Ed.; Oxford University Press: Oxford, UK, 1996.

27. Boyle, G. *Renewable Energy: Power for a Sustainable Future*; Oxford University Press: Oxford, UK, 1996; Volume 2.

28. Muradov, N.Z.; Veziroğlu, T.N. "Green" path from fossil-based to hydrogen economy: An overview of carbon-neutral technologies. *Int. J. Hydrogen Energy* **2008**, *33*, 6804–6839. [CrossRef]

29. Pei-dong, Z.; Guomei, J.; Gang, W. Contribution to emission reduction of CO_2 and SO_2 by household biogas construction in rural china. *Renew. Sustain. Energy Rev.* **2007**, *11*, 1903–1912. [CrossRef]

30. López Gómez, F.A.; Rodríguez, O.; Urien, A.; Lobato Ortega, B.; Álvarez Centeno, T.; Alguacil, F.J. Physico-chemical characteristics of the products derived from the thermolysis of waste abies alba mill. Wood. *J. Environ. Prot.* **2013**, *4*, 26–30. [CrossRef]

31. Chen, T.; Zhang, Y.; Wang, H.; Lu, W.; Zhou, Z.; Zhang, Y.; Ren, L. Influence of pyrolysis temperature on characteristics and heavy metal adsorptive performance of biochar derived from municipal sewage sludge. *Bioresour. Technol.* **2014**, *164*, 47–54. [CrossRef] [PubMed]

32. Oyarzún, B.; Bardow, A.; Gross, J. Integration of process and solvent design towards a novel generation of co2 absorption capture systems. *Energy Procedia* **2011**, *4*, 282–290. [CrossRef]

33. Pedroza, M.; Sousa, J.; Vieira, G.; Bezerra, M. Characterization of the products from the pyrolysis of sewage sludge in 1 kg/h rotating cylinder reactor. *J. Anal. Appl. Pyrolysis* **2014**, *105*, 108–115. [CrossRef]

34. Mesa-Perez, J.; Cortez, L.; Rocha, J.; Brossard-Perez, L.; Olivares-Gómez, E. Unidimensional heat transfer analysis of elephant grass and sugar cane bagasse slow pyrolysis in a fixed bed reactor. *Fuel Process. Technol.* **2005**, *86*, 565–575. [CrossRef]

35. Mesa Perez, J.M. Testes em Uma Planta de Pirólise Rápida de Biomassa em Leito Fluidizado: Critérios Para Sua Otimização. Ph.D. Thesis, University of Campinas, Campinas, Brazil, 2004.

36. Sousa, J.F.; Bezerra, M.B.; Almeida, M.B.; Moure, G.T.; Mesa-Perez, J.M.; Caramao, E.B. Characteristics of bio-oil from the fast pyrolysis of elephant grass (pennisetum purpureum schumach) in a fluidized bed reactor. *Am. Chem. Sci. J.* **2016**, *14*, 1–10. [CrossRef]

37. Meier, D.; Schoell, S.; Klaubert, H. New ablative pyrolyser in operation in Germany. *PyNe Newslett.* **2004**. Available online: https://www.researchgate.net/publication/237289951_New_Ablative_Pyrolyser_in_Operation (accessed on 9 November 2018).

38. Mohan, D.; Pittman, C.U.; Steele, P.H. Pyrolysis of wood/biomass for bio-oil: A critical review. *Energy Fuels* **2006**, *20*, 848–889. [CrossRef]

39. Demirbaş, A. Biomass resource facilities and biomass conversion processing for fuels and chemicals. *Energy Convers. Manag.* **2001**, *42*, 1357–1378. [CrossRef]

40. Demirbaş, A. Partly chemical analysis of liquid fraction of flash pyrolysis products from biomass in the presence of sodium carbonate. *Energy Convers. Manag.* **2002**, *43*, 1801–1809. [CrossRef]

41. Venderbosch, R.; Prins, W. Fast pyrolysis technology development. *Biofuels Bioprod. Biorefin.* **2010**, *4*, 178–208. [CrossRef]

42. Van de Velden, M.; Baeyens, J.; Brems, A.; Janssens, B.; Dewil, R. Fundamentals, kinetics and endothermicity of the biomass pyrolysis reaction. *Renew. Energy* **2010**, *35*, 232–242. [CrossRef]

43. Hosoya, T.; Kawamoto, H.; Saka, S. Pyrolysis behaviors of wood and its constituent polymers at gasification temperature. *J. Anal. Appl. Pyrolysis* **2007**, *78*, 328–336. [CrossRef]

44. McGrath, T.E.; Chan, W.G.; Hajaligol, M.R. Low temperature mechanism for the formation of polycyclic aromatic hydrocarbons from the pyrolysis of cellulose. *J. Anal. Appl. Pyrolysis* **2003**, *66*, 51–70. [CrossRef]

45. Scheirs, J.; Camino, G.; Tumiatti, W. Overview of water evolution during the thermal degradation of cellulose. *Eur. Polym. J.* **2001**, *37*, 933–942. [CrossRef]

46. Banyasz, J.; Li, S.; Lyons-Hart, J.; Shafer, K. Gas evolution and the mechanism of cellulose pyrolysis. *Fuel* **2001**, *80*, 1757–1763. [CrossRef]

47. Azeez, A.M.; Meier, D.; Odermatt, J. Temperature dependence of fast pyrolysis volatile products from european and african biomasses. *J. Anal. Appl. Pyrolysis* **2011**, *90*, 81–92. [CrossRef]

48. Wei, L.; Xu, S.; Zhang, L.; Zhang, H.; Liu, C.; Zhu, H.; Liu, S. Characteristics of fast pyrolysis of biomass in a free fall reactor. *Fuel Process. Technol.* **2006**, *87*, 863–871. [CrossRef]

49. Evans, R.J.; Milne, T.A. Molecular characterization of the pyrolysis of biomass. *Energy Fuels* **1987**, *1*, 123–137. [CrossRef]

50. Morf, P.; Hasler, P.; Nussbaumer, T. Mechanisms and kinetics of homogeneous secondary reactions of tar from continuous pyrolysis of wood chips. *Fuel* **2002**, *81*, 843–853. [CrossRef]

51. Neves, D.; Thunman, H.; Matos, A.; Tarelho, L.; Gómez-Barea, A. Characterization and prediction of biomass pyrolysis products. *Prog. Energy Combust. Sci.* **2011**, *37*, 611–630. [CrossRef]

52. Tilman, D.; Socolow, R.; Foley, J.A.; Hill, J.; Larson, E.; Lynd, L.; Pacala, S.; Reilly, J.; Searchinger, T.; Somerville, C. Beneficial biofuels—The food, energy, and environment trilemma. *Science* **2009**, *325*, 270–271. [CrossRef] [PubMed]

53. Demirbas, A. Pyrolysis of ground beech wood in irregular heating rate conditions. *J. Anal. Appl. Pyrolysis* **2005**, *73*, 39–43. [CrossRef]

54. Demiral, I.; Şensöz, S. The effects of different catalysts on the pyrolysis of industrial wastes (olive and hazelnut bagasse). *Bioresour. Technol.* **2008**, *99*, 8002–8007. [CrossRef] [PubMed]

55. Mohan, D.; Pittman, C.U., Jr.; Bricka, M.; Smith, F.; Yancey, B.; Mohammad, J.; Steele, P.H.; Alexandre-Franco, M.F.; Gómez-Serrano, V.; Gong, H. Sorption of arsenic, cadmium, and lead by chars produced from fast pyrolysis of wood and bark during bio-oil production. *J. Colloid Interface Sci.* **2007**, *310*, 57–73. [CrossRef] [PubMed]

56. Aho, A.; Kumar, N.; Eränen, K.; Salmi, T.; Hupa, M.; Murzin, D.Y. Catalytic pyrolysis of woody biomass in a fluidized bed reactor: Influence of the zeolite structure. *Fuel* **2008**, *87*, 2493–2501. [CrossRef]

57. Karaosmanoğlu, F.; Tetik, E. Fuel properties of pyrolytic oil of the straw and stalk of rape plant. *Renew. Energy* **1999**, *16*, 1090–1093. [CrossRef]

58. Jensen, P.A.; Sander, B.; Dam-Johansen, K. Pretreatment of straw for power production by pyrolysis and char wash. *Biomass Bioenergy* **2001**, *20*, 431–446. [CrossRef]

59. Pütün, E.; Uzun, B.B.; Pütün, A.E. Fixed-bed catalytic pyrolysis of cotton-seed cake: Effects of pyrolysis temperature, natural zeolite content and sweeping gas flow rate. *Bioresour. Technol.* **2006**, *97*, 701–710. [CrossRef] [PubMed]

60. Popp, J.; Lakner, Z.; Harangi-Rákos, M.; Fári, M. The effect of bioenergy expansion: Food, energy, and environment. *Renew. Sustain. Energy Rev.* **2014**, *32*, 559–578. [CrossRef]

61. Zafar, S. Importance of biomass energy. Available online: https://www.bioenergyconsult.com/tag/importance-of-biomass-energy/ (accessed on 9 November 2018).

62. Diniz, V.; Volesky, B. Biosorption of la, eu and yb using sargassum biomass. *Water Res.* **2005**, *39*, 239–247. [CrossRef] [PubMed]

63. Santos, J.; Nunes, L.; Melo, W.; Araújo, A. Tannery sludge compost amendment rates on soil microbial biomass of two different soils. *Eur. J. Soil Boil.* **2011**, *47*, 146–151. [CrossRef]

64. Karlen, D.L. *Cellulosic Energy Cropping Systems*; John Wiley & Sons: Hoboken, NJ, USA, 2014.

65. Laird, D.A.; Brown, R.C.; Amonette, J.E.; Lehmann, J. Review of the pyrolysis platform for coproducing bio-oil and biochar. *Biofuels Bioprod. Biorefin.* **2009**, *3*, 547–562. [CrossRef]

66. Luengo, C.; Felfli, F.; Bezzon, G. *Pirólise e Torrefação de Biomassa. Biomassa Para Energia. Cortez, LAB, Lora, EES, Gómez EO (Organizadores)*; Editora da Unicamp: Campinas, Brazil, 2008; pp. 333–351.

67. Goyal, H.; Seal, D.; Saxena, R. Bio-fuels from thermochemical conversion of renewable resources: A review. *Renew. Sustain. Energy Rev.* **2008**, *12*, 504–517. [CrossRef]

68. Gomes, M.D.S. *Produção de Bio-Óleo Através do Processo Termoquímico de Pirólise*; Fatec: Araçatuba, Brazil, 2010.

69. Huber, G.W.; Iborra, S.; Corma, A. Synthesis of transportation fuels from biomass: Chemistry, catalysts, and engineering. *Chem. Rev.* **2006**, *106*, 4044–4098. [CrossRef] [PubMed]

70. Strahan, G.D.; Mullen, C.A.; Boateng, A.A. Characterizing biomass fast pyrolysis oils by ^{13}C NMR and chemometric analysis. *Energy Fuels* **2011**, *25*, 5452–5461. [CrossRef]

71. Bridgwater, A.V. Review of fast pyrolysis of biomass and product upgrading. *Biomass Bioenergy* **2012**, *38*, 68–94. [CrossRef]

72. MOTA, A.M.A.; Lima, A.A.S.; Santos, F.F.P.; Caselli, F.D.T.R.; Viegas, R.A. Pirólise da biomassa lignocelulósica: Uma revisão. *Rev. GEINTEC-Gestão Inovação E Tecnol.* **2015**, *5*, 2511–2525. [CrossRef]

73. da Silva Mota, F.D.A.; Viegas, R.A.; da Silva Lima, A.A.; dos Santos, F.F.P.; Caselli, F.D.T.R. Pirólise da biomassa lignocelulósica: Uma revisão pyrolysis of lignocellulose biomass: A review. *Rev. GEINTEC* **2015**, *5*, 2511–2525. [CrossRef]

74. Bridgwater, T. Biomass for energy. *J. Sci. Food Agric.* **2006**, *86*, 1755–1768. [CrossRef]

75. Bridgwater, A. Principles and practice of biomass fast pyrolysis processes for liquids. *J. Anal. Appl. Pyrolysis* **1999**, *51*, 3–22. [CrossRef]

76. Czernik, S.; Bridgwater, A. Overview of applications of biomass fast pyrolysis oil. *Energy Fuels* **2004**, *18*, 590–598. [CrossRef]

77. Demirbas, A. The influence of temperature on the yields of compounds existing in bio-oils obtained from biomass samples via pyrolysis. *Fuel Process. Technol.* **2007**, *88*, 591–597. [CrossRef]

78. Demirbas, A. Combustion characteristics of different biomass fuels. *Prog. Energy Combust. Sci.* **2004**, *30*, 219–230. [CrossRef]

79. Lehmann, J.; Joseph, S. *Biochar for Environmental Management: Science, Technology and Implementation*; Routledge: Abingdon-on-Thames, UK, 2015.

80. Mašek, O.; Brownsort, P.; Cross, A.; Sohi, S. Influence of production conditions on the yield and environmental stability of biochar. *Fuel* **2013**, *103*, 151–155. [CrossRef]

81. Huang, G.; Chen, F.; Wei, D.; Zhang, X.; Chen, G. Biodiesel production by microalgal biotechnology. *Appl. Energy* **2010**, *87*, 38–46. [CrossRef]

82. Kanaujia, P.K.; Sharma, Y.; Agrawal, U.; Garg, M. Analytical approaches to characterizing pyrolysis oil from biomass. *TrAC Trends Anal. Chem.* **2013**, *42*, 125–136. [CrossRef]

83. Das, L.M.; Gulati, R.; Gupta, P.K. A comparative evaluation of the performance characteristics of a spark ignition engine using hydrogen and compressed natural gas as alternative fuels. *Int. J. Hydrog. Energy* **2000**, *25*, 783–793. [CrossRef]

84. Saidur, R.; Jahirul, M.I.; Moutushi, T.Z.; Imtiaz, H.; Masjuki, H.H. Effect of partial substitution of diesel fuel by natural gas on performance parameters of a four-cylinder diesel engine. *Proc. Inst. Mech. Eng. Part A J. Power Energy* **2007**, *221*, 1–10. [CrossRef]

85. He, M.; Xiao, B.; Liu, S.; Hu, Z.; Guo, X.; Luo, S.; Yang, F. Syngas production from pyrolysis of municipal solid waste (MSW) with dolomite as downstream catalysts. *J. Anal. Appl. Pyrolysis* **2010**, *87*, 181–187. [CrossRef]

86. Kantarelis, E.; Zabaniotou, A. Valorization of cotton stalks by fast pyrolysis and fixed bed air gasification for syngas production as precursor of second generation biofuels and sustainable agriculture. *Bioresour. Technol.* **2009**, *100*, 942–947. [CrossRef] [PubMed]

87. Tang, L.; Huang, H. Plasma pyrolysis of biomass for production of syngas and carbon adsorbent. *Energy Fuels* **2005**, *19*, 1174–1178. [CrossRef]

88. Ahmed, I.; Gupta, A.K. Syngas yield during pyrolysis and steam gasification of paper. *Appl. Energy* **2009**, *86*, 1813–1821. [CrossRef]

89. Fernández, Y.; Menéndez, J.A. Influence of feed characteristics on the microwave-assisted pyrolysis used to produce syngas from biomass wastes. *J. Anal. Appl. Pyrolysis* **2011**, *91*, 316–322. [CrossRef]

90. Basu, P. *Biomass Gasification, Pyrolysis and Torrefaction: Practical Design and Theory*; Academic Press: Cambridge, MA, USA, 2018.

91. Basu, P.; Kaushal, P. Modeling of pyrolysis and gasification of biomass in fluidized beds: A review. *Chem. Prod. Process Model.* **2009**, *4*. [CrossRef]

92. Basu, P.; Sadhukhan, A.K.; Gupta, P.; Rao, S.; Dhungana, A.; Acharya, B. An experimental and theoretical investigation on torrefaction of a large wet wood particle. *Bioresour. Technol.* **2014**, *159*, 215–222. [CrossRef] [PubMed]

93. Basu, P.; Acharya, B.; Dutra, A. Gasification in fluidized beds—Present status & design. In Proceedings of the 20th International Conference on Fluidized Bed Combustion, Xi'an, China, 18–21 May 2009; pp. 97–103.

94. Papadikis, K.; Gu, S.; Bridgwater, A. CFD modelling of the fast pyrolysis of biomass in fluidised bed reactors. Part b: Heat, momentum and mass transport in bubbling fluidised beds. *Chem. Eng. Sci.* **2009**, *64*, 1036–1045. [CrossRef]

95. Deo, M.D.; Fletcher, J.V.; Shun, D.; Hanson, F.V.; Oblad, A.G. Modelling the pyrolysis of tar sands in fluidized bed reactors. *Fuel* **1991**, *70*, 1271–1276. [CrossRef]

96. Zhang, S.; Yan, Y.; Li, T.; Ren, Z. Upgrading of liquid fuel from the pyrolysis of biomass. *Bioresour. Technol.* **2005**, *96*, 545–550. [CrossRef] [PubMed]

97. Bridgwater, A.; Peacocke, G. Fast pyrolysis processes for biomass. *Renew. Sustain. Energy Rev.* **2000**, *4*, 1–73. [CrossRef]

98. Lappas, A.; Samolada, M.; Iatridis, D.; Voutetakis, S.; Vasalos, I. Biomass pyrolysis in a circulating fluid bed reactor for the production of fuels and chemicals. *Fuel* **2002**, *81*, 2087–2095. [CrossRef]

99. Wang, X.; Kersten, S.R.; Prins, W.; van Swaaij, W.P. Biomass pyrolysis in a fluidized bed reactor. Part 2: Experimental validation of model results. *Ind. Eng. Chem. Res.* **2005**, *44*, 8786–8795. [CrossRef]

100. Kersten, S.R.; Wang, X.; Prins, W.; van Swaaij, W.P. Biomass pyrolysis in a fluidized bed reactor. Part 1: Literature review and model simulations. *Ind. Eng. Chem. Res.* **2005**, *44*, 8773–8785. [CrossRef]

101. De Filippis, P.; Borgianni, C.; Paolucci, M.; Pochetti, F. Gasification process of cuban bagasse in a two-stage reactor. *Biomass Bioenergy* **2004**, *27*, 247–252. [CrossRef]

102. Leung, D.Y.; Yin, X.; Wu, C. A review on the development and commercialization of biomass gasification technologies in china. *Renew. Sustain. Energy Rev.* **2004**, *8*, 565–580. [CrossRef]

103. Wang, L.; Weller, C.L.; Jones, D.D.; Hanna, M.A. Contemporary issues in thermal gasification of biomass and its application to electricity and fuel production. *Biomass Bioenergy* **2008**, *32*, 573–581. [CrossRef]

104. Chopra, S.; Jain, A. A review of fixed bed gasification systems for biomass. 2007. Available online: https://ecommons.cornell.edu/bitstream/handle/1813/10671/Invited?sequence=1 (accessed on 9 November 2018).

105. Martini, P.R.R. Conversão Pirolítica de Bagaço Residual da Indústria de Suco de Laranja e Caracterização Química dos Produtos. Ph.D. Thesis, UFSM, Santa Maria, Brazil, 2009.

106. Berton, R.P. Análise Teórica Comparativa de Eficiência Energética de Sistemas Integrados Para Pirólise Rápida de Biomassa. Master's Thesis, University of Campinas, Campinas, Brazil, 2012.

107. Santos, K.G. Aspectos Fundamentais da Pirólise de Biomassa em Leito de Jorro: Fluidodinâmica e Cinética do Processo. Ph.D. Thesis, Federal University of Uberlândia, Uberlândia, Brazil, 2011.

108. Aldaco, R.; Irabien, A.; Luis, P. Fluidized bed reactor for fluoride removal. *Chem. Eng. J.* **2005**, *107*, 113–117. [CrossRef]

109. Rao, M.; Singh, S.; Sodha, M.; Dubey, A.; Shyam, M. Stoichiometric, mass, energy and exergy balance analysis of countercurrent fixed-bed gasification of post-consumer residues. *Biomass Bioenergy* **2004**, *27*, 155–171. [CrossRef]

110. Warnecke, R. Gasification of biomass: Comparison of fixed bed and fluidized bed gasifier. *Biomass Bioenergy* **2000**, *18*, 489–497. [CrossRef]

111. Radmanesh, R.; Chaouki, J.; Guy, C. Biomass gasification in a bubbling fluidized bed reactor: Experiments and modeling. *AIChE J.* **2006**, *52*, 4258–4272. [CrossRef]

112. Sadaka, S.; Boateng, A. *Pyrolysis and Bio-Oil, Agriculture and Natural Resources*; FSA1052; University of Arkansas: Fayetteville, AK, USA, 2008.

113. Li, X.; Grace, J.; Lim, C.; Watkinson, A.; Chen, H.; Kim, J. Biomass gasification in a circulating fluidized bed. *Biomass Bioenergy* **2004**, *26*, 171–193. [CrossRef]

114. Corella, J.; Sanz, A. Modeling circulating fluidized bed biomass gasifiers. A pseudo-rigorous model for stationary state. *Fuel Process. Technol.* **2005**, *86*, 1021–1053. [CrossRef]

115. Sanz, A.; Corella, J. Modeling circulating fluidized bed biomass gasifiers. Results from a pseudo-rigorous 1-dimensional model for stationary state. *Fuel Process. Technol.* **2006**, *87*, 247–258. [CrossRef]

116. Helleur, R.; Popovic, N.; Ikura, M.; Stanciulescu, M.; Liu, D. Characterization and potential applications of pyrolytic char from ablative pyrolysis of used tires. *J. Anal. Appl. Pyrolysis* **2001**, *58*, 813–824. [CrossRef]

117. Jones, S.B.; Valkenburt, C.; Walton, C.W.; Elliott, D.C.; Holladay, J.E.; Stevens, D.J.; Kinchin, C.; Czernik, S. *Production of Gasoline and Diesel from Biomass via Fast Pyrolysis, Hydrotreating and Hydrocracking: A Design Case*; Pacific Northwest National Lab. (PNNL): Richland, WA, USA, 2009.

118. Jones, S.B.; Snowden-Swan, L.J. *Production of Gasoline and Diesel from Biomass via Fast Pyrolysis, Hydrotreating and Hydrocracking: 2012 State of Technology and Projections to 2017*; Pacific Northwest National Lab. (PNNL): Richland, WA, USA, 2013.

119. Verma, M.; Godbout, S.; Brar, S.; Solomatnikova, O.; Lemay, S.; Larouche, J. Biofuels production from biomass by thermochemical conversion technologies. *Int. J. Chem. Eng.* **2012**, *2012*, 542426. [CrossRef]

120. Nam, H.; Capareda, S.C.; Ashwath, N.; Kongkasawan, J. Experimental investigation of pyrolysis of rice straw using bench-scale auger, batch and fluidized bed reactors. *Energy* **2015**, *93*, 2384–2394. [CrossRef]

121. Phan, B.M.Q.; Duong, L.T.; Nguyen, V.D.; Tran, T.B.; Nguyen, M.H.H.; Nguyen, L.H.; Nguyen, D.A.; Luu, L.C. Evaluation of the production potential of bio-oil from vietnamese biomass resources by fast pyrolysis. *Biomass Bioenergy* **2014**, *62*, 74–81. [CrossRef]

122. Montoya, J.I.; Valdés, C.; Chejne, F.; Gómez, C.A.; Blanco, A.; Marrugo, G.; Osorio, J.; Castillo, E.; Aristóbulo, J.; Acero, J. Bio-oil production from Colombian bagasse by fast pyrolysis in a fluidized bed: An experimental study. *J. Anal. Appl. Pyrolysis* **2015**, *112*, 379–387. [CrossRef]

123. Boateng, A.A.; Daugaard, D.E.; Goldberg, N.M.; Hicks, K.B. Bench-scale fluidized-bed pyrolysis of switchgrass for bio-oil production. *Ind. Eng. Chem. Res.* **2007**, *46*, 1891–1897. [CrossRef]

124. Greenhalf, C.E.; Nowakowski, D.J.; Harms, A.B.; Titiloye, J.O.; Bridgwater, A.V. A comparative study of straw, perennial grasses and hardwoods in terms of fast pyrolysis products. *Fuel* **2013**, *108*, 216–230. [CrossRef]

125. Scott, D.S.; Majerski, P.; Piskorz, J.; Radlein, D. A second look at fast pyrolysis of biomass—The RTI process. *J. Anal. Appl. Pyrolysis* **1999**, *51*, 23–37. [CrossRef]

126. Tsai, W.T.; Lee, M.K.; Chang, Y.M. Fast pyrolysis of rice husk: Product yields and compositions. *Bioresour. Technol.* **2007**, *98*, 22–28. [CrossRef] [PubMed]

127. Garcia-Perez, M.; Chaala, A.; Roy, C. Vacuum pyrolysis of sugarcane bagasse. *J. Anal. Appl. Pyrolysis* **2002**, *65*, 111–136. [CrossRef]

128. Garcia-Nunez, J.A.; Pelaez-Samaniego, M.R.; Garcia-Perez, M.E.; Fonts, I.; Abrego, J.; Westerhof, R.J.M.; Garcia-Perez, M. Historical developments of pyrolysis reactors: A review. *Energy Fuels* **2017**, *31*, 5751–5775. [CrossRef]

129. Le Brech, Y.; Jia, L.; Cissé, S.; Mauviel, G.; Brosse, N.; Dufour, A. Mechanisms of biomass pyrolysis studied by combining a fixed bed reactor with advanced gas analysis. *J. Anal. Appl. Pyrolysis* **2016**, *117*, 334–346. [CrossRef]

130. Xu, Y.; Wang, T.; Ma, L.; Zhang, Q.; Chen, G. Technology of bio-oil preparation by vacuum pyrolysis of pine straw. *Trans. Chin. Soc. Agric. Eng.* **2013**, *29*, 196–201.

131. Schulzke, T.; Conrad, S.; Westermeyer, J. Fractionation of flash pyrolysis condensates by staged condensation. *Biomass Bioenergy* **2016**, *95*, 287–295. [CrossRef]

132. Henrich, E.; Dahmen, N.; Weirich, F.; Reimert, R.; Kornmayer, C. Fast pyrolysis of lignocellulosics in a twin screw mixer reactor. *Fuel Process. Technol.* **2016**, *143*, 151–161. [CrossRef]

133. Amutio, M.; Lopez, G.; Alvarez, J.; Olazar, M.; Bilbao, J. Fast pyrolysis of eucalyptus waste in a conical spouted bed reactor. *Bioresour. Technol.* **2015**, *194*, 225–232. [CrossRef] [PubMed]

134. Alvarez, J.; Lopez, G.; Amutio, M.; Bilbao, J.; Olazar, M. Bio-oil production from rice husk fast pyrolysis in a conical spouted bed reactor. *Fuel* **2014**, *128*, 162–169. [CrossRef]

135. Garcìa-Pérez, M.; Chaala, A.; Pakdel, H.; Kretschmer, D.; Roy, C. Vacuum pyrolysis of softwood and hardwood biomass: Comparison between product yields and bio-oil properties. *J. Anal. Appl. Pyrolysis* **2007**, *78*, 104–116. [CrossRef]

136. Sanginés, P.; Domínguez, M.P.; Sánchez, F.; Miguel, G.S. Slow pyrolysis of olive stones in a rotary kiln: Chemical and energy characterization of solid, gas, and condensable products. *J. Renew. Sustain. Energy* **2015**, *7*, 043103. [CrossRef]

137. Mohanty, A.K.; Misra, M.; Drzal, L. Sustainable bio-composites from renewable resources: Opportunities and challenges in the green materials world. *J. Polym. Environ.* **2002**, *10*, 19–26. [CrossRef]

138. Rogers, J.; Brammer, J. Estimation of the production cost of fast pyrolysis bio-oil. *Biomass Bioenergy* **2012**, *36*, 208–217. [CrossRef]

139. Lora, E.; Andrade, R. Biomass as energy source in brazil. *Renew. Sustain. Energy Rev.* **2009**, *13*, 777–788. [CrossRef]

140. Jackson, T. *Material Concerns: Pollution, Profit and Quality of Life*; Routledge: Abingdon-on-Thames, UK, 2013.

141. Demirbas, M.F.; Balat, M.; Balat, H. Potential contribution of biomass to the sustainable energy development. *Energy Convers. Manag.* **2009**, *50*, 1746–1760. [CrossRef]

142. Saxena, R.; Adhikari, D.; Goyal, H. Biomass-based energy fuel through biochemical routes: A review. *Renew. Sustain. Energy Rev.* **2009**, *13*, 167–178. [CrossRef]

143. Hill, J. Environmental costs and benefits of transportation biofuel production from food-and lignocellulose-based energy crops: A review. In *Sustainable Agriculture*; Springer: New York, NY, USA, 2009; pp. 125–139.

144. Brick, S.; Lyutse, S. *Biochar: Assessing the Promise and Risks to Guide Us Policy*; NRDC Issue Paper; Natural Resources Defense Council: New York, NY, USA, 2010.

145. Calonaci, M.; Grana, R.; Barker Hemings, E.; Bozzano, G.; Dente, M.; Ranzi, E. Comprehensive kinetic modeling study of bio-oil formation from fast pyrolysis of biomass. *Energy Fuels* **2010**, *24*, 5727–5734. [CrossRef]

146. Ahmad, M.; Rajapaksha, A.U.; Lim, J.E.; Zhang, M.; Bolan, N.; Mohan, D.; Vithanage, M.; Lee, S.S.; Ok, Y.S. Biochar as a sorbent for contaminant management in soil and water: A review. *Chemosphere* **2014**, *99*, 19–33. [CrossRef] [PubMed]

147. Hornung, A. *Transformation of Biomass: Theory to Practice*; John Wiley & Sons: Hoboken, NJ, USA, 2014.

148. Sattar, A.; Leeke, G.A.; Hornung, A.; Wood, J. Steam gasification of rapeseed, wood, sewage sludge and miscanthus biochars for the production of a hydrogen-rich syngas. *Biomass Bioenergy* **2014**, *69*, 276–286. [CrossRef]

149. Hornung, A.; Apfelbacher, A. Thermal Treatment of Biomass. U.S. Patent No. 8,835,704, 16 September 2014.

150. Brown, T.R.; Wright, M.M.; Brown, R.C. Estimating profitability of two biochar production scenarios: Slow pyrolysis vs. fast pyrolysis. *Biofuels Bioprod. Biorefin.* **2011**, *5*, 54–68. [CrossRef]

151. Vamvuka, D. Bio-oil, solid and gaseous biofuels from biomass pyrolysis processes—An overview. *Int. J. Energy Res.* **2011**, *35*, 835–862. [CrossRef]

152. Sanna, A. Advanced biofuels from thermochemical processing of sustainable biomass in Europe. *BioEnergy Res.* **2014**, *7*, 36–47. [CrossRef]

153. Rasul, M.; Jahirul, M.I. *Recent Developments in Biomass Pyrolysis for Bio-Fuel Production: Its Potential for Commercial Applications*; Central Queensland University, Centre for Plant and Water Science, Faculty of Sciences, Engineering and Health: Norman Gardens, Ausralia, 2012.

154. Brown, D.; Rowe, A.; Wild, P. A techno-economic analysis of using mobile distributed pyrolysis facilities to deliver a forest residue resource. *Bioresour. Technol.* **2013**, *150*, 367–376. [CrossRef] [PubMed]

155. Brown, D. Using Mobile Distributed Pyrolysis Facilities to Deliver a Forest Residue Resource for Bio-Fuel Production. Ph.D. Thesis, University of Victoria, Victoria, BC, Canada, 2013.

156. Li, D.; Briens, C.; Berruti, F. Improved lignin pyrolysis for phenolics production in a bubbling bed reactor–effect of bed materials. *Bioresour. Technol.* **2015**, *189*, 7–14. [CrossRef] [PubMed]

157. Babu, B.; Chaurasia, A. Pyrolysis of biomass: Improved models for simultaneous kinetics and transport of heat, mass and momentum. *Energy Convers. Manag.* **2004**, *45*, 1297–1327. [CrossRef]

158. Burton, A.; Wu, H. Mechanistic investigation into bed agglomeration during biomass fast pyrolysis in a fluidized-bed reactor. *Energy Fuels* **2012**, *26*, 6979–6987. [CrossRef]

159. Burton, A.H. Bed Agglomeration during Biomass Fast Pyrolysis in a Fluidised Bed Reactor. Ph.D. Thesis, Curtin University, Bentley, Australia, 2016.

160. Lin, C.-L.; Peng, T.-H.; Wang, W.-J. Effect of particle size distribution on agglomeration/defluidization during fluidized bed combustion. *Powder Technol.* **2011**, *207*, 290–295. [CrossRef]

161. Carrier, M.; Müller, N.; Grandon, H.; Segura, C.; Wilkomirsky, I.; Berg, A. Depolymerization of acetosolv lignin through in-situ catalytic fast pyrolysis. Available online: https://publicaciones.udt.cl/pub/depolymerization-of-acetosolv-lignin-through-in-situ-catalytic-fast-pyrolysis/ (accessed on 9 November 2018).

162. Babu, B.; Chaurasia, A. Modeling for pyrolysis of solid particle: Kinetics and heat transfer effects. *Energy Convers. Manag.* **2003**, *44*, 2251–2275. [CrossRef]

163. Solantausta, Y.; Lehto, J.; Oasmaa, A.; Kjäldman, L. Status of fast pyrolysis bio-oil technologies. In Proceedings of the 21st European Biomass Conference and Exhibition (EUBCE 2013), Copenhagen, Denmark, 3–7 June 2013.

164. Babu, B.; Chaurasia, A. Dominant design variables in pyrolysis of biomass particles of different geometries in thermally thick regime. *Chem. Eng. Sci.* **2004**, *59*, 611–622. [CrossRef]

165. Babu, B.; Chaurasia, A. Heat transfer and kinetics in the pyrolysis of shrinking biomass particle. *Chem. Eng. Sci.* **2004**, *59*, 1999–2012. [CrossRef]

166. Thunman, H.; Niklasson, F.; Johnsson, F.; Leckner, B. Composition of volatile gases and thermochemical properties of wood for modeling of fixed or fluidized beds. *Energy Fuels* **2001**, *15*, 1488–1497. [CrossRef]

167. Johansson, R.; Andersson, K.; Leckner, B.; Thunman, H. Models for gaseous radiative heat transfer applied to oxy-fuel conditions in boilers. *Int. J. Heat Mass Transf.* **2010**, *53*, 220–230. [CrossRef]

168. Boulard, D.C. *Bio-oil: The New Crude*; Ensyn RTPTM Bio-Refinery: Concord, NH, USA, 2002.

169. Brown, D. Continuous ablative regenerator system. In *Proceedings of the 2nd EU-Canada Workshop on Thermal Biomass Processing, 1996*; CPL Press: Newbury, UK, 1996; pp. 96–100.

170. Bridgewater, A.V.; Hogan, E.N. *Bio-Oil Production & Utilisation: Proceedings of the 2nd EU-Canada Workshop on Thermal Biomass Processing*; CPL Scientific Information Services Ltd.: Newbury, UK, 1996.

171. Babu, B.; Chaurasia, A. Optimization of pyrolysis of biomass using differential evolution approach. In Proceedings of the Second International Conference on Computational Intelligence, Robotics, and Autonomous Systems (CIRAS-2003), Singapore, 15–18 December 2003.

172. Thunman, H.; Leckner, B. Thermal conductivity of wood—Models for different stages of combustion. *Biomass Bioenergy* **2002**, *23*, 47–54. [CrossRef]

173. Thunman, H.; Davidsson, K.; Leckner, B. Separation of drying and devolatilization during conversion of solid fuels. *Combust. Flame* **2004**, *137*, 242–250. [CrossRef]

174. Niklasson, F.; Thunman, H.; Johnsson, F.; Leckner, B. Estimation of solids mixing in a fluidized-bed combustor. *Ind. Eng. Chem. Res.* **2002**, *41*, 4663–4673. [CrossRef]

Article

A Study of Sewage Sludge Co-Combustion with Australian Black Coal and Shiitake Substrate

Guan-Bang Chen [1,*], Samuel Chatelier [2], Hsien-Tsung Lin [3], Fang-Hsien Wu [1] and Ta-Hui Lin [1,2,*]

[1] Research Center for Energy Technology and Strategy, National Cheng Kung University, Tainan 70101, Taiwan; z10602031@email.ncku.edu.tw

[2] Department of Mechanical Engineering, National Cheng Kung University, Tainan 70101, Taiwan; n16057164@mail.ncku.edu.tw

[3] Department of Aeronautics and Astronautics, National Cheng Kung University, Tainan 70101, Taiwan; P48021027@mail.ncku.edu.tw

* Correspondence: gbchen@mail.ncku.edu.tw (G.-B.C.); thlin@mail.ncku.edu.tw (T.-H.L.);
Tel.: +886-6-275-7575 (ext. 51030) (G.-B.C.); +886-6-2757575 (ext. 51031) (T.-H.L.)

Received: 4 November 2018; Accepted: 4 December 2018; Published: 7 December 2018

Abstract: Co-combustion technology can be a gateway to sewage sludge valorization and net CO_2 reduction. In this study, combustion characteristics of sewage sludge, Australian black coal, shiitake substrate, and their blends were analyzed via thermogravimetric analysis (TGA) coupled with Fourier transform infrared spectroscopy. The ignition temperature, burnout temperature, flammability index (C), and combustion characteristics index (S) of the fuels and their respective blends were estimated. Kinetic parameters were also estimated using the Coats-Redfern method. The results showed that the oxidation of the blends had two distinct stages. Synergistic effects existed for all the blends, with negative ones occurring at temperatures between 300 and 500 °C and positive ones during the char oxidation period. In the first oxidation stage, both C and S indexes increased with sludge addition to the coal. However, they decreased with sludge addition in the final oxidation stage. The catalytic effect of the sludge and the shiitake was pronounced in the final oxidation stage and it resulted in a decrease of activation energy. As for the pollutant emissions, the results showed that NO_x and SO_2 emissions decreased for 25 wt.% sludge addition to the coal. For the sludge-shiitake blends, NO_x and SO_2 emissions decreased with increasing shiitake addition. The single-pellet combustion results showed that ignition delay time reduced with increasing sludge/coal ratio but increased with increasing sludge/shiitake ratio. The volatile combustion duration decreased with the addition of sludge and total combustion time decreased sharply with increasing sludge ratio.

Keywords: co-combustion; sewage sludge; thermogravimetric analysis; Fourier transform infrared spectroscopy; synergistic effect; single-pellet combustion

---

1. Introduction

Sewage sludge is a common by-product resulting from the treatment processes of industrial or municipal wastewaters. Its global production has increased considerably in the past years. This increase is expected to continue in the coming years as more and more people move to urban areas. In 2012, around 2.37 million tons of sludge were produced in Taiwan [1]. If not properly disposed of, sewage sludge can be a threat to the environment and human health because of the high presence of heavy metals and pathogens [2,3].

Prior to disposal, sewage sludge is usually processed in order to reduce smell, the quantity of organic solids, water content, and to eliminate some disease-causing bacteria. Therefore, at the wastewater treatment plant, the sludge goes through different processes such stabilization, conditioning and dewatering [4]. Traditionally, sewage sludge is disposed of through landfilling at sanitary sites and agricultural use (directly or indirectly). Currently, most of the sludge produced in Taiwan is disposed of through landfilling [1]. In the future, with the scarcity of available land, stricter environmental regulations, rising concerns regarding health and safety, and willingness to build and promote a circular economy model, it is expected that these practices will change. Sewage sludge can also be disposed of through incineration or thermally processed to recover its energy content.

Sewage sludge has been classified as biomass [5] due to its high calorific value and as semi-biomass [6] in the literature. It is even sometimes compared to low-rank coals such as lignite in terms of energy content [7,8]. However, sewage sludge differs from plant biomass in many aspects, which might pose technical challenges to its combustion. Describing the combustion process of sewage sludge is not an easy task due to the large heterogeneity in the physical and chemical properties. Cui et al. [9] studied the combustion process of pelletized raw sewage sludge and found that, as with biomass particle, sewage sludge combustion comprises different stages, such as drying, devolatilization, volatiles combustion, and char combustion. Each stage of combustion may last for a specific time depending on the fuel physical and chemical characteristics and some external parameters, such as the combustor configuration, heat and mass transfer with the surroundings, heating rate, and particle size [9,10]. At a relatively low heating rate and for a small particle, there is a clear distinction between the stages. However, at higher heating rates and for larger particles, there is a certain degree of overlap between the stages [11]. High inorganic matter contents in sludge might pose some challenges to its combustion process. Slagging (alkali-induced or silica-melt induced), ash accumulation in the furnace (fouling), corrosion, ash handling, heavy metals and emission levels are potential major issues for sludge combustion.

From a waste minimization viewpoint, sewage sludge incineration has the advantage that sludge with a wider range of water content may be used, and therefore requires less pretreatment. However, for sludge energetic valorization, such a process can be energy-deficient as supplementary fuel might be required to handle the wet sludge. Therefore, co-utilization of sludge with other fuels may be more feasible from an energy perspective. Sludge co-combustion also offers some economic advantages as there is no need to hire new experienced personnel and existing facilities and devices for gaseous emissions control can be used; consequently, no additional investments are needed for a new processing plant [12]. From an environmental viewpoint, co-combustion of sewage sludge and coal can help reduce net CO_2 emissions. Currently, sewage sludge is co-fired with coal in power plants, with other biomass, with municipal solid waste in incinerators, and in clay brick manufacture and cement kilns [13,14].

During co-combustion, dried sewage sludge replaces only a small fraction of the plant's feedstock. Werther and Ogada [4] reported that sludge should be milled and dried, leaving a water content of less than 10%, when co-fired with bituminous coal in pulverized coal power stations. Otero et al. [15] stated that dried sludge can be co-fired at a percentage of up to 50% with no major consequences regarding gaseous emissions. Thermogravimetric analysis (TGA) is widely used in studies on the thermal degradation of sludge co-combustion. Nimoya et al. [16] showed that sludge co-combustion with coal has better performance than sludge mono-combustion. An investigation by Folgueras et al. [17] using TGA of sewage sludge and bituminous coal blends revealed that blends with a sludge content of 10% behave similarly to coal, whereas blends of 50% clearly show two combustion stages (a clear volatile matter and more reactive structures stage and a second thermal decomposition stage of more complex structures).

However, few studies have elaborated on sludge co-combustion with plant biomass. Kijo-Kleczkowska et al. [18] studied the co-combustion of pelletized sewage sludge with coal and biomass, and reported that adding sludge to biomass (willow Salix viminalix) increases the fuel ignition temperatures, whereas increasing the sludge percentage in sludge-coal (lignite) blends decreases the ignition temperature due to the presence of volatiles. It is expected that mixing low-grade biomass with higher-quality biomass may reduce flame stability problems while minimizing corrosion due to the deposition of ash, which contains low-melting-point salts.

Although there are many studies on sludge co-combustion in the literature, further study is still essential to research the co-combustion behavior and interaction due to the complicated composition of sludge and the diversity of biomass species. Therefore, this study investigates sludge co-combustion with coal and biomass by means of TGA and single-pellet combustion experiments. The selected biomass is a mushroom growing substrate, which is one of the major agricultural wastes in Taiwan. Subsequently, the kinetic parameters and the emission characteristics are also investigated using the Coats-Redfern integral method and Fourier transform infrared spectroscopy (FTIR), respectively. The existence of synergistic effects with addition of sludge to the blends is also taken into account.

2. Materials and Methodology

2.1. Sample Pretreatment

Original sewage sludge was obtained from a WWTP in Tainan City, Taiwan (R.O.C.). In the as-received (ar) basis (Figure 1a), the sludge had a water content of more than 50%. Prior to composition analysis, the sludge was sun-dried to reduce its moisture content; the process is shown in Figure 1b. The final sun-dried sewage sludge is referred to as dried sewage sludge (DSS) and is considered as the as-received sludge in this study.

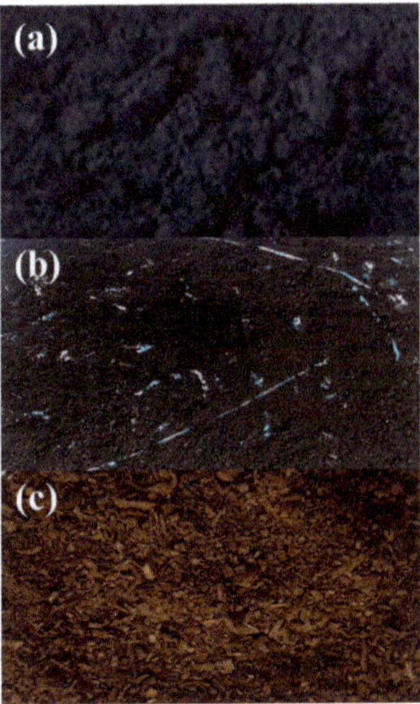

Figure 1. Photographs of different fuels: (**a**) original sewage sludge; (**b**) sun-drying process; (**c**) shiitake growing substrate.

Shiitake mushrooms, originally grown at high altitudes, are now widely grown throughout many areas in Taiwan. They are commonly grown in cultivation bottles or plastic bags, called "space bags" by farmers, filled with mushroom growing substrate (Figure 1c). Sawdust from the lumber industry and most recently chopped dried rice straw chips are used as the mushroom growing substrate [19]. Each year in Taiwan, an estimated 29.6 million bags and 7.5 million bottles are emptied of substrate [20]. The majority is reused to grow other types of mushrooms and eventually become agricultural waste. Shiitake substrate was obtained from a company that recycles agricultural waste. In this study, the term shiitake is used interchangeably with shiitake substrate. As for the coal used in this study (Australian black coal), it is a sub-bituminous coal and is often used for power generation in Taiwan. It was ground and provided by the Industrial Technology Research Institute.

To ensure homogeneity of the fuels, all the samples were crushed and sieved to a particle size of less than or equal to 96 μm (Mesh #160), and then dried in an oven at 103 °C for 10 h, in order to minimize water content, before composition analysis. To study the co-combustion behavior of sewage sludge mixed with coal and the shiitake substrate, two different blending ratios were defined; a biomass blending ratio (BBR) and a sludge-shiitake ratio (SSR). The two ratios were varied from 0% to 100% in increments of 25%. BBR = 0% represents the case of pure Australian black coal, SSR = 0% represents the case of a pure shiitake substrate, and BBR = SSR = 100% represents the case of pure sewage sludge. To prepare a mixture, a 10 g sample was prepared in a glass cylindrical container and well stirred using spherical steel balls for each case.

2.2. Fuel Properties

The ultimate analysis and proximate analysis of the samples in their as-received (ar) basis was determined through an elemental analyzer (Elementar vario EL III) and a thermal analyzer (PerkinElmer STA 8000), respectively. The proximate was obtained in accordance with the ASTM methods (ASTM D3302-12, ASTM D3175-11, ASTM D3174-11, and ASTM D7582-10e1) [21]. A bomb calorimeter (Parr 6200 Isoperibol Calorimeter) was used according to standardized procedures to estimate the gross calorific value of the samples and their blends. The proximate analysis results and heating values of the oven-dried sewage sludge and shiitake were used to calculate the values on a dry and ash-free (daf) basis.

2.3. Thermogravimetric Analysis

The weight loss history and exothermic phenomenon of the individual fuels and their blends in a linearly heated environment were studied using a thermal analyzer (PerkinElmer STA 8000) which was coupled with a differential scanning calorimetry (DSC) to measure the heat flow into or out of the sample over time. For each experiment, a sample weighing of about 9–10 mg was placed in a small crucible, and then put in a thermal analyzer to study its thermal degradation. The sample was heated from 30 °C to 1000 °C at a heating rate of 20 °C·min^{-1}. Air was used as a carrier gas at a flow rate of 50 mL·min^{-1}. Each experiment was repeated at least twice to ensure reproducibility. The recorded weight loss, namely, the TG curve, of the samples and its first derivative (DTG) were obtained using Pyris Manager software.

2.4. Synergistic Effect Analysis

In order to study whether there existed a coupling synergistic effect [22] between the samples, the calculated weight curves of the mixtures were estimated as the sum of the TGA curves of each individual sample. The calculated weight is denoted as W_{cal}, defined by the linear additive rule expressed by Equations (1) and (2) for the sludge-coal mixture and the blend of sludge and shiitake, respectively.

$$W_{cal} = \left(1 - \gamma_{sludge}\right)W_{coal} + \gamma_{sludge}W_{sludge} \tag{1}$$

$$W_{cal} = \left(1 - \gamma_{sludge}\right)W_{shiitake} + \gamma_{sludge}W_{sludge} \tag{2}$$

where γ_{sludge} represents the weight proportion of sewage sludge in the mixture, and W_{coal}, W_{sludge}, and $W_{shiitake}$ are the TG curves, at 20 °C·min^{-1}, of Australian black coal, sewage sludge, and shiitake, respectively. A positive synergistic effect may exist when the values of the experiment curves, W_{exp}, are lower than that of W_{cal} in the temperature range of the experiment. Equation (3) expresses the deviation between the experimental and the calculated curves, namely, ΔW, from which the degree of the synergistic effects was evaluated. Sadhukhan et al. [23] used the root mean square (RMS) values of the deviation to measure whether synergistic effects existed.

$$\Delta W = W_{exp} - W_{cal} \tag{3}$$

2.5. Combustion Characteristics Parameters

The intersection method was used in this study to estimate the ignition temperature (T_i) from the TGA results. For further readings regarding this method, one can refer to the studies of Lu et al. [24]. The burnout temperature (T_b) was estimated by the conversion method [24,25], where T_b is defined as the temperature at which the conversion of the fuel reaches 99%.

Knowing both T_i and T_b, two combustion characteristic parameters, namely, the combustion characteristics index (S) [25] and the flammability index (C) [26], were calculated. Typically, a fuel with a larger value of S and C has an improved combustion performance and combustion stability, respectively.

2.6. Kinetic Parameters Analysis

Kinetic parameters, obtained using the Coats and Redfern integral method [27], were estimated from the TGA results to further study the combustion characteristics of the samples. The fundamental rate equation of a heterogeneous solid phase reaction can be described by the following equation:

$$\frac{d\alpha}{dt} = k(T)f(\alpha) \tag{4}$$

where α is the conversion degree, expressed as:

$$\alpha = \frac{W_s - W}{W_s - W_e} \tag{5}$$

The temperature dependence of k is expressed by the Arrhenius equation:

$$k(T) = A exp\left(-\frac{E}{RT}\right) \tag{6}$$

where A is the pre-exponential factor (min^{-1}), E is the apparent activation energy ($kJ \cdot mol^{-1}$), T is the reaction temperature (K), and R is the gas constant ($8.314 \times 10^{-3} \, kJ \cdot mol^{-1} \cdot K^{-1}$). For non-isothermal and heterogeneous solid phase reactions at a constant heating rate, the temperature increasing rate could be described as $\beta = dT/dt$. Therefore, Equations (4) and (6) can be combined to become:

$$\frac{d\alpha}{dT} = \frac{1}{\beta}\frac{d\alpha}{dt} = \frac{1}{\beta} A exp\left(-\frac{E}{RT}\right) f(\alpha) \tag{7}$$

The specific form of $f(\alpha)$ represents the hypothetical model of the reaction mechanism or "model function", and can be written as $f(\alpha) = (1 - \alpha)^n$, in the form of an nth-order reaction. It is assumed that the conversion function, $f(\alpha)$, depends on the rate of weight loss.

Integration of Equation (7) yields:

For $n = 1$

$$ln\left[\frac{-ln(1-\alpha)}{T^2(1-n)}\right] = ln\left[\frac{AR}{\beta E}\left(1 - \frac{2RT}{E}\right)\right] - \frac{E}{RT} \tag{8}$$

For $n \neq 1$

$$ln\left[\frac{1 - (1-\alpha)^{1-n}}{T^2(1-n)}\right] = ln\left[\frac{AR}{\beta E}\left(1 - \frac{2RT}{E}\right)\right] - \frac{E}{RT} \tag{9}$$

For the tested temperature ranges and most values of E, $\frac{E}{RT} \geq 1$ and $1 - \frac{2RT}{E} \approx 1$. Thus, Equations (8) and (9) can be further simplified as:

$$ln\left[\frac{-ln(1-\alpha)}{T^2(1-n)}\right] = ln\left(\frac{AR}{\beta E}\right) - \frac{E}{RT}; \ (n=1) \tag{10}$$

$$ln\left[\frac{1-(1-\alpha)^{1-n}}{T^2(1-n)}\right] = ln\left(\frac{AR}{\beta E}\right) - \frac{E}{RT}; \ (n \neq 1) \tag{11}$$

The activation energy (E) and the pre-exponential factor (A) can be obtained from the slope, $-E/R$, and the intercept of the regression line, respectively, by plotting $ln\left[\frac{-ln(1-\alpha)}{T^2(1-n)}\right]$ versus $1/T$ and $ln\left[\frac{1-(1-\alpha)^{1-n}}{T^2(1-n)}\right]$ versus $1/T$. E and A can be used to evaluate the difficulty and intensity of combustion reactions of the fuels, respectively. Different regression lines are plotted for different values of reaction order (n = 0.3, 0.6, 0.9, 1, 1.2, 1.5, 2, and 3). The best fitting ones are used to estimate the activation energy and the pre-exponential factor.

2.7. Fourier Transform Infrared Spectroscopy

Evolved gas products in the TGA experiments were analyzed using a Fourier transform infrared spectrometer (PerkinElmer Spectrum Two FT-IR). To prevent the condensation of less-volatile products, the gas cell and the transfer line were both heated at 280 °C. The FTIR spectra were obtained in the scanning range of 4000 to 600 cm^{-1}, in terms of wavenumber. The resolution and number of accumulations were set to 1 cm^{-1} and 1, respectively. FTIR spectra of the gaseous products were collected continuously. The evolved gas products were obtained and analyzed using the software Spectrum TimeBase. In the study, the studied gases (CH_4, CO, CO_2, NO_x, SO_2) and their wavenumber intervals are listed in Table 1 [28–33]. Finally, the yields of gaseous products are an important indicator to understand the evolved gas from the combustion experiments. Because absorbance varies with species concentration during FTIR quantitative analysis, a normalized value of the gas yield was calculated for better comparison purpose. The gas yield was defined as the ratio of the integral values under the evolution curves of the different gaseous products and the sample initial weight; and is expressed as follow [34]:

$$\text{Gas yield} = \frac{\text{integral value of gaseous product}}{\text{initial sample weight}} \tag{12}$$

Table 1. Wavenumber intervals of different gaseous species.

Gas Species	Wavenumber Range (cm^{-1})	Functional Groups	Ref.
CH_4	3000–2700	C-H	[28]
CO_2	2400–2240	C=O	[29–31]
CO	2240–2060	C-O	[29–31]
NO_x	1795–1520	-NO_2 and -NO	[32,33]
SO_2	1374–1342	S=O	[29,33]

2.8. Single-Pellet Combustion

A dry pressing die set with a diameter of 12.7 mm was used to make the pellets. About 1 g of the sample powder was poured into a steel sleeve and compacted, using a load of about two tons on the pushing rod, with a hydraulic press. This technique was used to make the pure fuels and their respective blended pellets prior to the single-pellet combustion experiments.

Detailed measurement of the behavior of individual pellets of the different fuels might be useful for improving of the modeling and design of a boiler plant. In order to understand the combustion behavior of different biomass blends and to simulate real furnace conditions, each pellet was burned in a single-pellet free-drop furnace. Figure 2 shows a diagram of the single-pellet combustion system. Two electrically heated plates, controlled by a proportional-integral-derivative controller, provide heat to the furnace and a K-type thermocouple that sends feedback temperature measurements to the controller. Three temperatures (600, 700, and 800 °C) were used in this study. Air was used as the carrier gas for all experiments and the flow was controlled by an electronic flowmeter at 3.5 SLPM. Before entering the combustion chamber, the carrier gas was heated by a preheater. In the experiment, the pellet was dropped through the transmission tube toward the stainless mesh platform, which was connected to a quartz holder. Real-time weight measurements (one data point per second) of the pellet were recorded using an electronic scale with 0.001 g precision. A Sony FDR-AX100 4K camera (30 fps) was used to record video of the pellet combustion process through the observation window. Post-analysis of the video was performed using standard software that allows frame-by-frame viewing.

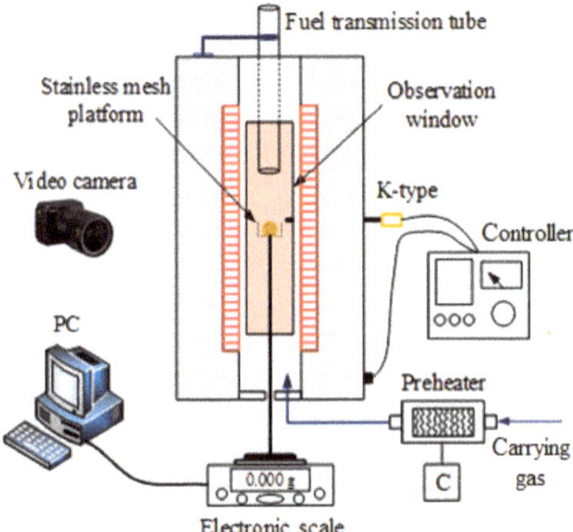

Figure 2. Single-pellet combustion system.

The moment when the scale showed the approximate recorded measurement after the pellet fell on the platform was taken as $t = 0$ s for the weight measurement data. The ignition delay time (t_{id}) was estimated from an analysis of recorded images. t_{id} is the time that it takes for the pellet to ignite after falling into the hot environment. Generally, volatile matter, flammability of volatiles, and transport from the particle determine how ignition of a particle occurs, either homogeneously (gas phase ignition) or heterogeneously. Homogeneous ignition is considered in this study for determining t_{id}. The first frame showing a glowing flame of the burning volatiles is defined as the moment of ignition. According to adiabatic thermal explosion theory, the homogeneous ignition delay time can be estimated as [35]:

$$t_{id} = \frac{c_{v,g}\left(\frac{T_0^2}{T_a}\right)}{q_c Y_{F,0} B exp\left(\frac{-T_a}{T_0}\right)} \tag{13}$$

where $c_{v,g}$ is the specific heat of the background gas (constant volume), T_0 is the initial temperature of the local volatiles and oxidizer mixture, q_c is the heat released by combustion per mass of volatiles, $Y_{F,0}$ is the initial mass fraction of the volatiles and $B exp(-T_a/T_0)$ is the reactivity of the mixture of volatiles and oxidizer. The released volatiles act like a shield, preventing O_2 from diffusing to the sludge particle surface. They ignite and burn in gas phase. After the pellet bursts into flame, the volatiles released are burned homogeneously until extinction. This length of time is demarcated as the volatile combustion duration t_f in this study. Then, the char keeps burning heterogeneously until the sample weight does not change. Finally, the end of the single-pellet combustion, namely t_{tot} was defined as the time when the measured weight showed no change in a period of 60 s. Each experiment was repeated five times. The average value of the parameters was calculated and reported in this study.

3. Results and Discussion

3.1. Fuel Properties

The ultimate analysis and the proximate analysis results of the samples used in this study are shown in Table 2. The ultimate analysis results of the fuels in their "ar" basis show that sewage sludge was 28.4 wt.% carbon, 5.29 wt.% hydrogen, and 25.58 wt.% oxygen. Compared to Australian coal (73.1 wt.% carbon) and shiitake (40.49 wt.% carbon), sewage sludge has the lowest carbon content. The oxygen content of Australian black coal and shiitake is 5.27 wt.% and 29.03 wt.%, respectively. The results also show that the nitrogen content, 4.65 wt.%, and the sulfur content, 2.66 wt.%, are the highest for sewage sludge. Consequently, the combustion of sludge should receive attention with respect to some pollution and corrosion issues.

Table 2. Ultimate analysis and proximate analysis of sewage sludge, shiitake, and Australian coal.

Sample	Ultimate Analysis (wt.%)				
	C	H	O	N	S
DSS_{ar}	28.40	5.29	25.58	4.65	2.66
$Shiitake_{ar}$	40.49	5.95	29.03	3.23	0.38
Australian coal	73.10	4.30	5.27	1.65	0.48

Sample	Proximate Analysis (wt.%)				HHV ($MJ \cdot kg^{-1}$)
	M	VM	ASH	FC *	
DSS_{ar}	8.07	48.90	39.61	3.42	11.37
DSS_{daf}	-	91.50	-	8.50	21.67
$Shiitake_{ar}$	13.28	69.05	13.52	4.15	15.48
$Shiitake_{daf}$	-	92.83	-	7.17	21.66
Australian coal	1.96	30.77	17.18	50.09	26.10

* Calculated from the difference. ar: as-received basis. daf: dry and ash-free basis.

From the proximate analysis results of the samples, the sun-dried sewage sludge (DSS) was 8.07 wt.% moisture, 48.9 wt.% volatile matter, 39.61 wt.% ash, and 3.42 wt.% fixed carbon. However, drying the sludge and estimating its proximate analysis on an ash-free basis yields 91.5 wt.% and 8.5 wt.% for volatile matter and fixed carbon content, respectively. Shiitake and Australian black coal's volatile matter content on an as-received basis is 69.05 wt.% and 30.77 wt.%, respectively. The fixed carbon content of the shiitake substrate and Australian black coal is 4.15 wt.% and 50.09 wt.%, respectively. The amount of volatiles strongly influences the thermal decomposition and combustion behavior of solid fuels, while the amount of fixed carbon influences the char oxidation process. Australian black coal has the lowest moisture and ash content of the samples; with sewage sludge ash content, 39.61 wt.%, being the highest because of its high inorganic content.

The measured heating value of the individual fuels are also shown in Table 2. Sewage sludge, with a heating value of 11.37 $MJ \cdot kg^{-1}$, has the lowest HHV of the studied fuels. Such low HHV is perhaps due to the fact that sewage sludge has high inorganic matter and low carbon content. However, taken on an ash-free basis, its HHV increased to 21.67 $MJ \cdot kg^{-1}$, similar to that of shiitake, 21.66 $MJ \cdot kg^{-1}$. The HHV of Australian black coal is 26.1 $MJ \cdot kg^{-1}$. The heating value of the sewage sludge and the shiitake is almost 80% of that of Australian black coal.

3.2. Thermogravimetric Analysis

The combustion profile of sewage sludge is shown in Figure 3. Sludge oxidation occurred in three stages. A dehydration stage, where moisture in the fuel is released, and a combustion stage that can be divided into main and final oxidation stages. The main oxidation stage is referred to as stage 1 in this study, and has a temperature range of 192.60 to 370.78 °C. It might be attributed to the reaction of air with the volatiles and some reactive structures in the sludge. During this stage, the volatiles released did not burn intensely, as shown by the heat flow curve. The maximum oxidation rate in this stage occurs at 286.61 °C. The weight loss during stage 1 is about 24.89 wt.%. The final oxidation stage, stage 2, started at about 370.78 °C, and showed a total weight loss of 25.86 wt.% at 800 °C. It might be attributed to the reaction of air with the heavier components or more complex structures and char oxidation. The maximum oxidation rate occurs at 481.55 °C. Stage 2 is associated with intense heat release, as shown by the DSC curve. Comparing the weight loss rates of the two stages, it is found that the burning rate (-4.80 wt.%·min^{-1}) of stage 1 was faster than that of stage 2 (-3.93 wt.%·min^{-1}), which is due to the high porosity [4] and high surface area [29] of the sludge.

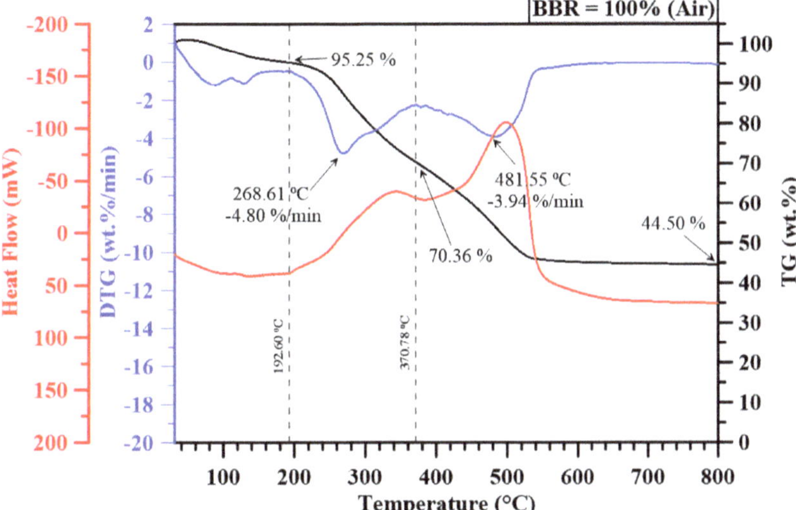

Figure 3. Thermogravimetric analysis results of sewage sludge under combustion condition.

The TGA results of Australian black coal are shown in Figure 4, and display a typical combustion profile of sub-bituminous coal with one-stage thermal decomposition, as shown by the DTG curve. The peak (-6.33 wt.%·min^{-1}) of the burning rate was at 580.73 °C and corresponded to intense heat release, as shown by the DSC curve. This stage is referred to as stage 2 in this study, and it is the result of the devolatilization process as well as char oxidation of the coal. The total weight loss of the oxidation stage at 800 °C is approximately 80.24 wt.%.

Shiitake oxidation occurred in two stages, as shown in Figure 5. Stage 1 showed one peak between 200 and 407 °C with a maximum value of -14.86 wt.%·min^{-1} at 317.54 °C. This peak can be attributed to the release and oxidation of the holocellulose (hemicellulose + cellulose) and a small amount of the lignin in the substrate, showing a total weight loss of about 51.77 wt.%. Further decomposition of the lignin and char oxidation occurred in the final oxidation stage, which showed a burning rate peak, at 439.13 °C, of -6.48 wt.%·min^{-1}. The total weight observed in the final oxidation stages at 800 °C is 27.89 wt.%. At higher temperatures (620–720 °C), a small weight loss can be observed. As this was not associated with any significant heat release, it is not considered as a stage itself and therefore is included in stage 2. It might be related to an intrinsic characteristic of the substrate, which is beyond the scope of this study.

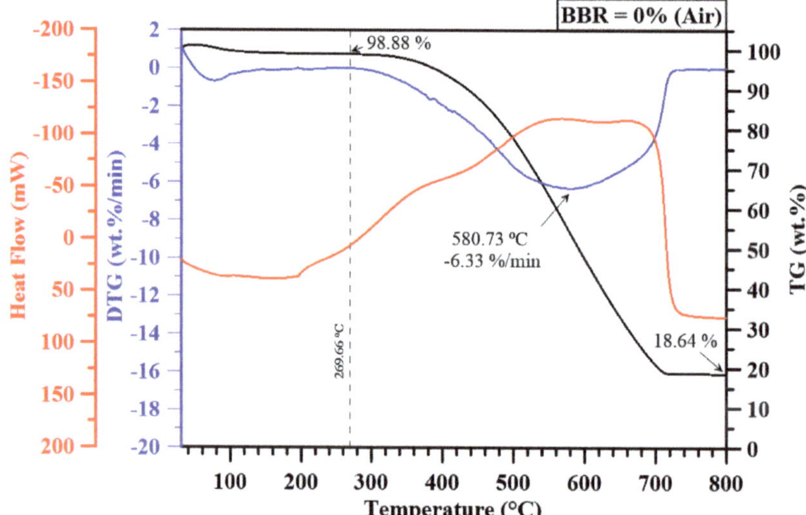

Figure 4. Thermogravimetric analysis results of Australian black coal under combustion condition.

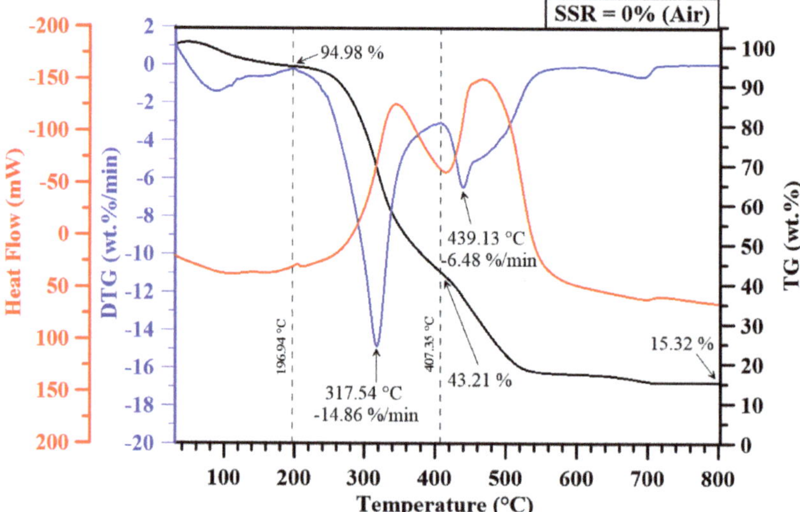

Figure 5. Thermogravimetric analysis results of shiitake substrate under combustion condition.

3.3. Synergistic Effect Analysis

The prediction of the TG curves in Figures 6 and 7 for the blends using Equations (1) and (2) revealed that they are non-additive and therefore might show some synergistic effects. A comparison of the weight loss from the experimental TGA curves for various BBRs to that calculated from the linear additive rule is shown in Figure 8 for the sludge-coal blends. The figure shows the various trends of W_{exp} and W_{cal} for the different blends for $\beta = 20\ °C \cdot min^{-1}$. For the BBR = 25% blend, the residual values of the experimental curve were lower than those of the calculated curve between 600 and 800 °C. This suggests that a positive synergistic effect may exist between coal and sludge for this blend in this temperature range. A similar tendency can be seen in Figure 9 for all sludge-shiitake blends at higher temperatures. In contrast, the residual values of the experimental curve for BBR = 50% and 75% were greater than the calculated values, implying negative synergy.

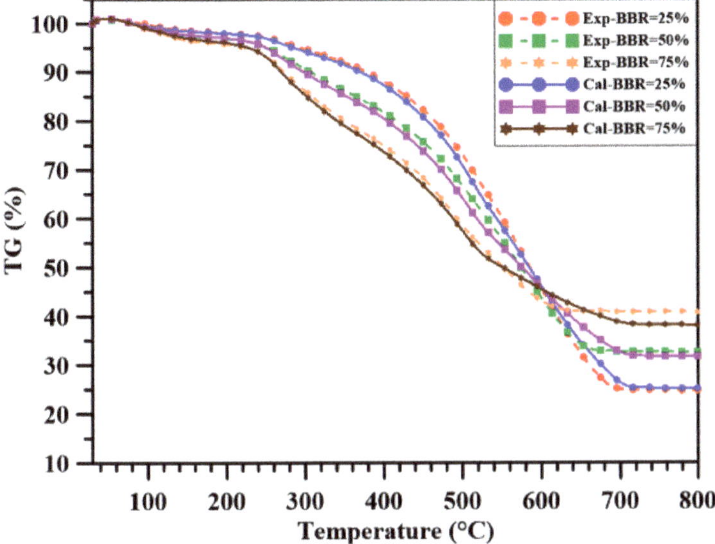

Figure 6. Synergistic effect for various sludge-coal ratios.

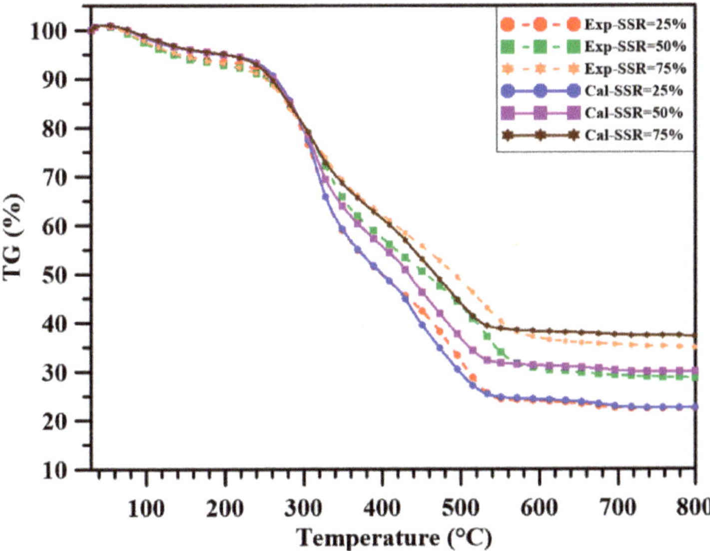

Figure 7. Synergistic effect for various sludge-shiitake ratios.

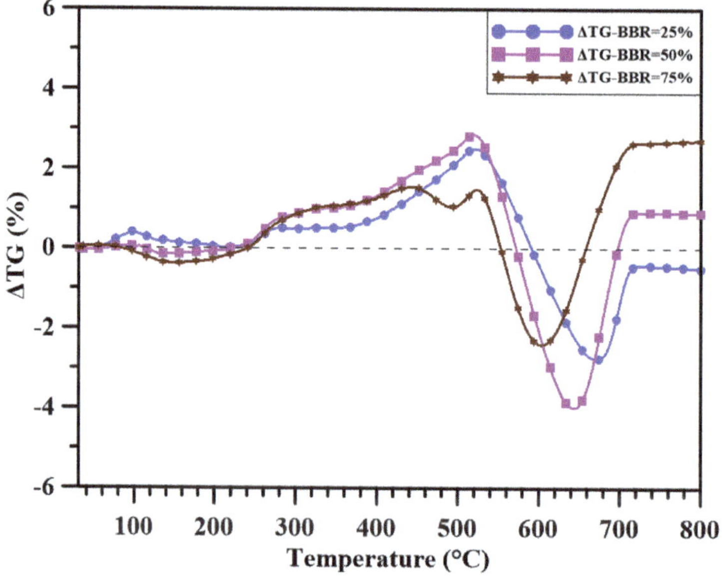

Figure 8. Degree of synergistic effect for various sludge-coal ratios.

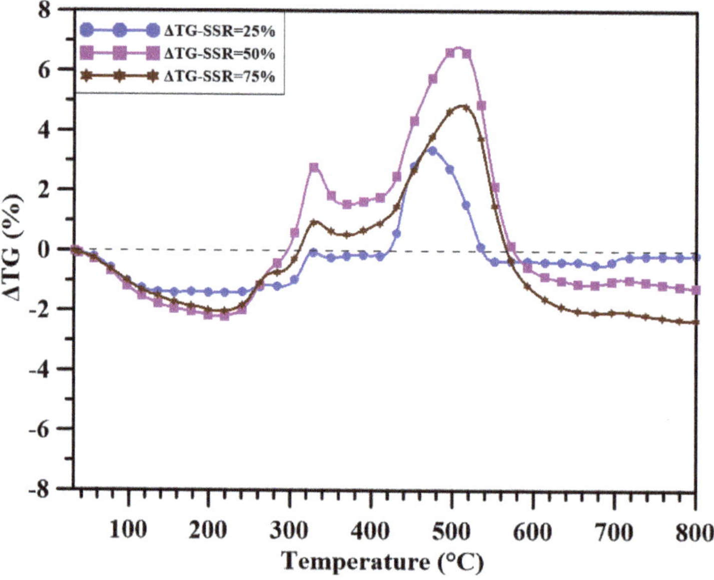

Figure 9. Degree of synergistic effect for various sludge-shiitake ratios.

Plots of the deviation between the experimental and calculated curves are shown in Figures 8 and 9. Those two figures clearly show that synergistic effects exist in the blends to a certain extent. For the BBR blends, positive synergy occurred during char oxidation. The catalytic effects of the sludge char might promote reactions at high temperature. However, at low temperature, the large quantity of released volatiles from the sludge may hinder the release of volatiles from the coal, which explains the negative synergistic effect. For the SSR blends, the addition of shiitake to the blends promoted reactions in the blend at both high and low temperatures. For all the blends, negative or low synergistic effects existed for temperatures between 300 and 500 °C. The values of the root mean square (RMS) of ΔW were all between 1.06 and 1.57 for the blends, higher than those used to measure synergistic effects and reported in previous studies [23,30], indicating that there were synergistic effects in the blends.

3.4. Combustion Characteristics Parameters

The combustion characteristics for the sludge-coal blends are shown in Table 3. Sludge high porosity and high surface area can explain why its devolatilization started at a lower temperature compared to the other fuels, and this led to a decrease in the ignition temperature with sludge addition to the blends in stage 1. Increasing sludge content in the blends resulted in an increase in burnout temperature in stage 1. Sewage sludge having higher devolatilization rate than Australian coal can explain why both the maximum and mean burning rates increased for the blends in the main oxidation stage (stage 1) with increasing sludge ratio, which also explained the increase in the values calculated for both the flammability index and the combustion characteristics index. The highest values for C and S in stage 1 for the blend were observed for BBR = 75% ($C_1 = 5.72 \times 10^{-5}$ and $S_1 = 3.91 \times 10^{-7}$, respectively). Sludge, having less material in the final decomposition stage (stage 2), showed the lowest maximum and mean burning rate values, and thus had the lowest values for both indexes. Positive synergistic effects in the blends decreased their burnout temperature. Although T_{b2} decreased with sludge addition to the blends, the flammability index and combustion characteristics index values decreased due to the decrease of the maximum and mean burning rates. The lowest values for both indexes in stage 2 were observed for BBR = 75%, and they are $C_2 = 2.15 \times 10^{-5}$ and $S_2 = 1.01 \times 10^{-7}$, respectively. On the other hand, sludge addition to the blends resulted in an increase in the ignition temperature, up to BBR = 50%, which then decreased with further sludge addition in stage 2.

The combustion characteristics for the sludge-shiitake blends are also shown in Table 3. As shown, the ignition and burnout temperatures of the blends both increased with shiitake addition in stage 1. The results also show that the flammability index and combustion characteristics index of the shiitake substrate have the highest values among SSRs for the main oxidation stage, namely, $C_1 = 19.30 \times 10^{-5}$ and $S_1 = 33.56 \times 10^{-7}$, respectively. The Shiitake high devolatilization rate may explain why both the maximum and mean burning rates of the blends in the main decomposition stage were higher for the blends with lower sludge ratio. As a result, both the flammability index and the combustion characteristics index in stage 1 decreased with increasing sludge ratio in the blends. C and S values increased from $C_1 = 9.12 \times 10^{-5}$ and $S_1 = 10.39 \times 10^{-7}$ for SSR = 25% to $C_1 = 14.61 \times 10^{-5}$ and $S_1 = 19.42 \times 10^{-7}$ for SSR = 75%. In the final oxidation stage (stage 2), the flammability index and the combustion characteristics index values of shitake are $C_2 = 3.56 \times 10^{-5}$ and $S_2 = 0.96 \times 10^{-7}$, respectively. A 25 wt.% addition of shiitake to the blend (SSR = 75%) increased both indexes of pure sludge, resulting from increases in the values of the maximum and mean burning rates of the blend. The values of C and S both decreased for SSR = 50%, with further shiitake addition, due to the synergistic effects between shiitake and sludge at higher temperatures. The highest value for the flammability index of the blends, $C_2 = 2.29 \times 10^{-5}$, was found for SSR = 25%. The highest values for the combustion characteristics index, $S_2 = 0.66 \times 10^{-7}$, was found for SSR = 75%.

Table 3. Combustion characteristics for different sludge-coal ratios (BBR) and sludge-shiitake ratios (SSR) (BBR = SSR = 100%: pure sludge).

BBR %	T_i		T_b		DTG_{max}		DTG_{mean}		$C\times10^5$		$S\times10^7$	
	T_{i1}	T_{i2}	T_{b1}	T_{b2}	Stage 1	Stage 2	Stage 1	Stage 2	C_1	C_2	S_1	S_2
0	-	438.29	-	709.15	-	6.33	-	5.15	-	3.30	-	2.39
25	247.30	458.42	306.90	693.62	1.13	5.68	0.96	4.68	1.85	2.70	0.58	1.82
50	240.97	462.41	357.28	665.57	2.11	4.74	1.77	3.93	3.64	2.22	1.81	1.31
75	238.31	446.67	367.44	629.45	3.25	4.29	2.51	2.95	5.72	2.15	3.91	1.01
100	239.40	439.32	368.57	633.29	4.80	3.94	3.48	1.65	8.37	2.04	7.90	0.53

SSR %	T_i		T_b		DTG_{max}		DTG_{mean}		$C\times10^5$		$S\times10^7$	
	T_{i1}	T_{i2}	T_{b1}	T_{b2}	Stage 1	Stage 2	Stage 1	Stage 2	C_1	C_2	S_1	S_2
0	277.43	426.59	404	680.82	14.86	6.48	7.02	1.83	19.30	3.56	33.56	0.96
25	272.38	457.96	417.81	666.85	10.84	4.80	5.56	1.68	14.61	2.29	19.42	0.58
50	266.82	469.67	413.85	688.17	6.95	3.52	4.49	1.67	9.76	1.59	10.59	0.39
75	253.93	436.20	379.99	620.62	5.88	4.14	4.33	1.89	9.12	2.18	10.39	0.66
100	239.40	439.32	368.57	633.29	4.80	3.94	3.48	1.65	8.37	2.04	7.90	0.53

3.5. Kinetic Parameters Analysis

According to the fitting results of linear regression using various kinetic mechanism equations, reaction orders of 1.2 and 1.5 were found to be the best model functions for stage 1 and stage 2, respectively; with corresponding correlation coefficients ranging from 0.950 to 0.994. The results of the calculated activation energy and frequency factor are shown in Table 4. As shown, the calculated E and A for stage 1 of sewage sludge are 80.42 kJ·mol^{-1} and 6.18×10^{12} s^{-1}, respectively. Those for shiitake are 97.07 kJ·mol^{-1} and 1.46×10^{14} s^{-1}, respectively. This shows that more energy is needed for the combustion of the volatiles in the shiitake substrate during stage 1. The kinetic parameters of the blends are non-additive, and therefore their values could not be predicted. BBR = 25% could be divided into two oxidation stages, and the calculated activation energy and frequency factor for stage 1 are 85.68 kJ·mol^{-1} and 2.80×10^{12}·s^{-1}, respectively. The addition of 50 wt.% sludge resulted in a decrease of both parameters due to the catalytic effect of sludge addition to the blend. Further sludge addition to the blend resulted in increases in both the activation energy and the frequency factor. This might be due to the occurrence of secondary reactions in the main oxidation stage. The activation energy and the pre-exponential factor are $E_1 = 80.90$ kJ·mol^{-1} and $A_1 = 3.83 \times 10^{12}$ s^{-1}, respectively, for BBR = 75%. For the sludge-shiitake blends, increases in both the activation energy and the frequency factor were observed in stage 1 with shiitake addition to the blends, except when 50 wt.% shiitake was added to the sludge, as shown in Table 4.

Table 4. Kinetic parameters for various sludge-coal ratios and sludge-shiitake ratios calculated using Coats-Redfern method (BBR = SSR = 100%: pure sludge).

BBR %	Temperature Range (°C)	E (kJ·mol^{-1})	A (s^{-1})	N	R^2
0	-	-	-	-	-
	278–730	105.74	1.06×10^{13}	1.20	0.971
25	210–308	85.68	2.80×10^{12}	1.50	0.991
	309–714	59.27	4.93×10^{9}	1.20	0.994
50	203–359	75.18	4.83×10^{11}	1.50	0.969
	360–692	54.70	3.48×10^{9}	1.20	0.993
75	202–369	80.90	3.83×10^{12}	1.50	0.950
	370–673	70.89	1.66×10^{11}	1.20	0.964
100	200–370	80.42	6.18×10^{12}	1.50	0.960
	371–624	64.89	1.07×10^{11}	1.20	0.981

SSR %	Temperature Range (°C)	E (kJ·mol^{-1})	A (s^{-1})	N	R^2
0	206–408	97.07	1.46×10^{14}	1.50	0.972
	408–579	62.93	8.23×10^{10}	1.20	0.989
25	206–432	87.00	1.55×10^{13}	1.50	0.956
	432–562	75.30	5.97×10^{11}	1.20	0.980
50	200–417	79.25	2.98×10^{12}	1.50	0.977
	417–632	61.80	4.38×10^{10}	1.20	0.983
75	194–382	84.33	1.25×10^{13}	1.50	0.982
	383–648	61.83	6.05×10^{10}	1.20	0.994
100	200–370	80.42	6.18×10^{12}	1.50	0.960
	371–624	64.89	1.07×10^{11}	1.20	0.981

A comparison of the activation energy values of the individual fuels in the final oxidation stage shows that Australian black coal has the highest value for E (105.74 kJ·mol^{-1}), as expected. The values for sludge and shiitake are 64.89 and 62.93 kJ·mol^{-1}, respectively. The catalytic effect of sludge addition to the blends was also noticed in the final oxidation stage. The activation energy and the pre-exponential factor both decreased with sludge addition, except for BBR = 75%, due to the occurrence of secondary reactions. The inorganic materials in the sludge catalytically promoted char formation and tar-cracking reactions in the sludge-coal blends, and the lignin in the substrate produced the same effect to the sludge-shiitake blends, which is shown by a decrease in the calculated E values for the sludge-shiitake blends. A previous study found that the presence of alkaline salts in biomass may lower the apparent activation energy of thermal reactions and promote the formation of char [36]. However, the opposite effect was observed with further addition of sludge and shiitake to the sludge-coal and sludge-shitake blends, respectively, due to further reactions between the heavier and complex compounds in the fuels.

3.6. Fourier Transform Infrared Spectroscopy

Figure 10 shows the CH_4 emission profiles for the sludge, the Australian black coal and their blends. The curve for sewage sludge shows two peaks, one at 280 °C and the other one at 474 °C, corresponding to two burning regions involving volatiles and fixed carbon. The release of CH_4 during the final decomposition stage was much lower than that in the main decomposition stage. The release of CH_4 in stage 1 is due to the primary pyrolysis of the volatiles and the cellulose in the sludge. A significant amount of the cellulose is converted into tar during the primary pyrolysis, and then some residual parts of the tar are converted into gases during the secondary reaction, which explains the occurrence of the second peak. The Australian coal, having low amount of volatile and one thermal decomposition stage, has one main peak around 471 °C. Adding sludge to the blends increased the amount of CH_4 released in the main decomposition stage, but decreased that of stage 2. The peaks for 25 wt.% sludge appeared to be the lowest. This was confirmed by taking the integral area under the curves and evaluating the gas yield using Equation (13). The results are shown in Table 5. They reveal that BBR = 25% yielded the lowest amount of CH_4. Figure 10 also shows the CH_4 emission profiles for the sludge-shiitake blends. Shiitake, having higher volatile and fixed carbon content than that of sludge, released a higher amount of CH_4 in both the main and final decomposition stages compared to that released by sludge. Adding shiitake to the blends reduced both peaks to approximately the same absorbance range, with SSR = 50% having the lowest peak for both stages. As shown in Table 5, adding sludge to the blends increased the amount of CH_4 emitted, and SSR = 25% showed the lowest methane emission.

The CO emission profiles for the individual fuels and their blends are shown in Figure 11. Only one absorption peak was observed for coal and the sludge-coal blends. In contrast, all profiles of the sludge-shiitake blends showed two peaks, in the temperature ranges of 275–400 °C and 425–575 °C, respectively. The first CO peak for sludge was observed at about 341 °C and the second one was observed at about 495 °C. The CO generation of stage 1 and stage 2 was attributed to the decomposition of the volatiles and decarbonylation reaction, respectively. Adding sludge to the blends reduced the absorption value of their peak, which is consistent with the results in Table 5. BBR = 75% yielded the lowest value (0.078). However, adding shiitake to the sludge-shiitake blends increased the CO emission in stage 1, which was compensated with a reduction in CO in stage 2. Overall, CO emission of the blends decreased with shiitake addition to the blends, with SSR = 25% having the lowest gas yield value (0.049).

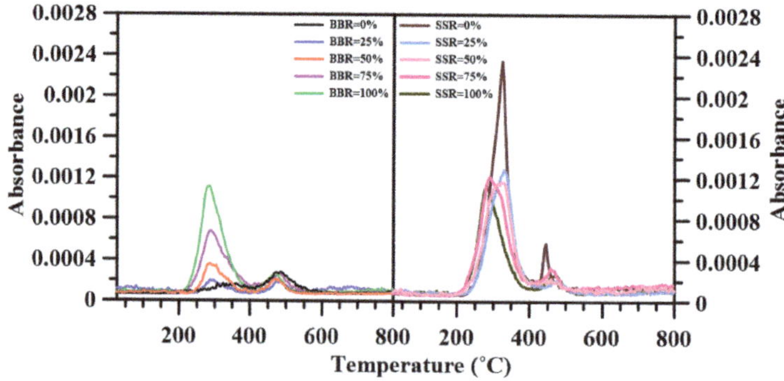

Figure 10. Evolution of CH_4 for various sludge-coal and sludge-shiitake ratios in TGA-FTIR.

Table 5. Evolved gas production rates of different sludge-coal ratios (BBR) and sludge-shiitake ratios (SSR) (BBR = SSR = 100%: pure sludge).

BBR %	CH$_4$	CO	CO$_2$	NO$_x$	SO$_2$
0%	0.010	0.186	5.035	0.275	0.180
25%	0.009	0.146	4.114	0.228	0.150
50%	0.010	0.124	3.757	0.271	0.180
75%	0.013	0.078	2.324	0.268	0.180
100%	0.017	0.052	2.042	0.337	0.227

SSR %	CH$_4$	CO	CO$_2$	NO$_x$	SO$_2$
0%	0.023	0.060	3.492	0.259	0.166
25%	0.018	0.049	2.466	0.203	0.131
50%	0.019	0.055	2.598	0.240	0.152
75%	0.021	0.061	2.639	0.296	0.195
100%	0.017	0.052	2.042	0.337	0.227

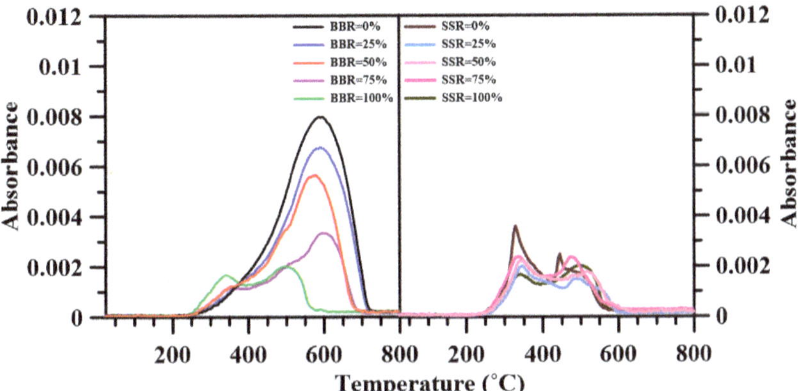

Figure 11. Evolution of CO for various sludge-coal and sludge-shiitake ratios in TGA-FTIR.

Figure 12 shows the CO$_2$ emission profiles of the respective fuels and their blends. All profiles displayed only one peak. The peak of the sludge-coal blends occurred at higher temperature, 425–700 °C, compared to that of the sludge-shiitake blends, which was observed between 300–600 °C. As shown in Figure 12, sludge addition to the blends decreased the amount of CO$_2$ emitted. BBR = 75% showed the lowest value of the yielded CO$_2$ of the blends, as shown in Table 5. In contrast, an addition of 25 wt.% shiitake to sludge led to an increase of about 30% for CO$_2$ emissions, as shown in Table 5. However, CO$_2$ emission decreased with further shiitake addition to the blends.

NO$_x$ formation during fuel combustion arises from three possible routes, thermal-NO$_x$, prompt-NO$_x$ and fuel-NO$_x$ [35]. Generally, thermal-NOx is generated from thermal oxidation of nitrogen at high temperature (>1500 °C). Although its emission increases with increasing oxygen concentration and combustion temperature, its role is negligible at temperatures below 1500 °C, as is the case in this study. The formation of prompt-NO$_x$ is more complex, involving CH radical intermediates that react with the nitrogen in air to form HCN, which further reacts to generate NO. The fuel-NO$_x$ formation mechanism is even more complicated. It is formed from the nitrogen content in the fuel via two routes: (1) by homogeneous reactions of the nitrogenous compounds of the volatiles and (2) via heterogeneous reactions of the nitrogen components bound to the char. For the experiments conditions in this study, most of the NO$_x$ came from fuel-NO$_x$. Figure 13 shows the NO$_x$ emission profiles of the individual fuels and their blends, showing one peak between 250 °C and 375 °C, which indicates significant NO emission produced by the volatile-N oxidation, for all the blends.

Sewage sludge, having the highest nitrogen content yielded the highest NO_x of the fuels. The yielded values of NO_x emission in Table 5 are all lower than that of pure coal, with NO_x emissions of BBR = 25% being the lowest. A similar observation was made for the yielded NO_x values for the sludge-shiitake blends in Table 5, with a decrease in NO_x emissions found with increasing shiitake ratio. The oxides in the char/ash play an active role in NO_x reduction. Lower oxygen concentration in the oxidizer was beneficial for reducing NO_x. As the volatiles in the blends increased, their combustion consumed the oxygen near the char particles, and therefore decreased the oxygen concentration around the char.

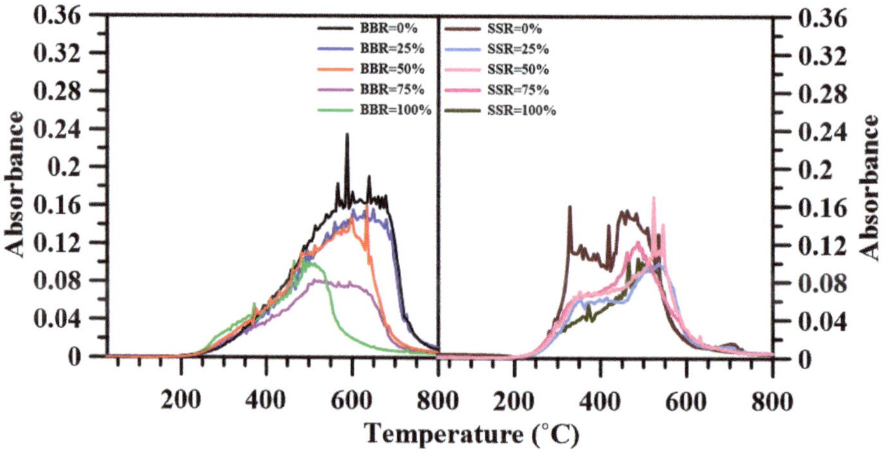

Figure 12. Evolution of CO_2 for various sludge-coal and sludge-shiitake ratios in TGA-FTIR.

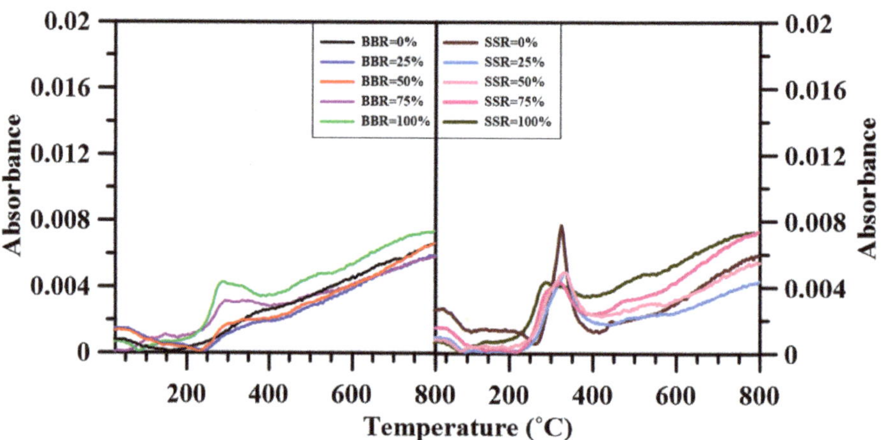

Figure 13. Evolution of NO_x for various sludge-coal and sludge-shiitake ratios in TGA-FTIR.

Figure 14 shows the SO_2 emission profiles for the individual fuels and their blends. As shown, all curves have smooth peaks, except the shiitake one. SO_2 generation was closely related to the sulfur content in the fuels, as shown by the results of Table 5. Sewage sludge, having the highest sludge content, showed the highest values of yielded SO_2. An addition of 25 wt.% sludge to the sludge-coal blends reduced the amount of yielded SO_2. An increase of SO_2 emission was observed with sludge further addition. The alkali and alkaline earth metal oxides in the sludge might act as a desulfurizing catalyst for such blends. However, further sludge addition of sludge increased the amount of yielded SO_2. For the sludge-shiitake blends, increasing the shiitake ratio decreased SO_2 emissions.

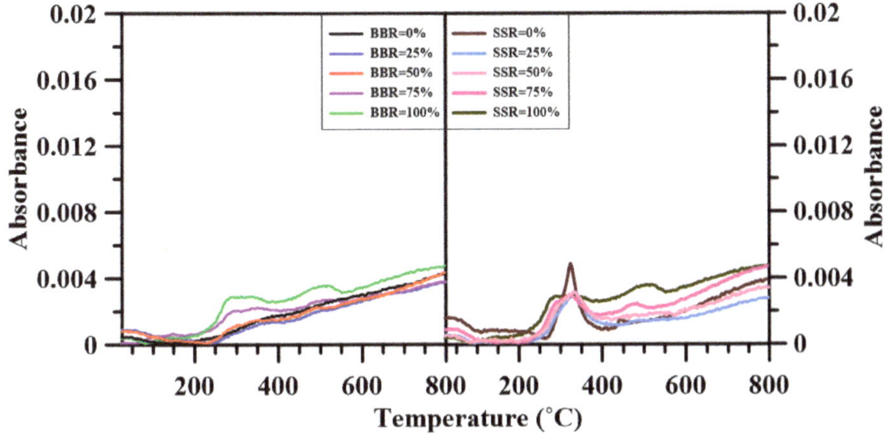

Figure 14. Evolution of SO_2 for various sludge-coal and sludge-shiitake ratios in TGA-FTIR.

In summary, BBR = 25% showed the lowest CH_4, NO_x and SO_2 emissions. The values of emitted CO and CO_2 decreased with addition of sludge to the blends. NO_x and SO_2 emissions both decreased with shiitake addition to the sludge-shiitake blends. An addition of 25 wt.% shiitake to the sludge increased the amount of CH_4, CO and CO_2. However, further shiitake addition decreased their respective yields.

3.7. Single-Pellet Combustion

The combustion process of a pure sludge pellet and the way used to demarcate some of the parameters are shown in Figure 15 for ambient temperatures of 700 °C. They show the volatiles, from the pellet, burning in the gas phase after the pellet fell into the hot environment before extinguishing. Then, char oxidation occurs homogeneously until burnout. The results of the studied parameters for the blends, i.e., ignition delay time (t_{id}), volatile combustion duration (t_f), and total combustion time (t_{tot}) for the single-pellet combustion experiments at 600 °C, 700 °C and 800 °C are shown in Tables 6 and 7. The weight loss history and its first derivative for pure sewage sludge pellet at 700 °C are presented in Figure 16 for illustration. Different temperatures have similar trends. An abrupt change in the pellet weight could be noticed during the devolatilization and volatiles combustion stage, whereas char combustion occurred at an almost constant weight loss rate until burnout.

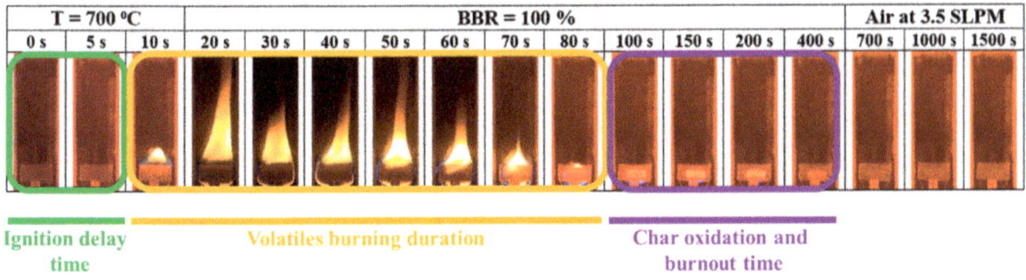

Figure 15. Graphic of stages in pure sewage sludge pellet combustion process.

Table 6. Single-pellet combustion characterization for various sludge-coal ratios (BBR = SSR = 100%: pure sludge).

BBR %	Single Pellet Combustion Characterization at 600 °C				HHV (MJ·kg⁻¹)
	t_{id} (s)	t_f (s)	t_{tot} (s)	Ash (wt.%)	
0	Indistinct	Indistinct	1611 ± 44.69	19.63 ± 0.87	26.10 ± 0.05
25	Indistinct	Indistinct	1371.80 ± 38.71	24.01 ± 0.79	22.48 ± 0.06
50	Indistinct	Indistinct	1174.20 ± 41.67	30.53 ± 0.19	18.64 ± 0.18
75	Indistinct	Indistinct	904.50 ± 27.93	36.46 ± 0.32	15.17 ± 0.03
100	26.27 ± 4.82	73.92 ± 2.07	615.6 ± 39.60	41.86 ± 0.82	11.91 ± 0.05

BBR %	Single Pellet Combustion Characterization at 700 °C				HHV (MJ·kg⁻¹)
	t_{id} (s)	t_f (s)	t_{tot} (s)	Ash (wt.%)	
0	9.04 ± 0.66	75.94 ± 1.83	1634.80 ± 41.89	18.23 ± 1.22	26.10 ± 0.05
25	7.22 ± 0.49	83.89 ± 2.38	1517.40 ± 100.34	23.94 ± 0.23	22.48 ± 0.06
50	6.71 ± 0.28	82.23 ± 0.10	1179.80 ± 82.01	30.19 ± 0.58	18.64 ± 0.18
75	7.14 ± 0.42	80.62 ± 2.06	894 ± 22.77	34.96 ± 0.94	15.17 ± 0.03
100	7.60 ± 0.39	69.03 ± 0.48	519.40 ± 19.70	41.76 ± 1.14	11.91 ± 0.05

BBR %	Single Pellet Combustion Characterization at 800 °C				HHV (MJ·kg⁻¹)
	t_{id} (s)	t_f (s)	t_{tot} (s)	Ash (wt.%)	
0	3.67 ± 0.32	73.30 ± 1.20	1534.60 ± 29.12	17.58 ± 0.51	26.10 ± 0.05
25	2.75 ± 0.21	76.88 ± 1.59	1223.40 ± 36.59	23.28 ± 0.46	22.48 ± 0.06
50	2.47 ± 0.18	75.88 ± 0.96	1031.80 ± 27.68	29.12 ± 1.08	18.64 ± 0.18
75	3.04 ± 0.19	68.47 ± 0.54	783.60 ± 41.21	34.39 ± 0.44	15.17 ± 0.03
100	3.47 ± 0.07	61.49 ± 2.09	458 ± 21.62	40.90 ± 1.15	11.91 ± 0.05

Table 7. Single-pellet combustion characterization for various sludge-shiitake ratios.

SSR %	Single Pellet Combustion Characterization at 600 °C				HHV (MJ·kg⁻¹)
	t_{id} (s)	t_f (s)	t_{tot} (s)	Ash (wt.%)	
0	19.52 ± 3.61	86.89 ± 3.01	689.6 ± 12.92	13.43 ± 1.06	16.65 ± 0.06
25	27.10 ± 0.96	77.64 ± 0.65	646.40 ± 19.74	22.95 ± 0.49	15.20 ± 0.01
50	34.94 ± 0.51	68.53 ± 1.00	618.40 ± 22.42	29.82 ± 0.63	14.03 ± 0.06
75	33.47 ± 6.22	67.12 ± 5.53	597.40 ± 29.21	37.40 ± 0.28	12.69 ± 0.02
100	26.27 ± 4.82	73.92 ± 2.07	615.6 ± 39.60	41.86 ± 0.82	11.91 ± 0.05

SSR %	Single Pellet Combustion Characterization at 700 °C				HHV (MJ·kg⁻¹)
	t_{id} (s)	t_f (s)	t_{id} (s)	Ash (wt.%)	
0	5.36 ± 0.59	84.91 ± 1.65	602.20 ± 21.01	13.70 ± 0.61	16.65 ± 0.06
25	5.48 ± 0.44	79.88 ± 2.43	599.40 ± 33.28	22 ± 0.43	15.20 ± 0.01
50	6.69 ± 0.54	78.76 ± 1.06	578.20 ± 30.72	28.53 ± 1.08	14.03 ± 0.06
75	7.54 ± 0.51	73.72 ± 1.06	572.80 ± 13.14	36.42 ± 1.10	12.69 ± 0.02
100	7.60 ± 0.39	69.03 ± 0.48	519.40 ± 19.70	41.76 ± 1.14	11.91 ± 0.05

SSR %	Single Pellet Combustion Characterization at 800 °C			HHV (MJ·kg⁻¹)	
	t_{id} (s)	t_f (s)	t_{tot} (s)	t_{id} (s)	t_f (s)
0	2.16 ± 0.18	77.53 ± 1.20	0	2.16 ± 0.18	77.53 ± 1.20
25	2.27 ± 0.50	76.60 ± 1.21	25	2.27 ± 0.50	76.60 ± 1.21
50	3.00 ± 0.04	72.41 ± 0.07	50	3.00 ± 0.04	72.41 ± 0.07
75	3.03 ± 0.34	67.17 ± 1.25	75	3.03 ± 0.34	67.17 ± 1.25
100	3.47 ± 0.07	61.49 ± 2.03	100	3.47 ± 0.07	61.49 ± 2.03

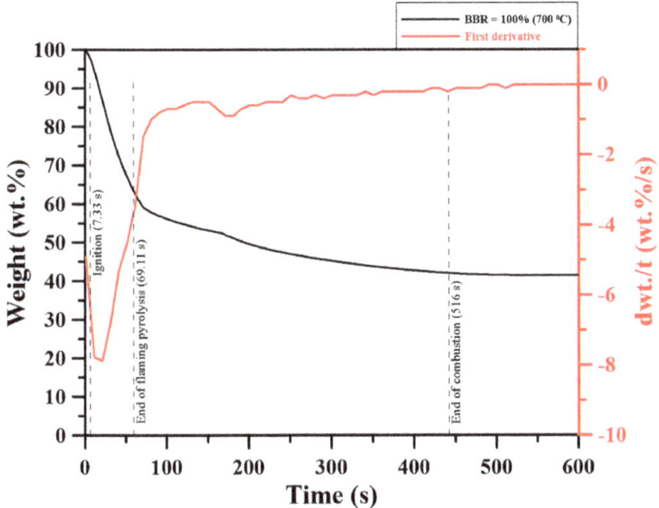

Figure 16. Weight loss history of pure sewage sludge pellet combustion process at 700 °C.

At 600 °C, the pure sludge pellets (BBR = 100%) took on average 26.27 s to ignite. No gas-phase ignition occurred for pure coal pellets and sludge-coal blends pellets with BBR = 25–75%. This might be because the volatiles released from Australian coal and the blends have an auto-ignition temperature that is higher than 600 °C. As shown from the FTIR experiments, the hydrocarbons released from the devolatilization process of the blends with BBR= 25–75% (peaks at 225–425 and 425–550 °C) turned into CO at higher temperature (peak between 550–650 °C), which then oxidized. Methane and carbon monoxide auto-ignition temperature was reported to be higher than 600 °C [37,38]. However, in contrast to the sludge-coal pellets, all the sludge-shiitake pellets ignited at 600 °C. This is consistent with the FTIR results, which shows that the hydrocarbons released from the devolatilization process of the blends with SSR= 25–75% (peaks at 200–400 and 400–500 °C) turned into CO at lower temperature (peaks between 250–475 and 475–575 °C) than those in sludge-coal blends, which then oxidized.

All the BBR pellets ignited in the gas phase at 700 °C. It took on average 7.60 s for the pellets made of sewage sludge to ignite, and pure coal pellets ignited after 9.04 s. Sludge pellets, having more volatiles content, took less time to ignite compared to pure coal ones; as predicted by Equation (15). The ignition delay time decreased for BBR = 25% and 50%. The addition of sludge to the blends increased their volatiles content, resulting in a decrease of their ignition delay time. The slight increase in ignition delay time, might be due to the endothermic evaporation of moisture delayed volatiles evolution. Adding more sludge to the blend might increase the required drying time. A similar trend was observed at 800 °C. Higher volatile content of shiitake explains why pellets made of pure shiitake took the lowest amount of time to ignite, as shown in Table 7. Adding shiitake to the blends decreased their ignition delay time.

The combustion of volatiles, t_f, is an important step in the combustion process of solid fuels, as the maximum weight loss takes place in this stage. Although the sludge volatile content is higher than that of coal, the pure coal pellet flame lasted longer than that of pure sludge pellets. A majority of the volatiles in the sludge pellet vaporized before homogeneous gas phase combustion took place. Adding 25 wt.% sludge to the blend increased the volatile combustion duration from 75.94 to 83.89 s at 700 °C, which is due to the increased volatiles content of the blend. A decrease in the volatile combustion duration was observed for blends with sludge content greater than or equal to 50 wt.%. Pellets with a higher sludge ratio have more volatiles and, in theory, should have a longer volatile combustion duration. However, conducting TGA experiments of sewage sludge under pyrolysis conditions shows that volatiles in sludge escape at a higher rate in sludge than in coal. Therefore, the volatile combustion duration reduced for BBR = 50–75% pellets. Moreover, the pure sludge pellet volatile flame lasted 69.03 s on average, the shortest time. On average, volatile combustion took about 13.36% of the total combustion time for the pure sludge pellets; for pure coal pellet, volatile combustion took about 4.71% of the total combustion time at 700 and 800 °C. Single-pellet combustion experiments conducted at 800 °C showed similar patterns to those observed at 700 °C. For the sludge-shiitake blends, t_f increased with shiitake addition to the blends. On average, volatile combustion took about 14.10% of the total combustion time for the pure shiitake pellets, 13.29% for the pure sludge pellets, and 4.65% for the pure coal pellets.

The total combustion (t_{tot}) time for various BBR pellets is also shown in Table 6. As expected, the total combustion time of the coal pellet was much greater than that of the sludge pellet. Since coal has more fixed carbon content than does sewage sludge, its char oxidation takes longer than that of sludge at the same ambient temperature, which explains why pure coal pellets had the longest total combustion time. A linear decrease in the total combustion time was observed with the addition of sludge to the blends, with the pure sludge pellets having the shortest total combustion time. Char formed from sludge is more reactive than that of coal and possesses more pores and thus takes less time to burnout. This explains why the addition of sewage sludge to the blends significantly decreased their total combustion time for all ambient temperatures. As shown in Table 7, the char of the shiitake pellets, having more fixed carbon, took more time to burnout than that of the sludge pellet. A linear increase in the total combustion time was observed with shiitake addition to the blend, due to the increase in fixed carbon content in the blends. Shiitake addition to the sludge can help with the issues related to ash, such as de-ashing, ash transport, storage, and disposal. Shiitake addition to the sludge may also minimize fouling, as the blend deposited ash might retain the low-melting-point salts in sludge.

Sewage sludge, having higher inorganic content than coal, left on average more residues after the pellet combustion experiments. About 41.50% of the initial weight of the sludge pellet was left on the stainless mesh platform. The pure coal pellets left, on average, a residual weight of about 18.48%. As shown in Table 6, ash formed in the single-pellet combustion experiments increased with the addition of sewage sludge to the blends. Experiments conducted at higher temperatures showed that a high temperature in the combustor enhanced the volatilization of some of the ash-forming heavy metals resulting in lower residues. However, it should be mentioned that such a decrease in bottom ash might cause an increase in the amount of fly ash carried with the flue gas. Shiitake addition to the sludge can help with the issues related to ash, such as de-ashing, ash transport, storage, and disposal. Shiitake addition to the sludge may also minimize fouling, as the blend deposited ash might retain the low-melting-point salts in sludge. The results in Table 7 show that shiitake addition to the blends decreased the amount of ash formed in the single-pellet combustion experiments.

The heating values of the blended fuels were measured, and the results are shown in the last column in Tables 6 and 7 for reference. As shown, adding sludge to the blends decreased their HHV, with the HHV for BBR = 25% and 75% being 22.48 and 15.17 MJ·kg^{-1}, respectively. Adding shiitake to the sludge also had a positive effect on the heating values of the blended fuels. As the heating value of the blended fuels increased with shiitake addition. The heating values for SSR = 75% and 25% being 12.69 and 15.20 MJ·kg^{-1}, respectively.

4. Conclusions

In this study, the combustion behavior of sewage sludge, Australian black coal, a shiitake substrate, and their blends was analyzed via thermogravimetric analysis and FTIR analysis. The findings can be summarized as follow:

1. Compared to both the shiitake substrate and Australian coal, dried sewage sludge has the lowest carbon content and the highest nitrogen and sulfur content. Furthermore, the sludge hydrogen and oxygen content levels are higher than those of Australian black coal, but lower than those of the shiitake substrate. Sewage sludge has the lowest HHV. However, on a dry and ash-free basis,

its HHV is similar to that of shiitake. The proximate analysis results show that sewage sludge has the lowest fixed carbon content and the highest ash content on an as-received basis of the fuels. The shiitake substrate has the highest amount of volatiles.

2. Both pure sludge and shiitake have higher flammability index and combustion characteristics index in the main decomposition stage compared to those of coal. Sludge addition to coal increased both indexes in stage 1 and a similar observation was made for the sludge-shiitake blends with shiitake addition to the sludge. In the final oxidation stage, adding sludge to the blends decreased their flammability index and combustion characteristics index, due to the low decomposition rate of the heavier and complex compounds in the sludge. For the sludge-shiitake blends, a 25 wt.% sludge addition to the sludge increases both indexes. The flammability index and the combustion characteristics index both decreased for SSR = 50%; and then increased with 75 wt.% shiitake addition to the sludge.

3. Synergistic effects occurred for both the sludge-coal blends and sludge-shiitake blends. For all the blends, negative or low synergistic effects existed in the temperature range of 300 and 500 °C. For higher temperature, during the char oxidation stage, positive synergistic effects occurred. The addition of shiitake to the blends had positive synergy for all the blending ratios. However, when the BBR $\geq$ 50 wt.%, negative synergy occurred at high temperature.

4. The results of the kinetic parameters showed that a reduction in both parameters occurred for BBR = 50% during the main oxidation stage. A similar observation was made for the sludge-shiitake blends. The catalytic effect of the sludge and the shiitake is pronounced in the final oxidation stage. A decrease in both the activation energy and the frequency factor occurred with sludge addition to the coal due to the catalytic effect of the inorganic materials in the sludge. A similar observation was made for the sludge-shiitake blends, as the lignin in the substrate catalytically promoted the reactions in the final oxidation stage. However, due to further reactions between the heavier and complex compounds in the samples, both the activation energy and the frequency factor increased with further sludge or shiitake addition to their respective blends (BBR = 75% and SSR = 25%).

5. The FTIR results show that adding sludge to the sludge-coal blends decreased both CO_2 and CO emissions. A decrease in the emitted CH_4, NO_x and SO_2 was found for BBR = 25%. However, further addition of sludge to the blends increased their respective emissions. For the sludge-shiitake blends, both yielded values of NO_x and SO_2 decreased with shiitake addition. A 25 wt.% shiitake addition to the sludge (SSR = 75%) increased the amount of emitted CH_4, CO and CO_2. However, further shiitake addition decreased their respective yields.

6. For single-pellet combustion experiments, adding sludge to coal decreased the ignition delay time of the pellet to a certain extent. However, adding sludge to shiitake increased the ignition delay time. Volatile combustion durations of the blends were longer than those of the individual fuels, with BBR = 25%. For other cases, they decreased with sludge addition to the blends. The volatile burning duration decreased with further sludge addition. Increasing the sludge ratio sharply decreased the total combustion time and increased the percentage of residual weight (ash) in the single-pellet combustion experiments.

Author Contributions: G.-B.C. and T.-H.L. contributed to the concept, results explanation and writing of the paper. S.C. performed the experiments and analyzed the data. H.-T.L. and F.-H.W. contributed to experimental method and paper review.

Funding: The Ministry of Science and Technology (MOST), Taiwan (R.O.C.), financially supported this research under contract number MOST 106-221-E-006-185.

Conflicts of Interest: The authors declare no conflicts of interest.

References

1. Environmental Protection Administration, R.O.C. *Review of Sludge Treatment Status and Response Strategies*; Environmental Protection Administration: Taipei, Taiwan, 2014.
2. Lundin, M.; Olofsson, M.; Pettersson, G.; Zetterlund, H. Environmental and economic assessment of sewage sludge handling options. *Resour. Conserv. Recycl.* **2004**, *41*, 255–278. [CrossRef]
3. Muchuweti, M.; Birkett, J.; Chinyanga, E.; Zvauya, R.; Scrimshaw, M.D.; Lester, J. Heavy metal content of vegetables irrigated with mixtures of wastewater and sewage sludge in Zimbabwe: Implications for human health. *Agric. Ecosyst. Environ.* **2006**, *112*, 41–48. [CrossRef]
4. Werther, J.; Ogada, T. Sewage sludge combustion. *Prog. Energy Combust. Sci.* **1999**, *25*, 55–116. [CrossRef]
5. European Union. Directive 2009/28/EC of the European Parliament and of the Council of 23 April 2009 on the promotion of the use of energy from renewable sources and amending and subsequently repealing Directives 2001/77/EC and 2003/30/EC. *Off. J. Eur. Union* **2009**, *5*, 2009.
6. Vassilev, S.V.; Baxter, D.; Andersen, L.K.; Vassileva, C.G.; Morgan, T.J. An overview of the organic and inorganic phase composition of biomass. *Fuel* **2012**, *94*, 1–33. [CrossRef]
7. García, G.; Arauzo, J.; Gonzalo, A.; Sánchez, J.L.; Ábrego, J. Influence of feedstock composition in fluidised bed co-gasification of mixtures of lignite, bituminous coal and sewage sludge. *Chem. Eng. J.* **2013**, *222*, 345–352. [CrossRef]
8. Vamvuka, D.; Sfakiotakis, S.; Saxioni, S. Evaluation of urban wastes as promising co-fuels for energy production—A TG/MS study. *Fuel* **2015**, *147*, 170–183. [CrossRef]
9. Cui, H.; Ninomiya, Y.; Masui, M.; Mizukoshi, H.; Sakano, T.; Kanaoka, C. Fundamental Behaviors in Combustion of Raw Sewage Sludge. *Energy Fuels* **2006**, *20*, 77–83. [CrossRef]
10. Miller, R.; Bellan, J. Analysis of reaction products and conversion time in the pyrolysis of cellulose and wood particles. *Combust. Sci. Technol.* **1996**, *119*, 331–373. [CrossRef]
11. Koppejan, J.; van Loo, S. *The Handbook of Biomass Combustion and Co-Firing*; Earthscan: London, UK, 2008.
12. Rulkens, W. Sewage sludge as a biomass resource for the production of energy: Overview and assessment of the various options. *Energy Fuels* **2007**, *22*, 9–15. [CrossRef]
13. Stasta, P.; Boran, J.; Bebar, L.; Stehlik, P.; Oral, J. Thermal processing of sewage sludge. *Appl. Therm. Eng.* **2006**, *26*, 1420–1426. [CrossRef]
14. Roy, M.M.; Dutta, A.; Corscadden, K.; Havard, P.; Dickie, L. Review of biosolids management options and co-incineration of a biosolid-derived fuel. *Waste Manag.* **2011**, *31*, 2228–2235. [CrossRef] [PubMed]
15. Otero, M.; Díez, C.; Calvo, L.F.; García, A.I.; Morán, A. Analysis of the co-combustion of sewage sludge and coal by TG-MS. *Biomass Bioenergy* **2002**, *22*, 319–329. [CrossRef]
16. Ninomiya, Y.; Zhang, L.; Sakano, T.; Kanaoka, C.; Masui, M. Transformation of mineral and emission of particulate matters during co-combustion of coal with sewage sludge. *Fuel* **2004**, *83*, 751–764. [CrossRef]
17. Folgueras, M.B.; Díaz, R.M.; Xiberta, J.; Prieto, I. Thermogravimetric analysis of the co-combustion of coal and sewage sludge. *Fuel* **2003**, *82*, 2051–2055. [CrossRef]
18. Kijo-Kleczkowska, A.; Środa, K.; Kosowska-Golachowska, M.; Musiał, T.; Wolski, K. Experimental research of sewage sludge with coal and biomass co-combustion, in pellet form. *Waste Manag.* **2016**, *53*, 165–181. [CrossRef] [PubMed]
19. Taiwan Agricultural Research Institute Council of Agriculture, Executive Yuan, Alternative Substrates for Growing Mushrooms. Available online: https://www.tari.gov.tw/english/form/index-1.asp?Parser=20,15,926,81,,,3007 (accessed on 22 July 2015).
20. Taiwan Business TOPICS, Fungus among Us—The History of Mushrooms in Taiwan. Available online: https://topics.amcham.com.tw/2017/01/fungus-among-us-history-mushrooms-taiwan/ (accessed on 16 January 2017).
21. Cassel, B.; Menard, K.; Shelton, C.; Earnest, C. *Proximate Analysis of Coal and Coke Using the STA 8000 Simultaneous Thermal Analyzer*; PerkinElmer Application Note; PerkinElmer: Waltham, MA, USA, 2012.
22. Wu, Z.; Wang, S.; Zhao, J.; Chen, L.; Meng, H. Synergistic effect on thermal behavior during co-pyrolysis of lignocellulosic biomass model components blend with bituminous coal. *Bioresour. Technol.* **2014**, *169*, 220–228. [CrossRef] [PubMed]
23. Sadhukhan, A.K.; Gupta, P.; Goyal, T.; Saha, R.K. Modelling of pyrolysis of coal–biomass blends using thermogravimetric analysis. *Bioresour. Technol.* **2008**, *99*, 8022–8026. [CrossRef]

24. Lu, J.J.; Chen, W.H. Investigation on the ignition and burnout temperatures of bamboo and sugarcane bagasse by thermogravimetric analysis. *Appl. Energy* **2015**, *160*, 49–57. [CrossRef]
25. Chen, G.-B.; Li, Y.-H.; Lan, C.-H.; Lin, H.-T.; Chao, Y.-C. Micro-explosion and burning characteristics of a single droplet of pyrolytic oil from castor seeds. *Appl. Therm. Eng.* **2017**, *114*, 1053–1063. [CrossRef]
26. Jin, Y.; Li, Y.; Liu, F. Combustion effects and emission characteristics of SO2, CO, NOx and heavy metals during co-combustion of coal and dewatered sludge. *Front. Environ. Sci. Eng.* **2016**, *10*, 201–210. [CrossRef]
27. Coats, A.W.; Redfern, J. Kinetic parameters from thermogravimetric data. *Nature* **1964**, *201*, 68. [CrossRef]
28. Gao, N.; Li, A.; Quan, C.; Du, L.; Duan, Y. TG–FTIR and Py–GC/MS analysis on pyrolysis and combustion of pine sawdust. *J. Anal. Appl. Pyrolysis* **2013**, *100*, 26–32. [CrossRef]
29. Xu, T.; Huang, X. Study on combustion mechanism of asphalt binder by using TG–FTIR technique. *Fuel* **2010**, *89*, 2185–2190. [CrossRef]
30. Oudghiri, F.; Allali, N.; Quiroga, J.M.; Rodríguez-Barroso, M.R. TG–FTIR analysis on pyrolysis and combustion of marine sediment. *Infrared Phys. Technol.* **2016**, *78*, 268–274. [CrossRef]
31. Fan, C.; Yan, J.; Huang, Y.; Han, X.; Jiang, X. XRD and TG-FTIR study of the effect of mineral matrix on the pyrolysis and combustion of organic matter in shale char. *Fuel* **2015**, *139*, 502–510. [CrossRef]
32. Mau, V.; Gross, A. Energy conversion and gas emissions from production and combustion of poultry-litter-derived hydrochar and biochar. *Appl. Energy* **2018**, *213*, 510–519. [CrossRef]
33. Wang, C.; Wu, Y.; Liu, Q.; Yang, H.; Wang, F. Analysis of the behaviour of pollutant gas emissions during wheat straw/coal cofiring by TG–FTIR. *Fuel Process. Technol.* **2011**, *92*, 1037–1041. [CrossRef]
34. Yao, Z.; Ma, X.; Wu, Z.; Yao, T. TGA–FTIR analysis of co-pyrolysis characteristics of hydrochar and paper sludge. *J. Anal. Appl. Pyrolysis* **2017**, *123*, 40–48. [CrossRef]
35. Law, C.K. *Combustion Physics*; Cambridge University Press: Cambridge, UK, 2006.
36. Nassar, M.M. Thermal Analysis Kinetics of Bagasse and Rice Straw. *Energy Sources* **1998**, *20*, 831–837. [CrossRef]
37. Robinson, C.; Smith, D.B. The auto-ignition temperature of methane. *J. Hazard. Mater.* **1984**, *8*, 199–203. [CrossRef]
38. Mc Dougall, I. Chapter 4—Ferroalloys Processing Equipment A2—Gasik, Michael. In *Handbook of Ferroalloys*; Butterworth-Heinemann: Oxford, UK, 2013; pp. 83–138.

Article

Effect of Accelerated High Temperature on Oxidation and Polymerization of Biodiesel from Vegetable Oils

Jae-Kon Kim [1],*,†, Cheol-Hwan Jeon [1],†, Hyung Won Lee [2], Young-Kwon Park [2], Kyong-il Min [1], In-ha Hwang [1] and Young-Min Kim [3],*

[1]   Research Institute of Petroleum Technology, Korea Petroleum Quality & Distribution Authority, Cheongju 28115, Korea; chjeon@kpetro.or.kr (C.-H.J.); muggu@kpetro.or.kr; (K.-i.M.); ihhwang@kpetro.or.kr (I.-h.H.)

[2]   School of Environmental Engineering, University of Seoul, Seoul 02504, Korea; adexhw@nate.com (H.W.L.); catalica@uos.ac.kr (Y.-K.P.)

[3]   Department of Environmental Sciences and Biotechnology, Hallym University, Chuncheon 24252, Korea

*   Correspondence: jkkim@kpetro.or.kr (J.-K.K.); analyst@hallym.ac.kr (Y.-M.K.); Tel.: +82-43-240-7932 (J.-K.K.); 7932 +82-33-248-332 (Y.-M.K.)

†   Co-first authors.

Received: 2 November 2018; Accepted: 13 December 2018; Published: 17 December 2018

Abstract: Oxidation of biodiesel (BD) obtained from the decomposition of biomass can damage the fuel injection and engine parts during its use as a fuel. The excess heating of vegetable oils can also cause polymerization of the biodiesel. The extent of BD oxidation depends on its fatty acid composition. In this study, an accelerated oxidation test of BDs at 95 °C was investigated according to ASTM D 2274 by applying a long-term storage test for 16 weeks. The density, viscosity, and total acid number (TAN) of BDs increased because of the accelerated oxidation. Furthermore, the contents of unsaturated fatty acid methyl esters (FAMEs), C18:2 ME, and C18:3 ME in BDs decreased due to the accelerated oxidation. The ^{1}H-nuclear magnetic resonance spectrum of BDs that were obtained from the accelerated high temperature oxidation at 180 °C for 72 h differed from that of fresh BDs. The mass spectrum obtained from the analysis of the model FAME, linoleic acid (C18:2) methyl ester, which was oxidized at high temperature, indicated the formation of dimers and epoxy dimers of linoleic acid (C18:2) methyl ester by a Diels-Alder reaction.

Keywords: biodiesel; fatty acid methyl ester; free fatty acids; oxidation stability; antioxidant

---

1. Introduction

The use of biomass for the production of biofuels has been encouraged because of the severe global issues that are related to climate change and energy security [1,2]. Biodiesel (BD), a mixture of fatty acid methyl esters (FAMEs) extracted from vegetables or animal fats, is a renewable fuel that can be used for diesel engines [3–5]. BD can be used directly or after blending with petroleum diesel (PD) without the need to significantly modify conventional diesel engines. The emissions of particulate matter and exhaust gas, such as hydrocarbons, sulfur dioxide, and carbon monoxide, can also be reduced by blending PD with BD with the decrease of smoke opacity [6–9].

Bai et al. [10] reported the relationship between the blending ratio of BDs in diesel fuel and their effects in engine and environmental aspects and recommended to mix 20–50% of BDs to provide the environmental and technical advantages. High pressure (e.g., 180 MPa) pumping is applied to modern light duty diesel engines to provide short injection, good spray, and air mixing of diesel fuels [11]. The excess fuel is returned to the fuel tank by the recirculation system, which houses the fuel injectors. Owing to fuel recirculation, the oil can be partially oxidized and mixed with fresh oil in the tank after it cools the fuel injection system. The temperature of the recirculating oil can rise to 60–150 °C, while it cools the hot injection system and makes contact with high temperature air [12–16].

BD, obtained from common biomass feedstocks, is a mixture of FAMEs consisting of fatty acid chains having different numbers of carbon atoms (16 or 18) and double bonds (0~3). The oxidation of FAMEs forms peroxy radicals and it is initiated by the formation of free radicals at the allylic position and the reaction with oxygen [17–19]. The peroxy radicals formed are converted to hydroperoxides and alkyl radicals via intramolecular isomerization or by the extraction of hydrogen from other fatty acid chains. The hydroperoxides can be converted to hydroxyl and alkoxy radicals, which can decompose to alkyl aldehydes and alkyl radicals, produce acids via oxidation, or undergo polymerization by reacting with other alkenes [20–25].

The content of unsaturated fatty acids in BD is directly related to the extent of oxidation of BD, because the rate of oxidation increases with the number of double bonds in fatty acid chains [26]. Owing to the importance of fuel stability during its use as a fuel [23], the use of BD with high contents of double bonds (e.g., linolenic acid (C18:3) or linoleic acid (C18:2); Figure 1) in FAMEs is being limited, even though they can provide excellent cold flow properties.

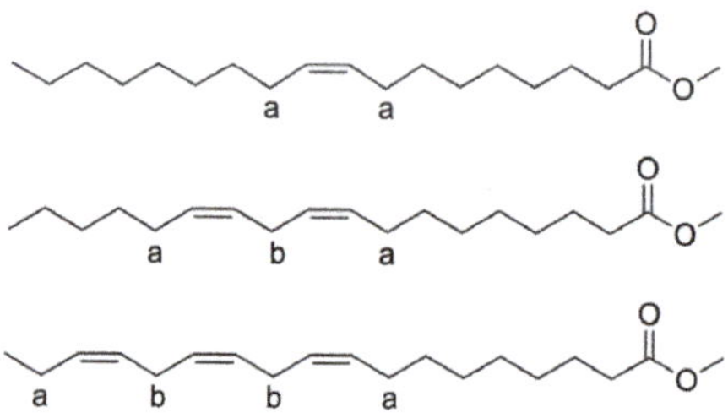

Figure 1. Common unsaturated C18 methyl esters in biodiesel: C18:1 ME, oleic acid methyl ester, C18:2 ME, linoleic acid methyl ester, and C18:3 ME, linolenic acid methyl ester, with (**a**) allylic positions and (**b**) bis-allylic positions.

In addition, the presence of oxygenates (e.g., acids, alcohols, aldehydes, and peroxides) in BD causes the accumulation of deposits and gum, as well as an increase in the fuel darkness, which decreases BD stability during long-term storage [27,28]. The presence of air, moisture, heat, and other contaminants can also decrease the stability of BD, because they can cause air oxidation, thermal oxidation, wet hydrolysis, microbial degradation, etc. [11–13], which can drastically change fuel properties [29,30]. Among various kinds of reactions, oxidation is considered to be the most significant contributor to the instability of BD, because of the ease with which it occurs during long-term storage of BD. Excessive degradation of BD can increase the contents of high molecular insoluble polymers, which can not only have negative effects (e.g., coking, clogging, and contamination deposits) on many vehicle parts (e.g., injector, fuel filter, fuel lines, and the engine), but also cause incomplete combustion [17,31].

Biodiesel stability tests can be classified by their purpose into the thermal stability test (ASTM D6468-08), storage stability test (ASTM D4625-16, ASTM D5304-15 and EN 15751), and oxidation stability tests (ASTM D2274, ASTM D7545-14, and EN 14112) [20,28–30,32].

In this study, the oxidation stability of commercial grade BDs was examined using ASTM D2274 by applying accelerated oxidation conditions normally at 95 °C during long-term storage for 16 weeks to provide the handling guidelines on the production, use, and storage of BDs. The accelerated high temperature oxidation at 180 °C for 72 h was also investigated to know the polymerization of fresh and oxidized unsaturated FAMEs.

2. Materials and Methods

Figure 2 shows the experimental procedure for the accelerated oxidation and polymerization of BD and linoleic acid methyl esters and product analysis methods. The change in total acid number (TAN), density, kinematic viscosity, and C18 FAME content during the accelerated oxidation were measured to evaluate the BD quality. Linoleic acid (C18:2) methyl ester was also used as a model FAME, because it can be polymerized at high temperatures. After polymerization at 180 °C (72 h), the polymerized linoleic acid methyl ester and the oxidized BDs were purified by flash chromatography to the polymerized FAME products and were evaluated using [1]H-nuclear magnetic resonance (NMR, Bruker, Billerica, MA, USA) spectroscopy and normal phase high performance liquid chromatography-atmospheric pressure chemical ionization mass spectrometry (HPLC-APCI MS, Thermo Fisher Scientific, Bannockburn, IL, USA).

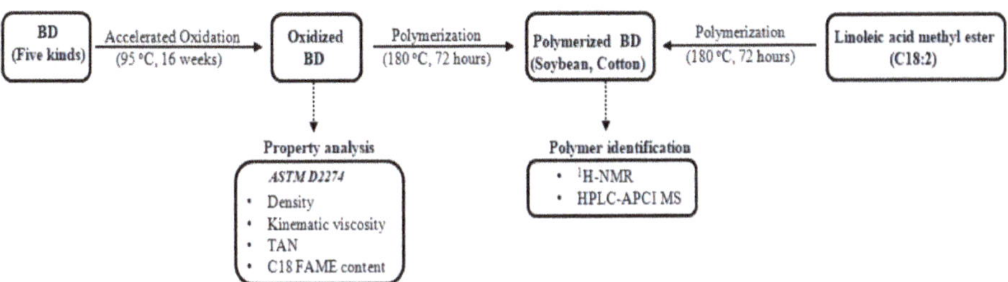

Figure 2. Experimental procedure for the accelerated oxidation and polymerization of biodiesel (BD) and linoleic acid methyl esters and product analysis methods.

2.1. Materials

Five kinds of commercial grade BD having no antioxidant, namely soybean oil, waste cooking oil, rapeseed oil, cottonseed oil, and palm oil, extracted from different types of biomass were purchased from Dansuck Co., Ltd. In Korea to give the guidelines on the production, use, and storage of commercialized BDs with their oxidation characteristics. High-purity linoleic acid methyl ester (C18:2 ME) was purchased from Sigma-Aldrich Co.

2.2. Fuel Properties of Biodiesels

Table 1 shows the properties of the BDs that were analyzed in this study before the accelerated oxidation. The physical and chemical properties of the fresh BDs, except oxidation stability, satisfy Korean and European standards for fuels that can be used for compression-ignition diesel engines [33]. The lower oxidation stability of the BDs can be explained by the absence of antioxidants [34,35]. Among the five BDs analyzed in this study, the BD that was extracted from palm revealed the longest oxidation stability time (4.48 h), followed by that from soybean (3.16 h), rapeseed (3.05 h), waste cooking oil (1.24 h), and cottonseed (1.03 h). The FAME compositions of five BDs are enumerated in Table 2. All BDs had FAMEs consisting of C18 as the main component. However, the unsaturated FAME contents differed depending on the biomass species. Rapeseed BD revealed the highest unsaturated FAME content (89.39%), followed by soybean BD (83.23%), waste cooking BD (79.02%), cottonseed BD (70.03%), and palm BD (49.16%). Owing to the high content of palmitic acid (16:0) methyl ester, palm BD had the lowest unsaturated FAME content.

Table 1. Physicochemical characteristics of BD samples.

Item	Korean Limit	EN 14214 Limit	Soybean	Waste Cooking	Rapeseed	Cottonseed	Palm
FAME content (wt. %)	96.5 min.	96.5 min.	98.20	97.59	97.35	98.56	97.68
Kinematic viscosity (40 °C, mm^2/S)	1.9~5.0	3.5~5.0	4.03	4.16	4.43	4.04	4.53
Sulfur content (mg/kg)	10 max.	10 max.	0.7	1.5	2.0	1.1	1.1
Flash point (°C)	120 min.	120 min.	177	177	183	173	171
Carbon residue (wt. %)	0.1 max.	0.3 max.	0.02	0.03	0.02	0.02	0.03
Total contamination (mg/kg) [a]	24 max.	24 max.	5.1	13.0	7.6	10.0	8.9
Density (15 °C, kg/m^3)	860~900	860~900	886	884	885	882	876
Water content (mg/kg)	500 max.	500 max.	133	105	183	80	272
Oxidation stability (110 °C, h) [b]	6 min.	6 min.	3.16	1.24	3.05	1.03	4.48
Total acid number (mg KOH/g)	0.50 max.	0.50 max.	0.24	0.32	0.36	0.15	0.30
Iodine number	-	120 max.	133	113	108	105	50
Total glycerol (wt. %)	0.24 max.	0.25 max.	0.13	0.13	0.11	0.10	0.15
CFPP (°C)	0	-	−3.0	−2.0	−11.0	7.0	>10.0
Cloud point (°C)	-	-	0.0	3.0	−3.0	9.0	16.0
Monoglyceride (wt. %)	0.80 max.	0.80 max.	0.45	0.40	0.10	0.24	0.45
Diglyceride (wt. %)	0.20 max.	0.20 max.	0.00	0.00	0.01	0.05	0.18
Triglyceride (wt. %)	0.20 max.	0.20 max.	0.00	0.00	0.00	0.00	0.04
Free glycerol (wt. %)	0.02 max.	0.02 max.	0.01	0.02	0.00	0.02	0.00
Phosphorus content (mg/kg)	10 max.	10 max.	0.08	0.35	0.07	0.06	0.19

[a] 10% of sample, [b] No addition of antioxidant.

Table 2. Fatty acid methyl ester (FAME) compositions of BD samples.

FAME (wt. %)	Soybean	Waste Cooking	Rapeseed	Cottonseed	Palm
C14:0	0.08	0.31	0.06	0.96	1.01
C14:1	0.01	0.05	-	-	-
C16:0	10.35	14.22	5.58	25.32	44.39
C16:1	0.12	0.93	0.25	0.59	0.22
C18:0	4.53	4.09	1.94	2.79	4.28
C18:1	21.39	30.37	55.11	15.91	38.48
C18:2	54.20	42.96	26.24	51.94	9.99
C18:3	7.22	4.31	6.76	1.05	0.29
C20:0	0.42	0.44	0.64	0.18	0.39
C20:1	0.21	0.34	1.01	0.10	0.16
C22:0	0.38	0.21	0.32	0.10	0.08
C24:0	0.16	0.09	0.17	0.03	0.09
C24:1	0.08	0.06	0.02	0.44	0.02
Not identified	0.85	1.62	1.9	0.59	0.6
Sat. FAME	15.92	19.36	8.71	29.38	50.24
Unsat. FAME	83.23	79.02	89.39	70.03	49.16

2.3. Accelerated Oxidation

The accelerated oxidation of BDs was performed according to the ASTM D 2274 method using the heat and aeration that were reported in previous studies [36,37]. For this, 1 L of BD in a brown glass tube was aerated at 95 °C in a thermal oven under 100 mL/min of zero-grade air for 16 weeks. The temperature of the oil condenser was maintained at 20 °C. The sampling of aged BD was performed once a week, and the sampled BD was kept in a dark container (4 °C) in a nitrogen atmosphere until analysis.

2.4. Property Analysis of Oxidized BD

The quality of the aged BDs was evaluated by measuring their FAME content, density, kinetic viscosity, and TAN values. The composition of FAMEs in BD was analyzed by a gas chromatography/flame ionization detector (7890 A, Agilent Technology) using a fused silica capillary column (30 m length $\times$ 0.25 mm inner diameter $\times$ 0.2 μm film thickness; SP-2380, Supelco) for the separation of FAMEs and methyl heptadecanoate-heptane as an internal standard. The density and kinematic viscosity of BD were evaluated according to KS M 2014 while using a digital density meter (DA300, Koto Electronics) and an automated kinematic viscometer (CAV-2100F, Cannon instrument), respectively. The TAN value of BD was measured according to KS M ISO 6245 using an 809 Titrando instrument (Brinkman Metrohm).

2.5. FAME Polymerization

Linoleic acid methyl ester (C18:2 ME; 3 g) and 10 g of oxidized soybean and cotton BDs during 16 weeks were heated in a sealed tube in air at 180 $^{\circ}$C for 72 h. After polymerization, they were purified by flash chromatography to yield the oxidation product as a brown oil (R_f 0.35, SiO_2, 10% $MeOH$-CH_2Cl_2).

2.6. Property Analysis of Polymerized Samples

The NMR spectra of polymerized soybean BD were obtained on a Bruker Avance 500 (Billerica, MA, USA) spectrometer operating at 500 MHz with $CDCl_3$ as the solvent. High speed LC connected to a LTQ orbitrap XL mass spectrometer equipped with APCI ionization (Thermo Fisher Scientific, Bannockburn, IL, USA) was performed using a HPLC-APCI-MS according to the method that was developed by Holcapek and Jandera [38,39]. For this, 5 μL of linoleic acid (C18:2) methyl ester was separated with a Hypersil GOLD (100 mm $\times$ 2.1 mm, 1.9 μm) using a mobile phase consisting of methanol (solvent A) and acetonitrile (Solvent B), pumped at 0.5 mL/min. A mobile phase flow gradient was applied to provide effective separation (75% A + 25% B to 5 min; 90% A + 15% B to 35 min; and 100% A to 45 min) using isocratic elution.

3. Results and Discussion

3.1. Fuel Properties of Accelerated Oxidation

Figures 3 and 4 show the change in the TAN, density, and kinematic viscosity of BDs during the accelerated oxidation. Although the TAN values of BDs gradually increased by increasing the oxidation time, the TAN values were below the threshold specified in the Korean and European standards (0.5 mg KOH/g until three weeks). Moreover, the TAN values of the BDs increased from four weeks to 16 weeks. After accelerated oxidation for 16 weeks, the TAN value was the highest (5.0 mg KOH/g) for cottonseed BD and lowest (2.3 mg KOH/g) for soybean BD. The sudden increase in TAN values can be explained by the oxidation of the FAMEs in the BDs. During the oxidation of FAMEs, free radicals are formed by the abstraction of hydrogen from the methylene groups in the allylic positions of the unsaturated FAMEs. By further reaction with atmospheric oxygen, these free radicals can be converted to hydroperoxides, causing the TAN value of the BDs to increase. Therefore, the rapid increase in TAN value after four weeks indicates that the rate of hydroperoxide formation is accelerated after four weeks during the accelerated oxidation test. The density values of BDs also increased gradually until four weeks; thereafter, they increased rapidly up to 16 weeks. However, except for the density values of cottonseed BD obtained after oxidation for 14 weeks, the BD density values remained below the thresholds that were specified in the Korean and European standards (900 kg/m^3 at 15 $^{\circ}$C) after 16 weeks of accelerated oxidation. The kinematic viscosity values of BDs also showed a change trend that was similar to that of the TAN and density during the accelerated oxidation test, exhibiting the rapid increase after four weeks. This indicates that the increase in density and kinematic viscosity are closely related to the change in TAN values during BD oxidation. The presence of unsaturated FAMEs causes the additional polymerization of BD, producing compounds with higher molecular weights and increasing the density and kinematic viscosity values of BDs during long-term accelerated oxidation. Interestingly, the rates of increase of TAN, density, and kinematic viscosity of cottonseed BD were much higher than those of the other BDs tested in this study. This indicates that the structural properties of FAMEs in cottonseed BD underwent more profound changes those in other BDs.

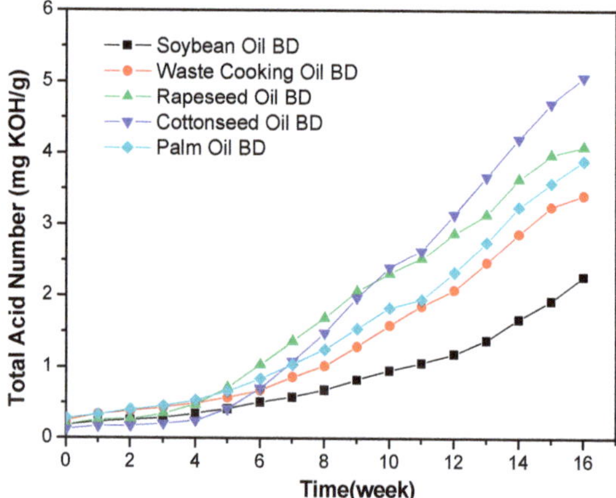

Figure 3. Change in total acid number (TAN) values of BDs during accelerated oxidation.

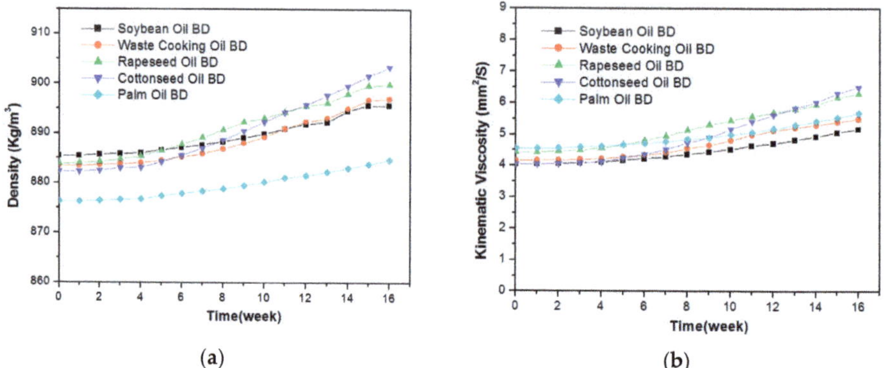

(a)   (b)

Figure 4. Change in (**a**) density and (**b**) kinematic viscosity of BDs during accelerated oxidation.

Figure 5 shows the change in contents of typical unsaturated FAMEs, C18:2 ME, and C18:3 ME, which accounted for the largest fraction of fresh BDs during accelerated oxidation. The contents of both unsaturated FAMEs decreased as the oxidation time increased, indicating the oxidation and/or polymerization of unsaturated FAMEs by oxidative aging. The C18:2 ME content in the palm BD, rapeseed BD, cottonseed BD, waste cooking BD, and soybean BD decreased to 51.6%, 40.7%, 27.7%, 19.9%, and 8.4%, respectively, after 16 weeks of accelerated oxidation. The content of C18:3 ME also decreased to 5.2% for soybean BD and to values lower than 2% for other BDs. Among the five BDs, the cottonseed BD, which exhibited the largest increase in TAN, density, and kinematic viscosity (Figure 5), showed the largest decrease in both unsaturated FAMEs. This confirms that a rapid decrease in unsaturated FAMEs increases the TAN, density, and kinematic viscosity of BDs by oxidation and polymerization. Frankel et al. [24] and Cosgrove et al. [40] also indicated that an auto-oxidation chain reaction is initiated from di-and tri-unsaturated fatty acids due to their high reactivity.

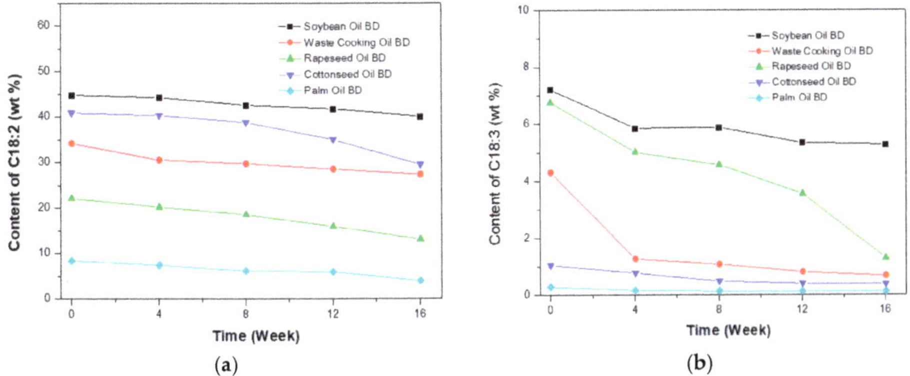

Figure 5. Change in C18 FAMEs contents in BDs during accelerated oxidation. (**a**) C18:2 ME; (**b**) C18:3 ME.

3.2. Polymerization of FAMEs

The FAMEs that are polymerized due to oxidation can be decomposed and they can be re-polymerized by additional polymerization reactions with other FAMEs and oxidized FAMEs. Owing to the negative effect of polymerized FAMEs in BD on the vehicle, the polymerization of BD is being regulated on its use. Although FAMEs can be polymerized by the thermal aging that accompanies the rapid decrease in unsaturated FAMEs, the thermal polymerization of BDs is difficult to monitor because they are not stored at high temperatures, which cause changes to the structure of FAMEs. However, polymerization of BDs in actual vehicles can occur by the repeated heating of the engine before combustion, which results in instability in the BDs [39].

To identify the chemical properties of polymerized FAMEs, the accelerated oxidation of soybean and cotton BDs were additionally performed at 180 °C for 72 h. For reference, the accelerated oxidation of linoleic acid (C18:2) methyl ester was also performed under the same condition. After the accelerated oxidation, the polymerized FAMEs were purified using flash chromatography and identified using ^{1}H-NMR and HPLC-APCI MS because it is difficult to study the polymerized chemicals using GC technologies. Figure 6 shows the ^{1}H-NMR spectra of purified soybean BD before and after the accelerated oxidation at 180 °C for 72 h. When compared to the non-oxidized BD, the oxidized BD revealed broader spectra and decreased intensities for the olefinic protons (-CH=CH-) of the double bonds at 5.45–5.49 ppm, suggesting the polymerization of FAMEs and the decrease in double bonds in these polymerized FAMEs [41]. Chuck et al. [13] also explained that the oxidation BDs had the decreased intensities of unsaturated fatty acids by increasing the oxidation temperature and time.

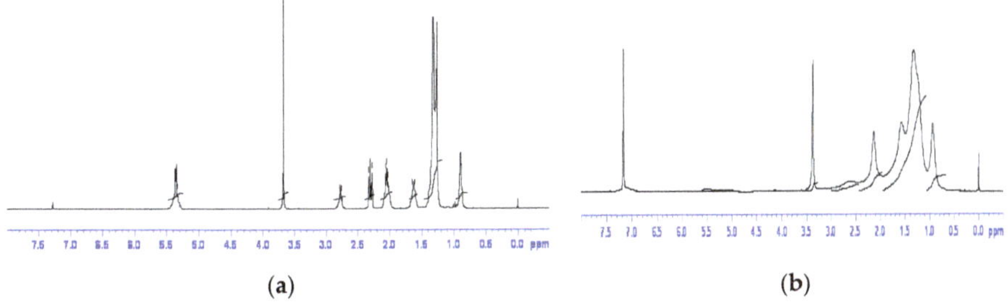

Figure 6. Change in the ^{1}H-NMR spectrum of soybean BD by accelerated oxidation at 180 °C for 72 h. (**a**) Fresh BD; (**b**) Oxidized BD.

HPLC-APCI MS analysis of the oxidized linoleic acid (C18:2) methyl ester at 180 °C for 72 h in a sealed tube in air was performed to study the polymerization of FAMEs in BDs, because linoleic acid (C18:2) methyl ester accounts for high fractions in soybean (54.20%) and cottonseed (51.94%) BDs, and it can be polymerized by the Diels-Alder reaction at high temperatures [32,42,43]. Figure 7 shows the mass spectra for the fresh and oxidized linoleic acid (C18:2) methyl ester. The fresh linoleic acid (C18:2) methyl ester, shown in Figure 7a, revealed typical ions, m/z 265.2526 and 297.2788. However, the oxidized one, as shown in Figure 7b, had large molecular ions, such as m/z 557.49, 587.50, 589.51, and 603.49, indicating its polymerization due to the accelerated oxidation. The possible structural information for the polymerized linoleic acid (C18:2) methyl ester that was suggested by the isotopic cluster for these ions (Figure 7c) is shown in Table 3. Figure 8 shows the possible reaction pathways for the formation of dimers and epoxy-dimers of linoleic acid (C18:2) methyl ester. For the thermal oxidation process, one of the conjugated diene groups in the chain can react with the olefinic group from the nearby fatty acid chain and form a substituted cyclo-ring, as part of the Diels-Alder reaction. The formation of these dimers and oxidized dimers of linoleic acid (C18:2) methyl ester can be explained by thermal polymerization during the Diels-Alder reaction [30,44]. When linoleic acid (C18:2) methyl ester was oxidized, the epoxy dimer of linoleic acid (C18:2) methyl ester could be formed by the Diels-Alder reaction between the oxidized and fresh linoleic acid (C18:2) methyl esters. These results indicate that the high temperature oxidation of BDs increase the contents of dimers and epoxy dimers of unsaturated FAMEs in BDs via their oxidation and polymerization.

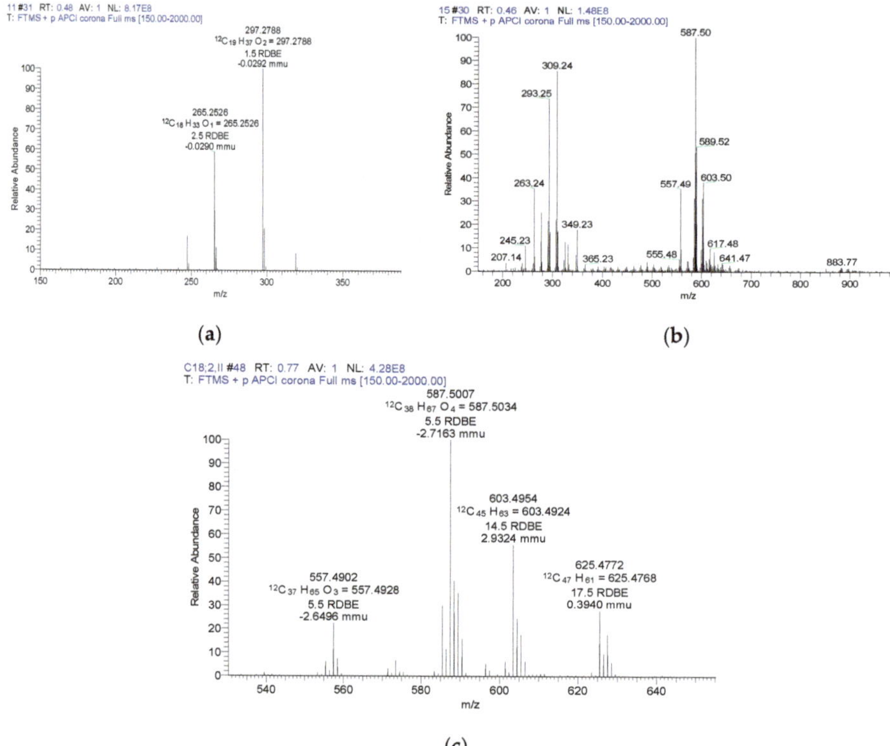

Figure 7. High performance liquid chromatography-atmospheric pressure chemical ionization mass spectrometry (HPLC-APCI MS) spectra for the fresh and polymerized linoleic acid (C18:2) methyl ester. (**a**) Fresh linoleic acid (C18:2) methyl ester; (**b**) Polymerized linoleic acid (C18:2) methyl ester; and, (**c**) Expanded spectrum for polymerized linoleic acid (C18:2) methyl ester.

Table 3. Identification by HPLC-APCI MS detection of FAME polymer from C18:2 ME.

Ion Fragment(m/z)	Theo. Mass	Elemental Composition	Proposed Structure
557.4902	557.4928	C37H65O3	-
587.5007	587.5034	C38H67O4	Dimer
589.5159	589.5190	C38H69O4	Dimer
603.4954	603.4957	C38H67O6	Epoxy-dimer

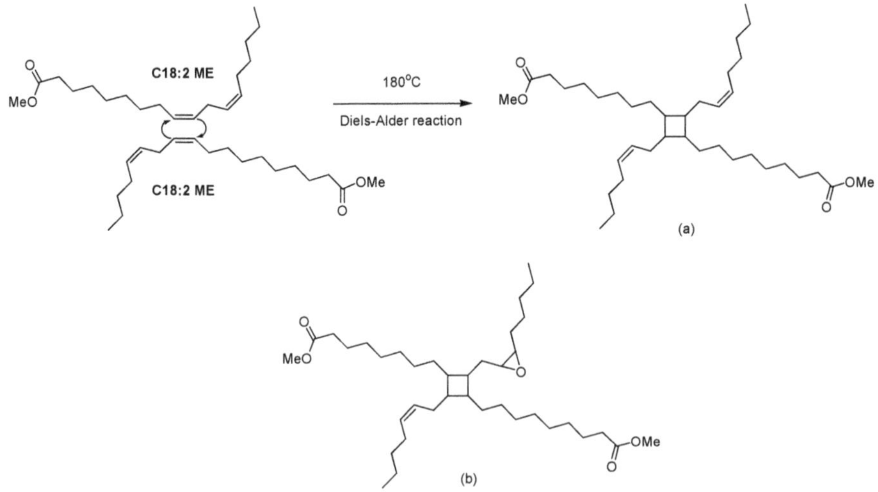

Figure 8. Diels-Alder reaction from C18:2 ME: (**a**) dimer of linoleic acid (C18:2) methyl ester; (**b**) epoxy-dimer of linoleic acid (C18:2) methyl ester.

4. Conclusions

The oxidation of BDs that were obtained from the decomposition of biomass can damage the fuel injection and engine parts during its use as a fuel. The excess heating of BDs can also cause polymerization of the biodiesel. In this study, the stability and polymerization of BDs by the accelerated oxidation were investigated. High unsaturated FAME contents in BDs led to a reduction in oxidation stability as compared to that of BDs with high saturated FAME contents. During the 16 week accelerated oxidation test at 95 °C, the TAN, density and kinematic viscosity of BDs decreased gradually until three weeks, and the decreased rapidly until 16 weeks. The unsaturated FAME contents in BDs showed similar change trends, decreasing slowly until three weeks and then rapidly until 16 weeks, indicating the occurrence of structural changes to the unsaturated FAMEs as a result of the oxidation. The accelerated high temperature oxidation at 180 °C for 72 h results in the polymerization of fresh and oxidized unsaturated FAMEs via the Diels-Alder reaction. The polymerized products that were obtained by the oxidation of unsaturated FAMEs at high temperature were identified as the dimers and epoxy dimers of FAMEs through the Diels-Alder reaction between the FAME and oxidized FAME molecules. The formation of these dimers and oxidized dimers of linoleic acid (C18:2) methyl ester can be explained by thermal polymerization. This study will provide the important basic data on the oxidation and polymerization of BDs, which can be used to know their effects on diesel vehicles and give the guideline on their production, use, and storage.

Author Contributions: Conceptualization, J.-K.K.; Data curation, formal analysis, investigation, C.-H.J., K.i-M., and I.-h.H.; Methodology, J.-K.K., Y.-M.K.; H.W.L., Supervision, Y.-K.P.; Validation, J.-K.K.; Writing—original draft, J.-K.K.; Writing—review & editing, Y.-M.K.

Acknowledgments: This work was supported by the Advanced Biomass R & D Center (ABC) of the Global Frontier Project, funded by the ministry of Science and ICT (ABC-2015M3A6A2076483).

Conflicts of Interest: The authors declare no conflict of interest.

References

1. Molino, A.; Larocca, V.; Chianese, S.; Musmarra, D. Biofuels production by biomass gasification: A review. *Energies* **2018**, *11*, 811. [CrossRef]
2. Zhou, Z.; Lei, F.; Li, P.; Jiang, J. Lignocellulosic biomass to biofuels and biochemicals: A comprehensive review with a focus on ethanol organosolv pretreatment technology. *Biotechnl. Bioeng.* **2018**, *115*, 2683–2702. [CrossRef] [PubMed]
3. Lee, H.; Kim, Y.M.; Lee, I.G.; Jeon, J.K.; Jung, S.C.; Chung, J.D.; Choi, W.G.; Park, Y.K. Recent advances in the catalytic hydrodeoxygenation of bio-oil. *Korean J. Chem. Eng.* **2016**, *33*, 3299–3315. [CrossRef]
4. Lee, Y.; Shafaghat, H.; Kim, J.K.; Jeon, J.K.; Jung, S.C.; Lee, I.G. Upgrading of pyrolysis bio-oil using WO_3/ZrO_2 and Amberlyst catalysts: Evaluation of acid number and viscosity. *Korean J. Chem. Eng.* **2017**, *34*, 2180–2187. [CrossRef]
5. Ali, O.M.; Mamat, R.; Abdullah, N.R.; Abdullah, A.A. Analysis of blended fuel properties and engine performance with palm biodiesel–diesel blended fuel. *Renew. Energy* **2016**, *86*, 59–67. [CrossRef]
6. Mohadesi, M.; Aghel, B.; Khademi, M.H.; Sahraei, S. Optimization of biodiesel production process in a continuous microchannel using response surface methodology. *Korean J. Chem. Eng.* **2017**, *34*, 1013–1020. [CrossRef]
7. Binnal, P.; Babu, P.N. Production of high purity biodiesel through direct saponification of wet biomass of *Chlorella protothecoides* in a low cost microwave reactor: Kinetic and thermodynamic studies. *Korean J. Chem. Eng.* **2017**, *34*, 1027–1036. [CrossRef]
8. Yaakob, Z.; Mohammad, M.; Mohammad, A.; Alam, Z.; Sopian, K. Overview of the production of biodiesel from waste cooking oil. *Renew. Sustain. Energy Rev.* **2013**, *18*, 184–193. [CrossRef]
9. Knothe, G.; Dunn, R.O. Biofuels derived from vegetables oils and fats. In *Oleo Chemical Manufacture and Applications*; Gunstone, F.D., Hamilton, R.J., Eds.; CRC Press: Sheffield, UK, 2001; pp. 106–163.
10. Bai, A.; Jobbágy, P.; Farkas, F.; Popp, J.; Grassell, G.; Szendre, J.; Balogh, P. Technical and environmental effects of biodiesel use in local public transport. *Transp. Res. Part D* **2016**, *47*, 323–335. [CrossRef]
11. Li, C.-H.; Ku, Y.-Y.; Lin, K.W.; SAE Technical Paper, No. *Experimental Study of Combustion Noise of Common Rail Diesel Engine Using Different Blend Biodiesel*, SAE Technical Paper No. 2014-01-1690; USA. 2014. [CrossRef]
12. Nwafor, O.M.I. The effect elevated fuel inlet temperature on performance of diesel engine running on neat vegetable oil at constant speed conditions. *Renew. Energy* **2003**, *28*, 171–181. [CrossRef]
13. Chuck, C.J.; Bannister, C.D.; Jenkins, R.W.; Lowe, J.P.; Davidson, M.G. A comparison of analytical techniques and the products formed during the decomposition of biodiesel under accelerated conditions. *Fuel* **2012**, *96*, 426–433. [CrossRef]
14. Westbrook, S.R.; Shoffner, B.A.; Blanks, M.G.; Brunner, K.; Wilson, G.R., III; Hansen, G.A., Sr. *Thermal and Oxidative Instability in Biodiesel Blends during Vehicle Use and Onboard Fuel Storage*; CRC Report No. AVFL-17c; Coordinating Research Council, Inc.: Alpharetta, GA, USA, 2016.
15. Stavinoha, L.L.; Howell, S. *Potential Analytical Methods for Stability Testing of Biodiesel and Biodiesel Blends*, SAE Technical Paper No. 1999-01-3520; USA. 1999; 108, 1566–1580. [CrossRef]
16. Petiteaux, M.; Monsallier, G. *Impacts on Biodiesel Blends on Fuel Filters Functions, Laboratory and Field Test Results*, SAE Technical Paper No. 2009-01-1876; USA. 2009. [CrossRef]
17. McCormick, R.L.; Westbrook, S.R. Storage stability of biodiesel and biodiesel blends. *Energy Fuels* **2010**, *24*, 690–698. [CrossRef]
18. Sugiyama, G.; Maeda, A.; Nagai, K. *Oxidation Degradation and Acid Generation in Diesel Fuel Containing 5% FAME*, SAE Technical Paper No. 2007-01-2027; USA. 2007. [CrossRef]
19. Waynick, J.A. *Characterization of Biodiesel Oxidation and Oxidation Products: Technical Literature Review*; Task 1 Results (CRC AVFL-2b); Coordinating Research Council, Inc.: Alpharetta, GA, USA, 2005.
20. Knothe, G.; Van Gerpen, J.H.; Krahl, J. *The Biodiesel Handbook*, 1st ed.; AOCS Press: Champaign, IL, USA, 2005.
21. Choe, E.; min, D.B. Mechanisms and factors for edible oil oxidation. *Compr. Rev. Food Sci. Food Saf.* **2006**, *5*, 169–186. [CrossRef]
22. Frankel, E.N. Lipid metabolism in the adipose tissue of ruminant animals. *Prog. Lipid Res.* **1980**, *19*, 1–22. [CrossRef]

23. Walling, C. Some properties of radical reactions important in synthesis. *Tetrahedron* **1985**, *41*, 3887–3900. [CrossRef]
24. Frankel, E.N. *Lipid Oxidation*, 2nd ed.; Oily Press: Cambridge, UK, 2005.
25. Russell, G.A. Deterium-isotope effects in the autoxidation of aralkyl hydrocarbons. Mechanism of the interaction of PEroxy radicals. *J. Am. Chem. Soc.* **1957**, *79*, 3871–3877. [CrossRef]
26. Knothe, G. Dependence of biodiesel fuel properties on the structure of fatty acid alkyl esters. *Fuel Process Technol.* **2005**, *86*, 1059–1070. [CrossRef]
27. Yamane, K.; Kawasaki, K.; Sone, K.; Hara, T.; Prakoso, T. Oxidation stability of biodiesel and its effect on diesel combustion and emission. *Int. J. Engine Res.* **2007**, *8*, 307–319. [CrossRef]
28. Dunn, R.O. Effect of temperature on the oil stability index (OSI) of biodiesel. *Energy Fuels* **2008**, *22*, 657–662. [CrossRef]
29. Jain, S.; Sharma, M.P. Stability of biodiesel and its blends: A review. *Renew. Sustain. Energy Rev.* **2010**, *14*, 667–678. [CrossRef]
30. Yaakob, Z.; Narayanan, B.N.; Padikkaparamil, S.; Unnu, K.S.; Akbar, P.M. A review on the oxidation stability of biodiesel. *Renew. Sustain. Energy Rev.* **2014**, *35*, 136–153. [CrossRef]
31. Fang, H.; McCormick, R.L. *Spectroscopic Study of Biodiesel Degradation Pathways*, SAE Technical Paper No. 2006-01-3300; USA. 2006. [CrossRef]
32. Ball, J.C.; Anderson, J.E.; Pivesso, B.P.; Wallington, T.J. Oxidation and polymerization of soybean biodiesel/petroleum diesel blends. *Energy Fuels* **2018**, *32*, 441–449. [CrossRef]
33. Østerstrøm, F.F.; Anderson, S.A.; Collings, T.; Ball, J.C.; Wallington, T.J. Oxidation stability of rapeseed biodiesel/petroleum diesel blends. *Energy Fuels* **2016**, *30*, 344–351. [CrossRef]
34. Anderson, J.E.; Collings, T.R.; Mueller, S.A.; Ball, J.C.; Wallington, T.J. Soy oxidation at vehicle fuel system temperature: Influence of aged fuel on fresh fuel degradation to simulate refueling. *SAE Int. J. Fuels Lubr.* **2017**, *10*, 296–303. [CrossRef]
35. Foglia, T.A.; Jones, K.C.; Phillips, J.G. Determination of biodiesel and triacylglycerols in diesel fuel by LC. *Chromatographia* **2005**, *62*, 115–119. [CrossRef]
36. Monteiroa, M.R.; Ambrozina, A.R.P.; Liaob, L.M.; Ferreirac, A.G. Critical review on analytical methods for biodiesel characterization. *Talanta* **2008**, *77*, 593–605. [CrossRef]
37. Ministry of Trade, Industry and Energy (MOTIE). Fuel Specification for Biodiesl. 2018. Available online: http://www.motie.go.kr/motie/ms/nt/gosi/bbs/bbsView.do?bbs_seq_n=62710&bbs_cd_n=5¤tPage=1&search_key_n=title_v&cate_n=&dept_v=&search_val (accessed on 25 October 2018).
38. Mittelbach, M.; Remschmid, C. *Biodiesel-The Comprehensive Handbook*; Boersedruck GmbH: Graz, Austria, 2004.
39. Kumar, N. Oxidation stability of biodiesel: Cause, effects and prevention. *Fuel* **2017**, *190*, 328–350. [CrossRef]
40. Cosgrove, J.P.; Church, D.F.; Pryor, W.A. The kinetics of the autoxidation of polyunsaturated fatty acids. *Lipids* **1987**, *22*, 299–304. [CrossRef]
41. Knothe, G. Analysis of oxidized biodiesel by ^{1}H-NMR and effect of contact area with air. *Eur. J. Lipid Sci. Technol.* **2006**, *108*, 493–500. [CrossRef]
42. De Araujo, T.A.; Barbosa, A.M.J.; Viana, L.H.; Ferreira, V.S. Electroanalytical determination of TBHQ, a synthetic antioxidant in soybean biodiesel samples. *Fuel* **2010**, *90*, 707–712. [CrossRef]
43. Flitsch, S.; Neu, P.M.; Schober, S.; Kienzl, N.; Ulmann, J. Quantitation of aging products formed in biodiesel during the rancimat accelerated oxidation test. *Energy Fuels* **2014**, *28*, 5849–5856. [CrossRef]
44. Velasco, J.; Berdeaux, O.; Márquez-Ruiz, G.; Dobarganes, M. Sensitive and accurate quantitation of monoepoxy fatty acids in thermoxidized oils by gas-liquid chromatography. *J. Chromatogr. A* **2002**, *982*, 145–152. [CrossRef]

Article

Hydrogen Production from Coffee Mucilage in Dark Fermentation with Organic Wastes

Edilson León Moreno Cárdenas [1,†], Arley David Zapata-Zapata [2,†] and Daehwan Kim [3,*]

[1] Laboratorio de Mecanización Agrícola, Departamento de Ingeniería Agrícolay de Alimentos, Universidad Nacional de Colombia-Sede Medellín, Carrera 64c # 63-120, Código Postal 050034, Colombia; elmorenoc@unal.edu.co

[2] Universidad Nacional de Colombia-Sede Medellín-Escuela de Química-Laboratorio de Procesos Biológicos-Carrera 65 # 59A-110, Medellín, Código Postal 050034, Colombia; adzapata@unal.edu.co

[3] Department of Biology, Hood College, 401 Rosemont Avenue, Frederick, MD 21701, USA

* Correspondence: kimd@hood.edu; Tel.: +1-765-637-8603

† Authors contributed equally to the study.

Received: 5 November 2018; Accepted: 27 November 2018; Published: 27 December 2018

Abstract: One of primary issues in the coffee manufacturing industry is the production of large amounts of undesirable residues, which include the pericarp (outer skin), pulp (outer mesocarp), parchment (endocarp), silver-skin (epidermis) and mucilage (inner mesocarp) that cause environmental problems due to toxic molecules contained therein. This study evaluated the optimal hydrogen production from coffee mucilage combined with organic wastes (wholesale market garbage) in a dark fermentation process. The supplementation of organic wastes offered appropriate carbon and nitrogen sources with further nutrients; it was positively effective in achieving cumulative hydrogen production. Three different ratios of coffee mucilage and organic wastes (8:2, 5:5, and 2:8) were tested in 30 L bioreactors using two-level factorial design experiments. The highest cumulative hydrogen volume of 25.9 L was gained for an 8:2 ratio (coffee mucilage: organic wastes) after 72 h, which corresponded to 1.295 L hydrogen/L substrates (0.248 mol hydrogen/mol hexose). Biochemical identification of microorganisms found that seven microorganisms were involved in the hydrogen metabolism. Further studies of anaerobic fermentative digestion with each isolated pure bacterium under similar experimental conditions reached a lower final hydrogen yield (up to 9.3 L) than the result from the non-isolated sample (25.9 L). Interestingly, however, co-cultivation of two identified microorganisms (*Kocuria kristinae* and *Brevibacillus laterosporus*), who were relatively highly associated with hydrogen production, gave a higher yield (14.7 L) than single bacterium inoculum but lower than that of the non-isolated tests. This work confirms that the re-utilization of coffee mucilage combined with organic wastes is practical for hydrogen fermentation in anaerobic conditions, and it would be influenced by the bacterial consortium involved.

Keywords: hydrogen; coffee mucilage; organic wastes; dark fermentation; anaerobic digestion

---

1. Introduction

Conventional fossil fuels, the main energy sources for industrial/technological development, have been meeting about 80% of the fundamental energy demand and supply in the world [1–3]. However, this fossil fuel-dependent energy system causes problems of limited resources, greenhouse gas emissions, and environmental pollution issues [4–6]. Among diverse alternative clean energy resources, hydrogen has been considered as a prospective future energy source for replacing the gradual depletion of fossil fuels and addressing the lack of sustainability. Hydrogen energy is not only unrestricted by greenhouse gas emissions but also produces more than 2.5 times higher energy than the energy from hydrocarbons [7,8]; this ability of hydrogen energy is proposed for an essential renewable and energy source for transportation. The current physico-chemical practice in most facilities achieves approximately 90% yields of hydrogen through steam refining of methane (40%), and gasification/partial oxidation of crude oil and coal (48%). However, these are still highly dependent upon fossil fuels because of cost effectiveness and the lack of a suitable alternative technique [1]. With these concerns about rapid depletion of petroleum stores and environmental problems, biological processes have been identified as a promising technology for hydrogen production. Its basic concept is to catalyze water decomposition or digest organic compounds in an environmentally friendly way using microorganisms, such as algae, *Cyanobacteria*, or photosynthetic bacteria. Biological methods can be classified into four groups: (1) direct bio-photolysis, (2) indirect bio-photolysis, (3) photo fermentation, and (4) dark fermentation. Although photodecomposition methods result in relatively higher hydrogen yields than the other approaches, the anaerobic fermentative process, in particular, dark fermentation, is widely thought to be an attractive approach. Since various organic wastes and wastewater can be used as carbohydrate-rich substrates in the anaerobic digestion process, and it is capable of transforming organic wastes into value-added molecules without light sources [9–11].

There is no doubt that coffee is one of the most largely consumed beverages along with water and tea worldwide; millions of people around the globe consume coffee each day, and the number of coffee-consuming people and nations are increasing. It is known that coffee is ranked number two as a traded commodity only after crude oils, with a worldwide production of coffee is estimated to be 152 million 60 kg bags [12]. Over the past decade, coffee production and its associated market have been rapidly growing with attractive research in functional foods, for example, the cognitive and physical behavior effects of caffeine. As coffee production and consumption increases, large amounts of undesirable byproducts (skin, parchment, pericarp, pulp, and mucilage) are also generated during the coffee separation process. In general, only coffee beans are used for brewed coffee, but the other components are separated and removed; they constitute more than 50% of an initial coffee fruit weight [13]. The residual coffee wastes after the wet separation process include 43.2% (w/w) skin and pulp, 6.1% (w/w) parchment, and 11.8% (w/w) mucilage and solubles [14]. Mussato et al. [15] reported the generation of residual wastes from the preparation of instant coffee were around 6 million tons per year worldwide, while more recent work has estimated the coffee residual byproducts would be approximately 15 million tons per year [16]. Coffee wastes can be utilized in animal feed [17–19], manure [20], antioxidant polyphenols [21,22], adsorption molecules [23–25], α-amylase [26], and ethanol production [17,19,27]; however, most coffee wastes are unutilized and dumped into land or water for economical and/or technical reasons [13,16]. Considering the current facts and issues, further investigations and practical applications with coffee residue by-products are required.

Coffee mucilage is a colorless thin layer, mainly composed of water, sugars, protein, and pectin that covers the parchment and outer skin (pericarp), and protects the inner fibrous pulp and endosperm (coffee bean) components. Due to the high carbohydrate and nitrogen content in mucilage, it is one of the direct resources for animal feeds after agricultural processing and can be a suitable source of value-added molecules, such as ethanol, lactic acid, and hydrogen. Previous works identified that the coffee mucilage contained 85–91% (w/w) water, 6.2–7.4% (w/w) sugars, 4–5% (w/w) protein, and 1% (w/w) pectin substances [13,20,28,29], which were relatively higher than those obtained from other coffee by-products such as husks, skin, and pulp [19,30]. Furthermore, the sugars in the coffee mucilage include a high portion of reducing sugars (63%, w/w) that facilitates the utilization of sugars to other molecules and commodities [17,31]. Orrego et al. [13,16] reported that coffee mucilage from the wet separation process had >50 g/L of fermentable sugars (mainly glucose and galactose), acetic acid, protein, and some minerals (calcium, iron, magnesium, potassium, phosphorus, and sodium), which could be directly transformed into other molecules (e.g., ethanol), without requiring any pretreatment and carbon or nitrogen source supplements.

In order to develop the hydrogen process, several research studies have been conducted in the anaerobic digestion process in the presence of pure culture medium using a specific microorganism such as *Clostridium*, *Bacillus*, and *Thermoanaerobacterium* [9,32,33]. However, hydrogen production under special conditions using defined/pretreated culture medium with a pure microbial inoculum restricts further understanding of the co-cultivation, substrate changes, biochemical and molecular interactions of the bacterial population into substrates and their patterns. Various seeding substrates, rich in carbohydrates, such as wastewater, sludge, compost, manure, and soil, are acceptable sources for fermentative hydrogen production, while to the best of our knowledge, there has been no work regarding optimization of hydrogen production from coffee mucilage in the dark fermentation method with organic wastes. This work reports that the use of coffee mucilage combined with supplemental organic wastes can be a potential approach for hydrogen production. The main objective of this study is to determine the effective practical conditions for fermentative digestion of organic compounds into hydrogen, which are evaluated at 30 L bioreactors for different ratios of coffee mucilage and organic wastes without inoculating any microorganism. The maximal cumulative hydrogen is achieved at a two-level factorial experimental design, and further tests are carried out and compared with different independent factors (ratio of coffee mucilage and organic wastes, chemical oxygen demand, pH, and temperature). Moreover, the impact of microbial consortiums (bacterial populations) on hydrogen production are tested with isolated bacteria under similar experimental conditions and those results were compared to non-isolated fermentation.

2. Materials and Methods

2.1. Raw Materials

Coffee mucilage was supplied by the San Rafael farm (Antioquia, Colombia), located at 1575 m above sea level with an average temperature of 21 °C. Organic wastes were collected from the Central Mayorista de Antioquia (Antioquia's Wholesale Market, Medellín, Colombia), any mainly contained fruit and vegetable wastes (lettuce, orange, guava, mango, and papaya), not suitable for human consumption (expired products). As soon as the raw materials were obtained, the coffee mucilage was autoclaved at 121 °C for 15 min and stored with intact organic wastes at 4 °C. The large solids in the mucilage sample were sieved over a 20-mesh screen (0.84 mm, Tyler USA standard testing sieve, VWR, Philadelphia, PA, USA), and the resulting slurry was centrifuged at 8000 rpm at 5 min in order to separate the remaining solids. Sugars and the acetic acid content of the mucilage liquid was determined by HPLC in our previous study, including glucose (37.1 g/L), galactose (14.7 g/L), lactose (0.8 g/L), and acetic acid (1.2 g/L), respectively [13]. All other chemicals and reagents in this study were purchased from Sigma Aldrich (St. Louis, MO, USA).

2.2. Experimental Design

Anaerobic fermentative digestion was prepared and evaluated by the Minitab 16 software program (Minitab 16, Minitab Inc., State College, PA, USA) with a two-level factorial experimental design. Initially, two prepared coffee mucilage and organic wastes were blended into three different ratios (w/w) of 8:2, 5:5, and 2:8, which were named level 1, level 2, and level 3, respectively. Two more levels with only coffee mucilage (level 0) or organic wastes (level 4) were added as control tests. Each anaerobic batch fermentation was carried out in a 30 L bioreactor with a working volume of 20 L under given experimental conditions: temperature range of 30–40 °C, chemical oxygen demand (COD) range of 20 g oxygen/L–60 g oxygen/L, and pH range of 5.0–8.0 until the hydrogen production was completed. The COD was determined by the 5220D Standard Chemical Oxygen Demand Method [34,35], and the initial pH was adjusted by adding 2 M of NaCl or NaOH. The quantity of total solids and volatile solids were determined following the previous work [36]. The bioreactor was operated with a helical ribbon impeller mixing at 100 rpm to avoid deposition of solids and to enhance the hydrogen turnover to the gas phase. The desirable temperature during the fermentation was kept by a heating jacket equipped with a main system, which was recorded with an automatic thermometer of 1 °C resolution and an accuracy of ±1 °C. The biogas samples from fermentative digestion were collected in 1 L gas sampling Tedlar bags (model number: 22,950, Restek, Los Angeles, CA, USA) every 24 h.

2.3. Isolation, Gram Staining, Biochemical, Sequence Analysis of Microorganisms

In some experiments with high hydrogen production, liquid samples were taken after fermentation performances for further analyses for microbial identification, biochemical tests, sequence analysis, acids, and fermentation tests with pure bacterium. The liquid broth samples were bottled in sterilized jars and stored at 4 °C prior to use. Each sample was shaken for 2 min in a shaking incubator at 200 rpm, and cell concentration was adjusted to a 10^{-8} (Colony-forming unit) CFU/mL by a serial dilution with sterilized water. Each diluted aliquot was spread out on nutrient agar medium (0.5% peptone, 0.3% yeast extract, 1.5% agar, and 0.5% sodium chloride) and cultured at room temperature for a week. Each grown colony was picked, diluted in 0.5 mL of distilled water, and kept with 50% glycerol solution at −80 °C prior to further use.

In order to a microbial identification, the colorimetric identification card method was prepared via a compact Vitek2 device (Biomerieux, Lyon, France) equipped with a reactive card for biochemical tests according to previous work [37]. Briefly, each isolated microorganism was suspended with a sterilized saline solution (containing 0.5% NaCl, pH 7.0) until turbidity between 0.5 and 0.63 units on the McFarland scale (approximately cell density between 1×10^8–1.89×10^8). The suspended cells (3 mL) on the identification card were installed in the Vitek2 device, and each sample was incubated at 35 °C for 12 h. After incubation, biochemical reactions were carried out with the values from the device's database, providing appropriate results based on the reactions.

In order to verify unknown microorganisms obtained from the best anaerobic fermentative digestion, the ribosomal DNA (16S rDNA) of each microorganism was prepared and isolated by the FastDNA spin kit for soil DNA (VWR catalog number: ICNA116560200, VWR Scientific, Bridgeport, NJ, USA). The 16S rDNA sequence was amplified through the polymerase chain reaction (PCR) with two designed primers: forward prime 5′-AGAGTTTGATCCTGGCTCAG-3′ and reverse prime: 5′-GGTTACCTTGTTACGACTT-3′. The polymerase chain reaction step was conducted in a Perkin-Elmer thermal cycler (GeneAmp PCR System 9700, Norwalk, CT, USA) 30 times. Each cycle included the denaturation step at 95 °C for 45 s, the annealing step at 56 °C for 2 min, and the extension step at 72 °C for 3 min [38]. The amplified PCR products were purified using the Wizard PCR preps DNA purification system (catalog number: A7231, Promega Corporation, Madison, WI, USA). The sequences of both directions of the DNA was confirmed via the ABI PROSM 3700 DNA analyzer (Applied Biosystem, Midland, ON, Canada) and the sequence alignment was carried out with BLAST at the NCBI.

For further dark fermentation with isolated pure bacterium, each isolated cell was grown overnight in a 500 mL Erlenmeyer flask (Belloco, Vineland, NJ, USA) in the presence of a YEPD medium (1% yeast extract, 1% peptone, and 2% glucose) at 30 °C with 200 rpm. The cells were harvested by centrifugation (5 min, 8000 rpm) then were suspended in YEP (no glucose) medium [39]. This liquid was utilized to inoculate the single cell fermentation with an initial cell concentration of 1 g dry cells/L. Each run was tested in a 30 L bioreactor (20 L working volume) under similar experimental conditions until the hydrogen production was completed and compared to the results from those from co-cultivations. All fermentation tests were conducted in duplicate.

2.4. Analysis

Hydrogen gas was analyzed using gas chromatography (3000 MicroGC system, Agilent, San Jose, CA, USA) equipped with a thermal conductivity detector (TCD) and capillary HP-PLOT U column (0.32 mm ID × 8 m length × 10 μm film). The temperatures of the injector, column, and detector were operated at 60 °C, 80 °C, and 300 °C respectively. The pressure was kept at 206.8 kPa. The carrier gas (argon gas) was utilized with a flow rate of 0.9 mL/min and G 2.5 volumetric gas meter (Metrex, Popayán, Cauca, Colombia) with a precision of 0.04 m^3/h, and a maximum working pressure of 40 kPa was used to register the gas. For statistical analysis of hydrogen production in different fermentative condition, the *t*-test was performed using the Minitab 16 program, with 95% significant differences.

3. Results and Discussion

3.1. Hydrogen Production by Dark Fermentation

In order to verify whether the supplementation of organic wastes to coffee mucilage is feasible for hydrogen production and to determine the best operating conditions, a two level factorial design was applied to different independent factors of organic wastes ratio, chemical oxygen demand, temperature, and pH. The light independent process (anaerobic dark fermentation) principally occurs with anaerobic bacteria, which are able to grow on sources abundant in carbohydrates but not requiring light energy [9]. The Embden-Meyerhof (glycolytic pathway) is a well-known metabolic process for glucose decomposition converted into pyruvate. In this metabolism, two hydrogen atoms are released from a glucose molecule by donating electrons in the redox reaction while an oxidized nicotinamide adenine dinucleotide (NAD+) molecule is converted into a reduced form of nicotinamide adenine dinucleotide (NADH) by accepting proton from the nicotinamide ring (Equation (1)). Due to the presence of hexose sugars in coffee mucilage and carbon sources in organic wastes, these are possibly capable of converting sugars into pyruvate through anaerobic glycolysis and generating two molecules of hydrogen as by-products:

$$C_6H_{12}O_6 + 2\,NAD^+ \rightarrow 2\,CH_3COCOOH + 2\,NADH + 2\,H^+ \tag{1}$$

The two-level factorial design with IV resolution generated a total of 26 experimental runs, including two control tests. To prevent lurking variations, all designed experiments were performed in random order, and a cumulative hydrogen production was measured in each fermentation. The experiment sets, independent factors, and results of hydrogen yields are summarized in Table 1. The total cumulative hydrogen from mixed substrates varied to each different ratios; some tests promoted the formation of hydrogen within 72 h. However, less to no hydrogen yields were observed in other experiments, associated with experimental parameters of substrate ratio, Chemical Oxygen Demand (COD), temperature, and pH.

The highest hydrogen concentration of 25.9 L was achieved within 72 h at 30 °C, pH 7.0, chemical oxygen demand 60 g O_2/L with level 1 preparation (8 coffee mucilage: 2 organic wastes), which was equal to 1.295 L hydrogen/L substrate (Run 8 in Table 1). On the other hand, most of the tests with level 3 (2 coffee mucilage: 8 organic wastes) were not suitable for anaerobic fermentation (Runs 19–25 in Table 1), and results from 2 (5 parts coffee mucilage: 5 parts organic wastes) resulted in lower hydrogen yields (Runs 10–17 in Table 1). Even though some tests produced hydrogen in level 2 (up to 11.4 L), the fermentative digestion in the presence of 20% (*w/w*) organic wastes (level 1) was more effective compared to the other ratios. The control run without organic wastes (level 0, run 1) and the test with only organic wastes (level 4, test 26) resulted in little to no hydrogen yield, respectively. Further variance analysis of hydrogen production with the random effects model analysis of variance (ANOVA), tests showed that the high concentration of chemical oxygen demand, low temperature, and low temperature considerably influenced the final yields (Figure 1A). To obtain precise variability response with higher than 0.95 probability worth (>95% coefficient correlation, R^2), the weak multilateral factors were ruled out, and the accurate model was fitted with a >95% two-sided confidence interval. Individual and interaction effects of each factors were depicted in Figure 1B (Pareto Chart), presenting the similar data analysis of ANOVA tests that organic wastes, chemical oxygen demand, and temperatures were the main contributors for anaerobic hydrogen fermentation. The fitted model was generated with the major parameters in response to the reciprocal interaction of independent factors at a given condition: hydrogen production (L) = −79.70 + 0.3062 COD + 1.5725 Temperature + 14.175 pH − 0.007375 COD × Temperature + 0.02438 COD × pH − 0.2675 Temperature × pH. The optimal experimental condition for the maximal anaerobic hydrogen fermentation was calculated and obtained though the contour plots in Figure 1C. This condition is used for anaerobic fermentative digestion for hydrogen production and is applied to fermentation with the isolated pure bacterium.

Table 1. A two level half-fractional factional factorial design and hydrogen production from waste substrates (coffee mucilage combined to organic wastes). Anaerobic fermentative digestion was carried out in 30 L bioreactor (20 L working volume) at provided conditions with a helical ribbon impeller mixing at 100 rpm until the hydrogen generation was completed. All runs were conducted in duplicate and provided statistical analysis of hydrogen production with 95% significant differences.

Run	Ratio[1]	COD (g O_2/L)	Temperature (°C)	pH	Total Solids (g/L)	Total Volatile Solids (g/L)	Hydrogen Production (L)	Yield (L H_2/L Substrate)	Yield (mol H_2/mol Hexose)
1	10:0	40	35	6.0	12.80	10.76	0	0	0
2	8:2	20	30	5.0	12.52	10.46	2.4	0.12	0.02
3	8:2	20	40	5.0	13.61	11.40	3.2	0.16	0.031
4	8:2	20	30	7.0	18.75	15.70	15.6	0.78	0.149
5	8:2	20	40	7.0	44.62	37.50	11.2	0.56	0.107
6	8:2	60	30	5.0	19.42	16.30	10.6	0.53	0.101
7	8:2	60	40	5.0	40.00	33.60	8.6	0.43	0.082
8	8:2	60	30	7.0	46.02	38.66	25.9	1.295	0.248
9	8:2	60	40	7.0	25.23	21.16	18.4	0.92	0.176
10	5:5	20	30	5.0	41.72	35.00	0	0	0
11	5:5	20	40	5.0	62.31	52.30	0	0	0
12	5:5	20	30	7.0	69.92	58.75	8.3	0.415	0.079
13	5:5	20	40	7.0	68.00	57.13	2.6	0.13	0.025
14	5:5	60	30	5.0	61.91	52.00	11.4	0.57	0.109
15	5:5	60	40	5.0	77.42	65.00	7.0	0.35	0.067
16	5:5	60	30	7.0	84.42	70.90	8.6	0.43	0.082
17	5:5	60	40	7.0	71.02	59.63	6.8	0.34	0.065
18	2:8	20	30	5.0	31.52	26.44	0	0	0
19	2:8	20	40	5.0	26.31	22.12	0	0	0
20	2:8	20	30	6.0	63.52	53.38	0	0	0
21	2:8	20	40	8.0	28.62	24.06	0	0	0
22	2:8	60	30	5.0	51.84	43.50	0	0	0
23	2:8	60	40	5.0	67.57	56.66	0	0	0
24	2:8	60	30	8.0	32.72	27.48	1.9	0.095	0.018
25	2:8	60	40	8.0	76.42	64.14	3.0	0.15	0.029
26	0:10	40	35	6.5	45.42	38.16	0.4	0.02	0.004

[1] Substrate ratio (mucilage: organic wastes).

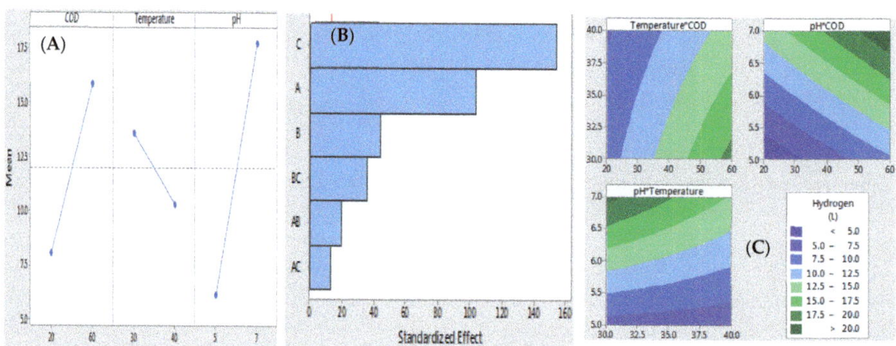

Figure 1. (**A**) The key effect of chemical oxygen demand, temperature, and pH on the hydrogen production from dark fermentation of complex substrates (level 1). (**B**) Individual and mutual interaction effects of chemical oxygen demand, temperature, and pH. A: chemical oxygen demand, B: temperature, C: pH. (**C**) Contour plots indicate influence of temperature and chemical oxygen demand, pH and COD, and pH and temperature, respectively. The different colors denote the different concentrations of hydrogen from dark fermentation. The dark green indicates a higher hydrogen yield while a dark blue presents a lower hydrogen yield.

3.2. Identification of Microorganisms

In order to identify the microorganisms which could dominantly grow and were associated with hydrogen production in complex waste substrates, a slurry sample from the best hydrogen production was spread out on a nutrient agar medium and anaerobically cultivated at room temperature. Each grown colony was suspended with a sterilized saline solution, cultivated, and biochemically reacted with provided reagents in the Vitek 2 analyzer (Biomérieux, Lyon, France). It is worthwhile to note that these isolation and identification methods would not provide all potential microorganisms in the mixed waste substrate. This work has considered isolating and identifying microorganisms that are able to primarily grow in the chosen agar medium under mesophilic and anaerobic experimental conditions (similar to bioreactors). The biochemical identification including morphology, gram staining, and species are summarized in Table 2.

Table 2. Isolation and biochemical analysis of microorganisms from the slurry sample from test 8.

Code	Morphology	Color	Gram Stain	Species	Certainty (%)
B1	Circular, entire edge, convex	Yellow	Positive	*Micrococcus luteus*	99
B2	Irregular, irregular edge, flat	Beige	Positive	*Kocuria kristinae*	87
B3	Circular, entire edge, flat	Beige	Positive	*Streptococcus uberis*	87
B4	Circular, entire edge, flat	Orange	Positive	*Leuconostoc mesenteroides* ssp. *cremosis*	94
B5	Irregular, irregular edge, convex	Beige	Positive	*Brevibacillus laterosporus*	94
B6	Irregular, entire edge convex,	Beige	Positive	*Bacillus farraginis/smithii/fordii*	97
B7	Circular, entire edge, convex	Beige	Positive	*Staphylococcus epudermidis*	95

Biochemical identification found that seven different bacteria existed, and they were involved in the dark hydrogen fermentation under the given conditions. All identified bacteria were Gram positive, anaerobic, mesophilic, and prefer to live in soil or organic agricultural wastes such as coffee mucilage. In particular, *Streptoccu uberis* (B3) and *Leuconostoc mesenteroides* ssp. *cremosis* (B4) are capable of producing lactic acid through glycolysis (Embden-Meyerhof-Parnas pathway) using glucose. Other bacteria containing *Brevibacillus laterosporus* (B5), *Bacillus farraginis/smithii/fordii* (B6), *Micrococcus luteus* (B1), and *Kocuria kristinae* (B2) are known to metabolize complex substrates by producing hydrolytic-protease and hydrolytic-glucosidase, which allow them to contribute toward hydrogen production [40–43]. Previous works observed that they tended to undergo an acetic or butyric pathway and produce hydrogen by consuming acetic or lactic acid as substrates [44,45]. An increase of lactic and acetic acid concentration was detected in the beginning of the fermentation (up to 13.76 g/L and 5.32 g/L, respectively) while none and a lower concentration (4.13 g/L) of both acids were determined at the end of fermentation. This observation supports that these strains possibly utilize intermediate molecules (lactic acid and acetic acid) for their growth, population, and metabolism that subsequently produce hydrogen. A similar study done by Hernández et al. [34] observed that the addition of swine manure content into coffee mucilage improved the hydrogen production and methanogenic process, which is considered a major limiting factor for the hydrogen metabolic pathway. They suggested that a high carbon/nitrogen ratio (C/N) contributed toward increasing hydrogen but decreasing the methane percentage by possibly changing the metabolic pathways through dominant microorganisms and their activity [34,46,47]. They also confirmed that the C/N ratio of 53.4 had a stable hydrogen production in the repetitive batch cultivation for 140 days, which may indicate that the methanogenic pathway was inhibited during the long fermentation times by changes in metabolic routes [32,33,48]. Further 16S rDNA and sequence analysis tests confirmed that B3, B5, and B6 strains were matched (>98% similarity) with *Bacillus firmus* (KT720243.1), *Bacillus simplex* (KT922035.1), and *Frigoritolerans* (KT719834.1), respectively. For the other strains it was not possible to match their sequences with the database from the Gene bank, BLASTN (National Center for Biotechnology Information, Bethesda, MD, USA).

3.3. Hydrogen Production from Single Bacterium Inoculum

The effect of single bacterial fermentation on hydrogen production was studied with seven different isolated bacteria under similar experimental conditions of test 8 in Table 1. All fermentation runs were completed within 48 h, and the resulting hydrogen yields are summarized in Table 3. When isolated pure bacteria were used, the final concentration of hydrogen was in the range of 0–9.3 L, which was significantly lower than the result from the initial test without inoculum (25.9 L) (Table 3).

Table 3. Anaerobic dark fermentation of the complex substrate using single isolated bacteria. All fermentation was carried out with the complex substrate (8 coffee mucilage: 2 organic wastes) at 30 °C, pH 7.0, chemical oxygen demand 60 g O_2/L with a helical ribbon impeller mixing at 100 rpm. All tests were in duplicate and provided statistical analysis of hydrogen yields with 95% significant differences.

Code	Species	H_2 Production (L)	Yield (L H_2/L Substrate)	Yield (mol H_2/mol Hexose)
B1	*Micrococcus luteus*	3.9	0.195	0.037
B2	*Kocuria kristinae*	9.3	0.465	0.089
B3	*Streptococcus uberis*	5.9	0.295	0.056
B4	*Leuconostoc mesenteroides* ssp. *cremosis*	0	0	0
B5	*Brevibacillus laterosporus*	5.6	0.28	0.054
B6	*Bacillus farraginis/smithii/fordii*	1.8	0.09	0.017
B7	*Staphylococcus epudermidis*	0.3	0.015	0.003
B2 and B3	*Kocuria kristinae, Streptococcus uberis*	14.7	0.735	0.14

Although anaerobic fermentative digestion was able to convert carbon sources into hydrogen, we hypothesized that co-cultivation (bacterial consortium) could be responsible for enhancing hydrogen production. To prove this hypothesis, two isolated pure bacteria having the highest yield, (*Kocuria kristinae* (B2) and *Steptococcus uberis* (B3)), were inoculated for a co-cultivation fermentation. As a result, >58% higher hydrogen production (14.7 ± 0.8 L, p-value < 0.05) was observed after 24 h, which was relatively higher than the results from single batch fermentation (Table 3). This result may suggest that these bacteria were significantly associated with hydrogen production. This study is in agreement with other earlier studies that microbial population shifts and enzymatic/metabolic shifts are critical factors for hydrogen production during dark fermentation, and these two parameters can independently or simultaneously affect the hydrogen yield [49,50]. For example, microbial sporulation, particularly in *Clostridium* sp., can be activated as a protection system when the microbial faced on un-favorite circumstance such as high temperature (>90 °C), low pH, dissolved oxygen concentration, and limitation of nutrient sources [51]. Addressing previous observations and current data, other combinations (of two or more bacteria) under different conditions could lead to better hydrogen yields via synergetic metabolisms and/or changing the pathways affecting different bacterial growth performances. The current study mainly focused on the utilization of complex substrates (coffee mucilage plus organic wastes) for hydrogen production and the determination of its optimal experimental conditions. However, the key outcome in this work is the observation that the bacterial population has a considerable effect on dark fermentation, and single batch fermentation is not suitable for efficient hydrogen production at the given conditions. This data indicates the need for further investigation with respect to the changes in bacterial growth, population, metabolic shift, microbial/product inhibition, and bioreactor classification, such as a continuously stirred tank reactor, anaerobic sequencing batch reactor, anaerobic membrane bioreactor, or immobilized bioreactor.

The current work is comparable with previous studies about the anaerobic hydrogen fermentation from other carbon sources, which include food wastes, apple, domestic wastewater, wastepaper, glycerol, and glucose; the detailed fermentation conditions and hydrogen yields are summarized in Table 4.

Table 4. Comparison of hydrogen yield from different carbon sources (wastes or hexose) in anaerobic batch fermentation using mixed culture or pure strain.

Organism	Carbon Source	Reactor	Hydrogen Yield	Reference
Mesophilic mixed culture	Coffee mucilage + organic wastes (20%, w/w)	Batch	0.248 mol H_2/mol hexose (1.295 L/L substrate)	Current work
Mesophilic mixed culture	Food wastes	Batch	0.05 mol H_2/mol hexose	[52]
Mixed culture	Apple (9 g COD/L)	Batch	0.9 L H_2/L substrate	[53]
Mixed culture	Domestic wastewater	Batch	0.01 L H_2/L substrate	[53]
Ruminococcus albus	Wastepaper	Batch	2.29 mol H_2/mol hexose (282.76 L/kg dry biomass)	[54]
Halanaerobium saccharolyticum	Glycerol	Batch	0.58 mol H_2/mol glycerol	[55]
Escherichia coli BW25113 (engineered)	Glucose	Batch	1.82 mol H_2/mol glucose	[56]

When the carbon substrates in waste sources were fermented under mixed batch culture conditions, the result from the present study was relatively higher than those from food waste (5 times), apple (1.44 times), and the domestic wastewater (130 times). It is possible that supplementation of organic wastes is positively effective for the growth and functional activity of hydrogen-producing bacteria to enhance the yield by providing essential nutrients such as nitrogen, phosphorous, ferrous, some mineral, and metals. Previous work demonstrated that additional nitrogen from organic wastes could enhance the hydrogen yield in anaerobic digestion by increasing the C/N ratios [34]. Since nitrogen is one of the vital sources for bacterial growth, the C/N ratio affects cell growth and the metabolic pathway, which suggests a range of 6.7–47 for optimal bacterial growth [57]. Another study with phosphorous (P) demonstrated that it was an essential component for adenosinetriphosphate (ATP) formation and could develop the metabolic pathway and hydrogen production by acting in enzyme linkage for its functions [57]. On the other hand, pure microbial fermentation in the presence of waste paper or hexose obtained 2.3–9.2 times higher hydrogen yields. It is assumed that differences in the composition of substrates and microbial consortium might be highly associated with their pathway during dark fermentation. Related work of the effect of carbohydrates (mainly glucose, fructose, sucrose, and cellobiose) elucidated that the hydrogen yield decreased from 1.82 to 1.38 (mol H_2/mol hexose) when the chains of carbohydrates increased due to the microbial population [58]. A similar study also found that a substrate rich in carbohydrate (sucrose) was more effective in producing hydrogen when the complex substrates were used under the same experimental conditions [57,59].

It is worthwhile to highlight that the carbohydrate degradation and hydrogen production from complex substrates like coffee mucilage and organic wastes are feasible without requiring aseptic condition, nutrients, or pure cell inoculum, suggesting potential and practical application for the real industrial field. The microbial diversity and its mechanisms for engaged management, however, still remain to be discovered. The proposed model and hydrogen yield in current study would be improved with the strategies for microbial population, their gene level resources, or performing conditions (culture operation, mixing condition, reactor type, engineered strain and others).

4. Conclusions

Anaerobic fermentative digestion enables the hexoses in coffee mucilage combined with organic waste to be metabolized and generate hydrogen. The optimal dark fermentation conditions determined via two level factorial designs were 30 °C, pH 7.0, chemical oxygen demand 60 g O_2/L in the presence of 20% (w/w) organic waste, which resulted in 25.9 L hydrogen yield. Moreover, this work found that seven different bacteria were involved in the best hydrogen production test, and their individual batch fermentations under similar experimental conditions, produced lower levels of hydrogen by the end of fermentations, suggesting that further knowledge of the microbial interactions with complex substrates and efficient anaerobic dark fermentation are required. In summary, hydrogen production from raw coffee mucilage and organic acid mixtures is feasible without any aseptic process and supplementation prior to anaerobic digestion. It can be considered a potential sources for practical hydrogen fermentation, requiring the extension of knowledge of microbial interaction to deal with complex substrates and efficient anaerobic dark fermentation.

Author Contributions: E.L.M.C. and A.D.Z.-Z. initiated current work at Universidad Nacional de Colombia and completed with D.K. at Hood College. As primary authors of this research, E.L.M.C. and A.D.Z.-Z. performed the designed dark fermentation tests, data collection, and literature research. D.K. assisted to review, data analysis and summary for this manuscript.

Funding: This research was funded by National University of Colombia, grant number 19522, 24025, and 30085.

Acknowledgments: The authors thank Donna Harrison and Craig Laufer at Hood College for their internal review of this work; the San Rafael farm and Central Mayorista de Antioquia for supporting coffee mucilage and organic wastes, respectively.

Conflicts of Interest: The authors declare no conflict of interest.

References

1. Brentner, L.B.; Jordan, P.A.; Zimmerman, J.B. Challenges in developing biohydrogen as a sustainable energy source: Implications for a research agenda. *Environ. Sci. Technol.* **2010**, *44*, 2243–2254. [CrossRef] [PubMed]
2. Kim, D. Physico-chemical conversion of lignocellulose: Inhibitor effects and detoxification strategies: A mini review. *Molecules* **2018**, *23*, 309. [CrossRef] [PubMed]
3. Brey, J.J.; Brey, R.; Carazo, A.F.; Contreras, I.; Hernández-Díaz, A.G.; Gallardo, V. Designing a gradual transition to a hydrogen economy in Spain. *J. Power Sources* **2006**, *159*, 1231–1240. [CrossRef]
4. Kim, D.; Orrego, D.; Ximenes, E.A.; Ladisch, M.R. Cellulose conversion of corn pericarp without pretreatment. *Bioresour. Technol.* **2017**, *245*, 511–517. [CrossRef] [PubMed]
5. Kim, D.; Ku, S. Beneficial effects of Monascus sp. KCCM 10093 pigments and derivatives: A mini review. *Molecules* **2018**, *23*, 98. [CrossRef] [PubMed]
6. Cao, G.; Ximenes, E.; Nichols, N.N.; Frazer, S.E.; Kim, D.; Cotta, M.A.; Ladisch, M. Bioabatement with hemicellulase supplementation to reduce enzymatic hydrolysis inhibitors. *Bioresour. Technol.* **2015**, *190*, 412–415. [CrossRef] [PubMed]
7. Jung, K.W.; Kim, D.H.; Kim, S.H.; Shin, H.S. Bioreactor design for continuous dark fermentative hydrogen production. *Bioresour. Technol.* **2011**, *102*, 8612–8620. [CrossRef] [PubMed]
8. Levin, D.B.; Pitt, L.; Love, M. Biohydrogen production: Prospects and limitations to practical application. *Int. J. Hydrogen Energy* **2004**, *29*, 173–185. [CrossRef]
9. Kapdan, I.K.; Kargi, F. Bio-hydrogen production from waste materials. *Enzyme Microb. Technol.* **2006**, *38*, 569–582. [CrossRef]
10. Vardar-Schara, G.; Maeda, T.; Wood, T.K. Metabolically engineered bacteria for producing hydrogen via fermentation. *Microb. Biotechnol.* **2008**, *1*, 107–125. [CrossRef]
11. Agler, M.T.; Wrenn, B.A.; Zinder, S.H.; Angenent, L.T. Waste to bioproduct conversion with undefined mixed cultures: The carboxylate platform. *Trends Biotechnol.* **2011**, *29*, 70–78. [CrossRef]
12. Coffee Production Worldwide from 2003/04 to 2017/18 (In Million 60 Kilogram Bags)*. Available online: https://www.statista.com/statistics/263311/worldwide-production-of-coffee/ (accessed on 1 October 2018).
13. Orrego, D.; Zapata-Zapata, A.D.; Kim, D. Optimization and scale-up of coffee mucilage fermentation for ethanol production. *Energies* **2018**, *11*, 786. [CrossRef]
14. Braham, J.E.; Bressani, R. *Coffee Pulp: Composition, Technology, and Utilization*; IDRC: Ottawa, ON, Canada, 1978.
15. Mussatto, S.I.; Machado, E.M.S.; Martins, S.; Teixeira, J.A. Production, Composition, and Application of Coffee and Its Industrial Residues. *Food Bioprocess Technol.* **2011**, *4*, 661–672. [CrossRef]
16. Orrego, D.; Zapata-zapata, A.D.; Kim, D. Bioresource Technology Reports Ethanol production from co ff ee mucilage fermentation by *S. cerevisiae* immobilized in calcium-alginate beads. *Bioresour. Technol. Rep.* **2018**, *3*, 200–204. [CrossRef]

17. Mussatto, S.I.; Machado, E.M.S.; Carneiro, L.M.; Teixeira, J.A. Sugars metabolism and ethanol production by different yeast strains from coffee industry wastes hydrolysates. *Appl. Energy* **2012**, *92*, 763–768. [CrossRef]

18. Pandey, A.; Soccol, C.R.; Nigam, P.; Brand, D.; Mohan, R.; Roussos, S. Biotechnological potential of coffee pulp and coffee husk for bioprocesses. *Biochem. Eng. J.* **2000**, *6*, 153–162. [CrossRef]

19. Brand, D.; Pandey, A.; Rodriguez-Leon, J.A.; Roussos, S.; Brand, I.; Soccol, C.R. Packed bed column fermenter and kinetic modeling for upgrading the nutritional quality of coffee husk in solid-state fermentation. *Biotechnol. Prog.* **2001**, *17*, 1065–1070. [CrossRef]

20. Murthy, P.S.; Naidu, M.M. Recovery of Phenolic Antioxidants and Functional Compounds from Coffee Industry By-Products. *Food Bioprocess Technol.* **2012**, *5*, 897–903. [CrossRef]

21. Borrelli, R.C.; Esposito, F.; Napolitano, A.; Ritieni, A.; Fogliano, V. Characterization of a New Potential Functional Ingredient: Coffee Silverskin. *J. Agric. Food Chem.* **2004**, *52*, 1338–1343. [CrossRef]

22. Napolitano, A.; Fogliano, V.; Tafuri, A.; Ritieni, A. Natural Occurrence of Ochratoxin A and Antioxidant Activities of Green and Roasted Coffees and Corresponding Byproducts. *J. Agric. Food Chem.* **2007**, 10499–10504. [CrossRef]

23. Franca, A.S.; Oliveira, L.S.; Nunes, A.A.; Alves, C.C.O. Microwave assisted thermal treatment of defective coffee beans press cake for the production of adsorbents. *Bioresour. Technol.* **2010**, *101*, 1068–1074. [CrossRef]

24. Oliveira, L.S.; Franca, A.S.; Alves, T.M.; Rocha, S.D.F. Evaluation of untreated coffee husks as potential biosorbents for treatment of dye contaminated waters. *J. Hazard. Mater.* **2008**, *155*, 507–512. [CrossRef]

25. Oliveira, W.E.; Franca, A.S.; Oliveira, L.S.; Rocha, S.D. Untreated coffee husks as biosorbents for the removal of heavy metals from aqueous solutions. *J. Hazard. Mater.* **2008**, *152*, 1073–1081. [CrossRef] [PubMed]

26. Murthy, P.S.; Naidu, M.M.; Srinivas, P. Production of α-amylase under solid-state fermentation utilizing coffee waste. *J. Chem. Technol. Biotechnol.* **2009**, *84*, 1246–1249. [CrossRef]

27. Brand, D.; Pandey, A.; Roussos, S.; Soccol, C.R. Biological detoxification of coffee husk by filamentous fungi using a solid state fermentation system. *Enzyme Microb. Technol.* **2000**, *27*, 127–133. [CrossRef]

28. Murthy, P.S.; Madhava Naidu, M. Sustainable management of coffee industry by-products and value addition—A review. *Resour. Conserv. Recycl.* **2012**, *66*, 45–58. [CrossRef]

29. Belitz, H.D.; Grosch, W.; Schieberle, P. *Food Chemistry*; Springer: Berlin, Germany, 2009; ISBN 9783540699330.

30. Ulloa Rojas, J.B.; Verreth, J.A.J.; Van Weerd, J.H.; Huisman, E.A. Effect of different chemical treatments on nutritional and antinutritional properties of coffee pulp. *Anim. Feed Sci. Technol.* **2002**, *99*, 195–204. [CrossRef]

31. Clifford, M.N. *Coffee: Botany, Biochemistry and Production of Beans and Beverage*; Springer: Berlin, Germany, 1985; ISBN 9781461566595.

32. Liu, Y.; Yu, P.; Song, X.; Qu, Y. Hydrogen production from cellulose by co-culture of Clostridium thermocellum JN4 and Thermoanaerobacterium thermosaccharolyticum GD17. *Int. J. Hydrogen Energy* **2008**, *33*, 2927–2933. [CrossRef]

33. Margarida, F. Temudo, Robbert Kleerebezem, M. van L. Determination of heat flux on dual bell nozzle by Monte carlo method. *J. Chem. Pharm. Sci.* **2007**, *98*, 69–79. [CrossRef]

34. Hernández, M.A.; Rodríguez Susa, M.; Andres, Y. Use of coffee mucilage as a new substrate for hydrogen production in anaerobic co-digestion with swine manure. *Bioresour. Technol.* **2014**, *168*, 112–118. [CrossRef]

35. Moreno Cardenas, E.L.; Cano Quintero, D.J.; Elkin Alonso, C.M. Generation of Biohydrogen by Anaerobic Fermentation of Organics Wastes in Colombia. *Liq. Gaseous Solid Biofuels-Convers. Tech.* **2013**, 378–400. [CrossRef]

36. Kim, M.S.; Lee, D.Y. Fermentative hydrogen production from tofu-processing waste and anaerobic digester sludge using microbial consortium. *Bioresour. Technol.* **2010**, *101*, S48–S52. [CrossRef]

37. Hata, D.J.; Hall, L.; Fothergill, A.W.; Larone, D.H.; Wengenack, N.L. Multicenter evaluation of the new VITEK 2 advanced colorimetric yeast identification card. *J. Clin. Microbiol.* **2007**, *45*, 1087–1092. [CrossRef] [PubMed]

38. Kim, D.; Ku, S. Bacillus cellulase molecular cloning, expression, and surface display on the outer membrane of Escherichia coli. *Molecules* **2018**, *23*, 503. [CrossRef] [PubMed]

39. Kim, D.; Ximenes, E.A.; Nichols, N.N.; Cao, G.; Frazer, S.E.; Ladisch, M.R. Maleic acid treatment of biologically detoxified corn stover liquor. *Bioresour. Technol.* **2016**, *216*, 437–445. [CrossRef] [PubMed]

40. Chang, J.J.; Chou, C.H.; Ho, C.Y.; Chen, W.E.; Lay, J.J.; Huang, C.C. Syntrophic co-culture of aerobic Bacillus and anaerobic Clostridium for bio-fuels and bio-hydrogen production. *Int. J. Hydrogen Energy* **2008**, *33*, 5137–5146. [CrossRef]

41. Goud, R.K.; Raghavulu, S.V.; Mohanakrishna, G.; Naresh, K.; Mohan, S.V. Predominance of Bacilli and Clostridia in microbial community of biohydrogen producing biofilm sustained under diverse acidogenic operating conditions. *Int. J. Hydrogen Energy* **2012**, *37*, 4068–4076. [CrossRef]

42. Zhang, K.; Ren, N.Q.; Wang, A.J. Fermentative hydrogen production from corn stover hydrolyzate by two typical seed sludges: Effect of temperature. *Int. J. Hydrogen Energy* **2015**, *40*, 3838–3848. [CrossRef]

43. Ratti, R.P.; Delforno, T.P.; Sakamoto, I.K.; Varesche, M.B.A. Thermophilic hydrogen production from sugarcane bagasse pretreated by steam explosion and alkaline delignification. *Int. J. Hydrogen Energy* **2015**, *40*, 6296–6306. [CrossRef]

44. Cordeiro, A.R.R. Biological Hydrogen Production Using Organic Waste and Specific Bacterial Species. Master's Thesis, University of Porto, Porto, Portugal, 2013.

45. Bala-Amutha, K.; Murugesan, A.G. Biohydrogen production using corn stalk employing Bacillus licheniformis MSU AGM 2 strain. *Renew. Energy* **2013**, *50*, 621–627. [CrossRef]

46. Saint-Amans, S.; Girbal, L.; Andrade, J.; Ahrens, K.; Soucaille, P. Regulation of Carbon and Electron Flow in Clostridium butyricum VPI 3266 Grown on Glucose-Glycerol Mixtures. *J. Bacteriol.* **2001**, *183*, 1748–1754. [CrossRef] [PubMed]

47. Jo, J.H.; Jeon, C.O.; Lee, D.S.; Park, J.M. Process stability and microbial community structure in anaerobic hydrogen-producing microflora from food waste containing kimchi. *J. Biotechnol.* **2007**, *131*, 300–308. [CrossRef] [PubMed]

48. Kim, I.S.; Hwang, M.H.; Jang, N.J.; Hyun, S.H.; Lee, S.T. Effect of low pH on the activity of hydrogen utilizing methanogen in bio-hydrogen process. *Int. J. Hydrogen Energy* **2004**, *29*, 1133–1140. [CrossRef]

49. Ye, N.F.; Lü, F.; Shao, L.M.; Godon, J.J.; He, P.J. Bacterial community dynamics and product distribution during pH-adjusted fermentation of vegetable wastes. *J. Appl. Microbiol.* **2007**, *103*, 1055–1065. [CrossRef] [PubMed]

50. Guo, X.M.; Trably, E.; Latrille, E.; Carrre, H.; Steyer, J.P. Hydrogen production from agricultural waste by dark fermentation: A review. *Int. J. Hydrogen Energy* **2010**, *35*, 10660–10673. [CrossRef]

51. Bastidas-Oyanedel, J.R.; Bonk, F.; Thomsen, M.H.; Schmidt, J.E. Dark fermentation biorefinery in the present and future (bio)chemical industry. *Rev. Environ. Sci. Biotechnol.* **2015**, *14*, 473–498. [CrossRef]

52. Shin, H.S.; Youn, J.H.; Kim, S.H. Hydrogen production from food waste in anaerobic mesophilic and thermophilic acidogenesis. *Int. J. Hydrogen Energy* **2004**, *29*, 1355–1363. [CrossRef]

53. Van Ginkel, S.W.; Oh, S.E.; Logan, B.E. Biohydrogen gas production from food processing and domestic wastewaters. *Int. J. Hydrogen Energy* **2005**, *30*, 1535–1542. [CrossRef]

54. Ntaikou, I.; Antonopoulou, G.; Lyberatos, G. Biohydrogen production from biomass and wastes via dark fermentation: A review. *Waste Biomass Valoriz.* **2010**, *1*, 21–39. [CrossRef]

55. Kivistö, A.; Santala, V.; Karp, M. Hydrogen production from glycerol using halophilic fermentative bacteria. *J. Eur. Ceram. Soc.* **2010**, *101*, 8671–8677. [CrossRef]

56. Mathews, J.; Li, Q.; Wang, G. Characterization of hydrogen production by engineered Escherichia coli strains using rich defined media. *Biotechnol. Bioprocess Eng.* **2010**, *15*, 686–695. [CrossRef]

57. Lin, C.Y.; Lay, C.H. Carbon/nitrogen-ratio effect on fermentative hydrogen production by mixed microflora. *Int. J. Hydrogen Energy* **2004**, *29*, 41–45. [CrossRef]

58. Quéméneur, M.; Hamelin, J.; Benomar, S.; Guidici-Orticoni, M.T.; Latrille, E.; Steyer, J.P.; Trably, E. Changes in hydrogenase genetic diversity and proteomic patterns in mixed-culture dark fermentation of mono-, di- and tri-saccharides. *Int. J. Hydrogen Energy* **2011**, *36*, 11654–11665. [CrossRef]

59. Lin, C.Y.; Lay, C.H. Effects of carbonate and phosphate concentrations on hydrogen production using anaerobic sewage sludge microflora. *Int. J. Hydrogen Energy* **2004**, *29*, 275–281. [CrossRef]

Article

Fatty Acids, Hydrocarbons and Terpenes of *Nannochloropsis* and *Nannochloris* Isolates with Potential for Biofuel Production

Alan Rodrigo López-Rosales [1], **Katia Ancona-Canché** [1], **Juan Carlos Chavarria-Hernandez** [1], **Felipe Barahona-Pérez** [1], **Tanit Toledano-Thompson** [1], **Gloria Garduño-Solórzano** [2], **Silvia López-Adrian** [3], **Blondy Canto-Canché** [4], **Erik Polanco-Lugo** [5] and **Ruby Valdez-Ojeda** [1,*]

[1] Unidad de Energía Renovable, Centro de Investigación Científica de Yucatán A.C., Parque Científico y Tecnológico del Estado de Yucatán, Carretera Sierra Papacal-Chuburná Puerto, Yucatán 97302, Mexico; alanrodrigolopez91@gmail.com (A.R.L.-R.); katiaancona17@gmail.com (K.A.-C.); jc.ch@cicy.mx (J.C.C.-H.); barahona@cicy.mx (F.B.-P.); tanit@cicy.mx (T.T.-T.)

[2] Facultad de Estudios Superiores Iztacala, UNAM. Tlalnepantla, Edo. de México 54090, Mexico; ggs@unam.mx

[3] Facultad de Veterinaria y Zootecnia, Universidad Autónoma de Yucatán. Mérida, Yucatán 97288, Mexico; ladrian@uady.mx

[4] Unidad de Biotecnología, Centro de Investigación Científica de Yucatán A.C., Parque Científico y Tecnológico del Estado de Yucatán, Carretera Sierra Papacal-Chuburná Puerto, Yucatán 97302, Mexico; blondy@cicy.mx

[5] Centro de Investigación y Asistencia en Tecnología y Diseño del Estado de Jalisco, Unidad Sureste, Parque Científico y Tecnológico del Estado de Yucatán, Carretera Sierra Papacal-Chuburná Puerto, Yucatán 97302, Mexico; erikpolanco@gmail.com

* Correspondence: dubi@cicy.mx; Tel.: +52-999-9428-330; Fax: +52-999-9813-900

Received: 29 October 2018; Accepted: 25 December 2018; Published: 31 December 2018

Abstract: Marine microalgae are a promising feedstock for biofuel production given their high growth rates and biomass production together with cost reductions due to the use of seawater for culture preparation. However, different microalgae species produce different families of compounds. Some compounds could be used directly as fuels, while others require thermochemical processing to obtain quality biofuels. This work focuses on the characterization of three marine microalgae strains native in Mexico and reported for the first time. Ultrastructure and phylogenetic analysis, suggested that they belong to *Nannochloropsis* sp. (NSRE-1 and NSRE-2) and *Nannochloris* sp. (NRRE-1). The composition of their lipid fractions included hydrocarbons, triacylglycerides (TAGs), free fatty acids (FFAs) and terpenes. Based on theoretical estimations from TAG and FFA composition, the potential biodiesels were found to comply with six of the seven estimated properties (ASTM D6751 and EN 14214). On the other hand, hydrocarbons and terpenes synthesized by the strains have outstanding potential as precursors for the production of other renewable fuels, mainly green diesel and bio-jet fuel, which are "drop-in" fuels with quality properties similar to fossil fuels. The validity of this theoretical analysis was demonstrated for the oxygenates of strain NSRE-2, which were experimentally hydrodeoxygenated, obtaining a high-quality renewable diesel as the reaction product.

Keywords: biodiesel; bio-jet fuel; triacylglycerides; Fatty Acid Methyl Ester; lipids; hydrodeoxygenation; drop-in fuel

---

1. Introduction

Microalgae offer an attractive way of generating renewable and sustainable biofuels [1] capable of helping to meet the global demand for transport fuels. However, a screening procedure to assess microalgae potential is a highly recommended prerequisite for biofuel feedstock production. Marine microalgae, which are abundant in seawater, are a particularly promising raw material due to the cost reductions resulting from using seawater for cultivation. The additional salinity level is also one of the methods of reducing contamination issues [2] in open-pond cultivation. For this reason, a number of companies currently operate commercial-scale mass cultivation of marine microalgae [3].

Proper characterization and adaptation to culture conditions of isolated native strains are the first steps in establishing any biotechnological process, including biofuel production. Strategies for improving microalgae growth include the variation of inoculum size, pH and culture medium composition. Inoculum size significantly affects microalgae cell growth characteristics, such as lag phase duration, maximal specific growth rate, biomass accumulation and metabolites production at the end of culturing [4]. An appropriate inoculum size can reduce cell mortality and increase biomass production and nutrient recovery [5]. With a proper culture pH, high cell densities and lipids accumulation can be obtained [6]. Culture medium composition affects the specific growth rate, biomass production and the biochemical composition of the resulting biomass and lipids [7].

Characterization of marine microalgae must include ultrastructure analysis of organelles and their disposition inside the cell, as well as light microscopy to elucidate reproduction mode, determination of the phylogenetic position of the strains under study, and analysis of their carotenoid and chlorophyll content in order to determine the taxonomic identity of each strain [8,9]. All of these analyses are particularly important in taxa such as *Nannochloropsis* sp. (Eustigmatophyceae) and *Nannochloris* sp. (Chlorophyceae), both extremely small unicellular marine eukaryotic algae with easily confused characteristics, which can even be impossible to distinguish due to their tiny cell size and simple and smooth cell wall morphology [10]. Among the most important marine microalgae for obtaining biofuels, precisely *Nannochloropsis* spp. stand out due to their high growth rates and lipid contents in the form of TAGs [11]. *Nannochloris*, meanwhile, is a species that has received less study for biofuel purposes, but has high lipid productivity [12]. To our knowledge, this is the first report on the isolation and characterization of *Nannochloropsis* and *Nannochloris* species in Mexico.

The lipidic fraction of the microalgal biomass is used for biofuel production. Triacylglycerides are converted to biodiesel, and this procedure has been extensively studied. However, other compounds produced by microalgae such as hydrocarbons and terpenes can also be used for the production of bio-jet fuel and green diesel. For example, phytol stands out as a natural diterpene alcohol that can be used as biofuel in diesel engines [13] or converted to gasoline by catalytic cracking [14]. Another terpene, neophytadiene, is one of the dominant ones in green microalgae [15] and it could also be converted to biofuel. Terpenes are hydrocarbons that are already important as bioactive compounds, with applications as functional compounds due to their broad spectrum of biological activity, such as antitumor and antiviral activity [16], among others. However, from more than 40,000 reported structures, specific terpenes (cyclic and acyclic forms) are recognized as special biofuels that can be used directly or blended with a range of fossil fuels, such as jet fuel, missile propellant, gasoline, or diesel fuels [17]. On the other hand, hydrocarbons produced from microalgae can be directly used in mixtures with the diesel fraction (C_{12}–C_{22}), or they can be subjected to a catalytic improvement stage (hydroisomerization or hydroisomerization plus hydrocracking process), through which paraffins with better fuel properties can be obtained [18]. The selection of the process will depend on the composition of the raw material and the type of fuel that is to be produced. For example, if the hydrocarbon fraction has a high content of paraffins with more than 21 carbon atoms, a hydroisomerization process with simultaneous hydrocracking would be recommended. This process would permit the formation of branched hydrocarbons with shorter carbon chains and products in the range of gasoline (bio-gasoline), jet fuel (bio-jet fuel) and diesel (green diesel).

The vast biodiversity available in Yucatan, Mexico, provides a valuable resource for the development of indigenous microalgal strains as biofuel feedstock. The objective of this research was to collect marine strains from coastal waters and analyze morphology, reproduction mode and organelle disposition by light microscopy and ultrastructure analysis to support phylogenetic studies and determine taxonomic identity. Additionally, the fatty acids and organic compounds synthesized by the strains under study were analyzed to determine their potential for biofuel production, either through catalytic processes for obtaining drop-in fuels (bio-jet fuel and renewable diesel) or biodiesel, or even using them without chemical modification in mixtures with fossil fuels. Finally, a catalytic hydrodeoxigenation reaction was carried out to experimentally demonstrate the potential of the extracted lipids as raw material to obtain high-quality renewable fuels.

2. Materials and Methods

2.1. Isolation and Cultivation of Microalgae

Marine microalgae were collected from the water column at three different sites in Progreso, Yucatan, located at 21°19′01.33″ N, 89°40′50.05″ W; 21°17′09.38″ N, 89°41′50.05″ W and 21°17′09.36″ N, 89°41′50.07″ W. Samples were collected at three different depths (0.0, 0.5 and 1 m) from water bodies during the mornings of May 2014 using a Van Dorn bottle (1120-G45, Wildco, Yulee, FL, USA). The pH, temperature and salinity values (Extech Stick EC500, Extech instruments, Boston, MA, USA) were registered [19]. The collected samples were allowed to settle for 3 to 4 h at 4 °C. The sedimented biomass was analyzed to proceed with the isolation of the microalgae species.

2.2. Obtaining Axenic Cultures

Microalgae cells were observed by inverted microscope (Axio Observer-5 ZEISS, Baden-Württemberg, Oberkoche, Germany) to select individual cells and inoculate in Guillard's f/2 (G9903, Sigma-Aldrich, San Luis, MO, USA) and saline BG-11 medium (2% NaCl) (C3061, Sigma). Then microalgae colonies (500 µL plated in solid Bold's basal medium (BBM) were transferred to both media for decontamination: *Centrifugation.* Microalgae cultures were centrifuged, pellets were washed in sterile seawater and centrifuged again, and were then inoculated [19]. *Serial dilution.* Dilutions of microalgae culture (1:10, 1:100 and 1:1000 (v/v)) were employed. *Antibiotic treatment.* After proliferation of bacteria in Müeller-Hinton media, antibiotic susceptibility testing disks (50295; Bio-Rad, Hercules, CA, USA) were used, giving maximum susceptibility to chloramphenicol, amikacin and streptomycin (100 mg/L) [20]. *Acid shock method.* Cultures in exponential phase were centrifuged at 2291× g for 7 min at 10 °C (Digicen 21R, OrtoAlresa, Madrid, Spain). The pellet was resuspended three times in 1 mL HCl (1N) by manual agitation [21] and re-suspended in both media and incubated. The axenic microalgal strains were cultivated (10 mL) in Guillard's f/2 (G9903, Sigma) and BG-11 (2% NaCl) (C3061, Sigma) under a 12:12 photoperiod at 25 ± 2 °C, 110 µmol m^{-2}s^{-1} light intensity and 120 rpm agitation. To improve cell densities, nutrient content (Guillard's f and f/2 medium) was evaluated at different pH (6, 7 and 8) and inoculum size (10, 20 and 30%). Cell concentration was determined every 48 h using a Neubauer hemocytometer [22].

2.3. Morphological Analysis. Light Microscopy and Transmission Electron Microscopy (TEM)

Morphological features (cell shape and reproductive features) observed through optical microscopy (Eclipse E200, Nikon instruments, Melville, NY, USA) were recorded. For TEM, the strains were preserved in 2% glutaraldehyde for 4 h and then post-fixed in 1% OsO$_4$ for 30 min, followed by progressive dehydration in 30 to 100% ethanol at 4 °C. The samples were then transferred to 100% propylene oxide and infiltrated into epon-propylene oxide (1-1) for 48 h, followed by embedding in epon and polymerization at 60 °C for 48 h [23]. Cross sections were performed with a Leica ultramicrotome (Leica, Wetzlar, Germany) and observed with the 1200-EX II TEM (JEOL, Peabody, MA, USA) located at the Institute of Cellular Physiology, UNAM.

2.4. DNA Extraction and 18S rRNA Gene Amplification

The microalgae cells were harvested at the stationary growth phase by centrifugation (30 mL) and washed three times in bidistilled sterile water. The cell (c.a. 200 mg) was stored at −20 °C until use.

For genomic DNA extraction, the Youssef protocol was used with minor modifications [24]. 18S rRNA gene amplification was implemented using forward primer A (5′-AACCTGGTTGATCCTGCCAG-3′) and reverse primer SSU-inR1 (5′-CACCAGACTTGCCCTCCA-3′) based on the conserved domain region of 18S rDNA [25].

18S rRNA region amplification was carried out with a 10 µL reaction mixture containing 1× PCR Buffer, 2 mM $MgCl_2$, 0.2 mM dNTPs, 1 mM primer, 1 U *Taq* DNA polymerase (Invitrogen, Waltman, MA, USA) and 10 ng of DNA. PCR conditions included 1 cycle of 1 min at 94 °C for initial denaturation, followed by 40 cycles of 30 s at 94 °C, 30 s at 57.5 °C, 1 min at 72 °C, and finally 7 min at 72 °C. Amplicons were checked by electrophoresis on 1% agarose gel in 1 × TAE buffer. The amplification products of approximately 500 pb were purified using NucleoSpin® Gel and PCR Clean-up commercial kit (Macherey-Nagel, Düren, Germany) to remove primer dimers according to the manufacturer's instructions. Direct sequencing of the amplification products was carried out by a two-directional (reverse and forward) procedure (Clemson Sequencing Service, Clemson, South Caroline, USA).

The 18S rRNA region sequences from DNA of NSRE-1, NRRE-1 and NSRE-2 of 540 pb (approx.) were used as "queries" in nucleotide Blast tools (NCBI website) to recover the ten best matches for each input sequence, based on percentage of sequence coverage, E-value and percentage of identity of 100%. After manual adjustment with MEGA version 2.1, sequences with their homologies in GenBank were deleted to a final sequence of 536 pb and aligned using CLUSTALX (EMBL, Heidelberg, Germany). Molecular phylogenetic analyses by the maximum likelihood method based on the Kimura 2-parameter model were conducted using MEGA version 2.1 [26]. One thousand replica samplings were analyzed for percent bootstrap values in a neighbor-joining tree. The percentage of trees in which the associated taxa clustered together is shown next to the branches. Evolutionary analyses were conducted in MEGA 7 [26]. Reference sequences retrieved from GenBank of 18S rRNA region sequences were: NSRE-1, MH921192; NRRE-1, MH921193; and NSRE-2, MH921194.

2.5. Biomass Production

Algal biomass was produced in a culture system consisting of three serially connected 250 mL Erlenmeyer flasks with 200 mL of culture medium and a 250 mL Erlenmeyer flask with sterile distilled water (with the purpose of acting as an air humidifier). The culture system was maintained under the following conditions: an aeration rate of 3.6 vvm, temperature of 24 ± 2 °C, and light intensity of 81 µmol m^{-2} s^{-1} with a 16:8 (light:dark) photoperiod [27]. The biomass was recovered at the beginning of the stationary phase (15 days of cultivation) by centrifugation at 5057× *g* for 7 min at 25 °C [28]. The pellet was washed with bidistilled sterile water to eliminate minerals and salts from the culture media. The total biomass was maintained at −80 °C for 24 h and freeze dried at 0.1 bar and −53 °C for 72 h [29].

2.6. Total Lipid Extraction

One g of dried biomass was subjected to solvent extraction with 15 mL of a 1:2 (*v/v*) chloroform-methanol mixture [30–33] for 3 h at 40 °C and 150 rpm (3×). The solvents were then eliminated by vacuum evaporation.

2.7. Thin Layer Chromatography

It was carried out on TLC Silica gel 60 F_{254} plates (Merck KGaA, New York, NY, USA). A 9:1:0.1 (*v/v*) mixture of hexane, ethyl acetate and acetic acid was used as the mobile phase. The plates were developed using a 10% (*w/v*) solution of phosphomolybdic acid in sulfuric acid [34].

2.8. Column Chromatography

The lipid extract was fractionated by column chromatography using silica gel (Kieselgel, 70–230 mesh) as the stationary phase. A solvent gradient elution was used to obtain five fractions from the lipid extract. The solvent system was hexane, then 9:1, 8:2 and 7:3 mixtures of hexane-ethyl acetate, and finally methanol [29,34,35].

2.9. Transesterification

The TAG fraction was transesterified with sodium metoxide (6% v/w) in a 25 mL vial with moderate agitation at 60 °C for 90 min. The reaction mixture was passed through silica gel to eliminate excess catalyst and glycerol. The solvent was then evaporated in a fume hood.

2.10. Lipid Profile Analysis (GC-MS)

The fractions were analyzed by gas chromatography–mass spectrometry (7890B and 5977A series GC/MSD, Agilent Technologies, Santa Clara, CA, USA), using a DB5-HT column (30 m × 250 μm ID × 0.25 μm film width) under the following conditions: 3 mL/min of helium flux as carrier gas. The detector and injector temperatures were set at 300 °C. The GC temperature program started at 70 °C/5 min, and was then raised to 200 °C using a rate of 10 °C/min with a 5 min hold at 200 °C and 20 °C/min until 290 °C, which was held for 22.5 min. The injection volume was 1 μL. The NIST 05 (National Institute of Standards and Technology, Gaithersburg, MD, USA) library was used for identification of the main peaks [36]. Six main groups of compounds were identified: hydrocarbons, FAMEs, organic acids, terpenes, sterols and amines.

2.11. Catalytic Hydrodeoxygenation of Oxygenates from Strain NSRE-2

The fraction of oxygenated compounds from strain NSRE-2 was subjected to a catalytic hydrodeoxygenation treatment (HDO) in a stainless steel (SS-316) fixed bed reactor (Microactivity Reference, Madrid, Sapin), with a length of 30.5 cm and a 0.9 cm internal diameter. The reaction mixture (31 mg lipid extract dissolved in 120 mL nonane) was fed using an HPLC pump (model 307 V 3.00, Gilson, Villiers-le-Bel, France) at a constant flow rate of 0.061 mL min-1 to attain a WHSV of 0.1 h^{-1}. Pt/HZSM-22-γ-Al$_2$O$_3$ synthesized in-house was used as catalyst (0.046 g) and activated in-situ under H$_2$ flow (50 mL min^{-1}) at 400 °C for 2 h with a heating rate of 5 °C min^{-1} and 20 bar with a pressure rate of 5 bar min^{-1}. The reaction conditions were 300 °C at a heating rate of 5 °C min^{-1} and a H$_2$ pressure of 40 bar. H$_2$ was fed at a gas flow rate of 55 mL min^{-1} and the total reaction time was 4 h. Samples of the reaction product were collected every hour and analyzed by GC-MS.

2.12. Statistical Analysis

Results of biomass and lipid production are presented as means ± standard deviations (SD) from three replicates. The statistical analysis was carried out using OriginPro (v 9.0, OriginLab, Northampton, MA, USA) to determine the degree of significance using one-way analysis of variance (ANOVA) at a probability level of $P < 0.05$.

3. Results and Discussion

3.1. Obtaining Axenic Cultures

The three isolated microalgae were named NSRE-1, NRRE-1 and NSRE-2, and placed under the conditions mentioned. Application of HCl proved to be an effective decontamination technique. This pH-shock treatment has been applied before for effective control of *V. chlorellavorus*, a predatory bacterium that can destroy a *Chlorella* culture in just a few days [37]. This effective treatment with high selectivity and low cost can be considered throughout the scale-up and production process [37]. The pH-shock treatment applied in this study is simpler than the one executed by Ganuza et al. [37] and patented by Ganuza and Tonkovich. Application of antibiotics has been used by other authors to obtain axenic cultures from freshwater bodies [38]. However, bacterial contamination persisted in all cultures when amikacin, active against aerobic gram-negative bacilli [19], was applied in this work.

The culturing of NSRE-1, NRRE-1 and NSRE-2 in Guillard's f/2 resulted in improved growth compared to BG-11 medium, as indicated by the cell densities (cell mL^{-1}) of NSRE-1, NRRE-1 and NSRE-2. In Guillard's f/2 they were 9.03×10^6, 5.46×10^6 and 6.28×10^6, while in BG-11 they were 1.48×10^6, 2.56×10^5 and 4.50×10^5 respectively. This result could be explained by the fact that Guillard's f/2 medium has a ratio of 16N:1P, which promotes higher cell density in shorter cultivation times [39]. Additionally, vitamins in Guillard's f/2 medium (cyanocobalamin, thiamine and biotin), absent in BG-11, contribute to the initial growth of microalgae species [40].

Improving microalgae culture parameters to increase biomass and growth is considered the most important factor in sustainable product development [41]. This includes the improvement of the medium composition and physical parameters [42]. In this study, modifications to pH were carried out to increase cell density (Figure 1). Inoculum size of microalgae is another important factor that significantly affects microalgae growth, specifically, the lag phase, maximum specific growth rate, biomass accumulation and the metabolite production at the end of culturing (Richmond 2004). In this study, an inoculum size of 10% *v/v*, yielded a higher cell density than 20% or 30%, probably because a greater availability of nutrients occurs at lower cell densities [43]. In this regard, inoculum size had an effect on final cell density. Selection of the appropriate inoculum size can help reduce the cell mortality ratio in microalgae and increase the biomass production and nutrient recovery [5].

The nutrient concentration of Guillard's medium employed in this study were half strength and full strength, denoted as f/2 and f respectively. The cell density double increased in all strains under study by using f Guillard's f media (Figure 1). Notably in NRRE-1, cell density was ten times greater than NSRE-1 and NSRE-2. These results are consistent with those obtained by Malakootian et al. (2015), in *Nannochloropsis oculata* with Walne medium, the composition of which is similar to Guillard's f medium, resulting in a highest cell density than with Guillard's f/2, Sato and TMRL media.

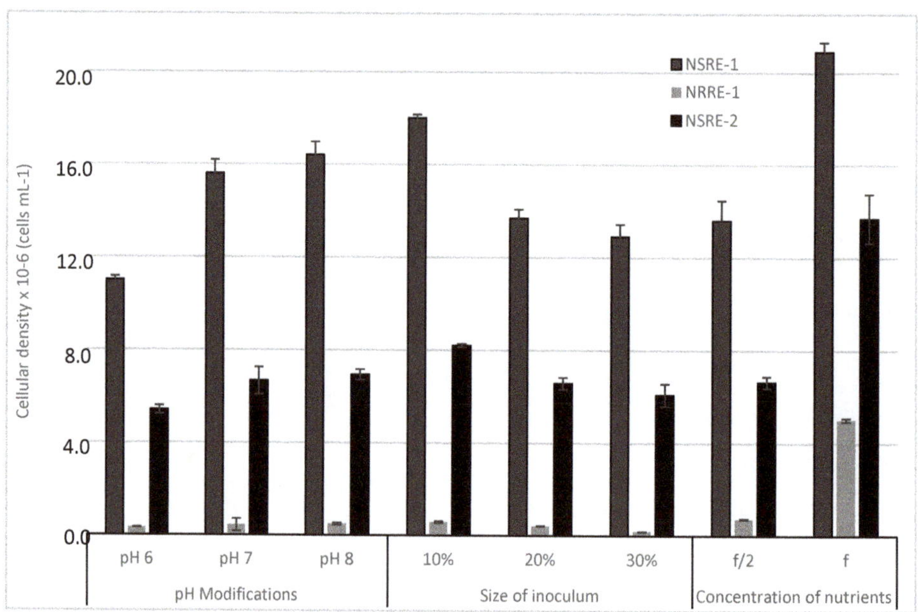

Figure 1. Mean cell density ± SE of NSRE-1, NRRE-1 and NSRE-2 strains at three different pH levels and inoculum sizes and two nutrient concentrations.

Based on the results obtained, stabilized microalgae strain culture conditions were as follows: pH 8, inoculum size 10% *v/v* and Guillard's f medium.

The three strains showed a short lag phase (2–4 days), followed by an exponential growth phase of 6 days. The early stationary phase began on day 18 of cultivation (Figure 2). The highest cell density was registered almost at the end of the stationary phase in NSRE-1 and NRRE-1. In NRRE-1 cell concentration decreased on day 20 of culture. Cell density was higher in NSRE-1 (2.16×10^7 cells mL^{-1}) than NSRE-2 (1.89×10^7 cells mL^{-1}), whereas NRRE-1 showed the lowest density (1.21×10^7 cells mL^{-1}).

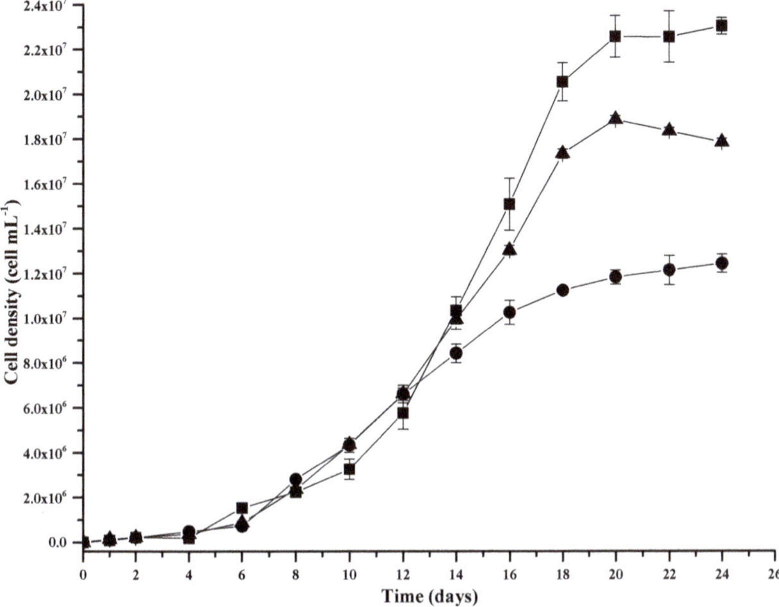

Figure 2. Growth curve of strains NSRE-1 (■), NRRE-1 (●) and NSRE-2 (▲).

3.2. Morphological Analysis (Light Microscopy and TEM)

A suite of analyses (morphological, ultrastructural, biochemical and molecular) may be required for the determination of some taxa [44]. Visualization by light microscopy mainly determines reproduction mode and other characteristics, while ultrastructure analysis determines the disposition of organelles and cellular details for taxa differentiation purposes. Under light microscopy (LM), NSRE-1 showed spherical or oval cells (1.0 to 3.0 μm in diameter) with rounded lobes and a very thin hyaline mucilage. In adult and young cells, a parietal chloroplast without pyrenoid covered three quarters of the cells. An eyespot was observed in some cells. Reproduction was asexual by 2 to 4 autospores, released by rupture and gelatinization of the mother cell wall. Mother cell wall remnants were observed in cell groups. The cells were observed mainly to be solitary vegetative cells or autospores clustered together. Our ultrastructure study through TEM analysis (Figure 3) showed lenticular ornamentations on the cell wall, some lipid bodies and the endoplasmic reticulum attached to the plastid and nuclear membrane. Four membranes were observed covering the plastid, characteristic of *Nannochloropsis* sp. (Eustigmatophyceae) [45].

NRRE-1 showed spherical or oval cells (2.0 to 4.0 μm in diameter), most of them clustered together (many dispersed solitary vegetative cells) or solitary cells of different size and phases. A cup-shaped parietal chloroplast with pyrenoid was observed. Under TEM, the cells were observed to be rounded with a multilayer cell wall (Figure 4). The plastid exhibited thylakoid membranes penetrating into the pyrenoid. Cells in division exhibited an evident cleavage furrow, characteristic of *Nannochloris* sp. (Chlorophyta). Under LM, reproduction was observed mainly by the formation of two autospores and tetrads of aplanospores, which are converted into colonies of sister cells.

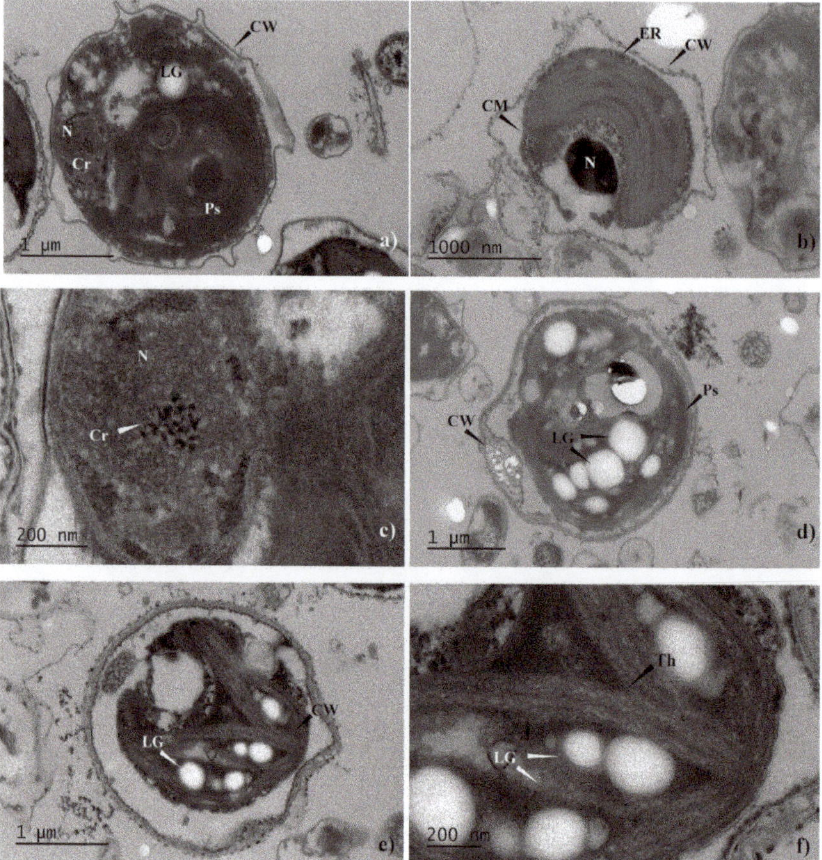

Figure 3. TEM of NSRE-1. (**a**) General appearance of *Nannochloropsis* sp. (**b**) Cell with detail of endoplasmic reticulum adjacent to the plastid, nucleus, membrane and ornamented wall. (**c**) Details of nucleus with chromatin. (**d**) Cell with lipid accumulation. (**e**) General aspect of cell in division. (**f**) Detail of plastids, thylakoids and lipids in dividing cells. N, nucleus; CM, cell membrane; CW, cell wall; ER, endoplasmic reticulum; Ps, plastid; Cr, chromatin; LG, lipid granules; Th, thylakoid.

Finally, NSRE-2 exhibited spherical or oval cells (1.4 to 2.8 μm in diameter), with a single parietal chloroplast, an eyespot, and with the pyrenoid surrounded by reserve substance. Solitary vegetative cells or cells clustered together were also observed in division into two with the formation of tetrads of aplanospores. Reproduction mode is mainly by autospores in two tetrads and by rupture of the autosporangium with the release of aplanospores. NSRE-2 showed large lipid bodies, a smooth cell wall, and the plastid attached to the endoplasmic reticulum and covered by four membranes, according to TEM (Figure 5). This characteristic evident in NSRE-1 and NSRE-2 strongly suggests that they belong to *Nannochloropsis* sp. (Eustigmatophyceae), in which the chloroplasts occupy most of the cell and contain a series of parallel lamellae composed of three thylakoids each, without lamella sheath [45]. Of the four membranes covering the chloroplast, the two external ones correspond to the endoplasmic reticulum, which apparently connects to the nuclear membrane [8].

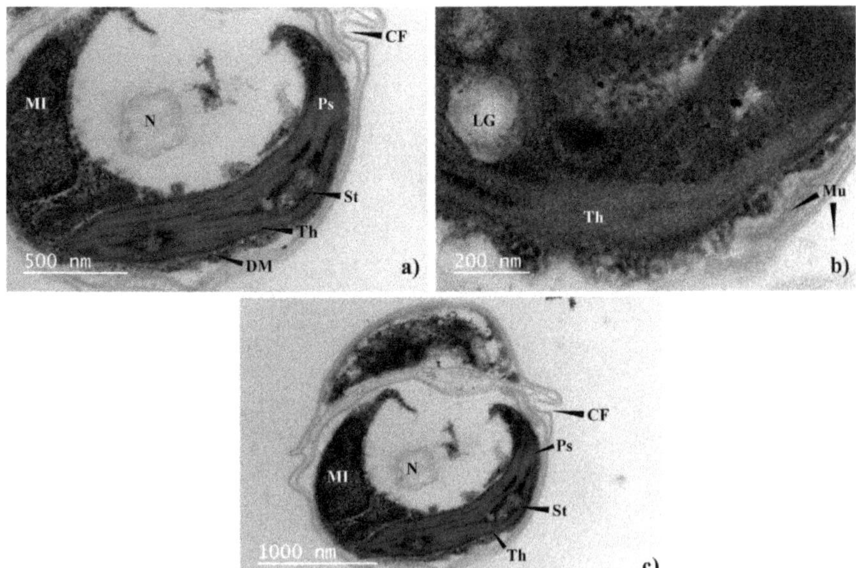

Figure 4. TEM of NRRE-1. (**a**) Cup-shaped plastid with starch granules and very broad, multilayer cell wall. The thylakoids penetrate the starch furrow and cup-shaped plastids with two membranes. (**b**) Detail of cell with nucleus and plastids with starch. (**c**) Cell in division with an evident furrow and cup-shaped chloroplast. MI, mitochondria; N, nucleus; Ps, plastid; Th, thylakoid; St, starch; CF, cleavage furrow; DM, double membrane; Mu, mucilage.

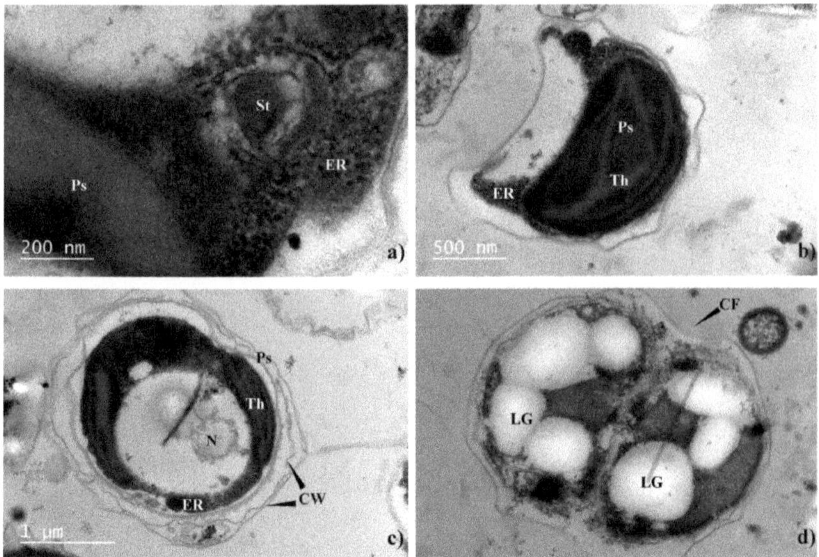

Figure 5. TEM of NSRE-2. (**a**) Detail of cells with reserve granules without the thylakoid membrane. (**b**) Cell with plastids, thylakoids, smooth-walled endoplasmic reticulum and chloroplast bounded by four envelopes. (**c**) Cell with nucleus, membrane and multilayer cell wall. (**d**) Dividing cell with large lipid granules. N, nucleus; Ps, plastid; Th, thylakoid; St, starch; CF, cleavage furrow; ER, endoplasmic reticulum; LG, lipid granules; CW, cell wall.

3.3. 18S rRNA Analysis

Morphological traits suggested the strains under study belong to *Nannochloropsis* (NSRE-1 NSRE-2) and *Nannochloris* species (NRRE-1). Molecular characterization was executed to support this classification based on morphological characteristics. The accession numbers, whose sequences showed homologies with 18S rDNA sequences of the strains under study, correspond to microorganisms with tiny cell sizes and from coastal marine environments. This confirms that the 18S rDNA lineages are consistent with the habitat [46]. The phylogenetic tree constructed with 18S rRNA region sequences supported splitting the strains into two groups (Figure 6), separating NSRE-1 and NRRE-1 from NSRE-2.

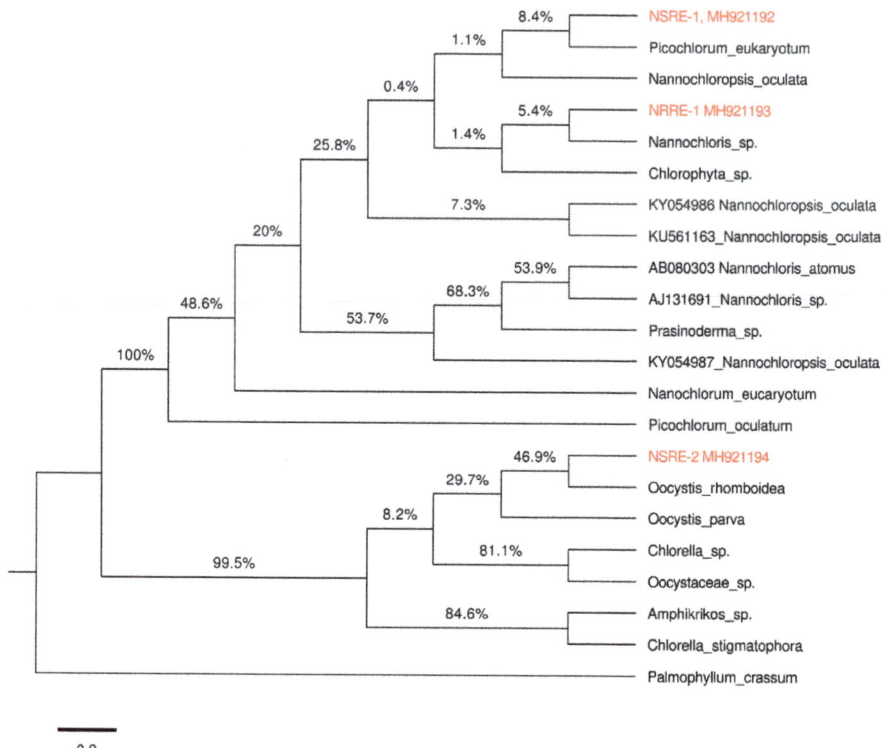

Figure 6. Molecular phylogenetic analysis by the maximum likelihood method inferred from 18S rDNA sequences. The percentage of trees in which the associated taxa clustered together is shown next to the branches.

Interestingly, NSRE-1 and NRRE-1 showed genetic relations with *Nannochloropsis oculata*, *Nannochloris* sp. and *Picochlorum eukaryotum*, all marine species [47] from coastal zones. The phylogenetic nearness of NSRE-1 and NRRE-1 to *Nannochloropsis* and *Nannochloris* must be revised in light of pigment composition as a taxonomic marker for the Eustigmatophyceae; e.g., astaxanthin and canthaxanthin present in authentic species of *Nannochloropsis* [48] and chlorophyll-*b*, to distinguish *Nannochloris* from *Nannochloropsis* [49]. Additionally, a larger fragment of the 18S rDNA, rbcL gene and ITS region must be considered to confirm their identity [8]. For now, NSRE-1 must be considered to belong to *Nannochloropsis* sp. and NRRE-1 to *Nannochloris* sp. pending further analysis.

On the other hand, NSRE-2 was clustered distantly from the rest of the strains under study. It falls into a freshwater clade dominated by *Oocystis* sp., characterized by synthesis of FAMEs and essential fatty acids. However, there are many morphologic details to reject this possibility. Oocystaceae (Trebouxiophyceae) have a characteristic cell wall substructure that is composed of several cellulose layers with perpendicular fibril orientations [50]. In fact, since the work of Komárek (1979) the definition of the family has been based upon the ultrastructure of the multilayered cell wall [51]. The phylogenetic position of the NSRE-2 strain must be revised, including a longer 18S rDNA fragment, although the morphologic details strongly indicate that it must be considered to be *Nannochloropsis* sp. In this context, it is worth mentioning that there are few *Nannochloropsis* spp. entry sequences in GenBank.

3.4. Biomass Concentration, Lipid Extraction and Analysis

The dry biomass weight for NSRE-1, NRRE-1 and NSRE-2 strains was 0.82 ± 0.03, 0.58 ± 0.01 and 0.56 ± 0.02 g L^{-1} respectively (Figure 7), with no significant differences (ANOVA, P < 0.05, n = 3) between NRRE-1 and NSRE-2 strains, in contrast with the NSRE-1 strain.

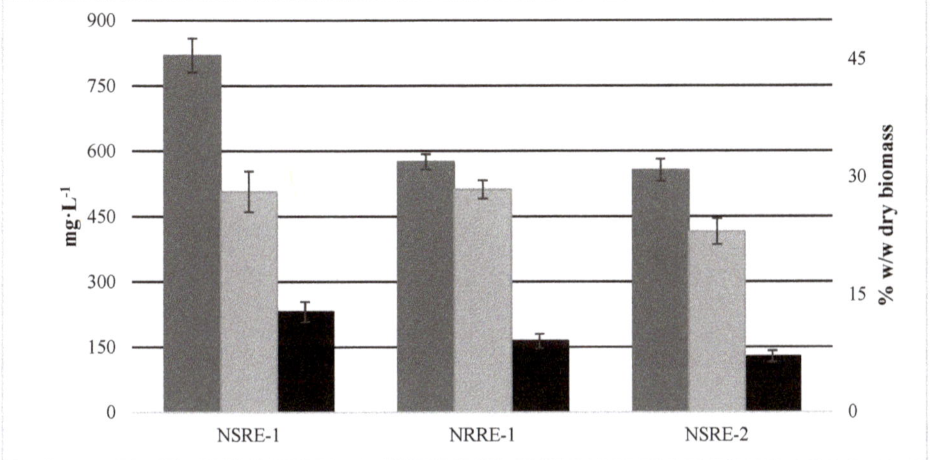

Figure 7. Dry biomass weight (dark gray bars), lipid extract (light gray bars) and lipid content (dark bars) of microalgae NSRE-1, NRRE-1 and NSRE-2.

Similar results to NSRE-1 were obtained in *Nannochloropsis salina* (0.9 g L^{-1}), *Nannochloropsis oleoabundans* (1.1 g L^{-1}) and *Dunaliella primolecta* (0.9 g L^{-1}), all marine microalgae. Meanwhile, for the NRRE-1 and NSRE-2 strains, dry biomass values were similar to those obtained in *Synechocystis salina* (0.6 g L^{-1}), *Microcystis aeroginosa* (0.8 g L^{-1}) and *Pseudokirchneriella subcapitata* (0.4 g L^{-1}) [52,53].

Total lipid extracts for NSRE-1, NRRE-1 and NSRE-2 were 307.91 ± 36, 218.02 ± 21 and 171.02 ± 16 mg L^{-1} (Figure 7) and lipid contents were 37.56, 37.91 and 30.75%, similar to the 25–56% lipid content found in several marine microalgae [52]. Meanwhile, lipid yields in seven marine isolated cultures of *Nannochloropsis* sp. (42.7–23.5% dry weight) [28] were similar to those obtained in this study from NSRE-1 and NSRE-2 (28.17 and 23.06% dry weight respectively).

Thin layer chromatography analysis indicated the presence of non-polar (terpenes and TAGs) and polar (fatty acids and phospholipids) lipids in the extracts. Therefore, to carry out their identification, the partitioning of the extract was executed by polarity gradient using column chromatography to obtain fractions of the extracts, which are summarized in Table 1.

Table 1. Fractions of the lipid extracts.

Fraction	% of Total Extract		
	NSRE-1	NRRE-1	NSRE-2
1	5.8	8.1	20.1
2	8.9	10.7	7.1
3	8.5	6	23.1
4	5.9	16.4	19.6
5	-	10	8.4
6	71	48.8	21.7

3.5. Lipid Profile Analysis (GC-MS)

The identification of the fractions by GC-MS generated numerous compounds grouped into six main types: hydrocarbons, FAMEs (after transesterification of TAGs), organic acids, terpenes, sterols and amines (Table 2).

Table 2. Compounds identified by GC-MS in strains NSRE-1, NRRE-1 and NSRE-2.

Compound (IUPAC)	Molecular Formula	% of Total Peak Area (wt % Fractions)		
		NSRE-1	NRRE-1	NSRE-2
Hydrocarbons				
Tetradecane	$C_{14}H_{36}$		5.64 (3.6)	
Pentadecane	$C_{15}H_{32}$	4.89 (8.3)		4.35 (4.5)
1-Heptadecene	$C_{17}H_{34}$	23.75 (59.2)		
Heptadecane	$C_{17}H_{36}$	6.91 (11.8)	8.48 (5.5)	5.97 (6.1)
Octadecane	$C_{18}H_{38}$	4.86 (8.3)		5.72 (5.9)
Nonadecane	$C_{19}H_{40}$		4.66 (3.0)	4.3 (1.1)
Eicosane	$C_{20}H_{42}$		7.8 (5.0)	8.78
Heneicosane	$C_{21}H_{44}$		7.28 (4.7)	8.31 (2.1)
Tetracosane	$C_{24}H_{50}$	3.95 (6.7)		7.76 (2.0)
Hexacosane	$C_{26}H_{54}$			4.86 (5.0)
Heptacosane	$C_{27}H_{56}$	3.37 (5.7)		
Neophytadiene	$C_{20}H_{38}$		64.22 (78.2)	64.98 (61.8)
Total (wt %)		33.8	59.9	46.9
FAMEs (TAG and FFA fractions)				
Hexadecanoic acid, methyl ester (Methyl palmitate, C16:0)	$C_{17}H_{34}O_2$	13.58 (4.16)	13.85 (28.48)	7.82 (10.82)
9,12-Octadecadienoic acid (Z,Z)-, methyl ester (Methyl linoleate, C18:2)	$C_{19}H_{34}O_2$		4.92 (10.11)	12.34 (26.35)
Octadecenoic acid. Methyl stearate (C18:0)			5.4 (10.4)	
(Z)-9-Octadecenoic acid, methyl ester (Methyl oleate, C18:1)	$C_{19}H_{36}O_2$		7.37 (15.15)	
Total (wt %)		9.5	24.7	23
Oxygenated compounds				
3-Hydroxy-butanoic acid	$C_4H_8O_3$	10.78 (2.2)		
Bis (2-ethylhexyl) hexadecanoic acid ester	$C_{22}H_{42}O_4$	45.19 (12.9)	4.93 (9.14)	39.03 (5.6)
Phytol	$C_{20}H_{40}O$		3.02 (10.0)	
6,10,14-Trimethyl-2-pentadecanone	$C_{18}H_{36}O$	10.54 (2.7)		
2-Palmitoylglycerol	$C_{19}H_{38}O_4$	9.94 (10.19)		
Campesterol	$C_{28}H_{48}O$		6.58 (12.20)	
Clionasterol	$C_{29}H_{50}O$		26.79 (49.68)	8.25 (2.4)
(Z)-13-Docosenamide	$C_{22}H_{43}NO$			61.35 (40.88)
2,2,4-Trimethyl-1,3-pentanediol 1-isobutyrate	$C_{12}H_{24}O_3$			5.01 (9.24)
2,4-Di-t-butylphenol	$C_{14}H_{22}O$		5.73 (18.97)	
Total (wt %)		56.8	15.3	30.1
% oxygen		19.2	6	12.5
H/C atomic ratio		1.9	1.7	1.9

3.5.1. Hydrocarbons

The hydrocarbons identified in the strains under study contained 14 to 27 carbon atoms (Table 2). Most common were even numbers (C_{12}, C_{16}, C_{18}, C_{22} and C_{24}), both saturated (alkanes) and unsaturated (alkenes). In all three strains, a higher percentage of saturated straight-chain hydrocarbons was detected (40, 80 and 90%, respectively). This result agrees with those reported in the literature, where e.g., heptadecane represented more than 80% of total hydrocarbons. Heptadecane, pentadecane, and 7- and 8-heptadecene were detected in several *Nannochloropsis* species (Eustigmatophyceae). In marine microalgae, odd-numbered hydrocarbons account for more than 60%, typically C15, C17 or C21, and are often saturated.

Dry weight percentages of hydrocarbons in all strains yielded 0.055, 0.041 and 0.044% for NSRE-1, NRRE-1 and NSRE-2 respectively. This result is consistent because the hydrocarbon content in microalgae is generally low (less than 0.07% in dry weight) and mostly straight-chain alkanes and alkenes. Alka(e)nes of various chain lengths are important targets for biofuel production because they are major components of gasoline (mainly C5–C9 hydrocarbons), jet fuels (C5–C16) and diesel fuels (C12–C20). Microalgae hydrocarbons represent a potential feedstock for fuel development, although their use is still under development because knowledge of physiological pathways of hydrocarbon synthesis in microalgae is necessary to improve synthesis yields. Additionally, knowledge is also required to attain an economically viable scale-up process.

3.5.2. FAMEs

The fatty acid methyl esters (FAMEs) identified correspond to saturated C_{16} and unsaturated, mono- and di-, C_{18} (Table 2). The NSRE-1 extract contained 10% more C16:0 than C18:1. The ratio of saturated/unsaturated FAMEs was 1:0.89. The NRRE-1 extract contained 11.2% more C16:0 than unsaturated C18:1 and C18:2, resulting in a saturated/unsaturated FAMEs ratio of 1:0.88. In contrast, the NSRE-2 extract showed 57.8% greater production of C18:2 than C16:0, with a saturated/unsaturated ratio of 1:1.57. Hexadecanoic fatty acid methyl ester (C16:0) was identified in all extracts at percentages higher than 15% compared to other fatty acids in marine and freshwater microalgae strains. In *Nannochloropsis* strains, C_{16} to C_{20} fatty acid chains were identified with a variable level of unsaturation: e.g., palmitic acid (16:0), stearic acid (18:0), oleic acid (18:1), linoleic acid (18:2) and eicosanoic acid (20:0) [53]. The same compounds were identified in NSRE-1 and NSRE-2, with additional compounds such as 9,12-octadecadienoic acid methyl ester and 9-octadecenoic acid. For biodiesel production, a low ratio of saturated to unsaturated fatty acids is preferred (less than 22% FAMEs by weight), because for high concentrations of saturated FAMEs, the CFPP increases above 0°C, and the biodiesel obtained cannot be used in cold climates [54].

3.5.3. Organic acids

Free fatty acids and unsaponifiable organic acids were identified in the strain extracts under study (Table 2). The 3-hydroxy butanoic acid in NSRE-1 is a precursor in the formation of homopolymers in marine cyanobacteria [55]. *n*-Hexadecanoic acid (palmitic acid) was more abundant in NSRE-2 (26.01%) than in NSRE-1 (7.98%). In marine microalgae species, hexadecanoic acid has been identified among major saturated fatty acids. Hexadecanoic acid bis(2-ethylhexyl) ester or di(2-ethylhexyl)adipate (DEHA), identified at a high percentage in NSRE-1 and NSRE-2, is a constituent of polymeric flexible materials (polyvinyl chloride, PVC) and films used in the flexible vinyl industry [53]. Commonly, this compound is used as a solvent and a component of aviation lubricants, among others uses [56]. DEHA is relatively insoluble in water and therefore it is likely to be distributed in sediment through biota of aquatic ecosystems or by effluents from plastic-producing plants [53]. Possible contamination of the sampling site is discarded due to the constant changes of culture and purification treatments applied in this study. As far as the authors know, this is the first study that reports its identification in the lipid composition of green microalgae.

3.5.4. Terpenes

Bis(2-ethylhexyl) phthalate (DEHP) detected in NSRE-2 (Table 2) is a plasticizer or solvent in a variety of industrial products. It is an environmental pollutant in marine sediments, although it has been identified in lipid extracts of *Tetraselmis suecica* and *Chlorella sorokiniana*, *Synechocystis* sp. and mixed cultures of *Streptomyces* sp. and *Chlorella sorokiniana* [32]. In NSRE-1,

6,10,14-trimethyl-2-pentadecanone (hexahydrofarnesyl acetone) or HHA was identified, derived from the diterpene phytol, a secondary metabolite synthesized by higher plants with the ability to esterify chlorophyll-*a* [53]. It is also characterized as being a natural fragrance in higher plants [57] and as a lipid component in green microalgae, such as *Mougeotia viridis, Cladophora rivularis* [58], *Botryococcus braunii* [52] and the marine brown macroalgae *Padina pavonia* [59].

3.5.5. Sterols

In NSRE-2, 25-oxo-25-norcholesterol and clionasterol were identified. 25-Oxo-25-norcholesterol has been identified in extracts of marine microalgae (*Skeletonema costatum*) [2], in cyanobacteria subjected to hydrocarbon biodegradation processes, and in the control of the invasive growth of diatom species [32]. The detection of campesterol and clionasterol in NRRE-1 and NSRE-2 can be explained because both derive from the conversion of 24-methylene cholesterol to fucosterol in the synthesis of isoprenoids in green microalgae (Chlorophyceae) [60]. Sterols, especially from green microalgae, differ from those of higher plants in the configuration of the C24 alkyl groups, these being 24*S*, in contrast to the 24*R* of sterols in higher plants. These sterols from marine microalgae have been used as nutrients for the growth of aquaculture organisms [61]. Alternatively, the use of sterols has been proposed as substitutes in edible oils [57]. On the other hand, 2-palmitoylglycerol was reported as a substrate of chloroplast (omega)-3 fatty acid desaturases isolated from plant cells [57].

3.5.6. Amines

(Z)-13-Docosenamide was identified in the NSRE-2 strain with 61.35% total peak area (Table 2) is an allelopathic substance in lipid extracts of *Microcystis aeruginosa* (Cyanobacteria) and *Quadrigula closterioides* (Chlorophyceae) [53].

3.6. Potential of Lipids as Feedstock for Biofuel Production

The products extracted from the three strains were grouped according to their nature: hydrocarbons, TAGs and FFAs, and oxygenated compounds (Table 2). The composition of each of these fractions is analyzed in the following sections with a view to their possible direct use as fuel or as a raw material to be chemically processed and obtain high-quality renewable fuels. The quality of the biodiesel that can be obtained from compounds from the three strains was estimated through the use of experimental data and the application of selected correlations taken from the literature, according to the procedures described in reference [62] with the following two exceptions:

i)   The equation reported by [63] was used in this work for the calculation of the cold filter plugging point (CFPP):

$$\text{CFPP } (^{\circ}\text{C}) = -26 + [5.76([\text{SAT}]\_(\text{C16-C24}))]^{\wedge}0.5 \tag{1}$$

where SATC16–C24 is the content (wt %) of saturated FAMEs having chains with 16 to 24 C atoms.

ii) Density at 15 °C, kinematic viscosity, higher heating value and oxidative stability of methyl oleate were taken from reference [64].

It is important to mention that no equation was found to estimate the CFPP for biodiesels with a content of saturated FAMEs as high as those of the three fractions of this work. However, Equation (1) was chosen because it is one of the most accurate equations found in the literature for estimating this property.

The potential of the hydrocarbon fractions to be used directly in mixtures with fossil fuel or to be fed to a catalytic improvement process for obtaining higher quality biofuels was also analyzed qualitatively, based on the lipid profiles obtained by GC-MS analyses. Finally, the fraction of oxygenated compounds was analyzed in a similar way to the hydrocarbon fraction, with the difference that in this case the composition of *Jatropha curcas* oil (JCO) was taken as a reference raw material for the oxygen elimination processes in the production of transportation biofuels.

3.6.1. Analysis of the Hydrocarbon, FAME and Oxygenated Fractions as Fuel Precursor

The composition of the hydrocarbon fractions obtained from the strains is shown in Table 2. Due to the high content (>59%) of mono-unsaturated compounds in each fraction, their direct use as diesel in mixtures with hydrocarbons of fossil origin is not recommended. This is because the presence of unsaturated molecules causes instability in fuels during storage [65]. For all strains, the shorter hydrocarbons are 14 or 15 C atoms. On the other hand, the fractions from NSRE-1 and NSRE-2 have hydrocarbons of more than 24 C atoms, although the concentration of these is relatively low (5.7 and 5% for strains NSRE-1 and NSRE-2 respectively). According to the above, the hydrocarbons of strain NRRE-1 could be improved through a single hydrogenation stage to saturate the double bonds in neophytadiene. This would produce diesel fuel with a high cetane number, although it would not have very good cold flow properties. It could, however, be mixed with diesel of fossil origin. Another alternative for the hydrocarbon fraction of the NRRE-1 strain would be to process it through a hydroisomerization stage. This could be more convenient, because the conditions of both processes are similar, with the isomerization process having the advantage of generating a given content of branched structures, which is necessary to improve the cold flow properties of the diesel obtained. In both cases, a diesel without the presence of heteroatoms would be obtained, which, when mixed with fossil diesel, would increase the yield of ultra-low-sulfur diesel.

On the other hand, the long hydrocarbons, which exceed the diesel range (>24 C atoms) of strains NSRE-1 and NSRE-2 could be separated by distillation and the remaining fraction could be processed in the manner described for strain NRRE-1. Another alternative for fractions one and three would be to transform the hydrocarbons without prior distillation in a hydroisomerization stage with simultaneous hydrocracking. This would make it possible to obtain high quality saturated and branched hydrocarbons as fuel, with chain sizes in the interval of diesel and bio-jet fuel.

For the analysis of oxygenated fractions, consider the following characteristic data of *Jatropha curcas* as a reference. According to in-house analysis, *J. curcas* oil (JCO), one of the raw materials that has been used to produce bio-jet fuel and green diesel through the HEFA process (HDO), has 11% by weight of oxygen and a H/C atomic ratio of 1.8. These values are close to others for different vegetable oils. Table 2 shows also the composition of the oxygenate fraction for the three strains, discarding TAGs and FFAs. The percentage of oxygen and the H/C atomic ratio for each fraction is also indicated.

As can be seen in Table 2, the oxygenated fraction of NSRE-1 has a component that has only four C atoms, which, when subjected to an HDO/DO process, will give rise to paraffin of three or four C atoms. While propane and butane are useful hydrocarbons as fuel or precursors for obtaining other compounds with higher added value, they are not part of liquid transportation fuels. The rest of the oxygenated components of NSRE-1 correspond to the diesel fraction according to the number of C atoms present in their structures. A simple HDO/DO stage would eliminate the oxygen heteroatoms and obtain linear paraffins with high cetane number. Alternatively, if the process is oriented towards the formation of linear and branched structures (hydrodeoxygenation plus hydroisomerization), a higher quality diesel could be obtained. Additionally, there is the option of favoring a moderate level of breakage of long molecules (hydrocracking), with which a certain yield of bio-jet fuel can also be obtained. However, it must be considered that the presence of cracking reactions and the consequent saturation of the resulting structures implies a higher net consumption of hydrogen in the process. The fractions from strains NRRE-1 and NSRE-2 can also be processed by HDO or DO to obtain renewable diesel. These fractions have the advantage of not having compounds with less than twelve carbon atoms. Therefore, a higher liquid yield would be expected in their processing. On the other hand, these fractions have structures with more than 22 C atoms, which would require the application of a certain level of hydrocracking in order to obtain hydrocarbons in the diesel range or even in the range of bio-jet fuel.

On the other hand, the oxygenated fractions of the three strains possess adequate characteristics to be transformed through the HEFA process to obtain bio-jet fuel as well as renewable diesel. When comparing the percentage of oxygen of the three fractions with that of JCO, that of NSRE-1 is observed to be considerably higher, while that of NRRE-1 corresponds to approximately half that of JCO. Finally, the oxygen content of NSRE-2 is similar to that of JCO. A greater or lesser amount of oxygen in the feed may imply the greater or lesser formation of CO and CO_2 byproducts, which, besides being polluting gases, are inhibiting agents of the bifunctional catalysts used in these processes. Moreover, the formation of CO and CO_2 is also directly related to the favored reaction routes which depend on the catalyst and the reaction conditions applied, since the hydrodeoxygenation route gives rise to the elimination of oxygen as water, while the decarbonylation/decarboxylation routes result in the formation of CO and CO_2. Whatever the catalytic route, the higher the oxygen content in the feed, the higher the catalytic capacity required to remove all oxygen present. Thus, hydroprocessing of the fraction from NSRE-1 may involve a slightly higher effort.

On the other hand, although the fraction from NRRE-1 has lower oxygen content, and this is convenient from the point of view of the removal of this heteroatom, this fraction has the highest content of structures with more than 24 C, which requires higher breaking levels to obtain equivalent yields of biofuels. A more significant presence of cracking reactions is also associated with a higher net consumption of hydrogen, as well as higher exposure to coke precursors, and a potential decrease in the lifetime of the catalysts used in the process. On the other hand, NSRE-1 requires a greater elimination of oxygen, but does not require favoring the breaking reactions. Regarding NSRE-2, it has an oxygen level equivalent to that of feeds processed by HEFA, and it has a low concentration of carbon chains with more than 24 C atoms, so its hydroprocessing would also be similar. As regards the H/C ratio of the three strains, the values are similar to the one present in JCO or other vegetable oils, so the net consumption of hydrogen would be comparable in orders of magnitude to those used in the processing of triglycerides by HEFA.

3.6.2. Catalytic Hydrodeoxygenation of Oxygenates from Strain NSRE-2

The reaction products obtained are listed in Table 3, and they were similar for the 4 h of time on stream, indicating a good stability of the catalyst used. The fact that the feed was composed of oxygenated C_{16} to C_{29} molecules, having 12.5 wt % oxygen and some unsaturations in their molecules (Table 2), indicates that certain level of hydrocracking occurred in the reaction, giving rise to hydrocarbons of smaller chain lengths. This, in turn, may imply the formation of some compounds shorter than C_{12}, that could not have been identified and quantified due to the large proportion of the C_9 used as solvent in the catalytic tests. However, these unidentified compounds would correspond to the gasoline fraction and would be free of oxygen, considering that the removal of this element is more difficult to achieve for larger compounds. In accordance with the above, it can be concluded that a complete elimination of the oxygen present in the feed was achieved in the HDO reaction. Likewise, a complete saturation of the identified reaction products was obtained, giving rise to a mixture of saturated hydrocarbons in the range of C_{12}-C_{20}.

Table 3. Compounds identified as HDO reaction products.

Compound (IUPAC)	Molecular Formula	% of Total Peak Area
Dodecane	$C_{12}H_{26}$	82.9
2,6,10-Trimethyldodecane	$C_{15}H_{32}$	1.8
Hexadecane	$C_{16}H_{34}$	1.8
Heptadecane	$C_{17}H_{36}$	5.9
Octadecane	$C_{18}H_{38}$	4.8
10-Methylnonadecane	$C_{20}H_{46}$	2.7

The product obtained is then an ultra-low sulfur renewable diesel with a high cetane number (conferred by the linear molecules of intermediate carbon chains lengths [C_{12}, C_{16}-C_{18}]), and good cold flow properties, due to the significant presence (4.5%) of structures with one or three branches. A diesel with this composition could probably be used in high proportions in mixtures with fossil diesel.

3.6.3. Estimation of Biodiesel Properties

Table 2 shows the composition of the TAG and FFA fraction expressed in terms of FAMEs resulting from their esterification and transesterification. The estimated properties of the biodiesels from the three strains studied are reported in Table 4. The following main fuel properties were considered in this analysis: density at 15 °C and 40 °C (ρ), cetane number (CN), kinematic viscosity at 40 °C (v), higher heating value (HHV), iodine value (IV), cold filter plugging point (CFPP) and oxidative stability (OSI). Also, the reference values of the estimated properties are indicated in the table when they are considered in the American (ASTM D6751) and European (EN 14214) biodiesel standards.

Table 4. Predicted biodiesel properties and specifications in U.S. and European standards.

Property (Units)	NSRE-1	NRRE-1	NSRE-2	ASTM D6751	EN 14214
ρ @15°C (kg/m^3)	886 [a]	880.4	883.0		860–900
ρ @40°C (kg/m^3)	850.8	855.5	856.3		
CN	74.5	66.6	65.9	47 min	51 min
v @40°C (mm^2/s)	4.7 [a]	4.5	4.2	1.9–6.0	3.5–5.0
HHV (MJ/kg)	38.4 [a]	39.7	39.5		
IV (g I/100 g biodiesel)	0.0	44.6	45.4		120 max
CFPP (°C)	31.6	17.9	23.4		
OSI (h)	27.4 [a]	5.4	7.1	3 h	6 h

[a] taken from [64].

It can be seen that the estimated densities of the three biodiesels at 15 °C fall within the limits established in EN 14214, while the U.S. biodiesel standard does not specify limits for this property.

Additional considerations on the density and the rest of the evaluated properties can be consulted elsewhere [62]. The CN of the three biodiesels is well above the minimum limit of 51 prescribed by the more stringent European standard. Moreover, the kinematic viscosity at 40 °C of the three biodiesels is also within the limits established by both standards. Furthermore, the higher heating value (HHV) estimated for biodiesels from the three strains lies between 38.4 and 39.7 MJ/kg, approximating the average value of 40 MJ/kg usually found for biodiesel. On the other hand, the estimated iodine values (IV), which are a measure of the unsaturation present in biodiesel, are well below the maximum limit of 120 g I/100 g biodiesel prescribed in the EN 14214 standard. Nevertheless, the estimated CFPP values for the three biodiesels are too high. The American and European standards for biodiesel do not consider defined limits for this property. However, a report is required depending on the season. Typical values of CFPP required in European countries are below 0 °C. For example, in Spain, the maximum limit of CFPP in summer is 0 °C, while the maximum in winter is −10 °C [66]. It is clear that the estimated CFPP values for the three biodiesels are very poor. This is explained by the high percentages of saturated species present in them, which range from 58 to 100%. For this reason, to be able to use these biodiesels it would be necessary to mix them with FAMEs having a higher content of unsaturated species to overcome the limitation on their applicability. Additionally, a warm climate would make their application more feasible. Contrary to what occurs for the CFPP parameter, oxidative stability (OSI) is favored by the presence of saturated fatty esters. For this reason, the biodiesels from the NSRE-1 and NSRE-2 strains, which possess higher percentages of saturated species, present OSI values higher than the minimum limit of 6 h indicated by the more stringent European standard. On the other hand, the OSI estimated for biodiesel from the NRRE-1 strain indicates that it complies with the American standard but not with the European one.

4. Conclusions

The collected strains were identified as belonging to *Nannochloropsis* (NSRE-1 and NSRE-2) and *Nannochloris* sp. (NRRE-1). To our knowledge, this is the first time that these species have been found in Mexico. The hydrocarbons extracted from the three strains studied fall within the range of diesel. However, their direct use is not recommended in pure form or mixed with fossil hydrocarbons due to the presence of mono-unsaturated components, as these confer instability to the fuel during storage. For hydrocarbons from NSRE-1 and NSRE-2, it would be advisable to apply a process of hydroisomerization plus simultaneous hydrocracking to obtain molecules with shorter and more branched carbon chains, useful as high-quality renewable diesel and even as bio-jet fuel. On the other hand, to properly take advantage of the hydrocarbon fraction from NRRE-1, a single catalytic improvement step would be adequate to saturate the double bonds and obtain a high cetane number diesel.

Among the oxygenated fractions, the one corresponding to NSRE-1 showed the highest oxygen content (19.2 wt %), which could represent a more significant challenge for its processing to obtain renewable transportation fuels. Nevertheless, the three fractions could be processed through a single HDO or DO stage to obtain high-quality renewable diesel as the main product. This was experimentally demonstrated by subjecting the fraction of oxygenated compounds of the NSRE-2 strain to an HDO reaction, which produced a high-quality, renewable diesel that was completely free of oxygen and unsaturated compounds.

On the other hand, if bio-jet fuel is the desired product, it would be more convenient to process the fractions through the HEFA process. The oxygenated fractions from strains NRRE-1 and NSRE-2 are of the higher quality required for processing by HEFA because, their oxygen content is lower than or similar to that of JCO taken as the reference feed for this process.

Regarding the analysis of FAMEs from the three strains studied, it was found that the three biodiesels satisfactorily comply with the majority of the evaluated properties according to the limits established in the American and European standards. The property that showed the poorest results was the cold filter plugging point (CFPP), with values of 17.9–31.6 °C. This is explained by the high content of saturated FAMEs, which ranges from 58 to 100%. For this reason, these biodiesels could only be used in warm climates and mixed with FAMEs with a higher content of unsaturated species.

According to this analysis, the lipids of the strains studied are very valuable precursors for obtaining high quality renewable fuels, since these can be processed through different catalytic routes that involve not only transesterification, but also (hydro)deoxygenation reactions, which have great potential to transform varied feeds and obtain fuels with properties similar or even superior to those of fossil fuels.

Author Contributions: Data curation, A.R.L.-R., J.C.C.-H., E.P.-L. and B.C.-C.; Formal analysis, A.R.L.-R., J.C.C.-H., F.B.-P. and S.L.-A.; Funding acquisition, R.V.-O.; Investigation, A.R.L.-R., G.G.-S. and R.V.-O.; Methodology, A.R.L.-R. an Data curation, A.R.L.-R., J.C.C.-H., E.P.-L. and B.C.-C.; Formal analysis, A.R.L.-R., J.C.C.-H., F.B.-P. and S.L.-A.; Funding acquisition, R.V.-O.; Investigation, A.R.L.-R., G.G.-S. and R.V.-O.; Methodology, A.R.L.-R. and Katia Ancona-Canché; Project administration, A.R.L.-R. and R.V.-O.; Resources, R.V.-O.; Supervision, T.T.-T. and R.V.-O.; Validation, A.R.L.-R., J.C.C.-H., F.B.-P., G.G.-S., E.P.-L., B.C.-C. and R.V.-O.; Visualization, A.R.L.-R. and S.L.-A.; Writing—original draft, A.R.L.-R., G.G.-S. and S.L.-A.; Writing—review & editing, A.R.L.-R., J.C.C.-H., F.B.-P. and R.V.-O. d Katia Ancona-Canché; Project administration, A.R.L.-R. and R.V.-O.; Resources, R.V.-O.; Supervision, T.T.-T. and R.V.-O.; Validation, A.R.L.-R., J.C.C.-H., F.B.-P., G.G.-S., E.P.-L., B.C.-C. and R.V.-O.; Visualization, A.R.L.-R. and S.L.-A.; Writing—original draft, A.R.L.-R., G.G.-S. and S.L.-A.; Writing—review & editing, A.R.L.-R., J.C.C.-H., F.B.-P. and R.V.-O.

Funding: The authors gratefully acknowledge the financial support granted by the Consejo Nacional de Ciencia y Tecnología (CONACyT, Mexico) through projects No. 166371 and No. 254667 (SENER –CONACYT), as well as the graduate grant for the first author No. 522627.

Acknowledgments: The authors also wish to thank José Luis Paredes (Instituto de Fisiología Celular, UNAM) for technical Support on TEM and Carlos López Salas for TEM figure editions.

Conflicts of Interest: The authors declare no conflict of interest.

References

1. Piloto-Rodríguez, R.; Sánchez-Borroto, Y.; Melo-Espinosa, E.A.; Verhelst, S. Assessment of diesel engine performance when fueled with biodiesel from algae and microalgae: An overview. *Renew. Sustain. Energy Rev.* **2017**, *69*, 833–842. [CrossRef]

2. Velasquez-Orta, S.B.; Lee, J.G.M.; Harvey, A.P. Evaluation of FAME production from wet marine and freshwater microalgae by in situ transesterification. *Biochem. Eng. J.* **2013**, *76*, 83–89. [CrossRef]

3. Maeda, Y.; Yoshino, T.; Matsunaga, T.; Matsumoto, M.; Tanaka, T. Marine microalgae for production of biofuels and chemicals. *Curr. Opin. Biotechnol.* **2018**, *50*, 111–120. [CrossRef] [PubMed]
4. Richmond, A. Biological principles of mass cultivation. *Handb. Microalgal Cult. Biotechnol. Appl. Phycol.* **2004**, 125–177.
5. Li, Z.; Haifeng, L.; Zhang, Y.; Shanshan, M.; Baoming, L.; Zhidan, L.; Na, D.; Minsheng, L.; Buchun, S.; Jianwen, L. Effects of strain, nutrients concentration and inoculum size on microalgae culture for bioenergy from post hydrothermal liquefaction wastewater. *Int. J. Agric. Biol. Eng.* **2017**, *10*, 194–204.
6. Bartley, M.L.; Boeing, W.J.; Dungan, B.N.; Holguin, F.O.; Schaub, T. pH effects on growth and lipid accumulation of the biofuel microalgae *Nannochloropsis salina* and invading organisms. *J. Appl. Phycol.* **2014**, *26*, 1431–1437. [CrossRef]
7. Sánchez, S.; Martínez, M.E.; Espinola, F. Biomass production and biochemical variability of the marine microalga *Isochrysis galbana* in relation to culture medium. *Biochem. Eng. J.* **2000**, *6*, 13–18. [CrossRef]
8. Cao, S.; Zhang, X.; Fan, X.; Qiao, H.; Liang, C.; Xu, D.; Mou, S.; Wang, W.; Ye, N. Phylogeny and characterisation of *Nannochloropsis oceanica* var. *sinensis* var. nov. (Eustigmatophyceae), a new oleaginous alga from China. *Phycologia* **2013**, *52*, 573–577.
9. Yamamoto, M.; Nozaki, H.; Kawano, S. Evolutionary relationships among multiple modes of cell division in the genus Nannochloris (Chlorophyta) revealed by genome size, actin gene multiplicity, and phylogeny. *J. Phycol.* **2001**, *37*, 106–120. [CrossRef]
10. Suda, S.; Atsumi, M.; Miyashita, H. Taxonomic characterization of a marine Nannochloropsis species, *N. oceanica* sp. nov. (Eustigmatophyceae). *Phycologia* **2002**, *41*, 273–279. [CrossRef]
11. Ma, X.-N.; Chen, T.-P.; Yang, B.; Liu, J.; Chen, F. Lipid production from Nannochloropsis. *Mar. Drugs* **2016**, *14*, 61. [CrossRef]
12. Mata, T.M.; Martins, A.A.; Caetano, N.S. Microalgae for biodiesel production and other applications: A review. *Renew. Sustain. Energy Rev.* **2010**, *14*, 217–232. [CrossRef]
13. Kumar, B.R.; Saravanan, S. Use of higher alcohol biofuels in diesel engines: A review. *Renew. Sustain. Energy Rev.* **2016**, *60*, 84–115. [CrossRef]
14. Tracy, N.I.; Crunkleton, D.W.; Price, G.L. Gasoline production from phytol. *Fuel* **2010**, *89*, 3493–3497. [CrossRef]
15. Abdel-Aal, E.I.; Haroon, A.M.; Mofeed, J. Successive solvent extraction and GC–MS analysis for the evaluation of the phytochemical constituents of the filamentous green alga *Spirogyra longata*. *Egypt. J. Aquat. Res.* **2015**, *41*, 233–246. [CrossRef]
16. Abrantes, J.L.; Barbosa, J.; Cavalcanti, D.; Pereira, R.C.; Fontes, C.L.F.; Teixeira, V.L.; Souza, T.L.M.; Paixão, I.C. *The Effects of the Diterpenes Isolated from the Brazilian Brown Algae Dictyota pfaffii and Dictyota menstrualis against the Herpes Simplex Type-1 Replicative Cycle*; Georg Thieme Verlag: Stuttgart, Germany, 2010.
17. Mewalal, R.; Rai, D.K.; Kainer, D.; Chen, F.; Külheim, C.; Peter, G.F.; Tuskan, G.A. Plant-derived terpenes: A feedstock for specialty biofuels. *Trends Biotechnol.* **2017**, *35*, 227–240. [CrossRef] [PubMed]
18. Rossetti, I.; Gambaro, C.; Calemma, V. Hydrocracking of long chain linear paraffins. *Chem. Eng. J.* **2009**, *154*, 295–301. [CrossRef]
19. Richmond, A. *Handbook of Microalgal Culture: Biotechnology and Applied Phycology*; John Wiley & Sons: Hoboken, NJ, USA, 2008.
20. Mühling, M.; Fuller, N.J.; Millard, A.; Somerfield, P.J.; Marie, D.; Wilson, W.H.; Scanlan, D.J.; Post, A.F.; Joint, I.; Mann, N.H. Genetic diversity of marine Synechococcus and co-occurring cyanophage communities: Evidence for viral control of phytoplankton. *Environ. Microbiol.* **2005**, *7*, 499–508. [CrossRef]
21. Ganuza, E.; Tonkovich, A.L. Method of Treating Bacterial Contamination in a Microalgae Culture with pH Shock. Google Patents US9181523B1, 10 November 2015.
22. Stewart, W.D.P. *Algal Physiology and Biochemistry*; University of California Press: Berkeley, CA, USA, 1974; Volume 10.
23. Holzinger, A.; Tschaikner, A.; Remias, D. Cytoarchitecture of the desiccation-tolerant green alga Zygogonium ericetorum. *Protoplasma* **2010**, *243*, 15–24. [CrossRef]
24. Youssef, M.; Valdez-Ojeda, R.; Ku-Cauich, J.R.; Escobedo-Gracia Medrano, R. Enhanced protocol for isolation of plant genomic DNA. *J. Agric. Environ. Sci.* **2015**, *4*, 172–180. [CrossRef]

25. Lee, S.-R.; Oak, J.H.; Chung, I.K.; Lee, J.A. Effective molecular examination of eukaryotic plankton species diversity in environmental seawater using environmental PCR, PCR-RFLP, and sequencing. *J. Appl. Phycol.* **2010**, *22*, 699–707. [CrossRef]

26. Kumar, S.; Stecher, G.; Tamura, K. MEGA7: Molecular evolutionary genetics analysis version 7.0 for bigger datasets. *Mol. Biol. Evol.* **2016**, *33*, 1870–1874. [CrossRef] [PubMed]

27. Le Rouzic, B.; Bertru, G. Phytoplankton community growth in enrichment bioassays: Possible role of the nutrient intracellular pools. *Acta Oecol.* **1997**, *18*, 121–133. [CrossRef]

28. Doan, T.T.Y.; Sivaloganathan, B.; Obbard, J.P. Screening of marine microalgae for biodiesel feedstock. *Biomass Bioenergy* **2011**, *35*, 2534–2544. [CrossRef]

29. Demirbaş, A. Biodiesel fuels from vegetable oils via catalytic and non-catalytic supercritical alcohol transesterifications and other methods: A survey. *Energy Conv. Manag.* **2003**, *44*, 2093–2109. [CrossRef]

30. Bligh, E.G.; Dyer, W.J. A rapid method of total lipid extraction and purification. *Can. J. Biochem. Physiol.* **1959**, *37*, 911–917. [CrossRef]

31. Wahlen, B.D.; Willis, R.M.; Seefeldt, L.C. Biodiesel production by simultaneous extraction and conversion of total lipids from microalgae, cyanobacteria, and wild mixed-cultures. *Biores. Technol.* **2011**, *102*, 2724–2730. [CrossRef]

32. Volkman, J.K.; Jeffrey, S.W.; Nichols, P.D.; Rogers, G.I.; Garland, C.D. Fatty acid and lipid composition of 10 species of microalgae used in mariculture. *J. Exp. Mar. Biol. Ecol.* **1989**, *128*, 219–240. [CrossRef]

33. Iverson, S.J.; Lang, S.L.C.; Cooper, M.H. Comparison of the bligh and dyer and folch methods for total lipid determination in a broad range of marine tissue. *Lipids* **2001**, *36*, 1283–1287. [CrossRef]

34. Kang, W.; Ji, Z.; Wang, J. Composition of the essential oil of *Adiantum flabellulatum*. *Chem. Nat. Compd.* **2009**, *45*, 575–577. [CrossRef]

35. Poirier, M.-A. Stable Alkoxylated Fatty Acid Alkyl Esters from Transesterification-Alkoxylation of Vegetable Oils. Google Patents US8523960B2, 3 September 2013.

36. Araujo, G.S.; Matos, L.J.; Fernandes, J.O.; Cartaxo, S.J.; Goncalves, L.R.; Fernandes, F.A.; Farias, W.R. Extraction of lipids from microalgae by ultrasound application: Prospection of the optimal extraction method. *Ultrason. Sonochem.* **2013**, *20*, 95–98. [CrossRef] [PubMed]

37. Ganuza, E.; Sellers, C.E.; Bennett, B.W.; Lyons, E.M.; Carney, L.T. A novel treatment protects chlorella at commercial scale from the predatory bacterium *Vampirovibrio chlorellavorus*. *Front. Microbiol.* **2016**, *7*, 848. [CrossRef] [PubMed]

38. Vidyashankar, S.; VenuGopal, K.S.; Swarnalatha, G.V.; Kavitha, M.D.; Chauhan, V.S.; Ravi, R.; Bansal, A.K.; Singh, R.; Pande, A.; Ravishankar, G.A. Characterization of fatty acids and hydrocarbons of chlorophycean microalgae towards their use as biofuel source. *Biomass Bioenergy* **2015**, *77*, 75–91. [CrossRef]

39. Ramos-Higuera, E.; Alcocer, J.; Ortega-Mayagoitia, E.; Camacho, A. Nitrógeno: Elemento limitante para el crecimiento fitoplanctónico en un lago oligotrófico tropical. *Hidrobiológica* **2008**, *18*, 105–113.

40. Chu, S. The influence of the mineral composition of the medium on the growth of planktonic algae: Part I. Methods and culture media. *J. Ecol.* **1942**, 284–325. [CrossRef]

41. Abirami, S.; Murugesan, S.; Sivamurugan, V.; Sivaswamy, S.N. Screening and optimization of culture conditions of *Nannochloropsis gaditana* for omega 3 fatty acid production. *J. Appl. Biol. Biotechnol.* **2017**, *5*, 13–17.

42. Santhanam, P.; Kumar, S.D.; Ananth, S.; Jeyanthi, S.; Sasirekha, R.; Premkumar, C. Effect of culture conditions on growth and lipid content of marine microalga Nannochloropsis sp. strain (PSDK11). *Indian J. Geo-Mar. Sci.* **2017**, *46*, 2332–2338.

43. Van den Hoek, C.; Stam, W.; Olsen, J. The emergence of a new chlorophytan system, and Dr. Kornmann's contribution thereto. *Helgoländer Meeresuntersuchungen* **1988**, *42*, 339–383. [CrossRef]

44. Shubert, E.; Gärtner, G. Nonmotile coccoid and colonial green algae. In *Freshwater Algae of North America*, 2nd ed.; Elsevier: Amsterdam, The Netherlands, 2014; pp. 315–373.

45. Lubián, L.M. *Nannochloropsis gaditana* sp. nov., una nueva *Eustigmatophyceae marina*. *Lazaroa* **1982**, *4*, 287–293.

46. Henley, W.J.; Hironaka, J.L.; Guillou, L.; Buchheim, M.A.; Buchheim, J.A.; Fawley, M.W.; Fawley, K.P. Phylogenetic analysis of the 'Nannochloris-like' algae and diagnoses of *Picochlorum oklahomensis* gen. et sp. nov. (Trebouxiophyceae, Chlorophyta). *Phycologia* **2004**, *43*, 641–652. [CrossRef]

47. Fawley, K.P.; Fawley, M.W. Observations on the diversity and ecology of freshwater Nannochloropsis (Eustigmatophyceae), with descriptions of new taxa. *Protist* **2007**, *158*, 325–336. [CrossRef] [PubMed]

48. Antia, N.J.; Cheng, J.Y. The keto-carotenoids of two marine coccoid members of the Eustigmatophyceae. *Br. Phycol. J.* **1982**, *17*, 39–50. [CrossRef]

49. South, G.; Whittick, A. Physiology and biochemistry. In *An Introduction to Phycology*; Blackwell Scientific Publications: Oxford, UK, 1987.

50. Liu, X.; Zhu, H.; Liu, B.; Liu, G.; Hu, Z. Phylogeny and morphology of genus Nephrocytium (Sphaeropleales, Chlorophyceae, Chlorophyta) from China. *Phytotaxa* **2017**, *319*, 84–92. [CrossRef]

51. Komárek, J. Änderungen in der taxonomie der chlorokokkalalgen. *Algolog. Stud. /Arch. Hydrobiol.* **1979**, *24*, 239–263.

52. Volkman, J.; Gllan, F.; Johns, R.; Eglinton, G. Sources of neutral lipids in a temperate intertidal sediment. *Geochim. Cosmochim. Acta* **1981**, *45*, 1817–1828. [CrossRef]

53. Yao, L.; Gerde, J.A.; Lee, S.-L.; Wang, T.; Harrata, K.A. Microalgae lipid characterization. *J. Agric. Food Chem.* **2015**, *63*, 1773–1787. [CrossRef] [PubMed]

54. Giakoumis, E.G. A statistical investigation of biodiesel physical and chemical properties, and their correlation with the degree of unsaturation. *Renew. Energy* **2013**, *50*, 858–878. [CrossRef]

55. Bondioli, P.; Della Bella, L.; Rivolta, G.; Chini Zittelli, G.; Bassi, N.; Rodolfi, L.; Casini, D.; Prussi, M.; Chiaramonti, D.; Tredici, M.R. Oil production by the marine microalgae Nannochloropsis sp. F&M-M24 and *Tetraselmis suecica* F&M-M33. *Biores. Technol.* **2012**, *114*, 567–572. [CrossRef]

56. El Shoubaky, G.A.; Salem, E.A. Terpenes and sterols composition of marine brown algae *Padina pavonica* (Dictyotales) and *Hormophysa triquetra* (Fucales). *Int. J. Pharmacogn. Phytochem. Res.* **2014**, *6*, 894–900.

57. Talebi, A.F.; Mohtashami, S.K.; Tabatabaei, M.; Tohidfar, M.; Bagheri, A.; Zeinalabedini, M.; Hadavand Mirzaei, H.; Mirzajanzadeh, M.; Malekzadeh Shafaroudi, S.; Bakhtiari, S. Fatty acids profiling: A selective criterion for screening microalgae strains for biodiesel production. *Algal Res.* **2013**, *2*, 258–267. [CrossRef]

58. Abomohra, A.E.-F.; El-Sheekh, M.; Hanelt, D. Screening of marine microalgae isolated from the hypersaline Bardawil lagoon for biodiesel feedstock. *Renew. Energy* **2017**, *101*, 1266–1272. [CrossRef]

59. Stevenson, R.J.; Bothwell, M.L.; Lowe, R.L.; Thorp, J.H. *Algal Ecology: Freshwater Benthic Ecosystem*; Academic Press: Cambridge, MA, USA, 1996.

60. Patterson, G.W.; Krauss, R.W. Sterols of Chlorella. I. The naturally occurring sterols of *Chlorella vulgaris, C. ellipsoidea*, and *C. saccharophila*. *Plant Cell Physiol.* **1965**, *6*, 211–220. [CrossRef]

61. Orcutt, D.; Patterson, G. Sterol, fatty acid and elemental composition of diatoms grown in chemically defined media. *Comp. Biochem. Physiol. Part B Comp. Biochem.* **1975**, *50*, 579–583. [CrossRef]

62. Valdez-Ojeda, R.; González-Muñoz, M.; Us-Vázquez, R.; Narváez-Zapata, J.; Chavarria-Hernandez, J.C.; López-Adrián, S.; Barahona-Pérez, F.; Toledano-Thompson, T.; Garduño-Solórzano, G.; Medrano, R.M.E.-G. Characterization of five fresh water microalgae with potential for biodiesel production. *Algal Res.* **2015**, *7*, 33–44. [CrossRef]

63. Serrano, M.; Oliveros, R.; Sánchez, M.; Moraschini, A.; Martínez, M.; Aracil, J. Influence of blending vegetable oil methyl esters on biodiesel fuel properties: Oxidative stability and cold flow properties. *Energy* **2014**, *65*, 109–115. [CrossRef]

64. Altaie, M.A.H.; Janius, R.B.; Taufiq-Yap, Y.H.; Rashid, U. Basic properties of methyl palmitate-diesel blends. *Fuel* **2017**, *193*, 1–6. [CrossRef]

65. Batts, B.; Fathoni, A.Z. A literature review on fuel stability studies with particular emphasis on diesel oil. *Energy Fuels* **1991**, *5*, 2–21. [CrossRef]

66. Fajar, R.; Sugiarto, B. Predicting fuel properties of partially hydrogenated jatropha methyl esters used for biodiesel formulation to meet the fuel specification of automobile and engine manufacturers. *Kasetsart J. Nat. Sci.* **2012**, *46*, 629–637.

Article

Optimization of Biodiesel Production Using Nanomagnetic CaO-Based Catalysts with Subcritical Methanol Transesterification of Rubber Seed Oil

Veronica Winoto and Nuttawan Yoswathana *

Department of Chemical Engineering, Faculty of Engineering, Mahidol University, Nakhon Pathom 73170, Thailand; vwinoto168@gmail.com or veronica.win@student.mahidol.edu
* Correspondence: egnyw8846@gmail.com or nuttawan.yos@mahidol.edu; Tel.: +662-889-2138 (ext. 6101)

Received: 20 November 2018; Accepted: 9 January 2019; Published: 12 January 2019

Abstract: The molar ratio of methanol to rubber seed oil (RSO), catalyst loading, and the reaction time of RSO biodiesel production were optimized in this work. The response surface methodology, using the Box–Behnken design, was analyzed to determine the optimum fatty acid methyl ester (FAME) yield. The performance of various nanomagnetic CaO-based catalysts—$KF/CaO-Fe_3O_4$, $KF/CaO-Fe_3O_4$-Li (Li additives), and $KF/CaO-Fe_3O_4$-Al (Al additives)—were compared. Rubber seed biodiesel was produced via the transesterification process under subcritical methanol conditions with nanomagnetic catalysts. The experimental results indicated that the $KF/CaO-Fe_3O_4$-Al nanomagnetic catalyst produced the highest FAME yield of 86.79%. The optimum conditions were a 28:1 molar ratio of methanol to RSO, 1.5 wt % catalyst, and 49 min reaction time. Al additives of $KF/CaO-Fe_3O_4$ nanomagnetic catalyst enhanced FAME yield without Al up to 18.17% and shortened the reaction time by up to 11 min.

Keywords: rubber seed oil; biodiesel production; nanomagnetic catalyst; subcritical methanol; FAME yield; Box-Behnken design

---

1. Introduction

The International Energy Agency (IEA) has reported that the global energy demand in 2020 would be 14,896 million tons of oil equivalent (Mtoe) and up to 18,048 Mtoe by 2035. The finite supply and depletion of fossil energy reserves, coupled with increased energy consumption is placing higher demands on energy production [1]. Renewable energy technology was introduced as an alternative energy to fulfill these needs. However, renewable energy contributes only 20% to the global energy demand and the other 80% is still supplied by fossil fuels [2]. Therefore, further studying and improving production technology is required [3]. Biodiesel—a liquid fuel produced from edible or non-edible vegetable oil or animal fats [4]—is an environmental friendly alternative energy, renewable, energy efficient, substitution fuel that does not sacrifice an engine's operational performance [5]. The increasing demand for biodiesel is also due to awareness of the environmental impact of emissions from conventional fossil fuel combustion and the decline in domestic oil production [6].

Rubber seed, the agriculture residues considered as waste in the rubber industry, has the potential to be a non-edible biodiesel production source in many country, such as Thailand [7], Malaysia [8], Indonesia [9], and Bangladesh [10]. The oil content in rubber seed is high, between 40 and 50% [11], so it has attracted attention as a biodiesel raw material. Thailand is an agricultural country and the largest rubber producer, producing 35% of the world's total production [12], with the rubber tree plantations covering around 3,548,274 ha or 35,482.7 km^2. This amount is continuously increasing each year [13].

For the conventional biodiesel technique, transesterification with methyl ester is performed with a homogeneous catalyst such as potassium methoxide or potassium hydroxide [14,15]. The catalysts have high activity and conversion ability, but they produce a large amount of chemical waste water [16], long-term consuming production, are expensive, and catalyst removal or recovery are difficult [14]. Therefore, heterogeneous catalysts were studied to replace homogeneous catalysts. Presently, heterogeneous nanocatalysts have exhibited good catalyst properties: high activity and large surface area. Most recent studies have shown that they demonstrated better performance and are more effective, easy to handle, and cost efficient [16]. CaO-based catalysts are widely used for biodiesel production due to their high efficiency in FAME enhancement, low cost, and availability [17,18]. For example, heterogeneous catalysts with CaO-based catalysts such as CaMgO, CaZnO [19], CaO-NiO, CaO-Nd$_2$O$_3$ [20], and CaO-La$_2$O$_3$ [21] improved biodiesel production and the resultant yields were 83%, 81% [19], 86.3%, 82.2% [20], and 86.5% [21], respectively. After combining the magnetic catalysts and nanocatalysts together, nanomagnetic catalysts were formed. The nanomagnetic catalyst has good catalyst properties conferred by the nanocatalyst and is easily removed due to its magnetic properties [14]. The heterogeneous CaO-based catalysts with Li additives reduced soaps formation due to the reaction of alkaline catalysts with free fatty acids or the saponification of triglycerides and enhanced methyl esters [22,23]. Similarly, the presence of Al additives in the form of sodium meta-aluminate improved transesterification in biodiesel production. Hence, both additives enhanced methyl esters production. The Al additives in CaO-based catalysts improved catalyst performance along with the presence of Na, an alkali metal that behaves similarly to Li [24].

The subcritical fluid technique was developed as a novel technique for biodiesel production. It is mostly considered environmental friendly and inexpensive due to the significant in of the production time [25]. The main advantages of transesterification in subcritical condition for biodiesel production was high conversion and reaction rates [26]. The operating temperature under subcritical methanol ranges from 150 to 250 °C [26,27]. Subcritical methanol at 220 °C achieved the highest FAME yield. For example, a 95.6% yield was obtained from soybean oil with K$_3$PO$_4$ [28], 94.7% from *Castor* oil, and 95.6% [29] from soybean oil with Na$_2$SiO$_3$ [30]. In the subcritical state, hydrogen bonding significantly decreases, which allows methanol to be a free monomer. The transesterification is completed via a methoxide transfer, whereby the fatty acid methyl ester and diglycerides are formed. Similarly, diglyceride is transesterified to form methyl ester and monoglycerides, which are further converted to methyl ester and glycerol in the last step [31].

Response Surface Methodology (RSM) with Box-Behnken experimental design have been effectively applied to determined optimum conditions. Mathematical and statistical methods have been used to predict the effect of input process parameters based on the output responses [32–34]. The three parameters that most influence biodiesel production are: methanol/oil ratio, catalyst loading, and reaction time [35,36].

In this research, rubber seed oil (RSO) was selected as a potential non-edible raw material for biodiesel production. Nanomagnetic catalysts were synthesized and applied in subcritical methanol biodiesel production to enhance biodiesel yield. The Box-Behnken experimental design, the most widely used for response surface methodology [37,38], was used to study the effect of each individual variable, such as molar ratio between methanol and RSO, catalyst amount, and reaction time. The response surface results were analyzed based on FAME yield percentage to determine the optimum conditions for rubber seed biodiesel production using subcritical methanol with nanomagnetic catalysts.

2. Materials and Methods

2.1. Materials

We used RSO and methanol (AR Grade, RCI Labscan, Bangkok, Thailand) as the raw materials for biodiesel production. The magnetic core (Fe$_3$O$_4$) was prepared from FeSO$_4$·7H$_2$O (AR Grade,

Ajax Finechem Pty Ltd., Victoria, Australia), $Fe(SO_4)_3$ (AR Grade, Ajax Finechem Pty Ltd.), and $NH_3 \cdot H_2O$ (25% ACS Reagent, J.T. Baker, Pittsburgh, PA, USA). The metal oxide base that was chosen was CaO (AR Grade, Ajax Finechem Pty Ltd.). In the impregnation method, calcined metal oxides mixed with a magnetic core were dipped in KF (AR Grade, Ajax Finechem Pty Ltd.). For Li and Al additives catalysts, $LiNO_3$ (AR Grade, Ajax Finechem Pty Ltd.) and Al with NaOH (AR Grade, Ajax Finechem Pty Ltd.), which formed sodium meta-aluminate, were added in nanomagnetic CaO-based catalysts.

2.2. Preparation of Rubber Seed Oil

Rubber seeds were planted and harvested in Phattalung, a southern province in Thailand. The RSO was obtained from Agricultural Occupation Promotion and Development Center in Cha-am district, Phetchaburi province, Thailand. Firstly, rubber seeds were inserted to the deshelling machine. Then, the deshelled rubber seeds were dried at 60 °C for 4 h before the extraction process. The mechanical hydraulic pressing machine was used to crush and squeeze the rubber seeds to obtain rubber seed oil. The oil was then analyzed using the ASTM D5555-95 (2011) standard test method [32]. The initial free fatty acids (FFA) content of rubber seed oil was 16.72%.

2.3. Preparation of Nanomagnetic Catalysts

The nanomagnetic catalyst preparation procedures were modified based on the literature [16,22–24]. Firstly, the magnetic core was prepared by dissolving 27.8 g $FeSO_4 \cdot 7H_2O$ and 79.6 g $Fe_2(SO_4)_3$ 79.6 g in deionized water. Then, 25% $NH_3 \cdot H_2O$ was added dropwise to the iron mixed solution. The mixed solution was vigorously stirred in a 60 °C water bath for 60 min. The final pH value of the aqueous solution was maintained at approximately 12.0. The mixed solution was then aged for 60 min. After aging, the black solid part that settled was then separated using a permanent magnet. The separated black solid was washed with distilled water until the final pH of the filtrate remained at 7.0. the wet magnetic core was dried at 60 °C for 24 h. After that, the black dried solid was pulverized into a fine black magnetic core powder. The metal oxide, CaO, was calcined at 600 °C for 3 h before the nanomagnetic solid base catalyst was prepared. Then, 100.0 g calcined metal oxides and a 5.0 g Fe_3O_4 magnetic core were mixed homogeneously. The impregnation method was performed next by completely dipping the mixture in 25 wt % KF aqueous solution, and then dried at 105 °C for 24 h. The dried mixture was calcined in a furnace at 600 °C [16]. As such, the synthesized nanomagnetic CaO-based catalyst, $KF/CaO-Fe_3O_4$, was prepared.

2.4. Additives

The heterogeneous CaO-based catalysts with lithium (Li) [22,23] or aluminum (Al) [24] additives were prepared as follows:

2.4.1. Lithium Additives (Li)

Calcined metal oxides (CaO; 100.0 g), 5.0 g Fe_3O_4 magnetic core, and 1.23% Li additives were mixed homogeneously. Then, metal oxides were mixed using aqueous $LiNO_3$ solutions of the appropriate concentration. The mixed slurry was stirred for 2 h and heated to 100 °C for 2 h [23]. Then, to continue with KF impregnation, drying and calcination were performed, as previously mentioned in Section 2.3. As a result, the $KF/CaO-Fe_3O_4$-Li nanomagnetic catalyst was synthesized.

2.4.2. Aluminum Additives (Al)

The molar ratio of NaOH, Al, and calcined metal oxides with Fe_3O_4 magnetic core was fixed at 2:2:3. Firstly, the NaOH solution was prepared by dissolving NaOH into the distilled water and then Al sheet was added into the solution. After that, the mixed slurry of sodium meta-aluminate solution (mixture of Al and NaOH) was stirred for 2 h and then heated to 100 °C for 2 h [24]. Then, we continued with KF impregnation, drying, and calcination as previously mentioned in Section 2.3. Finally, after the calcination process, the Al additives catalyst, KF/CaO-Fe$_3$O$_4$-Al, was prepared.

2.5. Transesterification Process

The transesterification reaction was carried out in a 40 mL tubular high-pressure reactor (20 MPa) using the subcritical fluid technique and nanomagnetic catalysts. The reaction temperature was fixed at subcritical methanol condition at 220 °C. The molar ratio between methanol and rubber seed oil was 10:1, 25:1, or 40:1. The synthesized nanomagnetic catalyst was used at 1.5, 3.0, or 4.5 wt % oil and the reaction time was 30, 45, or 60 min. After the reaction occurred completely, the catalyst was removed using a permanent magnet. Then, the two liquid phases, methyl ester and glycerol, were separated by separation funnel. We applied these transesterification process procedures to all 3 types of synthesized nanomagnetic catalysts; KF/CaO-Fe$_3$O$_4$, KF/CaO-Fe$_3$O$_4$-Li, and KF/CaO-Fe$_3$O$_4$-Al.

2.6. Experimental Design and Optimization

The three independent variables, the molar ratio between methanol and rubber seed oil, catalysts amount, and reaction time of rubber seed oil biodiesel production using subcritical methanol with various types of catalysts, were chosen for further optimization studies using RSM. These three variables were chosen because their effects are the most influential in biodiesel production [5,35]. The methanol to rubber seed oil ratio ranged from 10:1 to 40:1, catalysts loading ranged from 1.5 to 4.5 wt %, and reaction time ranged from 30 to 60 min [35,39–41].

A total of 15 runs of Box-Behnken Design (BBD) [36,38,42–44] with three variables at three levels were used in this study. The three levels of each variable examined were coded as –1 for the lower level, 0 for the middle level, and 1 for the higher level, as shown in Table 1. The molar ratio between methanol and rubber seed oil (the individual parameter) was defined as X_1, whereas the amount of catalysts and reaction time in subcritical condition were defined as X_2 and X_3, respectively. The response of each experiment was modeled as a second-order polynomial:

$$Y = \beta_0 + \beta_1 X_1 + \beta_2 X_2 + \beta_3 X_3 + \beta_{11} X_1^2 + \beta_{22} X_2^2 + \beta_{33} X_3^2 + \beta_{12} X_1 X_2 + \beta_{13} X_1 X_3 + \beta_{23} X_2 X_3 \quad (1)$$

where Y is the corresponding response variable and β_i ($i = 0, 1, 2, 3, 11, 12, 13, 22, 23,$ and 33) is the series of regression coefficients for the intercept, linearity, square, and interaction. The independent variables coded are of X_n ($n = 1, 2, 3$). The experimental design, data analysis, and response surface graphs were calculated and generated by combining regression and graphical analyses. The optimum operating conditions of rubber seed biodiesel production were determined by solving the regression equation and applying RSM. Design Expert Software (Version 11, Stat-Ease, Inc., Minneapolis, MN, USA) was used for data analyzation.

Table 1. The levels of independent variables examined and the range of dependent responses of the rubber seed oil biodiesel production using Box-Behnken design. RSO: Rubber seed oil.

Variable	Coded Level		
	−1	0	1
Molar ratio (X_1, methanol:RSO)	10:1	25:1	40:1
Amount of catalysts (X_2, wt %)	1.5	3	4.5
Reaction time in subcritical condition (X_3, min)	30	45	60

2.7. FAME Analytical Method

The FAME yield percentages were analyzed using gas chromatography (GC; Clarus 680 GC, PerkinElmer Inc., Waltham, MA, USA) with a Perkin Elmer Elite Series GC Column (PE-5 Phase, 60 mm × 0.250 mm, d_f = 0.25 µm) using the European Standard (EN) 14103 (European Standard) method.

The EN 14103 method specifies quality criteria for the biodiesel production via FAME content. The FAME yield from each experiment was calculated from its content in the composition as analyzed by GC.

2.8. Nanomagnetic Catalysts Characterization

The morphology of nanomagnetic CaO-based catalysts were analyzed using scanning electron microscopy (SEM, model JSM-6610LV, JEOL Ltd., Tokyo, Japan). The specific surface area, pore volume, and pore size were measured by nitrogen adsorption-desorption isotherms using a Brunauer-Emmett-Teller (BET) Model Autosorb-1 (Quantachrome Corporation, Boynton Beach, FL, USA).

3. Results and Discussion

3.1. Nanomagnetic Catalysts Properties

The morphologies of CaO powder and nanomagnetic CaO-based catalysts, KF/CaO-Fe$_3$O$_4$, KF/CaO-Fe$_3$O$_4$-Li, and KF/CaO-Fe$_3$O$_4$-Al, are shown in Figure 1a–d, respectively. The SEM images illustrate that nanomagnetic CaO-based catalysts particles were mostly agglomerated and formed into a larger crystal structure, especially KF/CaO-Fe$_3$O$_4$-Li, with Li additives. KF/CaO-Fe$_3$O$_4$ catalyst particles also agglomerated, but the overall crystal size was smaller than the Li additives catalysts. Although KF/CaO-Fe$_3$O$_4$-Al catalysts agglomerated, all small particles were identified.

Figure 1. The morphologies of CaO powder and various synthesized nanomagnetic CaO-based catalysts were analyzed using scanning electron microscopy (SEM): (**a**) CaO powder, (**b**) KF/CaO-Fe$_3$O$_4$, (**c**) KF/CaO-Fe$_3$O$_4$-Li, and (**d**) KF/CaO-Fe$_3$O$_4$-Al.

The BET specific surface area, pore volume, and Barrett-Joyner-Halenda (BJH) pore size of CaO powder and synthesized nanomagnetic CaO-based catalysts, KF/CaO-Fe$_3$O$_4$, KF/CaO-Fe$_3$O$_4$-Li and KF/CaO-Fe$_3$O$_4$-Al, are shown in Table 2. The properties of synthesized catalysts dramatically decreased compared to CaO powder due to magnetization and impregnation, due to the pore blockage caused by additives [45]. The specific surface area of synthesized nanomagnetic CaO-based catalyst was 27.84 m^2/g. The surface area increased to 34.93 and 95.70 m^2/g in the presence of Li and Al additives, respectively. The morphology of Al additive catalysts (Figure 1d) confirm that without the agglomeration effect, small particles were distributed independently, producing a higher specific surface area [46]. The larger surface area of the catalyst provided a larger area for adsorption on the surface, which enhanced the forward reaction [47] and resulted in higher FAME conversion. The pore sizes of the three types of synthesized nanomagnetic CaO-based catalyst ranged from 2.15 to 3.79 nm, which means that mesopores catalysts were produced [45]. The BET results illustrate that the Li additive catalysts had higher selectivity due to their smaller pore size and smaller pore volume compared to the other two synthesized catalysts. In contrast, the Al additive catalysts had the highest surface area, pore volume, and pore size, which led to better catalyst performance [46].

Table 2. Brunauer-Emmett-Teller (BET) specific surface area, pore volume and pore size data of synthesized nanomagnetic catalysts.

Nanomagnetic Catalyst	BET Surface Area (m^2/g)	Pore Volume (cm^3/g)	BJH Pore Size (nm)
CaO	121.00	0.1351	3.79
$KF/CaO\text{-}Fe_3O_4$	27.84	0.0510	2.16
$KF/CaO\text{-}Fe_3O_4\text{-}Li$	34.93	0.0308	2.15
$KF/CaO\text{-}Fe_3O_4\text{-}Al$	95.70	0.1140	3.79

3.2. Response Surface Methodology Analysis

A Box–Behnken design containing 15 trials was used as the experimental design. Table 3 summarizes the details of the experimental matrix and the corresponding results. The experimental FAME yield data that were analyzed from the rubber seed biodiesels using GC were used to calculate the coefficients of a second-order polynomial:

$$Y = 81.1533 + 7.6263X_1 - 6.3088X_2 - 6.1150X_3 - 11.1054X_1^2 - 1.6054X_2^2 - 8.2629X_3^2 + 3.7425X_1X_2 - 0.0550X_1X_3 - 10.5550X_2X_3 \qquad (2)$$

where Y denotes the predicted FAME yield of rubber seed biodiesel production, and X_1, X_2, and X_3 denote the molar ratio between methanol and rubber seed oil (methanol: RSO), $KF/CaO\text{-}Fe_3O_4\text{-}Al$ catalysts amount (wt %), and reaction time (min), respectively.

Table 3. Box-Behnken design conditions for optimizing the rubber seed oil biodiesel production with corresponding experimental results and predicted response values.

Variables	Molar Ratio		Catalyst Amount		Time		FAME Yield	
(unit)	(methanol:RSO)		(wt %)		(min)		(%)	
Run	X_1	x_1	X_2	x_2	X_3	x_3	Experimental	Predicted
1	10:1	−1	1.5	−1	45	0	72.11	70.87
2	40:1	1	1.5	−1	45	0	78.51	78.64
3	10:1	−1	4.5	1	45	0	50.89	50.77
4	40:1	1	4.5	1	45	0	72.26	73.50
5	10:1	−1	3.0	0	30	−1	60.54	60.22
6	40:1	1	3.0	0	30	−1	77.27	75.58
7	10:1	−1	3.0	0	60	1	46.41	48.10
8	40:1	1	3.0	0	60	1	62.92	63.24
9	25:1	0	1.5	−1	30	−1	71.59	73.15
10	25:1	0	4.5	1	30	−1	81.20	81.65
11	25:1	0	1.5	−1	60	1	82.48	82.03
12	25:1	0	4.5	1	60	1	49.87	48.31
13	25:1	0	3.0	0	45	0	83.32	81.15
14	25:1	0	3.0	0	45	0	81.01	81.15
15	25:1	0	3.0	0	45	0	79.13	81.15

Based on Equation (2), the predicted FAME yield of rubber seed biodiesel is shown in Table 3. The coefficient of determination (R^2) between the experimental and predicted values was determined to be 0.9898, indicating that approximately 98.98% of the variations for the response could be explained. The Analysis of Variance (ANOVA) summary of the RSM is presented in Table 4. The *p*-value of the model term was found to be significant at less than 0.05 as shown in Table 4.

Table 4. ANOVA for quadratic model of the rubber seed oil biodiesel production. D_f: Degrees of freedom; pred R^2: Predicted value of R-squared; adj R^2: Adjusted R-squared; adeq: Adequacy of R-squared.

Source	Sum of Squares	D_f	Mean Square	F-Value	*p*-Value Probability > F
Model	2243.92	9	249.32	53.90	0.0002
X_1	465.28	1	465.28	100.59	0.0002
X_2	318.40	1	318.40	68.84	0.0004
X_3	299.15	1	299.15	64.68	0.0005
X_1X_2	56.03	1	56.03	12.11	0.0177
X_1X_3	0.01	1	0.01	0.00	0.9612
X_2X_3	445.63	1	445.63	96.35	0.0002
X_1^2	455.37	1	455.37	98.45	0.0002
X_2^2	9.52	1	9.52	2.06	0.2109
X_3^2	252.10	1	252.10	54.50	0.0007
Residual	23.13	5	4.63		
Lack of fit	14.32	3	4.77	1.08	0.5129
Pure error	8.81	2	4.40		
SD	R^2	pred R^2	adj R^2	adeq precision	Mean
2.15	0.9898	0.8902	0.9714	19.3251	69.97

The response surface curves and contour plots of three independent variables illustrated in Figures 2–4 were generated based on the regression model. Both the influence and interaction of each independent variable were evaluated. Each of the response surfaces, as shown in Figures 2–4, produced a clear peak point that indicated that optimum condition and the maximum FAME yield in rubber seed biodiesel production, which were achieved inside the design boundary region for all three variables. The optimum values of X_1, X_2, and X_3 derived from Equation (2) were -0.1749, -0.7829 and 0.3974, respectively. Hence, the optimum point of the model includes a 28:1 molar ratio between methanol and RSO, 1.5 wt % of KF/CaO-Fe_3O_4-Al catalyst, and a 49 min reaction time. The optimum condition leads to the maximum predicted value of FAME yield of 86.79%.

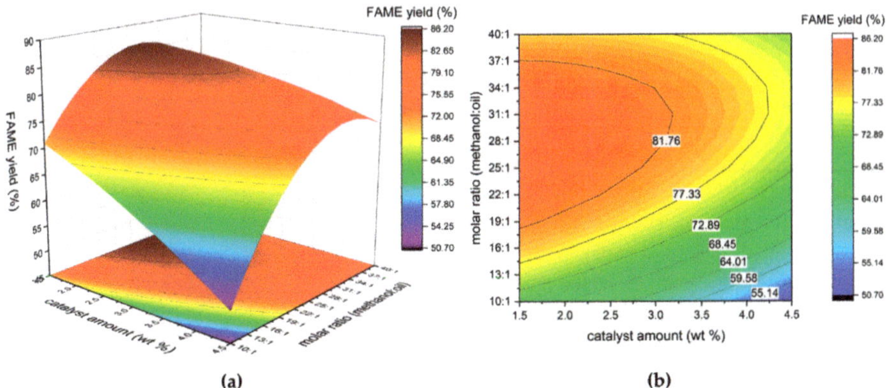

Figure 2. The interaction effect of molar ratio and KF/CaO-Fe_3O_4-Al catalyst amount on FAME yield for rubber seed biodiesel production: (**a**) response surface curve and (**b**) contour plot.

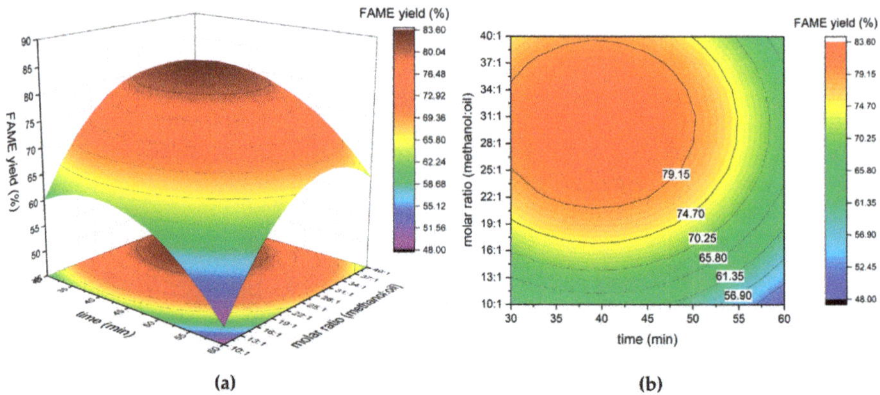

Figure 3. The interaction effect of molar ratio and reaction time on FAME yield for rubber seed biodiesel production: (**a**) response surface curve and (**b**) contour plot.

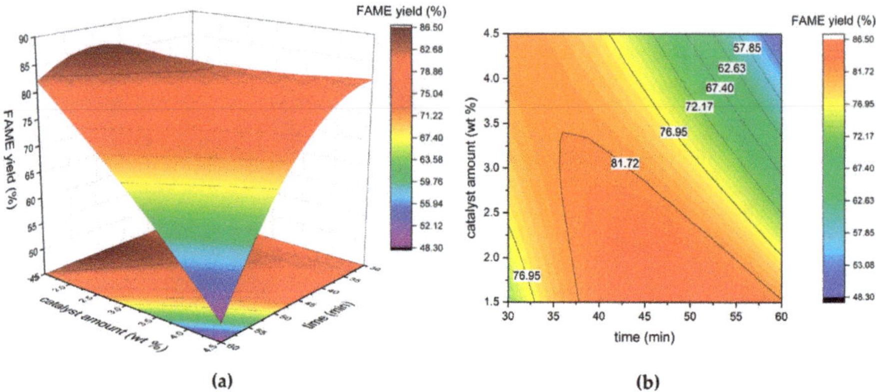

Figure 4. The interaction effect of KF/CaO-Fe$_3$O$_4$-Al catalyst amount and reaction time on FAME yield for rubber seed biodiesel production: (**a**) response surface curve and (**b**) contour plot.

The response surface illustrated in Figure 2 represents the interaction effect of molar ratio (X_1) from 10:1 to 40:1 and KF/CaO-Fe$_3$O$_4$-Al catalyst amount (X_2) from 1.5 to 4.5 wt % at a constant reaction time (X_3) of 45 min. It can be observed that the highest FAME yield percentage was produced at a 28:1 molar ratio and 1.5 wt % catalyst. As the molar ratio increased, the FAME yield increased [30]. The FAME yield continuously increased by introducing more methanol to shift the equilibrium to the right-hand side, but when the ratio was more than 37:1, then the excess methanol would cause the FAME yield to decrease. The low level of excess methanol in biodiesel production encourages the RSO to FAME conversion, but at some point (in this work, at 37:1) the excess methanol and rubber seed oil separated into two phases of fluids, which caused a shift in the biodiesel conversion step [48,49]. A high molar ratio of methanol to oil increased FAME conversion, which confirmed the data in Table 3 for runs 2 and 3. The contour in Figure 2 shows that a lower catalysts usage enhances FAME yield percentage. This was caused by excess catalysts remaining in the system, which resulted in lower FAME conversion [48]. In additional, excess catalyst caused lower FAME percentages, confirmed by the data in Table 3 for runs 11 and 12.

The response surface shown in Figure 3 represents the interaction effect of molar ratio (X_1) in the range of 10:1 to 40:1 and reaction time (X_3) from 30 to 60 min at constant KF/CaO-Fe$_3$O$_4$-Al catalyst amount (X_2) of 3 wt %. It can be observed that the highest FAME yield was produced by an approximately 30:1 molar ratio and 38 min reaction time. The FAME yield increased due to the increasing molar ratio. Similar to the previous paragraph, the FAME yield increased by introducing more methanol but with excess methanol, the FAME yield began to decrease [48,49]. The response surface also indicates that the advantage of subcritical methanol for biodiesel production is that it significantly reduces the reaction time compared to the KOH conventional method for biodiesel production, which requires more than 120 min to complete the process [50].

The response surface illustrated in Figure 4 represents the interaction effect when the KF/CaO-Fe$_3$O$_4$-Al catalyst amount (X_2) was varied from 1.5 to 3 wt % and reaction time (X_3) was increased from 30 to 60 min at a constant molar ratio (X_1) of 25:1. It can be observed that the highest FAME yield percentage was produced at 1.5 wt % of catalyst and 47 min reaction time. Similar to Figure 2, the amount of remaining excess catalysts led to lower FAME conversion due to rubber seed oil separating into two liquid phases, causing a shift in the biodiesel conversion step [48]. Using a small amount of catalysts is recommended. In general, the high FFA content in the feedstock oil should be esterified and run through two-step biodiesel production, which is time consuming and complicated [50]. In this work, subcritical methanol successfully produced biodiesel with the single-step transesterification of a high-FFA feedstock, rubber seed oil. The contour in Figure 4 shows that the FAME yield peaked at approximately 48 min of reaction time.

3.3. Comparison of Nanomagnetic Catalysts and Subcritical Methanol Performance

The optimum condition of three types of nanomagnetic catalysts, KF/CaO-Fe$_3$O$_4$, KF/CaO-Fe$_3$O$_4$-Li and KF/CaO-Fe$_3$O$_4$-Al for rubber seed biodiesel production were investigated in this work. The role of different components in various catalysts produces different FAME yields. Overall, the nanomagnetic CaO-based catalyst with additives appeared to perform better and produced higher FAME yields than the one without additives. The Li additives affected the transesterification in biodiesel production [22,23,51,52]. A high free fatty acids (FFA) content in RSO [7,10,53,54] significantly affected the catalysts performance compared to the other types of oil [16,23,24].

The roles of different types and components in nanomagnetic CaO-based catalysts were investigated through the optimum conditions and FAME yield of rubber seed biodiesel production, shown in Table 5. Overall, the synthesized catalysts significantly reduced the catalyst amount and reaction time required for biodiesel production and show the reasonable FAME yield as mentioned in the previous paragraph. In rubber seed biodiesel production, limestone-based catalysts produce a higher FAME yield [55], but the process is time consuming (by approximately more than four times) compared to the catalysts synthesized in this work. Limestone-based catalysts are complicated to prepare and have variable compositions such as Cao, SiO$_2$, Al$_2$O$_3$, Fe$_2$O$_3$, MgO, SO$_3$, K$_2$O, Na$_2$O, P$_2$O$_5$, and TiO$_2$ [55]. The acceptable FAME yield was determined to be 80%, and a high FAME yield is considered to be 85%. Hence, the FAME yield of KF/CaO-Fe$_3$O$_4$-Li is acceptable and the FAME yield for KF/CaO-Fe$_3$O$_4$-Al is considered high. The results indicate that both Li and Al additives significantly improved catalyst performance (from KF/CaO-Fe$_3$O$_4$). The reaction of alkali catalyst in Li additives or sodium meta-aluminate in Al additives catalysts with free fatty acids reduced saponification and enhanced methyl esters formation [22–24].

Various types of CaO-based catalyst are commonly used in *Jatropha curcas* biodiesel production. *Jatropha curcas* oil has a high FFA content (14–15%), which is similar to that of rubber seed oil [56–58]. The results showed that the nanomagnetic CaO-based catalysts synthesized in this work have better performance due to their faster reaction time, lower catalyst amount required, and easier removal with a permanent magnet [19–21]. Table 5 illustrated that Al-additive catalysts produced a yield of 86.79%, which was the highest FAME yield in this work and is competitive compared to other catalysts previously analyzed in the literature. The time required for biodiesel production using these catalysts was much shorter and used significantly less catalyst. Besides catalysts performance, the major advantages of all three types of synthesized nanomagnetic CaO-based catalysts are that they are easily removed after reaction and the produced biodiesel was clearly separated from the glycerol, which enhanced FAME yield.

Table 5. Comparison of optimum conditions with various catalysts for biodiesel production achieved during this work and other works.

Type of Catalyst	Biodiesel Source	Optimum Conditions				Reference
		Molar Ratio (Methanol:RSO)	Catalyst Amount (wt %)	Time (min)	FAME Yield (%)	
$KF/CaO\text{-}Fe_3O_4$	Rubber seed oil	34:1	1.6	60	68.62	This work
$KF/CaO\text{-}Fe_3O_4\text{-}Li$	Rubber seed oil	26:1	1.5	60	84.86	This work
$KF/CaO\text{-}Fe_3O_4\text{-}Al$	Rubber seed oil	28:1	1.5	49	86.79	This work
$KF/CaO\text{-}Fe_3O_4$	*Stillingia* oil	12:1	4	180	95.0	[16]
Limestone based (CaO, SiO_2, Al_2O_3, Fe_2O_3, MgO, SO_3, K_2O, Na_2O, P_2O_5 and TiO_2)	Rubber seed oil	5:1	5	240	96.9	[55]
CaO/Li	Karanja oil	12:1	2	480	94.9	[23]
$CaMgO$	*Jatropha curcas*	15:1	4	360	83	[19]
$CaZnO$	*Jatropha curcas*	15:1	4	360	81	[19]
$CaO\text{-}NiO$	*Jatropha curcas*	15:1	5	360	86.3	[20]
$CaO\text{-}Nd_2O_3$	*Jatropha curcas*	15:1	5	360	82.2	[20]
$CaO\text{-}La_2O_3$	*Jatropha curcas*	24:1	4	360	86.5	[21]
$CaO/Al/Fe_3O_4$	Rapeseed oil	15:1	6	180	98.71	[24]

4. Conclusions

Rubber seeds are agricultural residuals that have potential for use as a non-edible biodiesel raw material. Rubber seed oil successfully produced biodiesel through the transesterification process using subcritical methanol with nanomagnetic catalysts: $KF/CaO\text{-}Fe_3O_4$, $KF/CaO\text{-}Fe_3O_4\text{-}Li$, and $KF/CaO\text{-}Fe_3O_4\text{-}Al$. Response surface methodology with the Box-Behnken design was employed to determine a feasible experimental plan to optimize the rubber seed oil for the biodiesel conversion procedure due to the high FFA content of RSO (16.7%). The effect of all three independent parameters (methanol to oil molar ratio of 10:1–40:1, catalysts loading of 1.5–4.5 wt %, and reaction time of 30–60 min) were statistically investigated on the fatty acid methyl ester yield for rubber seed biodiesel production at 220 °C. The limitation of the developed response surface model is that all the p-values were less than 0.05. $KF/CaO\text{-}Fe_3O_4\text{-}Al$ is nanomagnetic catalyst with a BET surface area of 95.70 m^2/g, pore volume of 0.1140 cm3/g, BJH pore size of 3.79 nm, and the highest FAME yield. The optimal conditions for Al additives of $KF/CaO\text{-}Fe_3O_4$ nanomagnetic catalyst enhanced FAME yield without Al up to 18.17% and the process was 11 min faster, while Li additives enhanced FAME yield without Li up to 16.24%. The optimum conditions for rubber seed biodiesel production using the subcritical methanol with $KF/CaO\text{-}Fe_3O_4\text{-}Al$ as the nanomagnetic catalyst were optimized as a 28:1 molar ratio of methanol to RSO, 1.5 wt % catalyst, and 49 min reaction time. The optimum conditions led to the maximum predicted FAME yield of 86.79%.

Author Contributions: V.W. performed and developed the experiment, learned Design Expert software, analyzed the data and wrote the paper. N.Y. provided the research idea, developed the experiment, analyzed data, organized the research study including mentorship, and provided laboratory facilities and instruments for research.

Funding: The research was financially supported by The National Research Council of Thailand, Thesis Grant for Doctoral Degree Student Program of the fiscal year 2016.

Acknowledgments: The authors gratefully acknowledge the provision of the rubber seed oil processed by the Agricultural Occupation Promotion and Development Center in Cha-am district, Phetchaburi province, Thailand, and experimental facility at Faculty of Engineering, Mahidol University.

Conflicts of Interest: The authors declare no conflict of interest.

References

1. Martchamadol, J.; Kumar, S. Thailand's energy security indicators. *Renew. Sustain. Energy Rev.* **2012**, *16*, 6103–6122. [CrossRef]

2. Abas, N.; Kalair, A.; Khan, N. Review of fossil fuels and future energy technologies. *Futures* **2015**, *69*, 31–49. [CrossRef]

3. Refaat, A.A. Biodiesel production using solid metal oxide catalysts. *Int. J. Environ. Sci. Technol.* **2011**, *8*, 203–221. [CrossRef]

4. Siriwardhana, M.; Opathella, G.K.C.; Jha, M.K. Bio-diesel: Initiatives, potential and prospects in Thailand: A review. *Energy Policy* **2009**, *37*, 554–559. [CrossRef]

5. Takase, M.; Zhao, T.; Zhang, M.; Chen, Y.; Liu, H.; Yang, L.; Wu, X. An expatiate review of neem, jatropha, rubber and karanja as multipurpose non-edible biodiesel resources and comparison of their fuel, engine and emission properties. *Renew. Sustain. Energy Rev.* **2015**, *43*, 495–520. [CrossRef]

6. Alengaram, U.J.; Al Muhit, B.A.; bin Jumaat, M.Z. Utilization of oil palm kernel shell as lightweight aggregate in concrete—A review. *Constr. Build. Mater.* **2013**, *38*, 161–172. [CrossRef]

7. Samart, C.; Karnjanakom, S.; Chaiya, C.; Reubroycharoen, P.; Sawangkeaw, R.; Charoenpanich, M. Statistical optimization of biodiesel production from para rubber seed oil by SO_3H-MCM-41 catalyst. *Arabian J. Chem.* **2015**. [CrossRef]

8. Gimbun, J.; Ali, S.; Kanwal, C.C.S.C.; Shah, L.A.; Muhamad, N.H.; Cheng, C.K.; Nurdin, S. Biodiesel Production from Rubber Seed Oil using Activated Cement Clinker as Catalyst. *Procedia Eng.* **2013**, *53*, 13–19. [CrossRef]

9. Wibowo, A.D.K. Study on Production Process of Biodiesel from Rubber Seed (*Hevea Brasiliensis*) by in Situ (Trans)Esterification Method with Acid Catalyst. *Energy Procedia* **2013**, *32*, 64–73.

10. Morshed, M.; Ferdous, K.; Khan, M.R.; Mazumder, M.S.I.; Islam, M.A.; Uddin, M.T. Rubber seed oil as a potential source for biodiesel production in Bangladesh. *Fuel* **2011**, *90*, 2981–2986. [CrossRef]

11. Mukherjee, I.; Sovacool, B.K. Palm oil-based biofuels and sustainability in southeast Asia: A review of Indonesia, Malaysia, and Thailand. *Renew. Sustain. Energy Rev.* **2014**, *37*, 1–12. [CrossRef]

12. Ahmad, J.; Yusup, S.; Bokhari, A.; Kamil, R.N.M. Study of fuel properties of rubber seed oil based biodiesel. *Energy Conver. Manag.* **2014**, *78*, 266–275. [CrossRef]

13. Roschat, W.; Siritanon, T.; Yoosuk, B.; Sudyoadsuk, T.; Promarak, V. Rubber seed oil as potential non-edible feedstock for biodiesel production using heterogeneous catalyst in Thailand. *Renew. Energy* **2017**, *101*, 937–944. [CrossRef]

14. Liu, H.; Su, L.; Shao, Y.; Zou, L. Biodiesel production catalyzed by cinder supported CaO/KF particle catalyst. *Fuel* **2012**, *97*, 651–657. [CrossRef]

15. Reshad, A.S.; Tiwari, P.; Goud, V.V. Extraction of oil from rubber seeds for biodiesel application: Optimization of parameters. *Fuel* **2015**, *150*, 636–644. [CrossRef]

16. Hu, S.; Guan, Y.; Wang, Y.; Han, H. Nano-magnetic catalyst KF/CaO–Fe_3O_4 for biodiesel production. *Applied Energy* **2011**, *88*, 2685–2690. [CrossRef]

17. Bet-Moushoul, E.; Farhadi, K.; Mansourpanah, Y.; Nikbakht, A.M.; Molaei, R.; Forough, M. Application of CaO-based/Au nanoparticles as heterogeneous nanocatalysts in biodiesel production. *Fuel* **2016**, *164*, 119–127. [CrossRef]

18. Marwaha, A.; Rosha, P.; Mohapatra, S.K.; Mahla, S.K.; Dhir, A. Waste materials as potential catalysts for biodiesel production: Current state and future scope. *Fuel Process. Technol.* **2018**, *181*, 175–186. [CrossRef]

19. Taufiq-Yap, Y.H.; Lee, H.V.; Hussein, M.Z.; Yunus, R. Calcium-based mixed oxide catalysts for methanolysis of *Jatropha curcas* oil to biodiesel. *Biomass Bioenergy* **2011**, *35*, 827–834. [CrossRef]

20. Teo, S.H.; Rashid, U.; Taufiq-Yap, Y.H. Biodiesel production from crude *Jatropha Curcas* oil using calcium based mixed oxide catalysts. *Fuel* **2014**, *136*, 244–252. [CrossRef]

21. Taufiq-Yap, Y.H.; Teo, S.H.; Rashid, U.; Islam, A.; Hussien, M.Z.; Lee, K.T. Transesterification of *Jatropha curcas* crude oil to biodiesel on calcium lanthanum mixed oxide catalyst: Effect of stoichiometric composition. *Energy Convers. Manag.* **2014**, *88*, 1290–1296. [CrossRef]

22. MacLeod, C.S.; Harvey, A.P.; Lee, A.F.; Wilson, K. Evaluation of the activity and stability of alkali-doped metal oxide catalysts for application to an intensified method of biodiesel production. *Chem. Eng. J.* **2008**, *135*, 63–70. [CrossRef]

23. Meher, L.C.; Dharmagadda, V.S.S.; Naik, S.N. Optimization of alkali-catalyzed transesterification of *Pongamia pinnata* oil for production of biodiesel. *Bioresour. Technol.* **2006**, *97*, 1392–1397. [CrossRef] [PubMed]

24. Tang, S.; Wang, L.; Zhang, Y.; Li, S.; Tian, S.; Wang, B. Study on preparation of Ca/Al/Fe_3O_4 magnetic composite solid catalyst and its application in biodiesel transesterification. *Fuel Process. Technol.* **2012**, *95*, 84–89. [CrossRef]

25. Xie, W.; Peng, H.; Chen, L. Calcined Mg-Al hydrotalcites as solid base catalysts for methanolysis of soybean oil. *J. Mol. Catal. A Chem.* **2006**, *246*, 24–32. [CrossRef]

26. Sitthithanaboon, W.; Reddy, H.K.; Muppaneni, T.; Ponnusamy, S.; Punsuvon, V.; Holguim, F.; Dungan, B.; Deng, S. Single-step conversion of wet *Nannochloropsis gaditana* to biodiesel under subcritical methanol conditions. *Fuel* **2015**, *147*, 253–259. [CrossRef]

27. Micic, R.D.; Tomić, M.D.; Kiss, F.E.; Nikolić-Djorić, E.B.; Simikić, M.Đ. Optimization of hydrolysis in subcritical water as a pretreatment step for biodiesel production by esterification in supercritical methanol. *J. Supercrit. Fluids* **2015**, *103*, 90–100. [CrossRef]

28. Yin, J.-Z.; Ma, Z.; Shang, Z.-Y.; Hu, D.-P.; Xiu, Z.-L. Biodiesel production from soybean oil transesterification in subcritical methanol with K_3PO_4 as a catalyst. *Fuel* **2012**, *93*, 284–287. [CrossRef]

29. Sánchez, N.; Encinar, J.M.; Martínez, G.; González, J.F. Biodiesel Production from Castor Oil under Subcritical Methanol Conditions. *Int. J. Environ. Sci. Dev.* **2015**, *6*, 61.

30. Yin, J.-Z.; Ma, Z.; Hu, D.-P.; Xiu, Z.-L.; Wang, T.-H. Biodiesel Production from Subcritical Methanol Transesterification of Soybean Oil with Sodium Silicate. *Energy Fuels* **2010**, *24*, 3179–3182. [CrossRef]

31. Edem, D.O. Palm oil: Biochemical, physiological, nutritional, hematological and toxicological aspects: A review. *Plant Foods Hum. Nutr.* **2002**, *57*, 319–341. [CrossRef] [PubMed]

32. Jahirul, M.I.; Koh, W.; Brown, R.J.; Senadeera, W.; O'Hara, I.; Moghaddam, L. Biodiesel Production from Non-Edible Beauty Leaf (*Calophyllum inophyllum*) Oil: Process Optimization Using Response Surface Methodology (RSM). *Energies* **2014**, *7*, 5317–5331. [CrossRef]

33. Onumaegbu, C.; Alaswad, A.; Rodriguez, C.; Olabi, A.G. Optimization of Pre-Treatment Process Parameters to Generate Biodiesel from Microalga. *Energies* **2018**, *11*, 806. [CrossRef]

34. Bobadilla, M.C.; Martínez, R.F.; Lorza, R.L.; Gómez, F.S.; González, E.P.V. Optimizing Biodiesel Production from Waste Cooking Oil Using Genetic Algorithm-Based Support Vector Machines. *Energies* **2018**, *11*, 2995. [CrossRef]

35. Onoji, S.E.; Iyuke, S.E.; Igbafe, A.I.; Daramola, M.O. Transesterification of Rubber Seed Oil to Biodiesel over a Calcined Waste Rubber Seed Shell Catalyst: Modeling and Optimization of Process Variables. *Energy Fuels* **2017**, *31*, 6109–6119. [CrossRef]

36. Betiku, E.; Etim, A.O.; Pereao, O.; Ojumu, T.V. Two-Step Conversion of Neem (*Azadirachta indica*) Seed Oil into Fatty Methyl Esters Using a Heterogeneous Biomass-Based Catalyst: An Example of Cocoa Pod Husk. *Energy Fuels* **2017**, *31*, 6182–6193. [CrossRef]

37. Das, D.; Thakur, R.; Pradhan, A.K. Optimization of corona discharge process using Box–Behnken design of experiments. *J. Electrostat.* **2012**, *70*, 469–473. [CrossRef]

38. Yin, Y.; Wang, J. Optimization of Hydrogen Production by Response Surface Methodology Using γ-Irradiated Sludge as Inoculum. *Energy Fuels* **2016**, *30*, 4096–4103. [CrossRef]

39. Hebbar, H.R.H.; Math, M.C.; Yatish, K.V. Optimization and kinetic study of CaO nano-particles catalyzed biodiesel production from Bombax ceiba oil. *Energy* **2018**, *143*, 25–34. [CrossRef]

40. Khazaai, S.N.M.; Maniam, G.P.; Rahim, M.H.A.; Yusoff, M.M.; Matsumura, Y. Review on methyl ester production from inedible rubber seed oil under various catalysts. *Ind. Crop. Prod.* **2017**, *97*, 191–195. [CrossRef]

41. Mansir, N.; Teo, S.H.; Rashid, U.; Taufiq-Yap, Y.H. Efficient waste *Gallus domesticus* shell derived calcium-based catalyst for biodiesel production. *Fuel* **2018**, *211*, 67–75. [CrossRef]

42. Box, G.E.P.; Behnken, D.W. Some New Three Level Designs for the Study of Quantitative Variables. *Technometrics* **1960**, *2*, 455–475. [CrossRef]

43. Wu, Z.; Chen, C.; Wan, H.; Wang, L.; Li, Z.; Li, B.; Guo, Q.; Guan, G. Fabrication of Magnetic NH_2-MIL-88B (Fe) Confined Brønsted Ionic Liquid as an Efficient Catalyst in Biodiesel Synthesis. *Energy Fuels* **2016**, *30*, 10739–10746. [CrossRef]

44. Zhang, J.; Sun, H.; Pan, C.; Fan, Y.; Hou, H. Optimization of Process Parameters for Directly Converting Raw Corn Stalk to Biohydrogen by *Clostridium* sp. FZ11 without Substrate Pretreatment. *Energy Fuels* **2015**, *30*, 311–317. [CrossRef]

45. Rezayan, A.; Taghizadeh, M. Synthesis of magnetic mesoporous nanocrystalline KOH/ZSM-5-Fe_3O_4 for biodiesel production: Process optimization and kinetics study. *Process Saf. Environ. Protect.* **2018**, *117*, 711–721. [CrossRef]

46. Atadashi, I.M.; Aroua, M.K.; Aziz, A.R.A.; Sulaiman, N.M.N. The effects of catalysts in biodiesel production: A review. *J. Ind. Eng. Chem.* **2013**, *19*, 14–26. [CrossRef]

47. Levenspiel, O. Chemical Reaction Engineering. *Ind. Eng. Chem. Res.* **1999**, *38*, 4140–4143. [CrossRef]

48. Encinar, J.M.; Pardal, A.; Martinez, G. Transesterification of rapeseed oil in subcritical methanol conditions. *Fuel Process. Technol.* **2012**, *94*, 40–46. [CrossRef]

49. Go, A.W.; Nguyen, P.L.T.; Huynh, L.H.; Liu, Y.-T.; Sutanto, S.; Ju, Y.-H. Catalyst free esterification of fatty acids with methanol under subcritical condition. *Energy* **2014**, *70*, 393–400. [CrossRef]

50. Patil, P.D.; Gude, V.G.; Deng, S. Transesterification of *Camelina Sativa* Oil using Supercritical and Subcritical Methanol with Cosolvents. *Energy Fuels* **2009**, *24*, 746–751. [CrossRef]

51. Alonso, D.M.; Mariscal, R.; Granados, M.L.; Maireles-Torres, P. Biodiesel preparation using Li/CaO catalysts: Activation process and homogeneous contribution. *Catal. Today* **2009**, *143*, 167–171. [CrossRef]

52. Kaur, M.; Ali, A. Lithium ion impregnated calcium oxide as nano catalyst for the biodiesel production from karanja and jatropha oils. *Renew. Energy* **2011**, *36*, 2866–2871. [CrossRef]

53. Onoji, S.E.; Iyuke, S.E.; Igbafe, A.I.; Nkazi, D.B. Rubber seed oil: A potential renewable source of biodiesel for sustainable development in sub-Saharan Africa. *Energy Convers. Manag.* **2016**, *110*, 125–134. [CrossRef]

54. Ramadhas, A.S.; Jayaraj, S.; Muraleedharan, C. Biodiesel production from high FFA rubber seed oil. *Fuel* **2005**, *84*, 335–340. [CrossRef]

55. Gimbun, J.; Ali, S.; Kanwal, C.; Shah, L.A.; Ghazali, N.H.M.; Cheng, C.K.; Nurdin, S. Biodiesel Production from Rubber Seed Oil Using A Limestone Based Catalyst. *Adv. Mater. Phys. Chem.* **2012**, *2*, 138–141. [CrossRef]

56. Silitonga, A.S.; Ong, H.C.; Mahlia, T.M.I.; Masjuki, H.H.; Chong, W.T. Biodiesel Conversion from High FFA Crude *Jatropha Curcas, Calophyllum Inophyllum* and *Ceiba Pentandra* Oil. *Energy Procedia* **2014**, *61*, 480–483. [CrossRef]

57. Berchmans, H.J.; Hirata, S. Biodiesel production from crude *Jatropha curcas* L. seed oil with a high content of free fatty acids. *Bioresour. Technol.* **2008**, *99*, 1716–1721. [CrossRef]

58. Tiwari, A.K.; Kumar, A.; Raheman, H. Biodiesel production from jatropha oil (*Jatropha curcas*) with high free fatty acids: An optimized process. *Biomass Bioenergy* **2007**, *31*, 569–575. [CrossRef]

 *energies*

Review

A Review of Gasoline Compression Ignition: A Promising Technology Potentially Fueled with Mixtures of Gasoline and Biodiesel to Meet Future Engine Efficiency and Emission Targets

Yanuandri Putrasari [1,2] and Ocktaeck Lim [3,*

[1]   Graduate School of Mechanical Engineering, University of Ulsan, San 29, Mugeo2-dong, Nam-gu, Ulsan 44610, Korea; y.putrasari@gmail.com or yanu001@lipi.go.id
[2]   Research Centre for Electrical Power and Mechatronics—Indonesian Institute of Sciences, Jl. Cisitu No 154/21D, Bandung 40135, Indonesia
[3]   School of Mechanical Engineering, University of Ulsan, San 29, Mugeo2-dong, Nam-gu, Ulsan 44610, Korea
*   Correspondence: otlim@ulsan.ac.kr; Tel.: +82-10-7151-8218

Received: 25 November 2018; Accepted: 5 January 2019; Published: 14 January 2019

Abstract: Efforts have been made to develop efficient and alternative powertrains for internal combustion engines including combustion at low-temperature (LTC) concepts. LTC has been widely studied as a novel combustion mode that offers the possibility to minimize both nitrogen oxide (NOx) and particulate matter (PM) via enhanced air-fuel mixing and intake charge dilution, resulting in lower peak combustion temperatures. Gasoline compression ignition (GCI) is a new ignition method related to the extensive classification of combustion at low-temperature approaches. In this method of ignition, a fuel with high evaporation characteristics and low autoignition sensitivity, for instance gasoline, is burned in a high pressure process. Despite many research efforts, there are still many challenges related with GCI performance for compression ignition (CI) engines. Unstable combustion for idle- to low-load operation was observed because of the low reactivity characteristics of gasoline, and this will affect the efficiency and emissions of the engine. This paper contributes a detailed review of several topics associated with GCI engines and the effort to improve its efficiency and emissions, including its potential when using gasoline-biodiesel blends. Some recommendations are proposed to encourage GCI engines improvement and development in the near future.

Keywords: GCI; biodiesel; diesel; combustion; emission

---

1. Introduction

Energy savings, emission reduction, environmental friendliness and sustainable energy supplies are issues that attract much current attention. Investment in the research and development of alternative fuels and energy resources is growing rapidly at both national and international levels. The focus on the future energy supply is stimulated by anxieties associated with fossil fuel reserves. Global energy consumption forecasts continue to predict increasing request for liquid hydrocarbon fuels next shortly by internal combustion (IC) engines because these fuels are abundant, cheap and convenient. However, large-scale and extensive use of fossil fuels results in two bad situations: oil reserve depletion and environmental deterioration in the form of air pollution, increase in the average global temperature manifested as global warming, and climate change [1]. Therefore, it is important to promote and develop novel engine technologies and combustion strategies that focus on high efficiency and clean internal combustion engines.

Several attempts have been made to promote alternative and efficient powertrains for IC engines, including the combustion at low temperature (LTC) concept. LTC has been widely studied as a novel combustion mode, which offers the possibility to reduce both nitrogen oxide and particulate matter via enhanced air-fuel mixing and intake charge dilution resulting in lower peak combustion temperatures [2,3]. Compression ignition of gasoline (GCI) is a new combustion concept in the extensive classification of combustion at low-temperature approaches. For this new ignition type, a hydrocarbon fuel that has high evaporation property and less sensitivity to autoignition, for example gasoline, is burned under high pressure conditions [4–9]. Compression ignition of gasoline was first suggested by Kalghatgi to take advantage of the benefits of the high vapor and high autoignition of petroleum gasoline-like fuel and the high ratio of compression (CR) of CI engines to obtain maximum performance and near to zero emissions at the same time [4]. However, the lubricity of market gasoline is not adequate to protect today's fuel injection components, so either the engine components must become more robust or fuel lubricity additives will be needed. Furthermore, the major challenge for GCI is the very small cetane value of gasoline that is usually estimated to be no higher than about 15. This low value leads to long ignition delays and misfires. Therefore, the utilization of gasoline–diesel blends was suggested in engine ignition strategies. Besides the lengthened autoignition, their altered physical characteristics may promisingly influence the injection of fuel and characteristics of the jet spray, that are equally crucial for homogeneous mixture of fuel-air configuration and the combustion inside the cylinder of engine [10,11]. Later on, a certain portion of biodiesel was added to the gasoline fuel to achieve fuel properties that are appropriate for GCI engines and can overcome the autoignition problems observed in gasoline [12].

Despite many research efforts, many challenges related to GCI operation for CI engines still exist. Gasoline shows low reactivity characteristics, leading to unstable combustion for idle- to low-load operation, which will affect the efficiency and emissions of the CI engine operation. The efficiency of a GCI engine is estimated to be roughly equal that of a diesel engine and the emissions should be better than those of an SI engine. This research project deals with improving the efficiency and emission behaviors of CI engines using GCI mode and gasoline-biodiesel blend fuels. Understanding the main properties of fuel and quantifying the influences of several parameters on CI engines on compression ignition of gasoline mode, fueled with a biodiesel blended in gasoline are important for speeding up the contributions and theory to realize the utilization of gasoline in diesel engines and biofuels in the transportation area. Therefore, the aim of this paper is to contribute a detailed review of several topics associated with GCI engines and the efforts made to improve their efficiency and emissions, including their potential when using gasoline-biodiesel blends. Some recommendations are proposed to encourage GCI engine improvements and development in the near future.

2. Gasoline-Biodiesel Blends as Substitute Fuels for CI Engines

Gasoline with high octane number usually has high vaporization characteristics and is very difficult to ignite, and thus is called low reactivity fuel [13]. Spark ignition (SI) or gasoline engines are engines that are suitable for this gasoline fuel. However, a big issue of this fuel implementation in SI engines is the knocking phenomenon. This phenomenon allows gasoline engine to operate only at a low ratio of compression around 8 to 11, which is a limitation of the SI engine. Due to the low compression ratio, the limitation of gasoline engines is that they always display small efficiency. Furthermore, the gasoline engine application usually requires spark plugs to give a spark firing at the maximum of the stroke of compression as the trigger of combustion. Lastly, NOx emissions will be a final product because gasoline engines are usually operated with almost-stoichiometric mixtures which require high-temperature combustion [14].

Biodiesel is a substitute fuel for CI engines. It is a chemically modified biofuel made from either animal fats or vegetable oils. According to the ASTM D6751 biodiesel standard, biodiesel can be described as a "fuel containing monoalkyl esters of long-chain-fatty-acids processed from fats of animal or oils of vegetable, named B100". Biodiesel can be produced from many kind of seeds and crop plants, for example poppy seed, maize seed, camelina seed, seed of pumpkin, castor, grape, pea, beechnut, lupin, linseed, chestnut, rapeseed, peanut, olive, soybean, hemp, seed of sunflower, palm, Shea butter, and cottonseed [15,16]. Biodiesel is required to have the same chemical and physical features as diesel fuel. The biodiesel standards and properties of biodiesel around the globe are briefly presented in Table 1 [17], while, the superior features or deficiencies of biodiesel compared to diesel or petroleum fuels are as follows [17]:

Superiority of biodiesel compared to diesel fuel:

- Biodiesel is more portable, available and renewable.
- Emissions such as CO, CO_2, PM, SO_2, and HC from biodiesel are lower compared to petroleum-based diesel fuel.
- It is easier and faster to produce biodiesel than diesel petroleum-based fuel.
- The cetane value of biodiesel is higher than 100 that can results in better performance of CI engines than petroleum diesel fuel.
- Biodiesel's lubricity is better compared to petroleum-based diesel fuel which can reduce engine maintenance and improve the engine lifetime.
- Different from conventional diesel, biodiesel can be used without additional lubricants because of its purity and the clarity.
- Biodiesel can be used to solve energy security problems and has great potential to stimulate rural development and sustainability.
- Biodiesel can be obtained without mining, transporting, or refining activities like petroleum diesel fuel.
- Biodiesel can be produced locally, thus being cheaper than petroleum diesel fuel.
- The aromatic content, flash point, sulfur content, and biodegradability of biodiesel are better excellent than those of petroleum diesel fuel.
- Biodiesel is safer, much more non-toxic, and more biodegradable compared to petroleum-based diesel fuel.
- Biodiesel is non-flammable, however it has higher combustion efficiency because of its huge oxygen fraction if compared to petroleum diesel.
- Biodiesel can produce low emissions, less visible smoke, and smalles amounts of noxious fumes and odors.
- Blends of up to 20% biodiesel can be run without any modification of engines.

Table 1. Biodiesel standards and properties of biodiesel [Reprinted from Procedia Engineering, Vol. 56, Masjuki Hj. Hassan, Md. Abul Kalam, An Overview of Biofuel as a Renewable Energy Source: Development and Challenges, Copyrights (2013); with permission from Elsevier].

Properties (units)	Malaysia	Indonesia Republic of	Thailand	US	European Union	Brazil
		SNI	E 14214	ASTM D6751	E 14214	ANP 42
Minimum flash point (°C)	182	100	120	130	120	100
Viscosity at 40 °C (cSt)	4.415	2.3–6.0	3.5–5	1.9–6	3.5–5	-
Maximum of sulphated ash (%-mass)	0.01	0.02	0.02	0.02	0.02	0.02
Minimum of sulphur (%-mass)	0.001	0.001	0.001	0.001	0.001	-
Maximum of cloud point (°C)	15.2	18	-	-	-	-
Classification of copper corrosion (3 h, 50 °C)	1	3	1	3	1	1
Minimum of cetane number	-	51	51	47	51	-
Maximum of sediment and water content (volume %)	0.05	0.05	-	0.05	-	0.05
Maximum of CCR 100% (%-mass)	-	-	0.3	0.05	-	0.1
Neutralization value (mg, KOH/gm)	-	-	-	0.05	0.05	0.08
Maximum Free glycerin (%-mass)	0.01	0.02	0.02	0.02	0.02	0.02
Maximum Total glycerin (%-mass)	0.01	0.24	0.25	0.24	0.25	0.38
Maximum Phosphorus (%-mass)	-	10	0.001	0.001	0.001	-
Maximum distillation temperature (°C)	-	360	-	360	-	360
Oxidation stability (h)	-	-	6	3	6	6

Deficiencies of biodiesel compared to petroleum diesel fuel:

- It produces higher emissions of NOx than petroleum diesel fuel.
- Biodiesel has cold weather starting problems due to the higher cloud and pour point causing fuel freezing.
- Biodiesel is naturally corrosive if exposed to brass and copper.
- Biodiesel has around 11 to 17 times higher viscosity than diesel. Furthermore, pumping, atomization in the injector systems and combustion problems commonly happen in diesel engines due to the larger chemical structure and molecular mass of biodiesel than diesel.
- The engine speed and power of the diesel engine is lower when utilizing biodiesel.
- Biodiesel produces deposits in injectors, on the pistons, on the combustion chamber walls, and on the head of engines.
- Deposit formation inside injector and on rings and filter and line plugging will happen due to the gumming and sticking which are promoted by its high viscosity in long term operation using biodiesel.
- Biodiesel is incompatible with petroleum-based engine lubricating oils.
- Excessive engine wear can be caused by biodiesel.
- The cost-competitiveness of biodiesel cannot be compared with petroleum gasoline or diesel.

LTC is a famous combustion concept in CI engines, which proposes the possibility to reduce nitrogen oxide and particulate matter (PM). By using suitable modification of fuel properties, by the addition of gasoline [10], which has higher vaporization characteristics and lower cetane number is possible to achieve LTC in a CI engine with GCI mode [18,19]. The more retarded autoignition timing of gasoline because of its high octane number leads to adequate air-fuel mixing, thus, the enhanced thermal efficiency improves the combustion pre-mixing, resulting in decreased soot and NOx values [18]. The high thermal efficiency, low soot and NOx emissions, and low combustion temperature compared to gasoline engines are the main benefits of GCI engines [20]. However, the required high intake temperature, low lubricity, and requirement for a high ratio of compression like a diesel engine are the deficiencies of GCI engines. Biodiesel has high possibilities to solve many issues in GCI engine implementations when blended with gasoline, such as the low lubricity. Furthermore, because of the high content of oxygen in biodiesel, perfect combustion may also be achieved [21].

3. Challenges and Opportunities in GCI Engine Research

The common emission problems in CI engines are NOx and particulate matter (PM) formation [22,23]. Serious environmental and health problems might be caused by air pollution which is contributed by these emissions. Advanced technologies such as subsequent systems or after- treatment systems are being promoted to solve and control these engine emissions. However, these technologies are complicated, expensive, and reduce the primary benefit of CI engines. Additionally, emission regulations, particularly for diesel engine vehicles, are getting to be tighter everywhere throughout the globe. Therefore, researchers in the engine field have a motivation to study the necessity of implementation of high vapor fuels, for instance gasoline and other substitute fuels, for diesel engine combustion operations that afford high performance but near-zero exhaust emissions. A sustainable substitute fuel that is very suitable for diesel engines is biodiesel. Various renewable resources can be used to make biodiesel fuel [24,25]. Furthermore, it has proven that because the oxygen fraction in biodiesel has a main function of reducing soot formation during the combustion process [26], biodiesel has excellent benefits in reducing the soot emissions of CI engines [27,28].

The efficiency of diesel engines or diesel engines are much higher than that of gasoline engines, according to some considerable analysis [29]. First, CI engines can be run better at higher compression ratios because do not suffer deterioration from knocking at high loads compared to SI engines. Second, part load operation can be carried out in CI engines by reducing the injected fuel, instead of managing the air-mass compressed in the chamber of combustion. Third, performance near to an ideal cycle efficiency can be achieved due to the fact only air is trapped during the compression movement in a CI engine, rather than an air-fuel mixture. However, the huge emissions, especially soot/particulate matter/smoke and NOx, that are difficult to reduce by using after-treatments, are always produced by CI engines using diesel fuel. On the contrary, lower engine efficiency and lower exhaust emissions, especially nitrogen oxide and particulate matter are produced by SI engines using a petroleum-based gasoline fuel. Therefore, based on these realities, it is necessary to apply an advance combustion strategy to obtain an efficiency as high as an CI engine and emit less emissions like an SI engine [4,30]. Nowadays, consideration is being given to the GCI engine as the LTC method with the most potential because high thermal efficiency and low emission behavior can be produced using this concept [20,31–36]. GCI combustion is more practical to solve the issues of the complexity of combustion controllability than other LTC concepts for CI engines, for example premixed-charge-CI (PCCI) and homogeneous-charge-CI (HCCI), even though these concepts also offer interesting ignition phenomena under homogeneously lean air-fuel mixing conditions [11,12,37–40].

Engine experiments are very important for understanding the real phenomena of the combustion process and emission behaviors of an IC engine. However, these are usually costly and complicated. Therefore, only a few researchers have included engine experiments in their studies. An extensive search of the literature showed that there are only a few references for experimental studies of GCI engines, and studies on GCIs fueled with gasoline-biodiesel blends are even rarer. Most of the studies use simulations and numerical methods. Thus, sequential and complementary experimental studies are needed to achieve a better understanding of the process of combustion inside the cylinder and emission behaviors of GCI engines fueled using blend of biodiesel in gasoline fuel.

4. LTC Concept

The objective of advanced engine operating strategies is to increase the efficiency and decrease exhaust emissions in IC engines. The primary research focus is the development of LTC concepts. LTC methods offer the possibility to reduce both nitrogen oxide and particulate matter (PM) via enhanced air-fuel mixing and intake charge dilution, resulting in lower peak combustion temperatures [41]. Depending on the air and mixing of fuel distribution and temperatures of combustion in the combustion chamber at various operating points, there exist regions of excessive soot or NOx formation (also referred to as soot or NOx 'islands') [42]. The soot or NOx islands and LTC concept derivatives (HCCI) are shown in Figure 1 [2].

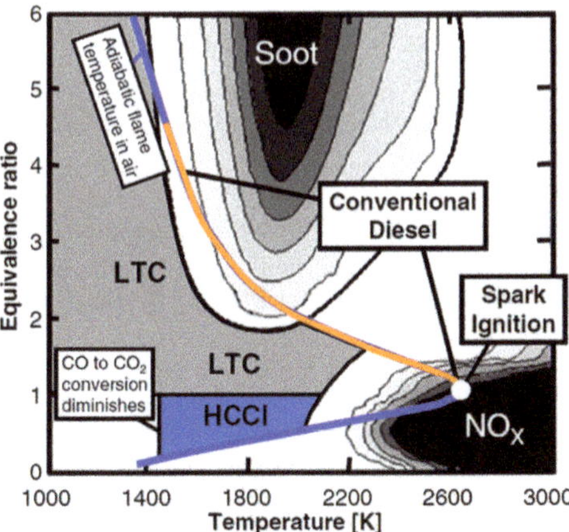

Figure 1. Soot or NOx islands, LTC and conventional CI combustion regimes in φ-T space [Reprinted from Proceedings of the Combustion Institute, Vol. 32, John E. Dec, Advanced compression-ignition engines—understanding the incylinder processes, Copyrights (2009); with permission from Elsevier].

According to Loeper et al. [42], reducing soot and NOx emissions simultaneously in conventional diesel combustion has long proved challenging due to the operating points that exist within "islands" of soot or NOx formation, as shown in Figure 1. This duality of achieving either low NOx or soot formation, but not both, is attributed to as the soot-nitrogen oxide trade-off. Low NOx formation rates can be achieved by reducing the combustion temperature, but this is traditionally accompanied by either high soot formation (due to poor air utilization and/or incomplete combustion) or significant restrictions in engine load (or power). Of course, soot formation can be reduced with superior air utilization, but this results in higher combustion temperatures and NOx formation rates. These compromises in conventional diesel combustion have spurred engine researchers to seek combustion strategies (e.g., LTC, HCCI, etc.) that avoid the NOx/soot islands altogether.

HCCI combustion, which operates on the similar basic fundamentals as a 4-stroke IC engine and employs the main components of gasoline and diesel engines is the most famous type of LTC concept, nowadays [43]. HCCI is the pioneer of one type of LTC, and is probably the most extensively researched [41]. In the HCCI concept, a uniform mixture of air and fuel is compressed in the engine cylinder until its pressure and temperature reach auto ignition. The clean soot characteristics of traditional fully-mixing gasoline engine combustion, homogeneous, high efficiencies (obtained by lean and unthrottled conditions) which is usual characteristics of diesel engine combustion, are combined in the HCCI method. Through the high amounts of mixture with air and/or fuels HCCI that keeps the NOx emissions at a low level, HCCI can achieve the lowest soot emissions and NOx at high efficiencies equal to, or higher than, traditional CI combustion because of these mentioned factors [2]. However, because of the minimum of straight control through the beginning and HRR, HCCI is only can be achieved over a limited range by operating a part-load area [44]. A preferable understanding of the principal combustion processes is needed to provide control through HR in HCCI. A more practical LTC strategy that can solve the problems of HCCI engines is GCI, in which a more difficult auto ignition fuel (i.e., gasoline) is directly injected into the chamber using a fueling system (common rail) in a diesel engine [4].

Even though the LTC concept is radically different from the conventional SI combustion or CI diffusion combustion concepts, LTC technology offers excellent benefits in terms of simultaneous reduction of both NOx and particulate matter or soot in addition to reducing specific fuel consumption and offering the flexibility of using various fuels. Furthermore, the main requirement of the LTC concept is the availability of a homogeneous fuel–air mixture before the start of combustion. This requirement is potentially fulfilled by using CI engines fueled with gasoline. The gasoline fuel can be injected directly into the cylinder with an early injection strategy, pilot and main injection, and also the application of EGR and intake boosting. With the naturally volatile characteristics of the gasoline, the gasoline can be mixed with air nearly homogenously before the combustion as required by the LTC concept. It cannot be denied that as long as the combustion in GCI concept is by auto ignition, the controlling ignition timing and heat release rate (HRR) are the main problems and challenges to be managed before LTC is achieved. However, since the combustion occurrs in the CI engines with gasoline fuel, the low temperature regime, NTC, and high temperature regime from the typical HRR curve of CI engines combustion can be predicted. Thus, GCI engines are relevant to the LTC concept.

5. GCI Engines and CI Engines with Gasoline Fuel

5.1. GCI Engines

To provide a review of previous research works associated to GCI engines, utilization of gasoline-biodiesel blends as internal combustion engine fuel and experimental studies of GCI combustion are the purposes of this section. This section also provides a review of other relevant research studies. To offer a perception into how previous research works have laid the groundwork for further works the review is organized chronologically. The new research efforts can be suitably tailored to correlate with the current body of literature because the review is detailed.

Gasoline compression ignition is a novel concept of ignition that belongs to the wider classification of LTC approaches. In this method, a poor reactivity fuel with high volatility, for example gasoline, is burned perfectly by using pressurized conditions [4,20,31,35,45,46]. The throttle-less, lean, and low-temperature combustion operation can lead GCI to achieve the same or better efficiencies than CI engines. Very low soot (because of the excellent mixing characteristics of gasoline) and less NOx emissions (because of the LTC and absence of flames) compared to conventional CI combustion, can also be emitted by this combustion strategy. Based on a reactivity stratification of fuel, it can be explained that the combustion in GCI is a result of sequential autoignition [47], without significant flame propagation. The aim of GCI is to manage the ignition delay period by managing stratification of air-fuel equivalence ratio concentrations inside the combustion chamber. A highly volatile and poor reactivity fuel (e.g., gasoline) is burned solely by compression (without any support from a spark plug) to prevent propagating flames and control the combustion temperatures and NOx emissions. This has been designated in the literature by various names/titles and terms, for example gasoline-direct-compression-ignition (GDCI), and gasoline-direct-injection-compression-ignition (GDICI). Throughout this study we will refer to this concept as GCI.

Compression ignition of gasoline was first studied by Kalghatgi [4] to take advantage of the benefits of the high vapor characteristics and high auto-burning resistance of petroleum gasoline fuel and the high ratio of compression of CI engines to simultaneously achieve almost-zero emissions and excellent efficiency. The GCI fuels can be categorized into two groups, low octane number fuels and high-octane number fuels [48]. A high-octane number fuel commonly refers to petroleum gasoline fuel with an octane number bigger than 90. Fuels with low octane number have higher auto burning and can be used in broader operating ranges with smaller load if compared with high octane fuels. The common approach to get a low octane number fuel is by blending gasoline and diesel, and there is also blending with biodiesel, bioethanol and other cetane improvers [11,12,49,50].

Based on Cracknell et al. [31], the potential advantages of fueling CI engines with gasoline in GCI combustion mode are as follows: First, CI engines have higher efficiency over SI/gasoline engines and have the possibility to use a broader fuels range. Second, the ability of GCI engine concepts to use available market gasoline would allow these concepts to enter engine applications quickly without fuel constraints. Third, more gasoline consumption in passenger cars would help to rebalance the gasoline/diesel fuel demand in refineries and reduce GHG emissions from the fuel supply in several countries. Fourth, a successful GCI vehicle could potentially compete in predominantly gasoline markets in other parts of the world.

5.2. CI Engines with Gasoline Fuel

Because of the high octane number of 100% gasoline fuel and its difficulty for auto-burning at the low level of temperatures of mixtures common in low-load conditions, GCI combustion faces challenges under low load operation. Thus, several methods are required to obtain optimal combustion and good exhaust emissions. The focus of this work is to improve the emission behaviors and efficiency of CI engines fueled with gasoline-biodiesel blends using GCI mode. We performed a literature study before attempts were made to enhance the efficiency and emission behaviors of CI engines fueled with gasoline-biodiesel blends. This study gives an overview of currently available research regarding the use of gasoline in CI engines using several combustion strategies.

Kalghatgi et al. [4] investigated the influence of fuel quality for its auto-burning characteristics among four fuels ranging from diesel towards gasoline fuels on combustion process at two different EGR levels and inlet pressures. The CI engine can be operated without problems using gasoline by single injection near TDC, although it difficult to operated with too early fuel injection in the HCCI combustion because of auto ignition failure or because of excessive heat release. The results of Kalghatgi et al. [4] showed the possibility of using the autoignition resistance of fuels with higher octane/lower cetane to obtain high IMEP with a low NOx, low smoke combustion system. In another study, Kalghatgi et al. [45], improved the GCI concept by proposing partially pre-mixed auto-burning of gasoline to achieve better smoke and better NOx at high load in a CI engine; they also compared this with a diesel fuel. The results indicated that pilot injection with gasoline decreases the peak of HRR for an adjusted IMEP, and allows HR to appear retarded with a small cycle variation compared to one-time injection. However, there are still challenges related to further improvements that may be realized by various EGR ratio schemes and boosting intake pressure and by optimization of the mixture preparation and strategy of injection (for example multiple injections), and the design of injectors (for example many holes injectors) [45].

Weall and Collings [51] conducted a study in a multi-cylinder, light duty type CI engine, using PPC mode fueled using a gasoline-diesel compound. Their results show that an upward content of gasoline decreased smoke emissions at higher working loads by an upward premixing of the in charge obtained from an increase in auto burning timing and higher vapor characteristics of the fuel. The results of Weall and Colling's work [51] confirmed that a blending of fuel features indicates higher vapor characteristics and prolonged auto ignition timing would widen the low emission-operating regime, but combustion stability at low operating loads must be considered [51]. Furthermore, engine idling was possible with a 50% of gasoline ratio test fuel, but cold start issues could emerge as a significant problem with such combustion regimes when using alternative fuels with present technology diesel cold start systems.

Experiments on the effects of gasoline by the port injection method have been conducted in a 1-cylinder CI engine with direct injection type by Sahin et al. [52]. The gasoline fuel was introduced through intake port using a carburetor, and there no other changes to the engine specifications are needed. The results showed that a 4 to 9% increase of power output resulted and efficiency was increased by around 1.5 to 4% and fuel burning reduced by about 1.5 to 4% due to gasoline volatilization. However, the exhaust gas emissions from the engine were not analyzed. Meanwhile, a parametric study of a CI engine running with gasoline was conducted in numerical simulation by Ra et al. [6]. The results indicated that the predictions concerning high pressure direct injection gasoline engine combustion and its emissions agreed with experimental results for various parameters. For the similar CA50, gasoline will have a much longer CA10 compared to diesel fuel; thus, nitrogen oxide and PM emissions were greatly decreased compared to the similar cases of diesel. The results also described that optimal injection timing with lowest UHC and CO emissions was produced by accelerating the timing of main injection in the combustion of gasoline. However, the timing of main injection has to be chosen as an balance between increasing of PRR levels and emissions reduction.

Another type of in-cylinder mixing fuel of diesel and gasoline has been studied by Curran et al. [40], who introduced gasoline in a port-fuel-injection system. The results from this study showed that the temperature of the intake charge has a high effect on cylinder PRR, while the thermal efficiency increased simultaneously with NOx and PM emission reductions for around 90%. However, the indicated thermal efficiency for the many-cylinder tests were lower than estimated from simulation and 1-cylinder measurements. The smaller indicated thermal efficiency of the experimental result compared to the modeling measurements showed that an improvement of cylinder-to-cylinder regulation and more optimization in dual-fuel mode were needed.

The emission behaviors of a CI engine fueled with in-cylinder dieseline blends has been studied by Prikhodko et al. [39]. The results from this work showed that 87% and 99% reductions in NOx and PM emissions were achieved in combustion of dual-fuel RCCI, and the thermal efficiency was increased by 1.7% as compared to combustion of normal diesel. However, CO, aldehyde, HC, and ketone emissions increased significantly for combustion of dual-fuel RCCI compared to conventional diesel and PCCI combustion of diesel. Shi and Reitz conducted a study on optimization of a heavy-duty CI engine running under middle-load and high-load operation conditions using petroleum diesel, petroleum gasoline, and an ethanol blend (E10) [19]. The study indicated that identical-gasoline fuels are potentially useful for heavy-duty CI engines with an optimized injection system because of their decreased soot and emissions of NOx and lower fuel consumption compared to traditional diesel petroleum-based fuels. However, the high PRR corresponding to the PPC of gasoline-like fuels might be a restriction for high load operation and a challenge for small-load operation.

Several tests were conducted in 2-single-cylinder engines, using a total of three various engine settings to show the advantages of using gasoline from 80 to 69 RON in a heavy-duty type CI engine [53]. The authors of this research concluded that if in the constant engine setting gasoline of around 70 octane is the most suitable for this combustion, and if higher octane gasoline is used in partial premix combustion, either a variable compression ratio or a cylinder heater should be used to operate under lower-load conditions. A blend of 20% gasoline and 80% diesel named G20, was tested in a 1-cylinder optical CI engine with two different pressures of injection set at 700 and 1400 bar, and injection timings from 11 °CA BTDC to 5 °CA ATDC [54]. The results showed that the G20 at retarded timing of injection and 50% EGR increased the auto ignition timing enabling the operation with retarded injection timing with premixed fuel in partial LTC mode where the fuel is fully injected prior to the auto ignition happening. In this method, great reductions of nitrogen oxide and smoke were achieved with deterioration of efficiency.

Ra et al. [7] conducted a study of a GDICI in the LTC method. The results exhibited good agreement between the modeling and experiments. Furthermore, because of the high vapor characteristics and low number of cetane of gasoline incorporated with the decreased temperature of combustion due to EGR application, either NOx or PM emissions could be decreased up to 100 mg/kg-f while experimental gross isfc was controlled at around 0.18 kg/kw-h. Zhang et al. [55]. conducted another PPCI study of dieseline in a CI engine. The results showed that diesel-gasoline had significant benefits as a combustion fuel of PPCI for emission minimizing purposes. It was also found that the total particle number concentration could be decreased by 90% by mixing 50% gasoline in the diesel, in addition, low NOx and higher thermal efficiency of approximately 30% were controlled for all variations of the loads. There is also another study on a DICI engine fueled with gasoline in the LTC mode with triple-pulse injection conducted by Ra et al. [8]. They showed that the stability of combustion and peak PRR might be adjusted by the pulse number two, while the pulse number three can be utilized to manage the engine load. Furthermore, by using the triple-injection strategy along with EGR, either NOx or PM could be decreased up to around 100 mg/kg-f, while obtaining isfc at around 0.173 kg/kW-h.

An experimental study has been done to compare the emission characteristics and combustion process of the late and early injection for highly premixed charge combustion (L-HPCC and E-HPCC, respectively) modes and the mixed fuel LTC mode [18]. Combustion with the L-HPCC and LTC may occur with local regions that are close to stoichiometric because of fuel stratification. Hence, they have relatively high NOx emissions, which could be lowered with increasing EGR rates. The soot emissions for the three modes of combustion are lower and below the Euro 6 regulation limits. The HC emissions of the E-HPCC and LHPCC regime are higher than those of the LTC because the premixed mixture of gasoline was trapped in the crevice area.

Another experimental investigation of a GCI engine has been conducted by Loeper et al. [46], to study the light to medium load operating sensitivity. The results revealed that input parameters can be used to reduce nitrogen oxide emissions below 600 g/kg-f with appropriate stability of combustion which is COV of IMEP below 3%, through a large intake temperature range. In addition, optimization refers to the efficiency of combustion and either CO or UHC emissions could be realized with the input parameter controllability.

Adams et al. [12] investigated the influences of biodiesel fraction in gasoline on GCI mode by injecting the fuel into the engine chamber directly. The main recommendation of this work is that the temperature of intake demands were decreased by 288 K and 303 K for the 5% and 10% biodiesel contents, respectively, compared to pure gasoline at a similar CA50. Biodiesel content at 5% and 10% significantly reduced CA10 and therefore advanced the CA50 compared with operation on pure gasoline. Pure gasoline produce higher average bulk gas temperatures resulting in higher NOx emissions as well as lower unburned hydrocarbon and carbon monoxide. Oxidations of carbon monoxide were not improved by the increased oxygen content of the biodiesel blends. Another experimental study was done to analyze the performance, combustion noise and emissions in PCCI regime using dieseline fuel [11]. The results showed that as the gasoline content increases, the CA10 increases, resulting in higher mixing time between the close of injection and the auto burning timing, thus, lower equivalence ratios are produced in local areas, and soot emissions are lowered. Finally, the performances and combustion noise are reduced. However, NOx emissions are a little bit increased.

A novel combustion method called multiple premixed compression ignition or MPCI was suggested by Yang [56]. The results revealed that the RON 66 MPCI regime decreased NO, CO, and soot, increasing thermal efficiency, but also increasing THC emissions compared to conventional CI engines. However, it is difficult to obtain the similar performance for RON 76 and RON 86 because of the lower auto-ignition ability compared to RON 66. Other spray and combustion phenomena of dieseline were investigated in a DI common rail CI engine [57]. The results showed that the combustion of gasoline produced higher HC, CO, and NOx but emitted less soot than combustion of pure diesel. However, the PCCI with early injection application significantly reduced emissions of nitrogen oxide compared to gasoline ignition.

Experimental and simulation studies of swirl effects at the intake port of a GCI engine have been carried out by Loeper [42]. A lowering in swirl rate from 2.2 to 1.5 caused a 6 °CA earlier combustion phasing, while rising swirl rate from 2.2 to 3.5 leads in a 2 °CA delay of combustion phasing. This earlier combustion phasing at the 1.5 swirl rate was followed by a huge increase of NOx emissions from 200 to 1600 g/kg-f. Another study testing various dieseline ratios in a CI engine was carried out by Zhang et al. [14]. The results indicated that the total PM of G50 was decreased by up to 50% and 90%, and the count median diameter was reduced by 25% and 75% at small and medium loads, respectively. Then, the G50 produced the minimum smoke 0.5 FSN for all tested load operations, while, compared with the diesel, NOx emissions of G50 were reduced by 50% at low speed, and increased by 20% at medium speed. Furthermore, the diesel-gasoline produces lower PRR and HRR peaks at low speeds, and it extended the CA10 by up to 7 °CA compared to diesel combustion.

One more characterization of the CA10 for a CI engine fueled with dieseline was conducted by Thoo et al. [58]. Higher gasoline ratio delayed the beginning of fuel injection up to 3 °CA, because of the changes in physical characteristics. The change in timing of injection influenced CA50, but did not affect CA10 directly. An analysis of the stability of GCI engine mode was conducted on idle operating speed and load operating condition on a many-cylinder CI engine using a gasoline with 87 index of anti-knock (AKI) by Kolodziej et al. [59]. Although hot EGR was successfully able to significantly increase the intake temperature, the simultaneous reduction of in-cylinder trapped mass and slight reduction of charge oxygen concentration had a more significant effect on reducing the mixture reactivity.

Three low octane number fuels: naphtha, dieseline G70D30, and the gasoline-*n*-heptane blend G70H30 were characterized for their combustion and emission phenomenon in MPCI regime by Wang et al. [60]. The results showed that the CA10 of the gasoline was longer with a higher pressure of injection. Then, the emissions of soot decreased at increased pressure of injection with a deterioration of increased CO and HC. Another experimental work on fuel consumption and exhaust emissions of DI premixed combustion engines using dieseline was conducted by Du et al. [61]. The results indicated that the CA10 was retarded and resulted in reduction of smoke. However, too high gasoline ratio might have the worst fuel economy.

Yang and Chou [62] numerically studied the engine performance and emission behaviors of a DI CI engine fueled with dieseline. The results showed that the CA10 was retarded by increasing of gasoline content. The 100% diesel operation achieved higher performance at low load. On the contrary, at middle and high loads of blended fuels, better performance could be realized, but slightly higher NOx emissions also occurred. Another work concerned the experimental study of the influence of fuel injection strategies on LTC fueled with gasoline on a 1-cylinder diesel engine was conducted by Yang et al. [63]. The results showed that the double-injection mode enabled expanding the high-efficiency and low emission combustion area, with higher soot, THC, and CO emissions at high loads and a little decrease in the efficiency of combustion and thermal efficiency. However, peak PRR and soot emissions were the predominant constraints on the load expansion of gasoline LTC, and they are related to their trade-off relationship.

Kodavasal [35] studied the effects of swirl ratio, injection parameters, and boost on GCI engine at idle and small-load operations by closed-cycle CFD simulations with an 87 anti-knock index fuel. The results showed that smaller nozzle angles lead to more reactivity or easier to ignite fuel and are appropriate over a larger range of injection timings. However, for low-load operation, smaller injection pressures did not improve ignitability, and this may be because of the reduced chemical residence time due to longer injection periods. Huang et al. [64] carried out an experimental study to investigate the process of combustion and the emissions behaviors of a many-cylinder diesel engine using neat diesel, dieseline, diesel mixed with n-butanol, and dieseline mixed with n-butanol blends. The results revealed that the addition of gasoline or n-butanol increased BSFC and NOx but decreased soot. The most interesting finding in this research was that the emissions of D70B30 achieved the optimum amounts at approximately 25%EGR. Another set of numerical simulation and experiments to enhance the efficiency of fuel in CI engines using dieseline and a technology of optimization have been conducted by Lee et al. [65]. The results showed that the CA10-CA90 duration and the IMEP dropped as the gasoline ratio increased. The uniformity of the A/F mixture was improved because CA10 was retarded. The emissions of soot were reduced by a huge amount of up to 90% compared to 100% diesel operation. The NOx emissions of the dieseline were a little bit increased when the SOI was retarded near to TDC.

Experiments in a CI engine fueled with 100% diesel, gasoline-diesel (GD) blends, and gasoline-diesel-polyoxymethylene dimethyl ether (GDP) blends were done by Liu et al. [66]. The study showed that either GDP or GD have one-stage premixed HR. GDP had an earlier CA10, lower peak of PRR and COV-IMEP compared to GD. Furthermore, GD had smaller amounts of emissions of soot than neat diesel, while the blends of GDP had the lowest emissions of soot and indicated the best trade-off between nitrogen oxide-soot emissions. Another study of dieseline with MPCI was done on a light-duty 1-cylinder CI engine by Wang et al. [67]. Their results showed that dieseline can simultaneously reduce emissions of NOx and soot and also maintain or obtain a higher indicated thermal efficiency than combustion of diesel fuel. However, the emissions CO and HC were increased for the dieseline.

Yu et al. [48] conducted an experimental investigation on combustion of PPCI and MPCI fueled with five type of gasoline-like fuels: gasoline, gasoline mixed with ethylhexyl-nitrate blends (EHN), blends of dieseline, blends of gasoline and polyoxymethylene dimethyl ether (PODEn), and blends of dieselin-PODEn, designated as G, GE, GD, GP, GDP, respectively. The results indicated that the thermal efficiency in PPCI regime from the biggest to the smallest was GP, G, GDP, GE, and GD, while in MPCI regime was GP, GDP, GE, G, and GD. In PPCI regime, nitrogen oxide and emissions of soot were lower than 40 mg/kWh and 10 mg/kWh, respectively, fulfilling the Euro six emission standards. Meanwhile, nitrogen oxide emissions in MPCI regime were acceptable for Euro six. CFD closed-cycle simulations have been done by Kodavasal [47] on GCI combustion regime with a sector mesh to study the influences of model scenarios on the simulation results. PRR and HRRs are exceeding the values predicted by the simulation, as are nitrogen oxide, carbon monoxide, and hydrocarbon emissions. The exceeding predicted HRR and nitrogen oxide might be because of the good mixing assumption for the simulation of combustion.

The most comprehensive study about application of gasoline fuel in the CI engines was conducted by Reitz and Duraisamy using the RCCI combustion strategy [68]. However, RCCI is a little different from GCI combustion. RCCI involves blending a low reactivity fuel (gasoline) and a high reactivity fuel (diesel) using a dual fuel strategy which uses two types of fuel injection system with different injectors and separate fuel tanks. The first fuel system is for gasoline fuel with a port injection method, the second fuel system is for diesel fuel with a direct injection system. Meanwhile, GCI uses only one fuel system that is direct injection, and blends the two types of fuel in the fuel tank. GCI is simpler and easier in fuel system configuration and real application.

5.3. Potential Methods to Obtain High Efficiency and Low Emission Targets of CI Engines in GCI Modes

The literature review has revealed studies on CI engines fueled with gasoline and/or blended with other fuels running in GCI mode or a similar method. Significant progress in the GCI research field in both experimental works and simulation studies has been achieved. There are however still many areas that need further study due to the complexity of the CI engines, which operate in GCI combustion mode. The information on the combustion characteristics of CI engines when gasoline and GCI act as the main fuel and combustion mode, respectively, are still limited. An extensive review showed that most research dealt with gasoline and biodiesel as a fuel supplement/additive in CI engines.

In general, there is an apparent lack of experimental studies of CI engines fueled with gasoline-biodiesel blends. The characteristics of emissions, ignition, and combustion process of the dieseline blends that greatly influence the efficiency of CI engine are well understood, however, the ignition, combustion, efficiency and emission characteristics are not well understood in a CI engine in which gasoline-biodiesel fuel is used.

Fuel properties and formation play an important function in the combustion process. An experimental and simulation study on the gasoline-diesel blend mixing strategy inside the cylinder would be a good contribution to elucidate combustion in GCI engines. Some simulation results on combustion processes display good agreement with the experimental ones. However, many areas remain unaddressed by the previous simulation and experimental studies. Studies on gasoline-biodiesel blends as alternatives for CI engines are very rare (especially for experimental work). Combustion process and emissions analysis on GCI combustion fueled using gasoline-biodiesel blends based on experimental work would give a better foundation for efforts to enhance CI engine efficiency.

To fill these gaps, some sequential experimental and/or simulation works can be conducted to identify the important parameters and phenomena related to the combustion process and emission behaviors of a CI engine fueled with gasoline-biodiesel blends to improve its efficiency. The literature review above can provide guidance in arranging, planning and proposing an experimental procedure and hypothesis to reach high efficiency and low emissions targets for CI engines fueled with gasoline-biodiesel blends [67,69–98] A brief explanation on how to obtain the targets is summarized in the effect flowchart shown in Figure 2.

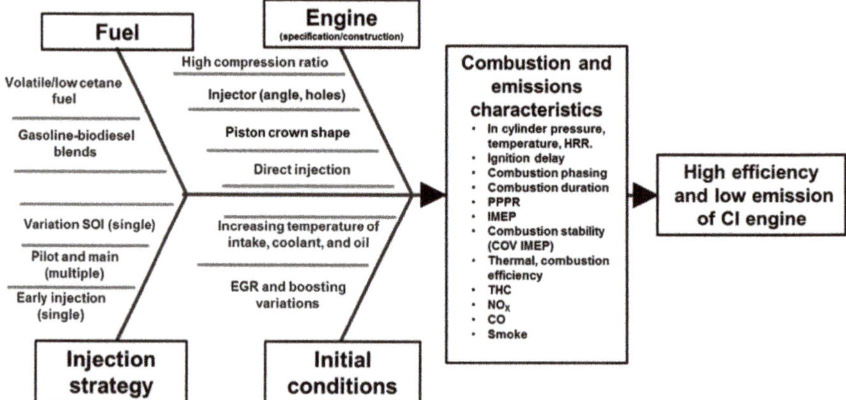

Figure 2. Flowchart of the potential strategies to obtain high efficiency and low emission CI engines fueled with gasoline-biodiesel blends.

5.4. Studies on GCI Engines Fueled Using Gasoline and Biodiesel Blends

Since early 2014, the Smart Powertrain Laboratory at the University of Ulsan has conducted several studies, including experimental and simulation activities, for potentially blending gasoline fuel with certain concentrations of biodiesel to obtain good combustion process and behaviors of emissions in a CI engine. Our specific achievements are detailed in the following subsections.

5.4.1. Gasoline-Biodiesel Preparation and Properties during Storage

Blending biodiesel and gasoline is known to have stability issues because of large density difference. In the case of low quality of biodiesel which has a high water content the stability issue should be solved. As studied previously emulsions of water and fuel (oil-based) display many phenomena related to their stability [99–104]. Therefore, further study of the properties of the blends, especially their stability due to the large density difference between gasoline and biodiesel is necessary. Currently, the maximum suitable biodiesel blend in petroleum fuel in term of temperature properties is 20% [21]. Other researchers have suggested that a petroleum based fuel consisting of 25% biodiesel was indicated to be the best-suited blend for an engine without heating and without any engine modification [76]. Meanwhile, Tinprabath [77] reported that fuel blends with biodiesel below 5% do not affect the cold flow properties.

As long as fuel grade (very low and near to zero water content) biodiesel is used, gasoline-biodiesel blends can be prepared by a simply mixing or shaking process for about 2–10 min to produce homogeneous fuel blends, which then must be immediately used in the engine experiments to avoid fuel stratification [105]. However, the fuel stratification during storage must be analyzed to understand the characteristics of the blends in real applications. Thongchai and Lim [106] prepared gasoline-biodiesel blends from 5% to 20% of biodiesel content and analyzed the phase stability and the properties during storage. Usually, phase separation occurred at low temperatures and long periods of storage. As shown in Figure 3, the ternary phase diagram indicates that no phase separation occurred for all gasoline-biodiesel blend ratios even though they were stored at a low ambient temperature from 20 °C to 10 °C, around 1 h, and for a long period of time at 25 °C, for around 45 days [106]. Figure 4 shows the clear appearance of the gasoline-biodiesel blends without phase separation [106]. The physical properties of gasoline, biodiesel, diesel, and gasoline-biodiesel blends are presented in Table 2. Meanwhile, the explanation that indicates that a 5% biodiesel content in gasoline-biodiesel blends fulfills the lubricity requirements for CI engine fuel systems can be seen in Figure 5.

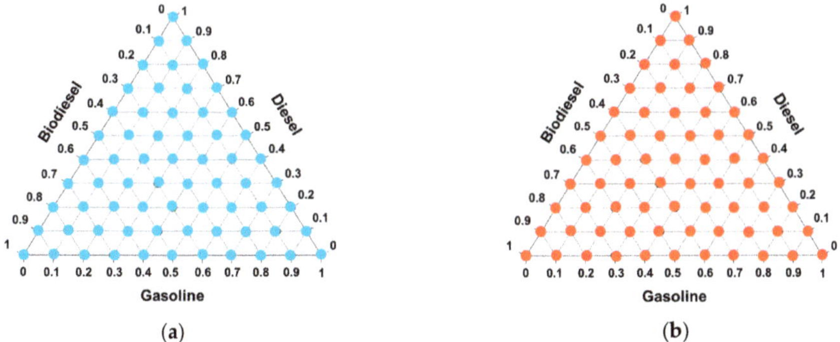

Figure 3. Ternary phase diagram of various ratio of gasoline-diesel-biodiesel blends stored at (**a**) 10–20 °C for 1 h and (**b**) 25 °C for 45 days (Adapted from [106]).

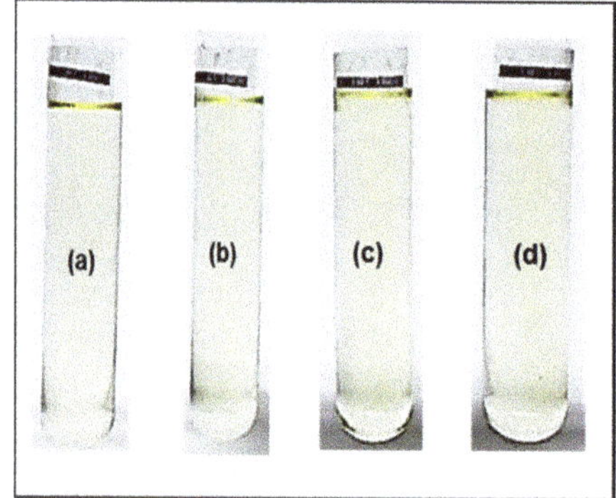

Figure 4. The appearance of gasoline-biodiesel blended fuels (GB) with biodiesel concentrations of (**a**) 20%, (**b**) 15%, (**c**) 10% and (**d**) 5% (Adapted from [106]).

Table 2. Physical properties of the fuels.

Test Item	Unit	Test Method	Gasoline	GB05	GB10	GB15	GB20	B100	Diesel
Heating value	MJ/kg	ASTM D240:2009	45.86	45.32	44.92	44.57	43.6	39.79	45.93
Kinematic Viscosity (40 °C)	Mm^2/s	ISO 3104:2008	0.735	-	-	-	-	4.229	2.798
Lubricity	μm	ISO 12156-1:2012	548	290	282	252	236	189	238
Cloud Point	°C	ISO 3015:2008	−57	−37	−32	−20	−16	3	−5
Pour Point	°C	ASTM D6749:2002	−57	−57	−57	−57	−57	1	−9
Density (15 °C)	kg/m^3	ISO 12185:2003	712.7	722.3	732.2	742.6	757.1	882.3	826.3

5.4.2. Spray Behaviors of Gasoline-Biodiesel Blending Fuel

In LTC and or GCI engine combustion, the main requirement of these concepts is the availability of a homogeneous fuel–air mixture before the start of combustion or prevention of locally fuel-rich mixture regions and the reduction of the in-cylinder combustion temperature. Because GCI uses the direct injection method, therefore the perfection of the spray breakup, evaporation, and mixing processes are most critical in obtaining the optimum combustion and emission performance. The increased surface area of a finely atomized spray enhances fuel evaporation and combustion rate, and the distribution and concentration of fuel vapor directly affect the combustion efficiency and emissions. In other word, the fuel atomization characteristics are very important parameters in a GCI engine because an increase in the number of small droplets means an increase of droplet surface area from the same volume of injected fuel spray, and optimum of the heat and energy transfer can be efficiently achieved through the droplet surface. Thus, how to control and obtain the efficiency of the atomization characteristics and process are very relevant to LTC and GCI concepts. Especially in GCI, due to the direct injection, controlling and managing spray characteristics to obtain efficient atomization is very relevant.

The spray behaviors of the non-vaporizing transient of gasoline and biodiesel blended into gasoline for 10%, 20% and 40% by volume under low-load conditions and various injection pressures were studied by using a constant volume chamber (CVC) [107]. The experimental CVC system can be seen in Figure 6 and a detailed explanation about the system can be found in the previous studies by Kanti et al. [107], and Thongchai and Lim [106]. The main finding is that 100% gasoline indicates significant atomization characteristics and increased gasoline fraction in gasoline-biodiesel blends lead to better quality of fuel and air mixes because of the atomization behaviors as indicated by the low Ohnesorge and high Reynolds number. The Ohnesorge diagram for the fuels in 1000 g/m^3 and 20,000 g/m^3 ambient gas density in 400, 800 and 1200 bar common-rail pressure can be seen in Figure 7.

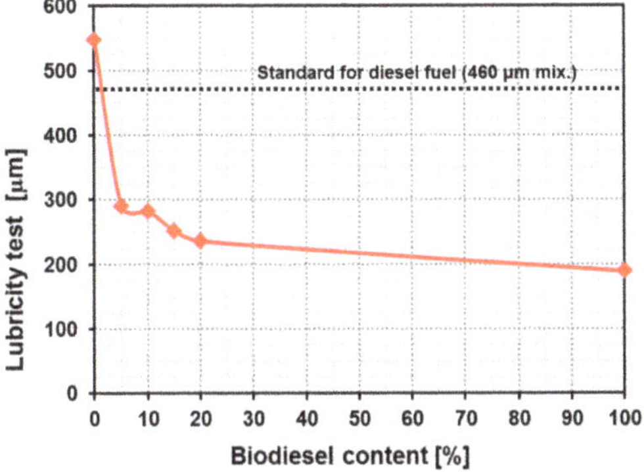

Figure 5. Effect of biodiesel concentration on gasoline- biodiesel lubricity.

The pressure of injection and period are also key combustion and behaviors of emission of GCI engines. The influence of the pressure of injection variations and period of injection on the fuel quantity when using different gasoline-biodiesel blends were investigated and compared [108]. The increase of injection pressure produces a bigger injection quantity for the fuels tested as seen in Figure 8 [108]. At each injection pressure, pure diesel (D100) results in a little bit smaller injection quantity than the gasoline-biodiesel blends. This condition indicates that a longer injection period is needed to deliver the same quantity of fuel when using conventional diesel fuel relative to the gasoline blends. Furthermore, when the quantity of biodiesel blended with gasoline is higher, the injection quantity is decreased (D100 < GB20 < GB10 < GB05). However, a longer injection period has no effect on the injection quantity for various fuels because a long injection duration enables to the work in a quasi-steady state flow duration in which the various densities or viscosities do not influence the amount of injected fuel [77].

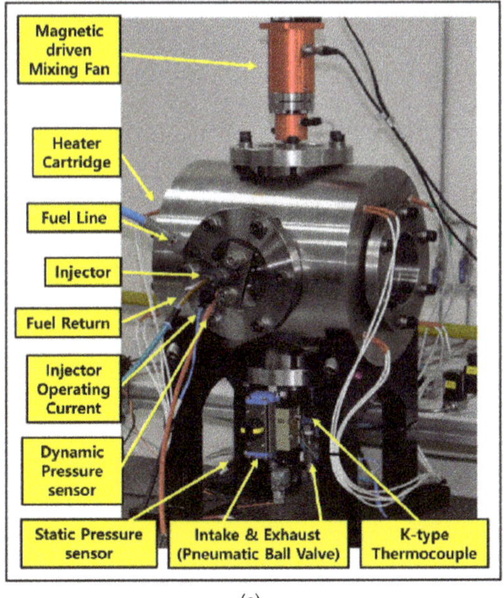

(**a**)

Figure 6. *Cont.*

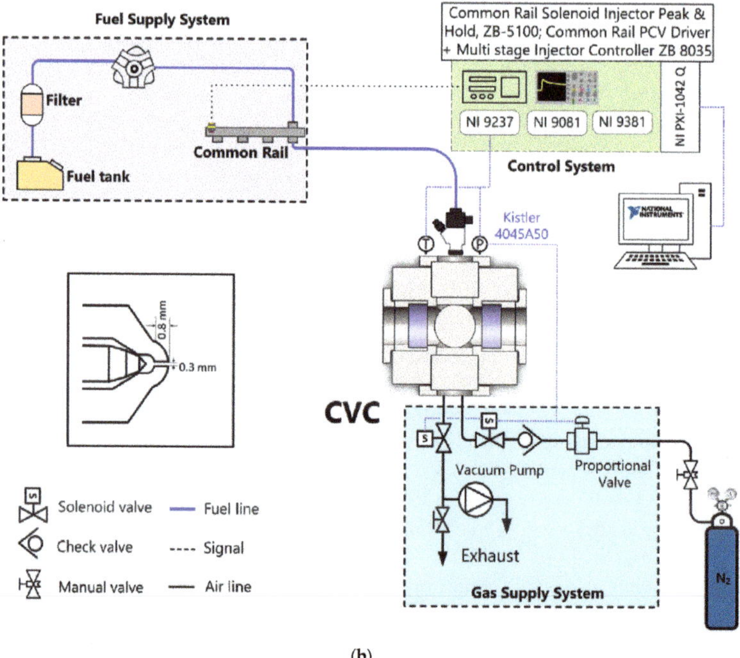

(b)

Figure 6. Experimental systems. (**a**) Pictorial representation of constant volume chamber (CVC), (**b**) Schematic diagram of the experimental system [Reprinted from Fuel Processing Technology, Vol. 178, Shubhra Kanti Das, Kihyun Kim, Ocktaeck Lim, Experimental study on non-vaporizing spray characteristics of biodiesel-blended gasoline fuel in a constant volume chamber, Copyrights (2018); with permission from Elsevier].

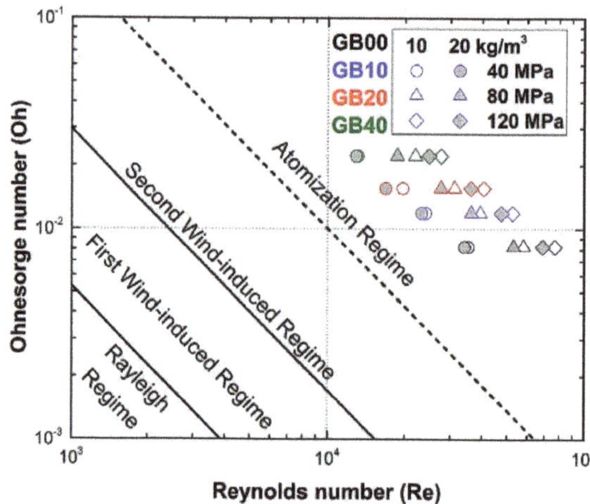

Figure 7. Ohnesorge diagram for all fuels under 10 kg/m^3 and 20 kg/m^3 ambient gas densities under 40, 80 and 120 MPa rail pressures [Reprinted from Fuel Processing Technology, Vol. 178, Shubhra Kanti Das, Kihyun Kim, Ocktaeck Lim, Experimental study on non-vaporizing spray characteristics of biodiesel-blended gasoline fuel in a constant volume chamber, Copyright (2018); with permission from Elsevier].

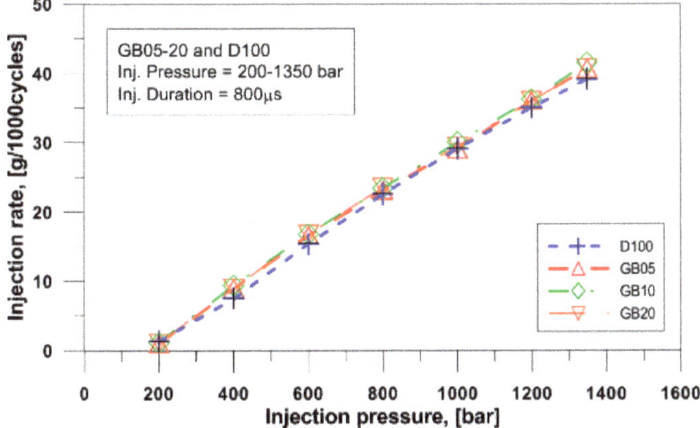

Figure 8. The effect of injection pressure on injection flow rate of a gasoline-biodiesel blend [Reprinted by permission from (the RightsLink Permissions Springer Nature Customer Service Centre GmbH): (Springer Nature) (Journal of Mechanical Science and Technology) [108], Copyright (2018)].

5.4.3. Auto Burning Behaviors of a Blended of Gasoline-Biodiesel

The biodiesel content effects of on the ignitability for gasoline–biodiesel blends were investigated using a rapid compression expansion machine (RCEM) [109]. The main finding was that a low fraction of biodiesel can enhance the autoignition behavior of gasoline as can be observed in Figure 9 [109]. The gasoline-biodiesel mixed fuel with bigger biodiesel content indicated a faster CA10 and no combustion at a lower temperature injection. Misfiring was detected at ambient temperatures at injection timings lower than 522 °C, 512 °C, 492 °C, 477 °C, and 559 °C for GB5%, GB10%, GB15%, GB20%, and G100%, respectively. A gasoline-biodiesel blended fuel with a bigger biodiesel ratio exhibited a smaller temperature of no combustion condition. This is the reason that biodiesel has a higher cetane number than gasoline and biodiesel.

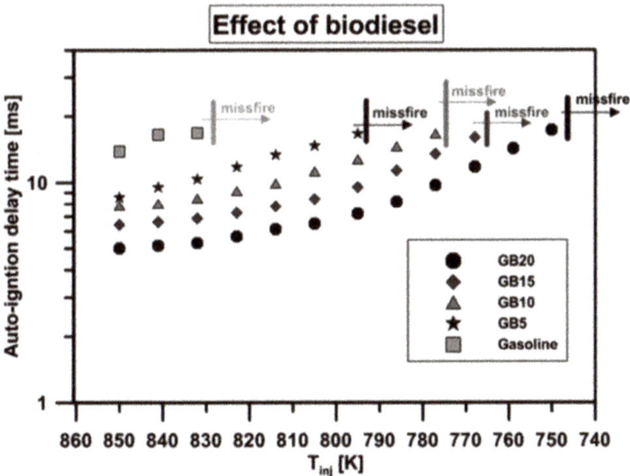

Figure 9. Effect of the biodiesel fraction on auto-ignition [Reproduced with permission from Vu, D.N.; Das, S.K.; Jwa, K.; Lim, O., Proc. Inst. Mech. Eng. Part D J. Authomob. Eng.; published by SAGE publication, 2018.].

5.4.4. Single Injection Mode of Gasoline-Biodiesel Blend

An experimental study on the engine combustion process and exhaust emission characteristics in an effort to simultaneously increase the efficiency of engines and reduce the exhaust emissions such as HC and CO, using a single injection strategy of GCI fueled with gasoline-biodiesel blends was conducted previously [105]. For every variation of start of injection of biodiesel blend in gasoline of the GCI engine, the IMEP was similar to 100% diesel. In addition, the combustion quality, which is represented by the variability coefficient of IMEP (COV-IMEP), indicated that all differences in start of injection of the biodiesel-gasoline blends reflected greater confidence compared to neat diesel. Lastly, the peak of combustion efficiency and smallest THC emission were achieved for GB20 with 40 °CA BTDC of injection timing, as can be observed in Figures 10 and 11.

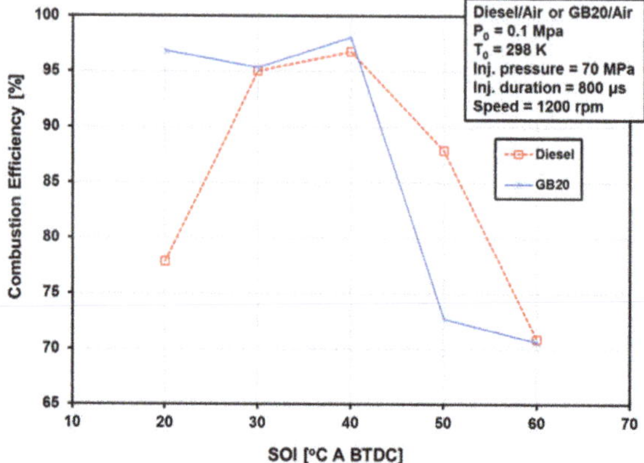

Figure 10. Combustion efficiency [Reprinted from Fuel, Vol. 189, Yanuandri Putrasari, Ocktaeck Lim, A study on combustion and emission of GCI engines fueled with gasoline-biodiesel blends, Copyright (2017); with permission from Elsevier].

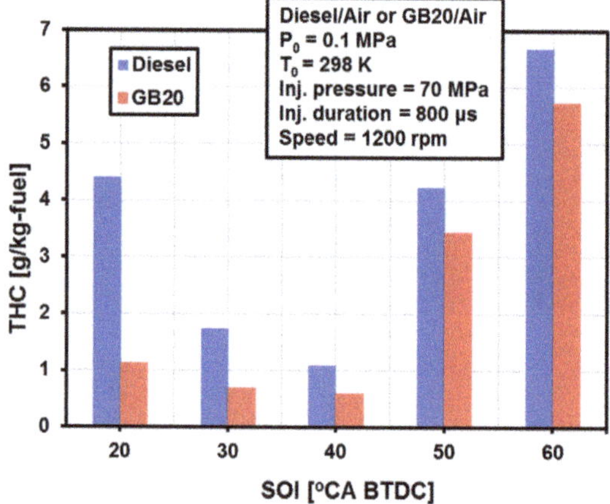

Figure 11. THC emissions [Reprinted from Fuel, Vol. 189, Yanuandri Putrasari, Ocktaeck Lim, A study on combustion and emission of GCI engines fueled with gasoline-biodiesel blends, Copyright (2017); with permission from Elsevier].

5.4.5. Double Injection Mode of Gasoline-Biodiesel Blends

GCI, where gasoline is used as a fuel in a CI engine without the use of a spark plug, combines the high efficiency associated with a CI engine with the low soot emissions associated with gasoline fuel. Further, combustion is a result of sequential auto burning without propagating flames, resulting in low temperature combustion which in turn significantly reduces NOx. The injection timing (the point in time when the fuel is injected into the engine chamber) is a main parameter determining the ignitability of the gasoline-air mixture, and thus engine operation stability. Double injection by early pilot injection and near top dead center of main injection in combination with high EGR rate is a potential way to extend the ignition delay, by causing the formation of a homogeneous mixture. As known the main concept of LTC combustion is formation of a homogeneous mixture which can be achieved by providing a sufficient premixing period. Lengthening the ignition timing is the most suitable method to obtain a homogeneous mixture because it enables sufficient time for mixing. Furthermore, extension of the ignition timing is possible by reducing the cetane number by adding a fuel with low-cetane and a high-octane number fuel such as gasoline and the relevant method is the GCI concept. Therefore, the proper combination of gasoline fuel, double injection strategy, and EGR in GCI engines can simultaneously reduce NOx and smoke emissions by enabling a sufficient air–fuel mixing period and the formation of a more homogeneous mixture.

A study on one- and many-injection strategies of GCI engine was successfully done to compare 100% gasoline with 5% added biodiesel with 100% diesel fuel to obtain the improvement of efficiency and decreased emission characteristics [110]. The findings were as follows: raising the inlet temperature of air, engine-oil, and engine-coolant and use of a multiple injections method could lead to enhancements of the combustion and thermal efficiencies of the GCI engine. Meanwhile, the quality of combustion study using coefficient of variation of indicated mean effective pressure and cyclic variations of maximum pressure of cylinder indicated a satisfactory result. However, the analysis of cycle to cycle variation for the ignition delay showed that these factors suffered with GB05 in many injection strategies (Figure 12). The earlier pilot injection of 5% biodiesel in gasoline fuel could be an argument for lowering carbon monoxide emissions. The biodiesel fraction and gasoline (as the very vapor fuel in the GB05) exhibited meaningful effects of reducing the emissions of carbon monoxide and total hydrocarbons. The emissions of nitrogen oxide from GB05 blend for either multiple or single injections seem to be higher than that of neat diesel fuel with many injections and even bigger than pure diesel fuel with a single injection method. It is believed to be because of the oxygen fraction in the gasoline-biodiesel blend fuel.

5.4.6. EGR and Intake Boosting of Gasoline-Biodiesel Blend

A study on a GCI engine was conducted by adding 5% biodiesel into gasoline and comparing the results with those from neat diesel in single-injection (PPCI) and multiple-injection (MPCI) modes combined with the application of EGR and intake boosting with the goal of obtaining high efficiency and low emissions.

The most important finding was that using 0.12 MPa of intake boosting reduced the NOx emissions with the gasoline-biodiesel blend by almost half compared to using 0.1 MPa intake boosting. Then, increasing the intake boost significantly reduced the smoke emissions with pure diesel fuel, but increasing the intake boost from ambient pressure to 0.12 MPa increased the smoke emissions for the gasoline-biodiesel blend. A summary of these finding can be seen in Figures 13 and 14.

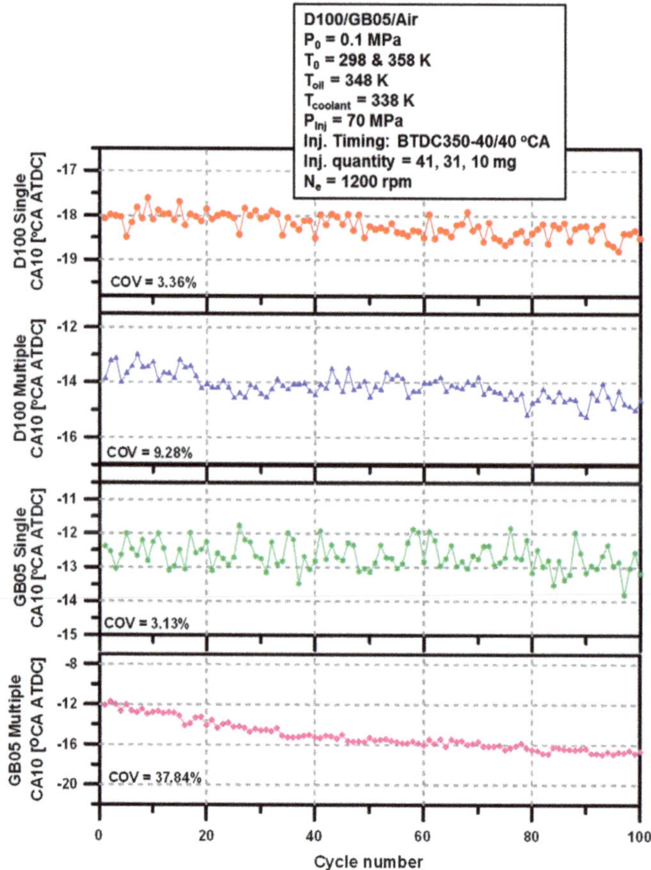

Figure 12. Cycle-by-cycle variations in CA10 under different conditions [Reprinted from Fuel, Vol. 221, Yanuandri Putrasari, Ocktaeck Lim, A study on combustion and emission of GCI engines fueled with gasoline-biodiesel blends, Copyright (2017); with permission from Elsevier.]

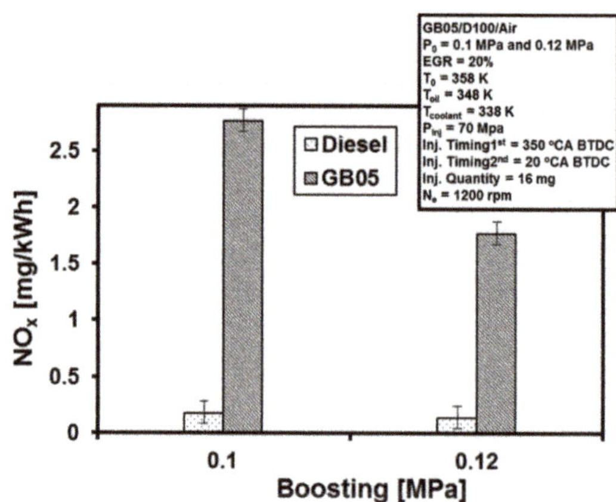

Figure 13. Effect of intake boosting on NOx emissions in MPCI mode.

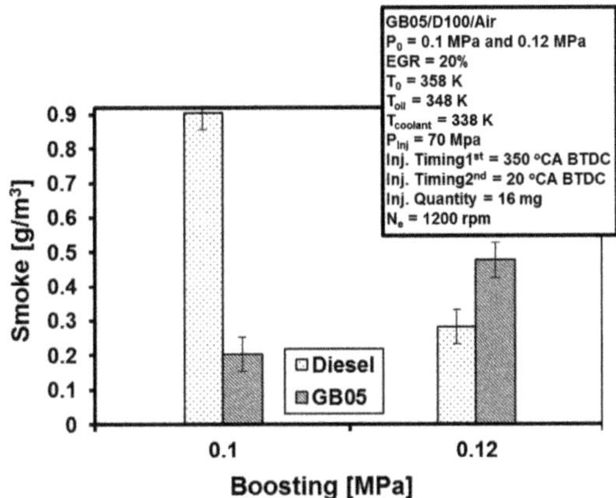

Figure 14. Effect of intake boosting on smoke emissions in MPCI mode.

6. Conclusions and Recommendations

6.1. Conclusions

This study has provided a review of studies on GCI engines fueled with biodiesel-gasoline blends in order to increase engine performance and decrease exhaust emissions. Based on a comprehensive review of the open literature, the present understanding of GCI engines fueled with biodiesel-gasoline blends research work includes the following contributions:

- A literature study of GCI engines potentially fueled with biodiesel-gasoline blends.
- The study on the combustion process and exhaust emission behaviors of gasoline CI engines using various injection timings in a single injection mode fueled with biodiesel-gasoline blends.
- The study about the effects of 5 and 20% biodiesel-gasoline blends on the efficiency and exhaust emissions in gasoline compression ignition engines under single injection mode.
- A study of the combustion and exhaust emissions of a gasoline CI engine fueled with biodiesel-gasoline through a double injection method which comprises of pilot and main injection.
- Experimental investigation of the combustion process and emission behaviors of CI engines fueled with biodiesel-gasoline blends in the early injection HCCI mode.
- An investigation on the effect of EGR and inlet boosting on the process of combustion and emission behaviors of a GCI engine using gasoline-biodiesel-blends.

6.2. Recommendations

Some valuable insights and new ideas were identified based on an experimental series and extensive investigation into GCI engines fueled with a mixture of gasoline and biodiesel. These have not been pursued more fully further within the current research. The following suggestions are provided for future research:

- Use of hot EGR to raise the inlet temperature and a smaller fuel injection pressure (to minimize too much mixing and cylinder wall-wet/impingement of fuel) is required for the GCI engine concept of volatile/vaporizable/less-reactivity and also low cetane-number fuels at smaller engine speeds or engine operating loads.
- Several issues related to the emission and combustion behaviors of gasoline CI engines using blends of biodiesel in gasoline need to be studied further, including stability under low and middle load conditions, cold start and idle conditions, operation in acceptable transient conditions, noise/engine sound quality and in-cylinder PRR at middle and big loads through injection of

fuel methods, exhaust emissions (especially HC, carbon monoxide, NOx and smoke emissions control), LTO and diesel particulate filters (DPFs).

- Several issues related to hardware optimization of gasoline CI engines using a mixing of gasoline-biodiesel also need to be studied further. In this case they include: the combustion chamber/cylinder-head/piston-crown design, injectors holes and angles, injection arrangement and injection method systems, cooled and uncooled EGR, turbochargers combined with superchargers or boosting to obtain higher intake pressures at big EGR ratio, subsequent/exhaust treatment, and quality of fuel (lubrication behavior, viscosity, and detergent-like properties). Thus, advanced additive technology has to be implemented for the various conditions come upon in gasoline compression ignition engines.

- More significant experimental, simulation and development studies and work are required to push gasoline compression ignition engine technology to the step of real application.

Author Contributions: Y.P. carried out the development of the concept of the paper, performed the literature study, prepare the paper structure, aim, objectives, research question, analyzed the data, and as well as writing the paper. O.L. supervised the research, advised on the research gap and objective, proofread the paper, and guided the writing process as well as reviewing the presented concepts and outcomes.

Funding: This research was supported by The Leading Human Resource Training Program of Regional Neo Industry through the National Research Foundation of Korea (NRF) funded by The Ministry of Science, ICT and Future Planning (2016H1D5A1908826).

Acknowledgments: Yanuandri Putrasari acknowledges support from the Program for Research and Innovation in Science and Technology (RISET-Pro) Ref. No. 61/RISET-Pro/FGS/III/2017, Ministry of Research, Technology and Higher Education, Republic of Indonesia

Conflicts of Interest: The authors declare no conflict of interest.

References

1. Tutak, W.; Lukacs, K.; Szwaja, S.; Bereczky, A. Alcohol-diesel fuel combustion in the compression ignition engine. *Fuel* **2015**, *154*, 196–206. [CrossRef]

2. Dec, J.E. Advanced compression-ignition engines—Understanding the in-cylinder processes. *Proc. Combust. Inst.* **2009**, *32*, 2727–2742. [CrossRef]

3. Lu, X.; Han, D.; Huang, Z. Fuel design and management for the control of advanced compression-ignition combustion modes. *Prog. Energy Combust. Sci.* **2011**, *37*, 741–783. [CrossRef]

4. Kalghatgi, G.T.; Risberg, P.; Ångström, H. *Advantages of Fuels with High Resistance to Auto-Ignition in Late-Injection, Low-Temperature, Compression Ignition Combustion*; SAE Technical Paper 2006-01-3385; SAE International: Warrendale, PA, USA, 2006. [CrossRef]

5. Won, H.W.; Peters, N.; Pitsch, H.; Tait, N.; Kalghatgi, G. *Partially Premixed Combustion of Gasoline Type Fuels Using Larger Size Nozzle and Higher Compression Ratio in a Diesel Engine*; SAE Technical Paper; SAE International: Warrendale, PA, USA, 2013; Volume 11. [CrossRef]

6. Ra, Y.; Yun, J.E.; Reitz, R.D. Numerical Parametric Study of Diesel Engine Operation with Gasoline. *Combust. Sci. Technol.* **2009**, *181*, 350–378. [CrossRef]

7. Ra, Y.; Loeper, P.; Reitz, R.; Andrie, M.; Krieger, R.; Foster, D.; Durrett, R.; Gopalakrishnan, V.; Plazas, A.; Peterson, R.; et al. Study of High Speed Gasoline Direct Injection Compression Ignition (GDICI) Engine Operation in the LTC Regime. *SAE Int. J. Engines* **2011**, *4*, 1412–1430. [CrossRef]

8. Ra, Y.; Loeper, P.; Andrie, M. Gasoline DICI Engine Operation in the LTC Regime Using Triple-Pulse Injection. *SAE Int. J. Engines* **2012**, 1109–1132. [CrossRef]

9. Sellnau, M.; Foster, M.; Hoyer, K.; Moore, W.; Sinnamon, J.; Husted, H. Development of a Gasoline Direct Injection Compression Ignition (GDCI) Engine. *SAE Int. J. Engines* **2014**, *7*, 835–851. [CrossRef]

10. Han, D.; Duan, Y.; Wang, C.; Lin, H.; Huang, Z. Experimental study on the two stage injection of diesel and gasoline blends on a common rail injection system. *Fuel* **2015**, *159*, 470–475. [CrossRef]

11. Benajes, J.; Broatch, A.; Garcia, A.; Monico Muñoz, L. *An Experimental Investigation of Diesel-Gasoline Blends Effects in a Direct-Injection Compression-Ignition Engine Operating in PCCI Conditions*; SAE Technical Paper 2013-01-1676; SAE International: Warrendale, PA, USA, 2013. [CrossRef]

12. Adams, C.A.; Loeper, P.; Krieger, R.; Andrie, M.J.; Foster, D.E. Effects of biodiesel-gasoline blends on gasoline direct-injection compression ignition (GCI) combustion. *Fuel* **2013**, *111*, 784–790. [CrossRef]

13. Zelenyuk, A.; Reitz, P.; Stewart, M.; Imre, D.; Loeper, P.; Adams, C.; Andrie, M.; Rothamer, D.; Foster, D.; Narayanaswamy, K.; et al. Detailed characterization of particulates emitted by pre-commercial single-cylinder gasoline compression ignition engine. *Combust. Flame* **2014**, *161*, 2151–2164. [CrossRef]

14. Zhang, F.; Rezaei, S.Z.; Xu, H.; Shuai, S.-J. Experimental Investigation of Different Blends of Diesel and Gasoline (Dieseline) in a CI Engine. *SAE Int. J. Engines* **2014**, *7*, 1920–1930. [CrossRef]

15. Putrasari, Y.; Praptijanto, A.; Budi, W.; Lim, O. Resources, policy, and research activities of biofuel in Indonesia: A review. *Energy Rep.* **2016**, *2*, 237–245. [CrossRef]

16. Salvi, B.L.; Panwar, N.L. Biodiesel resources and production technologies—A review. *Renew. Sustain. Energy Rev.* **2012**, *16*, 3680–3689. [CrossRef]

17. Hassan, M.H.; Kalam, M.A. An overview of biofuel as a renewable energy source: Development and challenges. *Procedia Eng.* **2013**, *56*, 39–53. [CrossRef]

18. Yang, B.; Li, S.; Zheng, Z.; Yao, M.; Cheng, W. *A Comparative Study on Different Dual-Fuel Combustion Modes Fuelled with Gasoline and Diesel*; SAE International: Warrendale, PA, USA, 2012; Volume 0694. [CrossRef]

19. Shi, Y.; Reitz, R.D. Optimization of a heavy-duty compression-ignition engine fueled with diesel and gasoline-like fuels. *Fuel* **2010**, *89*, 3416–3430. [CrossRef]

20. Rose, K.D.; Ariztegui, J.; Cracknell, R.F.; Dubois, T.; Hamje, H.D.C.; Pellegrini, L.; Rickeard, D.J.; Heuser, B.; Schnorbus, T.; Kolbeck, A.F. *Exploring a Gasoline Compression Ignition (GCI) Engine Concept*; SAE International: Warrendale, PA, USA, 2013. [CrossRef]

21. Misra, R.D.; Murthy, M.S. Blending of additives with biodiesels to improve the cold flow properties, combustion and emission performance in a compression ignition engine—A review. *Renew. Sustain. Energy Rev.* **2011**, *15*, 2413–2422. [CrossRef]

22. Kweon, C.; Okada, S.; Foster, D.; Bae, M.; Schauer, J. Effect of engine operating conditions on particle-phase organic compounds in engine exhaust of a heavy-duty direct-injection (DI) diesel engine. *SAE Trans.* **2003**, *112*, 460–476.

23. Kweon, C.; Okada, S.; Stetter, J.; Christenson, C.; Shafer, M.; Schauer, J.; Foster, D. *Effects of Fuel Composition on Combustion and Detailed Chemical/Physical Characteristics of Diesel Exhaust*; SAE International: Warrendale, PA, USA, 2003; Volume 22. [CrossRef]

24. Bae, C.; Kim, J. Alternative fuels for internal combustion engines. *Proc. Combust. Inst.* **2017**, *36*, 3389–3413. [CrossRef]

25. Tesfa, B.; Mishra, R.; Zhang, C.; Gu, F.; Ball, A.D. Combustion and performance characteristics of CI (compression ignition) engine running with biodiesel. *Energy* **2013**, *51*, 101–115. [CrossRef]

26. Wang, Z.; Li, L.; Wang, J.; Reitz, R.D. Effect of biodiesel saturation on soot formation in diesel engines. *Fuel* **2016**, *175*, 240–248. [CrossRef]

27. Cordiner, S.; Mulone, V.; Nobile, M.; Rocco, V. Impact of biodiesel fuel on engine emissions and Aftertreatment System operation. *Appl. Energy* **2016**, *164*, 972–983. [CrossRef]

28. Rakopoulos, C.D.; Rakopoulos, D.C.; Hountalas, D.T.; Giakoumis, E.G.; Andritsakis, E.C. Performance and emissions of bus engine using blends of diesel fuel with bio-diesel of sunflower or cottonseed oils derived from Greek feedstock. *Fuel* **2008**, *87*, 147–157. [CrossRef]

29. Pulkrabek Willard, W. *Engineering Fundamentals of Internal Combustion Engine*, 2nd ed.; Pearson Prentice Hall: Upper Saddle River, NJ, USA, 2003.

30. Kalghatgi, G. *Fuel/Engine Interactions*; Society of Automotive Engineers: Wallendale PA USA, 2014.

31. Cracknell, R.; Ariztegui Cortijo, J.; Dubois, T.; Engelen, B.; Manuelli, P.; Pellegrini, L.; Williams, J.; Deppenkemper, K.; Graziano, B.; Heufer, K.A.; et al. *Modelling a Gasoline Compression Ignition (GCI) Engine Concept*; SAE Technical Paper 2014-01-1305; SAE International: Warrendale, PA, USA, 2014; pp. 1–54. [CrossRef]

32. Lu, X.; Qian, Y.; Yang, Z.; Han, D.; Ji, J.; Zhou, X.; Huang, Z. Experimental study on compound HCCI (homogenous charge compression ignition) combustion fueled with gasoline and diesel blends. *Energy* **2014**, *64*, 707–718. [CrossRef]

33. Han, D.; Ickes, A.M.; Bohac, S.V.; Huang, Z.; Assanis, D.N. HC and CO emissions of premixed low-temperature combustion fueled by blends of diesel and gasoline. *Fuel* **2012**, *99*, 13–19. [CrossRef]

34. Feng, Z.; Zhan, C.; Tang, C.; Yang, K.; Huang, Z. Experimental investigation on spray and atomization characteristics of diesel/gasoline/ethanol blends in high pressure common rail injection system. *Energy* **2016**, *112*, 549–561. [CrossRef]

35. Kodavasal, J.; Kolodziej, C.P.; Ciatti, S.A. Effects of injection parameters, boost, and swirl ratio on gasoline compression ignition operation at idle and low-load conditions. *Int. J. Engine Res.* **2016**, *18*, 824–836. [CrossRef]

36. Sim, J.; Elwardany, A.; Jaasim, M. Numerical Simulations of Hollow-Cone Injection and Gasoline Compression Ignition Combustion With Naphtha Fuels. *J. Energy Resour. Technol.* **2017**, *138*, 052202. [CrossRef]

37. Zhong, S.; Wyszynski, M.L.; Megaritis, A.; Yap, D.; Xu, H. Experimental Investigation into HCCI Combustion Using Gasoline and Diesel Blended Fuels. *SAE Int. J. Engines* **2005**. [CrossRef]

38. Leermakers, C.A.J.; Van den Berge, B.; Luijten, C.C.M.; Somers, L.M.T.; de Goey, L.P.H.; Albrecht, B.A. *Gasoline-Diesel Dual Fuel: Effect of Injection Timing and Fuel Balance*; SAE Technical Paper 2011-01-2437; SAE International: Warrendale, PA, USA, 2011. [CrossRef]

39. Prikhodko, V.Y.; Curran, S.J.; Barone, T.L.; Lewis, S.A.; Storey, J.M.; Cho, K.; Wagner, R.M.; Parks, J.E. *Emission Characteristics of a Diesel Engine Operating with In-Cylinder Gasoline and Diesel Fuel Blending*; SAE International: Warrendale, PA, USA, 2010; Volume 2266, pp. 946–955.

40. Curran, S.; Prikhodko, V.; Cho, K.; Sluder, C.; Parks, J.; Wagner, R.; Kokjohn, S.; Reitz, R. *In-Cylinder Fuel Blending of Gasoline/Diesel for Improved Efficiency and Lowest Possible Emissions on a Multi-Cylinder Light-Duty Diesel Engine*; SAE Technical Paper 2010-01-2206; SAE International: Warrendale, PA, USA, 2010; pp. 1–20. [CrossRef]

41. Lawler, B.; Splitter, D.; Szybist, J.; Kaul, B. Thermally Stratified Compression Ignition: A new advanced low temperature combustion mode with load flexibility. *Appl. Energy* **2017**, *189*, 122–132. [CrossRef]

42. Loeper, P.; Ra, Y.; Foster, D.; Ghandhi, J. Experimental and computational assessment of inlet swirl effects on a gasoline compression ignition (GCI) light-duty diesel engine. In Proceedings of the SAE 2014 World Congress and Exhibition, Detroit, MI, USA, 8–10 April 2014; Volume 1. [CrossRef]

43. Agarwal, A.K.; Singh, A.P.; Maurya, R.K. Evolution, challenges and path forward for low temperature combustion engines. *Prog. Energy Combust. Sci.* **2017**, *61*, 1–56. [CrossRef]

44. Krasselt, J.; Foster, D.; Ghandhi, J.; Herold, R.; Reuss, D.; Najt, P. Investigations into the Effects of Thermal and Compositional Stratification on HCCI Combustion—Part I: Metal Engine Results. *SAE Int. J. Engines* **2009**, *2*, 1034–1053. [CrossRef]

45. Kalghatgi, G.T.; Risberg, P.; Angstrom, H.-E. *Partially Pre-Mixed Auto-Ignition of Gasoline to Attain Low Smoke and Low NOx at High Load in a Compression Ignition Engine and Comparison with a Diesel Fuel*; SAE Technical Paper 2007-01-0006; SAE International: Warrendale, PA, USA, 2007. [CrossRef]

46. Loeper, P.; Ra, Y.; Adams, C.; Foster, D.; Ghandhi, J.; Andrie, M.; Krieger, R.; Durrett, R. Experimental investigation of light-medium load operating sensitivity in a gasoline compression ignition (GCI) light-duty diesel engine. In Proceedings of the SAE 2013 World Congress and Exhibition, Detroit, MI, USA, 16–18 April 2013; Volume 2. [CrossRef]

47. Kodavasal, J.; Kolodziej, C.P.; Ciatti, S.A.; Sibendu, S. Computational Fluid Dynamics Simulation of Gasoline Compression Ignition. *J. Energy Resour. Technol.* **2017**, *137*, 032212. [CrossRef]

48. Yu, L.; Shuai, S.; Li, Y.; Li, B.; Liu, H.; He, X.; Wang, Z. An experimental investigation on thermal efficiency of a compression ignition engine fueled with five gasoline-like fuels. *Fuel* **2017**, *207*, 56–63. [CrossRef]

49. Zhou, L.; Boot, M.D.; De Goey, L.P.H. *Gasoline—Ignition Improver—Oxygenate Blends as Fuels for Advanced Compression Ignition Combustion*; SAE Technical Paper; SAE International: Warrendale, PA, USA, 2013; Volume 2. [CrossRef]

50. Doornbos, G.; Somhorst, J.; Boot, M. *Literature Study and Feasibility Test Regarding a Gasoline/EHN Blend Consumed by Standard CI-Engine Using a Non-PCCI Combustion Strategy*; SAE Technical Paper; SAE International: Warrendale, PA, USA, 2013. [CrossRef]

51. Weall, A.; Collings, N. *Investigation into Partially Premixed Combustion in a Light-Duty Multi-Cylinder Diesel Engine Fuelled Gasoline and Diesel with a Mixture of Gasoline and Diesel*; SAE Technical Paper 2007-01-4058; SAE International: Warrendale, PA, USA, 2007. [CrossRef]

52. Şahin, Z.; Durgun, O.; Bayram, C. Experimental investigation of gasoline fumigation in a single cylinder direct injection (DI) diesel engine. *Energy* **2008**, *33*, 1298–1310. [CrossRef]

53. Manente, V.; Johansson, B.; Cannella, W. Gasoline partially premixed combustion, the future of internal combustion engines? *Int. J. Engine Res.* **2011**, *12*, 194–208. [CrossRef]

54. Cnr, I.M.; Corcione, F.; Valentino, G.; Tornatore, C.; Merola, S.; Marchitto, L. *Optical Investigation of Premixed Low-Temperature Combustion of Lighter Fuel Blends in Compression Ignition Engines*; SAE Technical Paper 2011-24-0045; SAE International: Warrendale, PA, USA, 2011. [CrossRef]

55. Zhang, F.; Xu, H.; Zhang, J.; Tian, G.; Kalghatgi, G. Investigation into Light Duty Dieseline Fuelled Partially-Premixed Compression Ignition Engine. *SAE Int. J. Engines* **2011**, *4*, 2124–2134. [CrossRef]

56. Yang, H.; Shuai, S.; Wang, Z.; Wang, J. Fuel octane effects on gasoline multiple premixed compression ignition (MPCI) mode. *Fuel* **2013**, *103*, 373–379. [CrossRef]

57. Kim, K.; Kim, D.; Jung, Y.; Bae, C. Spray and combustion characteristics of gasoline and diesel in a direct injection compression ignition engine. *Fuel* **2013**, *109*, 616–626. [CrossRef]

58. Thoo, W.J.; Kevric, A.; Ng, H.K.; Gan, S.; Shayler, P.; La Rocca, A. Characterisation of ignition delay period for a compression ignition engine operating on blended mixtures of diesel and gasoline. *Appl. Therm. Eng.* **2014**, *66*, 55–64. [CrossRef]

59. Kolodziej, C.; Kodavasal, J.; Ciatti, S.; Som, S.; Shidore, N.; Delhom, J. *Achieving Stable Engine Operation of Gasoline Compression Ignition Using 87 AKI Gasoline Down to Idle*; SAE Technical Paper 2015-01-0832; SAE International: Warrendale, PA, USA, 2015. [CrossRef]

60. Wang, B.; Wang, Z.; Shuai, S.; Xu, H. Combustion and emission characteristics of Multiple Premixed Compression Ignition (MPCI) mode fuelled with different low octane gasolines. *Appl. Energy* **2015**, *160*, 769–776. [CrossRef]

61. Du, J.; Sun, W.; Guo, L.; Xiao, S.; Tan, M.; Li, G.; Fan, L. Experimental study on fuel economies and emissions of direct-injection premixed combustion engine fueled with gasoline/diesel blends. *Energy Convers. Manag.* **2015**, *100*, 300–309. [CrossRef]

62. Li, J.; Yang, W.M.; An, H.; Chou, S.K. Modeling on blend gasoline/diesel fuel combustion in a direct injection diesel engine. *Appl. Energy* **2015**, *160*, 777–783. [CrossRef]

63. Yang, B.; Yao, M.; Zheng, Z.; Yue, L. Experimental Investigation of Injection Strategies on Low Temperature Combustion Fuelled with Gasoline in a Compression Ignition Engine. *J. Chem.* **2015**, *2015*, 207248. [CrossRef]

64. Huang, H.; Zhou, C.; Liu, Q.; Wang, Q.; Wang, X. An experimental study on the combustion and emission characteristics of a diesel engine under low temperature combustion of diesel/gasoline/n-butanol blends. *Appl. Energy* **2016**, *170*, 219–231. [CrossRef]

65. Lee, S.; Jeon, J.; Park, S. Optimization of combustion chamber geometry and operating conditions for compression ignition engine fueled with pre-blended gasoline-diesel fuel. *Energy Convers. Manag.* **2016**, *126*, 638–648. [CrossRef]

66. Liu, H.; Wang, Z.; Wang, J.; He, X. Improvement of emission characteristics and thermal efficiency in diesel engines by fueling gasoline/diesel/PODEn blends. *Energy* **2016**, *97*, 105–112. [CrossRef]

67. Wang, B.; Wang, Z.; Shuai, S.; Wang, J. *Investigations into Multiple Premixed Compression Ignition mode Fuelled with Different Mixtures of Gasoline and Diesel*; SAE Technical Paper 2015-01-0833; SAE International: Warrendale, PA, USA, 2015. [CrossRef]

68. Reitz, R.D.; Duraisamy, G. Review of high efficiency and clean reactivity controlled compression ignition (RCCI) combustion in internal combustion engines. *Prog. Energy Combust. Sci.* **2015**, *46*, 12–71. [CrossRef]

69. Heywood, J.B. *Internal Combustion Engine Fundamentals*; McGraw-Hill: New York, NY, USA, 1988.

70. Kalghatgi, G.T. Fuel effects in CAI gasoline engines. In *HCCI and CAI Engines for the Automotive Industry*; Zhao, H., Ed.; Woodhead Publishing: Cambridge, UK, 2007; pp. 206–237.

71. Eng, J.A. *Characterization of Pressure Waves in HCCI Combustion Reprinted From: Homogeneous Charge Compression Ignition Engines*; SAE Technical Paper; SAE International: Warrendale, PA, USA, 2002; Volume 1, p. 15. [CrossRef]

72. Wang, Q.; Wang, B.; Yao, C.; Liu, M.; Wu, T.; Wei, H.; Dou, Z. Study on cyclic variability of dual fuel combustion in a methanol fumigated diesel engine. *Fuel* **2016**, *164*, 99–109. [CrossRef]

73. Maurya, R.K.; Agarwal, A.K. Experimental investigation of cyclic variations in HCCI combustion parameters for gasoline like fuels using statistical methods. *Appl. Energy* **2013**, *111*, 310–323. [CrossRef]

74. Wang, Y.; Xiao, F.; Zhao, Y.; Li, D.; Lei, X. Study on cycle-by-cycle variations in a diesel engine with dimethyl ether as port premixing fuel. *Appl. Energy* **2015**, *143*, 58–70. [CrossRef]

75. Christensen, M.; Johansson, B. *Supercharged Homogeneous Charge Compression Ignition (HCCI) with Exhaust Gas Recirculation and Pilot Fuel*; SAE Technical Paper 2000-01-1835; SAE International: Warrendale, PA, USA, 2000. [CrossRef]

76. Agarwal, A.K. Biofuels (alcohols and biodiesel) applications as fuels for internal combustion engines. *Prog. Energy Combust. Sci.* **2007**, *33*, 233–271. [CrossRef]

77. Sjöberg, M.; Dec, J.E.; Hwang, W. *Thermodynamic and Chemical Effects of EGR and Its Constituents on HCCI Autoignition*; SAE Technical Paper 2007-01-0207; SAE International: Warrendale, PA, USA, 2007; pp. 776–790. [CrossRef]

78. Iverson, R.J.; Herold, R.E.; Augusta, R.; Foster, D.E.; Ghandhi, J.B.; Eng, J.A.; Najt, P.M. *The Effects of Intake Charge Preheating in a Gasoline-Fueled HCCI Engine*; SAE Technical Paper 2005-01-3742; SAE International: Warrendale, PA, USA, 2005.

79. Andreae, M.M.; Cheng, W.K.; Kenney, T.; Yang, J. *Effect of Air Temperature and Humidity on Gasoline HCCI Operating in the Negative-Valve-Overlap Mode*; SAE Technical Paper 2007-01-0221; SAE International: Warrendale, PA, USA, 2007. [CrossRef]

80. Iida, M.; Aroonsrisopon, T.; Hayashi, M.; Foster, D.; Martin, J. *The Effect of Intake Air Temperature, Compression Ratio and Coolant Temperature on the Start of Heat Release in an HCCI (Homogeneous Charge Compression Ignition) Engine—Operation Ragin*; SAE International: Warrendale, PA, USA, 2014. [CrossRef]

81. Qiu, T.; Song, X.; Lei, Y.; Dai, H.; Cao, C.; Xu, H.; Feng, X. Effect of back pressure on nozzle inner flow in fuel injector. *Fuel* **2016**, *173*, 79–89. [CrossRef]

82. Desantes, J.M.; Payri, R.; Salvador, F.J.; Manin, J. *Influence on Diesel Injection Characteristics and Behavior Using Biodiesel Fuels*; SAE Technical Paper 2009-01-0851; SAE International: Warrendale, PA, USA, 2009; Volume 4970. [CrossRef]

83. Wislocki, K.; Pielecha, I.; Czajka, J.; Stobnicki, P. *Experimental and Numerical Investigations into Diesel High-Pressure Spray—Wall Interaction under Various Ambient Conditions*; SAE International: Warrendale, PA, USA, 2012; Volume 1662. [CrossRef]

84. Shehata, M.S.; Attia, A.M.A.; Abdel Razek, S.M. Corn and soybean biodiesel blends as alternative fuels for diesel engine at different injection pressures. *Fuel* **2015**, *161*, 49–58. [CrossRef]

85. Tirabnpath, P.; Hespel, C.; Chanchaona, S.; Foucher, F. Influence of biodiesel and diesel fuel blends on the injection rate under cold conditions. *Fuel* **2015**, *144*, 80–89. [CrossRef]

86. Anand, K.; Reitz, R.D. Exploring the benefits of multiple injections in low temperature combustion using a diesel surrogate model. *Fuel* **2016**, *165*, 341–350. [CrossRef]

87. Kim, D.; Bae, C. Application of double-injection strategy on gasoline compression ignition engine under low load condition. *Fuel* **2017**, *203*, 792–801. [CrossRef]

88. Jiang, X.; Deng, F.; Yang, F.; Zhang, Y.; Huang, Z. High temperature ignition delay time of DME/n -pentane mixture under fuel lean condition. *Fuel* **2017**, *191*, 77–86. [CrossRef]

89. Ying, W.; Li, H.; Jie, Z.; Longbao, Z. Study of HCCI-DI combustion and emissions in a DME engine. *Fuel* **2009**, *88*, 2255–2261. [CrossRef]

90. Dernotte, J.; Dec, J.E.; Ji, C. Investigation of the Sources of Combustion Noise in HCCI Engines. *SAE Int. J. Engines* **2014**, *7*. [CrossRef]

91. Ogunkoya, D.; Fang, T. Engine performance, combustion, and emissions study of biomass to liquid fuel in a compression-ignition engine. *Energy Convers. Manag.* **2015**, *95*, 342–351. [CrossRef]

92. Lapuerta, M.; Armas, O.; Rodríguez-Fernández, J. Effect of biodiesel fuels on diesel engine emissions. *Prog. Energy Combust. Sci.* **2008**, *34*, 198–223. [CrossRef]

93. Cairns, A.; Blaxill, H. *The Effects of Combined Internal and External Exhaust Gas Recirculation on Gasoline Controlled Auto-Ignition*; SAE International: Warrendale, PA, USA, 2005. [CrossRef]

94. Zhao, H.; Peng, Z.; Williams, J.; Ladommatos, N. *Understanding the Effects of Recycled Burnt Gases on the Controlled Autoignition (CAI) Combustion in Four-Stroke Gasoline Engines*; SAE International: Warrendale, PA, USA, 2001. [CrossRef]

95. Olsson, J.-O.; Tunestål, P.; Ulfvik, J.; Johansson, B. The effect of cooled EGR on emissions and performance of a turbocharged HCCI engine. *Soc. Autom. Eng.* **2003**, *2003*, 21–38. [CrossRef]

96. Yao, M.; Chen, Z.; Zheng, Z.; Zhang, B.; Xing, Y. *Effect of EGR on HCCI Combustion fuelled with Dimethyl Ether (DME) and Methanol Dual-Fuels*; SAE Technical Paper 2005-01-3730; SAE International: Warrendale, PA, USA, 2005. [CrossRef]

97. Sjöberg, M.; Dec, J.E. *EGR and Intake Boost for Managing HCCI Low-Temperature Heat Release over Wide Ranges of Engine Speed*; SAE International: Warrendale, PA, USA, 2007; pp. 776–790. [CrossRef]

98. Saxena, S.; Bedoya, I.D. Fundamental phenomena affecting low temperature combustion and HCCI engines, high load limits and strategies for extending these limits. *Prog. Energy Combust. Sci.* **2013**, *39*, 457–488. [CrossRef]

99. Suzuki, Y.; Harada, T.; Watanabe, H.; Shoji, M.; Matsushita, Y.; Aoki, H.; Miura, T. Visualization of aggregation process of dispersed water droplets and the effect of aggregation on secondary atomization of emulsified fuel droplets. *Proc. Combust. Inst.* **2011**, *33*, 2063–2070. [CrossRef]

100. Califano, V.; Calabria, R.; Massoli, P. Experimental evaluation of the effect of emulsion stability on micro-explosion phenomena for water-in-oil emulsions. *Fuel* **2014**, *117*, 87–94. [CrossRef]

101. Segawa, D.; Yamasaki, H.; Kadota, T.; Tanaka, H.; Enomoto, H.; Tsue, M. Water-coalescence in an oil-in-water emulsion droplet burning under microgravity. *Proc. Combust. Inst.* **2000**, *28*, 985–990. [CrossRef]

102. Strizhak, P.A.; Piskunov, M.V.; Volkov, R.S.; Legros, J.C. Evaporation, boiling and explosive breakup of oil–water emulsion drops under intense radiant heating. *Chem. Eng. Res. Des.* **2017**, *127*, 72–80. [CrossRef]

103. Kadota, T.; Yamasaki, H. Recent advances in the combustion of water fuel. *Prog. Energy Combust. Sci.* **2002**, *28*, 385–404. [CrossRef]

104. Watanabe, H.; Okazaki, K. Visualization of secondary atomization in emulsified-fuel spray flow by shadow imaging. *Proc. Combust. Inst.* **2013**, *34*, 1651–1658. [CrossRef]

105. Putrasari, Y.; Lim, O. A study on combustion and emission of GCI engines fueled with gasoline-biodiesel blends. *Fuel* **2017**, *189*, 141–154. [CrossRef]

106. Thongchai, S.; Lim, O. The influence of biodiesel blended in gasoline-based fuels on macroscopic spray structure from a diesel injector. *Int. J. Autom. Technol.* **2019**. accepted.

107. Kanti, S.; Kim, K.; Lim, O. Experimental study on non-vaporizing spray characteristics of biodiesel-blended gasoline fuel in a constant volume chamber. *Fuel Process. Technol.* **2018**, *178*, 322–335. [CrossRef]

108. Thongchai, S.; Lim, O. Investigation of the combustion characteristics of gasoline compression ignition engine fueled with gasoline-biodiesel blends. *J. Mech. Sci. Technol.* **2018**, *32*, 959–967. [CrossRef]

109. Vu, D.N.; Das, S.K.; Jwa, K.; Lim, O. Characteristics of auto-ignition in gasoline—Biodiesel blended fuel under engine-like conditions. *Proc. Inst. Mech. Eng. Part D J. Authomob. Eng.* **2018**, 1–13. [CrossRef]

110. Putrasari, Y.; Lim, O. A study of a GCI engine fueled with gasoline-biodiesel blends under pilot and main injection strategies. *Fuel* **2018**, *221*, 269–282. [CrossRef]

Article

A Novel Process for Production of *Calophyllum Inophyllum* Biodiesel with Electromagnetic Induction

Sri Kurniati [1], Sudjito Soeparman [2], Sudarminto Setyo Yuwono [3], Lukman Hakim [4] and Sudirman Syam [1,*]

1   Electrical Engineering Department, Science and Engineering Faculty, University of Nusa Cendana, Kupang 85228, Indonesia; sri_kurniati@staf.undana.ac.id
2   Mechanical Engineering Department, Engineering Faculty, Brawijaya University, Malang 65145, Indonesia; sudjitospn@yahoo.com
3   Department of Agricultural Product Technology, Faculty of Agricultural Technology, Brawijaya University, Malang 65145, Indonesia; ssyuwono2004@yahoo.com
4   Department of Chemistry, Engineering of Natural Sciences, Brawijaya University, Malang 65145, Indonesia; lukman.chemist@ub.ac.id
*   Correspondence: sudirman_s@staf.undana.ac.id; Tel.: +62-813-4264-6234

Received: 9 December 2018; Accepted: 17 January 2019; Published: 25 January 2019

Abstract: A novel method as proposed in the production of *Calophyllum inophyllum* biodiesel has been investigated experimentally. This study reports the results of biodiesel processing with electromagnetic induction technology. The applied method is aimed to compare the results of *Calophyllum inophyllum* biodiesel processing among conventional, microwave and electromagnetic induction. The degumming, transesterification, and esterification process of the 3 methods are measured by stopwatch to obtain time comparison data. Characteristics of viscosity, density, and fatty acid metil ester (FAME) are obtained from testing of a *Gas Chromatography-mass Spectrometry* (GCMS) at the Integrated Research and Testing Laboratory, Gadjah Mada University, Yogyakarta. The results present that the biodiesel produced by this method satisfies the biodiesel standards and their characteristics are better than the biodiesel produced by conventional and microwave methods. The electromagnetic induction method also offers a fast and easy route to produce biodiesel with the advantage of increasing the reaction rate and improving the separation process compared to other methods. This advanced technology has the potential to significantly increase biodiesel production with considerable potential to reduce production time and costs.

Keywords: renewable energy; microwave; free fatty acid; crude oil

---

1. Introduction

In the past two decades, the energy crisis has encouraged the development of alternative energy by seeking renewable energy resources. Alternative energy sources such as biodiesel have been developed to replace diesel or fuel oil. Generally, biodiesel is a liquid fuel processed from different sources such as palm oil [1–4], soybean oil [5–10], jatropha [11–18], cottonseed oil [19–25], soursop seed oil [26], recycled cooking oils [27–30], animal fats [31–35], and other potential triacylglycerol-containing feed-stocks [36]. Vegetable oils are mainly composed of triglycerides (98%) and a small proportion of diglycerides [37]. Biodiesel is a mixture of fatty acid alkyl esters obtained by treating triglycerides with methanol or ethanol [38,39].

One of the potential biodiesel plants is *Calophyllum inophyllum* or Nyamplung in Indonesia language. The benefits of *Calophyllum inophyllum* as biofuel are: having a higher yield than other crops (jatropha 40–60%, palm kernel 46–54%, and *Calophyllum inophyllum* (40–74%), and not having a chance to compete with food interests [40]. In addition, the productivity of *Calophyllum inophyllum* seeds of

20 tons/ha/year is higher than jatropha (5 tons/ha/year), and palm (6 ton/ha/year) [41]. *Calophyllum inophyllum* has a very high oil of 75% with an unsaturated fatty acid content of about 71% [42]. It is processed or pressed in the form of yellowish-greenish oil, similar to olive oil, which is aromatic and tasteless. It usually produces fruit twice a year around 100 kg in weight and with an oil content of about 18 kg [43]. *Calophyllum inophyllum* oil has a higher viscosity, but the capillary ability is lower than kerosene. *Calophyllum inophyllum* oil yields are forged or pressed between 40–70% of the dry seed mass [44], and the degumming process is 62.80–65.89% [45].

Generally, biodiesel is processed by thermal or heating system. Biodiesel processing with conventional heating systems is the most widely employed method. The conventional biodiesel processes are based on the use of high-power heating and magnetic stirring. Biodiesel processing is initiated with the process of dry seed pressing into *Crude Calophyllum Oil* (CCO). Through the degumming process, *Refined Crude Calophyllum Oil* (RCCO) is obtained, as followed by the esterification process. The next transesterification process is performed to produce *Crude Biodiesel* and, finally, biodiesel is obtained through washing and drying.

Due to the existence of several stages in the processing of biodiesel causing high cost of biodiesel production, the implementation of the production process is operationally inefficient. Such inefficiencies are marked from a degumming process which is heated at a temperature of 80 °C for 30 min, until the sediment appears, and an esterification process lasting for 1–2 h at temperature 60 °C and transesterification process lasting for 1 h at temperature 30–65 °C. In [46] describes, conventional biodiesel production is generally conducted at high temperatures with external heat sources. The heat transfer is considered less effective as it occurs as a conduction and convection system. In addition, conventional heating such as hotplate takes a long time and consumes high power. However, although, [47] has improved it by using a homogenizer system for transesterification reactions, the research has been only able to reduce the reaction time by 30 min (half of 60 min).

To address the challenge of high production cost, several efforts have been pursued to optimize the biodiesel production process. Microwave technology, has been selected as an alternative method with several advantages due to its quality issue [48,49], energy efficiency [50], and environmental impact [51,52]. The other technologies are less likely to be conducted as they produce more heats from the material and are not transferred from the outside materials as expected. The heating may also be selective, depending on the dielectric properties of the material. Moreover, microwave technology is preferred as it can propagate through the liquid by the more efficient heating process on the production of biodiesel, consuming a shorter time [50,53–55].

Regarding the processing biodiesel, a new method of electromagnetic induction heating is proposed by the researchers. Induction heating technology has been widely applied to the manufacture of induction cookers and is able to provide faster heat than the microwave. The induction heater (IH) is an alternating electrical current from the power unit flowing through a coil made of copper. This current will cause an electromagnetic field of varying magnitude. This field will generate an electric current on the inside metal material. This electric current is known as eddy current which generates heat to melt the metal.

2. Basic Principle

Induction heating is the process of heating objects that are electrically carried out (normally metal) with electromagnetic induction, where eddy currents are produced in the metal and the resistance leads to metal Joule heating.

The basic theory of electromagnetic-induction, however, is similar to a transformer. The transformer works due to the phenomenon of electromagnetic induction, when there is a current flowing in a closed circuit, it will produce various electromagnetic fields. As with transformers, electromagnetic fields (in primary coils) which affect secondary coils and secondary coils produce induced radiation and AC current flows (if the secondary coil is a closed circuit). An induction heater (IH) consists of three primary elements of electromagnetic-induction, skin effects, and heat-transfer [56].

Figure 1 displays the basic principle of induction heating, contain inductive-heating coils and current and illustrates electromagnetic-induction and skin-effects. Figure 1a indicates the simple structure of a transformer, where the secondary-current is in direct proportion to the main current in accordance with the turn-ratio.

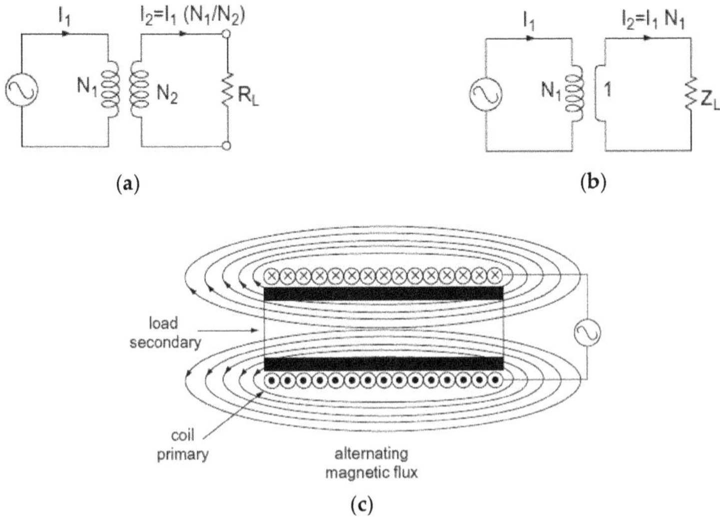

Figure 1. Basic induction: (**a**) equivalent-circuit of transformer; (**b**) secondary transformer of short circuit; (**c**) concept of induction heating.

When the secondary-coil is once turned on short-circuited, a major heat loss occurs to increase the secondary current (load-current), as shown in Figure 1b. The amount of current in the secondary coil (I_2) is determined from the magnitude of the current on the primary coil (I_1) and from the ratio of the windings between the primary and secondary coils (N_1/N_2). In this figure, when the secondary coil is replaced with 1 wire ($N_2 = 1$) and is made into a closed circuit, a large coherent ratio value of the primary and secondary coils is obtained causing a large secondary current (I_2).

This process will also be followed by a substantial increase in heat due to an increase in the load. The concept of induction heaters where energy is supplied from the same source of a combined number of primary and secondary losses is presented in Figure 1c. In this figure, the primary inductive windings have many turns, while the secondary windings are only one and are short-circuited. In IH design, it takes the maximum heat energy generated by the secondary windings, where the gap of the inductive heating coil is designed as small as possible and secondary is made with low-resistance and high-permeability. Nonferrous metals weaken energy efficiency due to their high resistance and low permeability.

2.1. Electromagnetic Induction

The output of IH flows an alternating current (AC) to supply an electric coil (induction-coil). The induction coil becomes a source of heat inducing an electric current to a work-piece or to a metal section to be heated. In addition, the induction coil as a heat source has no contact with the work-piece, and heat only occurs in the local area, especially in the zone inside the coil. This occurs because the presence of AC in the induction coil has an invisible electromagnetic force field (flux) around. When the work-piece is placed inside the induction-coil, the force lines concentrate in the air gap between the coil and the work-piece. The induction-coil functions as the major transformer with the work-piece to be heated into a secondary transformer. As a result, the force field in the induction coil causes the similar and opposite electric current in the work-piece. Further, the work-piece heats up due to the current flow resistance of the induced electric current. The level of heating of the work-piece is affected by the induction of current frequency and current intensity, material specific heat, material magnetic permeability, as well as by material resistance to current flow.

Based on Figure 1, when the AC current enters the coil, the magnetic-field is formed around the coil, according to Ampere Law as calculated in the following Equation (1):

$$\int Hdi = Ni = f$$
$$\varnothing = \mu HA$$

(1)

This formula states that: when the object is inserted into a magnetic field, they will cause changes in the velocity of magnetic motion.

The density of the magnetic field decreases as the object reaches close to the center of the surface. Faraday's Law states that the electric current produced on the surface of a conductive object has an inverse relationship with the current in an inductive-circuit as depicted in Equation (2).

$$E\frac{d\lambda}{dt} = \frac{d\varnothing}{dt}$$

(2)

Equation (3) describes an electrical energy caused by induced currents and eddy currents which changed into heat energy.

$$P = \frac{E^2}{R} = i^2 R$$

(3)

where,

R = Resistance is determined by resistivity (ρ) and permeability (μ) of conductive objects
i = Current is determined by the magnetic of field intensity.
E = Induction voltage
P = Power

2.2. Eddy Current

Induction heating arises when an electric current or eddy current is induced into the work-piece under conditions of poor electrical conductors. To produce the induction heating to be efficient and practical, the specific relationship between the frequency of an electromagnetic field that produces an eddy-current, and the properties of a work-piece, must be satisfied. The basis of induction heating is eddy currents which are produced outside the work-piece or commonly called as "skin effect" heating. The level of induction heating depends on the high frequency of the electric current, electrical resistance, and relative magnetic-permeability of the work-piece.

Eddy Current is the induction of alternating electric current in a conductive material by an alternating magnetic-field (as generated by alternating electric-current). The induced current inside the modified material causes a change in the value of the induced current through the material. The eddy current principle is based on Faraday's law which states that when a conductor is cut out the force lines of the magnetic field or electromotive force (EMF), it will be induced into the conductor. The amount of EMF depends on: (1) size, strength, and magnetic field density; (2) the speed at which the magnetic force lines are cut; and (3) quality of conductors. Since eddy current is an electric current in the conductor, it will also produce magnetic field. Lenz's law states that the magnetic field of the induced current has a direction opposite to the cause of the induced current. The magnetic field of eddy current is in opposite direction to the magnetic field of the coil as demonstrated in Figure 2.

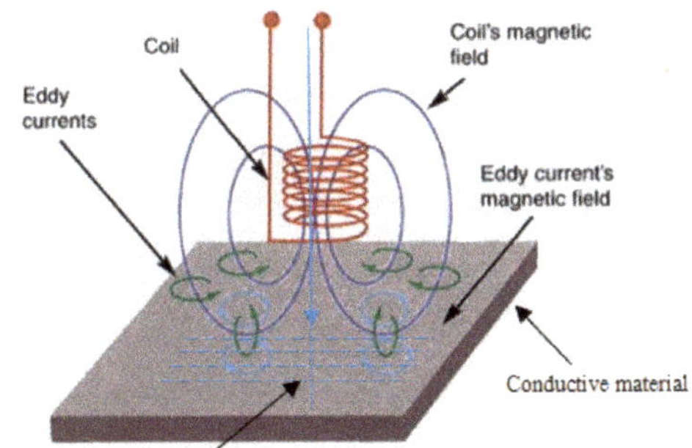

Figure 2. Direction of magnetic field.

2.3. Operating Fundamental of an IH

A block diagram of IH is shown in Figure 3. A high-frequency AC voltage is generated from the power module. Convert the AC current coming from the power source to DC using a rectifier. Then, connect this DC current to a high-frequency switching circuit to administer high-frequency current to the heating coil. This high frequency AC power is sent to the coil to cause flux, the size of the generated flux depends on the area of the induction coil used. This is because the induction heater utilizes losses that occur in the inducing coil. According to Ampere's Law, a high-frequency magnetic field is created around the heated coil. Eddy currents play a dominant role in the induction heating process, the heat generated in the material depends on the size of the eddy current induced by inducing windings. When the coil is flowed by AC, it creates a magnetic field around the conductive wire. The magnitude of the field changes according to the current flowing in the coil.

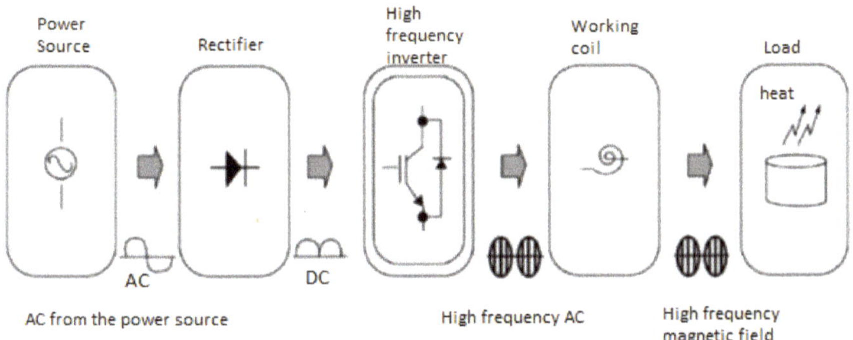

Figure 3. Block diagram of an induction heater.

3. Topology of IH

3.1. Power System of IH

Currently, IH becomes a popular heating technology which has been developed in industrial, domestic and medical applications, having advantages such as efficiency factor, fast-heating, safety, cleanliness and accurate control [57]. Since the discovery of semiconductor technology, the second major revolution of IH technology is developed. The use of high-power semiconductor devices and high-frequency power supply application has performed great progress. Among them are silicon-controlled rectifiers (SCR), and gate turn-off thyristor which have been widely utilized to implement a reliable power converter. The expansion of high-frequency power electronics for IH requires the metal-oxide-semiconductor field-effects (MOSFETs) and bipolar-junction-transistors (BJT), which allow higher efficiency power converter designs. Further progress in semiconductor technology is achieved by the discovery of an insulated-gate bipolar transistor (IGBT) that can improve the performance of IH design. IGBT has a low resistance and requires a very small power supply with a frequency of up to 100 kHz; however, a frequency of around 400 kHz is difficult to achieve. On the contrary, static induction thyristor (SIT) has defects such as high conduction loss compared to IGBT, complicated fabrication processes, high costs, and limited application. Therefore, switching to very high frequencies can be achieved by utilizing MOSFETs.

With regard to power systems, a new soft-switched converter circuit has been developed that integrates the advantages of both conventional pulse width modulation (PWM) and resonance converters. This soft-switched converter has a switching waveform that is identical to a conventional PWM converter, except that the rise and fall edges from the waveform, as 'smoothed' without a temporary surge. In contrast to resonant converters, new soft-switched converters usually apply a controlled resonance. In this case, efficient energy conversion at high-frequency switching is absolutely necessary by manipulating the voltage or current when switching to zero. The concept is to combine the resonant tank in the converter to create an oscillation voltage (generally sinusoidal) and/or the current waveform. As a result, the zero-voltage-switching (ZVS) or zero-current-switching (ZVS) conditions can be utilized for the power switch. The ZVS principle refers to the elimination of the turn-on switching loss having the zero voltage of the switching circuit before the circuit is turned "on". On the contrary, ZVS avoids turn-off switching-loss by allowing no current to pass through the circuit right before turning it "off". The voltage or current given to the switching circuit can be made zero by using resonance as generated by the inductance-capacitance (L-C) resonance circuit; by which this topology known as "resonant-converter". Resonance is allowed to occur before and during the turn-on and turn-off process to produce ZVS and ZCS conditions, like the conventional PWM converters.

3.2. Half-Bridge Resonance Inverter

In terms of the number of switching devices, the inverter topology is normally used in IH with a single switch [58], full bridge circuit [59], half bridge series resonant [60], and half bridge inverter [61]. In this paper, a power system with a half bridge resonance series of inverter circuit has been applied. In [60], explaining the advantages of half-bridge resonance series inverters is stable switching, low-cost, and a streamlined-design. Figure 4 shows the power operating system with a half-bridge-resonant inverter, consisting of AC power supply, main electrical circuit, control unit, input-current detection-circuit, a resonance-current detection circuit, and a gate operation circuit. All procedures are required to design and test the system as shown in the block diagram. The system for IH, however, does not need such large capacitors to produce DC flatter, as the major goal of this system is to generate heat energy. In contrast, a rough DC form helps increase the system's power factor.

AC power (220 V/50 Hz) passes through the rectifier diode to be transmitted to the capacitor (C). The capacitor in the available power system is unfortunately too small in its capacity to act in leveling work, which leads to the action of an increased current at 120 Hz, which is not the right level for DC operation. The system for IH, however, does not need large capacitors to produce DC flatter, because the major purpose of this system is to generate heat energy. In contrast, a rough DC form helps increase the system's power factor. Based on this circuit, the leveling capacitor functions as a filter that prevents high-frequency currents from flowing into the inverter and from entering the input part. Later, the input current becomes the average current of the inverter and acts as the rippling flow to the leveling capacitor.

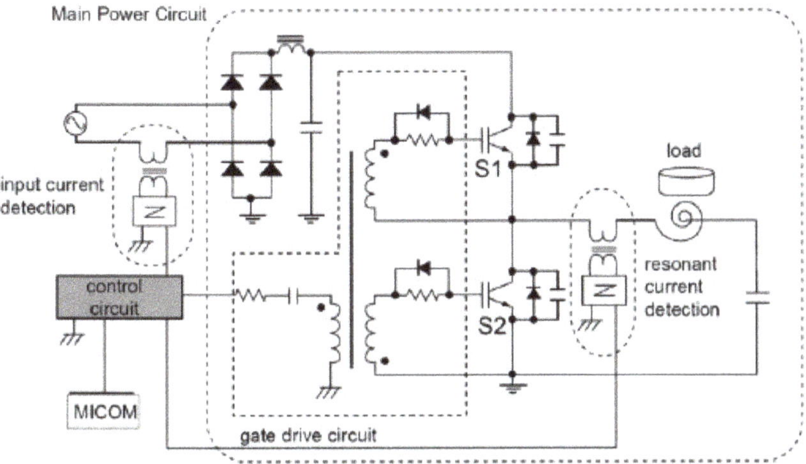

Figure 4. Half-bridge series resonant-inverter.

4. Experimental Set-up

4.1. Testing of Calophyllum-Inophyllum Oil

Calophyllum-inophyllum oil used in this study was obtained from Kebumen, East Java Province (Indonesia). This type of crude oil is an-inedible oil, which belongs to the Guttiferae family, usually called as "Nyamplung" oil in Indonesia. Nyamplung trees usually grow along coastal area and in adjacent lowlands forests, having no incubation in yielding, and seeds can be obtained throughout the year. Figure 5 shows the crude oil of *Calophyllum inophyllum.* The characteristics of Calophyllum inophyllum crude-oil have been tested using Gas Chromatography-Mass Spectrometry (GCMS) at the Integrated Research and Testing Laboratory, Gadjah Mada University, Yogyakarta.

Figure 5. Calophyllum inophyllum crude oil.

Based on this test the contents of palmitic, oleic, linoleic, and stearic and acid were obtained from the oil studied. This crude oil contains the content of saturated fatty acids including stearic acid (13.12%) and palmitic (13.92%). In contrast, high unsaturated fatty acids, representing 72.96% mainly consists of oleic acid (69.11%) and linoleic acid (3.85%).

4.2. Biodiesel Processing

A schematic process for producing biodiesel is illustrated in Figure 6. Firstly, biodiesel processing starts from the degumming process by utilizing the 3 heating methods at 80 °C to separate crude oil from its gum. The hotplates (conventional), microwaves, and electromagnetic induction heating are used in processing scheme as a comparison. A degumming process carried out using phosphoric acid (H_3PO_4) as much as 5% (*v/v*) of 400 mL *Calophyllum-inophyllum* oil, then it was heated to 80 °C. The duration of the degumming process depends on the heating method until the color changes to light brown. Table 1 shows a comparison of the time and degumming results of the three heating methods.

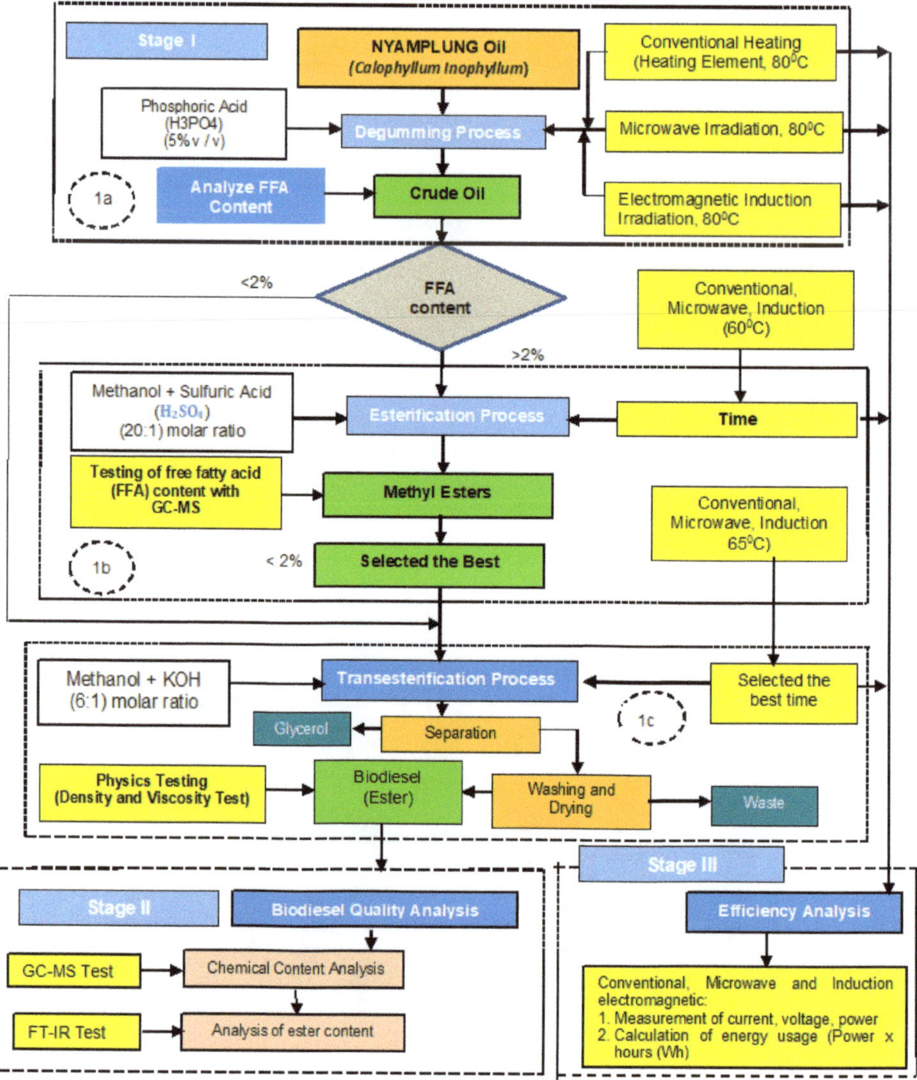

Figure 6. Schematic diagram of biodiesel process.

Table 1. A comparison of the time and degumming results of the three heating methods.

Methods	Temperature (°C)	Time (Min)	Power (W)	FFA (%) Before	FFA (%) After
Hotplate	80	15	600	20.27	17.66
Microwave	80	12	120	19.82	17.21
Electromagnetic Induction	80	3.21	145	19.7	17.03

Secondly, *Calophyllum-inophyllum* oil esterification reactions were carried out in the presence of H_2SO_4 (2 wt.%) at temperatures 60 °C with using molar ratio of methanol-to- oil (20:1, by mole basis). According to [30], the molar ratio and dosage of catalyst become one of the most important factors in biodiesel production, whereas the humidity is one of the least important factors. Briefly, the first catalyst is dissolved in methanol and added 5 g of *Calophyllum-inophyllum* oil is put in a 100 mL beaker. The mixture of methanol and catalyst H_2SO_4 was put into Erlenmeyer and stirred with a magnetic stirrer for 5 min. The beaker glass is filled with a mixture of methanol, *Calophyllum-inophyllum* oil, H_2SO_4, and it is heated using 3 heating methods by keeping the temperature up to 60 °C. After esterification, testing is carried out to determine the content of FFA. If the FFA is still high or in above 2%, the esterification process is repeated for 2 times until the FFA becomes precisely low (below 2%). After that, methanol must be removed and separated from oil product using a 100 mL separatory funnel for the purification stage. The product is washed with hot distilled water (>80 °C) to remove the remaining acid in the product, continuing until the washing water shows pH~7.

Further, the transesterification process is performed as a chemical reaction involving triglycerides and alcohol in the presence of a catalyst to form esters and glycerol. The transesterification process of Nyamplung oil is obtained at a methanol-to-oil molar ratio, 6:1 and KOH of 2% catalyst with a temperature of 65 °C. The time period of the transesterification process depends on the heating method used as seen in the oil color changes and the formed glycerol. Finally, the glycerol which is formed must be separated from biodiesel to be washed and dried.

4.3. Induction Heating Irradiation

In this study, an electromagnetic induction heater has been utilized in processing *Calophyllum inophyllum* biodiesel. The equipment specifications have been detailed in Table 2. The schematic diagram of the experimental set up is ills ustrated in Figure 7. The electromagnetic induction heat radiation induced by the metal mounted inside the coil is absorbed by the sample oil, which may lead to the appearance of warming in the sample. Heating with induced magnetic field radiation is faster and is evenly distributed along the metal mounted inside the coil, as it does not transfer heat from the outside. The duration of induction given to metals affects the rise in temperature/heat of the metal which affects the passage of the transesterification reaction. The greater the temperature given, the transesterification reaction runs faster and results in biodiesel conversion.

Table 2. Equipment specification.

No.	Name	Specification
1.	Dc Voltage	30 Volts
2.	Power Input	200 watts
3.	Current Output	1.2 Amps
4.	Temperature	0–1000 °C (adjustable)

Vibrations in molecules induced by induction heat radiation will produce an equal heat on the molecule, where the resulting induction heat penetrates and excites the molecules evenly, not solely in the surface. Induced heating radiation can speed up the reaction by vibrating the reactant molecule quickly. The longer the radiation time is given to the transesterification reaction, the heat generated by the reactant molecule's vibration will be greater until reaching its optimum state. The transesterification time at 65 °C has been controlled automatically by a microcontroller. Output data such as temperature, current, power, and voltage are displayed and stored on a laptop.

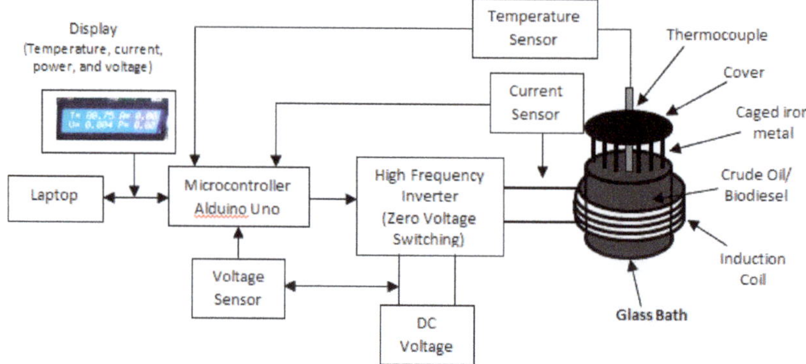

Figure 7. Schematic diagram of electromagnetic induction irradiation.

Eddy currents have the most dominant role in the induction of the heating process. The heat generated on the material depends on the number of eddy currents as induced by the inductor winding. When the winding is fed by alternating current, a magnetic field will occur around the conductor wire. The magnetic field varies according to the current flowing in the coil. According to [62], if there is conductive material around the changing magnetic field, the conductive material will flow a current called eddy current. The Eddy current principle is based on Faraday's law which states that when a conductor is cut out the force lines of the magnetic field or electromotive force (EMF), it will be inducted into the conductor. The amount of EMF depends on (1) the size, strength, and magnetic field density; (2) on the speed at which the magnetic force lines are cut; (3) and on the quality of conductors.

5. Result

5.1. Characteristics of IH Irradiation

One method of heating to allow the reaction to run faster is performed to achieve a transesterification reaction by using the electromagnetic induction of heat radiation. The electromagnetic IH radiation as induced by the metal mounted inside the coil is absorbed by the sample, which may lead to the appearance of warming in the sample. Heating with induced magnetic field radiation is faster and evenly distributed along the metal mounted inside the coil, rather than by transferring heat from the outside materials. The duration of induction given to metals affects the rise in temperature/heat of the metal which affects the passage of the transesterification reaction. The greater the temperature given, the transesterification reaction runs faster resulting in more biodiesel conversion.

Vibrations in molecules induced by IH radiation will produce an equal heat on the molecule, where the resulting induction heat penetrates and excites the molecules evenly, not solely in the surface. Induced heating radiation can speed up the reaction by vibrating the reactant molecule quickly. The longer the radiation time is given to the transesterification reaction; the heat generated by the reactant molecule's vibration will be greater, resulting the transesterification reaction will reach its optimum state at any time the trial. Eddy currents have the most dominant role in the induction heating process. The heat generated in the material depends on the amount of eddy current as induced by the inductor winding. When the windings are energized by alternating current, a magnetic field will occur around the conductor wire. Magnetic fields vary according to the current flowing in the coil.

Technically, induction heaters have characteristics that are able to release heat in a relatively short time due to high energy density. By the induction, it is possible to reach high temperatures. Heating can be performed at a specific location where the system can be made to work automatically. Induction heating, in general, has high energy efficiency, depending on the characteristics of the heated material as any heating losses can be minimized. According to [63], several factors determine the number of eddy currents in the metal, including: (1) the magnetic field inducing the metal; (2) and the metal materials used to generate heat (the smaller the resistance of the metal type, the metal is better as the object of heat); (3) the metal surface area (the more surface area of the metal, the more eddy currents will be on the metal surface); and (4) the greater the frequency, the more the magnetic field is generated. In addition, there are several advantages of using induction heater, including heat is generated directly inside the barrel wall; heat can be applied uniformly across the barrel; cold element operation is performed, having unlimited time; faster startup time; and more energy-efficient.

5.2. Advantages of Electromagnetic Induction Methods

The biodiesel production process must be refined to maximize material value and minimize costs. According to [64], if we can make low cost and high-acid-value oil available as resources, the cost of biodiesel production will be reduced significantly. However, these low-cost oils cannot be used to produce biodiesel directly because they usually contain a large amount of FFAs. They have to undergo a preparatory procedure to lower the acid value to a specific value. In addition, in industrialization processes, it not only saves time and effort but is also low-cost [65,66]. Therefore, low energy potential and lack of efficient technologies are the problems with these feedstocks for the commercial production of biodiesel [67].

Figure 8 shows a comparison of the testing result in biodiesel manufacturing processes among the conventional, microwave and electromagnetic induction. It is concluded that electromagnetic induction technology provides a great opportunity in the future production process. Compared to other methods, the proposed new method has a precisely short time at every stage of the production process such as in degumming (3.2 min), esterification (18.13 min), and transesterification (0.43 min). Overall, the time required for processing biodiesel is hotplate (130 min), microwave (112 min), and electromagnetic induction (21.93 min). Here, it appears that there is a significant time difference from the use of the induction method compared to the other two heating methods. With the achievement of faster time at each stage of biodiesel processing has advantages, such as reducing production costs, saving energy, and improving the quality of biodiesel. Another advantage is the cost of fabricating electromagnetic induction is cheaper compared to microwave and hotplate (in this study the cost of fabricating electromagnetic does not exceed $100 US).

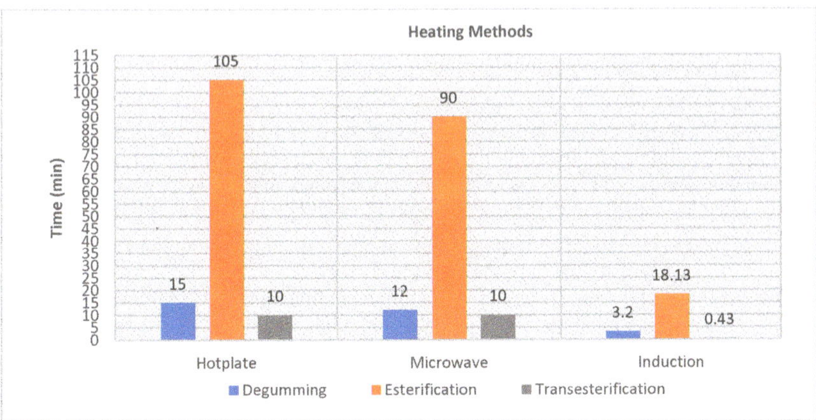

Figure 8. Time comparison of biodiesel production among conventional, microwave, and induction electromagnetic.

5.3. Analysis Free Fatty Acid (FFA)

5.3.1. Esterification Process

Esterification is the process of reacting FFA into esters with short-chain alcohols (methanol or ethanol) to produce fatty acid methyl ester (FAME) and water. The esterification process with an acid catalyst is required if vegetable oil contains FFA at above 2%. If oil with a high FFA level (>2%) is directly transesterified with a KOH catalyst, the FFA will react with the catalyst to form soap. The formation of large amounts of soap can inhibit the separation of glycerol from methyl esters and results in the formation of emulsions during the washing process. Esterification is employed as a preliminary process to convert FFA to methyl ester to reduce the levels of FFA in vegetable oils and is later transesterified with a base catalyst (to convert triglycerides to methyl esters).

Esterification is the conversion stage from FFA to esters. Esther reacts the fatty oils with alcohol. To encourage the reaction to proceed, the methanol reactant must be added in such excessive amounts (usually greater than 10 times of the stoichiometric ratio) and the water product following the reaction must be removed from the reaction phase, which is the oil phase. The 20:1 (methanol to oil) molar ratio between oil and methanol is utilized in this study. Heat treatment is carried out at an average temperature of 60 °C with a period of time depending on the characteristics of the applied heating method. To maximize the biodiesel yield from oils with high FFA levels, esterification must be performed to reduce the level of FFAs prior to transesterification [68,69]. Therefore, the esterification process is carried out twice [70], as the first esterification process obtains FFA content which is still high at above 2%. Table 3 presents the comparison of FFA values from the esterification results for the 3 (three) heating methods used.

Table 3. Comparison of Free Fatty Acid (FFA) values from esterification results among the three heating methods with a molar ratio of 20:1, and 2% *w*/*w* (of oil) H_2SO_4.

Method	Temperature (°C)	Esterfication I		Esterification II	
		Time (Min)	FFA (I) (%)	Time (Min)	FFA (II) (%)
Hotplate	60	75	6.8	30	1.7
Microwave	60	75	3.59	15	1.43
Induction Electromagnetic	60	12	5.8	6.13	1.6

5.3.2. Transesterification Process

This stage becomes the main stage in this study, where triglycerides are the main components of oil converted into biodiesel and glycerol. The product of pretreatment was heated to 65 °C in the three methods for transesterification with 6:1 M ratio of methanol and 2% *w*/*w* (of oil) KOH in the same setup. According to [71] the optimal temperature in the transesterification process with a KOH catalyst is 65 °C. Comparative data from the results of the transesterification of the three heating methods are illustrated in Table 4. Figures 9 and 10 show the transesterification process by using electromagnetic induction, respectively. The biodiesel production using electromagnetic induction has a higher FAME value than in conventional and microwave.

Table 4. A comparison of the transesterification process of Nyamplung oil with a molar ratio of 6:1, and 2% (*w*/*w*) KOH.

Method	Temperature (°C)	Time (Min)	FFA (%)	FAME (%)
Hotplate	65	10	0.56	35,1
Microwave	65	10	0.42	53.66
Induction Electromagnetic	65	0.43	0.4	65.96

Figure 9. The esterification process using electromagnetic induction.

(a)   (b)

Figure 10. Fatty acid metil ester (FAME): (**a**) esterification (**b**) biodiesel.

5.4. Analysis of Energy Use

Based on the measurement results, the amount of power utilized by all heating methods is the multiplication of voltage and output current. Energy consumption is calculated between the time (hour, h) multiplier and the power (Watt, W). Based on Table 5, it is apparent that energy with electromagnetic induction method is more efficient compared to that in conventional and microwave method, that is, 47.29 Wh. The amount of energy obtained is based on the multiplication of the overall time in the degumming, esterification and transesterification steps by the amount of heating equipment power.

Table 5. Comparison of energy consumption for *Calophyllum inophyllum* oil (200 mL).

Methods	Energy Consumption (Wh)			Total Energy (Wh)
	Degumming	Esterification	Transesterification	
Hotplate	150	900	96	1.146
Microwave	19.2	50	6.4	75.6
Electromagnetic Induction	7.68	38.7	0.91	47.29

5.5. Testing of Viscosity and Density

The viscosity of biodiesel plays an important role in the fuel injection process. A precisely small viscosity causes leaks in the fuel injection pump. Conversely, too high viscosity will produce large droplets of fuel and will have high momentum and collide with the cylinder wall, but the injection pump cannot fog properly. Density is related to the value of heat and to the energy produced by a diesel engine. Low density will produce high calorific value and vice versa. Density is a parameter marking the success of the transesterification reaction. Fulfillment of density parameters indicates that the biodiesel purification process is successful. The lack of biodiesel due to impurity can produce high-density values. Table 6 illustrates the Nyamplung biodiesel properties compared with those of SNI, ASTM D6751, ASTM PS 121, C1 Biodiesel, and EN 14214. The test results present that the obtained biodiesel viscosity is 5.54 Cst, meeting the requirements of the Indonesian National Standard (SNI) in the fuel injection process between 2.3 and 6.0 Cst.

Table 6. Nyamplung biodiesel properties compared with those of SNI, ASTM D6751, ASTM PS 121, C1 Biodiesel, and EN 14214.

Properties	ASTM D6751 (USA)	ASTM PS 121	EN 14214	C1 Biodiesel	SNI	Nyamplung Biodiesel (This Study)
Acid value (mg KOH/g)	<0.5	<0.5	<0.5	0.34	0.8	0.8
Density (20 C) (g/mL)	0.87–0.9	0.7328	No specific	0.877	0.850–0.890	0.882
Kinematic viscosity, 40 °C (mm^2/s)	1.9–6.0	1.9–6.0	3.5–5.0	5.6872	2.3–6.0	5.54

5.6. GC-MS Analysis

The biodiesel composition was quantified using Gas Chromatography - Mass Spectrometer detector (GCMS) testing at the Integrated MIPA Laboratory, State University of Malang. Based on the results of GC-MS testing that has been done, it can be identified the compound components of the Nyamplung oil biodiesel as shown in Table 7.

Table 7. Major Components Based on GC-MS Analysis.

Component	Peak	Detection Time	Identified Compounds	Molecular Formula	Percentage (%)
I.	6	35.006	Linoleic acid methyl ester	$C_{19}H_{34}O_2C_{19}H_{34}O_2$	16.78 +
	7	35.109			17.17 = 33.95
II.	8	35.388	Oleic acid methyl ester	$C_{19}H_{36}O_2$	20.06
III.	9	35.997	Stearic acid methyl ester	$C_{19}H_{38}O_2$	18.62
IV.	3	28.563	Palmitic acid methyl ester	$C_{17}H_{34}O_2$	17.12

As shown in Table 7, four fatty acids were clearly detected in the oil extracted by GC from the Nyamplung oil. The most major component was linoleic methyl ester 33.95% having a molecular mass of 294 m/z, at peaks 6 and 7 as shown in Figure 11. The percentage of the second major component of oleic methyl ester is 20.06% which has a molecular mass of 296 m/z at peak 8 (Figure 12). Figures 13 and 14 show stearic methyl ester with a mass of 298 m/z and palmitic methyl esters with a mass of 2970 m/z are the third and fourth most components at 18.62% and 17.12% respectively.

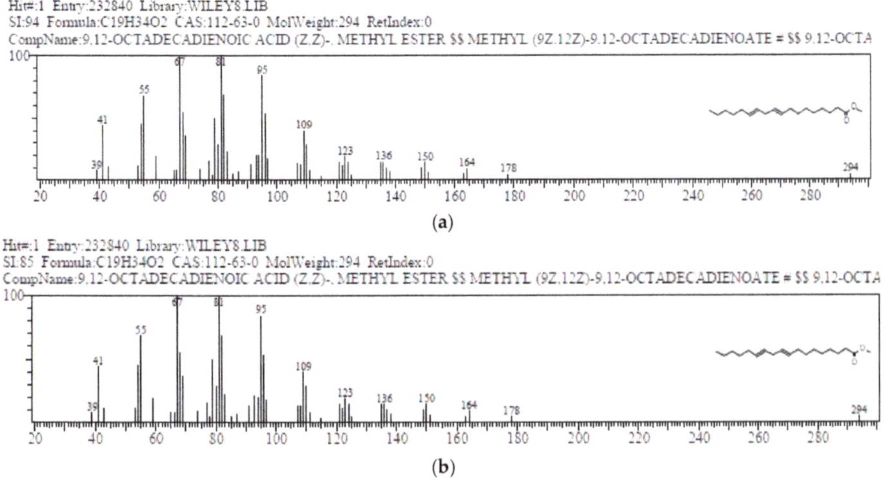

Figure 11. Fragmentation of linoleic acid methyl ester: (a) peak 6; (b) peak 7.

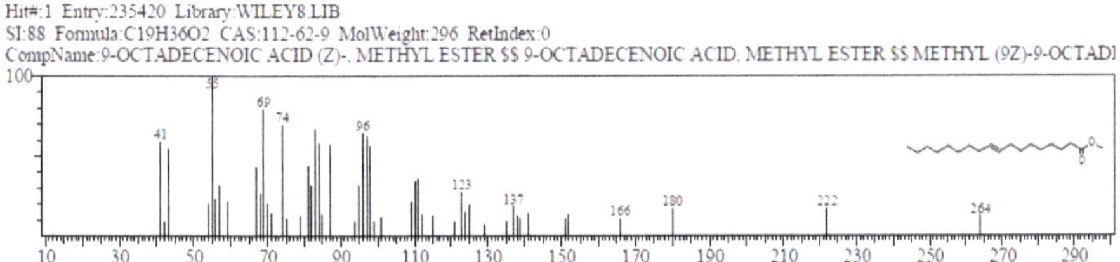

Figure 12. Fragmentation of oleic acid methyl ester.

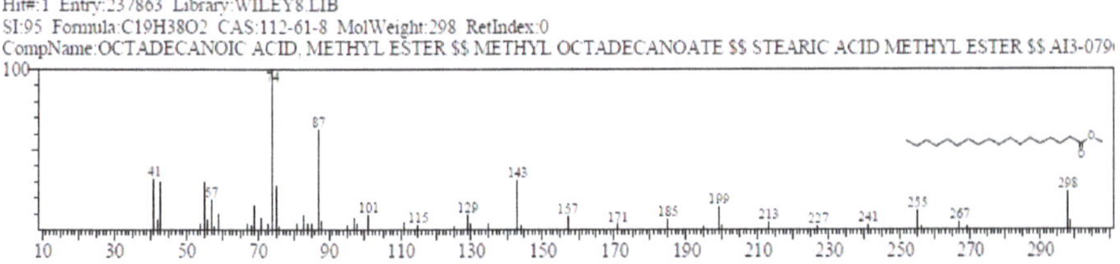

Figure 13. Fragmentation of stearic acid methyl ester.

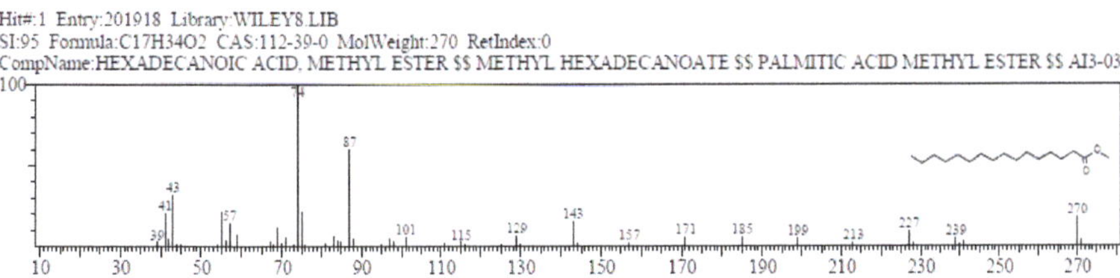

Figure 14. Fragmentation of palmitic acid methyl ester.

6. Discussion

Some ways to reduce the cost of biodiesel production are to find optimal parameters, including reaction temperature, reaction time, amount of catalyst to be added to the reaction, and the methanol-to-oil ratio. Previous researchers have experimented with base catalytic reactions, acid catalytic reactions, and enzymatic transesterification and have found that a base-catalytic reaction obtained the best result. Biodiesel can be produced under a lower temperature with a base catalyst, whereas a transesterification reaction with an acid catalyst requires higher temperatures and longer reaction time [72,73]. On the other hand, the use of longer reaction times in biodiesel processing has an impact on production costs. Generally, the biodiesel production process is performed by a heating device such as a hotplate or heating coil. With the development of high technology, the microwave radiation method in biodiesel making becomes a smart choice compared to conventional systems. An alternative heating system of "microwave irradiation" has been applied in transesterification reactions in recent years. Microwaves are considered as electromagnetic radiation that represents non-ionizing radiation influencing molecular motions such as ion migration or dipole rotation but that does not change the molecular structure [74,75].

According to [50], the advantage of using a microwave in the biodiesel process is that it efficiently accelerates the transesterification process through short reaction time. In addition, a drastic reduction in the number of products is obtained, a short separation time is obtained and a high yield of a very pure product is achieved in a short-time [76]. Compared to conventional systems, production costs are also reduced and fewer products occur. Therefore, microwave heating is more preferred compared to conventional methods, where heating time is very slow and is inefficient because the transfer of energy to the sample depends on the convection current and the thermal conductivity of the reaction mixture. However, a shorter heating reaction has been achieved through electromagnetic induction than in microwave technology.

Experimentally, a comparison of 3 methods in the production process of *Calophyllum inophyllum* biodiesel is attached. In the transesterification process, electromagnetic induction requires only 0.43 min, compared to both in microwave and conventional methods, lasting for 10 min. Likewise, other stages such as degumming and esterification, such as in Figure 8 present a significant time difference among the three methods. Some characteristics of the electromagnetic induction method show better progress than the others, such as the shorter time, more energy saving, more qualified FAME, better viscosity, and higher yield. Therefore, the findings of this method present promising expectations for biodiesel production.

7. Conclusions

An experimental investigation has been carried out to process *Calophyllum inophyllum* oil into biodiesel by a new method of electromagnetic induction. The most important conclusions obtained are summarized as follows:

1. Biodiesel produced from *Calophyllum inophyllum* oil with electromagnetic induction radiation generally meets ASTM D6751, C1 biodiesel, EN 14214, and SNI standards, so it can be used as an alternative for biodiesel processing.
2. Under optimal conditions, the energy consumption of electromagnetic induction is more efficient than the hotplate and microwave method. Compared to heating both hotplate and microwave, the reaction time is significantly reduced.
3. Due to having faster reaction time in the transesterification process, the FAME value obtained is higher than the hotplate and microwave. The optimal condition for this experiment is the molar ratio of methanol to oil 6:1, 2% (b/b) of KOH catalyst, a reaction temperature of 65 °C, the reaction time of 0.43 min, and FAME of 65.96%.
4. Compared to microwaves and hotplate, electromagnetic induction is achieved at a shorter time in all stages of degumming, esterification, and transesterification. Overall, the time needed to

process biodiesel is hotplate (130 min), microwave (112 min), and electromagnetic induction (21.93 min). As a result, the overall energy used from the biodiesel production stage is also more economical and efficient.

Author Contributions: Conceptualization, S.K., (S.S.) Soedjito Soeparman; Methodology, S.K., L.H.; Validation, S.K., S.S.Y. and (S.S.) Sudirman Syam; Formal Analysis, S.K., L.H.; Data Curation, (S.S.) Soedjito Soeparman; Writing-Original Draft Preparation, S.K.; Writing—Review & Editing, (S.S.) Sudirman Syam; Supervision, (S.S.) Sodjito Soeparman; Project Administration, S.K.; Funding Acquisition, S.K.

Funding: This research was funded by the Ministry of Research, Technology and Higher Education in the Doctoral Grant Research with contract number 42/UN15.19/LT/2018.

Acknowledgments: The authors would like to acknowledge the support of the Laboratory of Electrical Engineering, University of Nusa Cendana—Kupang and scholarship support from the Ministry of Research Technology and Higher Education.

Conflicts of Interest: The authors declare no conflict of interest.

References

1. Melero, J.A.; Bautista, L.F.; Morales, G.; Iglesias, J.; Sánchez-Vázquez, R. Biodiesel production from crude palm oil using sulfonic acid-modified mesostructured catalysts. *Chem. Eng. J.* **2009**, in press. [CrossRef]
2. Alkabbashi, A.N.; Alam, M.Z.; Mirghani, M.E.S.; Al-Fusaiel, A.M.A. Biodiesel production from crude palm oil by transesterification process. *J. Appl. Sci.* **2009**, *9*, 3166–3170. [CrossRef]
3. Crabbe, E.; Nolasco-Hipolito, C.; Kobayashi, G.; Sonomoto, K.; Ishizaki, A. Biodiesel production from crude palm oil and evaluation of butanol extraction and fuel properties. *Process Biochem.* **2001**, *37*, 65–71. [CrossRef]
4. Harsono, S.S. Biodiesel production from palm oil technology. *Res. J. Agric. Sci.* **2011**, *43*, 80–85.
5. Kaieda, M.; Samukawa, T.; Matsumoto, T.; Ban, K.; Kondo, A.; Shimada, Y.; Noda, H.; Nomoto, F.; Ohtsuka, K.; Izumoto, E.; et al. Biodiesel fuel production from plant oil catalyzed by Rhizopus oryzae lipase in a water-containing system biofuel's engineering process technology without an organic solvent. *J. Biosci. Bioeng.* **1999**, *88*, 627–631. [CrossRef]
6. Samukawa, T.; Kaieda, M.; Matsumoto, T.; Ban, K.; Kondo, A.; Shimada, Y.; Noda, H.; Fukuda, H. Pretreatment of immobilized candida antarctica lipase for biodiesel fuel production from plant oil. *J. Biosci. Bioeng.* **2000**, *90*, 180–183. [CrossRef]
7. Silva, C.C.C.M.; Ribeiro, N.F.P.; Souza, M.M.V.M.; Aranda, D.A.G. Biodiesel production from soybean oil and methanol using hydrotalcites as catalyst. *Fuel Process. Technol.* **2010**, *91*, 205–210. [CrossRef]
8. Cao, W.; Han, H.; Zhang, J. Preparation of biodiesel from soybean oil using supercritical methanol and co-solvent. *Fuel* **2005**, *84*, 347–351. [CrossRef]
9. Lee, J.H.; Kwon, C.H.; Kang, J.W.; Park, C.; Tae, B.; Kim, S.W. Biodiesel production from various oils under supercritical fluid conditions by candida antarctica lipase B using a stepwise reaction method. *Appl. Biochem. Biotechnol.* **2009**, *156*, 454–464. [CrossRef]
10. Yu, D.; Tian, L.; Wu, H.; Wang, S.; Wang, Y.; Ma, D.; Fang, X. Ultrasonic irradiation with vibration for biodiesel production from soybean oil by Novozym. *Process Biochem.* **2010**, *45*, 519–525. [CrossRef]
11. Chen, C.-H.; Chen, W.-H.; Chang, C.-M.J.; Lai, S.-M.; Tu, C.-H. Biodiesel production from supercritical carbon dioxide extracted Jatropha oil using subcritical hydrolysis and supercritical methylation. *J. Supercrit. Fluids* **2010**, *52*, 228–234. [CrossRef]
12. Chena, C.-R.; Cheng, Y.-J.; Ching, Y.-C.; Hsiang, D.; Chang, C.-M. Green production of energetic Jatropha oil from de-shelled *Jatropha curcas* L. seeds using supercritical carbon dioxide extraction. *J. Supercrit. Fluids.* **2012**, *66*, 137–143. [CrossRef]
13. Koh, M.Y.; Ghazi, T.I.M. A review of biodiesel production from *Jatropha curcas* L. oil. *Renew. Sustain. Energy Rev.* **2011**, *15*, 2240–2251. [CrossRef]
14. Syam, A.M.; Resul, M.F.M.G.; Yunus, R.; Ghazi, T.I.M. Reduction of free fatty acids in crude jatropha curcas oil via an esterification process. *Int. J. Eng. Technol.* **2008**, *5*, 92–98.
15. Shah, S.; Gupta, M.N. Lipase catalyzed preparation of biodiesel from Jatropha oil in a solvent free system. *Process Biochem.* **2007**, *42*, 409–414. [CrossRef]
16. Jain, S.; Sharma, M.P. Prospects of biodiesel from Jatropha in India: A review. *Renew. Sustain. Energy Rev.* **2010**, *14*, 763–771. [CrossRef]

17. Lu, H.; Liu, Y.; Zhou, H.; Yang, Y.; Chen, M.; Liang, B. Production of biodiesel from *Jatropha curcas* L. oil. *Comput. Chem. Eng.* **2009**, *33*, 1091–1096. [CrossRef]

18. Tiwari, A.K.; Kumar, A.; Raheman, H. Biodiesel production from jatropha oil (Jatropha curcas) with high free fatty acids: An optimized process. *Biomass Bioenergy* **2007**, *31*, 569–575. [CrossRef]

19. Berchmans, H.J.; Hirata, S. Biodiesel production from crude *Jatropha curcas* L. seed oil with a high content of free fatty acids. *Bioresour. Technol.* **2008**, *99*, 1716–1721. [CrossRef]

20. Köse, Ö.; Tüter, M.; Aksoy, H.A. Immobilized candida antarctica lipase-catalyzed alcoholysis of cotton seed oil in a solvent-free medium. *Bioresour. Technol.* **2002**, *83*, 125–129. [CrossRef]

21. He, C.; Baoxiang, P.; Dezheng, W.; Jinfu, W. Biodiesel production by the transesterification of cottonseed oil by solid acid catalysts. *Front. Chem. Eng. China* **2007**, *1*, 11–15.

22. Royon, D.; Daz, M.; Ellenrieder, G.; Locatelli, S. Enzymatic production of biodiesel from cotton seed oil using t-butanol as a solvent. *Bioresour. Technol.* **2007**, *98*, 648–653. [CrossRef]

23. Hoda, N. Optimization of biodiesel production from cottonseed oil by transesterification using NaOH and methanol. *Energy Sourcespart A Recoveryutilizationand Environ. Eff.* **2010**, *32*, 434–441. [CrossRef]

24. Azcan, N.; Danisman, A. Alkali catalyzed transesterification of cottonseed oil by microwave irradiation. *Fuel* **2007**, *86*, 2639–2644. [CrossRef]

25. Rashid, U.; Anwar, F.; Knothe, G. Evaluation of biodiesel obtained from cottonseed oil. *Fuel Process. Technol.* **2009**, *90*, 1157–1163. [CrossRef]

26. Su, C.; Nguyen, H.C.; Pham, U.K.; Nguyen, M.L.; Juan, H.-Y. Biodiesel production from a novel nonedible feedstock, soursop (*Annona muricata* L.) seed oil. *Energies* **2018**, *11*, 2562. [CrossRef]

27. Demirbaş, A. Biodiesel from waste cooking oil via base-catalytic and supercritical methanol transesterification. *Energy Convers. Manag.* **2009**, *50*, 923–927. [CrossRef]

28. Zhang, Y.; Dube, M.A.; McLean, D.D.; Kates, M. Biodiesel production from waste cooking oil: Process design and technological assessment. *Bioresour. Technol.* **2003**, *89*, 1–16. [CrossRef]

29. Issariyakul, T.; Kulkarni, M.G.; Meher, L.C.; Dalai, A.K.; Bakhshi, N.N. Biodiesel production from mixtures of canola oil and used cooking oil. *Chem. Eng. J.* **2008**, *140*, 77–85. [CrossRef]

30. Bobadilla, M.C.; Lorza, R.L.; García, R.E.; Gómez, F.S.; González, E.P.V. An improvement in biodiesel production from waste cooking oil by applying thought multi-response surface methodology using desirability functions. *Energies* **2017**, *10*, 130. [CrossRef]

31. Da Cunha, M.E.; Krause, L.C.; Moraes, M.S.A.; Faccini, C.S.; Jacques, R.A.; Almeida, S.R.; Rodrigues, M.R.A.; Caramão, E.B. Beef tallow biodiesel produced in a pilot scale. *Fuel Process. Technol.* **2009**, *90*, 570–575. [CrossRef]

32. Chung, K.H.; Kim, J.; Lee, K.Y. Biodiesel production by transesterification of duck tallow with methanol on alkali catalysts. *Biomass Bioenergy* **2009**, *33*, 155–158. [CrossRef]

33. Gürü, M.; Artukoğlu, B.D.; Keskin, A.; Koca, A. Biodiesel production from waste animal fat and improvement of its characteristics by synthesized nickel and magnesium additive. *Energy Convers. Manag.* **2009**, *50*, 498–502. [CrossRef]

34. Gürü, M.; Koca, A.; Can, Ö.; Cınar, C.; Şahin, F. Biodiesel production from waste chicken fat-based sources and evaluation with Mg based additive in a diesel engine. *Renew. Energy* **2010**, *35*, 637–643. [CrossRef]

35. Tashtoush, G.M.; Al-Widyan, M.I.; Al-Jarrah, M.M. Experimental study on evaluation and optimization of conversion of waste animal fat into biodiesel. *Energy Convers. Manag.* **2004**, *45*, 2697–2711. [CrossRef]

36. Öner, C.; Altun, Ş. Biodiesel production from inedible animal tallow and an experimental investigation of its use as alternative fuel in a direct injection diesel engine. *Appl. Energy* **2009**, *86*, 2114–2120. [CrossRef]

37. Knothe, G.; Razon, L.F.; Bacani, F.T. Kenaf oil methyl esters. *Ind. Crop. Prod.* **2013**, *49*, 568–572. [CrossRef]

38. Lam, M.K.; Lee, K.T. Mixed methanol-ethanol technology to produce greener biodiesel from waste cooking oil: A breakthrough for $SO_4\ 2-/SnO_2$–SiO_2 catalyst. *Fuel Process. Technol.* **2011**, *92*, 1639–1645. [CrossRef]

39. Paraschivescu, M.C.; Alley, E.G.; French, W.T.; Hernandez, R.; Armbrust, K. Determination of methanol in biodiesel by headspace solid phase microextraction. *Bioresour. Technol.* **2008**, *99*, 5901–5905. [CrossRef]

40. Joker, D. *Calophyllum inophyllum* L. *Seed Leaflet*; No 87; Forest & Landscape Denmark: Hørsholm, Denmark, 2004.

41. Anonim. Anonim. Nyamplung Plant-Based Alternative Energy Development Plan 2010–2014. Ministry of Forestry Republic of Indonesia, 2008. Available online: https://anzdoc.com/draft-rencana-aksi-pengembangan-energi-alternatif-berbasis-t.html (accessed on 30 December 2013). (In Indonesia)

42. Said, T.; Dutot, M.; Martin, C.; Beaudeux, J.L.; Boucher, C.; Enee, E. Cytoprotective effect against UV-induced DNA damage and oxidative stress: Role of new biological UV filter. *Eur. J. Pharm. Sci.* **2007**, *30*, 203–210. [CrossRef]

43. Dweek, A.C.; Meadows, T. Tamanu (*Calophyllum inophyllum* L.) the Africa, Asia Polynesia & Pasific Panacea. *Int. J. Cosmet. Sci.* **2002**, *24*, 341–348.

44. Kraftiadi, S. *Energy Analysis on Nyamplung Oil Making Process*; Department of Agricultural Engineering, Faculty of Agriculture, Bogor Agricultural Institute: Bogor, Indonesia, 2011. (In Indonesia)

45. Fathiyah, S. *Study Process of Oil Purification Nyamplung as Biofuels*; Department of Industrial Technology of Agriculture, Faculty of Agricultural Technology, Bogor Agricultural University: Bogor, Indonesia, 2010. (In Indonesia)

46. Motasemi, F.; Ani, F.N. A review on microwave-assisted production of biodiesel. *Renew. Sustain. Energy Rev.* **2012**, *16*, 4719–4733. [CrossRef]

47. Hsiao, M.-C.; Hou, S.-S.; Kuo, J.-Y.; Hsieh, P.-H. Optimized conversion of waste cooking oil to biodiesel using calcium methoxide as catalyst under homogenizer system conditions. *Energies* **2018**, *11*, 2622. [CrossRef]

48. Hernando, J.; Letón, P.; Matia, M.; Novella, J.; Alvarez-Builla, J. Biodiesel and FAME synthesis assisted microwave: Homogenous batch and flow process. *Fuel* **2007**, *86*, 1644–1646. [CrossRef]

49. Wu, L.; Zhu, H.; Huang, K. Thermal analysis on the process of microwave-assisted biodiesel production. *Bioresour. Technol.* **2013**, *133*, 279–284. [CrossRef]

50. Barnard, T.M.; Leadbeater, N.E.; Boucher, M.B.; Stencel, L.M.; Wilhite, B.A. Continuous flow preparation of biodiesel using microwave heating. *Energy Fuel* **2007**, *21*, 1777–1781. [CrossRef]

51. Saifuddin, N.; Chua, K.H. Production of ethyl ester (biodiesel) from used frying oil: Optimization of transesterification process using microwave irradiation. *Malays. J. Chem.* **2004**, *6*, 77–82.

52. Haryanto, A.; Silviana, U.; Triyono, S.; Prabawa, S. Biodiesel production from transesterification of waste cooking oil with the assistance of micro waves: The effect of power intensity and reaction time on rendement and biodiesel characteristics. *Agritech* **2015**, *35*. (In Indonesia)

53. Chen, K.-S.; Lin, Y.-C.; Hsu, K.-H.; Wang, H.-K. Improving biodiesel yields from waste cooking oil by using sodium methoxide and a microwave heating system. *Energy* **2012**, *38*, 151–156. [CrossRef]

54. Lertsathapornsuk, V.; Ruangying, P.; Pairintra, R.; Krisnangkura, K. Continuous Transethylation of Vegetable Oils by Microwave Irradiation. In Proceedings of the First Thai Energy Network Academic Conference Ambassador of Jomtien City, Chon Buri, Bangkok, 11–13 May 2005; pp. RE11-1–RE11-4.

55. Sherbiny, S.A.E.; Refaat, A.A.; Sheltawy, S.T.E. Production of biodiesel using the microwave technique. *J. Adv. Res.* **2010**, *1*, 309–314. [CrossRef]

56. Anonim. Induction Heating System Topology Review. 2000. Available online: www.fairchildsemi.com (accessed on 5 September 2016).

57. Lucia, O.; Maussion, P.; Dede, E.J.; Burdio, J. Induction heating technology and its applications: Past developments, current technology, and future challenges. *Ieee Trans. Ind. Electron.* **2013**, *6*, 2509–2520. [CrossRef]

58. Koertzen, H.W.; Ferreira, J.A.; van Wyk, J.D. A comparative study of single switch induction heating converters using novel component effectivity concepts. In Proceedings of the IEEE Power Electronics Specialists Conference, Toledo, Spain, 29 June–3 July 1992; pp. 298–305.

59. Dede, E.J.; Gonzalez, J.V.; Linares, J.A.; Jordan, J.; Ramirez, D.; Rueda, P. 25-kW/50-kHz generator for induction heating. *Ieee Trans. Ind. Electron.* **1991**, *38*, 203–209. [CrossRef]

60. Koertzen, H.W.; Wyk, J.D.V.; Ferreira, J.A. Design of the half-bridge series resonant converters for induction cooking. In Proceedings of the IEEE Power Electronics Specialist Conference Records, Atlanta, GA, USA, 18–22 June 1995; pp. 729–735.

61. Kamli, M.; Yamamoto, S.; Abe, M. A 50–150 kHz half-bridge inverter for induction heating applications. *IEEE Trans. Ind. Electron.* **1996**, *43*, 163–172. [CrossRef]

62. Rezon, A. *Half Bridge Inverter Design for Induction Heater Power Supply in Plastic Etruder Tool*; University of Diponegoro: Semarang, Indonesia, 2013. (In Indonesia)

63. Lozinski, M.G. *Industrial Application of Induction Heating*; Pergoman Press: London, UK, 1969.

64. Emil, A.; Yaakob, Z.; Kumar, M.S.; Jahim, J.M.; Salimon, J. Comparative evaluation of physicochemical properties of Jatropha seed oil from Malaysia, Indonesia and Thailand. *J. Am. Oil Chem. Soc.* **2010**, *87*, 689–695. [CrossRef]

65. Hsiao, M.-C.; Kuo, J.-Y.; Hsieh, P.-H.; Hou, S.-S. Improving biodiesel conversions from blends of high- and low-acid-value waste cooking oils using sodium methoxide as a catalyst based on a high-speed homogenizer. *Energies* **2018**, *11*, 2298. [CrossRef]
66. Huppertz, T. Homogenization of milk other types of homogenizer (high-speed mixing, ultrasonics, microfluidizers, membrane emulsification). *Encycl. Dairy Sci.* **2011**, 761–764.
67. Håkansson, A.; Trägårdh, C.; Bergenståhl, B. Studying the effects of adsorption, recoalescence and fragmentation in a high-pressure homogenizer using a dynamic simulation model. *Food Hydrocoll.* **2009**, *23*, 1177–1183. [CrossRef]
68. Ghorbani, A.; Rahimpour, M.R.; Ghasemi, Y.; Raeissi, S. The biodiesel of microalgae as a solution for diesel demand in Iran. *Energies* **2018**, *11*, 950. [CrossRef]
69. Hayyan, A.; Alam, M.Z.; Mirghani, M.E.; Kabbashi, N.A.; Hakimi, N.I.N.M.; Siran, Y.M.; Tahiruddin, S. Reduction of high content of free fatty acid in sludge palm oil via acid catalyst for biodiesel production. *Fuel Process. Technol.* **2011**, *92*, 920–924. [CrossRef]
70. Çayl, G.; Küsefoglu, S. Increased yields in biodiesel production from used cooking oils by a two-step process: Comparison with one step process by using TGA. *Fuel Process. Technol.* **2008**, *89*, 118–122. [CrossRef]
71. Refaat, A.A.; Attia, N.K.; Sibak, H.A.; El Sheltawy, S.T.; ElDiwani, G.I. Production optimization and quality assessment of biodiesel from waste vegetable oil. *Int. J. Environ. Sci. Technol.* **2008**, *5*, 75–82. [CrossRef]
72. Dmytryshyn, S.L.; Dalai, A.K.; Chaudhari, S.T.; Mishra, H.K.; Reaney, M.J. Synthesis and characterization of vegetable oil derived esters: Evaluation for their diesel additive properties. *Bioresour. Technol.* **2004**, *92*, 55–64. [CrossRef] [PubMed]
73. Vicente, G.; Martinez, M.; Aracil, J. Integrated biodiesel production: A comparison of different homogeneous catalysts systems. *Bioresour. Technol.* **2004**, *92*, 297–305. [CrossRef] [PubMed]
74. Özçimen, D.; Yücel, S. *Biofuel's Engineering Process Technology*; Yıldız Technical University, Bioengineering Department: Istanbul, Turkey, 1969; Volume 742, ISBN 978-953-307-480-1.
75. Kapilan, N.; Baykov, B.D. A review on new methods used for the production of biodiesel. *Coal Pet. Coal* **2014**, *56*, 62–73.
76. Sapiuddin, N.; Samiuddin, A.; Kumaran, P. A review on processing for biodiesel production. *Trends Appl. Sci. Res.* **2015**, *10*, 1–37.

Review

The Potential of Renewable Energy in Timor-Leste: An Assessment for Biomass

Lelis Gonzaga Fraga [1,2,*], José Carlos F. Teixeira [1] and Manuel Eduardo C. Ferreira [1]

[1] Department of Mechanical Engineering, University of Minho, 4800-058 Guimarães, Portugal; jt@dem.uminho.pt (J.C.F.T.); ef@dem.uminho.pt (M.E.C.F.)

[2] Department of Mechanical Engineering, Universidade Nacional Timor Lorosa'e, Rua Formosa, 10, Dili, Timor-Leste

* Correspondence: lelisfraga@hotmail.com; Tel.: +351-926946326

Received: 2 February 2019; Accepted: 10 April 2019; Published: 15 April 2019

Abstract: This paper assesses the potential of biomass energy resources in Timor-Leste (TL). Although other renewable energy sources are mentioned in this article, such as wind energy, solar energy, hydropower, bioenergy, including bioethanol and biogas, the main goal is to gather the data on biomass in TL and provide such data as useful information for a wide range of end-users. The current evaluation is based on various sources which include previous assessments on biomass and other renewable sources. The energy potential of biomass in TL apart that resulting from vegetation or flora and animals is also derived from agricultural waste, such as waste from rice, corn, and coffee. The analyses also include the contribution of agricultural waste, animal waste, and that from urban waste. The results from this article show that the potential of usable biomass energy in TL from forestry and agriculture is 1.68×10^6 toe/year, animal waste is 4.81×10^3 toe/year, and urban solid waste amounts to 9.55×10^3 toe/year. In addition, it is concluded that biomass alone can fully replace fossil fuels for electricity generation.

Keywords: renewable energy; biomass; waste

---

1. Introduction

Renewable energy is an important resource, which is widely available in nature. As fossil fuels are becoming scarce and are associated with climate change [1], a great opportunity is presented for the development and utilization of renewable energy [2]. Both government institutions and private sectors are starting investing in renewable energy development, and many developing and emerging countries start working and investing in renewables and related infrastructure. It was estimated that globally, renewable energy accounted for apparently 18.2% of the total final energy consumption in 2016 [3]. The International Energy Agency reported that in 2016, the worldwide total primary energy supply (TPES) was 13,761 Mtoe, of which 13.7%, or 1882 Mtoe (up from 1819 Mtoe in 2015), was produced from renewable sources [4]. The amount of global renewable energy supply, including solid biofuels/charcoal 62.4%, hydropower (2.5% of world TPES and 18.6% of renewables), and the rest, is speed over other sources, such as liquid biofuels, wind, geothermal, solar, biogases, renewable municipal waste, and tide. Depending on the sources and methods, other databases may estimate varying figures and indexes for renewable energy; nonetheless, the scenarios are in agreement with the previous statement. In the overall framework for the sustainable development goals, reported by the International Council for Science and International Social Science Council, amongst the 16 are included: end poverty, ensure healthy lives, ensure equitable quality education, ensure modern energy for all, reduce inequality, etc. These goals are targeted to be achieved in the year of 2030, named as 2030 Agenda. In this Agenda, Goal number seven aims to ensure universal access to affordable, reliable, sustainable, and modern energy for all. This is articulated with the goal of increasing the share of substantially renewable energy in the global energy mix by 2030 [5]. From this overview of renewable energy development, the policymaker has a key role in defining the process and routes to achieve the defined targets [3,4,6].

Biomass is one of the renewable energy sources, which contributes to the sustainable development of energy supply, with future development of bioenergy depending on the available biomass resources and their development [7]. The advantages of biomass, besides being a renewable and inexhaustible fuel source [1], is that it provides employment opportunities in rural areas [8]. Global utilization of biomass is increasing because being considered carbon neutral [9] can address the pressing need to reduce greenhouse gas emissions [10]. Biomass can be obtained from agricultural waste or fast-growing plants [11], industrial residues and process waste, sewage, and animal wastes [12]. Biomass is considered as an accessible and affordable source of energy, especially in rural areas [1]. However, the use of bioenergy for fuel applied in transportation and power sectors may have an impact on food availability and cost. This is a fact that the use of land for bioenergy production competes with alternative outputs from land and can affect agricultural land. The solution for this issue is that biofuel feedstocks must come from land that was neither forested before nor necessary for food crops (set aside land) as adopted by the U.S. Renewable Fuel Standard [13].

Timor-Leste (TL), a tropical country located on the south-east side of the Indonesia archipelago, is approximately 500 miles north of Australia, being washed by the Banda Sea and the Timor Sea to the north and south, respectively [14]. The country's size is 15,410 km^2, including the Oecusse enclave in the western part of the island and the islands of Atauro in the north and Jacob to the east. The territory of TL extends for about 265 km long (East-West) and 90 km wide (North-South) and reaches close to 3000 m in altitude. Administratively TL is divided into 13 districts, and the population of TL was about 1,183,643 inhabitants in 2015 [15]. Although statistical data is recent and unreliable, the total primary energy consumption has not varied substantially over the last years and was approximately 152 ktoe (as of 2014). The CO_2 emissions (resulting exclusively from petroleum products) were 0.36 ton/capita in 2014. This figure is less than an order of magnitude than that in developed countries, such as Western Europe. Nonetheless, this represents a three-fold increase over a decade.

Although TL is a net exporter of crude oil, there are no distilling facilities in the territory. Currently, the energy consumption in TL relies on the use of imported energy, mostly for electricity generation and transportation fuels, which leads to high expenses for the annual energy budget. As an example, in 2007, the government of TL spent approximately US $26 million per year on gasoline, which increased in subsequent years [16].

As a tropical country, TL has the potential for a wide variety of renewable energies, such as solar, wind, hydro, and bioenergy. Such resources of renewable energy, if properly managed, could make a useful contribution to the energy supply and consumption in TL in the future.

Wood biomass is mostly used for heating and cooking purposes in a household. Overall, it is estimated that up to 90% of the energy needs of TL citizens are provided by biomass as the electricity consumption is very low. In fact, the annual electricity consumption per capita is 97 kWh, well below the average in Europe (5500 kWh). Because of the growing interest in biomass as an energy source, many nations have dedicated a great effort into assessing correctly their potential, in order to support the development of policies to promote their use and increase their share of the energy unit. Due to multiple reasons, there is not a comprehensive assessment as such for TL. Although a brief overview of the renewable energy is presented here, the main objective of this paper is to collect the data and information regarding the potential of biomass resource in TL. The amount of biomass, including forestry and agriculture, animal, and urban waste, is calculated in terms of useful energy. It is a mostly scattered information regarding biomass resources, and this paper aims to organize such information into a coherent and structured body that can be useful for future use and to draw the likely impact that endogenous renewable energy resources can have in a small country that is taking its first steps into sustainable development.

2. Research Methods

As mentioned, this paper assesses the potential of renewable energy resources, mostly biomass, in TL. The data included the contribution from forestry, agriculture, urban, and animal waste and was collected by various methods at different times. This section details the various methods deployed.

2.1. Biomass Estimation from Forestry

The methodology applied to estimate the recoverable biomass resources follows various steps [14]: the land was organized in different categories, and for each one, the respective area was quantified; the expected average productivity was assumed; it was defined the period of rotation of species. The estimation of annual increment of biomass above ground was done by applying a simple calculation regarding the annual increment (I, ton/ha/year), average productivity (P, ton/ha), and age of rotation (R, years), with $I = 2P/R$. The usable biomass was estimated by taking into account its accessibility and the dynamics of the ecosystem. Meanwhile, the Shuttle Radar Mission Topography of National Aeronautics and Space Administration (NASA) was used to identify the different elevations of the land.

2.2. Biomass Estimation from Agricultural

The methodology used to estimate the biomass resources from agriculture included [14]: quantification of the annual production of the most representative large scale crops production and their waste for energy utilization in each district; selection of the empirical values for the ratio of production/waste produced, whether resulting in the field or from the crop processing. This data enables to mapping the availability of energy production from agriculture.

2.3. Biomass Collection from Urban Waste

The urban solid waste from Dili city was calculated based on the mean waste carried by trucks (m^3) multiplied by the number of trucks per day [17].

2.4. Biomass Collection from Animal Waste

Regarding the biomass from animal waste, this data is based upon the number of animals, organized by the most relevant species, which were registered in 2009. From this data, the waste was calculated based on the amount (kg or m^3) of biogas produced per animal on a daily basis [14,18].

3. Assessment of Renewable Energy Resources

3.1. Renewable Energy Resources in Timor-Leste

The renewable energy resources in TL, including wind, solar, hydro, and bioenergy, are briefly presented here as they may play an important role in the future energy mix.

In tropical and mountain areas, the potential of wind varies with time as it depends on winds driven by the thermal gradients and the local geographical conditions. In TL, information regarding the potential of wind power has been identified by a preliminary survey, suggesting that TL has several areas suitable for deployment of wind turbines [19]. The State Secretariat for Energy Policy (SEPE) during the term 2007–2012 expected an electricity production with a capacity of 11 MW in the area of Lariguto/Osso (Viqueque district) and Laleia (Manatuto district), combined with a capacity of 2 and 4 kW in the area of Turiscai/Foholau (Manufahi district) and Ataúro/Makadede (Dili district) [20].

TL, as a tropical country, has a huge potential for direct solar energy harvesting. A previous survey shows that the entire territory of TL has the potential to successfully produce solar energy. In terms of average annual range, it varies between 14.85 and 22.33 MJ/m^2 per day, which indicates that the whole territory of TL has the potential to successfully generate solar energy [19]. This compares favorably with Australia that has the highest average solar radiation per square meter of any continent in the world and whose potential annual average of solar radiation varies between 18 and 24 MJ/m^2

per day in the northwest and center of the country, the most favorable regions [21]. Due to the solar energy available in the country, a total amount of 1228 units of solar energy have been installed for family households in remote areas [20].

TL has a potential for developing the mini hydropower in specific districts. The analysis from the Strategic Development Plan (PED, in Portuguese) identified nearly forty locations that could generate power in the range from 1.2 MW to 50 MW [19]. The survey conducted by SEPE in the previous government describes that mini hydropower with the capacity of 36.1 GWh/year and 8.8 GWh/year have been installed in Atsabe (Ermera district) and Bulobo (Bobonaro district), respectively. Besides, a center for mini-hydro with a capacity of 10 kW capable of supplying 166 households, will still be identifying [20]. In addition, in the district of Lautem (Ira-lalaru), the potential of hydropower can be produced with the capacity of 12–28 MW [16], and also one mini hydroelectric power plant with the maximum capacity of 326 kW is located in Gariuai (Baucau district) [22].

3.2. Biomass Potential in Timor-Leste

Bioenergy is the general term regarding the energy derived from carbon-based materials, such as wood, straw or animal waste, where the living matter is relatively recent as opposed to fossil fuels. Such materials can be burned directly to produce heat or power, but can also be converted into biofuels. These (as charcoal and biodiesel) are mostly made from wood and plant seeds [14].

TL has a large potential of biomass as an energy resource. The analysis described in the PED [19] shows that the amount of biomass above the ground in TL is linked with the highest concentration of biomass plants in the tropical forests in upland areas, medium altitude areas, and lowland forest. Because of the terrain, the degree of accessibility of such resources varies significantly, and those from mountain areas are of difficult access.

Various types of biomass from agriculture available in TL, such as from forestry, include tree trunks, branches, leaves, shoots, bark, pine cones, etc. Other biomass sources are from agriculture plantations, such as coffee, rice fields, agricultural parcels with surrounding trees, meadows, etc. [14].

The economic structure of TL is mostly dependent upon the agriculture productivity. The 2001 Census identifies that the agricultural sector is the main activity for about 73% of the population and contributes to 25.4% to the total value of the GDP of TL [14]. This pattern is still valid today. For a large portion of the population, several agriculture plantations are their main livelihood, such as cassava, beans, sweet potatoes, and soybeans, while the most significant crops for life sustain are rice, corn, and coffee, which is the main commercial production. The biomass resources from agriculture waste can be obtained from those productivities. Other vegetable wastes that can contribute to power production include provenance tree, shrub, and herbaceous as a part of the forest and agricultural product. Table 1 summarizes the most important sources of biomass for energy purposes. The data include the total existing biomass for the various categories and the corresponding annual yield. The usable annual yield takes into account how easily the source can be accessed.

Being a mountainous island, some 30% of the amount of biomass actually produced in TL occurs at altitude (mostly varying between 1000 and 2000 m). Previously published data show that 73% of the East Timorese territory is categorized as forestry land, being some of that area as Tropical Monsoon Forest, such as the district of Lautem [14]. In such regions (tropical forest in the highlands and also at the middle and lower altitudes, provided it is dense forest), the annual yield varies between 90 and 160 tons/ha. Overall, the data show that the total stock of biomass in ecosystems suitable for the production of the electricity, including forestry, agriculture, and degraded pastures and forests, is 127,528,335 tons, with an annual increase of 5,222,999 tons/year (4%), much of which is usable (4,911,051 tons/year) [14]. Many species are endogenous to TL. One may take some for comparison purposes with other nations. Taking eucalyptus, in TL, the percentage of biomass in the ecosystem, based on the land occupation, is 15.99% [14]. While the percentage of biomass in the ecosystem in another country, such as Portugal, was 8.8% in the year 2010 [23]. From the data in Table 1, one can also estimate the annual yield (productivity) of eucalyptus. Considering a total area of 246,403 ha dedicated to eucalyptus, the yield rate is 4.5 tons/year/ha. The productivity depends on many factors (soil, water, and climate) but varies between 2.4 and 15 tons/year/ha [24].

Table 1. Amount of the stock biomass in ecosystems for power production. Data reproduced with the permission from the publisher [14]. Notes: a.g. = above-ground; s.o. = soil occupation.

categorized by Soil Occupation (s.o.)	Class of s.o.	Biomass a.g. (ton)	Annual yield of Biomass a.g (ton/year)	Usable Biomass (ton/year)
Forest	Mangrove forest	2893	386	0
	Coastal forest	1,021,854	68,209	68,209
	Humidity tropical forest of high zone, various species	10,608,297	265,207	79,562
	Casuarina	6,192,806	154,820	46,446
	Eucalyptus	22,225,526	1,111,276	1,111,276
	Humidity tropical forest of lower zone, various species	20,319,775	678,007	678,077
	Teak	305,296	15,265	15,265
	Dense humidity forest, of a lower and medium zone	40,985,126	1,639,405	1,639,405
	Sparse humidity forest, of a lower and medium zone	20,704,695	828,647	828,647
	Mount forest	940,339	18,765	1877
	Swamp	10,903	725	0
Agriculture field	Agriculture land (not irrigated)	1,003,663	134,491	134,491
	Coffee plantations	2,221,419	222,142	222,142
	Agriculture production of a coconut tree	470,793	37,663	37,663
	Abandoned agricultural areas	5147	690	690
	Irrigation areas	200,000	26,800	26,800
Degraded pastures and forests	Uncultivated	1594	105	105
	Savannah	82,091	5473	5473
	Pastures	222,097	14,658	14,658
	Shrubs and herbaceous species	4021	265	265
Total		**127,528,335**	**5,222,999**	**4,911,051**

In addition, the distribution through the territory of the amount of estimated biomass (stock) as the total existing biomass above-ground (a.g.), the annual growth and the usable biomass are depicted in Figure 1. The data show that the districts of Manatuto, Viqueque, Lautem, and Covalima are those with the largest resources. In addition, soil productivity does not vary significantly across the country, with an average yield spread of 3–4 tons/year per hectare.

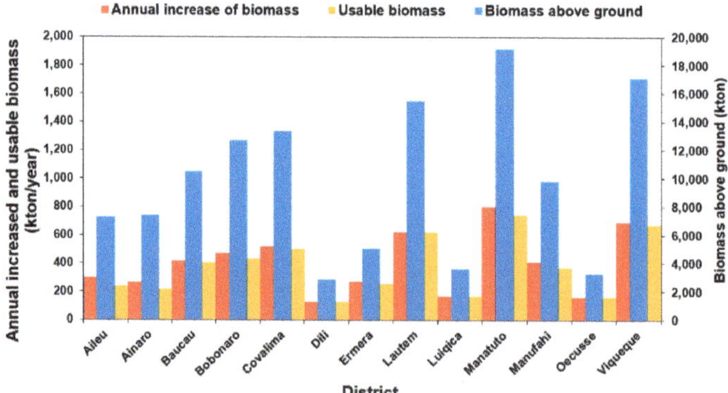

Figure 1. Estimation of the above-ground biomass stock at the district level. Based on data from [14]. Adapted with permission from the publisher.

3.3. Waste Biomass

Waste biomass in TL can be obtained from different sources, including animal waste, agriculture activities, forestry, and urban waste/solid waste (MSW).

3.3.1. Waste Biomass from Agriculture

The residues from agriculture crops, such as rice, corn, and coffee, are also available in the country. The total production of rice is 60,516 tons/year resulting in a waste production of 102,875 tons/year that includes straw and shell at 90,773 and 12,102 tons/year, respectively. From this, the usable waste from rice is 25,718 tons/year. The total productivity of corn is 65,202 tons/year, while the waste production is 84,760 tons/year with a usable waste of 21,190 tons/year. The total production of coffee is 9900 tons/year with a waste production of 9109 tons/year, and the usable waste is 2278 tons/year, as described in Figure 2 [14]. Taking the usable waste data, the waste biomass from corn in TL is approximately 95 GWh/year [14] compared to Portugal which is 3378 GWh/year [23].

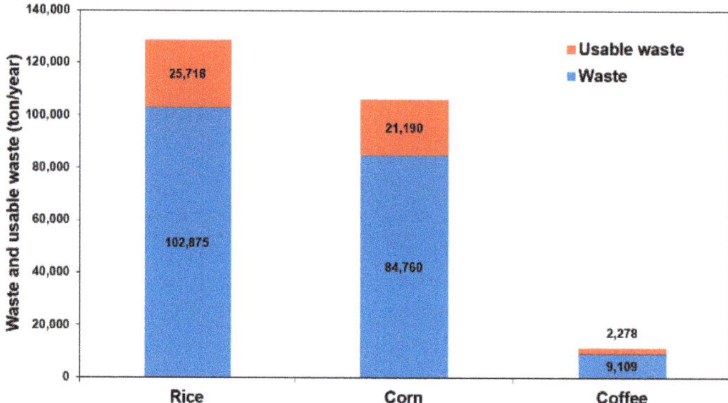

Figure 2. Waste from rice, corn, and coffee plantation. Based on data from [14]. Adapted with permission from the publisher.

The availability of the usable biomass mentioned above, which is composed of agricultural (rice, corn, and coffee), agroforestry, and animals, corresponds to 49,188, 4,901,483 and 60,636 tons/year, respectively. The districts with the highest potential of biomass energy and with usable waste biomass above 500 ktons/year are Manatuto, Viqueque, Lautem, and Covalima [14] (see Figure 3).

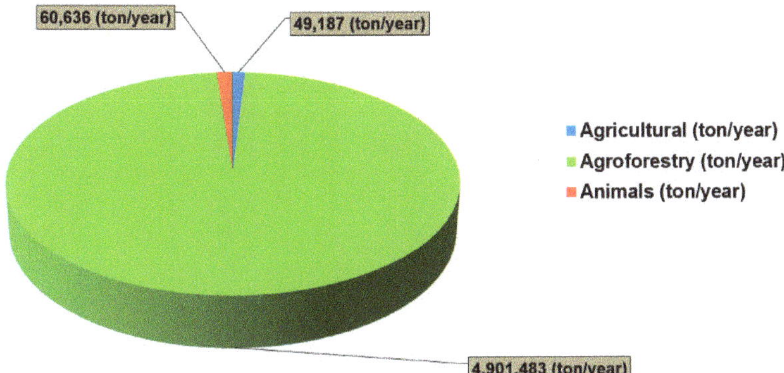

Figure 3. The availability of the usable biomass from agricultural, agroforestry, and animals in the country. Based on data from [14]. Adapted with permission from the publisher.

3.3.2. Municipal Solid Waste

TL has 13 districts; but from population Census of 2015, the three most populated districts are Dili (277,279), Ermera (125,702), and Baucau (123,203) as referred by Ferreira et al. [15]. In this paper, the information on the municipal solid waste (MSW) concerning Dili is presented because being the capital city of the country and also a center for various activities, it produces the maximum waste. Data from 2007 show that the collection of MSW in Dili is about 60 tons every day [16]. The waste in Dili is produced from households, markets, offices, schools, hospitals, and at the bus station. Amongst those, households and markets contribute to the largest share of waste. All the waste is collected and disposed to the final site, located 26 km from Dili city. It shows that the volume of the waste collected is around 158 tons/day in 2012 [17], assuming that for an economic developing country the bulk density of MSW is 660 kg/m^3 [25]. The increase in the amount of MWS from 2007 to 2012 is an indication of the increase in activity and population growth. The population growth in Dili during the period 2004 to 2010 and 2010 to 2015 is 33.17% and 18.48%, respectively, which is the largest among other districts [15,26]. Also, the stabilization of political life after the unrest following independence has contributed to such an evolution. The urban waste collection in Dili compared with another city, such as Kupang (East Nusa Tenggara, Indonesia), in the west of TL shows that in 2009 the waste collected was approximately 203 tons/day [27].

3.3.3. Waste Generated by Animal Husbandry

Based on a preliminary survey, it was estimated that the number of animals in the territory of TL in 2005 was 469,301 ruminants [14]. The animals mentioned in the survey include large species, such as buffalo, cow, and horse, mid-size animals, such as pig, goat, and sheep, and small livestock, such as chickens and ducks, with a total number of those animals estimated at 1,603,700 units (see Table 2). The waste products from those animals are approximately 505,317 tons/year, and the usable waste is 60,636 tons/year (see Table 3) [14].

Several districts, which include Viqueque, Oecusse, Baucau, and Bobonaro, have the potential of animal waste above 5000 tons/year, and from a total potential of approximately 60,000 tons/year, the districts of Viqueque and Baucau may contribute with 28% of the whole resource. The activities conducted by SEPE in the previous government resulted in the construction of two digesters with 20 and 10 m^3, capable of supplying biogas for 300 families in the Manufahi district [20]. Based on the information in TL, domestic and industrial waste can be processed in the factories, especially, designed to produce electricity [19]. The waste generated from animal husbandry in TL from different species mentioned in Table 3 is approximately 56 GWh/year [14] compared to Portugal from three different species, such as chickens, swine, and bovine, which is 725.4 GWh/year [23].

Table 2. Number of domestic animals in the country. Data reproduced with the permission from the publisher [14].

District	Cow	Buffalo	Horse	Goat/Sheep	Pig	Poultry
Aileu	5635	1342	1505	2261	3734	16,557
Ainaro	11,541	10,145	7985	4037	29,717	37,792
Baucau	8422	21,690	9577	32,279	39,608	73,280
Bobonaro	27,587	7142	3162	8396	31,404	85,149
Covalima	17,217	7608	1494	1765	20,049	27,405
Dili	6966	901	968	42,173	72,588	98,279
Ermera	10,768	4703	3389	2647	19,102	34,011
Lospalos	10,143	2682	1458	11,693	41,082	98,299
Luiqica	8967	13,161	3437	5203	21,134	42,279
Manatuto	8651	8588	2823	12,360	14,323	18,840
Manufahi	8664	6026	3655	1805	16,979	28,783
Oecusse	26,818	2037	1950	8652	26,698	129,101
Viqueque	22,714	22,644	6676	5189	48,118	60,088
Total	174,093	108,669	48,079	138,460	384,536	749,863
			1,603,700			

Table 3. Animal waste production and usable waste in the country. Data reproduced with the permission from the publisher [14].

District	Waste Production (ton/year)						Usable Waste	
	Cow	Buffalo	Horse	Goat/Sheep	Pig	Poultry	(ton/year)	(%)
Aileu	6086	1932	813	407	806	596	1273	2.1
Ainaro	12,464	14,609	4312	727	6419	1360	4883	8.1
Baucau	9096	31,234	5172	5810	8555	2638	7587	12.5
Bobonaro	29,794	10,284	1707	1511	6654	3065	6347	10.5
Covalima	18,594	10,956	807	318	4330	987	4436	7.3
Dili	7523	1297	523	7591	15,679	3538	4133	6.8
Ermera	11,629	6772	1830	477	4126	1224	3147	5.2
Lospalos	10,954	3862	787	2105	8874	3539	3369	5.6
Luiqica	9684	18,952	1856	937	4565	1522	4562	7.5
Manatuto	9343	12,367	1525	2225	3094	678	3619	6.0
Manufahi	9357	8677	1974	325	3667	1036	3042	5.0
Oecusse	28,963	2933	1053	1557	5767	4648	5104	8.4
Viqueque	24,531	32,607	3605	934	10,394	2163	9134	15.1
Total	188,020	156,483	25,964	24,925	82,930	26,995	60,636	-
			505,317				-	-

4. Future Development of Biomass Energy in Timor-Leste

Future development of renewable energy in TL is important as the energy demand is increasing, while the stock of energy in terms of fossil fuel is decreasing. In addition, the necessity for creating a good environment to foster the development of renewable resource is an important issue that needs to be considered.

4.1. The State's Policy on Biomass Energy

At approximately 745,174 hectares, the forest area of TL represents approximately 50% of the total land area. Biomass energy is available in certain parts of the country that are state-owned. TL has a law regulating all the natural resources in the country. According to that law, the state will hopefully show an interest in the development of the natural resources in the country, including biomass energy. Recently, the Government of TL has signed new global partnership for supporting the nation in its efforts to be effective in combating and diminishing the obstacle to the sustainability of biomass resources and to develop the energy access to the rural areas [28]. The partnership will be on the Production of Biomass as Sustainable Bioenergy (PBESB), and this project will be focused on the reduction of annual wood consumption for the traditional cooking stove. The previous SEPE has carried out many programs regarding the use of alternative energies, including solar panels, jatropha biofuels, and biogas. In addition, the law on alternative energy has been developed by SEPE [29].

4.2. International Role

International support is also a key element in the development of natural resources in the country. The Millennium Development Goal of achieving that 55% of territory be covered by forest remains an ambitious objective [19]. The consumption of energy from renewable energy sources in TL still represents a very small amount. Nonetheless, it is expected that the energy produced by wind, solar, hydropower, and other renewable sources may contribute to the mitigation efforts of TL to climate change and help to fulfill the obligations under international conventions on climate change [19].

4.3. Potential Consumers of Biomass

Firewood is an essential fuel for cooking in TL. According to the Japan International Cooperation Agency (JICA), it is estimated that in 2002 approximately 93% of households used biomass (firewood) to meet their energy needs, particularly for cooking and heating, which represents the consumption pattern of 1.3 million m^3 [14]. Such a large amount of users for firewood has an impact on the biomass resources in the country [28]. The very traditional method of cooking can result in severe air pollution (mostly airborne particles) and health problems. Although such route is of great relevance, there is a pressing issue and opportunity to develop and disseminate efficient energy devices for domestic use.

4.4. The Possibility of Future Development for Renewable Energy

For TL, as a new country, the development of every sector is important to enhance and improve its economy. Development of renewable energy is a key factor to support the country to secure the energy supply in a sustainable manner.

Nowadays, the government has extended the electricity grid to supply electric power to the entire country. There are two generating power stations to provide electric power for the whole of TL. One is at Hera (Dili district) with a capacity of 120 MW and another in Betano (Manufahi district) with a capacity of 137 MW. These two plants are using heavy fuel oil, but in the future, they can be retrofitted to use diesel or natural gas [30]. Taking into consideration the estimated usable biomass (Table 1) and assuming a 20% efficiency for a dedicated biomass power plant, this resource could guarantee the operation of at least 400 MW around the clock. Considering a typical biomass plant (approximately 10 MW electrical), this biomass could be funneled to approximately 40 plants dispersed throughout the country. This suggests that all the electricity generation could be provided by existing biomass resources, with spare capacity for future development. Such distributed generation capacity is mostly adequate for the country as the distribution network is limited in extension and unreliable in operation.

Energy consumption will certainly increase with population growth and with economic development. Back in 2011, it was expected that electricity consumption would increase from 160 GWh up to 800 GWh by 2020 [19]. In order to support the energy consumption on generator power, the requirement to develop renewable energy becomes significant. The potential renewable energy projects identified by the Strategic Development Plan exceeded 450 MW which is distributed by various technologies, as described in Figure 4 [19].

Based on the above information, the development of renewable energy, especially, in rural areas will benefit the people in remote areas. Several families in TL have benefited from the development of this renewable energy; approximately 8000 families in remote areas now have direct access to energy through the use of renewable energy resources. However, the development of more renewable energy resources is necessary to supply energy to other families and meet the expected future demand. There are still about 50,000 families who are not covered by the network distribution and do not have access to renewable energy systems [19]. In this scenario, biomass can provide a strong backbone for such development.

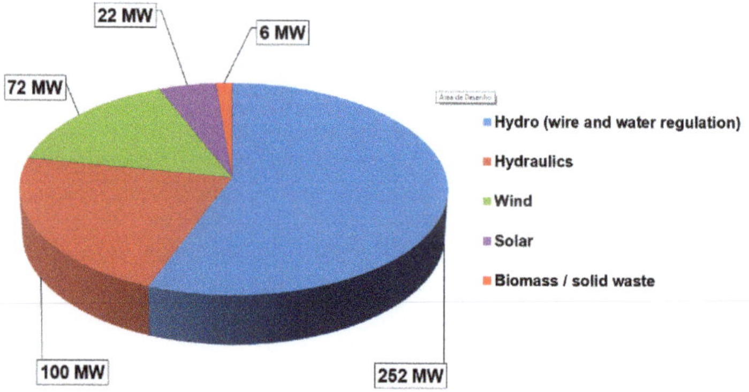

Figure 4. Potential of renewable energy projects technologies [19].

5. Conclusions

This paper presents an overall view of the energy scenario in TL regarding renewable resources. Specific attention is dedicated to biomass. As a tropical country, TL has a potential for developing endogenous renewable energy resources, such as wind, solar, hydropower, and bioenergy, including bioethanol, biogas, and biomass. Some initiatives to develop renewable energy have been implemented, but the absence of further development may prevent sustainable implementation in the future. If the biomass potential is properly managed, it can be developed to produce energy needs. Otherwise, this opportunity could be wasted.

The energy potential of biomass in TL, apart by vegetation or flora and animals, results also from the agricultural waste, such as waste from rice, corn, and coffee, animal, and urban solid waste. The potential of usable biomass energy in TL from forestry and agriculture is 1.68×10^6 toe/year, animal waste is 4.81×10^3 toe/year, and urban solid waste amounts to 9.55×10^3 toe/year.

Forestry is the largest source of biomass while agriculture waste results from rice, corn, and coffee plantations.

The data presented in this paper show that biomass could be used to generate electricity and displace costly fossil fuels either directly through dedicated biomass-powered plants or through co-firing (wood biomass blended with coal). This route has been successfully developed in the European Union [31].

Author Contributions: In this manuscript, all the authors have contributed to developing and finalizing the article: conceptualization, L.G.F. and M.E.C.F.; methodology, L.G.F.; validation, L.G.F., M.E.C.F. and J.C.F.T.; formal analysis, L.G.F.; investigation, L.G.F.; resources, M.E.C.F.; writing—original draft preparation, L.G.F.; writing—review and editing, L.G.F. and J.C.F.T.; visualization, L.G.F. and J.C.F.T.; supervision, J.C.F.T.; project administration, J.C.F.T.; funding acquisition, J.C.F.T.

Funding: This work is supported by FCT with the Reference Project UID/EEA/04436/2013, by FEDER Funds through the COMPETE 2020—Programa Operacional Competitividade e Internacionalização (POCI) with the

Reference Project POCI-01-0145-FEDER-006941. This work was financed by FCT, under the Strategic Project UID/SEM/04077/2013; PEst2015-2020 with the Reference UID/CEC/00319/2013.

Acknowledgments: Lelis Fraga acknowledges Universidade Nacional Timor Lorosa'e for granting the leave of absence to complete his Ph.D.

Conflicts of Interest: The authors declare no conflict of interest.

References

1. Saidur, R.; Abdelaziz, E.A.; Demirbas, A.; Hossain, M.S.; Mekhilef, S. A review on biomass as a fuel for boilers. *Renew. Sustain. Energy Rev.* **2011**, *15*, 2262–2289. [CrossRef]
2. Oyedeji, O.; Fasina, O. Impact of drying-grinding sequence on loblolly pine chips preprocessing effectiveness. *Ind. Crops Prod.* **2017**, *96*, 8–15. [CrossRef]
3. Renewable Energy Policy Network for the 21st Century (REN21). Renewables 2018 Global Status Report. 2018. Available online: http://www.ren21.net/wp-content/uploads/2018/06/17-8652_GSR2018_FullReport_web_final_.pdf (accessed on 11 April 2019).
4. International Energy Agency (IEA). *Renewables Information 2018: Overview*; OECD/IEA: Paris, France, 2018; pp. 1–492.
5. Ürge-Vorsatz, D.; Gomez-Echeverri, L.; Clair, A.L.S.; Jones, F.; Graham, P. Review of targets for the sustainable development goals: The science perspective. 2015. Available online: https://www.researchgate.net/publication/272238116_Ensure_access_to_affordable_reliable_sustainable_and_modern_energy_for_all (accessed on 11 April 2019).
6. FAO. OECD-FAO Agricultural outlook 2018-2027, Special focus: Middle East and North Africa. 2018. Available online: https://www.oecd-ilibrary.org/docserver/agr_outlook-2018-en.pdf?expires=1555005248&id=id&accname=guest&checksum=045BE3E9970080E767AE3736FF60DF09 (accessed on 11 April 2019).
7. Thrän, D.; Seidenberger, T.; Zeddies, J.; Offermann, R. Global biomass potentials—Resources, drivers and scenario results. *Energy Sustain. Dev.* **2010**, *14*, 200–205. [CrossRef]
8. González, J.F.; Ledesma, B.; Alkassir, A.; González, J. Study of the influence of the composition of several biomass pellets on the drying process. *Biomass Bioenergy* **2011**, *35*, 4399–4406. [CrossRef]
9. Euh, S.H.; Kafle, S.; Choi, Y.S.; Oh, J.H.; Kim, D.H. A study on the effect of tar fouled on thermal efficiency of a wood pellet boiler: A performance analysis and simulation using Computation Fluid Dynamics. *Energy* **2016**, *103*, 305–312. [CrossRef]
10. Alakoski, E.; Jämsén, M.; Agar, D.; Tampio, E.; Wihersaari, M. From wood pellets to wood chips, risks of degradation and emissions from the storage of woody biomass—A short review. *Renew. Sustain. Energy Rev.* **2016**, *54*, 376–383. [CrossRef]
11. Posom, J.; Shrestha, A.; Saechua, W.; Sirisomboon, P. Rapid non-destructive evaluation of moisture content and higher heating value of Leucaena leucocephala pellets using near infrared spectroscopy. *Energy* **2016**, *107*, 464–472. [CrossRef]
12. Fraga, L.; Teixeira, J.C.F.; Ferreira, M.E.C. Technical review on biomass resource for wood pellets in Timor-Leste. In Proceedings of the 2018 3rd International Conference, Production of Scientific Knowledge in Timor-Leste, Dili, Timor-Leste, 12–14 September 2018; pp. 1–13.
13. Wang, M.; Dunn, J.B. *Comments on Avoiding Bioenergy Competition for Food Crops and Land by Searchinger and Heimlich*; Argonne National Laboratory: Argonne, IL, USA, 2015; pp. 1–9. Available online: https://www.bing.com/search?q=comments+on+avoiding+bioenergy+competition+for+food+crops+and+land+by+searchinger+and+heimlich&form=EDGTCT&qs=PF&cvid=4476df8f29454458bb981e29bf736f84&refig=58af687fb9194d97cd4e4f1939c35142&cc=PT&setlang=en-GB (accessed on 11 April 2019).
14. Pires, A.M. *Disponibilidade de Biomassa em Timor Leste. Avaliação de potencial e de disponibilidade para aproveitamento para produção de energia eléctrica*; MEGAJOULE: Oporto, Portugal, 2009; pp. 1–70. (In Portuguese)
15. Ferreira, E.d.S. *Launch of the Main Results of the 2015 Census of Population and Housing*; Census presentation; La'o Hamutuk: Dili, Timor-Leste, 2016; Available online: https://www.laohamutuk.org/DVD/DGS/Cens15/Launch_2015_census_presentationEn.pdf (accessed on 11 April 2019).
16. MacDougall, J. *13591STL-Crise da Electricidade (2-Hotu). (Expand messages by Vicente Mau Bocy)*; Suara Timor Lorosae: Dili, Timor-Leste, 2007.

17. Costa, L.; Meitiniarti, V.I.; Mangimbulude, J.C. Manajemen sampah perkotaan di kota Dili, Timor Leste. *BioS Maj. Ilm. Semipopuler* **2012**, *5*, 1–7. (In Indonesian)

18. Esteves, T.C.d.J. Study of the bioenergy potential in the centre region of Portugal. Master Thesis, Univerdade Nova de Lisboa, Lisbon, Portugal, November 2010.

19. PED. *Timor-Leste plano estratégico de desenvolvimento 2011–2030*; Republic Democratic of Timor-Leste: Dili, Timor-Leste, 2011; pp. 1–279. (In Portuguese)

20. MF. *Orçamento geral do estado 2012 planos de acção anual livro 2*; Republic Democratic of Timor-Leste: Dili, Timor-Leste, 2012; pp. 1–384. (In Portuguese)

21. Geoscience Australia and Abare. Solar energy. In *Australian energy resource assessment*; 2010; pp. 261–284. Available online: https://www.arena.gov.au/assets/2013/08/Chapter-10-Solar-Energy.pdf (accessed on 11 April 2019).

22. Hoeiseth, J.; Klei, K. Gariuai mini HEP: The first hydroelectric plant in a new country. In Proceedings of the 2007 International Conference on Small Hydropower, Kandy, Sri Lanka, 22–24 October 2007; pp. 1–9. Available online: http://www.ahec.org.in/links/International%20conference%20on%20SHP%20Kandy%20Srilanka%20All%20Details/Papers/Policy,%20Investor%20&%20Operational%20Aspects-C/C22.pdf (accessed on 11 April 2019).

23. Ferreira, S.; Monteiro, E.; Brito, P.; Vilarinho, C. Biomass resources in Portugal: Current status and prospects. *Renew. Sustain. Energy Rev.* **2017**, *78*, 1221–1235. [CrossRef]

24. Forest plantation yields in the tropical and subtropical zone. Available online: http://www.fao.org/3/X8423E/X8423E08.htm (accessed on 22 March 2019).

25. Sebastian, R.M.; Kumar, D.; Alappat, B.J. A technique to quantify incinerability of municipal solid waste. *Resour. Conserv. Recycl.* **2019**, *140*, 286–296. [CrossRef]

26. Statistika, D.N. *ATLAS—Sensus ba populasaun ho uma-kain 2004*; Republic Democratic of Timor-Leste: Dili, Timor-Leste, 2006; pp. 1–92.

27. Jemaga, G. Pentingnya pengelolaan sampah di Kota Kupang. 2011. Available online: http://administrasikebijakankesehatan.blogspot.com/2011/03/pentingnya-pengelolaan-sampah-di-kota.html (accessed on 11 April 2019).

28. Press-Release. *Parseria foun entre MOP no PNUD atu kombate deflorestasaun, salva vida no kria kampu serbisu*; United Nations Development Programme (UNDP): Dili, Timor-Leste, 2014; (In Tétum).

29. Hamutuk, L. *Report Title*; Annual Report; La'o Hamutuk: Dili, Timor-Leste, 2011; pp. 1–35.

30. Hamutuk, L. Power plant and national electrical grid 2011-17. Available online: https://www.laohamutuk.org/Oil/Power/2011/11PowerPlant2011.htm (accessed on March 11, 2019).

31. Hoefnagels, R.; Junginger, M.; Faaij, A. The economic potential of wood pellet production from alternative, low-value wood sources in the southeast of the U.S. *Biomass Bioenergy* **2014**, *71*, 443–454. [CrossRef]

Article

Potential Protein and Biodiesel Sources from Black Soldier Fly Larvae: Insights of Larval Harvesting Instar and Fermented Feeding Medium

Chung-Yiin Wong [1,2,*], Siti-Suhailah Rosli [1,2], Yoshimitsu Uemura [2], Yeek Chia Ho [3], Arunsri Leejeerajumnean [4], Worapon Kiatkittipong [5], Chin-Kui Cheng [6], Man-Kee Lam [7] and Jun-Wei Lim [1,2,*]

[1] Department of Fundamental and Applied Sciences, Universiti Teknologi PETRONAS, Seri Iskandar 32610, Perak Darul Ridzuan, Malaysia; suhailah9875@gmail.com

[2] Centre for Biofuel and Biochemical Research, Institute of Self-Sustainable Building, Universiti Teknologi PETRONAS, Seri Iskandar 32610, Perak Darul Ridzuan, Malaysia; yoshimitsu_uemura@utp.edu.my

[3] Department of Civil and Environmental Engineering, Universiti Teknologi PETRONAS, Seri Iskandar 32610, Perak Darul Ridzuan, Malaysia; yeekchia.ho@utp.edu.my

[4] Department of Food Technology, Faculty of Engineering and Industrial Technology, Silpakorn University, Nakhon Pathom 73000, Thailand; LEEJEERAJUMNEAN_A@su.ac.th

[5] Department of Chemical Engineering, Faculty of Engineering and Industrial Technology, Silpakorn University, Nakhon Pathom 73000, Thailand; KIATKITTIPONG_W@su.ac.th

[6] Faculty of Chemical and Natural Resources Engineering, Universiti Malaysia Pahang, Lebuhraya Tun Razak, Gambang 26300, Pahang, Malaysia; chinkui@ump.edu.my

[7] Department of Chemical Engineering, Universiti Teknologi PETRONAS, Seri Iskandar 32610, Perak Darul Ridzuan, Malaysia; lam.mankee@utp.edu.my

[*] Correspondence: johnsonwcy@gmail.com (C.-Y.W.); junwei.lim@utp.edu.my (J.-W.L.)

Received: 10 January 2019; Accepted: 25 February 2019; Published: 25 April 2019

Abstract: Primarily produced via transesterification of lipid sources, fatty acid methyl ester (FAME) of biodiesel derived from insect larvae has gained momentum in a great deal of research done over other types of feedstock. From the self-harvesting nature of black soldier fly larvae (BSFL), research had, however, only concentrated on the harvest of BSFL on sixth instar. Through rearing BSFL on coconut endosperm waste (CEW), 100 BSFL were harvested at the fifth and sixth instar, then modification on CEW with mixed-bacteria powder was carried out. It was found that the fifth instar BSFL had 34% lipid content, which was 8% more than the sixth instar. Both instars had similar corrected protein contents around 35–38%. The sixth instar BSFL contained around 19% of chitin, which was about 11% more than the fifth instar. Biodiesel products from both instars showed no differences in terms of FAME content. With modification on CEW, at 0.5 wt% of mixed-bacteria powder concentration, the maximum waste-to-biomass conversion (WBC) and protein conversion (PC) were achieved at 9% and 60%, respectively. Moreover, even with the shorter fermentation time frame of CEW, it did not affect the development of BSFL in terms of its WBC and PC when fed with 14 and 21 days fermented medium. FAME from all groups set, which predominantly constituted about C12:0 at around 60%, followed by C14:0 at around 15%, C16:0, and C18:1 both at 10% on average. Lastly, the FAME yield from BSFL was improved from 25% (sixth instar) to 33% (fifth instar) and showed its highest at 38.5% with modification on raw CEW with 0.5 wt% mixed-bacteria powder and fermented for 21 days. Thus, harvesting BSFL at earlier instar is more beneficial and practical, as it improves the FAME yield from the BSFL biomass.

Keywords: black soldier fly larvae (BSFL); instar; lipid; fatty acid methyl ester (FAME); fermentation

1. Introduction

In recent years, investigations focusing on oleaginous microorganisms to produce biodiesel have initiated a third-generation biofuel into the renewable energy research realms [1–3]. However, oleaginous microorganism, which is defined as microbial (microalgae, bacterium, yeast, or fungi) with rich lipid content usually exceeding 20% [4–6] are often apt to be more buoyant or suspended in their cultivation medium, hence resisting settling in [5,7]. This unfortunate scenario has directly incurred high biomass harvesting expenses rising from the intensive time and energy required in handling such large microbial feedstock volume. At times, this accounts for up to approximately 20–30% of total production cost [8]. Thus, insect-derived biodiesel could be a better choice compared to oleaginous microorganisms to generate biodiesel. The rationale for this is that through insect farming, several biochemical products and by-products can be obtained, including insect-derived lipid, protein, and biodiesel [9]. In addition to that, solid waste could be treated and converted into value added products such as vermicomposting bio-fertilizer, which possess similar performances as conventional fertilizer [10].

Looking from an economical perspective, a rapid increment in global population has forced researchers to find alternative protein sources to sustain global demands. However, common protein from animal sources is not sustainable due to its higher energy input and global warming potential, high emission of green house gas (GHG), higher cost in terms of water and feed consumed, and huge land area requirement. Currently, over 1000 species of insects are used to fulfill protein demand; insects are more viable, as they have higher proficient feed conversion ratio, are able to consume numerous type of feed, and have swift growth rates and shorter life cycles compared to known poultry [11]. In general, insect possess up to 40 to 75% (dry weight basis) crude protein content, and this value solely depends on the insect species and the stage in their life cycle. Insects could be a food source either consumed raw, processed into fine powder, or integrated into animal feed. Insect diet consists of amino acids, which are greater than common poultry formulation and higher in protein content compared to soybean and conventional fish meal [12]. Of the thousand insect species in question, these demands are met mostly from applying black soldier fly larvae (BSFL) biomass into animal diet to replace common diet meal. Black soldier fly larvae are favored due to their higher protein content and greater amino acid composition [13–16].

Among the insect larvae, several species were investigated to measure their respective performances on biodiesel production, which were flesh fly, superworm, mealworm beetle, housefly, latrine blowfly, soldier fly, and ants [17–23]. Among these insect feedstocks, the usage of BSFL as the feedstock for lipid and protein content had been revealed. BSFL are saprophagous and polyphagous, as they are able to consume decaying matters and a wide variety of biomass [24]. Black soldier fly under the family of Stratiomyidae ranges from yellow, green black, or blue in color, and its appearance mimics the bees or wasps. BSFL stage consists of six instars and lasts for 14–22 days for biomass accumulation, yet the time frame depends on the feed medium and environmental factors as well. BSFL enter the last stage of the larvae stage, known as pre-pupae, where BSFL stop feeding and pupae at the dry surfaces. Pupation may take up to 14 days, and adults normally survive for about 10 days. Mating begins two days after adult emergence, and female BSFL oviposit the eggs at the crevices or cracks. The whole development process of BSF from birth to adulthood might take up to 40–43 days [25]. Lipid fractions of BSFL were determined in a previous study that showed BSFL contained mainly $C12:0$ (38.43 wt%), $C16:1$ (15.71 wt%), $C14:0$ (12.33 wt%), $C18:1$ (8.81 wt%) and $C18:0$ (2.95 wt%) [26].

Due to the natural self-harvesting ability of BSFL while reaching the pre-pupae stage, they have been widely studied for biodiesel production. During this stage, BSFL have their mouth transformed into a beak-like structure to crawl out from the feeding substrate and pupae on the dry surfaces. This phenomenon is known as the self-harvesting ability, and this process eases the separation of pupae from the feed medium [25]. However, separation and sorting processes are still required to separate pupae from feed residues, exuviae, and pre-pupae. Common ways to separate BSF pupae from the diet residue still mainly utilize fine mesh of bamboo baskets, and sorting solely relies on manpower done by hand [27].

Furthermore, BSFL biomass also contains high protein (~40%) and lipid (~30%) sources [28], which could plausibly be used as aquaculture, poultry, and broilers feed. For example, BSF lipid was introduced into aquaculture feed to promote growth, fatty acid, and lipid deposition in juvenile Jian carp. The results showed that there were no significant differences in growth rate, but the fatty acid composition (mainly C12:0 and C14:0) was visibly affected by the inclusion of BSF lipid. Hence, BSF lipid can be used to replace soybean oil in aquaculture feed since it can be introduced at 100% composition without any negative impacts on the growth of Jian carp [29]. Besides, alongside BSF lipid, BSF biomass can also be introduced into feed medium to replace common protein sources. Substitution of fish meal protein to defatted BSF biomass did not affect the growth performances of Jian carp, and it boosted the antioxidant status of Jian carp. However, substitution of BSF biomass greater than 75% in weight percentage caused intestinal damage and dietary stress. Hence, 50% substitution was recommended from the previous study [30]. Nevertheless, at this point in time, most of the research was performed utilizing BSFL pre-pupae biomass for biodiesel production [23,31,32]. Yet, during the pre-pupae stage, BSFL stopped feeding, and all the metabolic cost depended on the fat body tissues that stored during previous instars.

In Malaysia, the large areas of the palm oil plantation (3.87 million ha.), coconut plantation (147 thousand ha.), etc., have generated large amounts of cellulosic and non-cellulosic waste materials, which are causing environmental problems [33]. Yearly, 0.747 million tons of coconut shells, 0.374 million tons of husks, and 0.35 million tons of coconut copra are produced [34]. This huge amount of agricultural waste goes to landfill, causing soil deterioration and contamination of underground water. Through vermicomposting, these solid wastes can be reduced and transformed into value-added biochemical products. Compared to other types of feeding medium such as food waste, fruit waste, municipal waste, and animal manure, coconut endosperm waste (CEW) was selected as the feed medium in this BSFL experiment due to its year-long availability and easy handling necessities of homogenous quantity. Ultimately, the intention is to reduce waste accumulated in landfill. To that end, the primary objective of this study is to determine the impacts of harvesting periods and the modification of CEW as a feeding medium on BSFL's lipid, protein, and FAME content.

2. Materials and Methods

2.1. Acquisition of Neonate BSFL

The raw CEW was introduced into the opaque cylindrical plastic container with a diameter of 16.5 cm and height of 24.5 cm to serve as a bait. The moisture content of raw CEW was adjusted to 60–70% (wt/wt) of dry weight basis, and the thickness of the medium was about 3–5 cm. The corrugated boards were cut into 6 cm × 2 cm (length × width) and attached to the inner surface of the plastic container at about 4–6 cm above the medium, serving as a stand for female black soldier fly to oviposit eggs. The baits were placed at the natural vegetated area and checked every alternate day for BSF egg availability. The attached eggs on the corrugated boards were collected and examined under optical light microscope, as shown in Figure 1. The red dots indicate the head capsule part, and the blunt end indicates the bottom part. The eggs with live BSFL were then transferred to a sterile Petri dish, and the moisture was maintained by a wet filter paper with sterile distilled water and left to undergo eclosion.

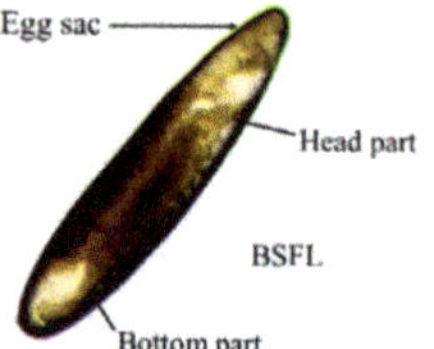

Figure 1. Black soldier fly larvae (BSFL) egg under magnification power of 40× through optical light microscope.

2.2. Acquisition of Coconut Endosperm Waste and Its Properties

Raw CEW was obtained from a local coconut milk seller in Seri Iskandar. Moisture content of raw CEW was determined by gravimetric method. Firstly, 10 g of raw CEW was weighed and dried at 105 °C until constant weight was obtained. The moisture content and dry matter were determined using Equations (1) and (2):

$$\text{Moisture content (\%)} = [(CEW_0 - CEW_f)/CEW_0] \times 100\% \tag{1}$$

$$\text{Dry matter (\%)} = (CEW_f/CEW_0) \times 100\% \tag{2}$$

where CEW_0 represents initial weight and CEW_f represents final weight of raw CEW, respectively. Next, the nitrogen content of raw CEW was determined using the Dumas method (Perkin Elmer 2400 series). Assuming proteins from coconut ranged from 18.5 to 19% nitrogen content [35], crude protein was determined by multiplying the total nitrogen content with the conversion factor of 5.30.

2.3. Preparation of Raw Coconut Endosperm Waste as Feeding Medium

Upon the determination of moisture content, the desired amount of distilled water was added to the raw CEW to achieve 70% moisture content (wt/wt) of dry weight basis. The amount of distilled water needed was calculated using Equation (3):

$$V_{H2O} \text{ (mL)} = \frac{(D_{H2O})(M_S)}{1 - D_{H2O}} - M_{H2O} \tag{3}$$

where V_{H2O} is the desired amount of distilled water to be added, D_{H2O} is the desired proportion of water in the medium, M_S is the oven dried weight of 100 g medium, and M_{H2O} is the mass of water in 100 g of oven dried medium [36]. The moisture adjusted CEW was used as a feeding medium to rear the 6 days old BSFL. The feed medium was introduced to 100 BSFL that were kept in a cylindrical container capped with a ventilated lid. The feeding was discontinued when the BSFL reached the fifth and the sixth instar by determining their head sizes and color changes, as indicated in Figure 2 [37]. The collected BSFL were then separated from the residue, water washed, and inactivated at 105 °C for 5 min before drying at 60 °C until constant weight was obtained [5,23]. The lipid content, protein content, and chitin content of BSFL at the fifth and the sixth instar were determined.

Figure 2. BSFL at the sixth (left) and the fifth (right) instar.

2.4. Fermented CEW Preparation and BSFL Rearing

After determining the best harvesting instar stage, a modification of raw CEW was done by introducing different concentrations of mixed-bacterial powder (Reckitt Benckiser, UPN:1920080310) at 0.02, 0.5, 0.1, and 2.5 wt% into CEW and mixed homogenously. The moisture of CEW was adjusted following Equation (3) and transferred into a polyethylene container and capped tightly to facilitate an anoxic fermentation at the sun-shaded area. The fermentation was carried out for 28 days. After fermentation, 10 g of CEW was administrated to 20 BSFL. The rearing was discontinued when the BSFL reached the fifth instar instead of the sixth instar. Then, once the performances of different mixed-bacteria powder concentration on BSFL growth were validated, the selected mixed-bacteria powder concentration was then mixed again with raw CEW with 70% moisture content and fermented for 7, 14, 21, and 24 days before it was administrated to 20 BSFL. The rearing was discontinued once the BSFL reached the fifth instar, and the samples were processed as described earlier. Next, the waste-to-biomass conversion (WBC) of BSFL was calculated using Equation (4) [38].

$$\text{WBC (\%)} = \frac{\text{BSFL}_{\text{DM}} \text{ (g)}}{\text{CEW}_{\text{DM}} \text{ (g)}} \times 100\% \tag{4}$$

where BSFL_{DM} and CEW_{DM} are the total dry matter in the BSFL biomass, and CEW is introduced as feed correspondingly.

2.5. Biochemical Products Analysis

2.5.1. Lipid Extraction

Lipid from BSFL biomass was extracted using petroleum ether (boiling point at 40–60 °C) solvent extraction. Approximately 0.1 g of grounded dried BSFL biomass was weighed, and 20 mL petroleum ether was added. The mixture was stirred for 24 h using a magnetic stirrer. The solvent layer was separated by filtration using filter paper, and the biomass residue was washed with 10 mL of petroleum ether twice. Then, the solvent layers were combined and dried under rotary evaporator. The extracted BSFL lipid was further dried in an oven at 105 °C for 1 h and cooled to room temperature in a desiccator. The weight of dried BSFL lipid was determined using the gravimetric method. The lipid content of BSFL was determined using Equation (5).

$$\text{Lipid content (\%)} = [\text{Lipid}_{\text{BSFL}} \text{ (g)}/\text{Biomass}_{\text{BSFL}} \text{ (g)}] \times 100\% \tag{5}$$

where $\text{Lipid}_{\text{BSFL}}$ is the total dried weight of BSFL, and $\text{Biomass}_{\text{BSFL}}$ is the total dried mass of BSFL used for lipid extraction.

2.5.2. Nitrogen Content and Protein Conversion of BSFL

Nitrogen content of BSFL was determined using the Dumas method (Perkin Elmer 2400 series). Assuming proteins from animal tissue contain 16% nitrogen content [35], crude protein was determined by multiplying the total nitrogen content with the conversion factor of 6.25. However, with the presence of chitin in BSFL, which accounted for about 6.89% of the nitrogen content, the direct protein conversion from total nitrogen might have led to an over-estimation of protein content. In order to recalculate the corrected protein content, chitin contents at the fifth and the sixth were determined [39], and the chitin-derived nitrogen content was subtracted from the total nitrogen content before protein conversion. Then, the amount of protein from CEW being converted to protein in BSFL, known as protein conversion (PC), was determined by Equation (6) [38].

$$\text{Protein conversion (\%)} = \frac{\text{BSFL}_{DM}\ (g)\ \times\ \text{Protein BSFL}_{DM}\ (\%)}{\text{CEW}_{DM}\ (g)\ \times\ \text{Protein CEW}_{DM}(\%)} \times 100\% \tag{6}$$

where BSFL_{DM} and CEW_{DM} are the dry matter of the BSFL and initial CEW introduced, and Protein BSFL_{DM} and Protein CEW_{DM} are the percentage of crude protein (dry basis) in BSFL and CEW, respectively.

2.5.3. Esterification and Trans-Esterification, FAME Content Analysis, and FAME Yield

The extracted lipid then underwent esterification reaction with 1% HCl in methanol at 75 °C for 60 min. The solution was then dried with anhydrous sodium sulphate to remove water produced. The ester and un-esterified lipid was recovered using petroleum ether and underwent trans-esterification reaction with 1% potassium hydroxide in methanol at 65 °C for 30 min [40]. The solution was then washed with 10% sodium chloride solution, dried with anhydrous sodium sulphate, and stored at 4 °C prior to gas chromatography (GC) analysis. Analysis of FAME in biodiesel was performed using a Shimadzu model GC-2010 plus system (Shimadzu, Kyoto, Japan) equipped with a polyethylene glycol capillary column BPX-BD20 (SGE, Melbourne, Australia) (30 m × 0.32 mm × 0.25 mm) and a flame ionization detector (Shimadzu, Kyoto, Japan). Helium was used as the carrier gas at a flow rate of 1.72 mL/min and a pressure of 83.9 kPa. The inlet was operated in split mode (50:1) at a temperature of 250 °C. The column temperature was programmed as follows: Holding at 150 °C for 1 min, increasing to 240 °C at 5 °C/min, and holding at 240 °C for 6 min. The methyl heptadecanoate (C17:0) was selected as the internal standard (ISTD) in determining the composition of FAME in biodiesel. Accordingly, 1.0 mL of heptane containing 1.001 mg/mL of methyl heptadecanoate (ISTD) was added to the vial containing biodiesel. Upon handheld reverse mixing, the mixture was filtered through a PTFE filter with a pore size of 0.20 μm before transferring into a GC vial. A sample volume of 1 μL was injected into the column, and FAME content in biodiesel was calculated based on Equation (7) [41].

$$\text{FAME content (\%)} = \frac{A_{FAME}}{A_{ISTD}} \times \frac{C_{ISTD} \times V_{ISTD}}{m} \times 100\% \tag{7}$$

where A_{FAME} is the peak area of specific FAME, A_{ISTD} is the peak area of C17:0, C_{ISTD} is the concentration of C17:0, which was 1.001 mg/mL, V_{ISTD} is the volume of C17:0, which was 1.0 mL, and m is the mass of the sample used to mix with C17:0. FAME yield from BSFL biomass was calculated using Equation (8).

$$\text{FAME yield (\%)} = [\text{FAME content} \times M_{Lipid}]/M_{BSFL} \tag{8}$$

where M_{Lipid} is total mass of the lipid and M_{BSFL} is total mass of BSFL biomass on dry weight basis.

2.6. Statistical Analysis

Throughout this study, all experiments were triplicated, and data for BSFL development under different treatment were analyzed with ANOVA followed by a Tukey Post-hoc pairwise comparison test at a level of significance, $\alpha = 0.05$, using the Minitab program (Version 17, Minitab Pty Ltd, NSW, Australia).

3. Results and Discussion

3.1. Impacts of Harvesting Instar on BSFL's Lipid and Protein Content

One hundred BSFL were collected from two different instar stages, which were the fifth and the sixth instar. As shown in Table 1, the fifth instar of BSFL contained higher amounts of lipid compared to the sixth instar, which were 34% and 26%, respectively. This gap potentially occurred because during the fifth instar, BSFL was still in an active ingestion state, allowing it to accumulate more body mass and lipid into its own. Meanwhile, when BSFL reached the sixth instar, its feeding mouthpart transformed into a claw shaped beak for it to migrate from the feeding medium and undergo pupation. At the same time, the digestive tract of BSFL was emptied out, and all of the metabolism cost solely depended on the fat body tissues that accumulated in the previous larvae forms [25]. Generally, the metamorphosis of BSFL from the fifth to the sixth instar can last for weeks; therefore, it would greatly influence the lipid content of BSFL, as parts of it would be used to maintain the BSFL at an active stage.

Table 1. Comparison of lipid, chitin, nitrogen, protein, and corrected protein content between the fifth and the sixth instars.

Properties	Fifth Instar	Sixth Instar
Lipid content (%)	34.23 ± 0.65	25.88 ± 0.36
Chitin content (%)	7.61 ± 0.93	18.62 ± 1.25
Nitrogen content (wt%)	6.07 ± 0.01	7.32 ± 0.06
Protein content (%)	37.94 ± 0.09	45.72 ± 0.40
Corrected protein content (%)	34.66 ± 0.31	37.70 ± 0.14

Moreover, when the nitrogen content was determined, it was found that the sixth instar BSFL contained 7.32% nitrogen, which was 1.25% more than the fifth instar. This phenomenon could be explained by the sixth instar containing higher chitin levels compared to the fifth instar, which were at around 19% and 8%, respectively. Chitin, a polymer of $(C_8H_{13}O_5N)_n$, contains 6.89% nitrogen, and the Dumas combustion method was practiced to determine the total nitrogen content of the substances. Hence, it was necessary to deduct the percentage of nitrogen from chitin from the total nitrogen content prior to protein conversion calculation to avoid overestimation of protein content. The calculated protein content is referred to as corrected protein content. The corrected protein content of BSFL between the fifth and the sixth instar was approximately 35% and 38%, with only a 3% minor difference.

After the lipid was extracted from BSFL biomass, the residual BSFL biomass could be further processed into animal feed or human consumption as a protein source. As the human population keeps increasing, the living standard is enhanced, and the need for better food, such as protein sources from livestock, is increased too. This greatly impacts the environment—for example, dairy farming generates most of the methane gases that affect the balance of nature and cause global warming. Hence, these BSFL biomass residuals are great candidates to fulfill this niche, since they could be a potential protein source replacement and can be used as feed meal in aquaculture, poultry, broilers, as well as for human consumption. However, chitin should be removed prior to feed administration, as chitin in BSFL biomass might reduce feed intake and nutrient availability in aquaculture (Turbot, African catfish), as it reduces growth performance and nutrient assimilation at lower or higher inclusion rates [42,43]. Yet, chitin from BSFL could be further processed into chitosan, which is widely used as biopolymers in biotechnology, pharmaceutical, food, cosmetic, textile, paper, and wastewater treatment industries [15,44].

3.2. FAME Content, Yield, and Profile of Fifth and Sixth Instar of BSFL

Extractable lipid from both instars was then esterified and trans-esterified into biodiesel, and the FAME content was determined. It was noted that sample products from both the fifth and the sixth instar had high FAME contents around 97% and 98%, respectively. FAME profiles from both sample products were analyzed, and it was found that BSFL-derived biodiesel contained mostly C12:0, which accounted for about 50–55%, followed by C14:0 at around 20%, C16:0 and C18:1 at around 10%, and little amounts of C6:0, C10:0, C14:1, C16:1, and C18:0. There were very small differences of FAME obtained between the fifth and the sixth instar of BSFL, as shown in Figure 3. However, in terms of the FAME yield, the fifth instar BSFL indicated around 33%, which was higher than the sixth instar with a difference of around 8%, which was due to the higher lipid content in the fifth instar and that heavier BSFL was harvested at this instar. The total dried biomass for 100 BSFL was 5.24 g for the fifth instar BSFL, and only 3.70 g for the sixth instar BSFL.

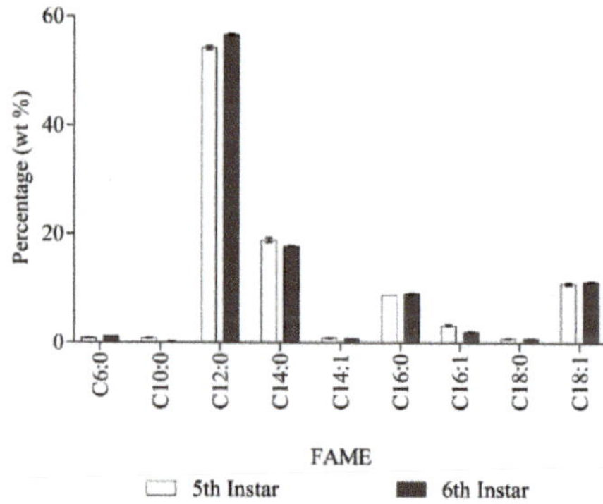

Figure 3. Fatty acid methyl ester (FAME) profile of biodiesel from the fifth and the sixth instar BSFL.

On an industrial scale, to increase yearly product batch and product yield, it is advised to harvest BSFL at the fifth instar since it is more beneficial than harvesting at the sixth instar. Most of the previous studies focused on the self-harvesting ability of BSFL, as they migrate away from the wet feeding medium and go through pupation on dry surfaces [45]. Yet, the current method practiced in the industry today still applies the manual sieving method to separate the larvae from the feeding medium [27]. Thus, it would be more profitable to harvest the BSFL at the fifth instar, as harvesting BSFL in this earlier instar does not affect the fatty acids present in the BSFL biomass much, and both the fifth and the sixth instar would still have to go through the sieving process for BSFL harvesting purposes, and harvesting the fifth instar could be more time and cost effective.

3.3. Effects of Different Concentrations of Mixed-Bacteria Powder on BSFL Development

The BSFL development in terms of waste-to-biomass conversion and protein conversion are presented in Figure 4. Under a controlled system with equal raw CEW, the BSFL attained a WBC at around 6%, and this value was increased to around 8% through a 0.02 wt% and 0.1 wt% increment of mixed-bacteria powder concentration. The highest value of WBC was reached only at a higher concentration of mixed-bacteria powder, which was at 0.5 wt% and showed around 9%. Further increment of mixed-bacteria powder to 2.5 wt% did not affect much on WBC, as it only dropped a little, to around 8.5%. From this result, it was assumed that at lower concentrations of mixed-bacteria powder, the fermented feed was not mature enough for BSFL consumption, resulting in a low waste-to-biomass conversion [9]. However, a higher mixed-bacteria concentration at 2.5 wt% led to a decrement in WBC due to the nutrient in fermented raw CEW being consumed by competitive microorganisms. In terms of protein conversion, at the control system, it showed 40% of PC. This value maintained even with increasing mixed-bacteria powder concentration to 0.02 wt% and 0.1 wt%. A significant boost of PC could be seen when 0.5 wt% of mixed-bacteria powder concentration was applied onto raw CEW for fermentation, as it showed around 60%. At the highest concentration of mixed-bacteria powder, the PC decreased with a small amount to around 55%. It was suggested that the low PC at the control with 0.02 wt% and 0.1 wt% of mixed-bacteria powder was due to the fact that the fermented feed was only enough for the BSFL to maintain the daily metabolism cost but did not boost growth in terms of protein content. At the concentration of 0.5 wt%, the PC increased 20% more compared to lower concentrations of mixed-bacteria powder. This could have been due to the fact that the fermented feed was mature and ready for BSFL rearing, which improved their development at the same time. With this, 0.5 wt% of mixed-bacteria powder was chosen as the optimal concentration for raw CEW fermentation for the following experiment.

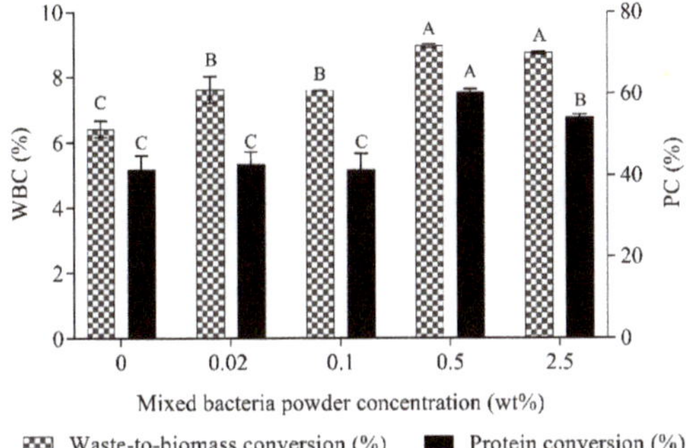

Figure 4. Performance of different concentrations of mixed-bacterial powder on waste-to-biomass and protein conversion when BSFL was fed with fermented CEW. Mean values indicated by same alphabetical letter were not significantly different.

3.4. Effects of Different Fermentation Time Frame on BSFL Development

After the validation of the optimal mixed-bacteria powder concentration, the next aim of this study was to shorten the fermentation time frame from the initial 28 days to the proposed time frames, which were 7, 14, or 21 days. Throughout fermentation, a significant number of acid-producing bacteria was found, and it had caused a drop in medium pH, indicating the presence of organic acid. In Figure 5, under the control system, the WBC was initially indicated at 6%, and the value increased to about 6.5% with an increment of fermentation time frame to seven days. Nevertheless, prolonging the fermentation time frame to 14 days also enhanced the WBC and boosted it to its highest value at around 9%. This value maintained for 21 and 28 days of the fermentation time frame. A similar pattern could be seen in terms of the PC of BSFL fed with different fermentation times of raw CEW under the control system. The PC was only around 35%, which was the lowest. The PC of BSFL increased to around 50% when it was fed with seven days fermented raw CEW, and the value lifted to around 60% and maintained at 14, 21, and 28 days of the fermentation time frame. Figure 5 shows that the WBC and PC both reached their peak conditions when fed with raw CEW fermented at 0.5 wt% for 14, 21, and 28 days, as the values of WBC and PC could no longer be improved after reaching the maximum point. This occurrence could be plausibly due to the presence of the organic acids that enhanced the gut health and development [46]. Also, through fermentation, part of fibers from CEW could be biotransformed into digestible organic acids, amino acids, and vitamins [27]. However, the nature of raw CEW also played a significant role in limiting the growth performance of BSFL, as it contained a high amount of polymer structures such as cellulose, which was hard to digest even with the introduction of mixed-bacteria powder as the aiding agent. It could also have been due to the low protein content of raw CEW, which was only measured to contain around 1.1% of nitrogen content, equivalent to 5.83% of protein content. Other diets such as animal manure and fruits and vegetables waste impact the developmental times of BSFL because the diets are lower in protein and energy with high fat content diet, and too much fat causes the BSFL to face difficulties in breaking down the fat during the metamorphosis [24]. From the previous study, the lipid and protein content of BSFL can be boosted with the introduction of protein sources into the feeding medium [41]. With the co-digestion technique of dairy manure topped with soybean curd residue, it was confirmed that nitrogen content reduction was more effective compared to only dairy manure [47]. Thus, a balanced diet in the perspective of energy, fat, protein, carbohydrates, amino acids, vitamins, etc. is required to enhance the development of BSFL as a sustainable source for energy security and food security purposes.

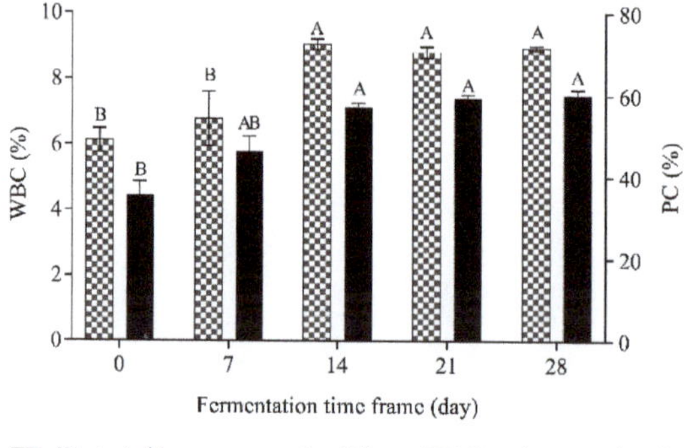

Figure 5. Performance of 0.5 wt% of mixed-bacterial powder at different fermentation time frames on waste-to-biomass and protein conversion when BSFL was fed with fermented CEW. Mean values indicated by same alphabetical letter were not significantly different.

3.5. Impacts of Raw CEW Modification on FAME Profile

Under treatment of raw CEW modification with different concentrations of mixed-bacteria powder and fermentation time frames, the lipid of each group set was trans-esterified into crude biodiesel, and FAME analysis was done. From Figure 6, disregarding the concentration of mixed-bacteria powder, the most abundant FAME was C12:0, followed by C14:0 and C16:0. At the same time, low amounts of C16:1, C18:2, and C10:0 could be observed in all concentrations of mixed-bacteria powder treatment. However, FAME including C6:0, C14:1, and C18:0 could only be found in certain specific concentrations of mixed-bacteria powder. The possible reason for these results might have been that these fatty acids were already low in concentrations and could not be detected by instrumentation. On the other hand, it could have been that the BSFL used up those types of fatty acids for metabolism cost, but this was a very low possibility compared to the previous reason. Further investigation would be required to determine the changes of fatty acids in BSFL at different instar or ages in order to have a detailed mapping of the changes.

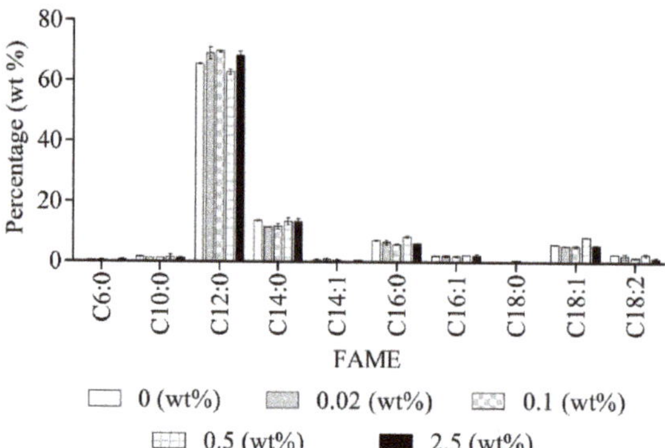

Figure 6. FAME profile of biodiesel from BSFL under different concentrations of mixed-bacterial powder.

Moving forward with the different fermentation time frames, the profile shown in Figure 7 shared a similar pattern with Figure 6. The majority of FAMEs were C12:0, C14:0, C16:0, and C18:1, followed by low amounts of C10:0, C16:1, and C18:2. In comparison, it showed that under modification on raw CEW, it did not change the FAME composition or profile of the crude biodiesel derived from BSFL, which could be seen in Figures 6 and 7. Overall, BSFL-derived biodiesel contained 87% of saturated fatty acids and 13% of unsaturated fatty acids. First and foremost, the high percentage of long chain saturated fatty acids resulted in poor cold flow property, and the low percentage of unsaturated fatty acids caused a low oxidative stability of the product. Several studies also supported that the BSFL's lipid could be an ideal feedstock for biodiesel production [23,47]. Meanwhile, with this modification, the biomass of BSFL was further improved and reached its ultimate when it was fed with raw CEW fermented at 0.5 wt% for 14 days onwards, as discussed earlier regarding its WBC and PC. In terms of FAME yield, it ranged from its lowest at 35.0% to its highest at 38.5%, as shown in Table 2. The FAME yield of BSFL-lipid was slightly influenced ($p > 0.05$) by the modification of raw CEW with a 3.5% boost, but it was affected at the fifth and the sixth instar, as discussed earlier.

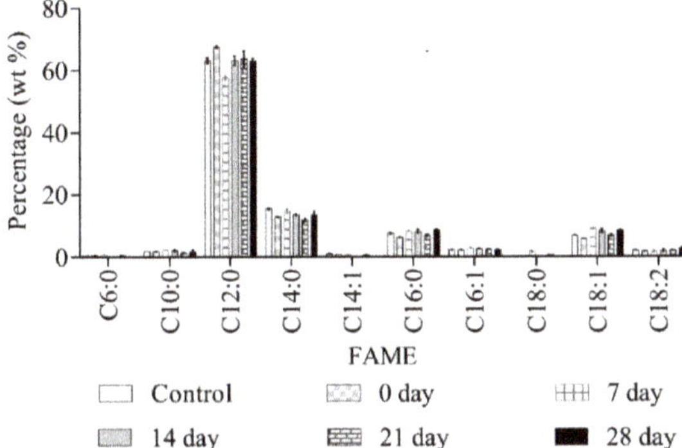

Figure 7. FAME profile of biodiesel from BSFL under 0.5 wt% of mixed-bacterial powder concentration at different fermentation time frames.

Table 2. FAME yield of BSFL biomass under treated raw coconut endosperm waste.

Parameter	Range	FAME Yield (%)
Mixed-bacteria powder concentration (wt%)	0	36.4
	0.02	35.6
	0.1	36.9
	0.5	37.8
	2.5	37.1
Fermentation time frame (day)	0	35.0
	7	35.0
	14	37.0
	21	38.5
	28	37.8

Moreover, several variables could impact the fatty acids composition in BSFL, including the type of feeding medium, feeding rate, and rearing duration. A detailed FAME composition of crude biodiesel derived from BSFL reared on 0.5 wt% fermented raw CEW for 14 days was tabulated in Table 3, and a comparison was made with other studies. In this study, the highest percentage of FAME was C12:0, which was 63.1%, and it was also the highest compared to other studies that fed BSFL with food waste, rice straw with restaurant waste, and dairy manure. Moreover, the second most abundant FAME was C14:0 at 13.5%, which was also the highest compared with food waste and rice straw with restaurant waste. However, it was not detected in the crude biodiesel derived from BSFL fed with dairy manure. On the other hand, BSFL fed with other types of waste had higher compositions of C16:0 and C18:1 with huge differences compared to the findings in this study. Surprisingly, when BSFL was fed with rice straw, restaurant waste, and dairy manure, odd numbered FAMEs such as C15:0, C17:1, C19:0, and C19:1 could be found. It can be suggested that BSFL might be able to assimilate the fatty acids present in the feeding medium and incorporate them into its own fatty acids for later use, but further investigation is still required to determine the changes of FAME composition in BSFL when fed with different feeding mediums.

Table 3. Comparison of FAME profile from BSFL-derived biodiesel fed with different feeding mediums.

Fatty Acid Methyl Ester	This Study	Food Waste [48]	Rice Straw & Restaurant Waste [49]	Dairy Manure [50]
C10:0	2.2	n/a	3.8	3.1
C12:0	63.1	44.9	27.8	35.6
C14:0	13.5	8.3	8.1	n/a
C14:1	n/a	n/a	n/a	7.6
C15:0	n/a	n/a	1.5	1.0
C16:0	8.2	13.5	14.2	14.8
C16:1	2.7	2.4	4.5	3.8
C17:0	n/a	n/a	0.8	n/a
C18:0	n/a	2.1	7.6	3.6
C18:1	8.3	12.0	22.5	23.6
C18:2	2.0	9.9	1.8	2.1
C18:3	n/a	0.1	2.1	n/a
C19:0	n/a	n/a	1.7	n/a
C19:1	n/a	n/a	n/a	1.4
C22:1	n/a	n/a	n/a	1.4

Note: The n/a stands for not reported.

4. Conclusions

To quicken the bioconversion process, harvesting the BSFL at the fifth instar did impact the lipid and total biomass gained from the larvae, since the sixth instar BSFL had to utilized the stored fat body tissues to sustain their metabolism. At the same time, harvesting the BSFL at the earlier instar did not affect the distribution of FAME, which was targeted for biodiesel production. Additionally, modification of raw CEW was carried out through the introduction of mixed-bacteria powder at different concentrations, and it showed that at a concentration of 0.5 wt%, the WBC and PC achieved their maximum points, which were around 9% and 60%, respectively. Next, the fermentation time frame was reduced from 28 days to 14 days at 0.5 wt% mixed-bacteria concentration and did not affect the development of BSFL in terms of its WBC and PC. With the modification of raw CEW, there was a minor impact on the FAME compositions, yet the most abundant FAME was C12:0 at around 60%, followed by C14:0 at around 15%, and both C16:0 and C18:1 at about 10% on average. Lastly, the FAME yield from BSFL was improved from 25% (sixth instar) to 33% (fifth instar) and showed its highest at 38.5% with the modification of raw CEW with 0.5 wt% mixed-bacteria powder and fermentation for 21 days.

Author Contributions: Conceptualization, C.-Y.W., S.-S.R.; Data curation, C.-Y.W., S.-S.R.; Formal analysis, C.-Y.W.; Methodology, C.-Y.W.; Supervision, Y.U., J.-W.L.; Writing—original draft, C.-Y.W.; Writing—review & editing, Y.C.H., A.L., W.K., C.-K.C., M.-K.L.

Funding: This research was funded by YUTP-FRG and The Murata Science Foundation.

Acknowledgments: The financial support from Universiti Teknologi PETRONAS via Graduate Assistantship Scheme (GA) to Chung-Yiin Wong is gratefully acknowledged. Also, Jun-Wei Lim wishes to thank the research fundings provided by YUTP-FRG and The Murata Science Foundation.

Conflicts of Interest: The authors declare no conflict of interest.

References

1. Allen, E.; Wall, D.M.; Herrmann, C.; Xia, A.; Murphy, J.D. What is the gross energy yield of third generation gaseous biofuel sourced from seaweed? *Energy* **2015**, *81*, 352–360. [CrossRef]
2. Srinophakun, P.; Thanapimmetha, A.; Rattanaphanyapan, K.; Sahaya, T.; Saisriyoot, M. Feedstock production for third generation biofuels through cultivation of arthrobacter ak19 under stress conditions. *J. Clean. Prod.* **2016**, *142*, 1259–1266. [CrossRef]
3. Lam, M.K.; Yusoff, M.I.; Uemura, Y.; Lim, J.W.; Khoo, C.G.; Lee, K.T.; Ong, H.C. Cultivation of *Chlorella vulgaris* using nutrients source from domestic wastewater for biodiesel production: Growth condition and kinetic studies. *Renew. Energy* **2017**, *103*, 197–207. [CrossRef]
4. Sitepu, I.R.; Sestric, R.; Ignatia, L.; Levin, D.; German, J.B.; Gillies, L.A.; Almada, L.A.G.; Boundy-Mills, K.L. Manipulation of culture conditions alters lipid content and fatty acid profiles of a wide variety of known and new oleaginous yeast species. *Bioresour. Technol.* **2013**, *144*, 360–369. [CrossRef]
5. Pinzi, S.; Leiva, D.; López-García, I.; Redel-Macías, M.D.; Dorado, M.P. Latest trends in feedstocks for biodiesel production. *Biofuel Bioprod. Biorefin.* **2014**, *8*, 126–143. [CrossRef]
6. Cea, M.; Sangaletti-Gerhard, N.; Acuña, P.; Fuentes, I.; Jorquera, M.; Godoy, K.; Osses, F.; Navia, R. Screening transesterifiable lipid accumulating bacteria from sewage sludge for biodiesel production. *Biotechnol. Rep.* **2015**, *8*, 116–123. [CrossRef]
7. Gerardo, M.L.; Den-Hende, S.V.; Vervaeren, H.; Coward, T.; Skill, S.C. Harvesting of microalgae within a biorefinery approach: A review of the developments and case studies from pilot-plants. *Algal Res.* **2015**, *11*, 248–262. [CrossRef]
8. Singh, M.; Shukla, R.; Das, K. Harvesting of microalgal biomass. In *Biotechnological Applications of Microalgae-Biodiesel and Value-Added Products*; Bux, F., Ed.; Taylor and Francis Group: Boca Raton, FL, USA, 2013; pp. 77–88.
9. Mohd-Noor, S.-N.; Wong, C.-Y.; Lim, J.-W.; Uemura, Y.; Lam, M.-K.; Ramli, A.; Bashir, M.J.; Tham, L. Optimization of self-fermented period of waste coconut endosperm destined to feed black soldier fly larvae in enhancing the lipid and protein yields. *Renew. Energy* **2017**, *111*, 646–654. [CrossRef]
10. Choi, Y.; Choi, J.; Kim, J.; Kim, M.; Kim, W.; Park, K.; Bae, S.; Jeong, G. Potential usage of food waste as natural fertilizer after digestion by *Hermetia illucens* (diptera: Stratiomyidae). *Int. J. Ind. Entomol.* **2009**, *19*, 171–174.
11. Gahukar, R. Chapter 4—Edible insects farming: Efficiency and impact on family livelihood, food security, and environment compared with livestock and crops. In *Insects as Sustainable Food Ingredients*; Elsevier: Amsterdam, The Netherlands, 2016; pp. 85–111.
12. Kelemu, S.; Niassy, S.; Torto, B.; Fiaboe, K.; Affognon, H.; Tonnang, H.; Maniania, N.; Ekesi, S. African edible insects for food and feed: Inventory, diversity, commonalities and contribution to food security. *J. Insects Food Feed* **2015**, *1*, 103–119. [CrossRef]
13. Maurer, V.; Holinger, M.; Amsler, Z.; Früh, B.; Wohlfahrt, J.; Stamer, A.; Leiber, F. Replacement of soybean cake by *Hermetia illucens* meal in diets for layers. *J. Insects Food Feed* **2016**, *2*, 83–90. [CrossRef]
14. Tomberlin, J.; Van Huis, A.; Benbow, M.; Jordan, H.; Astuti, D.; Azzollini, D.; Banks, I.; Bava, V.; Borgemeister, C.; Cammack, J. Protecting the environment through insect farming as a means to produce protein for use as livestock, poultry, and aquaculture feed. *J. Insects Food Feed* **2015**, *1*, 307–309. [CrossRef]
15. Veldkamp, T.; Van Duinkerken, G.; Van Huis, A.; Lakemond, C.; Ottevanger, E.; Bosch, G.; Van Boekel, T. *Insects as a Sustainable Feed Ingredient in Pig and Poultry Diets: A Feasibility Study*; Wageningen UR Livestock Research: Wageningen, The Netherlands, 2012.
16. Schiavone, A.; Cullere, M.; De Marco, M.; Meneguz, M.; Biasato, I.; Bergagna, S.; Dezzutto, D.; Gai, F.; Dabbou, S.; Gasco, L. Partial or total replacement of soybean oil by black soldier fly larvae (*Hermetia illucens* L.) fat in broiler diets: Effect on growth performances, feed-choice, blood traits, carcass characteristics and meat quality. *Ital. J. Anim. Sci.* **2017**, *16*, 93–100. [CrossRef]
17. Yang, S.; Li, Q.; Zeng, Q.; Zhang, J.; Yu, Z.; Liu, Z. Conversion of solid organic wastes into oil via boettcherisca peregrine (diptera: Sarcophagidae) larvae and optimization of parameters for biodiesel production. *PLoS ONE* **2012**, *7*, e45940. [CrossRef]
18. Leung, D.; Yang, D.; Li, Z.; Zhao, Z.; Chen, J.; Zhu, L. Biodiesel from zophobas morio larva oil: Process optimization and fame characterization. *Ind. Eng. Chem. Res.* **2012**, *51*, 1036–1040. [CrossRef]

19. Yang, S.; Liu, Z. Pilot-scale biodegradation of swine manure via chrysomya megacephala (fabricius) for biodiesel production. *Appl. Energy* **2014**, *113*, 385–391. [CrossRef]
20. Bowling, J.J.; Anderson, J.B.; Armbrust, K.L.; Hamann, M.T. Evaluation of potential biodiesel feedstock production from oleaginous insect *solenopsis* sp. *Fuel* **2014**, *117*, 5–7. [CrossRef]
21. Zheng, L.; Hou, Y.; Li, W.; Yang, S.; Li, Q.; Yu, Z. Exploring the potential of grease from yellow mealworm beetle (tenebrio molitor) as a novel biodiesel feedstock. *Appl. Energy* **2013**, *101*, 618–621. [CrossRef]
22. Yang, S.; Li, Q.; Gao, Y.; Zheng, L.; Liu, Z. Biodiesel production from swine manure via housefly larvae (musca domestica L.). *Renew. Energy* **2014**, *66*, 222–227. [CrossRef]
23. Li, Q.; Zheng, L.; Cai, H.; Garza, E.; Yu, Z.; Zhou, S. From organic waste to biodiesel: Black soldier fly, *Hermetia illucens* makes it feasible. *Fuel* **2011**, *90*, 1545–1548. [CrossRef]
24. Nguyen, T.T.X.; Tomberlin, J.K.; Vanlaerhoven, S. Influence of resources on *Hermetia illucens* (diptera: Stratiomyidae) larval deveopment. *J. Med. Entomol.* **2013**, *50*, 898–906. [CrossRef]
25. Diclaro, J.W.; Kaufman, P.E. Black soldier fly *Hermetia illucens* linnaeus (insecta: Diptera: Stratiomyidae). *EENY* **2009**, *461*, 1–3.
26. Ushakova, N.; Brodskii, E.; Kovalenko, A.; Bastrakov, A.; Kozlova, A.; Pavlov, D. *Characteristics of Lipid Fractions of Larvae of the Black Soldier Fly Hermetia Illucens*; Doklady Biochemistry and Biophysics; Springer: Berlin, Germany, 2016; pp. 209–212.
27. Caruso, D.; Devic, E.; Subamia, I.; Talamond, P.; Baras, E. *Technical Handbook of Domestication and Production of Diptera Black Soldier Fly (bsf), Hermetia Illucens, Stratiomyidae*; PT Penerbit IPB Press: Bogor, Indonesia, 2014.
28. Diener, S.; Lalander, C.; Zuebrueg, C.; Vinnerås, B. Opportunities and constraints for medium-scale organic waste treatment with fly larvae composting. In Proceedings of the 15th International Waste Management and Landfill Symposium, Cagliari, Italy, 5–9 October 2015.
29. Li, S.; Ji, H.; Zhang, B.; Tian, J.; Zhou, J.; Yu, H. Influence of black soldier fly (*Hermetia illucens*) larvae oil on growth performance, body composition, tissue fatty acid composition and lipid deposition in juvenile jian carp (cyprinus carpio var. Jian). *Aquaculture* **2016**, *465*, 43–52. [CrossRef]
30. Li, S.; Ji, H.; Zhang, B.; Zhou, J.; Yu, H. Defatted black soldier fly (*Hermetia illucens*) larvae meal in diets for juvenile jian carp (cyprinus carpio var. Jian): Growth performance, antioxidant enzyme activities, digestive enzyme activities, intestine and hepatopancreas histological structure. *Aquaculture* **2017**, *477*, 62–70. [CrossRef]
31. Leong, S.Y.; Kutty, S.R.M.; Malakamad, A.; Tan, C.K. Feasibility study of biodiesel production using lipids of *Hermetia illucens* larva fed with organic waste. *Waste Manag.* **2016**, *47*, 84–90. [CrossRef]
32. Nguyen, T.T.X.; Tomberlin, J.K.; Vanlaerhoven, S. Ability of black soldier fly (diptera: Stratiomyidae) larvae to recycle food waste. *Environ. Entomol.* **2015**, *44*, 406–410. [CrossRef]
33. Abdul Khalil, H.P.S.; Siti Alwani, M.; Mohd Omar, A.K. Chemical composition, anatomy, lignin distribution and cell wall structure of malaysian plant waste fibers. *BioResources* **2006**, *1*, 220–232.
34. Chuah, T.G.; Wan Azlina, A.G.K.; Robiah, Y.; Omar, R. Biomass as the renewable energy sources in malaysia: An overview. *Int. J. Green Energy* **2006**, *3*, 323–346. [CrossRef]
35. Jones, D.B. *Factors for Converting Percentages of Nitrogen in Foods and Feeds into Percentages of Proteins*; US Department of Agriculture: Washington, DC, USA, 1941.
36. Fathurochim, S.; Geden, C.J.; Axtell, R.C. Filth fly (diptera) oviposition and larval development in poultry manure of various moisture levels. *J. Entomol. Sci.* **1989**, *24*, 224–231. [CrossRef]
37. Kim, W.; Bae, S.; Park, H.; Park, K.; Lee, S.; Choi, Y.; Han, S.; Koh, Y.-H. The larval age and mouth morphology of the black soldier fly, *Hermetia illucens* (diptera: Stratiomyidae). *Int. J. Ind. Entomol.* **2010**, *21*, 185–187.
38. Lalander, C.; Diener, S.; Zurbrügg, C.; Vinnerås, B. Effects of feedstock on larval development and process efficiency in waste treatment with black soldier fly (*Hermetia illucens*). *J. Clean. Prod.* **2019**, *208*, 211–219. [CrossRef]
39. Draczyński, Z. Honeybee corpses as an available source of chitin. *J. Appl. Polym. Sci.* **2008**, *109*, 197–1981. [CrossRef]
40. Zheng, L.; Li, Q.; Zhang, J.; Yu, Z. Double the biodiesel yield: Rearing black soldier fly larvae, *Hermetia illucens*, on solid residual fraction of restaurant waste after grease extraction for biodiesel production. *Renew. Energy* **2012**, *41*, 75–79. [CrossRef]

41. Lim, J.-W.; Mohd-Noor, S.-N.; Wong, C.-Y.; Lam, M.-K.; Goh, P.-S.; Beniers, J.; Oh, W.-D.; Jumbri, K.; Ghani, N.A. Palatability of black soldier fly larvae in valorizing mixed waste coconut endosperm and soybean curd residue into larval lipid and protein sources. *J. Environ. Manag.* **2019**, *231*, 129–136. [CrossRef]

42. Makkar, H.P.S.; Tran, G.; Heuzé, V.; Ankers, P. State-of-the-art on use of insects as animal feed. *Anim. Feed Sci. Technol.* **2014**, *197*, 1–33. [CrossRef]

43. Spranghers, T.; Ottoboni, M.; Klootwijk, C.; Ovyn, A.; Deboosere, S.; De Meulenaer, B.; Michiels, J.; Eeckhout, M.; De Clercq, P.; De Smet, S. Nutritional composition of black soldier fly (*Hermetia illucens*) prepupae reared on different organic waste substrates. *J. Sci. Food Agric.* **2016**, *97*, 2594–2600. [CrossRef]

44. Waśko, A.; Bulak, P.; Polak-Berecka, M.; Nowak, K.; Polakowski, C.; Bieganowski, A. The first report of the physicochemical structure of chitin isolated from *Hermetia illucens*. *Int. J. Biol. Macromol.* **2016**, *92*, 316–320. [CrossRef]

45. Diener, S.; Zurbrügg, C.; Gutiérrez, F.R.; Nguyen, D.H.; Morel, A.; Koottatep, T.; Tockner, K. In Black soldier fly larvae for organic waste treatment-prospects and constraints. In Proceedings of the 2nd International Conference on Solid Waste Management in the Developing Countries, Khulna, Bangladesh, 13–15 February 2011.

46. Upadhaya, S.D.; Lee, K.Y.; Kim, I.H. Effect of protected organic acid blends on growth performance, nutrient digestibility and faecal micro flora in growing pigs. *J. Appl. Anim. Res.* **2016**, *44*, 238–242. [CrossRef]

47. Rehman, K.u.; Rehman, A.; Cai, M.; Zheng, L.; Xiao, X.; Somroo, A.A.; Wang, H.; Li, W.; Yu, Z.; Zhang, J. Conversion of mixtures of dairy manure and soybean curd residue by black soldier fly larvae (*Hermetia illucens* L.). *J. Clean. Prod.* **2017**, *154*, 366–373. [CrossRef]

48. Surendra, K.C.; Oliver, R.; Tomberlin, J.K.; Jha, R.; Khanal, S.K. Bioconversion of organic wastes into biodiesel and animal feed via insect farming. *Renew. Energy* **2016**, *98*, 197–202. [CrossRef]

49. Zheng, L.; Hou, Y.; Li, W.; Yang, S.; Li, Q.; Yu, Z. Biodiesel production from rice straw and restaurant waste employing black soldier fly assisted by microbes. *Energy* **2012**, *47*, 225–229. [CrossRef]

50. Li, Q.; Zheng, L.; Qiu, N.; Cai, H.; Tomberlin, J.K.; Yu, Z. Bioconversion of dairy manure by black soldier fly (diptera: Stratiomyidae) for biodiesel and sugar production. *Waste Manag.* **2011**, *31*, 1316–1320. [CrossRef]

Article

Safflower Biodiesel: Improvement of its Oxidative Stability by Using BHA and TBHQ

Sergio Nogales-Delgado [1],*, José María Encinar [1] and Juan Félix González [2]

[1] Department of Chemical Engineering and Physical Chemistry, University of Extremadura, Avda. De Elvas s/n, 06006 Badajoz, Spain; jencinar@unex.es
[2] Department of Applied Physics, University of Extremadura, Avda. De Elvas s/n, 06006 Badajoz, Spain; jfelixgg@unex.es
* Correspondence: senogalesd@unex.es

Received: 29 April 2019; Accepted: 18 May 2019; Published: 21 May 2019

Abstract: Biodiesel is gaining more and more importance due to environmental issues. This way, alternative and sustainable crops as new biofuel sources are demanded. Safflower could be a sustainable raw material for biodiesel production, showing one disadvantage (as many biodiesels from vegetable oils), that is, a short oxidative stability. Consequently, the use of antioxidants to increase this parameter is mandatory. The aim of this research work was to assess the effect of two antioxidants (butylated hydroxyanisole, BHA, and tert-butylhydroquinone, TBHQ) on the oxidative stability of safflower biodiesel, which was characterized paying attention to its fatty acid methyl ester profile. For oxidative stability, the Rancimat method was used, whereas for fatty acid profile gas chromatography was selected. For the remaining parameters, the methods were followed according to the UNE-EN 14214 standard. The overall conclusion was that safflower biodiesel could comply with the standard, thanks to the use of antioxidants, with TBHQ being more effective than BHA. On the other hand, the combined use of these antioxidants did not show, especially at low concentrations, a synergic or additive effect, which makes the mixture of these antioxidants unsuitable to improve the oxidative stability.

Keywords: Rancimat method; butylated hydroxyanisole; tert-butylhydroquinone; fatty acid methyl esters; viscosity; response surface

1. Introduction

Due to the environmental impact and the consequences related to the use of fossil fuels (especially on account of their contribution to greenhouse gases), the use of alternatives such as renewable energies is necessary. Indeed, there is a real concern about environmental conservation, and many countries and international agencies are promoting renewable energies, such as biofuels [1–3].

The main advantages related to biodiesel use are the zero-net CO_2 emissions, biodegradability, storage safety, efficient combustion, low sulphur content, lubricity, and good performance in diesel engines, among others [2,4,5]. Concerning the contribution for developing countries or emerging economies, the use of biodiesel could contribute to the energy independence of these countries, as many raw materials available in these areas might be suitable for biodiesel production, making the economic development more sustainable [6–9].

For biodiesel production, the use of oleaginous plants is usual, along with others, such as animal fats, fried oils, algae, bacteria, etc., and many research works about its production and performance in engines or tribology were carried out, which points out the importance that biodiesel has been gaining recently [10–18]. In the case of vegetable oils, such as canola, rapeseed, soya, or safflower oils, among others, they have been considered to produce biodiesel [7,19–22], with acceptable results.

Regarding safflower, which is a crop with many applications (for oil and natural dye production, among others), and widely used in many countries (most of them with dry climates) such as India, Ethiopia, the United States, Mexico, Australia, Argentina, Brazil, Romania, etc. [23–26], it could be a suitable biodiesel as it adheres to the abovementioned conditions and advantages. Moreover, the seed yield is around 800–900 kg ha^{-1}, which is an interesting production [27]. In Spain, it is becoming important, as it is usually a rotation crop, being an alternative to sunflower and other majority crops, due to the long primary root, which allows the surface soil to regain nutrients and use this plant under dry conditions [25,28,29]. Thus, the production of safflower in Spain was between 2 and 6 thousands of tons per year between 2011 and 2015 [30]. This way, it might be an alternative crop in some developed countries and in areas with seasonal rains [27]. In general, safflower biodiesel has been widely studied in the literature, paying attention to its performance in diesel engines [31–33], production, and oxidative stability [7,26,34].

The main drawback of biodiesel, including safflower biodiesel, is its short oxidative stability (which is usually determined by the Rancimat method and expressed in hours [35]), that is, its low storage stability [4,36], implying an important disadvantage compared to diesel. This fact is mainly due to the auto-oxidation of fatty acid methyl esters (FAMEs) that constitute biodiesel. Thus, depending on the molecular structure of these FAMEs (molecular branching or unsaturations, mainly), their degradation will be longer or shorter, implying a quality loss of biodiesel (FAME loss and increase in viscosity, mainly) [37,38] and not complying with the standards for its marketability, at least, as a pure biofuel [39].

Consequently, the mixture with other more stable biodiesel [40], the use of antioxidants (both natural and artificial) [41–45], and other chemical reactions [34] have been attempted in order to increase the oxidative stability of biodiesel samples, to revalue this product. Although the raw material usually contains natural antioxidants, during biodiesel production and purification they are usually missed. There are plenty of antioxidants to achieve this goal. For instance, butylated hydroxyanisole (BHA) and tert-butylhydroquinone (TBHQ) have been studied in many materials to assess their antioxidant activity, among other effects on biofuels [46–49], proving their effectiveness when it comes to keeping the oxidative stability of biodiesel. Thus, the use of this kind of antioxidants reduce the formation of unstable free-radicals during oxidation, as the presence of labile hydrogen in their molecular structure results in the formation of more stable free-radicals, which can also react each other and produce more stable molecules [38]. Moreover, the use of antioxidants could contribute to keep biodiesel from increasing NO_x emissions during combustion in diesel engines [38]. Although in some cases the combined use of antioxidants in biodiesel has been studied, there are controversial results about their synergetic or additive effect, depending on factors such as the raw material used [37,40,43,50,51]. In the case of safflower biodiesel, no studies about the improvement of its oxidative stability were found, making the compliment of this requirement an important aspect for the valorization of this biofuel.

To sum up, the use of safflower biodiesel could be an interesting energy source in developing regions, as it is becoming an important rotating crop. However, the use of antioxidants is required to make this product marketable, and the study of the effect of these antioxidants is necessary.

The aim of this research work was to assess the effectiveness of BHA and TBHQ on the oxidative stability of safflower biodiesel, in order to comply with the UNE-EN 14214 standard [39], especially concerning the oxidative stability (with a lower limit of 8 hours). For this purpose, a wide range of concentration (up to 1000 ppm) was used for each antioxidant, and the combined use of them was also studied, to check the additive, synergetic or inhibitory effect of both antioxidants when used together. Moreover, a thorough characterization of safflower biodiesel (paying attention to FAME profile and viscosity), along with the effect of the antioxidants on the most representative parameters (especially viscosity), was carried out.

2. Materials and Methods

2.1. Raw Material

The raw material was safflower seeds, collected in the Agricultural Research Center "La Orden" of Extremadura Government (CICYTEX) in 2018. Twenty liters of safflower oil was obtained by the use of a seed oil extractor, transporting it in covered containers and keeping it away from heat and light until the transesterification reaction took place. The reaction and the subsequent analysis were carried out as soon as possible (in a few days), to keep the properties of the raw material and the obtained biodiesel.

2.2. Transesterification Conditions

The vegetable oil obtained from safflower seeds, containing triglycerides, underwent transesterification (reacting with three moles of methanol, and KOH as a catalyst) to produce fatty acid methyl esters (FAMEs, considered as biodiesel) and glycerol. As this reaction is reversible, parameters such as temperature, catalyst concentration and molar ratio were selected to increase the FAME yield. Table 1 shows the reaction conditions for this research, according to previous research experiences, in order to assure the highest yield possible [20,21,52]. All the reagents were provided by Panreac (Germany).

Table 1. Transesterification reaction conditions for obtaining safflower biodiesel.

Reaction temperature (°C)	65
Reaction time (min)	60
Methanol/oil ratio	6:1
Catalyst concentration [1] (%)	1.5

[1] Sodium hydroxide, KOH.

The reaction took place in a 1-L reactor with three necks. The reaction temperature was selected according to the boiling point of the alcohol used (in this case methanol). In order to avoid evaporation of the alcohol, a condenser was connected to the reactor. Also, the temperature was continuously recorded and controlled.

Afterwards, the purification of FAMEs took place, by decanting (to remove glycerol) and washing (with distilled water). Once the sample was dried by heating at 100 °C, it was kept in topaz crystal bottles during storage and the experiments were carried out immediately.

2.3. Antioxidant Addition

In order to assess the effect of antioxidants on the oxidative stability of safflower biodiesel, butylated hydroxyanisole (BHA, Panreac Applichem GmbH, Germany) and tert-butylhydroquinone (TBHQ, Panreac Applichem GmbH, Germany) at a range concentration (from 250 to 1000 ppm) were used individually. A control sample (0 ppm) was used to compare the results obtained. In addition, the combined effect of both antioxidants was also considered. For the sample preparation, the suitable amount of antioxidant was added to 10 ml of sample, dissolving it by ultrasound for 1 min. The labeling of the samples was as follows: antioxidant abbreviation followed by the ppm value. For instance, for a sample treated with 500 ppm of butylated hydroxyanisole, it was labelled as BHA500. Table 2 shows a summary of the antioxidant addition. Furthermore, the structure of the antioxidants is shown in Figure 1.

Table 2. Antioxidant addition to safflower biodiesel.

Experiment	BHA Concentration (ppm)	TBHQ Concentration (ppm)
Effect of BHA addition	0, 100, 250, 500, 750 and 1000	0
Effect of TBHQ addition	0	0, 100, 250, 500, 750 and 1000
Effect of BHA and TBHQ addition	0, 100, 250, 500, 750 and 1000	0, 100, 250, 500, 750 and 1000

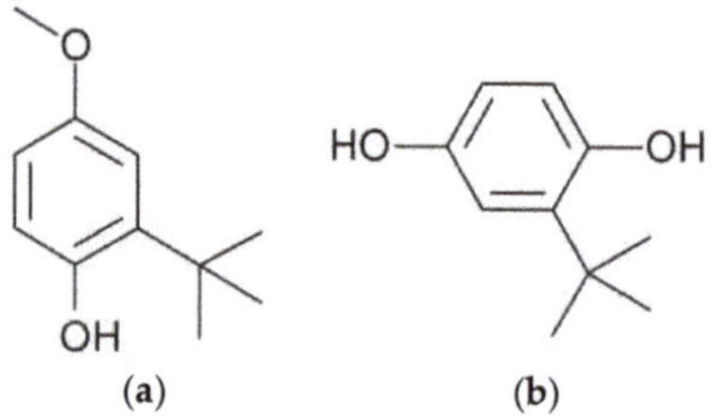

Figure 1. Molecular structure of (**a**) butylated hydroxyanisole (BHA) and (**b**) tert-butylhydroquinone (TBHQ).

2.4. Biodiesel Characterization

To assess the quality of the biodiesel obtained, its characterization was carried out according to standards [39]. Thus, density was obtained by using a pycnometer at room temperature. Viscosity was done according to the ISO 3104:1994 standard [53], by using an Ostwald viscosimeter at 40 °C. For cold filter plugging point (CFPP), the EN 166 standard was used [54]. Flash and combustion points were obtained by using the Cleveland open-cup method, according to EN 51023 standard [55]. For moisture, a Metrohm 870 trinitro plus equipment was used, using the Karl-Fischer method (EN-ISO-12937) [56]. Acid and iodine numbers were measured by using their corresponding standards [57].

2.5. FAME Characterization

For the characterization of FAMEs, the standards were followed [39]. A gas chromatograph (Varian 3900) coupled to a FID detector was used. A Zebron ZB-wax Plus capillary column (30 m long, 0.32 mm of inner diameter, 0.25 μm of film thickness and a maximum temperature of 260 °C) was used. The chromatography conditions are shown in Table 3.

Table 3. Chromatography conditions for fatty acid methyl ester (FAME) determination.

Oven temperature (°C)	220 for 23.5 min, 240 for 14 min
Injector temperature (°C)	270
Detector temperature (°C)	300
Column flow (cm^{3}min^{-1})	28
Carrier gas	Helium
Auxiliary gas	Nitrogen
Combustible gas (cm^{3}min^{-1})	Synthetic air (300)
Oxidizing gas (cm^{3}min^{-1})	Hydrogen (30)

For each FAME studied (methyl oleate, linoleate, palmitate, ricinoleate, linolenate, stearate, erucate, myristate, and palmitoleate), a calibration curve was done by using its corresponding standard (Sigma-Aldrich). The calibration was carried out by using an internal standard (methyl heptadecanoate, Sigma-Aldrich). All the gases used in this research were supplied by Linde (Munich, Germany).

2.6. Oxidative Stability Determination

The oxidative stability was obtained according to the Rancimat method [35]. Around 3 g of the sample was placed in a test tube, bubbling synthetic air (10 Lh^{-1}, Linde) and heating the tube at 100 °C. The resulting steam, after oxidizing the sample, passes through 50 mL of deionized water.

The conductivity of this amount of water was measured. As the sample was oxidized, some products were developed, dissolving in water and increasing its conductivity. Conductivity was recorded by a conductivity meter (Crison EC-Meter GLP31+, Spain). The experimental setup is shown in Figure 2a. Oxidative stability is expressed in hours, and for this purpose the induction point should be calculated. Thus, the conductivity plot, as it is shown in Figure 2b, has two clear stages. The first one, with a stationary conductivity evolution, is a line with a flat slope. The second one, when most by-products resulting from auto-oxidation are released, results in a line with a pronounced slope. The induction point is the intersection of both lines, as it can be seen in Figure 2b.

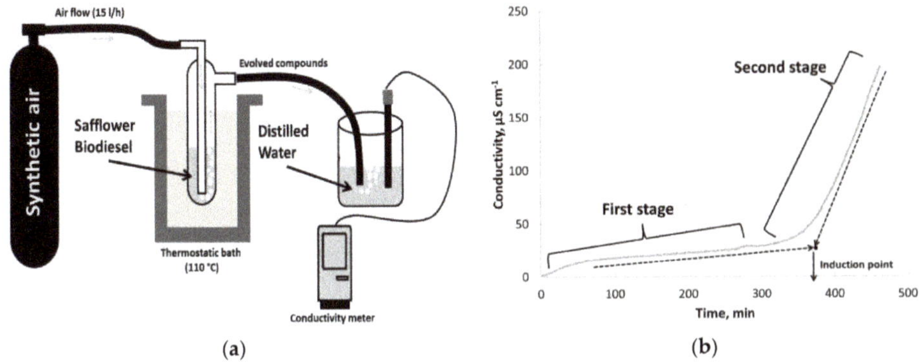

Figure 2. (**a**) Rancimat method and (**b**) induction point determination (example at 380 min).

3. Results and Discussion

3.1. Biodiesel Characterization

As it can be seen in Table 4, most characteristics of Safflower biodiesel complied with the EN-14214 [39].

Thus, it was a biofuel with a high yield in FAMEs and suitable characteristics for warm climates (acidity number, viscosity, and density showed intermediate values and cold filter plugging point was within the limits for warm climates, not for cold ones). Some properties were especially convenient, compared to diesel, such as flash and combustion points (well above the lower limit), whereas water content and iodine number (which is an indicator of the presence of unsaturations, sensitive to oxidation) were close to their corresponding upper limits.

However, the oxidative stability was well below the lower limit established by the standard. Consequently, the use of this pure biodiesel would not be possible, and a mixture with other biodiesel or the addition of antioxidants was required. Nonetheless, the abovementioned results are usual for biodiesel samples made by other authors, showing a similar behavior, that is, high flash and combustion points and short oxidative stabilities (from 2 to 6 h, depending on the raw material) [40,50]. Specifically, the results found for safflower biodiesel were similar to those found in the literature by other authors, although, there were some discrepancies (viscosity and density were slightly higher in some cases, for instance). Concerning the oxidative stability, it was even shorter than the results observed in Table 4, not achieving 1 h [26,58].

Table 4. Safflower biodiesel. Characterization and comparison with the EN-14214 standard.

Parameter	Value	EN-14214
Viscosity at 40 °C (cSt)	4.42	3.50–5.00
Density at 15 °C (g·dm^{-3})	880	860–900
Oxidative stability (h)	1.46	8 [1]
FAME content (%)	96.78	96.5 [1]
Flash point (°C)	180	120 [1]
Combustion point (°C)	190	Not included
CFPP (°C) [2]	−2	−20–+5
Water content (mg·Kg^{-1})	400	500 [3]
Acidity number (mg KOH·g^{-1})	0.35	0.5 [3]
Iodine number (g I$_2$·100 g^{-1})	115	120 [3]

[1] Lower limit. [2] For warm climates. [3] Upper limit.

3.2. FAME Profile

One of the most important characteristics of biodiesel is the fatty acid methyl ester profile. As many authors have pointed out, and from previous studies carried out by our research group, a strong influence of FAME proportion on many properties was found, especially concerning oxidative stability [20,50,59]. This way, majority FAMEs could play an important role in the global characteristics of biodiesel. The results obtained by gas chromatography are shown in Figure 3:

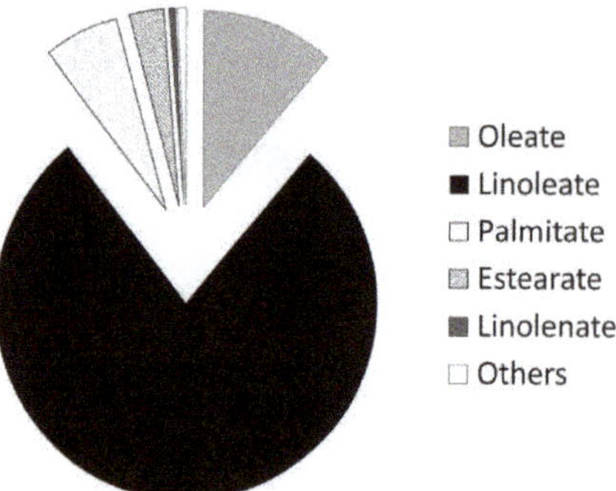

Figure 3. FAME profile for safflower biodiesel.

As it can be seen, there is a majority FAME, that is, methyl linoleate (over 75%), followed by methyl oleate (10%) and methyl palmitate (6%). The remaining FAMEs were under 5% or negligible. The high presence of methyl linoleate was slightly higher than the ones found in the literature, ranging from 61% to 70% [26,58].

Consequently, some characteristics of safflower biodiesel could be influenced by methyl linoleate, which is the majority FAME obtained. The molecular structure of this FAME is shown in Figure 4.

It should be pointed out the conjugated double bond of methyl linoleate, which is a reactive part of the molecular structure, being susceptible to oxidation to a larger extent when compared to mono or saturated FAMEs [2,60]. Indeed, the oxidative stability of methyl linoleate was 0.94 h, according to data [2]. This could explain the low oxidative stability of safflower biodiesel, compared to other compounds with lower methyl linoleate percentages. For instance, rapeseed biodiesel, usually with methyl linoleate percentages under 25% and high methyl oleate values (which is more stable), showed oxidative stability values (induction points) of 6 h [26]. Consequently, FAME determination is vital to assess the need of antioxidants and their approximate concentration.

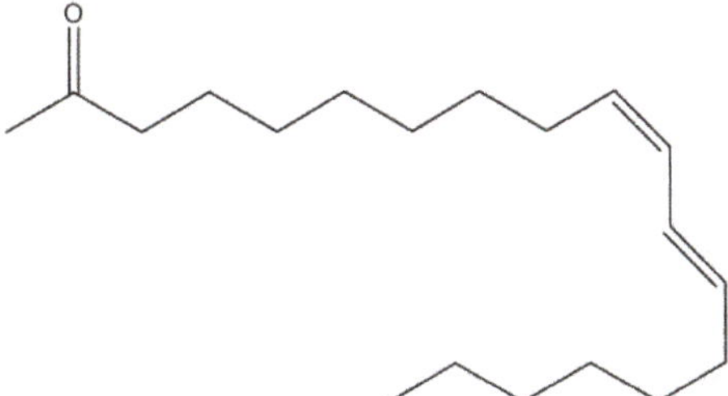

Figure 4. Molecular structure of methyl linoleate.

3.3. Antioxidant Addition

As mentioned previously, safflower biodiesel required antioxidant addition to increase its oxidative stability. For this purpose, BHA and TBHQ, two antioxidants with many uses in industry, were selected at the most usual concentrations found in the literature (from 100 to 1000 ppm) [38,42,61].

3.3.1. BHA Addition

The effect of BHA on oxidative stability of safflower biodiesel was shown in Figure 5. As expected, the addition of BHA increased the oxidative stability (induction point), from 1.5 h for the control sample to 5.75 h for BHA1000. However, the addition of BHA at these concentrations did not comply with the standard [39], requiring higher concentrations. According to the literature, BHA, compared to other antioxidants (such as pyrogallol, propyl gallate, or butylated hydroxyl toluene), showed shorter induction periods [49,62].

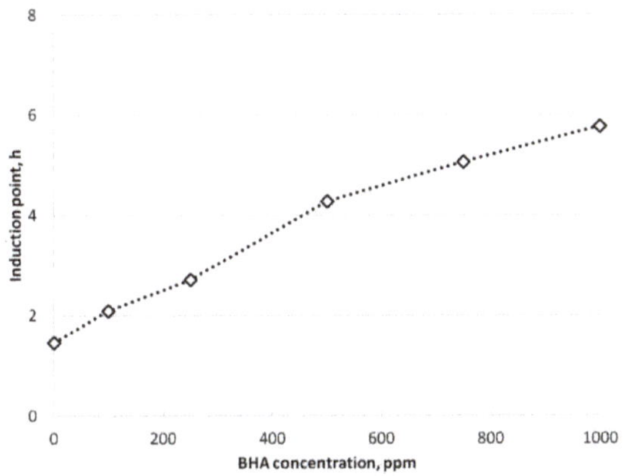

Figure 5. Effect of BHA on oxidative stability of safflower biodiesel.

Consequently, the addition of BHA would not be suitable for the treatment of safflower biodiesel, as the concentration required should be high, being a disadvantage in economic terms. Nonetheless, its use in other more stable biodiesel (with induction points longer than 6 h) could be feasible, as the concentration required in this case would be lower.

3.3.2. TBHQ Addition

Concerning TBHQ, the results are shown in Figure 6. The efficiency of TBHQ was higher than in the case of BHA. For the former, induction points over 8 h were found at 1000 ppm, whereas for the latter the induction point did not reach 6 h, well below the lower limit of the EN 14214 standard. The effectiveness of TBHQ was also observed by other authors, being higher, in most cases than in the case of BHA [37,46,48,62]. It could be due to the chemical structure of this antioxidant, with two hydroxyl groups available to neutralize free radicals in FAMEs whereas BHA has one hydroxyl group for this purpose (see Figure 1). However, other authors have found similar efficiency between both antioxidants [38].

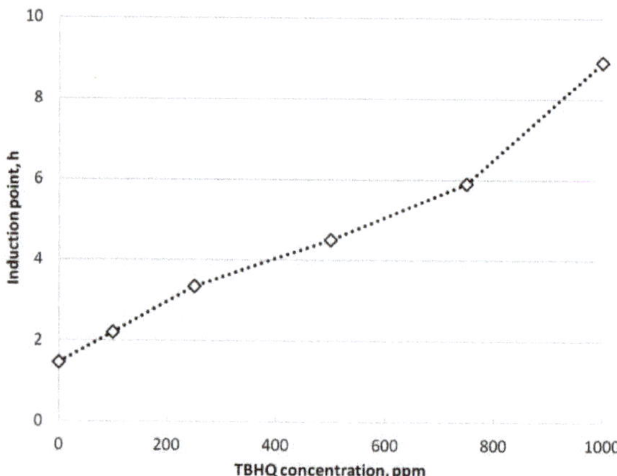

Figure 6. Effect of TBHQ on oxidative stability of safflower biodiesel.

This way, the use of TBHQ could be a suitable solution to the low oxidative stability of safflower biodiesel, recommending at least 1000 ppm of this antioxidant to comply with the standard, which is within the range of suitable concentrations found in the literature [48].

3.3.3. BHA and TBHQ Combined Addition

As it can be observed in Figure 7, the combined use of TBHQ and BHA did not show any additive or synergetic effect on the oxidative stability of safflower biodiesel. Indeed, there was an inhibitory effect especially at low concentrations, when the mixtures showed lower induction points than the addition of the corresponding results for BHA and TBHQ separately. For instance, in the case of BHA1000TBHQ1000, with the highest induction point increase compared to control sample (11.82 h), the addition of BHA1000 (4.3 h increase) and TBHQ1000 (7.45 h increase) was slightly lower (11.72 h). It could be considered an additive effect. On the contrary, the inhibitory effect was clear in the case of low concentrations. For instance, for BHA100TBHQ100 (0.89 h increase), the addition of BHA100 (0.63 h increase) and TBHQ1000 (0.74 h increase) was 50% higher (1.37 h increase). Indeed, this fact was also observed by other authors, which have not appreciated any improvement of the oxidative stability with the mixtures of antioxidants (TBHQ, BHA, and hydroxytoluene) [62]. Therefore, the "regeneration" effect (where some antioxidants are better conserved when combined with others) observed by other authors did not seem to apply in this case [51]. Indeed, there might be a pro-oxidation effect, as some authors have pointed out for natural antioxidants at certain concentrations [43].

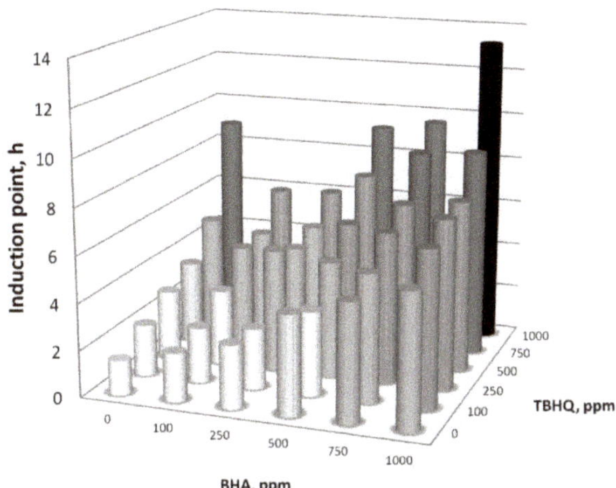

Figure 7. Effect of BHA and TBHQ mixtures on oxidative stability of safflower biodiesel.

This way, and taking into account that only an almost negligible synergetic effect (it can be considered an additive effect) was observed at high concentrations (which imply a great economic effort in industries), and there are clearly inhibitory effects when the antioxidants were mixed, the combined use of BHA and TBHQ is not recommended for safflower biodiesel, concerning the increase in oxidative stability.

3.3.4. Effect of Antioxidants on Other Biodiesel Characteristics

In this research work, the effect of antioxidants (BHA and TBHQ at 1000 and 2000 ppm) on safflower biodiesel viscosity was investigated. According to Figure 8, there was a slight increase in viscosity as the concentration of antioxidant added was higher. Due to the chemical structure of the antioxidants (see Figure 1), the possibility of forming crosslinks and hydrogen bonds with FAMEs could explain this variation, especially at high concentrations. Other authors, however, found that the addition of antioxidants (BHA among them) decreased slightly the viscosity of biodiesel [49]. It should be pointed out that the simple addition of antioxidants can vary viscosity of biodiesel (as it was observed in this research work), but during storage (when viscosity increases due to oxidation), as antioxidants keep from auto-oxidation, they help to keep viscosity values between the range of standards [37,61].

Moreover, the addition of TBHQ seemed to increase biodiesel viscosity (from 4.42 to 4.72 cSt) to a larger extent, compared to BHA (from 4.42 to 4.59 cSt). This fact could be due to the fact that the molecular structure of TBHQ (see Figure 1), with two hydroxyl groups, is prone to form more hydrogen bonds with FAMEs, increasing the resistance to flow (viscosity).

Concerning other parameters, such as density, the addition of BHA or TBHQ did not show any substantial changes (data not shown).

Consequently, the effect of the addition of antioxidants on some biodiesel characteristics, especially viscosity, should be taken into account, especially when the concentration of antioxidant required is high (because of the extreme low oxidative stability of biodiesel in certain cases) or the viscosity of biodiesel is close to the upper limit of the standard. For safflower biodiesel, the first case should be considered, as the antioxidant requirements, depending on pre-harvest conditions or the kind of antioxidant (and its effectiveness), could be increased. That is the case of BHA on safflower biodiesel, as higher concentrations are required to comply with the standard.

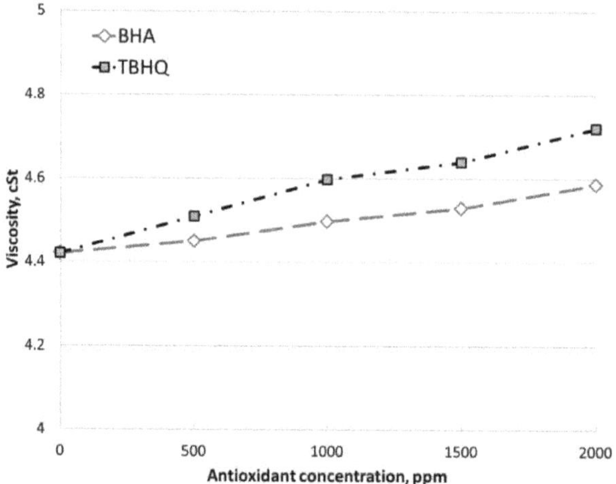

Figure 8. Effect of BHA and TBHQ on viscosity of safflower biodiesel.

4. Conclusions

The main findings of this research work were the following:

- The yield of biodiesel production from safflower oil was acceptable, making this product suitable for biodiesel production from an engineering point of view.
- Most characteristics of safflower biodiesel complied with the standard, except for oxidative stability. For the latter, the use of antioxidants is mandatory, as the oxidative stability was especially short and, therefore, not suitable for storage.
- As in many other parameters, the oxidative stability was strongly influenced by the FAME profile. In the case of safflower biodiesel, methyl linoleate was the majority compound, which is more unstable than mono or saturated FAMEs, due to its conjugated double bond (susceptible to self-oxidation).
- The use of BHA and TBHQ was effective at increasing the oxidative stability of safflower biodiesel. TBHQ was, in general and at the same concentrations, more effective than BHA. In fact, BHA did not comply with the lower limit of the standard (8 h) in the range studied.
- The combined use of BHA and TBHQ did not show any synergistic effect. Indeed, the effect of their combined use was inhibitory (especially at low concentrations of BHA and TBHQ), which makes the mixture of these antioxidants not desirable in this case.
- Apart from increasing the oxidative stability, the use of BHA and TBHQ influenced on viscosity. In general, an increase in viscosity was found, which could alter the compliance with standards. Therefore, both the initial viscosity and oxidative stability (and therefore the concentration of antioxidant required) of biodiesel are important to comply with the standard range.
- Consequently, as the oxidative stability of safflower biodiesel was so low, the use of an effective antioxidant, such as TBHQ at 1000 ppm, was recommended, not mixing it with other antioxidants. The mixture with other more stable biodiesel samples could be another alternative for safflower biodiesel revaluation.

Author Contributions: Conceptualization, S.N.-D. and J.M.E.; methodology, S.N.-D. and J.M.E.; formal analysis, S.N.-D. and J.M.E.; investigation, S.N.-D.; resources, J.M.E. and J.F.G.; data curation, S.N.-D.; writing—original draft preparation, S.N.-D.; writing—review and editing, S.N.-D., J.M.E., and J.F.G.; supervision, J.M.E. and J.F.G.; project administration, J.M.E. and J.F.G.; funding acquisition, J.M.E. and J.F.G.

Funding: This research was funded by the "Junta de Extremadura", GR 18150 and the FEDER "Fondos Europeos de Desarrollo Regional", IB 18028.

Acknowledgments: We thank the Agricultural Research Center "La Orden" of Extremadura Government (CICYTEX) for the supply of the raw material.

Conflicts of Interest: The authors declare no conflict of interest. The funders had no role in the design of the study; in the collection, analyses, or interpretation of data; in the writing of the manuscript, or in the decision to publish the results.

References

1. Manaf, I.S.A.; Embong, N.H.; Khazaai, S.N.M.; Rahim, M.H.A.; Yusoff, M.M.; Lee, K.T.; Maniam, G.P. A review for key challenges of the development of biodiesel industry. *Energy Convers. Manag.* **2019**, *185*, 508–517. [CrossRef]
2. Knothe, G.; Razon, L.F. Biodiesel fuels. *Prog. Energy Combust. Sci.* **2017**, *58*, 36–59. [CrossRef]
3. Huang, D.; Zhou, H.; Lin, L. Biodiesel: An alternative to conventional fuel. *Energy Procedia* **2011**, *16*, 1874–1885. [CrossRef]
4. Rodionova, M.V.; Poudyal, R.S.; Tiwari, I.; Voloshin, R.A.; Zharmukhamedov, S.K.; Nam, H.G.; Zayadan, B.K.; Bruce, B.D.; Hou, H.J.M.; Allakhverdiev, S.I. Biofuel production: Challenges and opportunities. *Int. J. Hydrog. Energy* **2017**, *42*, 8450–8461. [CrossRef]
5. Azadi, P.; Malina, R.; Barrett, S.R.H.; Kraft, M. The evolution of the biofuel science. *Renew. Sustain. Energy Rev.* **2017**, *76*, 1479–1484. [CrossRef]
6. Ji, X.; Long, X. A review of the ecological and socioeconomic effects of biofuel and energy policy recommendations. *Renew. Sustain. Energy Rev.* **2016**, *61*, 41–52. [CrossRef]
7. Cajamarca, F.A.; Lancheros, A.F.; Araújo, P.M.; Mizubuti, I.Y.; Simonelli, S.M.; Ida, E.I.; Guedes, C.L.B.; Guimarães, M.F. Evaluation of various species of winter oleaginous plants for the production of biodiesel in the State of Parana, Brazil. *Ind. Crop. Prod.* **2018**, *126*, 113–118. [CrossRef]
8. Cardoso, B.F.; Shikida, P.F.A.; Finco, A. Development of Brazilian biodiesel sector from the perspective of stakeholders. *Energies* **2017**, *10*, 399. [CrossRef]
9. Popp, J.; Harangi-Rákos, M.; Gabnai, Z.; Balogh, P.; Antal, G.; Bai, A. Biofuels and their co-products as livestock feed: Global economic and environmental implications. *Molecules* **2016**, *21*, 285. [CrossRef]
10. Poudel, J.; Karki, S.; Sanjel, N.; Shah, M.; Oh, S.C. Comparison of biodiesel obtained from virgin cooking oil and waste cooking oil using supercritical and catalytic transesterification. *Energies* **2017**, *10*, 546. [CrossRef]
11. Rahman, M.M.; Rasul, M.; Hassan, N.M.S. Study on the tribological characteristics of Australian native first generation and second generation biodiesel fuel. *Energies* **2017**, *10*, 55. [CrossRef]
12. Costarrosa, L.; Leiva-Candia, D.E.; Cubero-Atienza, A.J.; Ruiz, J.J.; Dorado, M.P. Optimization of the transesterification of waste cooking oil with mg-al hydrotalcite using response surface methodology. *Energies* **2018**, *11*, 302. [CrossRef]
13. Agarwal, A.K.; Gupta, J.G.; Dhar, A. Potential and challenges for large-scale application of biodiesel in automotive sector. *Prog. Energy Combust. Sci.* **2017**. [CrossRef]
14. Patel, P.D.; Lakdawala, A.; Chourasia, S.; Patel, R.N. Bio fuels for compression ignition engine: A review on engine performance, emission and life cycle analysis. *Renew. Sustain. Energy Rev.* **2016**, *61*, 113–149. [CrossRef]
15. Hamdan, S.H.; Chong, W.W.F.; Ng, J.H.; Ghazali, M.J.; Wood, R.J.K. Influence of fatty acid methyl ester composition on tribological properties of vegetable oils and duck fat derived biodiesel. *Tribol. Int.* **2017**, *113*, 76–82. [CrossRef]
16. Bobadilla, M.C.; Lorza, R.L.; Escribano-Garcia, R.; Gómez, F.S.; González, E.P.V. An improvement in biodiesel production from waste cooking oil by applying thought multi-response surface methodology using desirability functions. *Energies* **2017**, *10*, 130. [CrossRef]
17. Rahman, Z.; Rashid, N.; Nawab, J.; Ilyas, M.; Sung, B.H.; Kim, S.C. Escherichia coli as a fatty acid and biodiesel factory: Current challenges and future directions. *Environ. Sci. Pollut. Res.* **2016**, *23*, 12007–12018. [CrossRef]
18. Brigham, C. Perspectives for the biotechnological production of biofuels from CO2 and H2 using Ralstonia eutropha and other 'Knallgas' bacteria. *Appl. Microbiol. Biotechnol.* **2019**, *103*, 2113–2120. [CrossRef] [PubMed]

19. Abdulkhani, A.; Alizadeh, P.; Hedjazi, S.; Hamzeh, Y. Potential of Soya as a raw material for a whole crop biorefinery. *Renew. Sustain. Energy Rev.* **2017**, *75*, 1269–1280. [CrossRef]

20. Martínez, G.; Sánchez, N.; Encinar, J.M.; González, J.F. Fuel properties of biodiesel from vegetable oils and oil mixtures. Influence of methyl esters distribution. *Biomass Bioenergy* **2014**, *63*, 22–32. [CrossRef]

21. Encinar, J.M.; Pardal, A.; Sánchez, N. An improvement to the transesterification process by the use of co-solvents to produce biodiesel. *Fuel* **2016**, *166*, 51–58. [CrossRef]

22. Reeves, C.J.; Menezes, P.L.; Jen, T.C.; Lovell, M.R. The influence of fatty acids on tribological and thermal properties of natural oils as sustainable biolubricants. *Tribol. Int.* **2015**, *90*, 123–134. [CrossRef]

23. Chavoushi, M.; Najafi, F.; Salimi, A.; Angaji, S.A. Improvement in drought stress tolerance of safflower during vegetative growth by exogenous application of salicylic acid and sodium nitroprusside. *Ind. Crop. Prod.* **2019**, *134*, 168–176. [CrossRef]

24. Sabzalian, M.R.; Saeidi, G.; Mirlohi, A.; Hatami, B. Wild safflower species (*Carthamus oxyacanthus*): A possible source of resistance to the safflower fly (*Acanthiophilus helianthi*). *Crop Prot.* **2010**, *29*, 550–555. [CrossRef]

25. Landry, E.J.; Fuchs, S.J.; Bradley, V.L.; Johnson, R.C. The effect of cold acclimation on the low molecular weight carbohydrate composition of safflower. *Heliyon* **2017**, *3*, e00402. [CrossRef]

26. Mihaela, P.; Josef, R.; Monica, N.; Rudolf, Z. Perspectives of safflower oil as biodiesel source for South Eastern Europe (comparative study: Safflower, soybean and rapeseed). *Fuel* **2013**, *111*, 114–119. [CrossRef]

27. Khalid, N.; Khan, R.S.; Hussain, M.I.; Farooq, M.; Ahmad, A.; Ahmed, I. A comprehensive characterisation of safflower oil for its potential applications as a bioactive food ingredient—A review. *Trends Food Sci. Technol.* **2017**, *66*, 176–186. [CrossRef]

28. Sampaio, M.C.; Santos, R.F.; Bassegio, D.; de Vasconselos, E.S.; de Almeida Silva, M.; Secco, D.; da Silva, T.R. Fertilizer improves seed and oil yield of safflower under tropical conditions. *Ind. Crop. Prod.* **2016**, *94*, 589–595. [CrossRef]

29. Khalid, A.; Tamaldin, N.; Jaat, M.; Ali, M.F.M.; Manshoor, B.; Zaman, I. Impacts of biodiesel storage duration on fuel properties and emissions. *Procedia Eng.* **2013**, *68*, 225–230. [CrossRef]

30. Ministerio de Agricultura, Pesca y Alimentación. *Avances Mensuales de Superficies y Producciones Agrícolas*; Ministerio de Agricultura, Pesca y Alimentación: Madrid, Spain, 2018.

31. Çelebi, Y.; Aydın, H. Investigation of the effects of butanol addition on safflower biodiesel usage as fuel in a generator diesel engine. *Fuel* **2018**, *222*, 385–393. [CrossRef]

32. Işik, M.Z.; Aydin, H. Investigation on the effects of gasoline reactivity controlled compression ignition application in a diesel generator in high loads using safflower biodiesel blends. *Renew. Energy* **2019**, *133*, 177–189. [CrossRef]

33. Yang, R.; Zhang, L.; Li, P.; Yu, L.; Mao, J.; Wang, X.; Zhang, Q. A review of chemical composition and nutritional properties of minor vegetable oils in China. *Trends Food Sci. Technol.* **2018**, *74*, 26–32. [CrossRef]

34. Liu, W.; Lu, G.; Yang, G.; Bi, Y. Improving oxidative stability of biodiesel by cis-trans isomerization of carbon-carbon double bonds in unsaturated fatty acid methyl esters. *Fuel* **2019**, *242*, 133–139. [CrossRef]

35. Focke, W.W.; Van Der Westhuizen, I.; Oosthuysen, X. Biodiesel oxidative stability from Rancimat data. *Thermochim. Acta* **2016**, *633*, 116–121. [CrossRef]

36. Chan, C.H.; Tang, S.W.; Mohd, N.K.; Lim, W.H.; Yeong, S.K.; Idris, Z.; Kotwal, M.; Kumar, A.; Darbha, S.; Sharma, R.V.; et al. Tribological behavior of biolubricant base stocks and additives. *Renew. Sustain. Energy Rev.* **2018**, *93*, 145–157. [CrossRef]

37. Saluja, R.K.; Kumar, V.; Sham, R. Stability of biodiesel—A review. *Renew. Sustain. Energy Rev.* **2016**, *62*, 866–881. [CrossRef]

38. Kumar, N. Oxidative stability of biodiesel: Causes, effects and prevention. *Fuel* **2017**, *190*, 328–350. [CrossRef]

39. UNE-EN 14214:2013 V2+A1:2018. *Liquid Petroleum Products—Fatty Acid Methyl Esters (FAME) for Use in Diesel Engines and Heating Application—Requirements and Test Methods*; The National Standards Authority of Ireland (NSAI): Dublin, Ireland, 2018.

40. Canha, N.; Felizardo, P.; Correia, M.J.N. Controlling the oxidative stability of biodiesel using oils or biodiesel blending or antioxidants addition. *Environ. Prog. Sustain. Energy* **2018**, *37*, 1031–1040. [CrossRef]

41. França, F.R.M.; dos Santos Freitas, L.; Ramos, A.L.D.; da Silva, G.F.; Brandão, S.T. Storage and oxidation stability of commercial biodiesel using Moringa oleifera Lam as an antioxidant additive. *Fuel* **2017**, *203*, 627–632. [CrossRef]

42. Van der Westhuizen, I.; Focke, W.W. Stabilizing sunflower biodiesel with synthetic antioxidant blends. *Fuel* **2018**, *219*, 126–131. [CrossRef]
43. Varatharajan, K.; Pushparani, D.S. Screening of antioxidant additives for biodiesel fuels. *Renew. Sustain. Energy Rev.* **2018**, *82*, 2017–2028. [CrossRef]
44. Schober, S.; Mittelbach, M. The impact of antioxidants on biodiesel oxidation stability. *Eur. J. Lipid Sci. Technol.* **2004**, *106*, 382–389. [CrossRef]
45. Squissato, A.L.; Richter, E.M.; Munoz, R.A.A. Voltammetric determination of copper and tert-butylhydroquinone in biodiesel: A rapid quality control protocol. *Talanta* **2019**, *201*, 433–440. [CrossRef]
46. Yang, J.; He, Q.S.; Corscadden, K.; Caldwell, C. Improvement on oxidation and storage stability of biodiesel derived from an emerging feedstock camelina. *Fuel Process. Technol.* **2017**, *157*, 90–98. [CrossRef]
47. Dunn, R.O. Effect of antioxidants on the oxidative stability of methyl soyate (biodiesel). *Fuel Process. Technol.* **2005**, *86*, 1071–1085. [CrossRef]
48. Zhou, J.; Xiong, Y.; Liu, X. Evaluation of the oxidation stability of biodiesel stabilized with antioxidants using the Rancimat and PDSC methods. *Fuel* **2017**, *188*, 61–68. [CrossRef]
49. Kivevele, T.T.; Huan, Z. Effects of Antioxidants on the Cetane number, Viscosity, Oxidation Stability, and Thermal Properties of Biodiesel Produced from Nonedible Oils. *Energy Technol.* **2013**, *1*, 537–543. [CrossRef]
50. Serrano, M.; Martínez, M.; Aracil, J. Long term storage stability of biodiesel: Influence of feedstock, commercial additives and purification step. *Fuel Process. Technol.* **2013**, *116*, 135–141. [CrossRef]
51. Yahagi, S.S.; Roveda, A.C.; Sobral, A.T.; Oliveira, I.P.; Caires, A.R.L.; Gomes, R.S.; Trindade, M.A.G. An Analytical Evaluation of the Synergistic Effect on Biodiesel Oxidation Stability Promoted by Binary and Ternary Blends Containing Multifunctional Additives. *Int. J. Anal. Chem.* **2019**, *2019*, 6467183. [CrossRef]
52. Encinar, J.M.; González, J.F.; Pardal, A.; Martínez, G. Rape oil transesterification over heterogeneous catalysts. *Fuel Process. Technol.* **2010**, *91*, 1530–1536. [CrossRef]
53. UNE-EN ISO 3104/AC:1999. *Petroleum Products. Transparent and Opaque Liquids. Determination of Kinematic Viscosity and Calculation of Dynamic Viscosity (ISO 3104:1994)*; International Organization for Standarization (ISO): Geneva, Switzerland, 1999.
54. UNE-EN 116:2015. *Diesel and Domestic Heating Fuels—Determination of Cold Filter Plugging Point—Stepwise Cooling Bath Method*; British Standards Institution (BSI): London, UK, 2015.
55. UNE-EN 51023:1990. *Petroleum Products. Determination of Flash and Fire Points. Cleveland Open cup Method*; Asociacion Espanola de Normalizacion: Madrid, Spain, 1990.
56. UNE-EN-ISO-12937:2000. *Productos Petrolíferos. Determinación de Agua. Método de Karl Fischer por Valoración Culombimétrica*; The Spanish Association for Standardization and Certification (AENOR): Madrid, Spain, 2001.
57. UNE-EN 14111:2003. *Fat and Oil Derivatives. Fatty Acid Methyl Esters (FAME). Determination of Iodine Value*; International Organization for Standarization (ISO): Geneva, Switzerland, 2003.
58. Eryilmaz, T.; Yesilyurt, M.K. Influence of blending ratio on the physicochemical properties of safflower oil methyl ester-safflower oil, safflower oil methyl ester-diesel and safflower oil-diesel. *Renew. Energy* **2016**, *95*, 233–247. [CrossRef]
59. Jose, T.K.; Anand, K. Effects of biodiesel composition on its long term storage stability. *Fuel* **2016**, *177*, 190–196. [CrossRef]
60. García, M.; Botella, L.; Gil-Lalaguna, N.; Arauzo, J.; Gonzalo, A.; Sánchez, J.L. Antioxidants for biodiesel: Additives prepared from extracted fractions of bio-oil. *Fuel Process. Technol.* **2017**, *156*, 407–414. [CrossRef]
61. Dunn, R.O. Antioxidants for improving storage stability of biodiesel. *Biofuels Bioprod. Biorefining* **2008**, *2*, 304–318. [CrossRef]
62. Borsato, D.; de Moraes Cini, J.R.; Da Silva, H.C.; Coppo, R.L.; Angilelli, K.G.; Moreira, I.; Maia, E.C.R. Oxidation kinetics of biodiesel from soybean mixed with synthetic antioxidants BHA, BHT and TBHQ: Determination of activation energy. *Fuel Process. Technol.* **2014**, *127*, 111–116. [CrossRef]

Article

Biogas Potential from the Anaerobic Digestion of Potato Peels: Process Performance and Kinetics Evaluation

Spyridon Achinas [1],*, Yu Li [1], Vasileios Achinas [2] and Gerrit Jan Willem Euverink [1]

[1]   Faculty of Science and Engineering, University of Groningen, 9747 AG Groningen, The Netherlands;
      yu.li@rug.nl (Y.L.); g.j.w.euverink@rug.nl (G.J.W.E.)
[2]   Union of Agricultural Co-operatives of Monofatsi, Heraklion 700 16 Crete, Greece; vas.achinas@gmail.com
*   Correspondence: s.achinas@tug.nl

Received: 17 May 2019; Accepted: 11 June 2019; Published: 17 June 2019

Abstract: This article intends to promote the usage of potato peels as efficient substrate for the anaerobic digestion process for energy recovery and waste abatement. This study examined the performance of anaerobic digestion of potato peels in different inoculum-to-substrate ratios. In addition, the impact of combined treatment with cow manure and pretreatment of potato peels was examined. It was found that co-digestion of potato peel waste and cow manure yielded up to 237.4 mL CH_4/g VS_{added}, whereas the maximum methane yield from the mono-digestion of potato peels was 217.8 mL CH_4/g VS_{added}. Comparing the co-digestion to mono-digestion of potato peels, co-digestion in PPW/CM ratio of 60:40 increased the methane yield by 10%. In addition, grinding and acid hydrolysis applied to potato peels were positively effective in increasing the methane amount reaching 260.3 and 283.4 mL CH_4/g VS_{added} respectively. Likewise, compared to untreated potato peels, pretreatment led to an elevation of the methane amount by 9% and 17% respectively and alleviated the kinetics of biogas production.

Keywords: anaerobic treatment; biogas; kinetic study; potato peels; cow manure

---

1. Introduction

Potatoes are one of the largest crops worldwide and steadily grow hitherto as a staple food crop [1]. In the Netherlands, potatoes production is one of the major agricultural activities with more than 10 million tons produced in 2016 and approximately 50% of the potatoes undergo processing [2]. Potato peels are the main by-product of processing, producing about 8% waste by weight and its abatement is becoming a major issue. The potato industry produces on average 100 ktons of peels worldwide annually [3]. Current research is focused on the potato peel waste (PPW) recycling pathways for pharmaceutical and/or energy industries resulting in an enhanced potato peel waste management [4]. PPW is used for the production of low-value animal feed or as fertilizer in agricultural activities [5]. PPW has been previously characterized and contains starch (15–25%), non-starch polysaccharide (25–30%), acid insoluble and acid soluble lignin (15–20%), protein (18%), lipids (1%), and ash (6–10%) [6–9].

Its diverse composition can facilitate the production of bio-based products and fuels [10,11]. Peschel et al. [12] found an industrial approach for polyphenols extraction from potato peel waste. Recent studies also reported a promising method for lactic acid production with mixed microbial consortia in batch fermentation mode [13,14]. Moreover, its high content of carbon and nutrients renders it good feedstock for biogas production via the anaerobic digestion (AD) process [15]. AD is a biochemical procedure that degrades organic material and produces biogas (approximately 55–65% CH_4—35–45% CO_2) by facultative anaerobes and anaerobes bacteria in the absence of oxygen. AD mainly comprises three phases (hydrolysis, acidogenesis and methanogenesis) which illustrate biochemical events during the AD process [16–18]. AD technology has been broadly facilitated in the treatment of agricultural residuals, wastewaters, and animal slurries with primary objectives the energy production and waste reduction [19–21]. Versatile utilization of biogas, such as heat and power generation or vehicle fuel production, is an advantage among other gaseous biofuels [22,23]. Currently, more than 250 AD plants are operated in the Netherlands providing energy to approximately 2 million households and targeting to reduce more than 25 million metric tons CH_4 equivalent of greenhouse gas emissions by 2030 [24,25]. The inoculum to substrate ratio (ISR) has been regarded as an intrinsic factor influencing the AD performance. Notable lessening of methane yield was reported from the degradation of wastes, when ISRs were over a certain level [26,27].

Simultaneous treatment of solid and/or liquid organic waste also strengthens the process performance and increases the degradation efficacy, and this is attributed to a synergistic phenomena which occurs within the digester [28,29]. Comparing with mono-digestion, co-digestion of organic wastes with different animal slurries has positively affected the performance as it enhances the buffer capacity to maintain an optimal pH for methanogenic archaea. It also provides a better carbon-to-nitrogen (C/N) ratio, and utilizes microorganisms and macro and micronutrients which exist in various wastes [30]. Several reports refer to an optimal range for the C/N ratio as 20–30 which is far from the C/N ratio of 40–70 provided by the agricultural waste [31–34]. Furthermore, substrates with too high C/N ratio would not provide the sufficient nitrogen required for the microbial growth, while substrates with low C/N ratio would inhibit the process because of the NH_3 accumulation. Animal slurries contain a high amount of nitrogen that may hinder the bioreactor performance [35,36]. Pretreatment is also a parameter for further investigation as the degradation rate is affected by the composition and the characteristics of each substrate. The recalcitrance is a pivotal feature that impedes the decomposition of feedstocks and may hinder the process performance [16,37].

The present work aimed to add knowledge to the anaerobic digestion of PPW. The investigation of the anaerobic treatment of potato peels is crucial for the smooth operation and stability of continuous systems [38]. In the present report, AD performance was investigated by conducting batch experimental tests. The effect of homogenizing/grinding and acid hydrolysis as pretreatment methods are studied. The first-order and cone kinetic models were also applied for the prediction of the biogas production rate and to assess the fitting with the experimental data. This study aims to: (1) Investigate the effect of ISR on the biogas yield; (2) clarify how the addition of cow manure enhances the AD performance; and (3) provide insight of the kinetics of the PPW degradation.

2. Materials and Methods

2.1. Inoculum and Substrates

Microbial inoculum (sludge) was collected from a mesophilic operating anaerobic bioreactor from the wastewater treatment installation of Garmerwolde in Groningen, Netherlands. Cow manure was collected from a local farmer in the province of Groningen, Netherlands. Prior to their characterization and use, sludge and manure were stored at 6 °C to avoid undesirable fermentation processes. Potatoes were purchased in a local store and the peels were removed using a sharp knife and were cut into 0.5–1 cm pieces prior to use.

2.2. Experimental Design

The digestion tests were facilitated in batch mode using 500 mL glass bottles with a working volume of 400 mL operated at 36 °C for 55 days. The experimental conditions are given in Table 1. Three experimental studies were carried out to examine the influence of ISR (R1→5), co-digestion ratio (R6→8), and pretreatment method (R9→10) respectively on biogas yield by anaerobically treating in batch mode. The concentrations of the substrates were based on organic loading mass [39,40]. The amount of volatile solids (VS) of the inoculum and the substrate were calculated based on the predetermined ISRs. For all experiments, distilled water was added, and no additional nutrients/ trace elements were added to the reactors as it was assumed that they are provided by the inoculum (anaerobic sludge).

Table 1. Experimental conditions of the batch tests.

Particular	Experimental Design			
	ISR (based on VS)	Co-Digestion Ratio (based on VS)	Organic Load (g VS added per L)	Pretreatment
Experiment 1				
R1	0.25	100:0	10	-
R2	0.5	100:0	10	-
R3	1	100:0	10	-
R4	2	100:0	10	-
R5	4	100:0	10	-
Experiment 2				
R6	2	90:10	10	-
R7	2	75:25	10	-
R8	2	60:40	10	-
Experiment 3				
R9	2	60:40	10	Grinding
R10	2	60:40	10	Acid hydrolysis

Glass reactors were sealed with rubber stoppers and being flushed for 5 min with nitrogen to maintain an anoxic environment. Thereafter, they were incubated at temperature (36 °C) and rotating speed (150 rpm). Blank trials containing only anaerobic sludge were conducted to correct the biogas produced by the inoculum itself. Batch experiments were ended when no biogas was observed. In all experiments, triplicate serum bottles were used and results were means of triplicates ± standard deviation. Samples were taken at the beginning, the end, and in four predetermined time spots during the experimental period to determine the variation of pH, volatile fatty acids (VFAs and alkalinity. The experiments were terminated when no biogas was observed.

For the first batch experiment (R1, R2, R3, R4, and R5) of studying the influence of the ISR on biogas production, potato peels were used as the sole substrate and inoculum was added based on the set ISRs of 0.25, 0.5, 1, 2, and 4. For the second study (R6, R7, and R8) of investigating the impact of simultaneous digestion on biogas production, all experimental conditions were similar as those of the first experiment, apart from the type of substrate. Briefly, ISR was fixed at 2, and potato peels were digested with cow manure (CM) according to the mixing ratios (Table 1) to reach a total organic load per reactor of 4 g VS. The last experiment (R9 and R10) was conducted to examine the impact of potato peels pretreatment on biogas production, and all experimental conditions were the same as those of the second experiment, except that potato peel waste were underwent grinding and acid hydrolysis.

2.3. Analytical Methods

The amount of total solids (TS; g/kg) and volatile solids (VS; g/kg) was determined according to the recommendations of the Standard Methods of APHA et al. (2005) [41]. PH was estimated off-line using a pH meter (HI991001, Hanna Instruments, Woonsocket, RI, USA). The titration method was applied to determine the total alkalinity (TA; mg $CaCO_3$/L) and total volatile fatty acids (TVFAs; mg acetate/L) [42]. Chemical oxygen demand (COD; g/kg) was calculated using a test kit (Hach Lange GmbH, Germany) according to the manufacturer's guide and was quantified with a spectrophotometer (DR/2010, Hach, USA). For the acid hydrolysis pretreatment (AHP), 20 g of PPW was boiled in 200 mL of 0.9% (*w/w*) H_2SO_4 solution for 1 h and for the grinding, a commercial grinder was used to homogenize 30 g PPW with 90 mL water. The pH of the reactor with the acid-based treated PPW was adjusted to 7 by using 1 M NaOH solution. The lignin amount of substrates was determined according to the procedures established by the National Renewable Energy Laboratory (NREL) [43].

The method used to determine the VS removal was based on the previous study [44]. The method used to estimate the biochemical biogas potential was based on a volumetric test, which considered the displacement of a liquid into the measure of biogas production [45]. The water displacement equipment used in this work can provide biogas data within the 5% accuracy [46]. Biogas composition and its correction was carried out by the procedure followed in a previous study [31]. The methane potential assay can also be used to estimate the synergistic effect of co-substrates The synergistic effect can be expressed as the additional biogas yield for co-digestion feedstocks over the weighted average of the individual feedstock's biogas yield [47,48]. All the results were the means of replicates of triplicate ± standard deviation (SD) with an accuracy of 5%. Technical digestion time (TDT) (time needed to reach 80% of the maximal biogas yield) was used to assess the AD performance [49].

2.4. Kinetic Study

Biogas produced during the study was modeled by fitting the experimental data with two classical kinetic models to identify the kinetic parameters in order to enhance the full-scale application of the process. A regression analysis was conducted in Matlab R2015b (Mathworks, Natick, MA, USA) and Microsoft Office Excel (Microsoft Office 2010, Microsoft Corporation, Redmond, WA, USA) and the first order and cone models were used for the hydrolysis of organic matter and are described by the equations from previous studies [50,51]:

$$G(t) = G_O \times \left(1 - e^{(-Kt)}\right) \tag{1}$$

$$G(t) = \frac{G_O}{1 + (Kt)^{-n}} \tag{2}$$

$G(t)$ is the cumulative biogas yield at digestion time t days (mL biogas/g VSadded), G_O is the maximum biogas potential of the substrate the biogas potential (mL biogas/g VSadded), n is the shape factor, K is the hydrolysis constant (1/day), t is the time (days).

2.5. Statistical Analysis

The standardized residual is the residual divided by its standard deviation and is given by the equation:

$$\text{Standardized residual}\ (i) = \frac{residual\,(i)}{standard\ deviation\,(i)} \tag{3}$$

Single-factor analysis of variances (ANOVA) in Excel software 2010 was facilitated and statistical significance was established at a P-value less than 0.05 level.

3. Results and Discussion

3.1. Characterization of Substrates and Inoculum

The characteristics of the sludge, potato peel waste and cow manure were summarized in Table 2. It is noted that the characteristics of inoculum and cow manure were different for the three experimental periods.

3.2. Effect of ISRs on the Mono-Digestion PPW

The cumulative biogas yield (mL/g VSadded), daily biogas production rate (mL/g VSadded/day), pH, and total organic acids (%) from the anaerobic digestion of PPW at five ISRs were investigated, and the results are given in Figure 1. For the biogas correction, control reactors were operated treating only anaerobic sludge.

Biogas production rapidly began on the first day of digestion in all of the digesters (Figure 1A). Biogas production did not show interdependence with the increase of the ISR. The highest daily biogas production rate reached 47.4 and 45.5 mL/g VSadded/day on the second day of digestion at the ISRs of 2 and 4 (Figure 1A). When compared with those obtained at these two ISRs, a decrease of biogas production was observed at a lower ISR of 0.25, 0.5 and 1. The daily biogas production for the ISR 0.25, 0.5 and 1 fluctuated during the time period of 5–20 day of digestion within the range of 4–12 mL/g VS_{added}/day, and then dropped to a lower level. Biogas production was maintained in the period 5–20 days with the highest level of 25.6 and 19.5 for ISR of 2 and 4. It is notable that cumulative biogas production yields at the ISRs of 0.25 and 0.5 gradually augmented from day 3 until approximately day 25 and thereafter, it is gradually decreased (Figure 1B). In the ISR of 2 and 4, the production rate started to increase slowly until day 5. After 53 days of digestion, the biogas yields of PPW at the ISRs of 2 and 4 reached 383.7 and 361.2 mL/g VS_{added} respectively against 332.2 mL/g VS_{added} achieved at the ISR of 1 (Table 3). At the ISRs of 0.25 and 0.5, biogas yield reached 165.8 and 228.1 mL/g VS_{added} respectively, indicating the importance of sufficient availability of sludge for the organic mass degradation.

Table 2. Physical and chemical characteristics of the substrates used in the batch tests.

Parameter	Inoculum	Potato Peels Waste (PPW)	Cow Manure (CM)
TS (g/kg)	50.9 (0.2)	144.0 (4.3)	144.6 (1.5)
VS (g/kg)	29.1 (1.2)	139.4 (4.8)	119.4 (1.1)
VS/TS	0.60	0.97	0.78
COD (g/kg)	47.2 (1.0)	123.8 (4.2)	138.4 (1.9)
pH	7.36	NA	NA
Insoluble lignin (%TS)	NA	19.1 (0.7)	24.6 (2.5)
Soluble lignin (%TS)	NA	1.59 (0.1)	1.2 (0.4)
Ash (%)	NA	0.58 (0.1)	5.5 (1.1)

The methane contents at the ISRs of 0.25, 0.5, 1, 2, and 4 were 50.23%, 51.12%, 53.40%, 56.75%, and 52.8%, respectively (Table 3). The methane yields were calculated to be 83.3, 116.6, 177.4, 217.8, and 190.7 mL CH_4/g VS_{added} respectively. The methane content at the ISRs of 0.25 and 0.5 indicated that the degradation was lower than those obtained at the ISRs of 1, 2, and 4 (30.8% vs. 37.1–39.3%). The ISR of 4 showed obviously the shortest technical digestion time (TDT) of 17 days indicating the rapid degradation of the organic matter.

As shown in Table 3, the TDT at the ISRs of 0.25 and 0.5 were shorter than that obtained in the case of the ISR of 1 and 4 (22 days vs. 24 and 26 days). The TDT did not follow the tendency of degradation efficiency. Parawira et al. [52] studied the two-stage digestion of potato peels and observed high biogas yield. This might be due to the segregation of the digestion stages resulting in higher conversion efficiencies. Similar studies on the digestion of solid potato waste showed higher cumulative biogas yield than in our study, which might be due to the higher lignin percentage (approx. 20%) of the substrate, which burdens the degradation of hemicellulose components from the enzymes [53,54]. The complex structure of lignin prevents the carbohydrate-degrading microbes from attaching to the structural carbohydrates, i.e., cellulose and hemicellulose [55].

Table 3. Results from the mono-digestion of PPW.

Parameter	ISR				
	0.25 (R1)	0.5 (R2)	1 (R3)	2 (R4)	4 (R5)
Biogas yield (mL/g VS_{added})	165.8 ± 7.9	228.1 ± 6.8	332.2 ± 7.6	383.7 ± 10.7	361.2 ± 14.4
Methane content (%)	50.2 ± 0.5	51.1 ± 1.7	52.4 ± 0.5	56.8 ± 0.3	53.8 ± 0.3
Methane yield (mL/g VS_{added})	83.3 ± 0.9	116.6 ± 3.9	177.4 ± 2.3	217.7 ± 1.2	190.7 ± 2.9
VS_{added} degradation (%)	15.6 ± 1.2	19.7 ± 3.0	28.5 ± 2.1	33.1 ± 2.7	32.1 ± 3.4
Technical digestion time (d)	22	22	26	17	24

As depicted in Figure 1 and Table 3, both daily production and total biogas yield lowered when the ISRs were decreased from 1 to 0.25. Low efficiency of biogas production might be due to the insufficient methanogens, which could cause the accumulation of the volatile fatty acids in the bioreactor and thus the pH dropping. PH values in the range of 6.5–7.5 favor the growth and activity of methanogens [56,57].

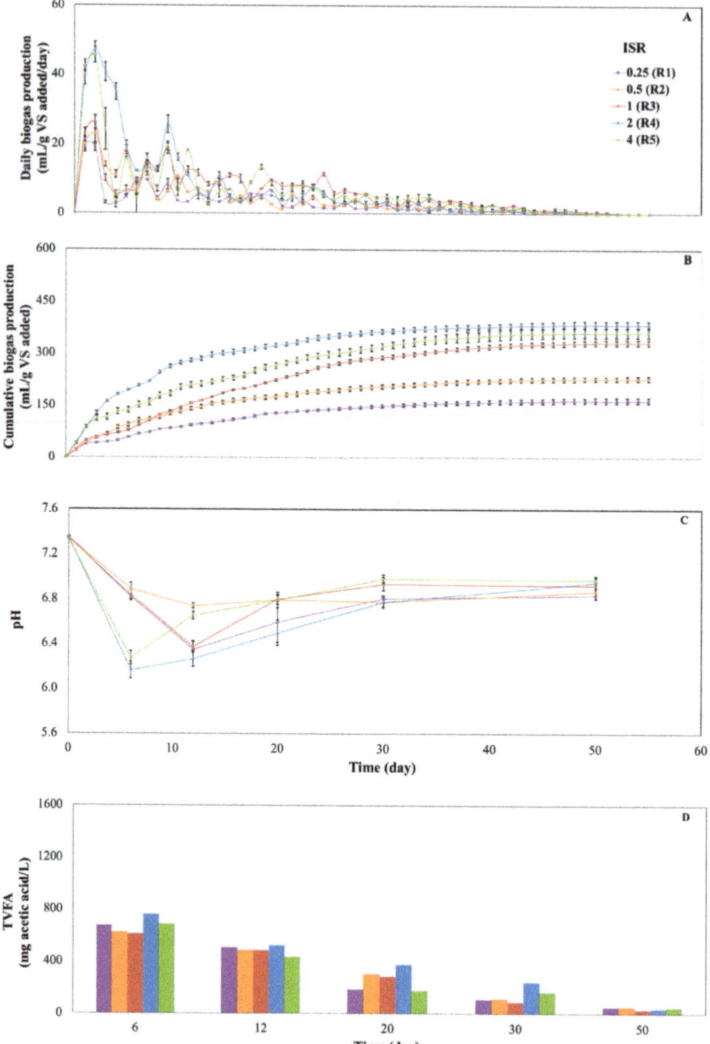

Figure 1. Effect of the inoculum-to-substrate ratio on the daily biogas production (**A**), the cumulative biogas production (**B**), the pH (**C**), and total volatile fatty acids (TVFA) accumulation (**D**), from the anaerobic digestion of potato peels.

The changes in VFA and pH were also recorded, and the outcomes are given in Figure 1C,D. As observed in the digestion of PPW at the ISR of 2 and 4, the pH in those digesters decreased to 6.16 and 6.27 on day 6 and slightly increased until day 12, corresponding well with higher VFA concentrations (Figure 1D) during acidogenesis of the substrate in this period. According to TVFA analysis, R1 and R2 reached 675.8 and 625.4 mg acetate/L on day 6. The digesters with ISR of 0.25 and 0.5, showed a slow decrease of the pH, which corresponds to the low availability of inoculum reaching day 12 pH of 6.35 and 6.74 respectively. TVFA levels were found to vary significantly in the two reactors at different sampling times. The AD process is regarded as stable when TVFA/TA ratio ranges between 0.23 and 0.3. In addition, the ISR of 2 showed significant higher buffer capacity maintaining an optimal pH for methanogenic archaea (Figure S1A).

VFAs are produced in anaerobic digesters and have different impacts on microbial activity. Liotta et al. referred that VFA concentration of 2100 mg/L does not inhibit the AD process stability [58]. Several studies have determined the effect of different VFAs on the dynamics of the methanogenic consortia during the anaerobic digestion resulting in variations of the process stability and efficiency [59,60]. Previous studies also reported a reduction of methane yield produced food wastes and/or green wastes at the ISR of over 4. Pellera and Gidarakos [61] verified that lower methane yields and TDT were observed in the treatment of olive pomace at the ISR of 4 than those obtained at ISRs of 0.5–1. The peak of daily biogas production at ISR of 4 was slightly higher and TDT was relatively shorter (17 days on ISR of 4, Table 3) than those obtained at the ISR = 0.5 and 1. The functional relationship between biogas yield and the inoculum-to-substrate ratio was non-linear and followed a bell curve (Figure S1B). Based on that, the maximum biogas yield was achieved at the ISR of 2 and this ratio was chosen for the following tests.

3.3. Co-Digestion of Potato Peel Waste with Cow Manure: Influence on Process Performance

To ameliorate the performance of anaerobic digestion by treating potato peel waste with cow manure, the mixing ratios of the two substrates were varied to produce different C/N ratios, while operating the digesters at the ISR of 2. The outcomes of mono-digestion of potato peel waste (R4) were used as the control. As in the anaerobic mono-digestion of potato peel waste, biogas was produced on day 6 in the reactors with the mixtures. Biogas production rapidly began on the first day of digestion in all of the digesters. Biogas production did show interdependence on the increase of cow manure fraction. The highest daily biogas production rate of R7 and R8 reached 31.5 and 25.5 mL/g VS$_{added}$/day on the second day of digestion and reactor R6 reached a daily production rate of 28.3 mL/g VS$_{added}$ on the first day of the digestion (Figure 2A).

For the determination of weighted methane yield, an additional control reactor RC2 was operated treating cow manure with inoculum and the methane yield is given in Table S1. Synergistic effects were noticed in PPW/CM ratio of 75:25 and 60:40 (Table 4). When compared with that obtained without cow manure, a significant decrease was noticed in the daily biogas production rate. As seen in Figure 2A, biogas production was maintained in high level the second week of the digestion with the highest level of 16.7, 29.1, and 23.4 mL/g VS$_{added}$ for the combined treatment ratios of 90:10, 75:25, and 60:40 respectively.

At PPW/CM ratio of 75:25 and 60:40, daily biogas production almost reached the former peaks on day 8 (29.14 and 23.4 mL/g VS$_{added}$/day) before decreasing to a low level on day 9. At a PPW/CM ratio of 90:10, R8 reached almost the half of the maximum biogas production (16.7 mL/g VS$_{added}$/day) on day 9 and thereafter it gradually remained at a lower level. Daily production fluctuated between day 15 and day 25 within the range of 6–11 mL/g VS$_{added}$/day, and then dropped to a low level.

After 53 days of digestion, the cumulative biogas amount in R6, R7, and R8 reached 395.46, 403.65 and 423.1 mL/g VS$_{added}$ respectively against 383.7 mL/g VS$_{added}$ achieved at the R4 without cow manure (Figure 2B). The cumulative biogas production observed adding cow manure was higher than the anaerobic mono-digestion of PPW.

Discrepancies in biogas and methane yields between mono-digestion and co-digestion of PPW were statistically significant. The methane percentages in R6, R7, and R8 were 52.28%, 53.40%, and 56.11% respectively, were lower than 56.75% in the mono-digestion of potato peels. The TDT at the three PPW/SM ratios was 29, 21, and 25 days, respectively, which is longer than that in R4 and the final methane yields at the three OP/SM ratios were calculated to be 206.8, 215.5, and 237.4 mL

CH$_4$/g VS$_{added}$ (Table 4). Opinions vary regarding if TVFA/TA is a sufficient parameter to evaluate the AD stability, with several authors advocating ratios below 0.3 as optimal for the smooth operation of the digesters. PH variation and changes in the TVFA concentration were therefore estimated, and the outcomes are depicted in Figure 2C,D. There was a positive correlation between pH variation and TVFA concentration within the three reactors. Elevated TVFA concentrations were detected from the co-digestion reaching 1069.2, 1152.8, and 1028.1 mg acetate/L on day 6 in ratios of 90:10, 75:25, and 60:40 respectively. However, TVFAs of R7 reduced slower than those on R6 and R8 reaching 939.6 mg acetate/L on day 12. The high bicarbonate concentration most likely contributed to the system buffering (Figure S2A).

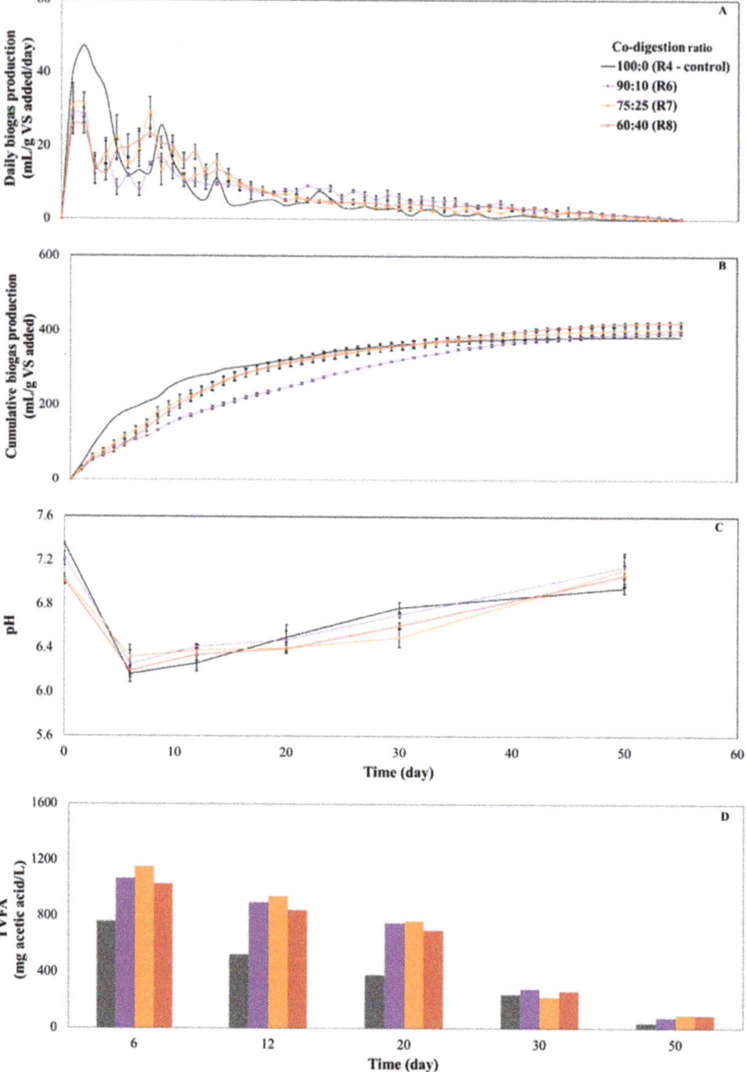

Figure 2. The effect of the co-digestion ratio on the cumulative biogas production (**A**), the daily biogas production (**B**), the pH (**C**), and the TVFA accumulation (**D**).

Table 4. Results from the co-digestion of potato peels with cow manure.

Parameter	Co-Digestion Ratio		
	90:10 (R6)	**75:25 (R7)**	**60:40(R8)**
Biogas yield (mL/g VS_{added})	395.4 ± 4.3	403.7 ± 9.8	423.1 ± 5.2
CH_4 content (%)	52.3 ± 1.6	53.4 ± 2.3	56.1 ± 0.8
CH_4 yield (mL/g VS_{added})	206.7 ± 3.6	215.5 ± 7.6	237.4 ± 3.1
Weighted methane yield (mL/g VS_{added})	211.6	202.5	193.3
Differential (EMY-weighted MY) (mL/g VS_{added})	-4.9	13.1	44.2
Synergistic effect	No	Yes	Yes
VS_{added} degradation (%)	34.5 ± 2.8	35.2 ± 4.3	37.2 ± 4.7
Technical digestion time (day)	29	21	25

From Figure 2D, the TVFAs concentration of all the reactors dropped below 300 mg acetate/L on day 30 and remained in the range of 225–283 mg acetate/L. The TVFA showed similar performance with those in the case of R4, and at the end of digestion did not exceed the level of 100 mg/L. Combined treatment provided a better carbon/nitrogen ratio to the system. Considering the above results, the part addition of cow manure typifies an alternative method for farm-scale digestion units. The functional relationship between biogas yield and cow manure fraction added at a constant ISR was also assessed (Figure S2B). The results indicated that the biogas yield increased linearly [y = 12.637x + 369.89 (R2 = 0.9685)] with respect to an increase of cow manure fraction in the reactors. Similarly, a positive effect of cow manure in this paper was consistent with previously reported results [62,63]. Mixtures at the PPW/CM ratio of 60:40 were prepared thereafter for examining the impact of grinding and acid hydrolysis as pretreatment method on the anaerobic digestion of potato peels.

3.4. Effect of Pretreatment on Biogas Production in Co-Digestion of Potato Peels and Cow Manure

To elevate the AD efficiency, the combined treatment of PPW with CM, potato peels underwent pretreatment in a co-digestion ratio of 60:40. Grinding/homogenizing and dilute acid pretreatments were chosen to represent physical and chemical pretreatment.

The results of combined digestion of potato peels and cow manure (experimental set R8) were used as a control. Biogas production quickly began on the first day for all digesters (Figure 3A). The highest daily biogas production rate of R9 and R10 reached 35.8 and 35.1 mL/g VS_{added}/day on the second and the first day respectively. The two pretreatments effectively enhanced the biogas production yields of PPW. Significant high biogas yield was achieved from R9 and R10 reaching 453.2 and 485.4 mL/g VS_{added} respectively (Table 5). The biogas production was quicker than R8, with R10 showing a technical digestion time of 19 days (Figure 3B). The changes of VFA and pH during digestion of R9 and R10 were monitored to examine the relationship between the improvement of the degradation efficiency and the corresponding high biogas yields, and the results are shown in Figure 3C,D.

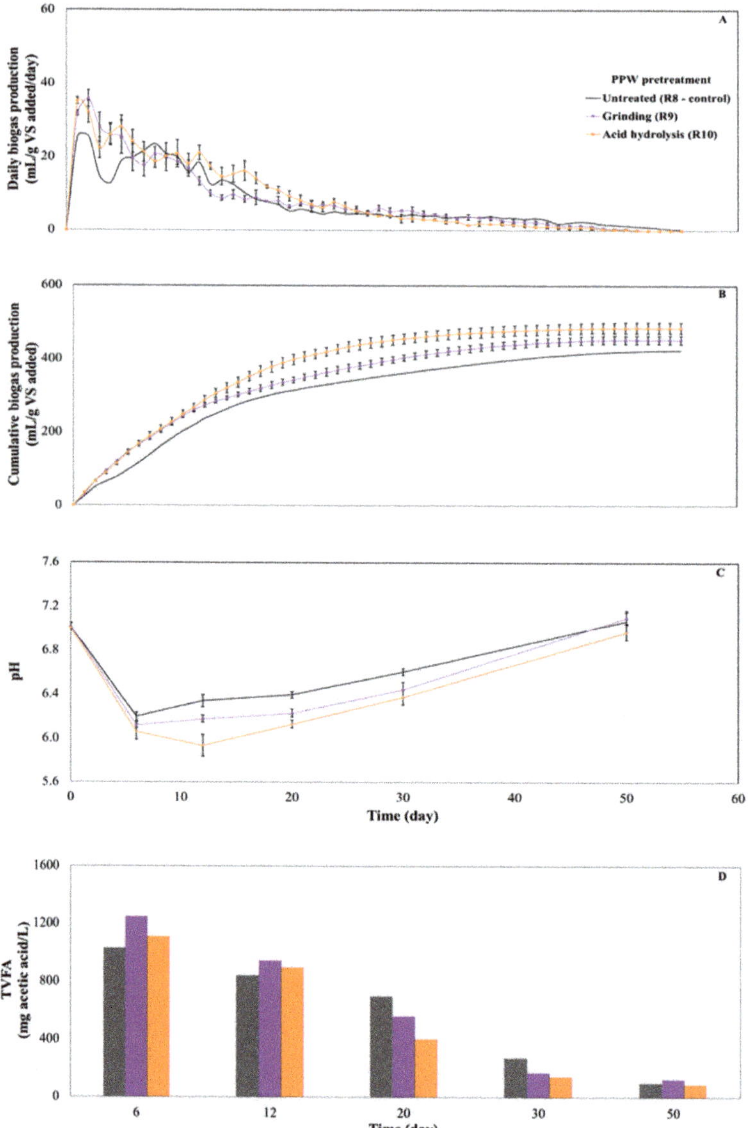

Figure 3. The effect of the pretreatment on the cumulative biogas production (**A**), the daily biogas production (**B**), the pH (**C**), and the TVFA accumulation (**D**).

Table 5. Results from the combined treatment of pretreated PPW and CM.

Parameter	Pretreatment	
	Grinding/Homogenizing (R9)	Acid Hydrolysis (R10)
Biogas yield (mL/g VS_{added})	453.2 ± 11.0	485.4 ± 14.6
Methane content (%)	57.5 ± 1.4	58.3 ± 0.8
Methane yield (mL/g VS_{added})	260.3 ± 3.2	283.4 ± 2.7
VS_{added} degradation (%)	39.5 ± 2.5	40.5 ± 3.2
Technical digestion time (day)	23	19

The pH in R9 followed the same trend with R8 and slightly increased until day 12, which was in line with the higher VFA concentrations (Figure 3C) during hydrolysis of the substrate in this period. In R10, pH decreased from 6.06 on day 6 to 5.94 on day 12 which might reflect an agglomeration of fatty acids which is usually associated with a lessening of the pH and lack of sufficient reactor buffering. However, on day 12, the pH increased as the methanogens quickly converted the VFAs to biogas indicating the sufficient response of the process to a possible imbalance between acidogens and methanogens. Volatile fatty acids are important intermediates of carbohydrates, lipids, and proteins degradation, and affect the efficiency of the AD process [64]. The methane production content (Table 5) in the experiment was increased by pretreatment of potato peels, such as acid hydrolysis and grinding treatments. The fast production of VFAs can decrease the pH and, thus impede the methanogenesis step.

Therefore, a smooth and stable reactor operation depends on the balance between VFAs and alkalinity (Figure S3). The performance of VFAs generation, conversion, and accumulation in reactor R9 and R10 are given in Figure 3D. The main difference between the R9 and R10 with R8 was the rapid response to TVFAs production and accumulation. From day 6 to day 12, the concentration of TVFA in R9 and R10 decreased from 1248.7 and 1109.6 mg acetate/L to 945.3 and 896.5 mg acetate/L respectively. These two reactors had a similar TVFA profile. However, the produced and increased TVFAs were converted rapidly without causing inhibition because of sufficient alkalinity (Figure S3). Higher biogas yield upon pretreatment is envisaged due to the reduced cellulose crystallinity and the lessening of lignin content [65].

The dilute acid pretreatment revealed an increment in the amount of biogas produced and justifies the enhanced biomass reactivity, a fact that is attributed to the incremental release of the sugars (monomers and oligomers). Depending on the pretreatment poignancy (time, acid concentration and temperature), sugars can be converted to aldehydes, such as furfural and hydroxymethylfurfural. Under controlled conditions, removal of hemicellulose from biomass displays the cellulose to microorganisms and increases the glucose yield [66]. Mancini et al. [67] reported the positive effect of chemical pretreatment on biogas production of wheat straw. This step enhances the susceptibility of the cellulosic components. In addition, other pretreatment methods would be useful to be applied to determine the effect on anaerobic digestion performance as they have advantages and limitations, and their application cannot be based on their influence on the AD process [68]. However, an economic analysis would be interesting to be conducted in order to evaluate which pretreatment has the best potential for full-scale applications.

3.5. VS Removal

From another view, the relationship of cumulative biogas yield and VS% removal was plotted (Figure 4). Based on the obtained data in this study, a linear regression equation was established ($y = 0.085x + 1.6183$).

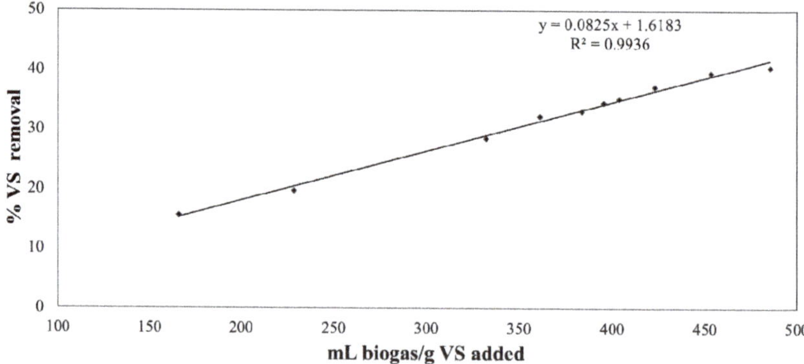

Figure 4. The relationship between biogas yield and volatile solids (VS) removal.

As expected, the increment of cumulative biogas yield follows similar to the raised degradation percentage of the organic material. The high correlation coefficient (R2) indicated that there was a positive correlation between biogas yield and VS reduction.

3.6. Kinetics Results and Correlation with Process Parameters

Outcomes of the kinetic study using the first order and cone models are summarized in Tables 6 and 7. Both models were found to have a good fit with the experimental data. In the first experiment, R1 and R2 showed good degradation rates, however, the biogas yields remained at a lower level. One possible reason for poor biogas yield could be methanogens growth inhibition due to the fast PPW's acidification and slow methanogenesis rate. The R4 with the ISR of 2 showed efficient performance providing the highest K value (0.911 d^{-1}) reaching 383.7 mL/g VS$_{added}$ of biogas yield.

Table 6. Results from the 1-order model.

Reactor	1-Order				
	K (1/day)	R^2	RMSE	Measured (mL/g VS$_{added}$)	Predicted (mL/g VS$_{added}$)
R1	0.0741	0.9934	4.05	165.8	163.3
R2	0.0733	0.9977	6.46	228.1	224.6
R3	0.067	0.9786	16.57	332.2	324.9
R4	0.0911	0.0323	15.49	383.7	381.6
R5	0.0667	0.0323	15.30	361.2	353.1
R6	0.0537	0.9924	14.15	395.4	376.9
R7	0.0767	0.999	5.49	403.7	398.6
R8	0.066	0.9943	8.89	423.1	413.3
R9	0.0702	0.9982	10.47	453.2	444.9
R10	0.0921	0.9909	19.38	485.4	482.9

During the second experiment, all reactors showed lower K values than the mono-digestion of PPW (R4) because the manure has a higher percentage of complex carbohydrates [35]. Although the hydrolysis was slower, microbial interactions from inoculum and manure favored the whole process performance degradation. The pretreatment positively affected the process performance and increased the biogas yield. Acid hydrolysis pretreated PPW showed high hydrolysis rate and increased the biogas yield. The difference between experimental and predicted values was low (0.52%) in the case of acid hydrolysis.

Table 7. Results from the cone model.

| Reactor | Cone | | | | | |
	K (1/day)	n	R^2	RMSE	Measured (mL/g VS_{added})	Predicted (mL/g VS_{added})
R1	0.1108	1.69	0.9657	8.273	165.8	158.4
R2	0.1258	1.59	0.9762	9.187	228.1	217.9
R3	0.0817	2.03	0.9724	16.51	332.2	317.3
R4	0.176	1.51	0.9818	12.54	383.7	371.6
R5	0.1155	1.55	0.9481	21.59	361.2	341.7
R6	0.0798	1.79	0.965	21.17	395.4	368.9
R7	0.1098	1.78	0.9871	12.61	403.7	387.8
R8	0.0958	1.80	0.9886	12.77	423.1	402.7
R9	0.1129	1.68	0.9779	18.13	453.2	432.9
R10	0.1111	1.98	0.9838	17.56	485.4	472.2

To assess the soundness of the results, the predicted values were plotted against the experimental results (Figures S4–S6). The low values of RMSE reflect the model's high ability to accurately predict the bioactivities. The statistical indicators (RMSE and R2) are given in Tables 6 and 7 providing the picture of kinetics study. In addition, to evaluate the accuracy of all replicates, the standardized residuals were calculated and plotted in Figure S7.

According to the kinetic study, this study concluded that the kinetic parameters can be affected by the process conditions. Therefore, the effects of these process parameters (ISR, co-digestion ratio, VS removal rate and pH) on kinetic parameters (G and K) were examined based on Pearson's correlation analysis (Table S2). In this case, the biogas production potential was significantly increased by a co-digestion ratio ($r = 0.742$, $P < 0.05$) and the VS removal rate ($r = 0.994$, $P < 0.01$), but decreased by pH ($r = -0.753$, $P < 0.05$). The hydrolysis constant (k) did not show any relevance to the process parameters or to biogas production potential.

4. Conclusions

This study investigated the impact of inoculum-to-substrate ratios, co-digestion ratios and pretreatment methods on the anaerobic digestion performance of potato peels. The overall results showed that co-digesting of acid based pretreated PPW with cow manure in a ratio of 60:40 reinforces biogas production and increases methane content. Specifically, biogas yields in the ISR of 2 showed the highest biogas yield and the lowest technical digestion time in comparison to other ISRs indicating that an appropriate amount of sludge is required for efficient operation. Moreover, the addition of cow manure in general improved the process efficiency with the PPW:CM ratio of 60:40 being ideal and reaching 423.1 mL/g VS_{added} of biogas yield. Notwithstanding this, the pretreatment methods applied to PPW affected positively the degradation yielding of 453.2 and 485.4 mL/g VS_{added} for grinding and acid hydrolysis as pretreatments respectively. It is noteworthy that pretreated potato peels show good potential for enhancement of biogas production and increment of the energy output by co-digesting with cow manure while providing a stable anaerobic digestion process.

Supplementary Materials: The following are available online at http://www.mdpi.com/1996-1073/12/12/2311/s1, Additional tables and figures are given in the supplementary file.

Author Contributions: Conceptualization, S.A.; writing—original draft, S.A.; software, Y.L.; writing—review & editing, V.A. and G.J.W.E.

Funding: This research received no external funding.

Conflicts of Interest: The authors declare no conflict of interest.

References

1. FAO. FAOSTAT Agricultural Data. Food and Agricultural Commodities Production. Available online: http://faostat.fao.org/site/339/default.aspx (accessed on 12 June 2014).
2. VAVI. Association for the Potato Processing Industry. Available online: http://www.vavi.nl/index.shtml (accessed on 26 November 2018).
3. Chang, K.C. Polyphenol antioxidants from potato peels: Extraction, optimization and application to stabilizing lipid oxidation in foods. In Proceedings of the National Conference on Undergraduate Research (NCUR), Ithaca College, NY, USA, 31 March–2 April 2011.
4. Wu, D. Recycle Technology for Potato Peel Waste Processing: A Review. *Procedia Environ. Sci.* **2016**, *31*, 103–107. [CrossRef]
5. Nelson, M.L. Utilization and application of wet potato processing coproducts for fishing cattle. *J. Anim. Sci.* **2010**, *88*, 133–142. [CrossRef] [PubMed]
6. Mader, J.; Rawel, H.; Kroh, L. Composition of phenolic compounds and glycoalkaloids a-solanine and a-chaconine during commercial potato processing. *J. Agric. Food Chem.* **2009**, *57*, 6292–6297. [CrossRef]
7. Schieber, A.; Stintzing, F.C.; Carle, R. By-products of plant food processing as a source of functional compounds—Recent developments. *Trends Food Sci. Technol.* **2001**, *12*, 401–413. [CrossRef]
8. Camire, M.E.; Violette, D.; Dougherty, M.P.; McLaughlin, M.A. Potato peel dietary fiber composition: Effects of peeling and extrusion cooking processes. *J. Agric. Food Chem.* **1997**, *45*, 1404–1408. [CrossRef]
9. Liang, S.; McDonald, A.G. Chemical and thermal characterization of potato peel waste and its fermentation residue as potential resources for biofuel and bioproducts production. *J. Agric. Food Chem.* **2014**, *62*, 8421–8429. [CrossRef] [PubMed]
10. Liang, S.; Han, Y.; Wei, L.; McDonald, A.G. Production and characterization of bio-oil and bio-char from pyrolysis of potato peel wastes. *Biomass Convers. Biorefin.* **2015**, *5*, 237–246. [CrossRef]
11. Arapoglou, D.; Varzakas, T.; Vlyssides, A.; Israilides, C. Ethanol production from potato peel waste (PPW). *Waste Manag.* **2010**, *30*, 1898–1902. [CrossRef]
12. Peschel, W.; Sanchez-Rabaneda, F.; Diekmann, W.; Plescher, A.; Gartzia, I.; Jimenez, D.; Lamuela-Raventos, R.; Buxaderas, S.; Codina, C. An industrial approach in the search of natural antioxidants from vegetable and fruit wastes. *Food Chem.* **2006**, *97*, 137–150. [CrossRef]
13. Liang, S.; Gliniewicz, K.; Mendes-Soares, H.; Settles, M.L.; Forney, L.J.; Coats, E.R.; McDonald, A.G. Comparative analysis of microbial community of novel lactic acid fermentation inoculated with different under fined mixed cultures. *Bioresour. Technol.* **2015**, *179*, 268–274. [CrossRef]
14. Liang, S.; McDonald, A.G.; Coats, E.R. Lactic acid production from potato peel waste by anaerobic sequencing batch fermentation using undefined mixed culture. *Waste Manag.* **2015**, *45*, 51–56. [CrossRef] [PubMed]
15. Chiumenti, A.; Boscaro, D.; da Borso, F.; Sartori, L.; Pezzuolo, A. Biogas from fresh spring and summer grass: Effect of the harvesting period. *Energies* **2018**, *11*, 1466. [CrossRef]
16. Achinas, S.; Achinas, V.; Euverink, G.J.W.M. A technological overview of biogas production from biowaste. *Engineering* **2017**, *3*, 299–307. [CrossRef]
17. Li, W.W.; Yu, H.Q. Advances in energy-producing anaerobic biotechnologies for municipal wastewater treatment. *Engineering* **2016**, *2*, 438–446. [CrossRef]
18. Chen, J.F. Green chemical engineering. *Engineering* **2017**, *3*, 283–284. [CrossRef]
19. Sahajwalla, V. Green Processes: Transforming Waste into Valuable Resources. *Engineering* **2018**, *4*, 309–310. [CrossRef]
20. Matsakas, L.; Gao, Q.; Jansson, S.; Rova, U.; Christakopoulos, P. Green conversion of municipal solid wastes into fuels and chemicals. *Electron. J. Biotechnol.* **2017**, *26*, 69–83. [CrossRef]
21. RedCorn, R.; Fatemi, S.; Engelberth, A.S. Comparing end-use potential for industrial food-waste sources. *Engineering* **2018**, *4*, 371–380. [CrossRef]
22. Solarte-Toro, J.C.; Chacón-Pérez, Y.; Cardona-Alzate, C.A. Evaluation of biogas and syngas as energy vectors for heat and power generation using lignocellulosic biomass as raw material. *Electron. J. Biotechnol.* **2018**, *33*, 52–62. [CrossRef]
23. Achinas, S.; Achinas, V. Biogas combustion: An introductory briefing. In *Biogas: Production, Applications and Global Developments*; Vico, A., Artemio, N., Eds.; Nova Science Publishers, Inc.: New York, NY, USA, 2017; pp. 179–193.

24. Chen, P.; Anderson, E.; Addy, M.; Zhang, R.; Cheng, Y.; Peng, P.; Ma, Y.; Fan, L.; Zhang, Y.; Lu, Q.; et al. Breakthrough technologies for the biorefining of organic solid and liquid wastes. *Engineering* **2018**, *4*, 574–580. [CrossRef]

25. EBA. *Biogas Report 2014—Biogas Production in Europe*; European Biogas Association: Bruxelles, Belgium, 2015.

26. Fabbri, A.; Serranti, S.; Bonifazi, G. Biochemical methane potential (BMP) of artichoke waste: The inoculum effect. *Waste Manag. Res.* **2014**, *32*, 207–214. [CrossRef] [PubMed]

27. Valenti, F.; Porto, S.M.C.; Cascone, G.; Arcidiacono, C. Potential biogas production from agricultural by-products in Sicily. A case study of citrus pulp and olive pomace. *J. Agric. Eng.* **2017**, *48*, 196–202. [CrossRef]

28. Scaglione, S.; Caffaz, S.; Ficara, E.; Malpei, F.; Lubello, C. A simple method to evaluate the short-term biogas yield in anaerobic codigestion of WAS and organic wastes. *Water Sci. Technol.* **2008**, *58*, 1615–1622. [CrossRef] [PubMed]

29. Dinuccio, E.; Gioelli, F.; Cuk, D.; Rollè, L.; Balsari, P. The use of co-digested solid fraction as feedstock for biogas plants. *J. Agric. Eng.* **2013**, *44*, 153–159. [CrossRef]

30. Achinas, S.; Euverink, G.J.W. Feasibility study of biogas production from hardly degradable material in co-inoculated bioreactor. *Energies* **2019**, *12*, 1040. [CrossRef]

31. Gomez, X.; Cuetos, M.J.; Cara, J.; Moran, A.; Garcia, A.I. Anaerobic co-digestion of primary sludge and the fruit and vegetable fraction of the municipal solid wastes: Conditions for mixing and evaluation of the organic loading rate. Renew. *Energy* **2006**, *31*, 2017–2024.

32. Bolzonella, D.; Innocenti, L.; Cecchi, F. BNR wastewater treatments and sewage sludge anaerobic mesophilic digestion performances. *Water Sci. Technol.* **2002**, *46*, 199–208. [CrossRef]

33. Perazzolo, F.; Mattachini, G.; Tambone, F.; Calcante, A.; Provolo, G. Nutrient losses from cattle co-digestate slurry during storage. *J. Agric. Eng.* **2016**, *47*, 94–99. [CrossRef]

34. Pontoni, L.; Panico, A.; Salzano, E.; Frunzo, L.; Iodice, P.; Cavinato, F. Innovative parameters to control the efficiency of anaerobic digestion process. *Chem. Eng. Trans.* **2015**, *43*, 2089–2094.

35. Cavinato, C.; Fatone, F.; Bolzonella, D.; Pavan, P. Thermophilic anaerobic co-digestion of cattle manure with agro-wastes and energy crops: Comparison of pilot and full-scale experiences. *Bioresour. Technol.* **2010**, *101*, 545–550. [CrossRef]

36. Naik, L.; Gebreegziabher, Z.; Tumwesige, V.; Balana, B.B.; Mwirigi, J.; Austin, G. Factors determining the stability and productivity of small-scale anaerobic digesters. *Biomass Bioenergy* **2014**, *70*, 51–57. [CrossRef]

37. Mancini, G.; Papirio, S.; Lens, P.N.L.; Esposito, G. Solvent pretreatments of lignocellulosic materials to enhance biogas production: A review. *Energy Fuels* **2016**, *30*, 1892–1903. [CrossRef]

38. Nelson, M.J.; Nakhla, G.; Zhu, J. Fluidized-bed bioreactor applications for biological wastewater treatment: A review of research and developments. *Engineering* **2017**, *3*, 330–342. [CrossRef]

39. Bolzonella, D.; Pavan, P.; Battistoni, P.; Cecchi, F. Mesophilic anaerobic digestion of waste activated sludge: Influence of the solid retention time in the wastewater treatment process. *Process Biochem.* **2005**, *40*, 1453–1460. [CrossRef]

40. Fabbri, A.; Bonifazi, G.; Serranti, S. Micro-scale energy valorization of grape MARCS in winery production plants. *Waste Manag.* **2015**, *36*, 156–165. [CrossRef] [PubMed]

41. APHA-AWWA-WEF. *Standard Methods for the Examination of Water and Wastewater*; APHA-AWWA-WEF: Washington, DC, USA, 2005.

42. Lossie, U.; Pütz, P. *Targeted Control of Biogas Plants with the Help of FOS/TAC*. Practice Report Hach-Lange. 2008. Available online: https://www.semanticscholar.org/paper/Targeted-control-of-biogas-plants-with-the-help-of-Lossie-Putz/b71fa7e5578e02348b65d22f7be11844e646c7c8 (accessed on 16 December 2018).

43. Sluiter, A.; Hames, B.; Ruiz, R.; Scarlata, C.; Sluiter, J.; Templeton, D.; Crocker, D. Determination of structural carbohydrates and lignin in biomass. *Lab. Anal. Proced.* **2008**, *1617*, 1–16.

44. Wickham, R.; Galway, B.; Bustamante, H.; Nghiem, L.D. Biomethane potential evaluation of co-digestion of sewage sludge and organic wastes. *Int. Biodeterior. Biodegrad.* **2016**, *13*, 3–8. [CrossRef]

45. Morosini, C.; Conti, F.; Torretta, V.; Rada, E.C.; Passamani, G.; Schiavon, M.; Cioca, L.I.; Ragazzi, M. Biochemical methane potential assays to test the biogas production from the anaerobic digestion of sewage sludge and other organic matrices. *WIT Trans. Ecol. Environ.* **2016**, *205*, 235–244.

46. WRC. *Equipment for Measurement of Gas Production at Low Rates of Flow*; Technical Memorandum TM104—Water Research Centre: Swindon, UK, 1975.

47. Dinuccio, E.; Balsari, P.; Gioelli, F.; Menardo, S. Evaluation of the biogas productivity potential of some Italian agro-industrial biomasses. *Bioresour. Technol.* **2010**, *101*, 3780–3783. [CrossRef]
48. Labatut, R.A.; Angenent, L.T.; Scott, N.R. Biochemical methane potential and biodegradability of complex organic substrates. *Bioresour. Technol.* **2011**, *102*, 2255–2264. [CrossRef]
49. Luna-del Risco, M.; Normak, A.; Orupõld, K. Biochemical methane potential of different organic wastes and energy crops from Estonia. *Agron. Res.* **2011**, *9*, 331–342.
50. Stone, R.J. Improved statistical procedure for the evaluation of solar radiation estimation models. *Sol. Energy* **1993**, *51*, 289–291. [CrossRef]
51. Bhattarai, S.; Oh, J.H.; Euh, S.H.; Kafle, G.K.; Kim, D.H. Simulation and model validation of sheet and tube type photovoltaic thermal solar system and conventional solar collecting system in transient states. *Sol. Energy Mater. Sol. Cells* **2012**, *103*, 184–193. [CrossRef]
52. Parawira, W.; Murto, M.; Read, J.S.; Mattiasson, B. A study of two-stage anaerobic digestion of solid potato waste using reactors under mesophilic and thermophilic conditions. *Environ. Technol.* **2007**, *28*, 1205–1216. [CrossRef] [PubMed]
53. Parawira, W.; Murto, M.; Zvauya, R.; Mattiasson, B. Anaerobic batch digestion of solid potato waste alone and in combination with sugar beet leaves. *Renew. Energy* **2004**, *29*, 1811–1823.
54. Muhondwa, J.P.; Martienssen, M.; Burkhardt, M. Feasibility of anaerobic digestion of potato peels for biogas as mitigation of greenhouse gases emission potential. *Int. J. Environ. Res.* **2015**, *9*, 481–488.
55. Cater, M.; Zorec, M.; Logar, R.M. Methods for improving anaerobic lignocellulosic substrates degradation for enhanced biogas production. *Springer Sci. Rev.* **2014**, *2*, 51–61. [CrossRef]
56. Franchi, O.; Rosenkranz, F.; Chamy, R. Key microbial populations involved in anaerobic degradation of phenol and p-cresol using different inocula. *Electron. J. Biotechnol.* **2018**, *35*, 33–38. [CrossRef]
57. Świątek, M.; Lewicki, A.; Szymanowska, D.; Kubiak, P. The effect of introduction of chicken manure on the biodiversity and performance of an anaerobic digester. *Electron. J. Biotechnol.* **2019**, *37*, 25–33. [CrossRef]
58. Liotta, F.; Esposito, G.; Fabbricino, M.; van Hullebusch, E.D.; Lens, P.N.L.; Pirozzi, F.; Pontoni, L. Methane and VFA production in anaerobic digestion of rice straw under dry, semi-dry and wet conditions during start-up phase. *Environ. Technol.* **2016**, *37*, 505–512. [CrossRef]
59. Franke-Whittle, I.H.; Walter, A.; Ebner, C.; Insam, H. Investigation into the effect of high concentrations of volatile fatty acids in anaerobic digestion on methanogenic communities. *Waste Manag.* **2014**, *34*, 2080–2089. [CrossRef] [PubMed]
60. Wang, Y.; Zhang, Y.; Wang, J.; Meng, L. Effects of volatile fatty acid concentrations on methane yield and methanogenic bacteria. *Biomass Bioenergy* **2009**, *33*, 848–853. [CrossRef]
61. Pellera, F.M.; Gidarakos, E. Effect of substrate to inoculum ratio and inoculum type on the biochemical methane potential of solid agro-industrial waste. *J. Environ. Chem. Eng.* **2016**, *4*, 3217–4229. [CrossRef]
62. Cestonaro, T.; de Mendonça Costa, M.S.S.; de Mendonça Costa, L.A.; Rozatti, M.A.T.; Pereira, D.C.; Lorin, H.E.F.; Carneiro, L.J. The anaerobic co-digestion of sheep bedding and >50% cattle manure increases biogas production and improves biofertilizer quality. *Waste Manag.* **2015**, *46*, 612–618. [CrossRef] [PubMed]
63. Alvarez, R.; Liden, G. Low temperature anaerobic digestion of mixtures of llama, cow and sheep manure for improved methane production. *Biomass Bioenergy* **2009**, *33*, 527–533. [CrossRef]
64. Wang, Q.; Kuninobu, M.; Ogawa, H.I.; Kato, Y. Degradation of volatile fatty acids in highly efficient anaerobic digestion. *Biomass Bioenergy* **1999**, *16*, 407–416. [CrossRef]
65. Achinas, S.; Euverink, G.J.W. Consolidated briefing of biochemical ethanol production form lignocellulosic biomass. *Electron. J. Biotechnol.* **2016**, *23*, 44–53. [CrossRef]
66. Johnson, D.K.; Elander, R.E. Treatment for enhanced digestability of feedstocks. In *Biomass Recalcitrance: Deconstructing the Plant Cell Wall for Bioenergy*; Himmel, M.E., Ed.; Blackwell Pub.: Oxford, UK, 2008; pp. 436–453.
67. Mancini, G.; Papirio, S.; Lens, P.N.L.; Esposito, G. Increased biogas production from wheat straw by chemical pretreatments. *Renew. Energy* **2018**, *119*, 608–614. [CrossRef]
68. Galbe, M.; Zacchi, G. Pretreatment: The key to efficient utilization of lignocellulosic materials. *Biomass Bioenergy* **2012**, *46*, 70–78. [CrossRef]

 *energies*

Article

Investigation of Temperature Effect on Start-Up Operation from Anaerobic Digestion of Acidified Palm Oil Mill Effluent

Muhammad Arif Fikri Hamzah [1], Jamaliah Md Jahim [1,2,*], Peer Mohamed Abdul [1,2] and Ahmad Jaril Asis [3]

[1] Research Centre for Sustainable Process Technology (CESPRO), Faculty of Engineering and Built Environment, Universiti Kebangsaan Malaysia (UKM), Bangi 436000 UKM, Selangor, Malaysia
[2] Chemical Engineering Program, Faculty of Engineering and Built Environment, Universiti Kebangsaan Malaysia (UKM), Bangi 436000 UKM, Selangor, Malaysia
[3] Sime Darby Research Sdn Bhd, Lot 42700, Pulau Carey, Banting 42960, Selangor, Malaysia
* Correspondence: jamal@ukm.edu.my
† This paper is an extended and revised article presented at the International Conference on Sustainable energy and Green Technology 2018 (SEGT 2018) on 11–14 December 2018 in Kuala Lumpur, Malaysia.

Received: 23 April 2019; Accepted: 31 May 2019; Published: 27 June 2019

Abstract: Malaysia is one of the largest palm oil producers worldwide and its most abundant waste, palm oil mill effluent (POME), can be used as a feedstock to produce methane. Anaerobic digestion is ideal for treating POME in methane production due to its tolerance to high-strength chemical oxygen demand (COD). In this work, we compared the culture conditions during the start-up of anaerobic digestion of acidified POME between thermophilic (55 °C) and mesophilic (37 °C) temperatures. The pH of the digester was maintained throughout the experiment at 7.30 ± 0.2 in a working volume of 1000 mL. This study revealed that the thermophilic temperature stabilized faster on the 44th day compared to the 52nd day for the mesophilic temperature. Furthermore, the thermophilic temperature indicated higher biogas production at 0.60 L-CH_4/L·d compared to the mesophilic temperature at 0.26 L-CH_4/L·d. Results from this study were consistent with the COD removal of thermophilic temperature which was also higher than the mesophilic temperature.

Keywords: thermophilic; mesophilic; palm oil mill effluent; acclimatization

---

1. Introduction

Malaysia is the main producer of palm oil in the world, with ~19.9 million tons of crude palm oil generated in 2017 [1]. However, palm oil mills create waste such as palm oil mill effluent (POME), which must be treated because it can pollute the environment. POME appears as brownish liquid, has high chemical oxygen demand (COD), and is usually discharged directly at high temperatures [2]. To solve this problem, anaerobic digestion is a method that can be used to treat organic wastes for the production of methane [3]. In the future, methane can potentially replace fossil fuel as a source of energy. Methane is an inexpensive fuel source that can be produced continuously, while fossil fuels will be depleted someday. Palm oil mills in Malaysia have generally used close tank in POME treatment to maximize methane collection. Until 2013, about 50% of palm oil mills still applied the pond system to treat POME without capturing methane gas [4] because a considerable amount of land is available in Malaysia [5]. The challenge in the construction of biogas plants is the high investment needed compared to the open pond system, because investors consider this project a high risk due to the utilization of POME as a new method of biogas production.

In theory, many parameters can affect anaerobic digestion. For example, operating temperature has a remarkable impact on methanogen reactions. High temperatures allow mixed microflora, including microorganisms responsible for the degradation of organic matter, to flourish and produce methane at a high rate. Generally, methanogens are active under two temperature categories: mesophilic (35–37 °C) and thermophilic (55–60 °C) [6]. Thermophilic digesters have higher biogas production and a faster biochemical reaction rate, whereas mesophilic digesters require lower input energy for operation [6,7]. Jeong et al. [8] compared the performance between mesophilic and thermophilic anaerobic reactors and reported that thermophilic reactors could produce 32.5% more biogas than mesophilic reactors at an organic loading rate (OLR) of 15 kg COD/m^3/d.

The start-up is the key to achieving successful anaerobic digestion [9]. The study of acclimatization includes monitoring the operational conditions and bacterial community involved in the production of methane [10]. The mixed culture must comprise a stable community between methanogenic and non-methanogenic bacteria to attain a stable digestion process. Methanogens need a long time to activate. Therefore, anaerobic digestion cannot begin with a high OLR during the early period of acclimatization [9] because methanogen growth will be disrupted, thereby affecting its performance. Typically, the period of acclimatization can be shortened by utilizing stepped-loading OLR to ensure that methanogens can grow efficiently with sufficient food. Anaerobic digestion of POME requires approximately 3 months for completion, to achieve high biogas production and efficient COD removal [9].

Running a larger system is more complicated than operating lab or pilot scales, due to the bigger number of aspects that require control. Overall, a common problem that happens in a large digester is when a sample is not mixed well and the reaction culminates in the lack of organic matters that can eventually inhibit methanogens [11]. Mixing properly in the digester is important to avoid the feed from being concentrated only in certain areas of the reactor. Low volatile suspended solid (VSS) reduction and COD removal occur due to poor mixing on the level of a large-scale operation, leading to low biogas production. For instance, Massé et al. [12] reported that an anaerobic bench-scale reactor (ABSR) with a 12 m^3 volume yielded higher COD removal (76.7% ± 4.3%) than 8 m^3 of its working volume counterpart (76.9 ± 4.2%). In addition, before constructing a large-scale biogas plant, its operating conditions must be verified first on a laboratory scale because optimization from small scale can reduce operating cost and allow multiple experiments to operate simultaneously. Factors that need to be considered in upscaling digesters include hydraulic retention time (HRT), OLR, mixing condition, and temperature [13].

To obtain successful anaerobic digestion, the effect of temperature must be fully understood. However, limited publications are available on the comparison of acclimatization between mesophilic and thermophilic conditions. By contrast, the focus has been only on the specific temperature condition during anaerobic digestion. For instance, Alrawi et al. [14] only investigated the start-up of methane production under mesophilic condition. Understanding the temperature effect on anaerobic digestion can improve its performance and productivity. Hence, this study aims to compare the performance between mesophilic and thermophilic conditions during the start-up period. The stability of digesters was monitored by analyzing parameters such as pH, total alkalinity, and volatile fatty acid (VFA). Any disturbance, especially the accumulation of VFA, during this period must be solved rapidly to prevent system failure.

2. Material and Methods

2.1. Seed Sludge and Substrate

The seed sludge was obtained from the anaerobic digestion pond located at Sime Darby palm oil mill in Tennamaram, Selangor, Malaysia. The seed sludge had pH and VSS of 7.25 and 10.5 g/L, respectively. This sludge was acclimatized by using acidified POME as the sole substrate. The acidified POME was collected from the effluent based on the previous study by Maarof et al. [2], wherein a biohydrogen fermenter was fed with raw POME. Table 1 summarizes the overall characteristics of the acidified POME. The acidified POME was characterized by its yellowish brown color and sour smell. Its VFA concentration was higher than that of raw POME, thereby allowing more degradation substrates to generate higher biogas production. It was stored in the chiller at 4 °C to reduce bacterial activity. Prior to feeding, it was heated in a water bath to maintain the temperature of the culture in the digester [15]. The volume of the acidified POME that fed into the digester was measured using a measuring cylinder, and was stirred homogeneously before feeding into the digester.

Table 1. Characteristic of substrate.

ID.	Parameter	Concentration Range	
		Acidified POME	**Raw POME [2]**
1	pH	5.14 ± 0.1	5.90 ± 0.2
2	Chemical oxygen demand (g/L)	44.3 ± 3.7	50.1 ± 1.4
3	Total solid (g/L)	31.2 ± 2.3	61.5 ± 1.5
4	Total suspended solid (g/L)	24.1 ± 4.4	42.3 ± 1.1
5	Volatile suspended solid (g/L)	20.4 ± 2.9	39.5 ± 1.3
6	Volatile fatty acid (mg/L as CH_3COOH)	6383.5 ± 1348.3	4600.0 ± 1300.0
7	Total nitrogen (mg/L)	254 ± 27	N/D

2.2. Experimental Set-Up and Digester Operation

In this experiment, anaerobic digestion was performed in Schott Duran bottles under sequencing batch mode with 1000 mL of working volume in mesophilic and thermophilic conditions. Figure 1 shows the experimental setup, which comprises Schott Duran bottles, a gas collector, and a water bath shaker (model SW22, 230 V/50e60 Hz). A rubber tubing was utilized for the purposes of liquid sampling and providing a pathway for the biogas to flow into the gas collector. Tubing inspection was performed daily to prevent any outside air from entering the bottle. The existing biogas in the gas collector was stored inside the acidified water (pH = 2.50 ± 0.5), as suggested by Ergüder et al. [16], to avoid dissolution into water. Prior to digestion, nitrogen gas was purged to allow the process to operate under anaerobic conditions. The bottles were then incubated in a separated water bath shaker with temperatures of 55 °C and 37 °C, and agitated at 80 rpm. Anaerobic digestion began with a HRT of 30 days and OLR ranged between 1.1–1.3 g COD/L·d by diluting the feed as done by Badiei et al. [17]. Feeding took place every two days. During each feeding, 7% of the medium was removed and replaced with the same volume of the new substrate. Liquid and gas samples were obtained prior to feeding and subjected to analysis. Mesophilic and thermophilic operations stopped upon reaching the steady state wherein biogas volume and volatile suspended solids (VSS) had less than 10% in variation.

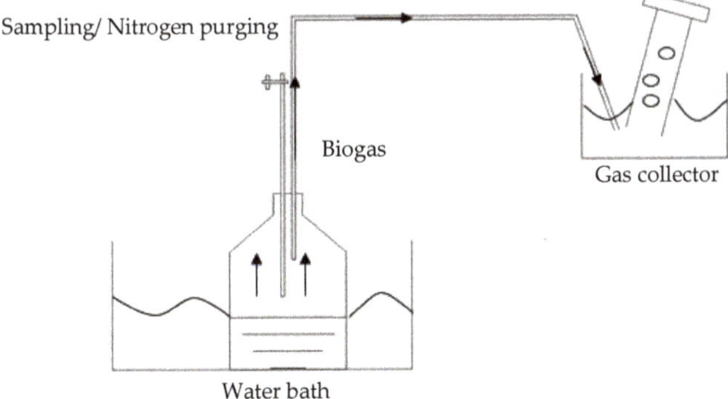

Figure 1. Schematic diagram of the anaerobic digestion process.

2.3. Analytical Method

The liquid samples were subjected to pH, total alkalinity, and VSS analysis based on American Public Health Association (APHA) [18]. Total nitrogen in the acidified POME was determined using the method suggested by HACH. VFAs were analyzed via high-performance liquid chromatography (Agilent 1200, California, USA). The HPLC system was run with a REZEX ROA column (Phenomenex, USA) and operated at a fixed flow rate of 0.6 mL/min. The mobile phase used in the system was 5 mmol/L sulfuric acid. Furthermore, chemical oxygen demand was analyzed using the dichromate method operating with a COD analyzer (DR 2800, HACH). The biogas flow rate was measured at a specific time interval using Equation 1 [19]. The samples were collected and analyzed via gas chromatography (GC, model SRI 8600C, USA) by a helium ionization detector equipped with a thermal conductivity detector. The sample was injected into the GC at an oven temperature and pressure of 40 °C and 2.7 psi for 5 min. Helium gas, with a purity of 99.99%, was used as the carrier gas and set at a flow rate of 25 mL/min. The temperature was ramped at 30 °C per minute and maintained for 10 min once the temperature reached 220 °C. Liquid and gas samples were analyzed in triplicate.

$$V_{H,i} = V_{H,i-1} + C_{H,i}\left(V_{g,i} - V_{g,i-1}\right) + V_H(C_{H,i} - C_{H,i-1}) \tag{1}$$

where $V_{H,i}$ = the cumulative methane gas volume (mL); $V_{g,i}$ = the cumulative biogas volume (mL); and $C_{H,i}$ = methane gas composition fraction and i denote at the current time, while $i-1$ denote previous time interval.

3. Results and Discussion

3.1. Start-Up Experiments

Start-up duration is vital in anaerobic digestion for methanogens to adapt well in a new environment so that the duration of a start-up can be shortened and anaerobic digestion can be operated smoothly. Abd Nasir et al. [20] recommend that the experiment should start with a low OLR to avoid organic matter overload and enable rapid growth of methanogens in the new environment. In addition, the biomass could washout when functioning at a high OLR and disturb the start-up operation. Productive acclimatization was demonstrated by Abd Nasir et al. [20], wherein acclimatization was operated at low OLR (0.5 g COD/L·d) and successfully attained 3.59 L-CH4/d of biogas production. Moreover, Seadi et al. [21] reported that HRT should be long enough to ensure a remarkably higher rate of methanogen reproduction than the death rate so that methanogens could fully utilize the substrate. Methanogens differ from other microorganisms involved in anaerobic digestion because of their slower regeneration time, which takes approximately 5–16 days to reproduce [22]. Few hydrogenotrophic species such as *Methanococcus maripaludis* only take 2 hours to regenerate [23]. Thus, low OLR (1.1–1.3 g COD/L·d) was applied in this study to allow methanogens to grow effectively. As proof, both mesophilic and thermophilic digestions were positively tested and obtained satisfactory biogas production and COD removal at the end of the start-up period. Subsequently, the digesters also revealed that no VFA accumulation occurred during this period.

3.2. Biogas Productivity

The production profile of biogas expressed in L-CH$_4$/L·d for both mesophilic and thermophilic conditions as illustrated in Figure 2 indicates that the biogas production increased from the initial anaerobic digestion until the biogas production stabilized at the end of the acclimatization period. Mesophilic and thermophilic trials had a significant increase after the 16th and 12th days, respectively, and corresponded to VSS, which also increased during the same period. Biogas production under thermophilic conditions increased from 0.05 L-CH$_4$/L·d to 0.28 L-CH$_4$/L·d. By contrast, the mesophilic digester increased from 0.08 L-CH$_4$/L·d to 0.13 L-CH$_4$/L·d. Wong et al. [24] reported that increased biogas production could be clarified by the methanogens that started to act in this period and began to consume VFAs to produce methane. This trend was consistent with the study by Yacob et al. [25] who proposed that increased biogas production during the start-up operation could be attributed to high microbial activity and stability in the system, thereby accelerating the breakdown rate. To achieve steady biogas production during the start-up period, anaerobic digestion should run under optimal conditions during hydrolysis, acidogenesis, acetogenesis, and methanogenesis phases [26]. Hence, monitoring operating conditions such as temperature, pH, and HRT is crucial to maintaining the system's stability.

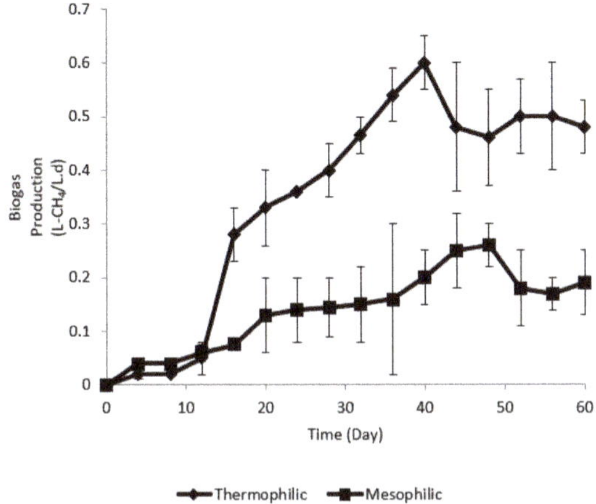

Figure 2. The range of biogas volume for thermophilic and mesophilic conditions during the acclimatization period. Notes: error bars show standard deviation of means (n = 3).

Biogas production under mesophilic and thermophilic conditions then began to decline by 20% and 30% on the 48th and 40th days, respectively. Both mesophilic and thermophilic conditions reached their peak (0.60 L-CH$_4$/L·d and 0.26 L-CH$_4$/L·d for thermophilic and mesophilic conditions, respectively) before biogas production started to decline. Alrawi et al. [14] demonstrated that decreased biogas production during acclimatization could be due to the low availability of organic matters to produce additional methane. Based on the experimental values, the result was corroborated by the low COD effluent in the final period of acclimatization in this study for both thermophilic and mesophilic conditions. Methanogens must be supplied with sufficient substrate to achieve stable biogas production because a low concentration of substrate will limit the performance and activity of methanogens.

At the end of the acclimatization period, thermophilic and mesophilic conditions registered stable biogas production with less than 10% variation as observed from the 44th and 52nd days onwards, respectively. The thermophilic digester recorded biogas production of 0.50 ± 0.04 L-CH$_4$/L·d while the mesophilic one was 0.20 ± 0.08 L-CH$_4$/L·d during the steady state. The results were consistent with the observation by Alrawi et al. [14], who stated that biogas production increased until substrate insufficiency became the limiting factor for methanogens to thrive. However, Alrawi et al. [14] reported only the mesophilic condition (37 °C) that took nearly 45 days to acclimatize completely. By contrast, the results from the present study showed that the mesophilic condition took a much longer time (52 days) to acclimatize compared to Alrawi et al.'s [14] study, which was 45 days due to the difference in operational conditions (OLR, HRT, and working volume) that may affect methanogens activity. OLR represents the organic matter inside the digester, while HRT influences the contact time between microorganism and feedstock. Feeding that occurs above sustainable OLR and HRT can upset the anaerobic digestion due to the production of inhibiting substances inside the digester that can disturb methanogen activity.

During the experiment, we observed that biogas production was 2.5 times higher under thermophilic condition than under mesophilic condition, wherein a larger difference of biogas production can be seen at the end of the start-up period, particularly on the 40th day. An underlying reason for this observation could the significant degradation rate of organic matters at high temperatures [6]. This can be enlightened by the high COD removal that allowed the degradation of more substrates into methane. High temperatures cause rapid reaction rates, leading to the removal of large amounts of organic matters.

3.3. Substrate Removal

Yu et al. [27] indicated that COD removal could signify the amount of substrate converted to methane. High COD removal indicates the successful conversion of substrates into methane. Figure 3 outlines the variation of COD removal during the acclimatization period. In the early stage, COD removal under thermophilic and mesophilic conditions decreased: the former declined from 40.7% to 12.4% and the latter from 43.6% to 17.3%. This is due to the fact that methanogens did not appear to be active during this period as COD removal for both mesophilic and thermophilic conditions continued to decrease until the 12th and 16th days, respectively. Isik et al. [28] illustrated that the production rate of VFA was remarkably higher than the utilization of VFA to methane at the early stage of anaerobic digestion of methane, resulting in VFA accumulation. The excess VFA concentration retarded the anaerobic digestion and eventually increased the COD effluent. Methanogens were less active at an early stage of start-up and required some time before VFA consumption started. This statement was also supported by the elevation in VFA concentrations under mesophilic and thermophilic conditions in the same interval.

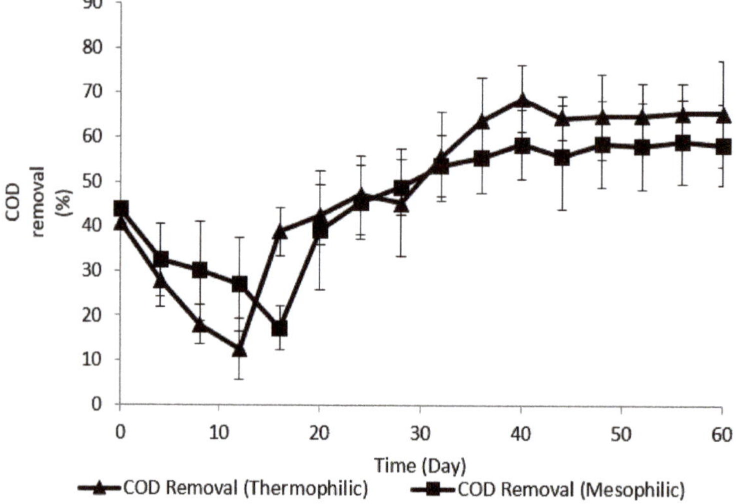

Figure 3. The range of COD removal for thermophilic and mesophilic effluents during the acclimatization period. Notes: error bars show standard deviation of means (n = 3).

Subsequently, COD removal further increased from 12.4% to 38.8% and 17.3% to 39.2% on the 12th and 16th days onward under thermophilic and mesophilic conditions, respectively. It is believed that methanogens adapted to the new environment and started to consume the substrate that was proven by the reduction in VFA effluent. Once methanogens already adjusted to the new environment, they could effectively consume the feedstock, thereby increasing COD removal continuously. The period wherein microorganisms acclimatize to the new environment is known as lag phase [29]. A long lag phase is usually related to the substrate with high fiber content in feedstock [30]. In addition, pH values under thermophilic and mesophilic conditions increased on the 12th and 16th days, respectively, thereby reflecting the ongoing production of methane by the methanogens. Corroborative findings were reported by Wong et al. [24] and Teng et al. [31] based on their studies on the start-up operation. For Wong et al. [24], an increased of COD removal started on the 10th day, whereas for Teng et al. [31], it began to rise on the 21st day. COD removal in the study by Wong et al. [24], took a shorter time to increase, indicating that methanogens adapted earlier to the new environment.

At the end of anaerobic digestion, COD removal reached the steady state. COD removal of 64.5% ± 0.2% was registered under the thermophilic condition, whereas COD removal of 58.5% ± 0.1% was recorded under the mesophilic condition. The higher COD removal under the thermophilic condition suggested better methanogen activity compared to the mesophilic condition, thereby enabling the conversion of more substrates into methane. For biochemical reactions, increment in temperature tends to escalate the rate of degradation. According to the van't Hoff equation, increasing the temperature by 10 °C doubles the reaction rate [32]. Similarly, this outcome is in agreement with Jeong et al. [8] who found that COD removal in the thermophilic digester was slightly higher than that in the mesophilic digester. Notably, the good result in COD removal also led to high biogas production in their study.

3.4. Variation of Volatile Suspended Solid and Total Suspended Solid

The results of VSS and total suspended solid (TSS) were plotted over time as shown in Figure 4. At the beginning of the experiment, VSS and TSS under thermophilic conditions decreased gradually with time. Mesophilic condition trends were also the same under the thermophilic condition. The VSS under mesophilic and thermophilic conditions decreased from 14.9 g/L to 12.0 g/L and from 16.7 g/L to 11.7 g/L, respectively, consistent with the results by Wong et al. [3] and Alrawi et al. [12], who stated that VSS declined because methanogens still attempted to adjust to the new environment before starting to grow. During this period, the methanogen colony was lower than the acidogenic bacteria that were responsible for the production of hydrogen, carbon dioxide, and organic acid [33]. Thus, a high concentration of acidogenic bacteria inhibited methanogen growth. The study by Wong et al. [24] experienced that VSS decreased from 34,190 mg/L to 8940 mg/L during the first 7 days of operation.

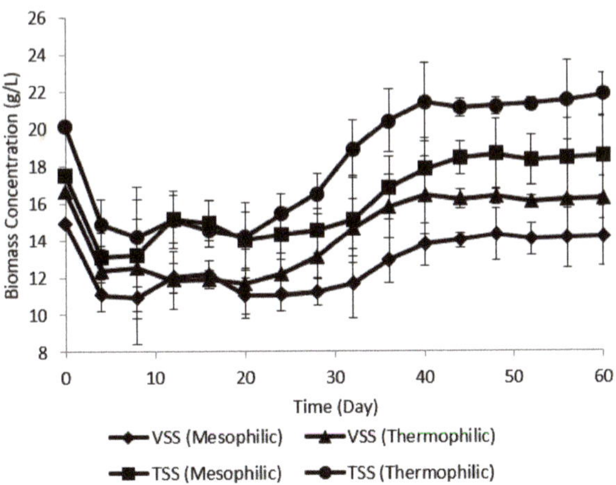

Figure 4. The range of VSS and TSS for thermophilic and mesophilic conditions during the acclimatization period. Notes: error bars show standard deviation of means (n = 3).

As can be observed in Figure 4, once anaerobic digestion reached the 40th day, VSS demonstrated a constant reading with less than 5% variation under thermophilic and mesophilic conditions, thereby indicating the final stage of the acclimatization period. Mesophilic and thermophilic digesters obtained

VSS of 14.0 and 16.2 g/L before reaching the steady state, respectively. Teng et al. [31] found a similar trend of VSS during acclimatization, in which VSS decreased at the beginning of acclimatization before starting to increase. These results showed that VSS decreased by 21.5% after the 21st day of operation. The high concentration of VSS at the end of the anaerobic digestion was found by Teng et al. [31], indicating that good anaerobic digestion occurred. This result was also supported by high COD removal in a similar duration. Mesophilic and thermophilic conditions showed a VSS:TSS ratio up to 0.78 and 0.79, respectively. The VSS:TSS ratio is essential for determining the rate of degradation of suspended solids during anaerobic digestion. A high VSS:TSS ratio indicates a high degradation rate, while a low ratio represents minimal consumption of organic matters in the digester. Sperling et al. [34] propose that the VSS:TSS ratio should be typically between 0.70–0.85 during the anaerobic process.

Overall, higher VSS concentration was observed under the thermophilic condition with 11.1% difference compared to that under the mesophilic condition, indicating better methanogen performance. Microorganisms' growth is dependent on aspects such as pH, temperature, and osmotic pressure. As anticipated, enzyme reaction was slow at low temperatures, thereby halting their activity. As the temperature increased, the chemical reaction also occurred at a higher rate and caused rapid cell growth. Similar to high biogas production in the thermophilic digester, high VSS concentration was compatible with the utilization of VFAs to produce methane. The same trend can be seen by Trisakti et al. [35] who claimed that mesophilic digestion obtained lower VSS in the system with a 10.3% difference compared to that in thermophilic digestion. The results in this study were consistent with those from Song et al. [36], such that low VSS associated with the mesophilic condition corresponded to high residual VFAs due to the low degradation rate.

3.5. Volatile Fatty Acids Production

Paritosh et al. [37] suggested that 65% of organic matter energy is in the form of VFA that can be converted into methane. Figure 5 shows that acetic acid, butyric acid, and propionic acid are three keys in VFAs that are responsible for methane production. The results were dominantly acetic acid (3493.9 ± 926 mg/L), followed by butyric acid (640.9 ± 277 mg/L), and propionic acid (446.6 ± 260 mg/L) in the mesophilic digester. The thermophilic digester also showed the same trend. The produced VFA was dependent on the pH change. Acetic acid and butyric acid were produced at low pH. On the other hand, high pH led to the production of propionic acid and acetic acid [38]. At an early start-up period, thermophilic and mesophilic conditions showed high VFA concentrations due to inactivity of the methanogen. Anaerobic digestion can be divided into four main phases (hydrolysis, acidogenesis, acetogenesis, and methanogenesis). During acidogenesis, fermentative bacteria are responsible for breaking down amino acids, fatty acids, and sugar into VFA. Therefore, high VFA concentrations were observed at this interval. Yacob et al. [25] also achieved a high VFA concentration at this period, and stated that VFAs accumulated from the initial start-up until they reached 970.0 mg/L as CH_3COOH. The high concentration of VFAs was unsuitable for anaerobic digestion because it may lead to low pH that could hinder anaerobic digestion. Furthermore, washing out VSS in the system was another likely cause of VFA accumulation that can lead to acidification [39]. Washout inside the digester could be prevented by applying long enough HRT to retain methanogen inside the digester. Hence, HRT does not affect the microorganism's growth limit. Given the increased of biogas production and COD removal with time, it was possible that the anaerobic digestion system could tolerate the concomitantly-produced VFAs. Further observation revealed that VFAs under thermophilic and mesophilic conditions subsequently started to decline as can be seen on the 12th and 16th days onward, thereby indicating the increase in methanogens and their consumption of VFAs. COD removal also began to increase simultaneously.

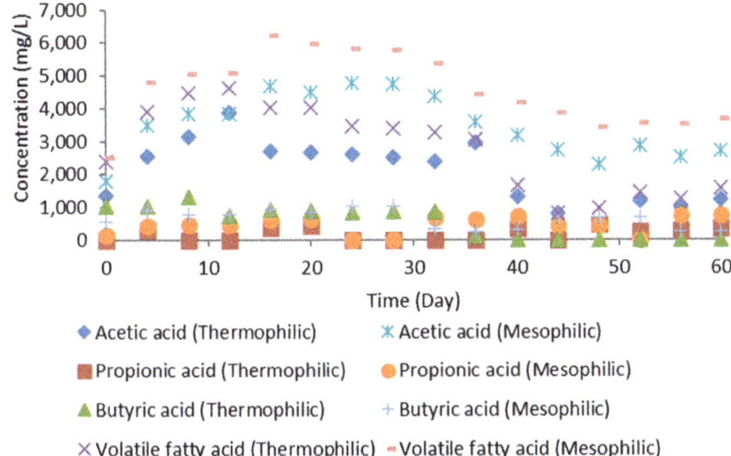

Figure 5. The range of volatile fatty acid of thermophilic and mesophilic effluents during the acclimatization period.

Moreover, it is noteworthy that the accumulation of the propionic acid could upset the biogas production as it culminated at a lower conversion rate of methane than acetic acid and butyric acid. It is known that low concentrations of propionic acid correlate with stable anaerobic digestion, whereas high concentrations of propionic acid are associated with digestion failure [40]. Work done by Demirul and Yanigun [41] recommends that propionic acid above 951.0 mg/L is not viable for anaerobic digestion. Propionic acid with a concentration of 2.0 g/L inhibited cellulose degradation, whereas a concentration of 4.0 g/L critically affected the degradation of glucose [42]. In this study, the concentration of propionic acid did not affect the system—446.6 ± 260 and 174.5 ± 180 mg/L under mesophilic and thermophilic conditions, respectively. The concentration of propionic acid must be controlled so that it does not inhibit the anaerobic digestion process.

The results display that the mesophilic condition yielded nearly twice more VFAs than its thermophilic counterpart. This result was supported by Kumar et al. [43] who discovered that VFAs in the mesophilic digester were 30% higher than those in the thermophilic digester. Consequently, a significant concentration of VFAs appeared to reduce biogas production and its stability due to the reduction in pH that can cause toxicity to anaerobic digestion. It is believed that the high concentration of VFAs under the mesophilic condition was due to the significantly lower conversion rate of VFA to methane than the VFA production. In this process, the low temperature resulted in the low reaction rate of methanogens to consume VFA.

3.6. Stability of The Digesters

The variation of pH over time for both thermophilic and mesophilic digestions is shown in Figure 6. At the early stage, thermophilic pH dropped from 7.33 to 6.70. Similarily, under the mesophilic condition, it declined from 7.40 to 6.60. Wong et al. [24] showed that in an early stage of anaerobic digestion (1st to 10th day), the pH decreased from 6.95 to 5.11 due to the prevalence of VFAs. Considerable amounts of VFA could upset the anaerobic digestion because they inhibit methanogenesis. A problem occurs when methanogens do not have sufficient time to remove organic acids, resulting in acid accumulation. Nevertheless, the anaerobic digestion system for both studies remained stable because of the high total alkalinity that acted as a buffer to neutralize the resultant VFAs. The reduction of pH at this interval did not critically affect the anaerobic digestion because biogas production continued to increase with time.

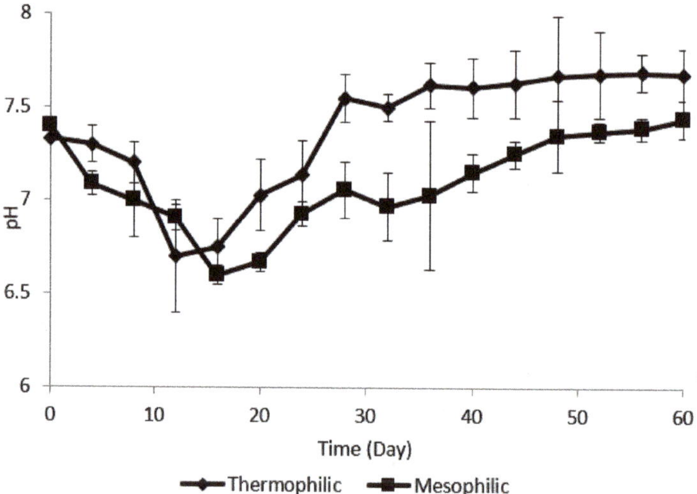

Figure 6. pH range of thermophilic and mesophilic effluents during the acclimatization period. Notes: error bars show standard deviation of means (n = 3).

Later, pH of the mesophilic condition escalated on the 16th day while that of the thermophilic condition increased on the 12th day. Thermophilic pH increased from 6.70 to 6.75, whereas the mesophilic pH increased from 6.60 to 6.67. This indicates that the thermophilic digester took a shorter time for the methanogens to utilize the VFAs. This outcome could be clarified by the observation of lower VFAs and higher COD consumption within this duration. The consumption of VFAs by methanogens in the production of biogas likely contributed to the increased pH in the system. At the final stage of acclimatization, thermophilic and mesophilic pH only slightly fluctuated on the 36th and 48th days, respectively. Thus, anaerobic digestion reached a stable phase. Teng et al. [31] discovered a pH of 8.47 at the steady state, which was slightly similar to that of this study, thereby reflecting high COD removal efficiency. It is believed that the high COD removal represented the high removal of VFAs that could lead to a high resultant pH.

As shown in Figure 6, the range of pH for both digesters was close to the neutral condition's pH. However, the recorded pH in the mesophilic digester was lower than that of the thermophilic digester with a 3.8% difference. The same observation was also illustrated by Labatut et al. [44] who stated that pH in the thermophilic condition was higher than that in mesophilic condition, corresponding to higher total alkalinity. Likewise, Kardos et al. [45] who studied sewage sludge revealed a higher pH (7.90 ± 0.6) in thermophilic digestion than in mesophilic digestion (7.60 ± 0.4) at the end of anaerobic digestion. Song et al. [36] proposed that the high thermophilic pH could result from the high degradation rate of nitrogenous compounds. Operating at high temperatures contributed to free ammonia formation. The production of free ammonia led to increased pH during anaerobic digestion. The authors reported a pH of 7.67 registered under the mesophilic digester, which was equivalent to 5.1% difference between mesophilic and thermophilic digesters.

It can be observed from this research that the final pH attained from mesophilic and thermophilic digestions was suitable for anaerobic digestion, in that it could yield satisfactory biogas production and COD removal. The results were consistent with findings obtained by Alrawi et al. [14] and Teng et al. [31], proving that methanogens were only active when pH was suitable for anaerobic digestion (7.35 ± 0.2) due to their sensitivity to pH changes. Non-ideal pH out of the neutral condition will impair biogas production. An excessively low pH is initiated by the accumulation of VFAs while an unreasonably high one will result in the production of additional ammonia, which is not conducive to the anaerobic process.

3.7. Total Alkalinity

Total alkalinity and VFA are two effective parameters that can be utilized for monitoring the progress of anaerobic digestion [46]. Total alkalinity acts as a pH buffer since it prevents rapid drastic changes in pH. Buffering capacity in the system is a result of bicarbonate ion that responsible for maintaining the pH between 6.50–7.60 [47]. Interestingly, the concentration of bicarbonate ion is dependent on the concentration of carbon dioxide in the gas phase. The reduction of total alkalinity can be an early sign of additional VFAs in the system, and for that reason the concentration of VFAs must be controlled to avoid anaerobic digestion failure. Based on Figure 7, the thermophilic condition exhibited a total alkalinity interval of 11,617 ± 4246 mg/L as $CaCO_3$ which was 15.5% higher than that under the mesophilic condition. Total alkalinity of thermophilic and mesophilic digesters increased from the beginning of the anaerobic digestion until the 48th day before stabilization started. Both pH and total alkalinity showed similar trends during the anaerobic digestion such that the increment of total alkalinity was followed by the increment of pH. Kugelman et al. [48] pointed out that high temperature could increase the breakdown of organic nitrogen into ammonium bicarbonate, which could eventually produce high total alkalinity. Girardi et al. [49] described that high total alkalinity was a result of successful VFA conversion into methane by methanogens. Total alkalinity and pH exhibited a strong correlation with VFAs. Since successful consumption of substrate will result in a low concentration of VFAs in the effluent, the pH and total alkalinity in the system will be high when the concentration of VFAs is low. In direct comparison, the total alkalinity in this study was lower than that in the study by Teng et al. [31], which was at 11,610 mg/L as $CaCO_3$. Nevertheless, total alkalinity in this study still led to superior biogas production and achieved high COD removal as the VFAs were successfully balanced by total alkalinity in the system without resulting in VFA accumulation.

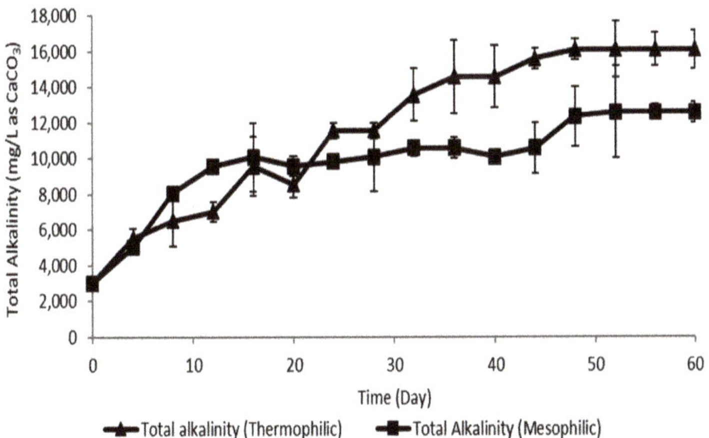

Figure 7. The range of total alkalinity for thermophilic and mesophilic effluents during the acclimatization period. Notes: error bars show standard deviation of means (n = 3).

3.8. Overall Performance

Table 2 outlines the comparison of previous studies in the start-up period of anaerobic digestion of methane with POME as the substrate except that this study uses acidified POME. The use of acidified POME as feedstock for methane production has been explored by numerous researchers [20,50,51]. According to Krishnan et al. [52], acidified POME contains abundant acetogenic and methanogenic bacteria, which were beneficial for the acceleration of anaerobic digestion. Acidified POME was an outcome of hydrogen fermentation containing large energy that remained as VFAs. Thus, further digestion of this effluent allowed the maximum recovery of energy from VFAs [51]. Moreover, it can be clearly seen from the study conducted by Alrawi et al. [14], who deployed the highest OLR (3.4 g

COD/L·d) and also the shortest HRT (20 days). The selection of OLR is vital for determining the stability of anaerobic digestion. Operating at excessively high OLR can cause overloading of organic matter inside the digester and eventually lead to anaerobic digestion failure. Meanwhile, the OLR used by Alrawi et al. [14] was nearly three times higher than that in current study (1.2 g COD/L·d). Compared to this study, the low OLR resulted in low methane production due to the insufficient substrate for the methanogens to produce additional methane. In this case, the lack of substrate became a limiting factor affecting methanogen growth. However, the production rate can still be enhanced by increasing OLR gradually so that additional substrate can be used to improve methane production. Increasing OLR slowly into the digester is crucial since methanogens require time to consume the substrate. Furthermore, it is can be noticed that despite operating at different HRT and OLR levels, the pH values of all studies were in a suitable range for methanogen growth (7.35 ± 0.2). The near-neutral pH value indicated that no accumulation of VFA occurred during digestion. The build-up of VFAs can upset anaerobic digestion because it leads to the reduction of pH and lowers biogas production. The pH obtained was the highest at 7.30 achieved by Alrawi et al. [14] and Teng et al. [31] indicating a better conversion of organic matter into methane compared to this study. As mentioned previously, high total alkalinity significantly influences pH. The high pH achieved by Teng et al. [31] corresponded to the highest total alkalinity among the studies. It was found that the results of the present study obtained higher VFA concentration in the mesophilic condition than in thermophilic condition but negatively affected anaerobic digestion as proven by the satisfactory methane production rate. Nevertheless, the high concentration of VFA resulted in low total alkalinity and pH in the mesophilic digester.

Table 2. Comparison of previous studies on acclimatization anaerobic digestion of methane from palm oil mill effluent.

ID	Parameter	[31]	[14]	Present Study	
1	Substrate	Palm oil mill effluent	Palm oil mill effluent	Acidified palm oil mill effluent	
2	Hydraulic retention time (days) & Organic loading rate (g COD/L·d)	Hydraulic retention time = 40 Organic loading rate = 1.4	Hydraulic retention time = 20 Organic loading rate = 3.4	Hydraulic retention time = 30 Organic loading rate = 1.22	
3	pH	Mesophilic (7.30)	Mesophilic (7.30)	Mesophilic (7.00)	Thermophilic (7.20)
4	Total alkalinity (mg/L as $CaCO_3$)	Mesophilic (11,610)	N/D	Mesophilic (7792)	Thermophilic (9540)
5	Volatile fatty acid (mg/L as CH_3COOH)	N/D	N/D	Mesophilic (4360.3)	Thermophilic (2714.2)
6	Chemical oxygen demand effluent (g/L)	Mesophilic (14.4)	Mesophilic (47.5)	Mesophilic (15.2)	Thermophilic (14.3)
7	Production rate (L-CH_4/L·d)	N/D	N/D	Mesophilic (0.14)	Thermophilic (0.34)

3.9. Statistical Analysis

Table 3 depicts the t-test results for biogas production under mesophilic and thermophilic conditions by using IBM SPSS Statistics 23 and $p < 0.05$ was considered significant. The analysis was carried out to determine whether the difference in temperatures (mesophilic and thermophilic) had a significant effect on biogas production during the acclimatization period. Biogas productivity is a beneficial indicator for determining the performance of the anaerobic digestion process. High biogas productivity indicated that anaerobic digestion had been operated successfully under optimum conditions. Based on the results, significance (2-tailed) was at 0.001. The obtained value was below 0.05, thereby indicating that a statistically significant difference exists between mesophilic and thermophilic conditions in biogas production. On the basis of the table, the thermophilic digester (0.34 L-CH_4/L·d) successfully produced more than twice the amount of biogas than the mesophilic digester (0.14 L-CH_4/L·d). This result verified that high thermophilic COD removal resulted in low effluent VFAs.

Table 3. Summary of t-test for the measured biogas production between mesophilic and thermophilic conditions.

Independent Samples Test		Levene's Test for Equality of Variances		t-Test for Equality of Means			
		F	Sig.	t	df	sig. (2-tailed)	Mean Difference
Productivity	Equal Variances Assumed	13.920	0.001	3.745	30	**0.001**	0.206
	Equal variances not assumed			3.745	19	0.001	0.206
Group Statistics							
	Condition	N	Mean	Std Deviation	Std Error Mean		
Productivity	Thermophilic	16	0.34	0.21	0.05		
	Mesophilic	16	0.14	0.08	0.02		

4. Conclusions

In the current study, we evaluated the significance of understanding the operation monitoring and acclimatization behavior during the anaerobic digestion of methane. Overall, the thermophilic digester obtained better biogas production than the mesophilic digester, with a difference of approximately 50% and required much shorter time to reach steady state during acclimatization compared to the mesophilic digester on the 44th day for the former and the 52nd day for the latter. High temperatures allowed enlarged substrate consumption, which was validated by high COD removal under the thermophilic condition. It is often assumed that the high biomass concentration under the thermophilic condition indicates good methanogenic activity to allow significant biogas production. This study emphasizes the role of VFAs and total alkalinity in monitoring the stability of the anaerobic digestion process. Total alkalinity in the system must be balanced with VFA concentration to avoid the accumulation of VFA inside the digester, which could upset the anaerobic digestion performance. In this study, the high concentration of VFAs under mesophilic and thermophilic conditions (4581.437 and 2765.106 mg/L as CH_3COOH, respectively) were well balanced with the total alkalinity of the system. As a result, no significant decrease in pH was observed in the experiment. Precautionary steps must be considered to avoid failure of the digestion process such as maintaining pH at neutrality during feeding for effective methanogen growth. In conclusion, acclimatization must be operated under thermophilic conditions to reach the steady state in a shorter amount of time and achieve higher biogas productivity compared to mesophilic conditions. This paper demonstrates that the production of biogas inside the thermophilic digester is almost twice that of the mesophilic digester, thereby justifying the operation at a larger scale. Moreover, the operation of thermophilic digestion was more stable than that of mesophilic digestion because it resulted in lower amounts of VFA. Commonly, raw POME was discharged at high temperatures. In practical applications, operating under thermophilic conditions eliminated the necessity for a cooling pond prior to feeding in the anaerobic digester.

Author Contributions: Conceptualization, J.M.J.; Methodology, J.M.J. and P.M.A.; Software, M.A.F.H. and P.M.A.; Validation, J.M.J. and P.M.A.; Formal Analysis, M.A.F.H., J.M.J. and P.M.A.; Investigation, M.A.F.H.; Resources, J.M.J. and P.M.A.; Data Curation, M.A.F.H. and J.M.J.; Writing-Original Draft Preparation, M.A.F.H.; Writing-Review & Editing, M.A.F.H., J.M.J. and P.M.A.; Visualization, J.M.J. and A.J.A.; Supervision, J.M.J. and P.M.A.; Project Administration, J.M.J. and A.J.A.; Funding Acquisition, J.M.J.

Acknowledgments: We gratefully acknowledge the financial and technical support provided by the Universiti Kebangsaan Malaysia-Yayasan Sime Darby (UKM-YSD) Chair for Sustainable Development: Zero Waste Technology. We would also like to extend our gratitude to the Government of Malaysia and Universiti Kebangsaan Malaysia for funding this work through KK-2015-002.

Conflicts of Interest: The authors declare no conflict of interest.

References

1. Production of Crude Palm Oil for the Month of January–December 2016 & 2017. 2018. Available online: http://bepi.mpob.gov.my/index.php/en/statistics/production/177-production-2017/792-production-of-crude-oil-palm-2017.html (accessed on 30 October 2018).
2. Maaroff, R.M.; Md Jahim, J.; Azahar, A.M.; Abdul, P.M.; Masdar, M.S.; Nordin, D.; Abd Nasir, M.A. Biohydrogen production from palm oil mill effluent (POME) by two stage anaerobic sequencing batch reactor (ASBR) system for better utilization of carbon sources in POME. *Int. J. Hydrogen Energy* **2018**, *44*, 3395–3406. [CrossRef]
3. Hanum, F.; Yuan, L.C.; Kamahara, H.; Aziz, H.A.; Atsuta, Y.; Yamada, T.; Daimon, H. Treatment of sewage sludge using anaerobic digestion in Malaysia: Current state and challenges. *Front. Energy Res.* **2019**, *7*. [CrossRef]
4. Chin, M.J.; Poh, P.E.; Tey, B.T.; Chan, E.S.; Chin, K.L. Biogas from palm oil mill effluent (POME): Opportunities and challenges from Malaysia's perspective. *Renew. Sustain. Energy Rev.* **2013**, *26*, 717–726. [CrossRef]
5. Wu, T.Y.; Mohammad, A.W.; Jahim, J.M.; Anuar, N. Pollution control technologies for the treatment of palm oil mill effluent (POME) through end-of-pipe processes. *J. Environ. Manag.* **2010**, *91*, 1467–1490. [CrossRef] [PubMed]
6. Gebreeyessus, G.; Jenicek, P. Thermophilic versus mesophilic anaerobic digestion of sewage sludge: A comparative review. *Bioengineering* **2016**, *3*, 15. [CrossRef]
7. İnce, E.; İnce, M.; Önkal Engin, G. Comparison of thermophilic and mesophilic anaerobic treatment for potato processing wastewater using a contact reactor. *Glob. NEST J.* **2017**, *19*, 318–326.
8. Jeong, J.-Y.; Son, S.-M.; Pyon, J.-H.; Park, J.-Y. Performance comparison between mesophilic and thermophilic anaerobic reactors for treatment of palm oil mill effluent. *Bioresour. Technol.* **2014**, *165*, 122–128. [CrossRef]
9. Sulaiman, A.; Tabatabaei, M.; Yusoff, M.Z.M.; Ibrahim, M.F.; Hassan, M.A.; Shirai, Y. Accelerated start-up of a semi-commercial digester tank treating palm oil mill effluent with sludge seeding for methane production. *World Appl. Sci. J.* **2010**, *8*, 247–258.
10. Burak, D.; Orhan, Y. Changes in microbial ecology in an anaerobic reactor. *Bioresour. Technol.* **2006**, *97*, 1201–1208. [CrossRef]
11. Hansen, T.L.; Schmidt, J.E.; Angelidaki, I.; Marca, E.; Jansen, J.C.; Mosbæk, H.; Christensen, T.H. Measurement of methane potentials of solid organic waste. *Waste Manag.* **2004**, *24*, 393–400. [CrossRef]
12. Massé, D.I.; Croteau, F.; Masse, L.; Danesh, S. The effect of scale-up on the digestion of swine manure slurry in psychrophilic anaerobic sequencing batch reactors. *Trans. ASAE* **2004**, *47*, 1367–1373. [CrossRef]
13. Mir, M.A.; Hussain, A.; Verma, C. Design considerations and operational performance of anaerobic digester: A review. *Cogent Eng.* **2016**, *3*. [CrossRef]
14. Alrawi, R.A.; Ahmad, A.; Norli, I.; Mohd Omar, A.K. Methane production during start-up phase of mesophilic semi-continues suspended growth anaerobic digester. *Int. J. Chem. React. Eng.* **2010**, *8*. [CrossRef]
15. Sidik, U.; Razali, F.; Alwi, S.; Maigari, F. Biogas production through co-digestion of palm oil mill effluent with cow manure. *Niger. J. Basic Appl. Sci.* **2013**, *21*, 79–84. [CrossRef]
16. Ergüder, T.; Tezel, U.; Güven, E.; Demirer, G. Anaerobic biotransformation and methane generation potential of cheese whey in batch and UASB reactors. *Waste Manag.* **2001**, *21*, 643–650. [CrossRef]
17. Badiei, M.; Jahim, J.M.; Anuar, N.; Sheikh Abdullah, S.R. Effect of hydraulic retention time on biohydrogen production from palm oil mill effluent in anaerobic sequencing batch reactor. *Int. J. Hydrogen Energy* **2011**, *36*, 5912–5919. [CrossRef]
18. APHA. *Standard Methods for the Examination of Water and Wastewater*; American Public Health Association: Washington, DC, USA; New York, NY, USA, 2005.
19. Logan, B.E.; Oh, S.-E.; Kim, I.S.; Van Ginkel, S. Biological hydrogen production measured in batch anaerobic respirometers. *Environ. Sci. Technol.* **2002**, *36*, 2530–2535. [CrossRef]
20. Abd Nasir, M.A.; Jahim, J.M.; Abdul, P.M.; Silvamany, H.; Maaroff, R.M.; Mohammed Yunus, M.F. The use of acidified palm oil mill effluent for thermophilic biomethane production by changing the hydraulic retention time in anaerobic sequencing batch reactor. *Int. J. Hydrogen Energy* **2018**. [CrossRef]
21. Seadi, T.A.; Rutz, D.; Prassl, H.; Kottner, M.; Finsterwalder, T.; Volk, S.; Janssen, R. *Biogas Handbook*; University of Southern Denmark: Esbjerg, Denmark, 2008.

22. Meegoda, J.; Li, B.; Patel, K.; Wang, L. A review of the processes, parameters, and optimization of anaerobic digestion. *Int. J. Environ. Res. Public Health* **2018**, *15*, 2224. [CrossRef]

23. Richards, M.A.; Lie, T.J.; Zhang, J.; Ragsdale, S.W.; Leigh, J.A.; Price, N.D. Exploring hydrogenotrophic. methanogenesis: A genome scale metabolic reconstruction of *Methanococcus maripaludis. J. Bacteriol.* **2016**, *198*, 3379–3390. [CrossRef]

24. Wong, Y.S.; Ong, S.A.; Lim, K.K.; Lee, H.C. Acclimatization and performance study of acidogenesis anaerobic degradation process for palm oil mill effluent. In Proceedings of the 2011 International Conference on Environment and Industrial Innovation (IPCBEE), Singapore, 26–28 February 2011.

25. Yacob, S.; Shirai, Y.; Hassan, M.A.; Wakisaka, M.; Subash, S. Start-up operation of semi-commercial closed anaerobic digester for palm oil mill effluent treatment. *Process. Biochem.* **2006**, *41*, 962–964. [CrossRef]

26. Enzmann, F.; Mayer, F.; Rother, M.; Holtmann, D. Methanogens: Biochemical background and biotechnological applications. *AMB Express* **2018**, *8*. [CrossRef] [PubMed]

27. Yu, H.-Q.; Fang, H.H.; Gu, G.-W. Comparative performance of mesophilic and thermophilic acidogenic upflow reactors. *Process. Biochem.* **2002**, *38*, 447–454. [CrossRef]

28. Işik, M.; Sponza, D.T. Substrate removal kinetics in an upflow anaerobic sludge blanket reactor decolorising simulated textile wastewater. *Process. Biochem.* **2005**, *40*, 1189–1198. [CrossRef]

29. Echiegu, E.A. Kinetic Models for Anaerobic Fermentation Processes—A Review. *Am. J. Biochem. Biotechnol.* **2015**, *11*, 132–148. [CrossRef]

30. Nielfa, A.; Cano, R.; Fdz-Polanco, M. Theoretical methane production generated by the co-digestion of organic fraction municipal solid waste and biological sludge. *Biotechnol. Rep.* **2015**, *5*, 14–21. [CrossRef] [PubMed]

31. Teng, T.T.; Wong, Y.-S.; Ong, S.-A.; Norhashimah, M.; Rafatullah, M. Start-up operation of anaerobic degradation process for palm oil mill effluent in anaerobic bench scale reactor (ABSR). *Procedia Environ. Sci.* **2013**, *18*, 442–450. [CrossRef]

32. Switzenbaum, M.S.; Jewell, W.J. Anaerobic attached film expanded bed reactor treatment. *J. Water Poll. Control. Fed.* **1980**, *52*, 1953–1965.

33. Idris, N.; Lutpi, N.A.; Wong, Y.S.; Izhar, T.N.T. Acclimatization study for biohydrogen production from palm oil mill effluent (pome) in continuous-flow system. *E3S Web Conf.* **2018**, *34*, 2054. [CrossRef]

34. Sperling, M.V. *Activated Sludge and Aerobic Biofilm Reactors*; IWA Publishing: London, UK, 2007.

35. Trisakti, B.; Manalu, V.; Taslim, I.; Turmuzi, M. Acidogenesis of palm oil mill effluent to produce biogas: Effect of hydraulic retention time and pH. *Procedia Soc. Behav. Sci.* **2015**, *195*, 2466–2474. [CrossRef]

36. Song, Y.-C.; Kwon, S.-J.; Woo, J.-H. Mesophilic and thermophilic temperature co-phase anaerobic digestion compared with single-stage mesophilic- and thermophilic digestion of sewage sludge. *Water Res.* **2004**, *38*, 1653–1662. [CrossRef] [PubMed]

37. Paritosh, K.; Kushwaha, S.K.; Yadav, M.; Pareek, N.; Chawade, A.; Vivekanand, V. Food waste to energy: An overview of sustainable approaches for food waste management and nutrient recycling. *Biomed. Res. Int.* **2017**, 1–19. [CrossRef] [PubMed]

38. Rabii, A.; Aldin, S.; Dahman, Y.; Elbeshbishy, E. A review on anaerobic co-digestion with a focus on the microbial populations and the effect of multi-stage digester configuration. *Energies* **2019**, *12*, 1106. [CrossRef]

39. Diamantis, V.; Aivasidis, A. Performance of an ECSB reactor for high-rate anaerobic treatment of cheese industry wastewater: Effect of pre-acidification on process efficiency and calcium precipitation. *Water Sci. Technol.* **2018**, *78*, 1893–1900. [CrossRef] [PubMed]

40. Bonk, F.; Popp, D.; Weinrich, S.; Sträuber, H.; Kleinsteuber, S.; Harms, H.; Centler, F. Ammonia inhibition of anaerobic volatile fatty acid degrading microbial communities. *Front. Microbiol.* **2018**, *9*. [CrossRef] [PubMed]

41. Demirel, B.; Yenigün, O. Two-phase anaerobic digestion processes: A review. *J. Chem. Technol. Biotechnol.* **2002**, *77*, 743–755. [CrossRef]

42. Siegert, I.; Banks, C. The effect of volatile fatty acid additions on the anaerobic digestion of cellulose and glucose in batch reactors. *Process. Biochem.* **2005**, *40*, 3412–3418. [CrossRef]

43. Kumar, G.; Sivagurunathan, P.; Park, J.-H.; Kim, S.-H. Anaerobic digestion of food waste to methane at various organic loading rates (OLRs) and hydraulic retention times (HRTs): Thermophilic vs. mesophilic regimes. *J. Environ. Eng.* **2016**, *21*, 69–73. [CrossRef]

44. Labatut, R.A.; Angenent, L.T.; Scott, N.R. Conventional mesophilic vs. thermophilic anaerobic digestion: A trade-off between performance and stability? *Water Res.* **2014**, *53*, 249–258. [CrossRef]

45. Kardos, L.; Juhász, A.; Palkó, G.Y.; Oláh, J.; Barkács, J.; Záray, G.Y. Comparing of mesophilic and thermophilic anaerobic fermented sewage sludge based on chemical and biochemical tests. *Appl. Ecol. Environ. Res.* **2011**, *9*, 293–302. [CrossRef]

46. Raposo, F.; Borja, R.; Martín, M.A.; de la Rubia, M.A.; Rincón, B. Influence of inoculum–substrate ratio on the anaerobic digestion of sunflower oil cake in batch mode: Process stability and kinetic evaluation. *Chem. Eng. J.* **2009**, *149*, 70–77. [CrossRef]

47. Labatut, R.A.; Gooch, C.A. Monitoring of anaerobic digestion process to optimize performance and prevent system failure. Proceeding Got Manure Enhancing Environment. *Econ. Sustain.* **2012**, *14*, 209–225.

48. Kugelman, I.J.; Guida, V.G. *Comparative Evaluation of Mesophilic and Thermophilic Anaerobic Digestion: Phase II—Steady State Studies*; National Service Centre for Environmental Publication: Cincinnati, OH, USA, 1989.

49. Girardi, G.; Berman, J.; Redecha, P.; Spruce, L.; Thurman, J.M.; Kraus, D.; Hollmann, T.J.; Casali, P.; Caroll, M.C.; Wetsel, R.A.; et al. Complement C5a receptors and neutrophils mediate fetal injury in the antiphospholipid syndrome. *J. Clin. Investig.* **2003**, *112*, 54. [CrossRef]

50. Mamimin, C.; Prasertsan, P.; Kongjan, P.; O-Thong, S. Effects of volatile fatty acids in biohydrogen effluent on biohythane production from palm oil mill effluent under thermophilic condition. *Electron. J. Biotechnol.* **2017**, *29*, 78–85. [CrossRef]

51. Hans, M.; Kumar, S. Biohythane production in two-stage anaerobic digestion system. *Int. J. Hydrogen Energy* **2018**. [CrossRef]

52. Krishnan, S.; Singh, L.; Sakinah, M.; Thakur, S.; Wahid, Z.A.; Sohaili, J. Effect of organic loading rate on hydrogen (H2) and methane (CH4) production in two-stage fermentation under thermophilic conditions using palm oil mill effluent (POME). *Energy Sustain. Dev.* **2016**, *34*, 130–138. [CrossRef]

Article

Application of Biochar Derived from Different Types of Biomass and Treatment Methods as a Fuel Source for Direct Carbon Fuel Cells

Lithnes Kalaivani Palniandy [1], Li Wan Yoon [1,*], Wai Yin Wong [2,*], Siek-Ting Yong [3] and Ming Meng Pang [1]

[1] School of Engineering, Faculty of Innovation and Technology, Taylor's University Lakeside Campus, Jalan Taylor's, Subang Jaya 47500, Selangor, Malaysia

[2] Fuel Cell Institute, Universiti Kebangsaan Malaysia, Bangi 43600, Selangor, Malaysia

[3] School of Engineering, Monash University Malaysia, Jalan Lagoon Selatan, Bandar Sunway 47500, Selangor, Malaysia

* Correspondence: LiWan.Yoon@taylors.edu.my (L.W.Y.); waiyin.wong@ukm.edu.my (W.Y.W.); Tel.: +603-56295098 (L.W.Y.); +603-8911-8588 (W.Y.W.)

Received: 15 May 2019; Accepted: 19 June 2019; Published: 27 June 2019

Abstract: The direct carbon fuel cell (DCFC) is an emerging technology for energy production. The application of biomass in DCFCs will be a major transition from the use of coal to generate energy. However, the relationship between biomass or biochar composition and the electrochemical performance of a DCFC is yet to be studied. The performance of a DCFC using fuel sources derived from woody and non-woody biomass were compared in this study. The effect of pyrolysis temperature ranges from 550 °C to 850 °C on the preparation of biochar from rubber wood (RW) and rice husk (RH) were evaluated for power generation from DCFCs. In addition, the effect of applying chemical pre-treatment and post-treatment on biochar were further investigated for DCFC performance. In general, the power density derived from rubber wood biochar is significantly higher (2.21 mW cm^{-2}) compared to rice husk biochar (0.07 mW cm^{-2}). This might be due to the presence of an oxygen functional group, higher fixed carbon content, and lower ash content in rubber wood biochar. The acid and alkaline pre-treatment and post-treatment have altered the composition with a lower ash content in rubber wood biochar. The structural and compositional alterations in alkaline pre-treatment bring a positive effect in enhancing the power density from DCFCs. This study concludes that woody biochar is more suitable for DCFC application, and alkaline pre-treatment in the preparation of biochar enhances the electrochemical activity of DCFC. Further investigation on the optimization of DCFC operating conditions could be performed.

Keywords: direct carbon fuel cell; biochar; pyrolysis; power density; pre-treatment; post-treatment

---

1. Introduction

Fuel cell technologies have recently attracted great attention owing to their advantages in producing clean energy using renewable energy sources [1,2]. These technologies use simple electrochemical processes to convert chemical energy into electrical energy. They differ from the fossil fuel energy generation, which requires a complex conversion from heat energy to mechanical energy and to electrical energy with lower efficiency that is limited by the Carnot efficiency [3]. Several types of fuel cells have been investigated and targeted for different applications. In view of the utilisation of biomass energy, direct carbon fuel cells (DCFC) have shown their potential as energy generation devices using carbon fuel in high temperature operations. The success of this system can benefit the chemical industry, in which waste heat energy can be integrated into the DCFC system to generate extra energy on-site. With a proper design of the DCFC system, it possesses the highest electrical efficiency of almost 100% among all types of fuel cells [4,5].

In the current situation, commercial carbon black and coals have been studied as DCFC carbon fuel for energy generation, with high energy efficiency obtained [4,5]. Nonetheless, these carbon sources are non-renewable and, hence, they do not contribute to the carbon-neutral cycle. Recent efforts have been focusing on searching for renewable carbon sources derived from biomass as a more sustainable option. Several works have investigated different types of biomass to produce biochar as a fuel source for DCFCs. The literature shows that the maximum power density achieved by different types of biochars ranged from 12 to 185 mW cm^{-2}, with the system operating at temperatures ranging between 700 and 800 °C [6]. It can be observed that biomass categorised as woody-type, such as almond shell [7], olive wood [8], and Acacia wood chips [9], obtain a relatively high power density of nearly 100 mW cm^{-2} and above, as compared to non-woody biomass, which only produce a power density of approximately 30 mW cm^{-2} and below. Among the many factors which would affect energy production in the DCFC system, the type of fuel source plays a vital role. The conversion of biomass into biochar using pyrolysis would alter the chemical composition of the fuel source and retain the highest energy value in the form of carbon content [10]. For example, the carbon content of rubber wood biochar after pyrolysis is approximately 97.3% [11], as compared to the raw rubber wood with a carbon content of only 43.98% [12].

Besides this, the presence of impurities and different functional groups in the biochar was believed to affect the DCFC performance as well [13,14]. It has been suggested that the differences in the chemical composition between woody and non-woody biomass has led to a major difference in the DCFC power density obtained. This is in line with the good cell performance of coal, with high power density of 165.4 mW cm^{-2} due to the presence of high fixed carbon, 81.5%, which contributed to the high chemical energy [15]. Besides the application of heat treatment, chemical treatment on the biomass or biochar can alter the chemical composition of the biochar fuel source. Acid pre-treatment of coal enhances the electrochemical activity in DCFCs, with an increment on oxygen-containing surface functional groups increasing the current density [13]. In another study using oak sawdust treated with nitric acid, the surface oxygen functional group increases and ash content reduces [16]. Thus, chemical treatment could be required for the enhancement of the carbonaceous fuel for DCFCs. However, a lack of study was observed on the comparison of pre-treatment and post-treatment in the preparation of biochar for the application of DCFCs.

Therefore, this study aims to explore the potential of woody (rubber wood) and non-woody (rice husk) biomass for DCFC application. In this research, the effect pyrolysis temperature on the preparation of biochar from rubber wood and rice husk on DCFC performance was investigated. Initial screening was conducted to choose the pyrolysis temperature that produces the biochar with the best DCFC performance. Besides this, the effect of different pre-treatment and post-treatment methods in biochar preparation on DCFC performance was investigated. Through this, the relationship between the structural composition of biochar and the performance of the DCFC can be further discovered.

2. Materials and Methods

2.1. Pyrolysis of Biomass

The rice husk was collected from PLS Marketing (M) Sdn Bhd, Sekinchan, Selangor, and the rubber wood biomass was collected from SYF Resources Bhd, Semenyih, Selangor. The biomass was washed using distilled water at room temperature. Then, it was dried in an oven at 110 °C for 24 h. The dried biomass was ground and sieved into particle size smaller than 500 μm. The pyrolysis process was conducted using a horizontal split tube furnace (Carbolite, HST 1200) under a nitrogen feed flow of 150 mL min^{-1}. The untreated biochar was produced at different pyrolysis temperatures at 550 °C, 650 °C, 750 °C, and 850 °C with a heating rate of 10 °C min^{-1} for 60 min. All the samples were labelled as RW for rubber wood (RW550, RW650, RW750 and RW850) and RH for rice husk (RH550, RH650, RH750 and RH850).

2.2. Direct Carbon Fuel Cell Performance

The DCFC setup used in this study was the same as that employed in previous studies [14]. Button cell was placed on the sample holder and solid biochar was loaded on the anode. The biochar loading in this study was 100 mg per run. Silver wire was used as the current collector at both the anode and the cathode. Both the anode and the cathode chambers were compressed mechanically. Nitrogen flowed through the anode at 200 mL min^{-1}, whereas air flowed through cathode at 200 mL min^{-1}. The DCFC system was heated to 850 °C at a heating rate of 10 °C min^{-1}. The electrochemical performance was studied using potentiostat (Interface 1000E, Gamry Instrument). Upon reaching the target temperature, open circuit potential (OCP) tests was performed at a scan rate of 1 mV s^{-1} using a potentiostat (Gamry, Interface 1000E). The internal resistance was tested at a high frequency of 1 kHz [15].

2.3. Chemical Treatment in Biochar Preparation

The results of DCFC performance in Section 2.2 show that rice husk generally produced a much lower power density compared to rubber wood. Hence, the effect of chemical treatment was further investigated for rubber wood raw biomass and biochar only. The structure and chemical composition of rubber wood biomass and biochar were further modified by applying acid and alkali treatments under different conditions. Pre-treatments of the biomass and post-treatment of rubber wood biochar, pyrolysed at 850 °C, were conducted using 2 mol dm^{-3} 50 cm^3 of HNO$_3$ and 2 mol dm^{-3} 50 cm^3 NaOH for 24 h. The experiment was conducted at room temperature with periodic stirring. A total of 2 g of rubber wood were added into 25 mL of the acid and alkaline solution. The samples were thoroughly washed with distilled water at room temperature until neutral pH was obtained [17]. The chemically treated samples were subjected for DCFC performance testing.

2.4. Biochar Fuel Characterization

Biochar derived from rice husk and rubber wood were further subjected for characterization. Proximate analysis of the raw rubber wood, raw rice husk raw, and biochar from these two biomasses was carried out using the Thermogravimetric Analyzer (TGA800, Perkin Elmer). A total of 10 mg of the sample was subjected to heating from room temperature to 110 °C under nitrogen gas flow and was held for 20 min. Then, the temperature was ramped from 110 °C to 950 °C and was held for 20 min. Gas was switched to air flow at 950 °C and was held for another 20 min [18].

The presence of a surface functional group of biochar was analysed using Fourier transform infrared (FTIR) (Spectrum100, Perkin Elmer) transmission analysis. The samples were analysed using the attenuated total reflection (ATR) from 4000 cm^{-1} to 650 cm^{-1} at a resolution of 4 cm^{-1} [19].

The surface morphology of untreated and chemically treated biochar samples was further studied via a variable pressure scanning electron microscope (VP-SEM) system (Hitachi S3400N-II). Samples were applied with platinum coating prior to the imaging.

3. Results and Discussion

3.1. Effect of Pyrolysis Temperature on the Biochar Characteristics

3.1.1. Biochar Yield at Different Pyrolysis Temperatures

Biochar yield for both woody and non-woody biomass shows a similar decreasing trend as the pyrolysis temperature was increased. The biochar yield decreased from 28% to 21.3% for rubber wood and 33.9% to 31.4% for rice husk. A similar trend was reported for the biochar production using corn straw pellets [1]. The results in Table 1 also show that different types of biomass generate a different biochar yield.

Table 1. Biochar yield from woody and non-woody biomass pyrolysed under different temperature.

Pyrolysis Condition	Rice Husk (%)	Rubber Wood (%)
550 °C	33.9 ± 0.26	28.0 ± 0.35
650 °C	33.6 ± 0.05	25.4 ± 0.10
750 °C	33.1 ± 0.05	24.6 ± 0.34
850 °C	31.4 ± 0.01	21.3 ± 1.40

3.1.2. Proximate Analysis for Biochar Characterization

The proximate analysis and weight loss of the biochar generated from different biomass sources are represented in Table 2 and Figure 1, respectively. The first mass loss observed at temperatures below 110 °C for all samples in Figure 1 was due to the removal of moisture. An increasing trend in the percentage of moisture was observed with the increase in pyrolysis temperature. The higher moisture content might be due to the hygroscopic characteristic of biochar pyrolysed at higher temperature, reabsorbing moisture from the surroundings [20]. The pyrolysed biochar for both rice husk and rubber wood show a significant decrease in volatile matter and increase in fixed carbon content as pyrolysis temperature increased. The production of volatile matter is due to the breakdown of carbohydrate fraction at higher temperatures [21]. Rubber wood biochar contains a higher fixed carbon content compared to rice husk in general. This might be attributed to the higher lignin and cellulose contents in raw rubber wood compared to rice husk. A maximum fixed carbon content of 67.0% was obtained from rubber wood, as compared to only 43.4% from rice husk.

Rubber wood-derived biochar has a significantly reduced ash content compared to rice husk. Fuel sources with lower ash content are reported to produce a better performance in DCFCs [16]. Generally, the ash content in rubber wood biochar reduces as the pyrolysis temperature increases. This might be attributed to the volatilization of inorganic compounds into gas or liquids at higher pyrolysis temperatures [22]. Conversely, pyrolysis temperature did not have a significant impact on the ash content in rice husk derived biochar. The high ash content may be attributed to the presence of a high silica content in the raw biomass. From Figure 1, it can be observed that the thermal stability of pyrolysed biochar increases compared to the raw biomass. The weight losses start at 550 °C for rice husk biochar, whilst significant weight losses of rubber wood start at 650 °C, showing that rubber wood possesses a higher thermal stability compared to rice husk.

Table 2. Proximate analysis of raw biomass and biochar.

Type of Biochar	Moisture (%)	Volatile Matter (%)	Fixed Carbon (%)	Ash (%)
Raw RH	11.5	58.5	10.3	19.7
RH550	3.5	40.9	19.5	36.1
RH650	5.2	18.2	41.6	35.1
RH750	6.7	12.9	42.7	37.7
RH850	7.3	11.9	43.4	37.4
Raw RW	10.4	59.9	11.1	18.6
RW550	7.5	23.8	62.0	6.7
RW650	8.1	22.4	67.0	2.5
RW750	10.8	24.0	62.7	2.4
RW850	14.2	21.8	62.0	2.0

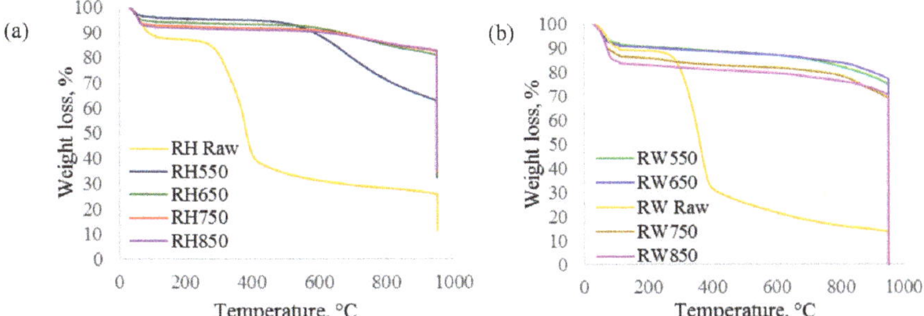

Figure 1. Thermogravimetric analysis curves for (**a**) rice husk derived biochar, (**b**) rubber wood derived biochar.

3.1.3. Surface Functional Analysis of Untreated Biochar

FTIR analysis allows the detection of the functional groups in biochar prepared at different pyrolysis temperatures. Figure 2 shows the FTIR spectra for rubber wood biochar and rice husk biochar. Prominent changes in the rubber wood biochar compared to the raw rubber wood were observed. A peak was observed at 1567 cm^{-1} for RW550, and this was broadened with the increase in pyrolysis temperature. This represents the aromatic C=C and C=O stretching of conjugated ketones and quinones, which suggests the presence of phenolic and carboxylic compounds in lignin increases in rubber wood biochar as pyrolysis temperature increases [22]. The peaks in the range of 1230 cm^{-1} to 1032 cm^{-1} (symmetric C=O stretching that presents in cellulose, hemicellulose, and lignin) [23] of raw rubber wood, decomposed further after an increase in pyrolysis temperature. This shows that the volatile matter starts to decompose after a pyrolysis temperature of 550 °C, and a broadened peak at approximately 1030 cm^{-1}, which possibly shows the trait of a C–C–O asymmetric stretch, was formed [19]. The formation of peaks in the range of 3000–2900 cm^{-1} when rubber wood biochar was pyrolysed at a temperature higher than 650 °C shows the presence of aliphatic C–H groups.

In addition, new peaks formed at approximately 1800 cm^{-1} (ester group) after the biomass was pyrolysed. This can be attributed to the interaction between cellulose and lignin during the heating of the rubber wood samples [24]. The changes of the bands at the region of 1000–1400 cm^{-1} show the band for C–O, the oxygen-containing functional groups, whereas O–H stretching vibrations are represented at the range of 3000–3445 cm^{-1} [25]. These peaks increased in intensity with the increase in pyrolysis temperature. This phenomenon might be caused by the increased moisture content in biochar produced at higher pyrolysis temperatures [26]. This activity may have resulted from the cracking and reforming reactions of aromatic hydrocarbons in the biochar [27].

Rice husk biochar has significant peaks in the range of 1090 cm^{-1} (Si–O–Si) and 788 cm^{-1} (Si–H), which represent the silica functional group [28]. This has a close similarity with the proximate analysis in Section 3.1.2 of rice husk biochar, which shows a higher ash content, which might be attributed to the high silica content in all the range of pyrolysed biochar.

As shown by the FTIR results, rubber wood biochar has a higher amount of surface oxygen functional groups compared to the rice husk. In a separate study with treated coal as the fuel source, more oxygen functional groups were produced, which contributed to higher electrochemical activity of 26 mW cm^{-2} compared to pyrolysed untreated coal at 750 °C, which produces only 8 mW cm^{-2}. This shows that the oxygen functional group facilitates the electrochemical kinetics [13] and provides a large number of reactive sites for the anode reaction [4].

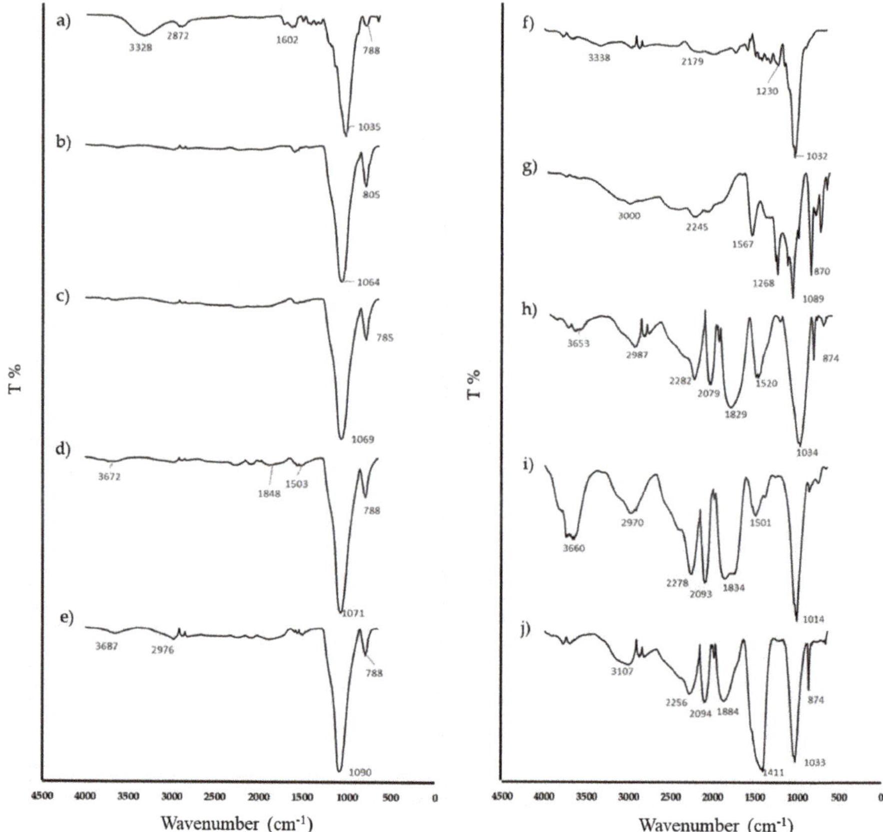

Figure 2. FTIR spectra for rice husk derived biochar (**a**) RH Raw, (**b**) RH550, (**c**) RH650, (**d**) RH750, (**e**) RH850; rubber wood derived biochar (**f**) RW Raw, (**g**) RW550, (**h**) RW650, (**i**) RW750, (**j**) RW850.

3.2. DCFC Performance Test with Biochar

DCFC performance tests were carried out by applying the solid carbonaceous fuel directly onto the button cell of the DCFC. Figure 3a,b shows the polarization curve of both rice husk and rubber wood biochar samples from various pyrolysis temperatures. From the figures, a rapid fall in the OCP for all rice husk derived biochar might be attributed to the activation resistance, as observed in other studies using activated carbon as the solid fuel [5]. However, a fast decrease with unstable potential change at a higher current density might show that the fuel consumption is faster than it is supplied to the electrode, which refers to the mass transport limitation [5]. By referring to the maximum power density, as shown Table 3, rubber wood biochar was shown to possess higher electrochemical activity (1.49–2.21 mW cm^{-2}) than rice husk biochar (0.05–0.07 mW cm^{-2}), disregard to the pyrolysis temperature in the biochar preparation. This indicates that woody biomass possesses a better electrochemical oxidation ability than non-woody biochar.

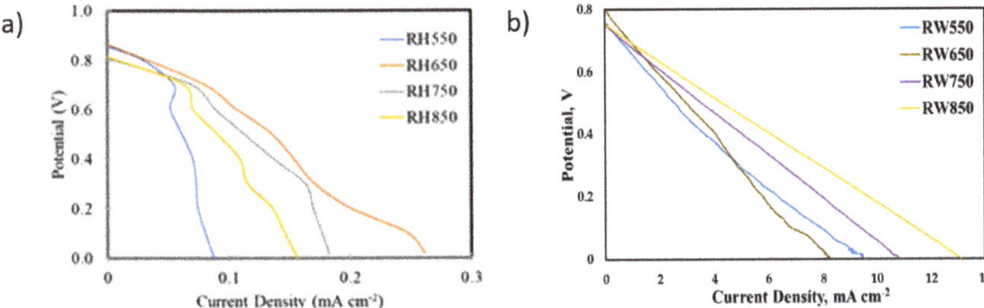

Figure 3. Polarization curves of (**a**) rice husk derived biochar, (**b**) rubber wood derived biochar from DCFC at 850 °C.

Table 3. Electrochemical data for untreated rubber wood and rice husk derived biochar from DCFC at 850 °C.

Parameter	RW550	RW650	RW750	RW850	RH550	RH650	RH750	RH850
OCP (V)	0.77 ± 0.03	0.78 ± 0.02	0.75 ± 0.01	0.76 ± 0.01	0.86 ± 0.01	0.86 ± 0.01	0.80 ± 0.02	0.82 ± 0.01
i at 0.7 V (mA cm^{-2})	0.69 ± 0.19	0.79 ± 0.17	0.54 ± 0.16	0.75 ± 0.13	0.04 ± 0.01	0.09 ± 0.01	0.07 ± 0.01	0.07 ± 0.01
i at 0.4 V (mA cm^{-2})	3.64 ± 0.01	3.95 ± 0.13	4.19 ± 1.16	5.44 ± 0.86	0.09 ± 0.02	0.19 ± 0.04	0.17 ± 0.04	0.13 ± 0.03
i at 0.1 V (mA cm^{-2})	7.60 ± 0.41	7.31 ± 0.75	8.57 ± 1.09	10.65 ± 1.03	0.11 ± 0.04	0.26 ± 0.01	0.21 ± 0.04	0.18 ± 0.05
P_{max} (mW cm^{-2})	1.49 ± 0.03	1.59 ± 0.05	1.74 ± 0.42	2.21 ± 0.33	0.05 ± 0.01	0.08 ± 0.01	0.07 ± 0.01	0.06 ± 0.01

The relatively low power density and quick potential drop observed in rice husk biochar could be attributed to the high ash content (ca. 36–37%). It was believed that the ash, which is mainly from silica content, acted as a barrier for the carbon to be in contact with the anode surface, and thus severely blocked the active sites of the anode for oxidation [29]. This is consistent with the presence of a silica functional group in the FTIR analysis, as depicted in Figure 2. In contrast, a lower ash content in woody biochar contributes to a better electrochemical activity, with a maximum reported power density value of 2.21 mW cm^{-2} at RW850. In addition, the higher composition of fixed carbon in rubber wood biochar samples could also possibly contribute to their better performance, due to a better fuel utilisation per active surface area. Rubber wood pyrolysed at 850 °C shows a slightly higher reactivity with the highest power density. In general, the electrochemical reactivity for rubber wood-derived biochar increases to the order of the rising pyrolysis temperature, from 550 °C to 850 °C.

The resistance of the cell was characterized by impedance spectra, as depicted in Figure 4. A small polarization arc was observed for all the rubber wood derived biochar. This suggests that the lower the resistance is, the more it contributes to better electrochemical activity, having a higher power density (1.49–2.21 mW cm^{-2}). Similarities were observed in another study with carbon black as the solid fuel, producing a power density output of 13.0 mW cm^{-2} with a lower resistance value (fuel cell operated at 900 °C) compared to the output of 3.6 mW cm^{-2} (fuel cell operated at 700 °C), which possessed a higher resistance value [30]. In contrast, a different trend was observed for the rice husk derived biochar, with all the samples showing a high polarization arc. Rice husk had a lower power density (0.05–0.07 mW cm^{-2}) with impedance spectra of a larger polarization, showing the presence of a charge transfer limit which is unfavorable for DCFC operation. A high polarization arc might have resulted from the difficulties in the transportation of the oxidants at the current collector and the electrode boundary, thus limiting the electrochemical activity [15].

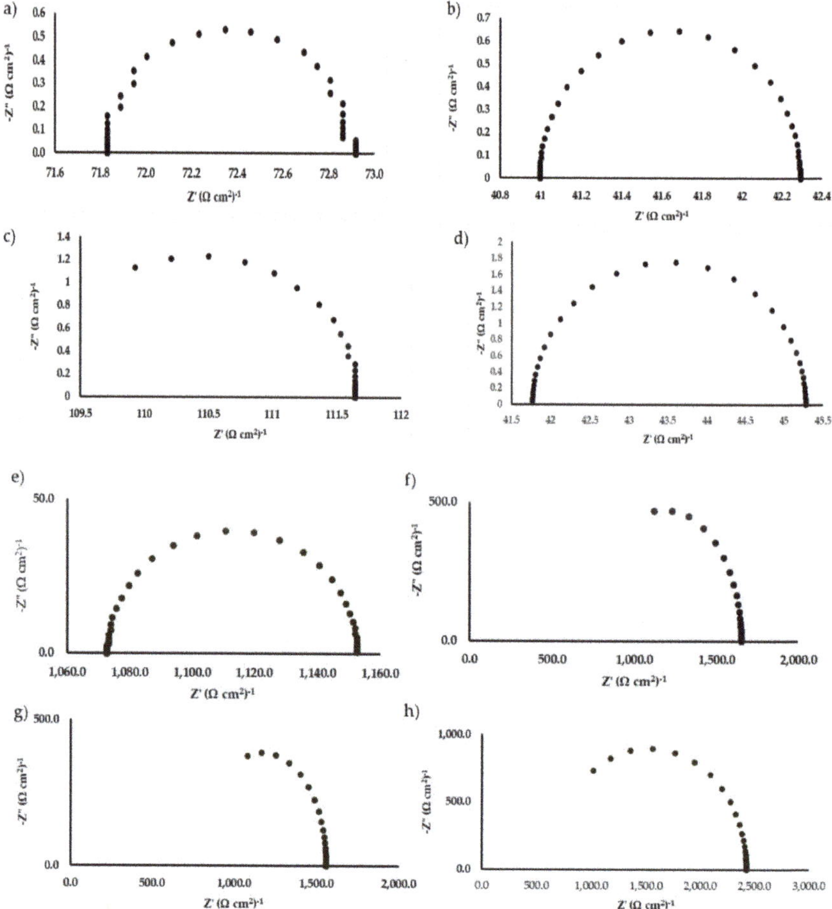

Figure 4. Electrochemical impedance spectra of DCFC at 850 °C for untreated rubber wood and rice husk biochar: (**a**) RW550, (**b**) RW650, (**c**) RW750, (**d**) RW850, (**e**) RH550, (**f**) RH650, (**g**) RH750, (**h**) RH850.

3.3. Effect of Pre-Treatment and Post-Ttreatment on Rubber Wood Biochar Production

From Section 3.1, rubber wood biochar pyrolysed at 850 °C recorded the best DCFC performance with the highest power density. The results show that the structural and chemical compositions of the biochar might have a significant impact on the DCFC performance. Hence, the structural and chemical composition of biochar was further altered with different chemical pre-treatment and post-treatment methods to study its effect on DCFC performance. Chemical treatments were conducted for the rubber wood biomass only as it possessed a higher electrochemical activity compared to the rice husk.

3.3.1. Biochar Yield of Pre-Treated and Post-Treated Woody Biochar

Table 4 shows the biochar yield when rubber wood was subjected to acid and alkaline pre- and post-treatment. In general, a slightly higher biochar yield was obtained from pre-treated alkali RW. In general, pre-treated samples gave a higher biochar yield compared to post-treated samples. The excessive washing imposed on biochar from the post-treatment might have caused the reduced yield.

Table 4. Biochar yield for chemically treated rubber wood biochar, pyrolysed at 850 °C.

Pyrolysis Condition	Biochar Yield, %
Post acid RW	22.0 ± 0.85
Post alkali RW	24.9 ± 0.49
Pre-treated acid RW	25.5 ± 4.44
Pre-treated alkali RW	31.8 ± 0.32

3.3.2. Proximate Analysis of Chemically Modified Rubber Wood Biochar

Table 5 presents the proximate analysis of post-treated and pre-treated rubber wood. The post-treated acid biochar possessed a lower moisture content among the treated samples. Besides this, chemical treatment aids in ash content reduction, as reported in all the treated samples. Pre-treated acidic RW shows a better demineralization effect compared to the alkali application in the pre-treatment technique. This is supported by a study using oak sawdust, which produced a lowered ash content after acid pre-treatment. The substantial reduction of ash removal via nitric acid removed the alkaline earth metals, such as ferric oxide and calcium oxide [16].

However, not much difference was observed for the content of fixed carbon. The major differences were observed in acid pre-treated rubber wood, which provides higher fixed carbon, 76.5%, and lower volatile matter, 11.3% compared to the untreated and other treated samples. Higher fixed carbon may contribute by the further disintegration of volatile fractions into a smaller molecular weight [15]. Post-treated acid biochar leads to higher volatile matter content, and this may contribute to nitration and oxidation effects during the process nitric acid treatment [13].

Table 5. Proximate analysis of chemical treated carbon fuel.

Type of Biochar	Moisture (%)	Volatile Matter (%)	Fixed Carbon (%)	Ash (%)
Post acid RW	9.4	32.9	56.2	1.5
Post alkali RW	16.4	31.3	63.4	-
Pre-treated acid RW	10.4	11.3	76.5	0.8
Pre-treated alkali RW	11.9	25.3	55.1	2.3

3.3.3. Surface Functional Study of Chemically Modified Rubber Wood Biochar

Figure 5 shows the spectra for chemically treated rubber wood for both pre-treatment and post-treatment techniques. It is evident that chemical treatment on biochar and biomass aids in the surface chemistry modification and disruption of structural components. The reduction of peaks at the range of 1500 cm^{-1}, which represents aromatic C=C and C=O stretching, shows that there is lignin decomposition after the chemical treatment. The peak reduction was greatly noticed in alkali pre-treatment and post-treatment. As discussed earlier in Section 3.1.3, the peak at the range of 1230 cm^{-1} to 1032 cm^{-1}, which contributes to symmetric C=O, was greatly reduced for pre-treated biochar compared to post-treated biochar. This suggests that the pre-treatment by acid and alkaline contributed to a major disruption of cellulose, hemicellulose, and lignin. Peak intensity at the range of 2300 cm^{-1} to 1900 cm^{-1} increased compared to untreated RW850. This peak may be attributed to the carboxyl and carbonyl functional group, observed for all the chemical-treated biochar samples. In addition, a new peak formation was observed at the range of 1800 cm^{-1}, which represents C=C for alkanes and aromatics [31]. A broad peak at 700 cm^{-1} was observed in pre-treated alkali RW, suggesting the presence of -OH out of plane bending modes [19]. It was observed that pre-treated alkali RW possessed higher oxygen surface functional groups compared to other chemically treated rubber wood derived biochar.

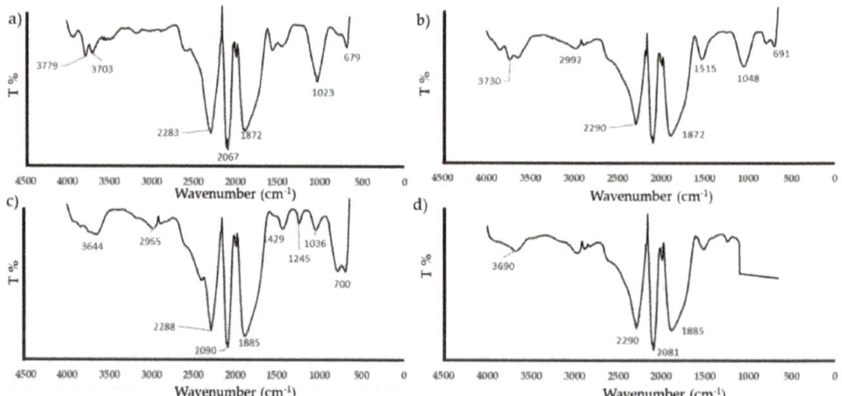

Figure 5. FTIR spectra for chemically treated rubber wood derived biochar: (**a**) post alkali RW, (**b**) post acid RW, (**c**) pre-treated alkali RW, (**d**) pre-treated acid RW.

3.3.4. Surface Morphology of Untreated Biochar and Modified Woody Biochar

Figure 6 shows the SEM morphology comparison between the untreated biochar derived from rubber wood, post-treated rubber wood biochar, and pre-treated rubber wood biochar at two different magnifications. The SEM images in Figure 6 show that there are no significant structural changes observed between the chemically treated biochar samples. Despite high temperature pyrolysis and chemical treatments, biomass structures and tissue morphology still remain intact.

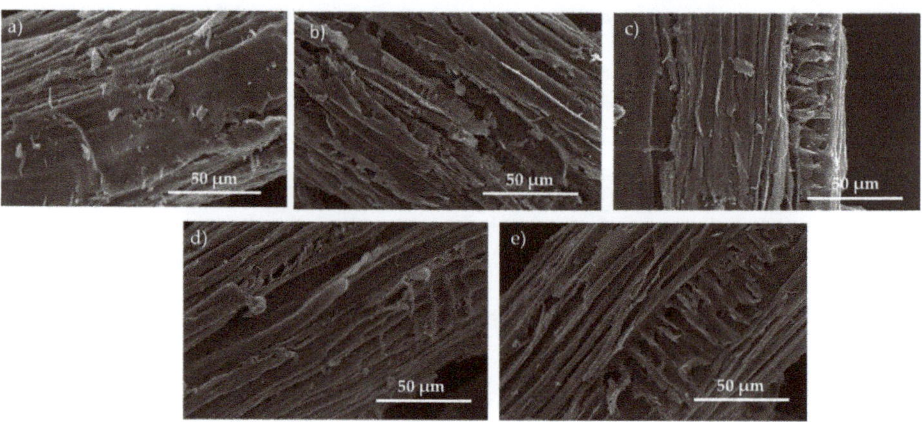

Magnification at 700×.

Figure 6. *Cont.*

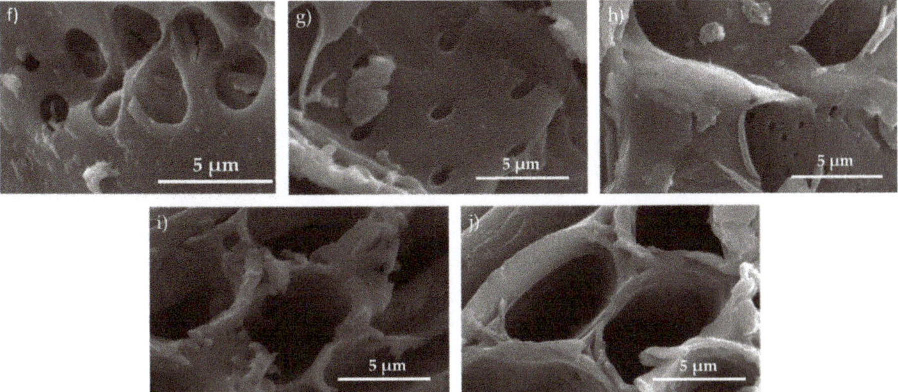

Magnification at 6500×.

Figure 6. Scanning Electron Microscope (SEM) images of untreated and chemical treated RW biochar pyrolysed at 850 °C. (**a**) untreated RW850; (**b**) post acid RW; (**c**) post alkali RW; (**d**) pre-treated acid RW; (**e**) pre-treated alkali RW; (**f**) untreated RW850; (**g**) post acid RW; (**h**) post alkali RW; (**i**) pre-treated acid RW; (**j**) pre-treated alkali RW.

3.3.5. DCFC Performance Test with Direct Solid Fuel of Modified Woody Biochar

From the results shown in Table 6, rubber wood biochar produced from the alkali pre-treatment and alkali post-treatment have given higher OCP values of 0.77 V and 0.76 V, respectively, compared to acidic post-treatment and acidic pre-treatment. A rapid potential reduction at lower current densties was observed for all the other chemically treated biochar samples compared to the pre-treated alkali RW, as shown in Figure 7. This further explains that the biochar samples were affected by activation polarization at lower current densities [7]. This might be attribute to the application of the solid fuel biochar directly to the button cell, causing the lack of activation energy needed for the electron movement at both the cathode and anode [15].

Based on Table 6, the power density of alkali pre-treated biochar (power density of 2.94 mW cm^{-2}) shows a better result compared to the post-treatment method. Based on the DCFC performance, alkali treatment in both conditions showed its suitability as an agent to modify the structure of rubber wood chemically, as it produced a higher power density compared to acid-treated biochar. This might be attributed to the function of alkaline in lignin removal and decomposition during treatment, which exposes more surface oxygen functional groups, that aid in better electrochemical activity. On the other hand, the power density was still higher in untreated rubber wood biochar compared to the acid-treated biochar in both post and pre-treated techniques. This suggests that the acid treatment on rubber wood did not enhance the electrochemical activity, further demonstrating that excessive degradation to the structure of biochar caused by acid treatment might not be favorable for rubber wood biochar for DCFC application. Besides, acid pre-treatment causes demineralization, which reduces the ash content [32]. However, the reduction of minerals from the acid treatment may leach out some minerals, that may act as a catalyst in enhancing DCFC performance.

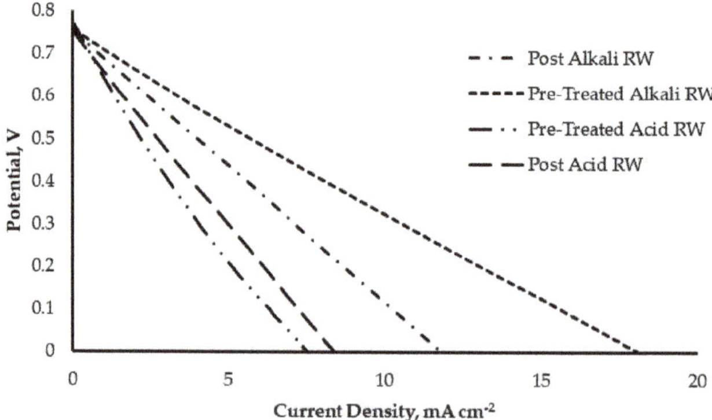

Figure 7. Polarization curve of chemical treated rubber wood derived biochar at pyrolysis temperature of 850 °C.

Table 6. Electrochemical data for chemical treatment of rubber wood derived biochar at pyrolysis temperature of 850 °C.

Parameter	Post Acid RW	Post Alkali RW	Pre-Treated Acid RW	Pre-Treated Alkali RW
OCP (V)	0.74 ± 0.00	0.76 ± 0.00	0.74 ± 0.03	0.77 ± 0.02
i at 0.7 V (mA cm^{-2})	0.51 ± 0.04	0.79 ± 0.06	0.41 ± 0.20	1.15 ± 0.01
i at 0.4 V (mA cm^{-2})	3.54 ± 0.43	5.27 ± 0.50	3.36 ± 0.36	7.23 ± 1.18
i at 0.1 V mA cm^{-2})	6.81 ± 0.65	9.75 ± 0.73	6.66 ± 0.48	14.36 ± 1.75
P_{max} (mW cm^{-2})	1.44 ± 0.18	2.13 ± 0.19	1.36 ± 0.15	2.94 ± 0.47

Figure 8 presents the electrochemical impedance spectra for chemically-treated rubber wood biochar samples. Alkali treatment (both pre and post) shows a greatly lowered polarization compared to other chemically-treated biochar samples. Higher polarization may be attributed to the presence of the activation resistance and concentration resistance that occurs during the cell reaction mechanism [20]. This might be the reason for a higher DCFC performance by alkali pre-treated biochar.

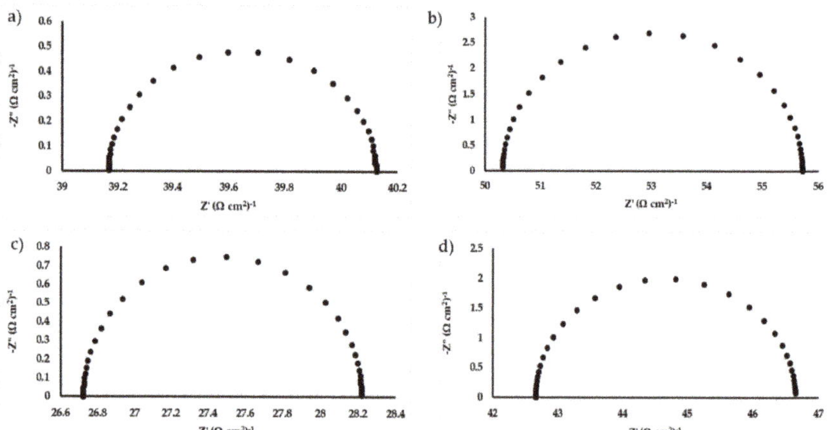

Figure 8. Electrochemical impedance spectra of DCFC for chemically treated rubber wood biochar: (**a**) pre-treated alkali RW; (**b**) pre-treated acid RW; (**c**) post alkali RW; (**d**) post acid RW.

4. Conclusions

This study explored the potential of biochar derived from woody and non-woody biomass as a fuel source for direct carbon fuel cells. The effect of pre-treatment and post-treatment on biochar was further evaluated through DCFC performance. A higher power density was generated by rubber wood biochar as compared to rice husk-derived biochar. The higher fixed carbon, lower ash content, and the presence of a surface oxygen functional group in rubber wood biochar might contribute to the better performance. The pre-treated alkali rubber wood biochar pyrolysed at 850 °C produced the highest power density, 2.94 mW cm^{-2}. Rubber wood biochar obtained from acidic pre-treatment and post-treatment generally produced a lower power density from DCFCs. Acid solution might leach out the minerals from biochar that are potential catalysts for the oxidation reaction. Conversely, alkali pre-treatment and post-treatment would retain the minerals and increase the surface functional groups in biochar that contribute to the improved DCFC performance. Further study on optimizing the operating conditions of DCFCs can be conducted to fully understand the significant factors affecting DCFC performance.

Author Contributions: Conceptualization, W.Y.W., L.W.Y. and S.-T.Y.; Methodology, W.Y.W., L.W.Y., and L.K.P.; Formal analysis and investigation, L.K.P.; writing—original draft preparation, L.K.P.; writing—review and editing, L.W.Y. and M.M.P.; supervision, L.W.Y., W.Y.W. and M.M.P.; project administration, W.Y.W., L.W.Y.; funding acquisition, W.Y.W., L.W.Y.

Funding: The research was funded by Taylor's University research grant scheme (TRGS/ERFS/2/2016/SOE/003) and Universiti Kebangsaan Malaysia's research grant (GUP-2018-013).

Acknowledgments: The author would like to thank the financial support by Taylor's University research grant scheme (TRGS/ERFS/2/2016/SOE/003) and Universiti Kebangsaan Malaysia's research grant (GUP-2018-013). This paper is an extended and revised article presented at the International Conference on Sustainable Energy and Green Technology 2018 (SEGT 2018) on 11–14 December 2018 in Kuala Lumpur, Malaysia.

Conflicts of Interest: The authors declare no conflict of interest. The funders had no role in the design of the study; in the collection, analyses, or interpretation of data; in the writing of the manuscript, or in the decision to publish the results.

References

1. Mahmud, L.S.; Muchtar, A.; Somalu, M.R. Challenges in fabricating planar solid oxide fuel cells: A review. *Renew. Sustain. Energy Rev.* **2017**, *72*, 105–116. [CrossRef]
2. Shaikh, S.P.S.; Muchtar, A.; Somalu, M.R. A review on the selection of anode materials for solid-oxide fuel cells. *Renew. Sustain. Energy Rev.* **2015**, *51*, 1–8. [CrossRef]
3. O'Hayre, R.; Cha, S.W.; Colella, W.; Prinz, B.F. *Fuel Cell Fundamentals*, 2nd ed.; John Wiley & Sons: Hoboken, NJ, USA, 2009; pp. 3–20.
4. Li, X.; Zhu, Z.H.; Marco, R.D.; Dicks, A.; Bradley, J.; Liu, S.; Lu, G.Q. Factors that determine the performance of carbon fuels in the direct carbon fuel cell. *Ind. Eng. Chem. Res.* **2008**, *47*, 9670–9677. [CrossRef]
5. Li, X.; Zhu, Z.H.; Chen, J.; Marco, R.D.; Dicks, A.; Bradley, J.; Lu, G.Q. Surface modification of carbon fuels for direct carbon fuel cells. *J. Power Sources* **2009**, *186*, 1–9. [CrossRef]
6. Jafri, N.; Wong, W.Y.; Doshi, V.; Yoon, L.W.; Cheah, K.H. A review on production and characterization of biochars for application in direct carbon fuel cells. *Proc. Saf. Environ. Prot.* **2018**, *118*, 152–166. [CrossRef]
7. Elleuch, A.; Boussetta, A.; Yu, J.; Halouani, K.; Li, Y. Experimental investigation of direct carbon fuel cell fueled by almond shell biochar: Part II. Improvement of cell stability and performance by a three-layer planar configuration. *Int. J. Hydrog. Energy* **2013**, *38*, 16605–16614. [CrossRef]
8. Elleuch, A.; Halouani, K.; Li, Y. Investigation of chemical and electrochemical reactions mechanism in a direct carbon fuel cell using olive wood charcoal as sustainable fuel. *J. Power Sources* **2015**, *281*, 350–361. [CrossRef]
9. Dudek, M.; Socha, R. Direct electrochemical conversion of the chemical energy of raw waste wood to electrical energy in tubular direct carbon solid oxide fuel cell. *Int. J. Electrochem. Sci.* **2014**, *9*, 7414–7430.

10. Tripathi, M.; Sahu, J.N.; Ganesan, P. Effect of process parameters on production of biochar from biomass waste through pyrolysis: A review. *Renew. Sustain. Energy Rev.* **2016**, *55*, 467–481. [CrossRef]

11. Ghani, W.A.W.A.K.; Mohd, A.; Silva, G.D.; Bachmann, R.T.; Taufiq, Y.; Yun, H.; Raashid, U.; Al-Muhtaseb, A. Biochar production from waste rubber-wood-sawdust and its potential use in C sequestration: Chemical and physical characterization. *Ind. Crops Prod.* **2013**, *44*, 18–24. [CrossRef]

12. Srinivasaakannan, C.; Bakar, M.Z.A. Production of activated carbon from rubber wood sawdust. *Biomass Bioenergy* **2004**, *27*, 89–96. [CrossRef]

13. Li, X.; Zhu, Z.; Marco, R.D.; Bradley, J.; Dicks, A. Modification of coal as a fuel for the direct carbon fuel cell. *J. Phys. Chem. A* **2010**, *114*, 3855–3862. [CrossRef] [PubMed]

14. Palniandy, L.K.; Wong, W.Y.; Yap, J.J.; Doshi, V.; Daud, W.R.W. Effect of alkaline pre-treatment on rice husk-derived biochar for direct carbon fuel cell. *J. Eng. Sci. Technol.* **2017**, 84–100.

15. Chien, A.C.; Arenillas, A.; Jiang, C.; Irvine, J.T.S. Performance of direct carbon fuel cells operated on coal and effect of operation mode. *J. Electrochem. Soc.* **2014**, *161*, F588–F593. [CrossRef]

16. Zhang, J.; Zhong, Z.; Zhao, J.; Yang, M.; Li, W.; Zhang, H. Study on the preparation of activated carbon for direct carbon fuel cell with oak sawdust. *Can. J. Chem. Eng.* **2012**, *90*, 762–768. [CrossRef]

17. Cao, D.; Wang, G.; Wang, C.; Wang, J.; Lu, T. Enhancement of electrooxidation activity of activated carbon for direct carbon fuel cell. *Int. J. Hydrog. Energy* **2010**, *35*, 1778–1782. [CrossRef]

18. Lim, S.H.; Yong, S.T.; Ooi, C.W.; Doshi, V.; Daud, W.R.W. Pyrolysis of palm waste for the application of direct carbon fuel cell. *Energy Proc.* **2014**, *61*, 878–881. [CrossRef]

19. Ghani, W.A.W.A.K. Sawdust-derived biochar: Characterization and CO2 adsorption/desorption study. *J. Appl. Sci.* **2014**, *14*, 1450–1454. [CrossRef]

20. Adam, C.R.; Giddey, S.; Kulkarni, A.; Badwal, S.P.S.; Bhattacharya, S.; Ladewig, B. Direct carbon fuel cell operation on brown coal. *Appl. Energy* **2014**, *120*, 56–64.

21. Park, J.; Hung, I.; Gan, Z.; Rojas, O.J.; Lim, K.H.; Park, S. Activated carbon from biochar: Influence of its physicochemical properties on the sorption characteristics of phenanthrene. *Bioresour. Technol.* **2013**, *149*, 383–389. [CrossRef]

22. Zhao, S.X.; Ta, N.; Wang, X. Effect of temperature on structural and physicochemical properties of biochar with apple tree branches as feedstock material. *Energies* **2017**, *10*, 1293. [CrossRef]

23. Ahmad, M.; Lee, S.S.; Dou, X.; Mohan, D.; Sung, J.; Yang, J.E.; Ok, Y.S. Effects of pyrolysis temperature on soybean stover- and peanut shell-derived biochar properties and TCE adsorption in water. *Bioresour. Technol.* **2012**, *118*, 536–544. [CrossRef] [PubMed]

24. Worasuwannarak, N.; Sonobe, T.; Taanthapanichakoon, W. Pyrolysis behaviors of rice straw, rice husk, and corncob by TG-MS technique. *J. Anal. Appl. Pyrolysis* **2007**, *78*, 265–271. [CrossRef]

25. Leng, L.; Yuan, X.; Zheng, G.; Shao, J.; Chen, X.; Wu, Z.; Wang, H.; Peng, X. Surface characterization of rice husk bio-char produced by liquefaction and application for cationic dye (Malachite green) adsorption. *Fuel* **2015**, *155*, 77–85. [CrossRef]

26. Claoston, N.; Samsuri, A.W.; Ahmad Husni, M.H.; Mohd Amran, M.S. Effects of pyrolysis temperature on the physicochemical properties of empty fruit bunch and rice husk biochars. *Waste Manag. Res.* **2014**, *32*, 331–339. [CrossRef] [PubMed]

27. Xiong, S.; Zhuo, J.; Zhang, B.; Yao, Q. Effect of moisture content on the characterization of products from the pyrolysis of sewage sludge. *J. Anal. Appl. Pyrolysis* **2013**, *104*, 632–639. [CrossRef]

28. Wei, L.; Huang, Y.; Li, Y.; Huang, L.; Mar, N.N.; Huang, Q.; Liu, Z. Biochar characteristics produced from rice husks and their sorption properties for the acetanilide herbicide metolachlor. *Environ. Sci. Pollut. Res.* **2017**, *24*, 4552–4561. [CrossRef]

29. Ahn, S.Y.; Seong, Y.E.; Young, H.R.; Yon, M.S.; Cheor, E.M.; Gyung, M.C.; Duck, J.K. Utilization of wood biomass char in a direct carbon fuel cell (DCFC) system. *App. Energy* **2013**, *105*, 207–216. [CrossRef]

30. Nabae, Y.; Pointon, K.D.; Irvine, J.T.S. Electrochemical oxidation of solid carbon in hybrid DCFC with solid oxide and molten carbonate binary electrolyte. *Energy Environ. Sci.* **2008**, *1*, 148–155. [CrossRef]

31. Angin, D. Effect of pyrolysis temperature and heating rate on biochar obtained from pyrolysis of safflower seed press cake. *Bioresour. Technol.* **2013**, *128*, 593–597. [CrossRef]
32. Dudek, M. On the utilization of coal samples in direct carbon solid oxide fuel cell technology. *Solid State Ion.* **2015**, *271*, 121–127. [CrossRef]

Article

Modification of a Direct Injection Diesel Engine in Improving the Ignitability and Emissions of Diesel–Ethanol–Palm Oil Methyl Ester Blends [†]

Norhidayah Mat Taib, Mohd Radzi Abu Mansor * and Wan Mohd Faizal Wan Mahmood

Department of Mechanical Engineering, Faculty of Engineering and Built Environment, Universiti Kebangsaan Malaysia, UKM Bangi 43600, Selangor, Malaysia
* Correspondence: radzi@ukm.edu.my
† This paper is an extended and revised article presented at the International Conference on Sustainable Energy and Green Technology 2018 (SEGT) on 11–14 December 2018 in Kuala Lumpur, Malaysia.

Received: 13 May 2019; Accepted: 6 July 2019; Published: 10 July 2019

Abstract: Blending diesel with biofuels, such as ethanol and palm oil methyl ester (PME), enhances the fuel properties and produces improved engine performance and low emissions. However, the presence of ethanol, which has a small cetane number and low heating value, reduces the fuel ignitability. This work aimed to study the effect of injection strategies, compression ratio (CR), and air intake temperature (T_i) modification on blend ignitability, combustion characteristics, and emissions. Moreover, the best composition of diesel–ethanol–PME blends and engine modification was selected. A simulation was also conducted using Converge CFD software based on a single-cylinder direct injection compression ignition Yanmar TF90 engine parameter. Diesel–ethanol–PME blends that consist of 10% ethanol with 40% PME (D50E10B40), D50E25B25, and D50E40B10 were selected and conducted on different injection strategies, compression ratios, and intake temperatures. The results show that shortening the injection duration and increasing the injected mass has no significant effect on ignition. Meanwhile, advancing the injection timing improves the ignitability but with weak ignition energy. Therefore, increasing the compression ratio and ambient temperature helps ignite the non-combustible blends due to the high temperature and pressure. This modification allowed the mixture to ignite with a minimum CR of 20 and T_i of 350 K. Thus, blending high ethanol contents in a diesel engine can be applied by advancing the injection, increasing the CR, and increasing the ambient temperature. From the emission comparison, the most suitable mixtures that can be operated in the engine without modification is D50E25B25, and the most appropriate modification on the engine is by increasing the ambient temperature at 350 K.

Keywords: combustion characteristics; injection strategies; compression ratio; intake temperature

1. Introduction

Palm oil methyl ester (PME) produced from edible sources, such as palm oil, is known as the best biodiesel in Malaysia [1]. PME has a performance that is almost similar to diesel fuel, but with good engine emissions [2]. However, the high density and viscosity of PME require much fuel consumption in the real-application engine. Therefore, blending PME in diesel was able to reduce the harmful emission from the diesel and improve the engine performance due to its high cetane number. However, the high density and viscosity of PME need the ethanol presence to prevent the injection system problem in the engine and reduce the fuel consumption. The reason is that ethanol has the lowest viscosity amongst diesel and PME, as explained in Table 1. Besides, ethanol is another renewable energy fuel, promising good engine emissions. Along with the energy efficient vehicle (EEV) concept, the PME and ethanol presence can reduce the emissions, improve the engine performance, and reduce petroleum dependency.

Table 1. Thermophysical properties of diesel, ethanol, and palm oil methyl ester (PME) [3–6].

Properties	Diesel	Ethanol	PME
Molecular formula	NC_7H_{16}	C_2H_5OH	$C_{18}H_{34}O_2$
Viscosity (cSt)	3.49	1.5	4.8
Density (kg/m3)	834	788	881
Low heating value (MJ/kg)	43.2	27	37.8
Cetane number	50	8	56
Octane number	25	107	-
Oxygen contents (%)	0.0	34.8	11
Auto-ignition temperature (K)	483	638	469

Ethanol was recommended to be used in commercial diesel engines due to its low production cost for biofuels. Ethanol is a low viscous liquid that is able to reach good air–fuel mixture homogeneity. However, ethanol that contains a deficient cetane number reduced the engine performance and delayed the combustion [7]. The poor ethanol ignitability is probably due to its higher evaporation enthalpy compared with diesel. Thus, the temperature was reduced during combustion due to the withdrawn heat during combustion [8]. The long delay in combustion causes the fuel to be less combustible and forms the excess fuel in the engine. However, a perfectly blended composition should be identified to have excellent fuel combustion.

Meanwhile, high ethanol contents in blends also affect the ignitability of the fuel blends, and the excess fuel leads to fuel deposition inside the cylinder. Therefore, modification of the injection parameter was suggested to improve the ignitability of blends. Modifications of injection parameters include the injection mass, injection pressure, injection duration, and injection timing. Modification of the injection timing can improve engine performance. However, not all the applied injection strategies can solve the ignitability problem of diesel–ethanol–PME blends.

Few researchers have studied the effect of advancing the injection timing for high oxygenated alternative fuels, such as ethanol and PME. The studies found that advancing the injection timing increased the in-cylinder pressure and heat release rate (HRR). However, advancing the injection timing can increase NO_x emission [9]. The reason is that advancing the injection timing allows the fuel to mix early and increase the temperature. Thus, the NO_x reaction takes part during the combustion easily.

Meanwhile, the presence of ethanol and PME also is another reason for the NO_x formation due to the high oxygen contents in both fuel molecules [10]. Therefore, suitable blends of diesel, ethanol, and PME should be formed to reduce the NO_x formation. Another study was conducted in improving the ethanol ignitability by increasing the engine compression ratio. Increasing the compression ratio of the engine increases the pressure and temperature to enhance the fuel ignitability and engine efficiency [11]. Considering that the auto-ignition temperature for the blends was high, another approach, such as increasing the ambient temperature at the intake, also helped improve the ignitability [12]. A study from Kuszewski (2018) shows that improving the ambient gas improved the auto-ignition of diesel–ethanol blends. The results showed that the ignition delay for a high ethanol percentage is longer than that for diesel. The maximum percentage of ethanol use in this experiment is 14% [13].

This research works aimed to study the ignitability of diesel–ethanol–PME blends and optimize the best solution in solving the non-combustible fuel blends. The non-combustible fuel blends are recognized when no ignition occurs, which usually happened on the high ethanol content blends when operating at high engine speeds. The simulation was conducted using Converge CFD software based on the direct injection Yanmar TF90 diesel engine parameter. The simulation was performed for low, medium, and high ethanol contents, containing 10% ethanol and 40% PME in 50% diesel (D50E10B40), D50E25B25, and D50E40B10 blends at 900 RPM, 1600 RPM, and 2400 RPM. The first work was to identify the non-combustible blends. The injection parameter, compression ratio, and intake temperature were modified due to the ignitability problem of high ethanol contents in the blends. The optimization of the best solution was identified from the combustion characteristic progress to solve the ignitability problem faced by high ethanol content blends.

2. Methodology

In this study, blends were simulated to analyze their ignitability and emission. Grid independence and emission tests were also conducted to determine the blend composition. The emission results from the simulation are verified with the results obtained from experimental work.

2.1. Simulation of Blends Ignitability and Emission

In this study, simulation work was conducted to analyze the combustion characteristics and emissions of diesel–ethanol–PME. The combustion characteristics analysis was used to study the combustion behavior of the fuel when entering the combustion chamber. This process includes the ignition delay, thermal efficiency, in-cylinder pressure and temperature, and heat release during combustion. This work focused on the ignitability optimization of high ethanol presence in diesel–PME blends. The simulation was conducted based on the single cylinder direct injection compression ignition Yanmar TF90 engine parameter by using Converge CFD software.

Converge CFD software is a unique software equipped with adaptive mesh refinery (AMR) with the ability to refine the mesh automatically during the combustion. A grid independence test was conducted to identify the most suitable grid size for meshing. This test aims to eliminate any unnecessary grid meshing sizes to improve simulation outcomes. The grid independence test was performed on different grid sizes of 0.003 m, 0.004 m, 0.005 m, and 0.006 m. The diesel fuel results were verified with the experiment results obtained by Ibrahim (2015) [14]. The most suitable grid applied on the engine model is 0.004 m by activating the AMR function to allow small automatic meshing occurring during the injection and combustion.

In this software, an engine model was prepared based on the engine specification, as shown in Figure 1. The three boundaries, which consist of a piston, head, and liner of the cylinder, were assigned as the initial conditions of each boundary. The cylinder wall (liner) and head were set in a fixed position with the wall temperature of 363 K and 319 K. Meanwhile, the piston was set as a moving boundary with the wall temperature of 403 K [15]. The parameter of the engine and injection system based on the real engine parameter for the Yanmar TF90 direct injection diesel engine was also explained in Table 2.

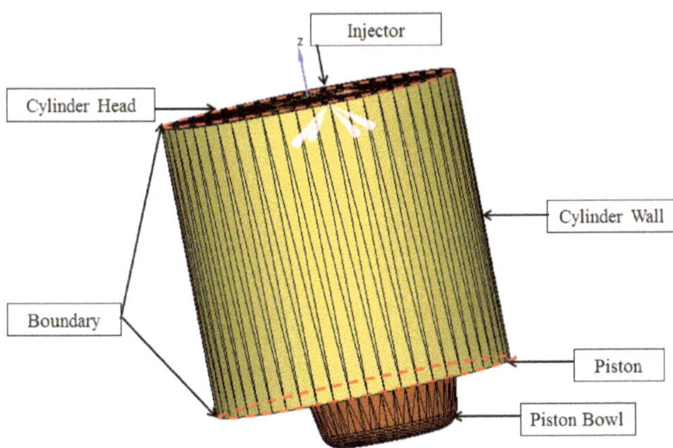

Figure 1. Yanmar TF90 combustion cylinder model.

Table 2. Engine and injection system specification [15].

Engine Specification		Injection System Specification	
Engine model	Yanmar TF90	Number of nozzles	4
Bore (m) × stroke (m)	0.085 × 0.087	Nozzle diameter (mm)	0.22
Connecting rod length (m)	0.13	Injection pressure (MPa)	19.613
Piston bowl diameter (m)	0.0463	Injection duration (°CA)	16
Piston bowl depth (m)	0.016	Injection timing (°CA BTDC)	−18
Compression ratio	18	Cone angle (°)	10

The Yanmar TF90 engine is a natural aspirated air single cylinder engine with a direct injection system. The air is entered the combustion chamber at the atmospheric pressure and temperature before the piston compressed it. The simulation is conducted for a cycle starting right after the intake valve closes at −168 °CA BTDC until the exhaust valve opens at 138 °CA ATDC. The initial temperature and pressure of all the boundaries were set at the atmospheric pressure and temperature, which is 101 kPa and 300 K. The real injection timing of this engine was started at −18 °CA BTDC located at the center of the cylinder head. The combustion was simulated using a SAGE combustion model in a closed-system chamber. The mathematical correlation was solved with the text data. In this simulation, the chemical reaction, gas transport, and thermodynamic data for all the fuel blends were obtained from the Lawrence Livermore National Laboratory. Meanwhile, the chemical kinetic reaction of biodiesel underwent a mechanism reduction process to reduce the number of reactions in the data. The emission model used in this simulation was the Zeldovich NO_x model, which utilizes the Hiroyasu soot model to identify emissions from combustion.

2.2. Blend Composition

A few different diesel–ethanol–PME blend compositions were applied to the engine running at 900 RPM, 1600 RPM, and 2400 RPM. The blends were injected through the injector nozzle into the engine combustion chamber. From all the blends, three blends were selected to run at 1600 RPM and 2400 RPM. The blends are D50E10B40 (50% diesel, 10% ethanol, and 40% PME), D50E25B25, and D50E40B10. Table 3 presents the mass fraction of these blends. These blends were chosen to study the effect of ethanol and the PME presence in the engine through combustion analysis.

Table 3. Blend composition. D50E10B40: 50% diesel, 10% ethanol, and 40% PME, etc.

No	Blends	Mass Fraction		
		Diesel	Ethanol	PME
1	Diesel	1.0	0.0	0.0
2	D50E10B40	0.5	0.1	0.4
3	D50E10B0	0.83	0.27	0.0
4	D50E25B25	0.5	0.25	0.25
5	D50E0B25	0.67	0.0	0.33
6	D50E40B10	0.5	0.4	0.1
7	D50E0B10	0.83	0.0	0.27

As explained in Table 3, another diesel–ethanol blend named as D50E10B0 was operated and compared with the D50E10B40 blend to study the effect of PME presence in diesel–ethanol blends. Meanwhile, two diesel–PME blends such as D50E0B25 and D50E0B10 were compared with D50E25B25 and D50E40B10 to study the effect of ethanol in these two diesel–ethanol–PME blends. However, due to the presence of ethanol, which has a very low cetane number, the non-combustible blends were identified from the combustion analysis at each engine speed.

From the analysis, the non-combustible blends are then selected for a few modifications to improve its ignitability. The modifications, such as the injection system parameter, engine compression ratio, and air intake temperature, were selected. These modifications were conducted to find the most suitable modification that can be applied to improve the ignitability of the fuels. The simulation was performed at two engine speeds: namely, 1600 RPM and 2400 RPM. Table 4 shows the modification applied to the engine by comparing the combustion characteristics from the in-cylinder pressure and HRR during combustion. From the results, the ignitability of each blend is investigated by identifying its ignition delay. Ignition delay is defined as the duration between the start of fuel injection and the start of ignition. Ignition results from the rapid rise in pressure and the occurrence of HRR. In addition, the purposes of blending the ethanol and PME in diesel are to reduce the greenhouse emission. Therefore, carbon footprint and NO_x emission analysis were also conducted from the simulation to identify the emission efficiency of each fuel blend and its modification.

Table 4. Injection system and engine modification parameter.

Injection Modification			Compression Ratio Modification	Ambient Temperature, T_i
Injection Timing	Injection Duration	Injected Mass		
−25 °CA BTDC	10 °CA	25 mg	18	300 K
−18 °CA BTDC	16 °CA	19 mg	20	350 K
−10 °CA BTDC	-	16 mg	24	400 K
−8 °CA BTDC	-	8 mg	-	-

Note: underlined parameter is the normal parameter for the Yanmar TF90 engine.

2.3. Validation of Emissions from Blend Combustion through Experiments

The simulation for the D50E10B40 and D50E25B25 blends was experimentally validated on the YanmarTf90 Engine. Table 2 presents the specifications. Without any further modifications, the engine ran at 900 RPM, 1600 RPM, and 2400 RPM at a 2-kW load. Figure 2 shows a schematic of the experiment setup. The emission of the diesel–ethanol–PME blend combustion was measured using a KANE gas analyzer, which assessed the CO, CO_2, and NO_x emissions. The limitation in this experiment is that the combustion characteristics could not be measured for validation. Therefore, the emission results from the experiment were compared with the simulation results.

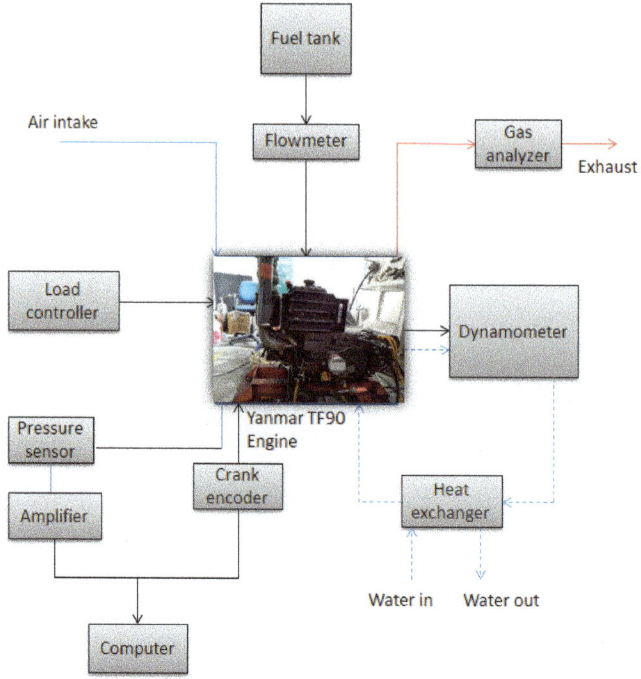

Figure 2. Schematic of the experimental engine setup.

3. Results and Discussion

The results of this study are focused on the ignitability of diesel–ethanol–PME blends at various compositions and engine speeds. The ignition abilities of the blends are discussed briefly for different cases especially when the blends contained high ethanol percentage. Thus, another method was applied to improve the ignition ability by working on the injection parameter change, compression ratio, and ambient temperature improvement.

3.1. Grid Independence Test

A grid independence test was conducted to identify the optimum grid size for meshing. The results of combustion characteristics were compared with the experiment results to determine the best grid size for meshing. Figure 3 shows the in-cylinder pressure of diesel fuel from a simulation with different grid sizes compared with the experiment results when the engine is running at 1600 RPM. The grid size comparison shows that 0.004 m is the most suitable grid size for meshing, which obtains the minimum error value compared with the experiment results at the peak pressure. The comparison of data sizes and operation working durations are also part of the primary consideration. Table 5 shows the comparison of data sizes, simulation durations, and error percentages obtained from the pressure results. A grid size of 0.004 m is the most suitable grid for further simulations.

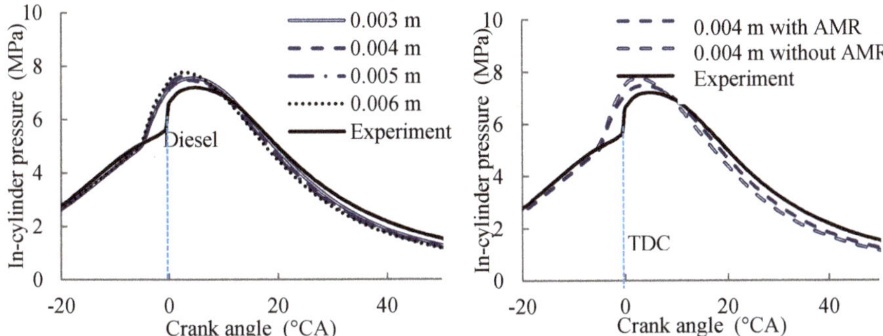

Figure 3. Pressure of diesel fuel from the simulation with different grid sizes and with and without adaptive mesh refinery (AMR) function when the engine is running at 1600 RPM compared with the experiment.

Table 5. Comparison of data sizes, simulation durations, and error percentages from the pressure during combustion with different grid sizes.

Grid Size	Error Percentage (%)		Number of Cells	File Size (GB)	Duration (min)
	At TDC	Peak Pressure			
0.003	8.424	5.120	1441 390	14.0	1557
0.004	8.190	4.024	614 589	7.76	996
0.004 without AMR	14.318	7.817	614 874	6.14	439
0.005	10.848	6.145	504 128	4.87	664
0.006	13.244	8.022	500 624	4.32	454

3.2. Ignitability and Emission of Diesel-ethanol-PME Blends with High Ethanol Contents

Figure 4 shows the comparison of in-cylinder pressure and HRR between the D50E10B40, D50E25B25, and D50E40B10 blends running at high engine speeds of 1600 RPM and 2400 RPM. The graph shows that the operating engine at high engine speed influenced the fuel ignitability. The result shows that at high engine speed, the ignitability of the blends weakened due to the ethanol presence that has low heating value and low cetane number. The reason is that the cetane number of ethanol was approximately five to 10 times lower than that of diesel and PME [16]. The low heating value of the fuel also indicates that the blends release a very low heat during the compression. This condition leads to a lack of energy and delays the ignition.

Figure 5 shows the ignition delay of the blends influenced by ethanol contents. The results show that the ignition condition is worsened when the high ethanol contents were used. Ignition delay timing in the crank angle at high engine speed is relatively lengthened, and shows opposite results in an ignition delay time in milliseconds due to the fast engine piston motion. When the engine speed moved rapidly, the engine duration to complete a cycle was shortened. This condition results in insufficient air into the cylinder and reduces the pressure in the cylinder. As presented in the graph, the D50E25B25 blend was not able to ignite when running at 2400 RPM. Meanwhile, the ignitability potential of the D50E40B10 blend was worsened at all speeds except for at 900 RPM.

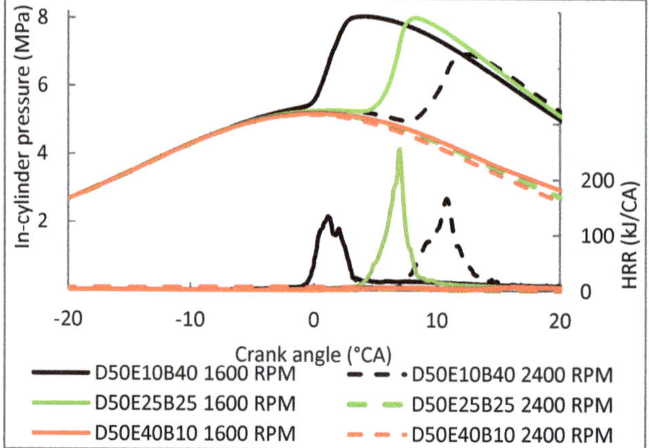

Figure 4. Graph of in-cylinder pressure and heat release rate (HRR) of D50E10B40, D50E25B25, and D50E40B10 blends at 1600 RPM and 2400 RPM.

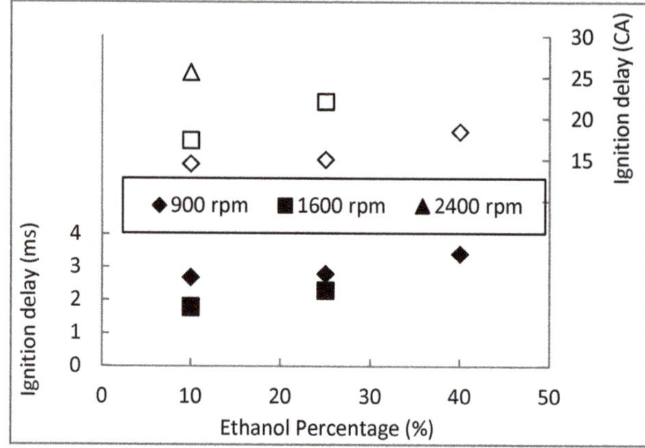

Figure 5. Ignition delay of the combustible blends with different ethanol percentages.

Figure 6 shows the three-dimensional (3D) visual temperature distribution for D50E10B30, D50E25B25, and D50E40B10 blends at 1600 RPM running at normal Yanmar TF90 engine specification without modification. Evidently, the high ethanol percentage in blends cannot be ignited at this engine specification due to the low ambient temperature. The maximum temperature for the blends without ignition is 850 K. This temperature is not high enough to ignite the blends with high ethanol contents. Therefore, the ignitability of ethanol present in the fuel decreases at high engine speed. However, the high ethanol contents in blends have low ignition potential and have small peak pressure due to the delay. In addition, it also has a rapid rise in HRR and small peak pressure that will cause the ignition duration. The results show that the maximum ethanol contents in the blend composition are 25% when running at 1600 RPM and are decreased when running at a high engine speed.

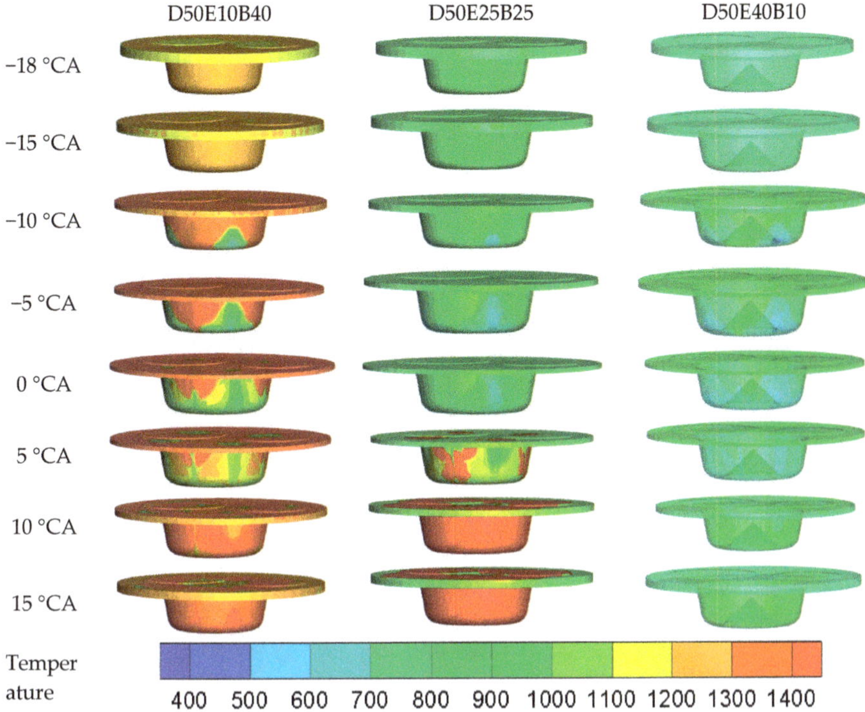

Figure 6. Three-dimensional (3D) view of ignitability difference between different diesel–ethanol–PME blends at 1600 RPM.

The emission of simulation results obtained from the combustion of all the blends is validated with the emission results from the experiment. However, due to the ignition problem faced by the D50E40B10 blend, the experiment is conducted only for D50E10B40 and D50E25B25. The average emission of CO_2, CO, and NO_x are calculated and compared to the experiment. Figure 7 shows the comparison of CO, CO_2, and NO_x emissions of D50E10B40, D50E25B25, and D50E40B10 blends.

The simulation and experiment results show that the NO_x and CO_2 emissions of the D50E10B40 blend are insignificantly different. Meanwhile, the NO_x, CO, and CO_2 emissions of D50E25B25 are unstable, especially at a high engine speed. This phenomenon is due to the high ethanol contents with unstable combustion due to insufficient oxygen. The D50E25B25 blend also showed evident separation after a certain time during the experiment, which is an indication for unstable combustion. Separation occurs due to the high water presence from ethanol. Hence, diesel is probably the only fuel that can be injected into the combustion chamber. The emission results show that the simulation accuracy of the experiment has a minimum error of 5% to 30%, except for unstable combustion blends. Therefore, the simulation for the combustion characteristics and emissions of diesel–ethanol–PME blends is validated for further study.

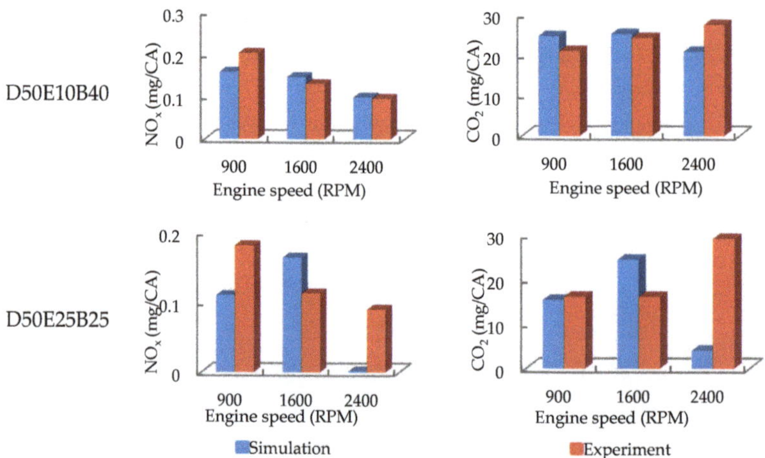

Figure 7. Comparison of emissions from simulations and experiments for D50E10B40 and D50E25B25 blends.

Figure 8 shows the graph of tri-fuel blends and dual-fuel blends to compare the difference of in-cylinder pressure and HRR when ethanol or PME is added. The results obviously show that the blend without any PME presence (D50E10B0) has slightly higher pressure and HRR compared with other blends. Meanwhile, the D50E0B10 blend that does not consist of any ethanol presence has shorter ignition delay and is obviously easily ignited compared with D50E40B10. This outcome is due to the biodiesel, which has a high cetane number and increased temperature to early ignite the fuel. D50E0B25 has a shorter ignition delay than D50E25B25. The diesel–PME blends without ethanol presence show a positive result to the ignition delay. Although the ignition delay of the blends with high PME contents is shortened, the HRR of the blends is very low. This condition is influenced by the properties of PME, which has high viscosity, a high molecular weight, and low burning velocity [16]. Therefore, the presence of ethanol in D50E25B25 helps the blends obtain high HRR for improved engine thermal efficiency.

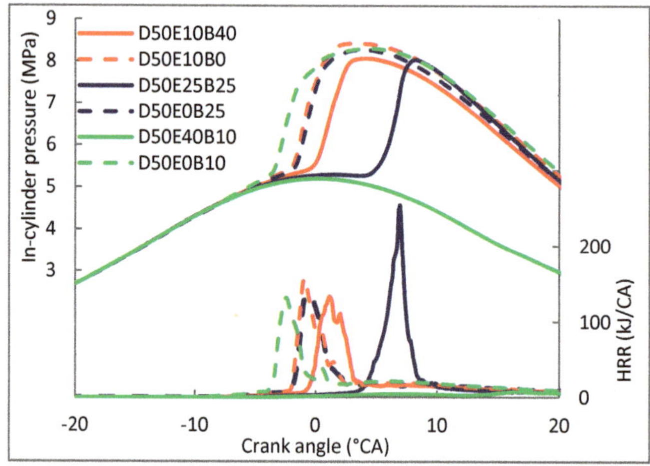

Figure 8. Pressure and HRR graphs of the blends with and without the addition of ethanol and PME.

3.3. Ignitability of Diesel–Ethanol–PME Blends with Injection Parameter Modification

Injection parameters, such as injection duration, injected mass, and injection timing, were modified due to the ignition problem of the D50E25B25 blend at 2400 RPM and the D50E40B10 blend at 1600 RPM and 2400 RPM. Figure 9 shows the graph of in-cylinder pressure and HRR against the crank angle of the D50E25B25 blend by using different injection durations. Figures 10 and 11 show the graph of pressure and HRR for the D50E25B25 and D50E40B10 blends by using different injected masses and injection timings at 1600 RPM. The result shows that modification of the injection duration has no significant change on pressure and HRR. The study from Adnan et al. (2012) also compared the injection duration of diesel but with hydrogen. They found the same trend: that no significant change is observed on the pressure and HRR for different injection durations [17]. Furthermore, reducing the injected fuel mass also reduces fuel ignitability. Consuming a great amount of fuel to the engine for high ethanol blends does not give positive results. Therefore, the injection duration and fuel mass should be maintained with 16 °CA and 19 mg of fuel, considering that increasing the fuel amount will not meet the EEV concept.

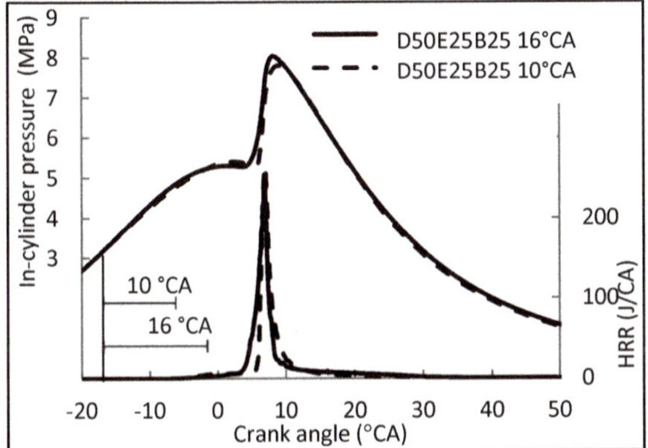

Figure 9. In-cylinder pressure and HRR graph of D50E25B25 blend against crank angle by using different injection durations.

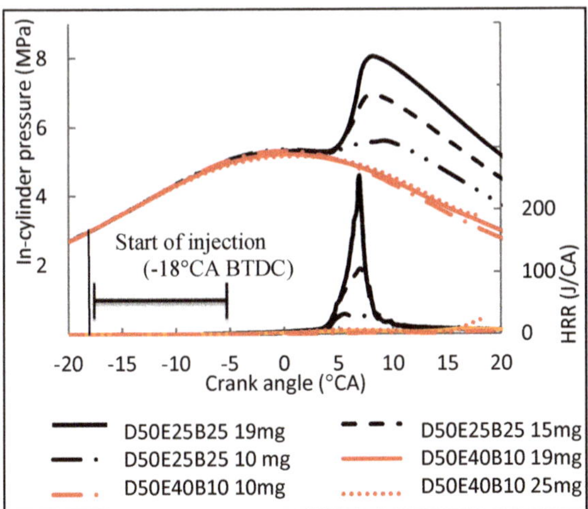

Figure 10. In-cylinder pressure and HRR graph of D50E25B25 and D50E40B10 blends against crank angle by using different injected mass.

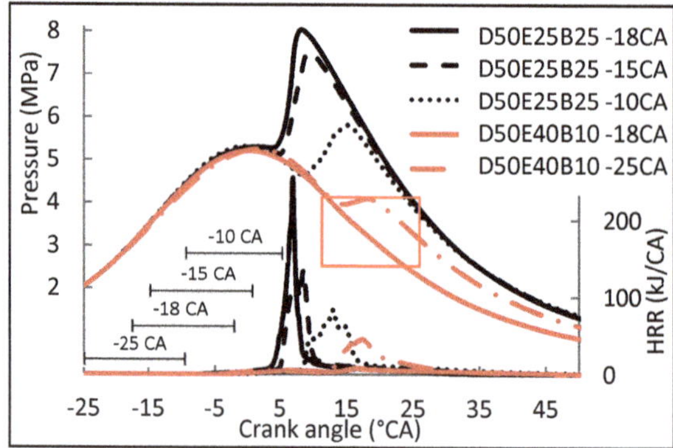

Figure 11. In-cylinder pressure and HRR graph of D50E25B25 and D50E40B10 blends against a crank angle by using different injection times.

Figure 11 illustrates the pressure and HRR graph of D50E25B25 and D50E40B10 blends running at 1600 RPM with different injection times. The graph shows that advancing the injection time from −18 °CA BTDC to −25 °CA BTDC for D50E40B10 blends exhibited positive results in solving the ignition problem. However, the ignition of this fuel is very weak, as marked in the red box in Figure 11. Meanwhile, retarding the injection timing of D50E25B25 reduced the pressure and HRR. This phenomenon proves that only advancing the injection timing can improve the blend ignitability, especially for the blends with high ethanol contents. The reason is that the advance injection can allow the fuel to be mixed early with air and increase the temperature during the premixed phase and ignite the fuel [18]. Compared with the previous study, Mendes Guedes et al. (2018) also found that an advanced injection timing increases the pressure and HRR during combustion. In addition, the advanced ignition has a short ignition duration that is able to reduce the carbon emission problems. However, high temperature may lead to NO_x emission [8].

However, advancing the injection timing at the early time before the top dead center (TDC) may lead to another problem. Although advancing the injection timing improves the fuel ignitability, injecting the fuel too early can lead to the incomplete fuel combustion. Thus, the situation will increase the hydrocarbon, carbon monoxide, and soot emission. As observed in the emission graphs in Figure 12, emissions of CO, HC, and C_2 are high with the advancement of the injection timing for the D50E25B25 blend. A few studies found that advancing the injection timing reduced the carbon emissions, while considering that the ignition started early. However, the transition of liquid fuel turning into gas is very slow, especially at high engine speeds, and leads to incomplete combustion due to the ethanol presence [19]. Figure 12 shows the emission graphs for D50E40B10 blends with two different injection timings. Injection at −18 °CA BTDC does not have any ignition due to the high ethanol presence. Advancing the injection timing with −25 °CA BTDC was able to ignite the fuel, as shown in Figure 8.

As illustrated in the graph, CO_2, CO, and NO_x emissions still occur, because the reaction between the hydrocarbon and air occurs in the combustion chamber. However, the reaction itself is insufficiently high to release additional energy for igniting the fuel, which results in a low temperature. Therefore, the ignition hardly occurs. The graph shows that the CO, CO_2, C_2, and NO_x emissions are too low for the D50E40B10 blend without modification. The hydrocarbon (HC) emission is higher than that of the other modifications, because C and H do not completely react with other elements upon injection. No NO_x emission is observed because the temperature in the cylinder is insufficiently high to trigger a reaction between nitrogen and oxygen to form NO_x.

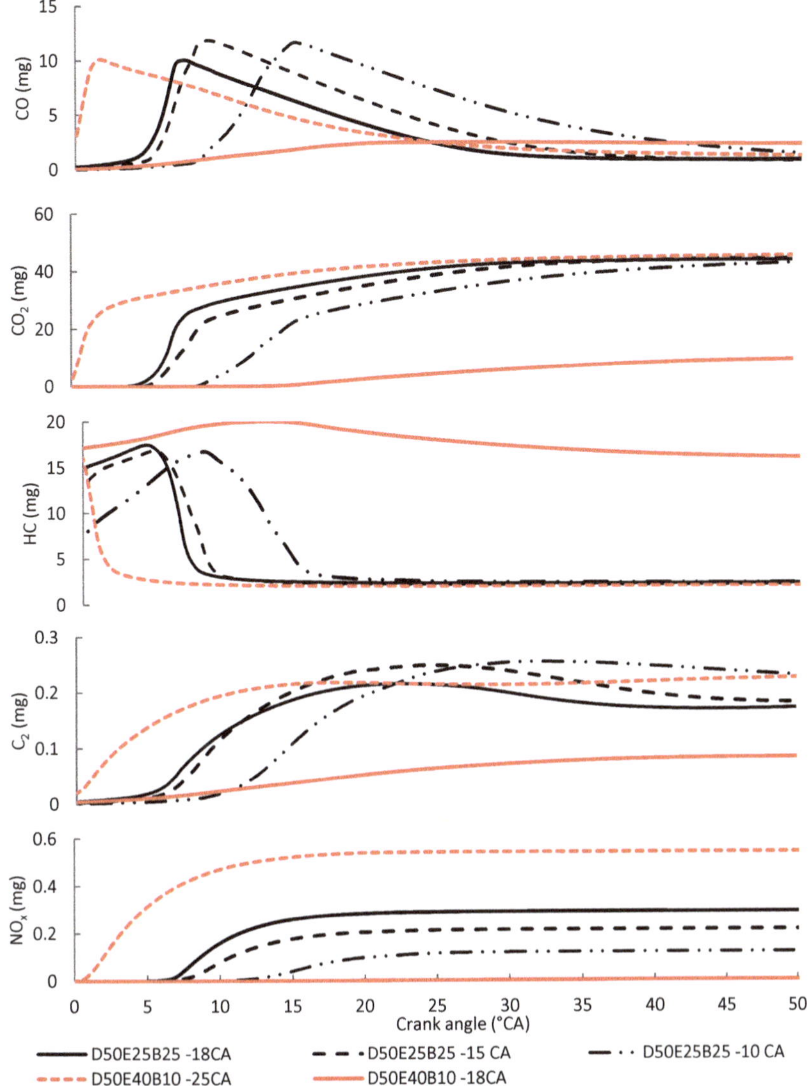

Figure 12. Emission graphs of D50E25B25 blends running by using different injection timing at 1600 RPM.

However, advancing the injection timing too early before TDC increased the emission of CO, CO_2, C_2, and NO_x. The reason is that the fuel is injected too early before the temperature in the cylinder reaches fuel auto-ignition. In addition, high ethanol blends have a very high auto-ignition temperature. Therefore, the fuel mixture in the combustion chamber is hard to ignite. This condition will lead to a long ignition delay. Therefore, the unburned fuel in the cylinder increases and leads to incomplete combustion. The trend of emissions between D50E25B25 and D50E40B10 blends is significantly different due to the ignitability of blends at a high ethanol ratio. Moreover, advancing the injection timing earlier than −25 °CA BTDC increases the temperature before the TDC and increases the NO_x emission.

3.4. Ignitability of Blends with Compression Ratio and Intake Temperature Modification

Diesel–ethanol–PME blends with high ethanol percentage faced the ignitability problem. Due to this reason, the compression ratio and intake temperature of the engine is modified.

3.4.1. Compression Ratio

The D50E40B10 blend has significantly shown that the ignitability of this blend is very weak due to the low cetane number and delay of ignition. Therefore, compression ratio modification has been selected to improve the ignitability. Figure 13 shows the graph of in-cylinder pressure and HRR of D50E40B10 blends operated at two different engine speeds of 1600 RPM and 2400 RPM by using various cylinder compression ratios of 18, 20, and 24. The standard compression ratio of 18 of the engine was not able to combust the fuel due to the low amount of air at high engine speed operation. Therefore, a modification of cylinder compression ratio was proved to be able to increase the in-cylinder pressure. Meanwhile, HRR decreases with the compression ratio, which is probably due to the decrease of specific energy caused by the low heating value of ethanol [20,21]. The reason is that the large cylinder volume is able to increase the temperature and pressure in the cylinder. High temperature and pressure in the cylinder help ignite the blends with high ethanol contents, considering that ethanol has a high auto-ignition temperature [19].

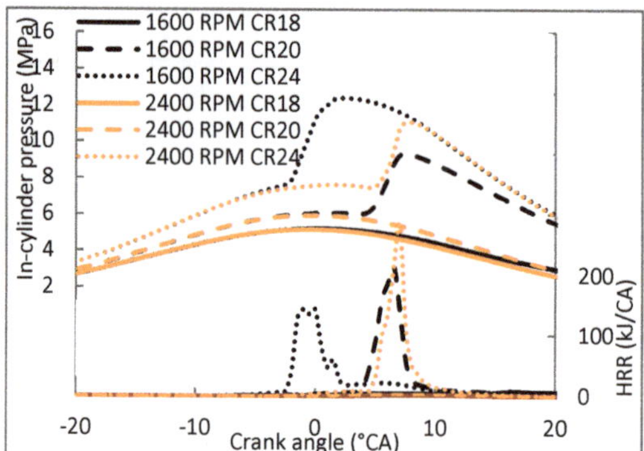

Figure 13. In-cylinder pressure and HRR graph of the D50E40B10 blend against a crank angle by using different engine compression ratios at different engine speeds.

The high engine compression ratio is also able to solve the problems of the D50E40B10 blends at high engine speed. In addition, the minimum CR allowed for the D50E40B10 blend operated at 1600 RPM is CR20, and at 2400 RPM is CR 24. A high compression ratio should be applied on high engine speed because the reaction rate is very fast and rapidly increases the pressure [11]. However, as the CR increases, the knock and misfiring phenomenon also occur, as observed on the HRR curve. Knock occurs when the unburned fuel in the cylinder is ignited by itself before the piston reached the TDC and forced the piston to go down. Therefore, increasing the compression ratio may solve the ignition problem of the blends with high ethanol contents. However, this phenomenon leads to unwanted phenomena, such as knocking and misfiring.

In addition, increasing the compression ratio can lead to another pollution problem. Figure 14 shows the emission graphs of D50E40B10 blends running at 1600 RPM at different compression ratios. The results show that increasing the compression ratio increases the emission of CO_2 and NO_x emissions and reduces the CO emissions during the expansion stroke. The graphs also show that the emissions of the blends running at a normal compression ratio are too low for CO, CO_2, C_2, and NO_x emissions. This situation proves that the fuel ignition at this condition is too weak, and the reaction of hydrocarbon with air is too low. The heat release from the reaction is insufficiently high to increase the temperature for ignition.

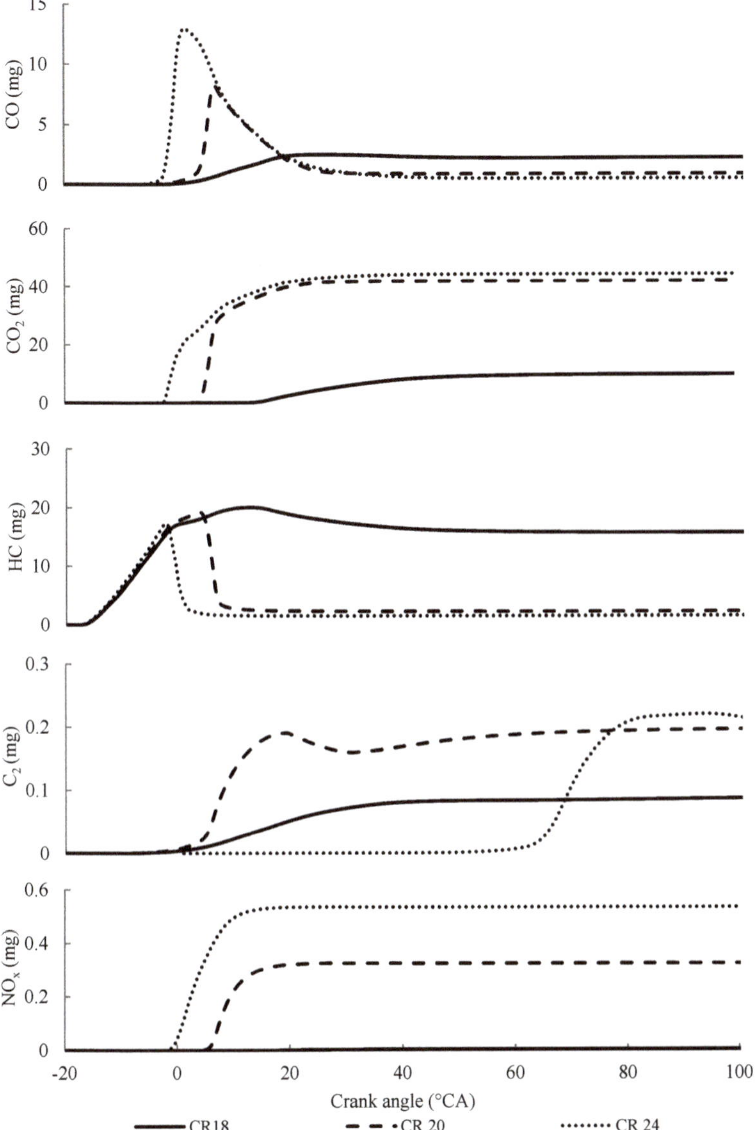

Figure 14. Emission graphs of the D50E40B10 blend at different compression ratios running at 1600 RPM.

The emission difference of CO between CR 20 and CR 24 was really significant, with a difference of approximately 55%. Meanwhile, the percentage difference of CO_2 emissions between CR 20 and CR 24 was only 5%. However, the emissions of hydrocarbon and soot for CR 24 was found to be lower than those for CR 20, because increasing the compression ratio shortened the ignition delay, and the temperature in the cylinder is high enough to combust the fuel and reduce the emissions of HC and CO. However, high temperature in the cylinder also increases the NO_x emissions [22]. The previous study also found similar changes in emissions, wherein the high compression ratio reduces the emissions of carbon but increases the NO_x emission [23,24]. From the comparisons of the blends on the combustion characteristics and emissions, CR 20 is considered the best for the D50E40B10 blend at 1600 RPM, and CR 24 works well with the blend at 2400 RPM.

3.4.2. Intake Temperature

Another method for improving the ignitability of high ethanol content blends is by increasing the in-cylinder temperature that can be applied by increasing the temperature or air at the intake T_i. Figure 15 shows the in-cylinder pressure and HRR of D50E40B10 blends operated at 2400 RPM by applying different T_i. The results show that increasing the air temperature increases the temperature in the cylinder and is able to improve the fuel ignitability. Increased T_i ignites the fuel early, has wide combustion, and produces a great amount of heat power. However, early ignition at high engine speed can cause misfiring. In addition, burning the diesel–ethanol–PME blends at 400 K has lower in-cylinder pressure compared with that at 350 K. Studies from Akashah et al. (2015) also found that a high ambient temperature has the weakest energy and low in-cylinder pressure for ethanol combustion. This phenomenon proves that the high ethanol presence has this effect whereby it causes the in-cylinder pressure at the highest ambient temperature to decrease [25]. Meanwhile, using 350 K of intake temperature increases the pressure rapidly and exhibits the highest peak pressure. Therefore, 350 K is the most suitable intake temperature needed to ignite the D50E40B10 blend when operating at high engine speed.

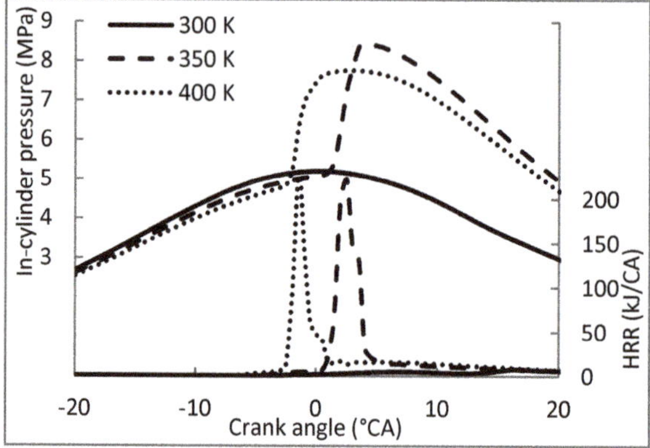

Figure 15. In-cylinder pressure and HRR graph of the D50E40B10 blend against a crank angle at 2400 RPM by using a different intake temperature, T_i.

Normally, to increase the intake temperature in the real engine, an intake heating system or exhaust gas recirculation (EGR) system can be applied to the engine with the temperature rise of approximately 350 K [7]. However, implementing an EGR should be reconsidered, because it can reduce engine efficiency. However, this process is very useful in reducing NO_x emissions. Therefore, applying the air-heating intake system is another solution in the future to control the temperature of air intake with the minimum temperature of 350 K if more than 40% ethanol in the blends is used. A patent from Linkenhoger (2005) designed an air-heating intake system to heat up the air and vaporize the fuel to improve the engine efficiency, improve the fuel economy, increase the engine power, reduce the carbon emission, and increase the engine life [26].

Figure 16 shows the emission results of the D50E40B10 blend at different intake temperatures running at 1600 RPM. The study found that increasing the intake temperature emits increased CO and CO_2 emissions. The NO_x emissions also increased with the increase of the temperature in the cylinder and contributed to NO_x reactions. The soot formation was also increased with the increment of intake temperature due to the shortened ignition delay [27].

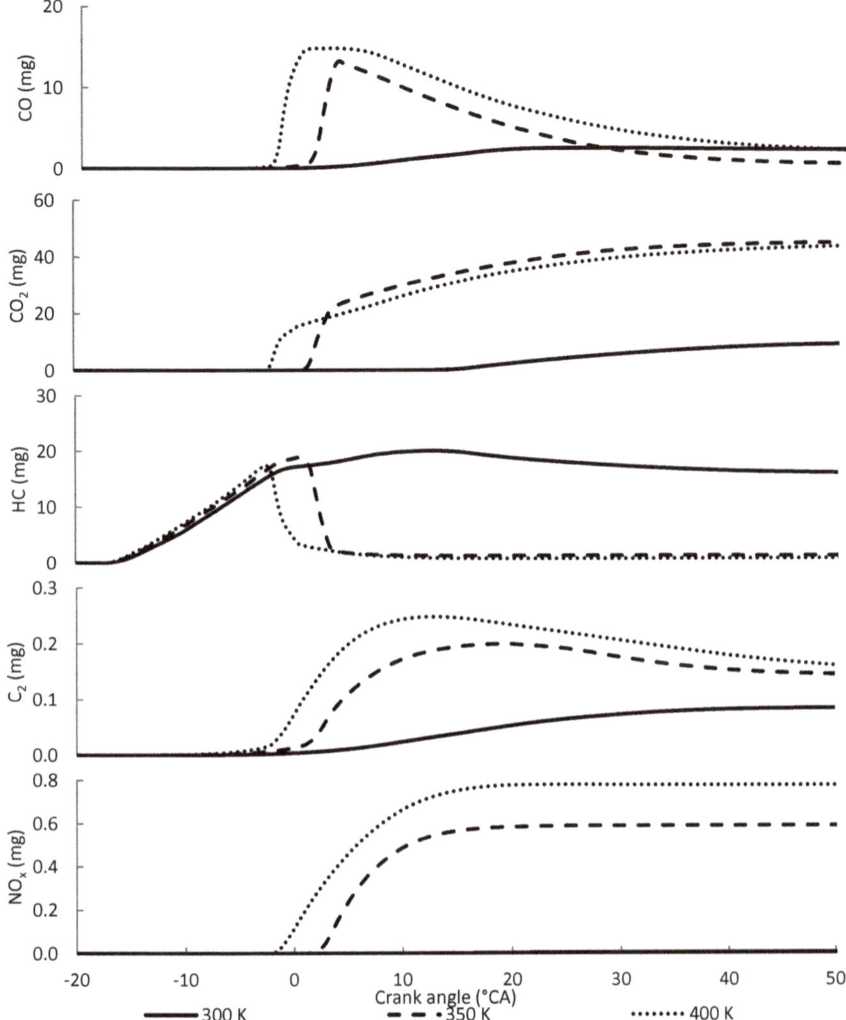

Figure 16. Emission graphs of D50E40B10 blend running at 1600 RPM at different intake temperature.

However, the hydrocarbon emissions decreased with the increased intake temperature, because the increased temperature has a good tendency to burn all the fuels that contain high ethanol contents due to the high auto-ignition temperature of hydrogen and long ignition delay. From the combustion characteristics and emission analysis, T_i = 350 K is the most suitable intake temperature to ignite the D50E40B10 blends with low emission and low effect of abnormal combustion.

3.5. Effect of Modification on Combustion Characteristics and Emissions

Here, we compare all the injection modifications, compression ratios, and intake temperatures in order to determine the best combustion efficiency and emissions. Figure 17 shows the pressure and HRR graph of the D50E40B10 blends running at 1600 RPM. The results show that the compression ratio modification has a dominant effect on in-cylinder pressure and HRR. Besides, injection timing modification also gives a very small effect on the combustion, since the ignition of the blend at this condition is very weak. However, the heat rate released from the combustion for ambient temperature modification and compression ratio modification shows no significant difference except the ignition delay. Operating the D50E40B10 blend at an ambient temperature of 350 K has a shorter ignition delay than that operated at a compression ratio of 20.

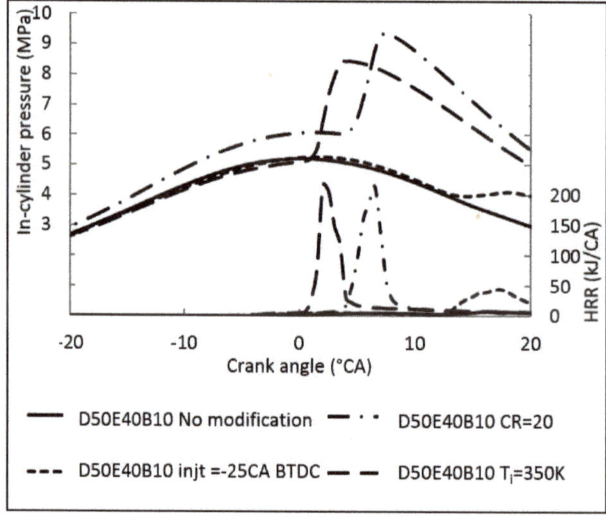

Figure 17. Pressure and HRR graphs of the D50E40B10 blend running at 1600 RPM at different modification comparisons.

Figure 18 shows the comparison of emissions between the injection timing advancement, compression ratio, and intake temperature modification. Although the modification of engine intake temperature and compression ratio release has high HRR and pressure, the emission of this modification is severed compared with the other modifications. The graphs show that increasing the engine intake temperature increases the CO and NO_x emissions. The reason is that the high temperature from the combustion and high oxygen contents from the blends leads to the NO_x formation. Therefore, intake temperature modification was chosen as the best engine modification to solve the blend ignition problem, improve the combustion efficiency, shorten the ignition delay, and reduce the carbon emissions.

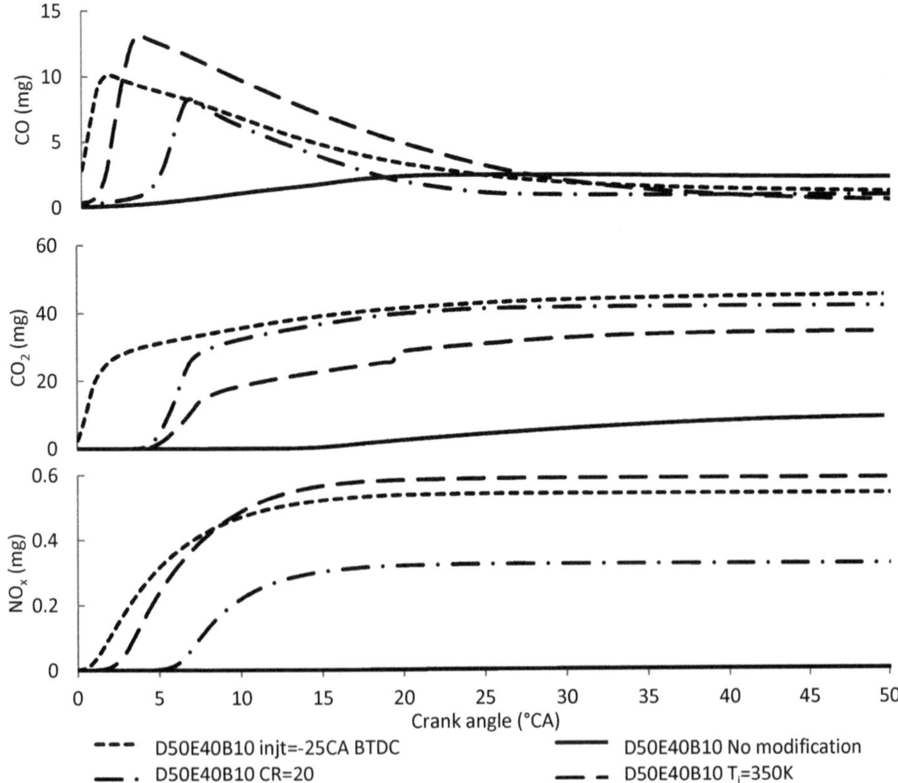

Figure 18. Comparison of CO, CO$_2$, and NO$_x$ emissions of D50E40B10 blends at the best injection modification, compression ratio, and intake temperature.

4. Conclusions

Blending diesel, ethanol, and PME in a diesel engine with different compositions results in a different ignitability behavior. Ethanol with a low heating value and low cetane number reduces the ignition ability due to the long ignition delay. A simulation study has been conducted for different diesel–ethanol–PME compositions. From the simulation, the following conclusions are drawn:

1. High ethanol contents reduce the ignitability of the blends, and the maximum speed achieved by D50E25B25 only at 1600 RPM and D50E40B10 was not ignited at both engine speeds.
2. Changing the injection duration and increasing the injected mass was not helpful enough to ignite the D50E40B10 blends at 1600 RPM and 2400 RPM. Meanwhile, advancing the injection timing at approximately −25 °CA increased the in-cylinder pressure and its HRR. Advancing the injection timing successfully ignites the fuel blends. However, the combustion was so weak, and the ignition delay was too long. Moreover, advancing the injection earlier than −25 °CA increases the NO$_x$ emission.
3. Therefore, modification of the compression ratio and ambient temperature produces good results in improving the ignitability of D50E40B10 blends with minimum CR20 at 1600 RPM and CR24 at 2400 RPM. Increasing the ambient temperature by implementing an air-heating or EGR system to the engine can improve the ignitability. The most suitable intake temperature for the D50E40B10 blend is 350 K, producing high peak pressure and HRR.

4. From the comparison between all the modifications of injection, compression ratio, and intake temperature, the most suitable modification for diesel–ethanol–PME blends is increasing the intake temperature. This procedure produces increased in-cylinder pressure and HRR, but has a short ignition delay. Although the NO_x emission of intake temperature modification is slightly high, the abnormal combustion phenomena can be avoided compared with the compression ratio modification.

In the future, the heating system should be mounted to the intake to increase the air intake temperature in the real engine application. Further studies on the diesel–ethanol–PME blend ignitability on the engine performance, combustion characteristics, and emissions should be investigated and another approach to reduce the emission focusing on NO_x emission should be applied.

Author Contributions: Conceptualization and Methodology N.M.T and M.R.A.M; Software, Validation, Analysis and Writing—Original Draft Preparation N.M.T; Writing-Review and Editing M.R.A.M; Supervision M.R.A.M and W.M.F.W.M.

Funding: This research was funded by Universiti Kebangsaan Malaysia [Grant number: GUP-2018-099].

Acknowledgments: The authors thank the Ministry of Higher Education Malaysia and Universiti Kebangsaan Malaysia for supporting and funding this research.

Conflicts of Interest: The authors declare no conflict of interest.

References

1. Monirula, I.M.; Masjukia, H.H.; Kalama, M.A.; Mosarofa, M.H.; Zulkiflia, N.W.M.; Teohab, Y.H.; How, H.G. Assessment of performance, emission and combustion characteristics of palm, jatropha and calophyllum inophyllum biodiesel blends. *Fuel* **2016**, *181*, 985–995. [CrossRef]
2. Ganjehkaviri, A.; Jaafar, M.N.M.; Hosseini, S.E.; Musthafa, A.B. Performance evaluation of palm oil-based biodiesel combustion in an oil burner. *Energies* **2016**, *9*, 97. [CrossRef]
3. Taib, N.M.; Mansor, M.R.A.; Mahmood, W.M.F.W.; Abdullah, N.R. Simulation Study of Combustion Characteristics of Diesel- Ethanol-Palm Oil Methyl Ester Blends in Diesel Engine. *J. Adv. Res. Fluid Mech. Therm. Sci.* **2018**, *44*, 149–156.
4. Pedrozo, V.B.; May, I.; Nora, M.D.; Cairns, A.; Zhao, H. Experimental analysis of ethanol dual-fuel combustion in a heavy-duty diesel engine: An optimisation at low load. *Appl. Energy* **2016**, *165*, 166–182. [CrossRef]
5. Yu, S.; Gao, T.; Wang, M.; Li, L.; Zheng, M. Ignition control for liquid dual-fuel combustion in compression ignition engines. *Fuel* **2017**, *197*, 583–595. [CrossRef]
6. Dong, S.; Cheng, X.; Ou, B.; Liu, T.; Wang, Z. Experimental and numerical investigations on the cyclic variability of an ethanol/diesel dual-fuel engine. *Fuel* **2016**, *186*, 665–673. [CrossRef]
7. Kumar, B.R.; Saravanan, S. Use of higher alcohol biofuels in diesel engines: A review. *Renew. Sustain. Energy Rev.* **2016**, *60*, 84–115. [CrossRef]
8. Guedes, A.D.M.; Braga, S.L.; Pradelle, F. Performance and combustion characteristics of a compression ignition engine running on diesel-biodiesel-ethanol (DBE) blends—Part 2: Optimization of injection timing. *Fuel* **2018**, *225*, 174–183. [CrossRef]
9. Sharma, A.; Murugan, S. Combustion, performance and emission characteristics of a di diesel engine fuelled with non-petroleum fuel: A study on the role of fuel injection timing. *J. Energy Inst.* **2015**, *88*, 364–375. [CrossRef]
10. Rahim, N.A.; Jaafar, M.N.M.; Sapee, S.; Elraheem, H.F. Effect on particulate and gas emissions by combusting biodiesel blend fuels made from different plant oil feedstocks in a liquid fuel burner. *Energies* **2016**, *9*, 659. [CrossRef]
11. Gnanamoorthi, V.; Devaradjane, G. Effect of compression ratio on the performance, combustion and emission of di diesel engine fueled with ethanol e Diesel blend. *J. Energy Inst.* **2015**, *88*, 19–26. [CrossRef]
12. Shahir, S.A.; Masjuki, H.H.; Kalam, M.A.; Imran, A.; Fattah, I.M.R.; Sanjid, A. Feasibility of diesel-biodiesel-ethanol/bioethanol blend as existing CI engine fuel: An assessment of properties, material compatibility, safety and combustion. *Renew. Sustain. Energy Rev.* **2014**, *32*, 379–395. [CrossRef]

13. Kuszewski, H. Experimental investigation of the effect of ambient gas temperature on the autoignition properties of ethanol–diesel fuel blends. *Fuel* **2018**, *214*, 26–38. [CrossRef]
14. Ibrahim, F.; Mahmood, W.M.F.W.; Abdullah, S.; Mansor, M.R.A. A Review of Soot Particle Measurement in Lubricating Oil. *Def. S T Tech. Bull.* **2015**, *8*, 141–152.
15. Ibrahim, F.; Mahmoof, W.M.F.W.; Abdullah, S.; Mansor, M.R.A. Comparison of Simple and Detailed Soot Models in the Study of Soot Formation in a Compression Ignition Diesel Engine. *SAE Int.* **2017**. [CrossRef]
16. Pradelle, F.; Braga, S.L.; Fonseca de Aguiar Martins, A.R.; Turkovics, F.; Nohra Chaar Pradelle, R. Performance and combustion characteristics of a compression ignition engine running on diesel-biodiesel-ethanol (DBE) blends—Potential as diesel fuel substitute on an Euro III engine. *Renew. Energy* **2019**, *136*, 586–598. [CrossRef]
17. Adnan, R.; Masjuki, H.H.; Mahlia, T.M.I. Performance and emission analysis of hydrogen fueled compression ignition engine with variable water injection timing. *Energy* **2012**, *43*, 416–426. [CrossRef]
18. Raeie, N.; Emami, S.; Sadaghiyani, O.K. Effects of injection timing, before and after top dead center on the propulsion and power in a diesel engine. *Propuls. Power Res.* **2014**, *3*, 59–67. [CrossRef]
19. Labeckas, G.; Slavinskas, S.; Mažeika, M. The effect of ethanol-diesel-biodiesel blends on combustion, performance and emissions of a direct injection diesel engine. *Energy Convers. Manag.* **2014**, *79*, 698–720. [CrossRef]
20. Hariram, V.; Shangar, R.V. Influence of compression ratio on combustion and performance characteristics of direct injection compression ignition engine. *Alex. Eng. J.* **2015**, *54*, 807–814. [CrossRef]
21. Muralidharan, K.; Vasudevan, D. Performance, emission and combustion characteristics of a variable compression ratio engine using methyl esters of waste cooking oil and diesel blends. *Appl. Energy* **2011**, *88*, 3959–3968. [CrossRef]
22. Hoseini, S.S.; Najafi, G.; Ghobadian, B.; Mamat, R.; CheSidik, N.A.; Azmi, W.H. The effect of combustion management on diesel engine emissions fueled with biodiesel-diesel blends. *Renew. Sustain. Energy Rev.* **2017**, *73*, 307–331. [CrossRef]
23. Bora, B.J.; Saha, U.K.; Chatterjee, S.; Veer, V. Effect of compression ratio on performance, combustion and emission characteristics of a dual fuel diesel engine run on raw biogas. *Energy Convers. Manag.* **2014**, *87*, 1000–1009. [CrossRef]
24. Amarnath, H.K.; Prabhakaran, P.; Bhat, S.A.; Paatil, R. A comparative analysis of thermal performance and emission characteristics of methyl esters of karanja and jatropha oils based on a variable compression ratio diesel engine. *Int. J. Green Energy* **2014**, *11*, 675–694. [CrossRef]
25. Akasyah, M.K.; Mamat, R.; Abdullah, A.; Aziz, A.; Yassin, H.M. Effect of ambient temperature on diesel-engine combustion characteristics operating with alcohol fuel. *Int. J. Automot. Mech. Eng.* **2015**, *11*, 2373–2382. [CrossRef]
26. Linkenhoger, T. Intake Air Pre-Heated Assembly for Automotive Gasoline Engines. U.S. Patent 20050257781 A1, 24 November 2005.
27. Pan, W.; Yao, C.; Han, G.; Wei, H.; Wang, Q. The impact of intake air temperature on performance and exhaust emissions of a diesel methanol dual fuel engine. *Fuel* **2015**, *162*, 101–110. [CrossRef]

Article

Product Characteristics of Torrefied Wood Sawdust in Normal and Vacuum Environments

Yi-Kai Chih [1], Wei-Hsin Chen [1,2,3,*], Hwai Chyuan Ong [4] and Pau Loke Show [5]

[1] Department of Aeronautics and Astronautics, National Cheng Kung University, Tainan 701, Taiwan; chihyikai@gmail.com
[2] Department of Mechanical Engineering, National Chin-Yi University of Technology, Taichung 411, Taiwan
[3] Research Center for Energy Technology and Strategy, National Cheng Kung University, Tainan 701, Taiwan
[4] Department of Mechanical Engineering, Faculty of Engineering, University of Malaya, Kuala Lumpur 50603, Malaysia; onghc@um.edu.my
[5] Department of Chemical and Environmental Engineering, Faculty of Engineering, University of Nottingham Malaysia Campus, Jalan, Broga, Semenyih 43500, Selangor Darul Ehsan, Malaysia; PauLoke.Show@nottingham.edu.my
* Correspondence: weihsinchen@gmail.com or chenwh@mail.ncku.edu.tw; Tel.: +886-6-2004456

Received: 1 September 2019; Accepted: 4 October 2019; Published: 11 October 2019

Abstract: To investigate the efficacy of torrefaction in a vacuum environment, wood sawdust was torrefied at various temperatures (200–300 °C) in different atmospheres (nitrogen and vacuum) with different residence times (30 and 60 min). It was found that the amount of biochar reduced at the same rate—regardless of atmosphere type—throughout the torrefaction process. In terms of energy density, the vacuum system produced biochar with better higher heating value (HHV, MJ/kg) than the nitrogen system below 250 °C. This was the case because the moisture and the high volatility compounds such as aldehydes diffused more easily in a vacuum. Over 250 °C, however, a greater amount of low volatility compounds evaded from the vacuum system, resulting in lower higher heating value in the biochar. Despite the mixed results with the solid products, the vacuum system increased the higher heating value of its liquid products more significantly than did the nitrogen system regardless of torrefaction temperature. It was found that 23% of the total energy output came from the liquid products in the vacuum system; the corresponding ratio was 19% in the nitrogen system. With liquid products contributing to a larger share of the total energy output, the vacuum system outperformed the nitrogen system in terms of energy density.

Keywords: torrefaction; vacuum; biomass pretreatment; bioenergy; energy yield; biochar

1. Introduction

Anthropogenic carbon dioxide emissions have a notable impact on both human society and the ecosystem [1–3]. As part of an ongoing effort to mitigate carbon dioxide emissions worldwide, bioenergy has achieved wide applicability [4,5]. Compared to other sources of renewable energy, bioenergy has major advantages in storage and applications because biomass can be stored and extracted in different phases [6]. In the solid phase, biomass can be processed into powder, pellets, blocks, and briquettes through thermal decomposition [7]. In the liquid phase, on the other hand, it can be processed into biodiesel, bioethanol, and bio-oil through thermal decomposition or liquefaction [8–10]. All of these products have wide industrial applicability [11–13].

Torrefaction is a type of pyrolysis, where biomass is thermally degraded in an inert system under atmospheric conditions at temperatures of 200–300 °C. The final product is called torrefied biomass (i.e., biochar) [14]. The process can be categorized into light, mild, and severe types according to the temperature ranges of 200–235 °C, 235–275 °C, and 275–300 °C, respectively [13]. It is also used as a method of pretreatment aimed at improving the physical, chemical, and biochemical characteristics of raw biomass [15]. Through torrefaction, raw biomass is upgraded to biochar, which exhibits better hydrophobicity, lower moisture content, better grinding property, and better higher heating value (HHV) [16]. Biochars are also easier to handle and store and more suitable for combustion and gasification than raw biomass. According to Santos et al. [17], biochars produced at high torrefaction temperature have an energy density comparable to low-rank coal, which is better than raw biomass. These biochars have less toxicity associated with emissions from combustion, more homogeneous combustion behavior, higher grinding ability, and lower hygroscopicity [18]. The said characteristics make biochar an ideal alternative to coal and solid fuels [19,20]. Thus, biochars have been dubbed as "solid fuels" [21,22].

Aside from solid biochars, torrefaction also generates liquid and gaseous products [23,24]. During torrefaction, the release and thermal decomposition of high volatility compounds from biomass produce non-condensable gases such as CO, CO_2, and H_2 in addition to small amounts of CH_4 [25,26], toluene, benzene, and low molecular weight hydrocarbons [27]. Depending on the torrefaction temperature, brown or black-colored liquid products are generated, consisting of condensable components such as water, acetic acid, alcohols, aldehydes, and ketones. Elliott's [28] analysis of the liquid products from gas chromatography-mass spectrometry (GC–MS) suggests that the main components in the liquid are monoaromatics.

The vacuum—also known as "low pressure"—technology has been used for the manufacturing of semiconductors, flat-panel displays, solar power panels, and scientific instrumentation. In the food and pharmaceutical industries, several manufacturing processes require gas pressure well below the atmospheric pressure. However, such a technology has yet to be applied to torrefaction. Only a few studies have tackled torrefaction in a vacuum. Lin et al. [29] investigated the thermal degradation of various wood species under vacuum and reported how different lignocellulosic components behave under different temperature conditions (200–230 °C, 0.2 °C min^{-1}, 200 hPa). Garcia-Perez et al. [30] used softwood and hardwood species in a vacuum pyrolysis pilot plant and characterized the resulting products. Murwanayashka et al. [31] carried out vacuum pyrolysis (8–15 kpa, 500 °C, 12 °C min^{-1}) of birch derived biomass and observed the evolution of phenols during the process. They reported that vacuum minimized the extent of secondary reactions during pyrolysis (1.3 kPa, 450 °C), facilitating a more uniform product. Pakdel and Roy [32] used vacuum pyrolysis to extract steroids from common lignocellulosic materials during thermal treatment, facilitating an easier extraction and further modification of steroids.

Although torrefaction has been performed conventionally in a nitrogen environment as a medium of convection for torrefied compounds, this study sets out to explore the potential advantages of torrefaction in a vacuum environment. One of the objectives is to assess whether a vacuum environment is able to increase the energy density of torrefied materials through diffusion rather than atmosphere-based convection.

2. Materials and Methods

2.1. Parameters for Torrefaction with a Vacuum and a Nitrogen System

The biomass used in this study was rubber wood obtained from Chinese Petroleum Corporation (CPC) in Taiwan. Every batch of the wood sawdust used in our experiments was ground by a crusher and shredded in a 100 mesh. After shredding and sieving, the raw material was dried in a convection oven at 105 °C for 24 h to eliminate the moisture and provide an experimental basis. This drying process was aimed at minimizing experimental error. The dried material was then kept in air-tight plastic sample boxes to prevent moisture absorption before samples were taken and used in experiments on the same day. A sample weight (30 g) of wood sawdust was placed in a sample glass tube mounted to the reactor. The furnace was preheated to the torrefaction temperature (200–300 °C), then the sample glass tube was rapidly pushed into the reactor using a glass rod [33]. The wood sawdust was torrefied at 200, 225, 250, 275, and 300 °C for periods of 0.5 or 1 h in the nitrogen (N_2) and the vacuum systems, respectively, as shown in Figure A1 (in Appendix A). Nitrogen gas was blown through the reactor at a flow rate of 0.1 L/min, as shown in Figure 1a. The vacuum pressure was controlled at −590 torrs (0.22 atm) in vacuum systems, as shown in Figure 1b. The temperature of the sample in the reactor was recorded at the end of each torrefaction period. Then the furnace was opened to cool the reactor for further sample analysis.

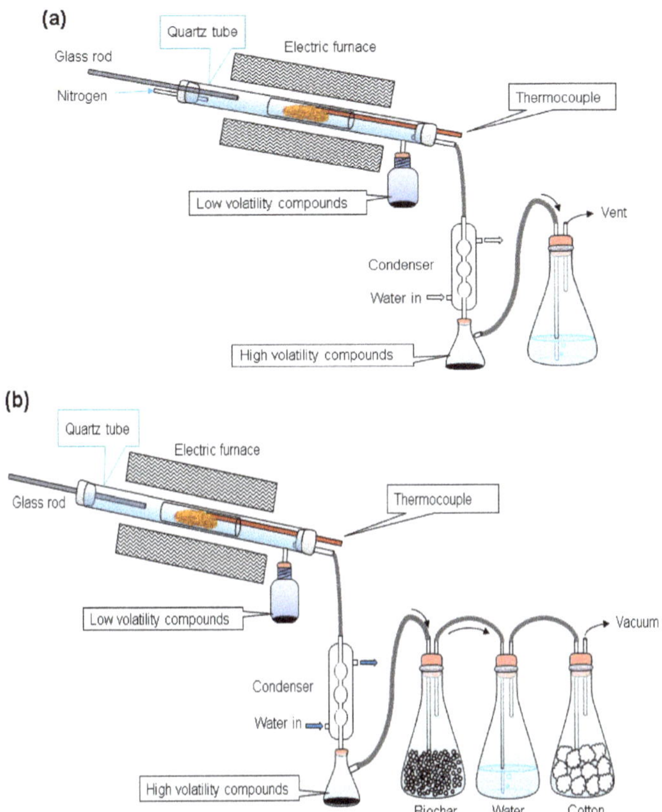

Figure 1. The experimental setups for the (**a**) nitrogen and the (**b**) vacuum systems.

During torrefaction, liquid compounds were released from the samples. The lower volatility compounds moved into a glass bottle, while the higher volatility compounds into an Erlenmeyer flask. Non-condensable gases were released into an exhaust ventilation system. The product left in the sample bottles after torrefaction was collected and stored at −4 °C for further analyses.

2.2. Sample Analysis

In this study, wood sawdust was torrefied at various temperatures (200–300 °C), under atmospheric condition (~1 atm) and vacuum for two residence times (30 and 60 min). The resulting products were characterized by proximate and elemental analysis, HHV and GC/MS methodologies.

The proximate, elemental, fiber and calorific analyses of the biomass materials were performed to figure out their basic properties [34]. The weight percentages of C, H, and N for biochar and for the low and the high volatility compounds were measured by an elemental analyzer. The proximate analysis of biomass was carried out in accordance with standard procedures of American Society for Testing and Materials (ASTM).

The weight percentages of C, H, and N in the biomass were measured by an elemental analyzer (PerkinElmer 2400 Series II CHNS/O Elemental Analyzer). The weight percentage of O was obtained by difference, that is, O (wt%) = 100 – C–H–N-Ash. The hemicellulose, cellulose, and lignin contents of biomass were analyzed following the fiber analysis in a previous study. The HHVs of the samples were measured by a bomb calorimeter (IKAC5000). The definitions of product yield and energy yield were used as follows:

$$\text{Product yield } (\%) = \frac{\text{Weight}_{\text{product}}}{\text{Weight}_{\text{raw}}} \times 100 \tag{1}$$

$$\text{Energy yield } (\%) = \frac{\text{Weight}_{\text{product}} \times \text{HHV}_{\text{product}}}{\text{Weight}_{\text{raw}} \times \text{HHV}_{\text{raw}}} \times 100 \tag{2}$$

$$\text{Gas energy yield } (\%) = 100 - \text{solid energy yield-liquid energy yield } (\%) \tag{3}$$

The subscripts "raw" and "product" represent raw and torrefied biomass, respectively.

The condensed low and high volatility compounds were measured using a gas chromatography-mass spectrophotometer (GC 6890N, MS 5973N, Autosample 7683B, GC–MS). GC column and solvent are GSBP 0125-3001HT (length 30 m, inner diameter 0.25 mm, film thickness 0.1 um) and hexane/acetone. The oven temperature was set as follows: (1) 50 °C for 2 min; (2) from 50 to 270 °C at a heating rate of 9 °C min^{-1}; and (3) the capillary column was maintained at 270 °C for about 5 min; (4) from 270 to 320 °C at a heating rate of 13 °C min^{-1}; and (5) the capillary column was maintained at 320 °C for about 2.5 min; (6) from 320 to 350 °C at a heating rate of 10 °C min^{-1}; and (7) the capillary column was maintained at 350 °C for about 2.5 min. High-purity helium at a flow rate of 1.0 mL min^{-1} was used as the carrier gas and sent into the reaction system. In the GC/MS, the mass-to-charge range used for the mass selective detector was between 40 and 550 m/z. For the purpose of this research, the existence of a volatile product was confirmed if it showed a qualification percentage of 60% or higher according to the MS database. The calibration curves as well as limit of detection and limit of quantification values were programmed in the MS instrument.

As for the Karl Fisher device, the water content of the low and high volatility compounds was measured by Karl Fischer titration (TitroLine® 7500 KF) method using the standard procedure of the American Society for Testing and Materials (ASTM D1744) [35]. This method uses Karl Fischer reagent, which reacts quantitatively and selectively with water, to measure moisture content. Karl Fischer reagent consists of iodine, sulfur dioxide, a base, and a solvent such as alcohol. The fundamental principle behind it is based on the Bunsen Reaction between iodine and sulfur dioxide in an aqueous medium [36]. The measurement quality of the GC/MS and Karl Fischer titration systems was reliable since the instruments had been calibrated periodically. It should be noted that the data presented in the tables and figures were average figures derived from multiple tests wherein the differences were found to be smaller than 5%.

3. Results and Discussion

3.1. Proximate Analysis for Wood Sawdust

The proximate analysis (dry ash-free) of raw and torrefied wood sawdust is shown in Table 1. Cellulose was the major component in the biomass contributing 46.32 wt%, followed by hemicellulose (27.58 wt%) and lignin (8.17 wt%). The raw wood sawdust contained moisture, VM, fixed carbon (FC), and ash in portions of 7.72 wt%, 74.57 wt%, 16.41 wt%, and 1.30 wt%, respectively. The C, H, N, and O contents in the raw wood sawdust were found to be 47.07 wt%, 6.10 wt%, 0.43 wt%, and 46.40 wt%, respectively. The HHV of the biomass was 17.68 MJ kg^{-1}.

Table 1. Proximate, fiber, elemental, and calorific analyses of raw wood sawdust.

Wood Sawdust	
Proximate Analysis (wt%)	
Volatility matter (VM)	74.57 ± 3.89
Fixed carbon (FC)	16.41 ± 3.51
Moisture	7.72 ± 1.25
Ash	1.30 ± 0.73
Fiber Analysis (wt%)	
Hemicellulose	27.58 ± 4.72
Cellulose	46.32 ± 4.10
Lignin	8.17 ± 2.33
Other	16.71 ± 3.47
Elemental Analysis (wt%, Dry-Ash-Free)	
C	47.07 ± 2.55
H	6.10 ± 1.82
N	0.43 ± 0.21
O (by difference)	46.40 ± 4.44
HHV (MJ kg^{-1}, dry basis)	17.68 ± 1.35

3.2. The Effect of Vacuum on Product Yields

The yield of solid and liquid products at various operating conditions is displayed in Figures 2 and 3. The solid yield decreased with torrefaction temperature between 200 and 300 °C for both systems (see Figure 2). Between 200 and 300 °C for 60 min, the biochar yield decreased from 93.19 to 56.08 wt% in the vacuum system, while it decreased from 94.29 to 56.91 wt% in the nitrogen system [30,37]. This demonstrates that the vacuum approach can produce a similar amount of biochar in comparison to the conventional approach.

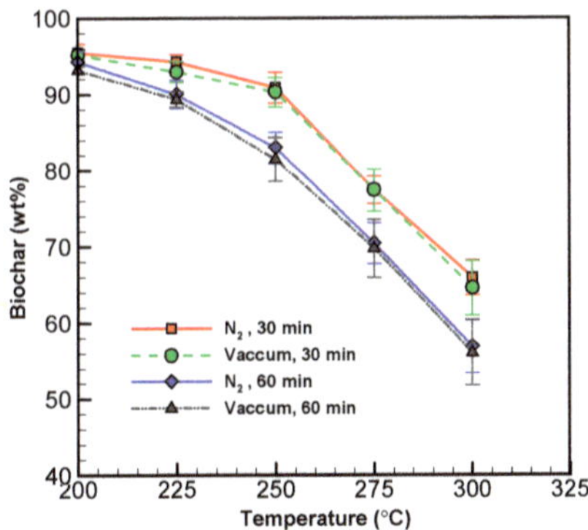

Figure 2. Profiles of solid (biochar) yield from wood sawdust torrefaction.

Besides the solid products, liquid yields from both systems were both found to be around 5 wt% at 250 °C with a torrefaction time of 30 min (see Figure 3a,b). According to the law of thermal decomposition, as torrefaction temperature increases, the formation of liquid products also increases. As anticipated, the effect of temperature picked up at 275 °C for both systems alike, where the liquid yield reached 15 wt%. With a torrefaction time of 60 min, liquid yield became 9–10 wt% below 250 °C and went above 30 wt% at 300 °C in both systems (see Figure 3b). The fact that the samples did not reached the target temperature of 300 °C during much of the 30 min torrefaction time provides a plausible explanation for the result that the vacuum system yielded less liquid than the nitrogen system. With a longer torrefaction time of 60 min, however, the advantage of the vacuum system was fully realized while a lot of volatile compounds were released from the solid components at 300 °C.

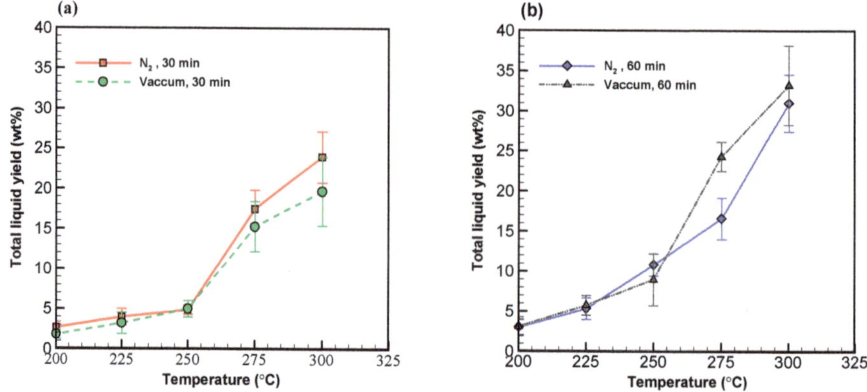

Figure 3. Profiles of liquid yields from wood sawdust torrefaction for (**a**) 30 min and (**b**) 60 min.

Table 2 shows the overall solid and liquid yields formed at various temperatures in both systems. The yields of low and high volatility liquid compounds are also available in the table. Overall, the total liquid yield increased from 2 to 33 wt% while temperature increased in both systems alike. It is noteworthy, nonetheless, that the vacuum system surpassed the nitrogen system in yielding liquid products with a peak liquid yield rate of 33.21 wt% at 300 °C with a torrefaction time of 60 min.

Table 2. Profiles of liquid and solid products from wood sawdust torrefaction.

Operating Conditions	Duration (min)	Temp. (°C)	Solid Yield Biochar (wt%)	Liquid Yield Low Volatility Compounds (wt%)	High Volatility Compounds (wt%)	Total (wt%)
N$_2$ (0.1 L/min)	30	200	95.47 ± 1.22	1.92 ± 0.21	0.72 ± 0.32	2.64 ± 0.72
		225	94.26 ± 1.01	2.34 ± 0.24	1.64 ± 0.43	3.98 ± 1.02
		250	90.92 ± 2.11	2.50 ± 0.35	2.37 ± 0.15	4.88 ± 0.57
		275	77.46 ± 1.82	6.11 ± 1.61	11.36 ± 1.25	17.48 ± 2.32
		300	65.97 ± 2.31	8.01 ± 2.37	15.93 ± 2.52	23.94 ± 3.19
N$_2$ (0.1 L/min)	60	200	94.29 ± 1.42	0.32±0.38	2.69 ± 0.42	3.01 ± 0.93
		225	90.02 ± 1.83	2.92 ± 0.82	2.41 ± 0.45	5.33 ± 1.34
		250	83.03 ± 2.03	3.26 ± 0.87	7.55 ± 0.64	10.81 ± 1.38
		275	70.44 ± 2.67	3.82 ± 1.32	12.79 ± 2.21	16.61 ± 2.58
		300	56.91 ± 3.47	9.65 ± 2.53	21.38 ± 3.52	31.03 ± 3.57
Vacuum	30	200	95.20 ± 0.73	-	1.83 ± 0.54	1.83 ± 0.54
		225	92.97 ± 1.34	-	3.24 ± 1.32	3.24 ± 1.32
		250	90.33 ± 1.93	0.54 ± 0.92	4.48 ± 0.74	5.02 ± 1.02
		275	77.45 ± 2.78	1.65 ± 0.72	13.63 ± 1.84	15.28 ± 3.13
		300	64.61 ± 3.67	4.98 ± 2.77	14.70 ± 3.52	19.68 ± 4.28
Vacuum	60	200	93.19 ± 0.50	-	3.09 ± 1.19	3.09 ± 1.19
		225	89.34 ± 1.02	-	5.74 ± 1.21	5.74 ± 1.21
		250	81.47 ± 3.88	0.65 ± 0.21	8.32 ± 3.31	8.97 ± 3.25
		275	69.76 ± 3.78	5.11 ± 0.92	19.17 ± 0.58	24.28 ± 1.82
		300	56.08 ± 4.33	8.61 ± 3.53	24.61 ± 3.82	33.21 ± 4.95

In further analysis breaking down liquid yields into low and high volatility compounds, low molecular weight (i.e., high volatility) products were absent at lower temperatures (i.e., below 250 °C) in the vacuum system. This was because—in a vacuum—most of the low molecular weight products diffused to the Erlenmeyer flask and left a considerably smaller amount of such products in the glass bottle [38,39]. By subtracting the weight percentage of the solid and the liquid yields from the total weight percentage of the original sample (i.e., 100%), the mass of losses in gaseous products can be obtained. In the nitrogen system, 2%–10% of the gaseous compounds and moisture were lost, while 3%–15% of such were lost in the vacuum system between 200 and 300 °C.

3.3. GC/MS Analysis of Liquid Compounds

Gas chromatography-mass spectrometry (GC/MS) is an instrumental technique consisting of a gas chromatograph (GC) coupled with a mass spectrometer (MS). It allows complex mixtures of chemicals to be separated, identified, and quantified and is thus ideal for the analysis of low molecular weight compounds (LMC). Recently, GC/MS has been employed for the in-depth analysis of thermal decomposition and the products of pyrolysis from biomass [40–42]. Table 3 and Figure A2 (in Appendix A) show the relative mass contents of sawdust torrefied in the vacuum and the nitrogen systems at 300 °C for 60 min. The liquid products were largely composed of condensable components such as phenols, acids, alcohols, aldehydes, and ketones (see Table 3) [43–45]. All of the results reported in Table 3 are mutually comparable because they were obtained under the same operating conditions as programed in the MS instrument.

Table 3 shows that—in the nitrogen system—low molecular weight compounds (LMC) and high molecular weight compounds (HMC) can be found in both the glass bottle (where low volatility compounds concentrated) and the Erlenmeyer flask (where high volatility compounds concentrated). Compared to torrefaction performed in nitrogen, torrefaction conducted in vacuum drastically reduced the furfural, 2-furancarboxaldehyde, and 2,6-dimethoxy-phenol in the products. Apparently, torrefaction in vacuum altered the ratio between LMC and HMC. In a vacuum, most of the LMC, especially furfural, were removed from both the glass bottle and the erlenmeyer flask. The LMC evaporated and mixed in the gas stream, which vented out of the system. On the other hand, HMC could only be found in the erlenmeyer flask in the vacuum system [46,47]. It is clear that the moving of torrefied liquid compounds through diffusion played a major role in creating such a diverged pattern [48,49]. These findings suggest that a vacuum environment is better able to stratify liquid compounds by their molecular weight.

3.4. Elemental Analysis of Solid and Liquid Products

Atomic H/C and O/C ratios as well as HHV are important indices for a material used as a source of energy [22,50]. The atomic H/C and O/C ratios for the solid and liquid products are reported in Table 4.

With regards to the liquid products, Figure 4a shows the H/C ratios for the low volatility compounds in both systems, which dropped along with torrefaction temperature by a similar pattern. At 275 °C and higher, the same figures came down to below 10%. In much the same way, Figure 4b shows uniformly low O/C ratios across the board [51]. However, it is noteworthy that higher torrefaction temperature (i.e., carbonization) appeared to have affected hydrogen to a greater extent than it did oxygen [26,52,53].

The covariation of the H/C and the O/C ratios is presented in a van Krevelen diagram (see Figure 4c). Both ratios decreased as torrefaction temperature increased, resulting in the shift of data points from the upper right to the bottom left corner. The H/C and O/C ratios of the sample torrefied at 200 °C for 60 min in vacuum were 25.36 and 11.49; such figures decreased to 3.64 and 1.29, respectively, when the sample was torrefied at 300 °C.

The outcome of the elemental analysis, as well as atomic H/C and O/C ratios for biochar, low volatility, and high volatility compounds in both systems, is reported in Table 4. The atomic H/C and O/C ratios for biochar were in the intervals 1.0–1.6 and 0.5–0.75 in vacuum and nitrogen, respectively. For the liquid products, data for high volatility compounds were unavailable for the vacuum system at lower torrefaction temperatures. Such compounds presumably had molecular weights that were too small to be captured in the process of diffusion. In other words, high volatility compounds could only be collected in the erlenmeyer flask at lower torrefaction temperatures in vacuum.

Table 3. Relative peak area distribution of the main products for biomass in GC/MS with vacuum and nitrogen systems.

Bio-Low Volatility Compounds Fraction	Formula	N_2 (300 °C, 60 min)				Vacuum (300 °C, 60 min)			
		Low Volatility Compounds		High Volatility Compounds		Low Volatility Compounds		High Volatility Compounds	
		RT	%	RT	%	RT	%	RT	%
Furfura	$C_5H_4O_2$	1.699	1.38	1.612	9.32	-	-	-	-
2-Furancarboxaldehyde	$C_5H_4O_2$	2.276	1.08	2.274	3.13	-	-	2.272	2.18
1,2-Cyclopentanedione	$C_5H_6O_2$	2.908	1.65	2.872	1.64	2.859	1.00	2.875	1.17
2-methoxy-Phenol	$C_7H_8O_2$	3.674	2.03	3.650	2.70	3.649	0.74	3.650	2.06
Maltol	$C_6H_6O_3$	3.903	1.04	3.858	0.81	3.856	0.55	3.856	0.71
Creosol	$C_8H_{10}O_2$	5.219	1.12	5.185	1.78	5.186	0.72	5.183	1.43
1,2-Benzenediol	$C_6H_6O_2$	6.407	1.38	6.159	0.63	6.236	0.67	6.215	0.42
4-ethyl-2-methoxy-Phenol	$C_9H_{12}O_2$	6.604	0.97	6.565	1.09	6.564	0.51	6.563	0.63
2-Methoxy-4-vinylphenol	$C_9H_{10}O_2$	7.106	1.10	7.059	0.76	7.066	1.28	7.061	0.82
2,6-dimethoxy-phenol	$C_8H_{10}O_3$	7.719	4.48	7.559	3.05	-	-	-	-
Eugenol	$C_{10}H_{12}O_2$	7.847	0.52	7.792	0.56	7.798	0.67	7.791	0.69
2-methoxy-4-propyl-phenol	$C_{10}H_{14}O_2$	8.028	0.55	7.988	0.43	-	-	-	-
Vanillin	$C_8H_8O_3$	8.354	1.09	8.133	0.43	8.172	0.83	8.137	0.55
trans-isoeugenol	$C_{10}H_{12}O_2$	8.600	0.56	8.564	0.47	8.566	0.53	8.563	0.52
3,5-Dimethoxy-4-hydroxytoluene	$C_9H_{12}O_3$	9.253	3.83	9.102	1.74	9.136	1.97	9.104	1.30
2-methoxy-4-(1-propenyl)-Phenol	$C_{10}H_{12}O_2$	9.295	1.55	9.201	1.31	9.217	1.28	9.204	1.36
Apocynin	$C_9H_{10}O_3$	9.724	1.12	9.503	0.44	9.547	0.64	9.511	0.53
2,6-dimethoxy-4-(2-propenyl)-Phenol	$C_{11}H_{14}O_3$	-	-	11.389	0.69	11.413	1.22	11.395	0.83
4-hydroxy-3,5-dimethoxy-Benzaldehyde	$C_9H_{10}O_4$	-	-	11.911	0.71	11.985	1.17	11.936	1.17
2,6-Dimethoxy-4-(prop-1-en-1-yl)phenol	$C_{11}H_{14}O_3$	12.893	4.72	12.727	2.29	12.781	3.32	12.744	2.99
4-Hydroxy-2-methoxycinnamaldehyde	$C_{10}H_{10}O_3$	-	-	12.901	0.24	-	-	12.935	1.21
2,9-Dimethyl-2,3,4,5,6,7-hexahydro-1H-2-benzazonine	$C_{14}H_{21}N$	15.877	0.33	-	-	15.817	0.32	-	-
3,5-Dimethoxy-4-hydroxycinnamaldehyde	$C_{11}H_{12}O_4$	16.181	0.85	15.966	0.50	16.058	1.27	-	-
gamma-phenyl-carbonic acid	$C_{10}H_{10}O_2$	-	-	16.424	0.17	-	-	-	-
n-Hexadecanoic acid	$C_{16}H_{32}O_2$	16.522	0.65	-	-	16.452	0.51	-	-
4,4'-(1-methylethylidene)bisphenol	$C_{15}H_{16}O_2$	-	-	18.166	0.05	18.235	0.37	-	-
Oleic Acid	$C_{18}H_{34}O_2$	18.446	0.54	-	-	18.385	0.49	-	-
Octadecanoic acid	$C_{19}H_{36}O_2$	18.727	0.14	18.687	0.10	18.695	0.13	-	-
Dehydroabietic acid	$C_{20}H_{28}O_2$	20.998	0.06	-	-	20.981	0.14	-	-

Table 4. Elemental analysis of raw and torrefied samples (dry-ash-free) in vacuum and nitrogen systems.

| Condition | Duration (min) | Temp. (°C) | Solid | | | | | Liquid | | | | | | | | | |
| | | | Biochar(wt%, Dry Basis) | | | | | Low Volatility Compounds (wt%) | | | | | High Volatility Compounds (wt%) | | | | |
			C (%)	H (%)	O (%)	H/C Atom Ratio	O/C Atom Ratio	C (%)	H (%)	O (%)	H/C Atom Ratio	O/C Atom Ratio	C (%)	H (%)	O (%)	H/C Atom Ratio	O/C Atom Ratio
Raw	-	-	47.70	6.19	45.67	1.54	0.71	-	-	-	-	-	-	-	-	-	-
N_2 (1 L/min)	30	200	47.67	6.60	46.17	1.65	0.73	3.06	9.44	86.64	36.75	21.25	3.60	10.13	85.48	33.53	17.83
		225	47.9	6.48	45.58	1.61	0.71	6.56	9.07	83.65	16.47	9.57	4.39	10.44	84.48	28.34	14.45
		250	50.71	6.53	43.54	1.53	0.64	7.55	8.08	83.40	12.75	8.29	5.03	10.76	83.66	25.49	12.49
		275	53.75	5.81	39.41	1.29	0.55	27.96	8.58	61.85	3.65	1.66	16.76	9.14	73.42	6.50	3.29
		300	53.75	5.48	40.25	1.21	0.56	34.16	8.49	55.91	2.96	1.22	17.69	7.98	73.01	5.38	3.10
N_2 (1 L/min)	60	200	50.01	6.27	43.17	1.49	0.64	3.50	11.35	84.22	38.64	18.06	3.41	11.78	84.31	41.16	18.56
		225	49.48	5.92	44.02	1.42	0.66	7.58	9.95	81.77	15.64	8.10	6.37	10.27	82.74	19.21	9.75
		250	49.06	5.66	44.79	1.37	0.68	30.62	8.22	59.94	3.20	1.47	14.81	7.80	76.47	6.28	3.88
		275	55.22	5.59	38.52	1.20	0.52	49.90	7.79	40.54	1.86	0.61	19.23	9.73	70.53	6.03	2.75
		300	52.20	4.90	42.19	1.11	0.60	50.39	7.45	40.84	1.76	0.61	16.84	7.44	75.16	5.26	3.35
Vacuum	30	200	48.67	6.37	44.46	1.56	0.69	-	-	-	-	-	1.75	10.46	87.39	71.22	37.48
		225	48.24	6.48	45.16	1.60	0.70	-	-	-	-	-	2.22	12.79	84.49	68.65	28.57
		250	48.21	6.04	45.37	1.49	0.71	-	-	-	-	-	8.18	11.53	79.76	16.80	7.32
		275	54.21	6.36	38.92	1.40	0.54	52.14	7.21	39.16	1.64	0.56	18.88	8.81	70.87	5.56	2.81
		300	54.45	5.75	39.34	1.26	0.54	53.83	7.41	37.24	1.64	0.51	25.73	7.00	67.05	3.24	1.95
Vacuum	60	200	47.36	5.78	46.24	1.45	0.73	-	-	-	-	-	5.39	11.47	82.49	25.36	11.49
		225	49.87	5.95	43.65	1.42	0.66	-	-	-	-	-	7.71	7.01	83.89	10.83	8.17
		250	51.65	5.76	42.22	1.33	0.61	55.51	7.12	34.87	1.53	0.47	18.14	9.50	71.06	6.24	2.94
		275	54.30	5.42	39.72	1.19	0.55	52.53	7.00	39.12	1.59	0.56	18.62	7.84	72.53	5.02	2.92
		300	56.35	5.15	38.00	1.09	0.51	53.09	7.01	38.28	1.57	0.54	32.66	9.97	55.99	3.64	1.29

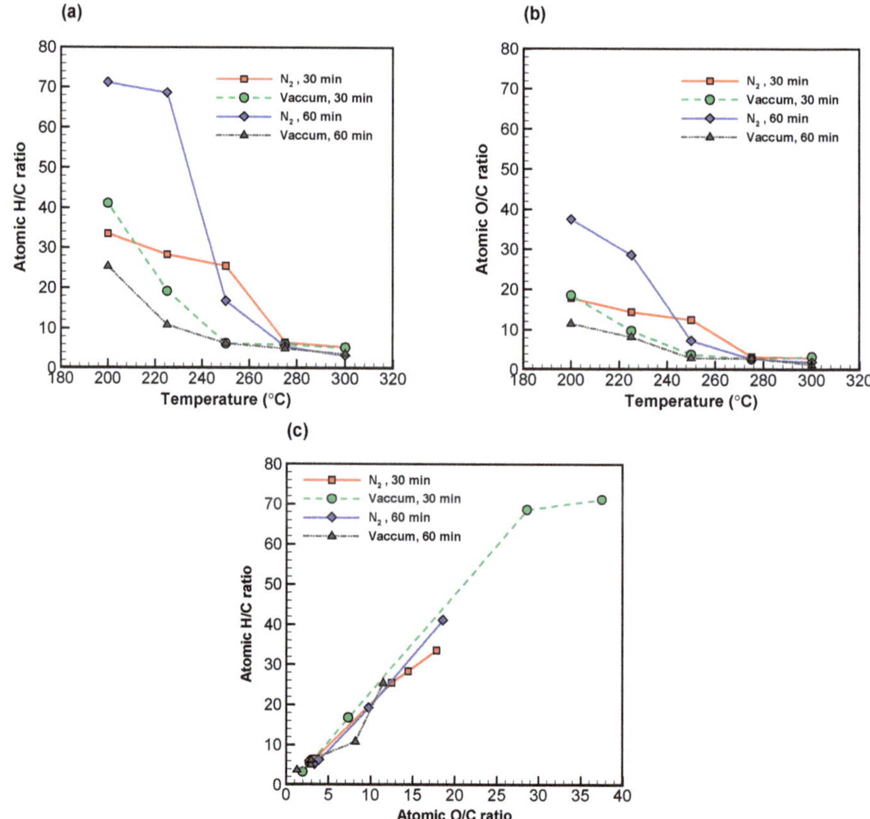

Figure 4. Profiles of (**a**) atomic H/C ratio versus temperature, (**b**) atomic O/C ratio versus temperature, and (**c**) van Krevelen diagram. All of the data presented were of low volatility compounds.

In contrast, in the nitrogen system, the atomic H/C and O/C ratios for low volatility compounds fell in the intervals 38.0–1.7 and 21.0–0.6, respectively. This indicated that both high and low volatility compounds were condensed into the glass bottle, resulting in mixed components therein.

With respect to high volatility compounds with a torrefaction time of 30 min in a vacuum, the C content increased from 1.75 to 25.4% while torrefaction temperature increased from 200 to 300 °C. At the same time, the H and O contents reduced from 6.25 to 5.92% and from 40.37 to 37.68%, respectively. With a torrefaction time of 60 min in a vacuum, the C content increased from 5.39 to 32.66% while temperature went from 200 °C to 300 °C. Meanwhile, the H and O contents reduced from 11.47 to 9.97% and from 82.49 to 55.99%, respectively. Similar patterns were found on the high volatility compounds in nitrogen. In fact, the atomic H/C and O/C ratios for high volatility compounds in vacuum with a torrefaction time of 60 min were lower than those in the nitrogen system. This indicated that—at a temperature of 275 °C or above—the H and O elements in the liquid compounds resulting from the vacuum system reduced by the same rate as they did in nitrogen.

3.5. Higher Heating Value Analysis

The HHV of torrefied products is a function of temperature [54]. All HHV and temperature figures from the experiments can be found in Figure 5 and Table 5. Across all materials, HHV increased along with the temperature. The increase in HHV was positively associated with temperature and residence time. This is due to an increase in the carbon content of the torrefied materials as a result of reduced H and O. With increased temperature, the total mass of the solid decreased, while the HHV increased. The HHVs of the solid materials from both systems were 27–35%, which were higher than that of the original biomass.

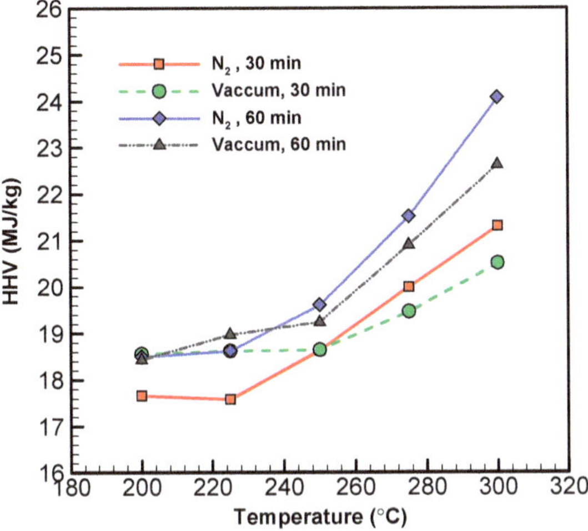

Figure 5. Profiles of the HHV of torrefied wood sawdust in vacuum and nitrogen systems with different temperatures and times.

Table 5. HHVs of untreated and torrefied biomasses.

			Solid	Liquid	
Condition	Time (min)	Temp. (°C)	Biochar HHV (Dry Basis) (MJ/kg)	Low Volatility Compounds HHV (MJ/kg)	High Volatility Compounds HHV (MJ/kg)
Raw	-	-	17.687	-	-
N$_2$ (0.1 L/min)	30	200	17.662	2.902	1.160
		225	17.583	4.403	1.653
		250	18.637	4.978	4.429
		275	19.995	10.721	6.628
		300	21.308	13.762	6.839
N$_2$ (0.1 L/min)	60	200	18.496	3.533	1.021
		225	18.624	5.197	1.026
		250	19.605	9.644	5.499
		275	21.518	17.685	6.425
		300	24.071	19.205	6.816
Vacuum	30	200	18.561	-	1.455
		225	18.624	-	1.533
		250	18.647	15.600	1.776
		275	19.465	19.034	6.724
		300	20.512	19.101	9.131
Vacuum	60	200	18.431	-	2.801
		225	18.970	-	5.047
		250	19.235	-	5.804
		275	21.698	19.726	6.921
		300	22.521	19.450	9.333

As for the liquid products, the HHV of low volatility compounds from the vacuum system did not go up until temperature reached 275 °C [55]. In contrast, the HHV of the low volatility compounds from the nitrogen system only started to pick up above 300 °C. Overall, the HHV of the high volatility compounds from both systems were lower—to a similar degree—than that of their respective solid compounds (17.68 MJ kg^{-1}). However, the high volatility compounds from the vacuum system exhibited higher HHVs than those from the nitrogen system. These findings suggest that torrefaction in vacuum improves the energy density of biomass more significantly than does a nitrogen system. Thus, the application of the vacuum technique to torrefaction has the potential for improving combustion efficiency.

3.6. Biomass Energy Conversion

In converting raw biomass into various forms of energy such as heat, solid, liquid, and gaseous fuels, there are four approaches: direct combustion, physical conversion, biochemical, and thermochemical conversions [56,57]. Energy yield, which stands for the ratio of total energies between processed and raw biomass samples, is a measure of preserved energy in the processed materials, corresponding to their solid, liquid, and gaseous states [58].

Torrefaction, a method of thermochemical conversion, registers a high energy yield of 90% for solid products when performed at 250 °C or below. While higher torrefaction temperature increases the energy density of the resultant biochar, it does so at the expense of reduced solid yield. When this is the case, lower energy yield means energy transfer from the solid product to liquid or gaseous products [59]. Figure 6a shows the percentage of energy yield for the solid, liquid, and gaseous products in a nitrogen system with a torrefaction time of 60 min at 300 °C. The resulting portions are 77%, 19%, and 4%, respectively. In comparison, the energy yield for the solid, liquid, and gaseous products from a vacuum system under the same conditions become 71%, 23%, and 6%, respectively. These suggest a 6% energy transfer from the solid biochar to the liquid and gaseous products in a vacuum environment.

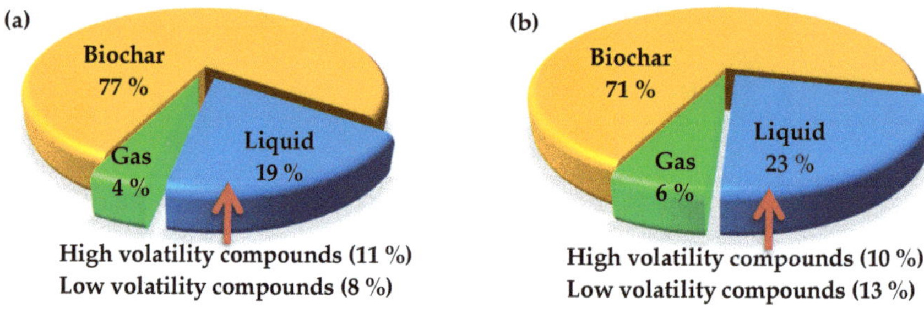

Figure 6. Energy distribution of biomass sources torrefied at 300 °C for 60 min in (**a**) nitrogen and (**b**) vacuum.

In addition, the moisture content was evaluated. The moisture contents in the liquid products at the different torrefaction temperatures in vacuum and nitrogen with different torrefaction times are presented in Table A1 (in Appendix A). In both systems, the moisture contents exceeded 85 wt% at 200 °C. The moisture content decreased along with temperature [26]. At 300 °C, the moisture contents of low and high volatility compounds were around 15% and 45%, respectively, in the vacuum system. At the same temperature, the moisture content of the low volatility compounds in the nitrogen system was 25% after 60 min of torrefaction and 49% after 30 min, while the high volatility compounds exhibited a consistent level of moisture content around 62% regardless of torrefaction time. In contrast, the moisture contents of the low volatility products from the vacuum system at 300 °C were between 14% and 16% after 60 and 30 min, respectively, while the high volatility compounds hovered around 44% and 46% regardless of time. In sum, the moisture contents of the liquid products from the vacuum system are lower than those of the nitrogen-system products across the board. It is thus clear that torrefaction in vacuum serves to improve the HHV of liquid products by bringing down their moisture level, making liquid and gaseous products of torrefaction qualified as fuels and other functional materials [58].

4. Conclusions

With the vacuum technique applied to torrefaction, it was found that the amount of biochar (solid products) resulting from the vacuum system was reduced at the same rate as the nitrogen system. In terms of the energy density, the vacuum system was able to produce biochar with considerably higher HHV than did the nitrogen system below 250 °C. This is because the moisture and the high volatility compounds (i.e., aldehydes) dispersed more smoothly in a vacuum. At a temperature higher than 250 °C, however, most of the low volatility compounds evaporated, resulting in lower HHV in the biochar produced in the vacuum.

Despite such mixed results with the solid products, the vacuum system increased the HHV of its liquid products more significantly than did the nitrogen system regardless of torrefaction temperature. As the GC/MS analysis revealed, the vacuum system separated a greater amount of smaller molecules (such as water) from bigger ones and thus produced high-volatility compounds that contained considerably less moisture. Another proof of reduced moisture in the liquid products of the vacuum system was a Van Krevelen plot of the H/C and O/C ratios, which demonstrated a linear relationship between torrefaction temperature and the hydrophobic property of the liquid products.

In light of the above, the vacuum system outperformed the nitrogen system in terms of the HHV, owing largely to the former's less moist liquid products. Unlike the case with the solid products, this pattern exhibited by the liquid products was unaffected by torrefaction temperature. Since the vacuum system converted more energy from solid biomass to liquid products, it was able to enhance the total energy output of torrefaction.

Author Contributions: Y.-K.C. performed the experiments, analyzed data, and wrote the paper. W.-H.C. created the research concept, organized the work, designed the experiments, analyzed data, and provided facilities and instruments for the research. H.C.O. analyzed data and wrote the paper. P.L.S. analyzed data and wrote the paper.

Funding: This research was funded by the Ministry of Science and Technology, Taiwan, R.O.C., under the contracts MOST 106-2923-E-006-002-MY3 and MOST 108-3116-F-006-007-CC1. This research was also funded in part by Higher Education Sprout Project, Ministry of Education to the Headquarters of University Advancement at NCKU.

Acknowledgments: The authors acknowledge the financial support of the Ministry of Science and Technology, Taiwan, R.O.C., under the contracts MOST 106-2923-E-006-002-MY3 and MOST 108-3116-F-006-007-CC1 for this research. This research is also supported in part by Higher Education Sprout Project, Ministry of Education to the Headquarters of University Advancement at NCKU.

Conflicts of Interest: The authors declare no conflict of interest.

Appendix A

Table A1. The moisture content of high and low volatility compounds at various operating conditions.

Operating Conditions	Temp. (°C)	Low Volatility Compounds		High Volatility Compounds	
		60 min	30 min	60 min	30 min
N$_2$	200	95.03	85.55	97.94	95.11
	300	24.6	49.12	62.06	61.32
Vacuum	200	-	-	98.65	89.54
	225	-	-	82.34	91.91
	300	14.3	15.68	45.33	44.28

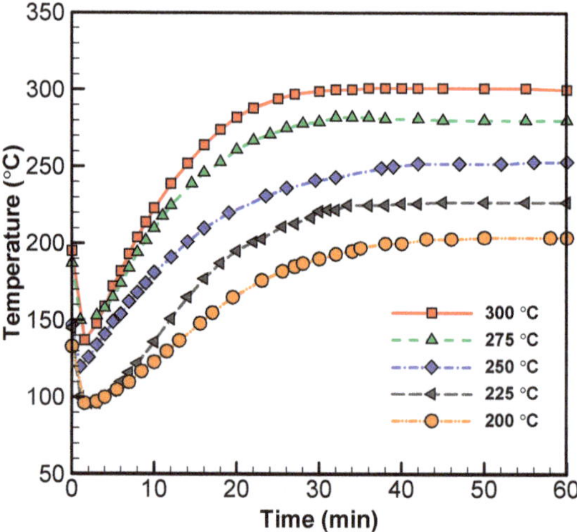

Figure A1. Temperature vs. time plots for the center sample in reactor trials (vacuum and nitrogen systems) for wood sawdust.

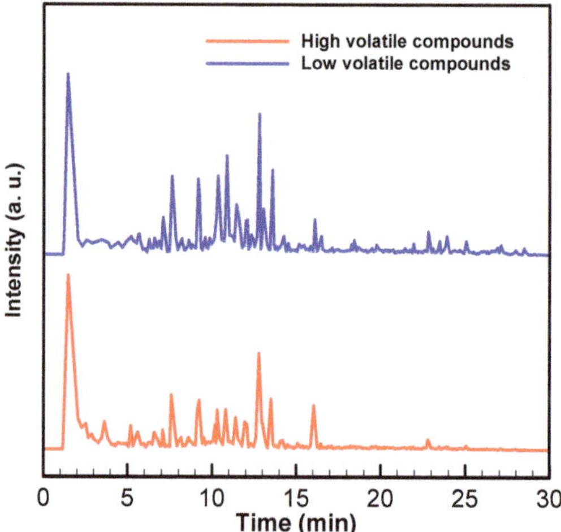

Figure A2. GC/MS spectra of high and low volatility compounds at 300 °C in vacuum systems.

References

1. Zachos, J.; Pagani, M.; Sloan, L.; Thomas, E.; Billups, K. Trends, Rhythms, and Aberrations in Global Climate 65 Ma to Present. *Science* **2001**, *292*, 686. [CrossRef]
2. Hansen, J.; Ruedy, R.; Sato, M.; Lo, K. Global surface temperature change. *Rev. Geophys.* **2010**, *48*. [CrossRef]
3. Dai, A. Increasing drought under global warming in observations and models. *Nat. Clim. Chang.* **2012**, *3*, 52. [CrossRef]
4. John, R.P.; Anisha, G.S.; Nampoothiri, K.M.; Pandey, A. Micro and macroalgal biomass: A renewable source for bioethanol. *Bioresour. Technol.* **2011**, *102*, 186–193. [CrossRef]
5. Parikka, M. Global biomass fuel resources. *Biomass Bioenergy* **2004**, *27*, 613–620. [CrossRef]
6. Long, H.; Li, X.; Wang, H.; Jia, J. Biomass resources and their bioenergy potential estimation: A review. *Renew. Sustain. Energy Rev.* **2013**, *26*, 344–352. [CrossRef]

7.   Prins, M.J.; Ptasinski, K.J.; Janssen, F.J.J.G. Torrefaction of wood. Part 2. Analysis of products. *J. Anal. Appl. Pyrolysis* **2006**, *77*, 35–40. [CrossRef]

8.   Grilc, M.; Likozar, B.; Levec, J. Kinetic model of homogeneous lignocellulosic biomass solvolysis in glycerol and imidazolium-based ionic liquids with subsequent heterogeneous hydrodeoxygenation over NiMo/Al$_2$O$_3$ catalyst. *Catal. Today* **2015**, *256*, 302–314. [CrossRef]

9.   Grilc, M.; Likozar, B.; Levec, J. Simultaneous liquefaction and hydrodeoxygenation of lignocellulosic biomass over NiMo/Al$_2$O$_3$, Pd/Al$_2$O$_3$, and Zeolite Y catalysts in hydrogen donor solvents. *ChemCatChem* **2016**, *8*, 180–191. [CrossRef]

10. Grilc, M.; Likozar, B.; Levec, J. Hydrotreatment of solvolytically liquefied lignocellulosic biomass over NiMo/Al$_2$O$_3$ catalyst: Reaction mechanism, hydrodeoxygenation kinetics and mass transfer model based on FTIR. *Biomass Bioenergy* **2014**, *63*, 300–312. [CrossRef]

11. Chen, W.-H.; Hsu, H.-J.; Kumar, G.; Budzianowski, W.M.; Ong, H.C. Predictions of biochar production and torrefaction performance from sugarcane bagasse using interpolation and regression analysis. *Bioresour. Technol.* **2017**, *246*, 12–19. [CrossRef]

12. Song, X.; Yang, Y.; Zhang, M.; Zhang, K.; Wang, D. Ultrasonic pelleting of torrefied lignocellulosic biomass for bioenergy production. *Renew. Energy* **2018**, *129*, 56–62. [CrossRef]

13. Chen, W.-H.; Zhuang, Y.-Q.; Liu, S.-H.; Juang, T.-T.; Tsai, C.-M. Product characteristics from the torrefaction of oil palm fiber pellets in inert and oxidative atmospheres. *Bioresour. Technol.* **2016**, *199*, 367–374. [CrossRef]

14. Lu, J.-J.; Chen, W.-H. Product yields and characteristics of corncob waste under various torrefaction Atmospheres. *Energies* **2014**, *7*, 13. [CrossRef]

15. Chen, W.-H.; Cheng, W.-Y.; Lu, K.-M.; Huang, Y.-P. An evaluation on improvement of pulverized biomass property for solid fuel through torrefaction. *Appl. Energy* **2011**, *88*, 3636–3644. [CrossRef]

16. Xing, X.; Fan, F.; Jiang, W. Characteristics of biochar pellets from corn straw under different pyrolysis temperatures. *R. Soc. Open Sci.* **2018**, *5*. [CrossRef]

17. Santos, L.B.; Striebeck, M.V.; Crespi, M.S.; Capela, J.M.V.; Ribeiro, C.A.; De Julio, M. Energy evaluation of biochar obtained from the pyrolysis of pine pellets. *J. Therm. Anal. Calorim.* **2016**, *126*, 1879–1887. [CrossRef]

18. Wang, Z.; Dunn, J.B.; Han, J.; Wang, M.Q. Effects of co-produced biochar on life cycle greenhouse gas emissions of pyrolysis-derived renewable fuels. *Biofuels Bioprod. Biorefining* **2014**, *8*, 189–204. [CrossRef]

19. Sajdak, M.; Muzyka, R.; Hrabak, J.; Różycki, G. Biomass, biochar and hard coal: Data mining application to elemental composition and high heating values prediction. *J. Anal. Appl. Pyrolysis* **2013**, *104*, 153–160. [CrossRef]

20. Bridgeman, T.G.; Jones, J.M.; Shield, I.; Williams, P.T. Torrefaction of reed canary grass, wheat straw and willow to enhance solid fuel qualities and combustion properties. *Fuel* **2008**, *87*, 844–856. [CrossRef]

21. Sun, J.; He, F.; Pan, Y.; Zhang, Z. Effects of pyrolysis temperature and residence time on physicochemical properties of different biochar types. *Acta Agric. Scand. Sect. B Soil Plant Sci.* **2017**, *67*, 12–22. [CrossRef]

22. Bach, Q.-V.; Chen, W.-H.; Chu, Y.-S.; Skreiberg, Ø. Predictions of biochar yield and elemental composition during torrefaction of forest residues. *Bioresour. Technol.* **2016**, *215*, 239–246. [CrossRef]

23. Chen, W.-H.; Chu, Y.-S.; Lee, W.-J. Influence of bio-solution pretreatment on the structure, reactivity and torrefaction of bamboo. *Energy Convers. Manag.* **2017**, *141*, 244–253. [CrossRef]

24. Couhert, C.; Salvador, S.; Commandré, J.M. Impact of torrefaction on syngas production from wood. *Fuel* **2009**, *88*, 2286–2290. [CrossRef]

25. Dhanavath, K.N.; Bankupalli, S.; Bhargava, S.K.; Parthasarathy, R. An experimental study to investigate the effect of torrefaction temperature on the kinetics of gas generation. *J. Environ. Chem. Eng.* **2018**, *6*, 3332–3341. [CrossRef]

26. Chen, D.; Gao, A.; Ma, Z.; Fei, D.; Chang, Y.; Shen, C. In-depth study of rice husk torrefaction: Characterization of solid, liquid and gaseous products, oxygen migration and energy yield. *Bioresour. Technol.* **2018**, *253*, 148–153. [CrossRef]

27. Fisher, T.; Hajaligol, M.; Waymack, B.; Kellogg, D. Pyrolysis behavior and kinetics of biomass derived materials. *J. Anal. Appl. Pyrolysis* **2002**, *62*, 331–349. [CrossRef]

28. Elliott, D.C. Historical developments in hydroprocessing bio-oils. *Energy Fuels* **2007**, *21*, 1792–1815. [CrossRef]

29. Lin, B.-J.; Colin, B.; Chen, W.-H.; Pétrissans, A.; Rousset, P.; Pétrissans, M. Thermal degradation and compositional changes of wood treated in a semi-industrial scale reactor in vacuum. *J. Anal. Appl. Pyrolysis* **2018**, *130*, 8–18. [CrossRef]

30. Garcìa-Pérez, M.; Chaala, A.; Pakdel, H.; Kretschmer, D.; Roy, C. Vacuum pyrolysis of softwood and hardwood biomass: Comparison between product yields and bio-oil properties. *J. Anal. Appl. Pyrolysis* **2007**, *78*, 104–116. [CrossRef]

31. Murwanashyaka, J.N.; Pakdel, H.; Roy, C. Step-wise and one-step vacuum pyrolysis of birch-derived biomass to monitor the evolution of phenols. *J. Anal. Appl. Pyrolysis* **2001**, *60*, 219–231. [CrossRef]

32. Pakdel, H.; Roy, C. Separation and characterization of steroids in biomass vacuum pyrolysis oils. *Bioresour. Technol.* **1996**, *58*, 83–88. [CrossRef]

33. Chen, W.-H.; Kuo, P.-C. Torrefaction and co-torrefaction characterization of hemicellulose, cellulose and lignin as well as torrefaction of some basic constituents in biomass. *Energy* **2011**, *36*, 803–811. [CrossRef]

34. Chen, W.-H.; Tu, Y.-J.; Sheen, H.-K. Impact of dilute acid pretreatment on the structure of bagasse for bioethanol production. *Int. J. Energy Res.* **2010**, *34*, 265–274. [CrossRef]

35. *ASTM D 1744-13—Standard Test Method for Determination of Water in Liquid Petroleum Products by Karl Fischer Reagent (Withdrawn 2016)*; ASTM International: West Conshohocken, PA, USA, 2013. [CrossRef]

36. Scholz, E. Karl Fischer titration: determination of water. In *Springer Science & Business Media*; Springer: Berlin/Heidelberg, Germany, 2012; ISBN 978-3-642-69991-7.

37. Islam, M.A.; Asif, M.; Hameed, B.H. Pyrolysis kinetics of raw and hydrothermally carbonized Karanj (Pongamia pinnata) fruit hulls via thermogravimetric analysis. *Bioresour. Technol.* **2015**, *179*, 227–233. [CrossRef]

38. Konopel'ko, N.A.; Shakhov, E.M.J.F.D. Rarefied gas flow into a vacuum from a plane long channel closed at one end. *Fluid Dyn.* **2016**, *51*, 552–560. [CrossRef]

39. ElhusseinŞahin, S.J.H.; Transfer, M. Drying behaviour, effective diffusivity and energy of activation of olive leaves dried by microwave, vacuum and oven drying methods. *Heat Mass Transf.* **2018**, *54*, 1901–1911. [CrossRef]

40. González Martínez, M.; Dupont, C.; Thiéry, S.; Meyer, X.-M.; Gourdon, C. Impact of biomass diversity on torrefaction: Study of solid conversion and volatile species formation through an innovative TGA-GC/MS apparatus. *Biomass Bioenergy* **2018**, *119*, 43–53. [CrossRef]

41. Zhang, L.; Li, K.; Zhu, X. Study on two-step pyrolysis of soybean stalk by TG-FTIR and Py-GC/MS. *J. Anal. Appl. Pyrolysis* **2017**, *127*, 91–98. [CrossRef]

42. Xin, X.; Pang, S.; de Miguel Mercader, F.; Torr, K.M. The effect of biomass pretreatment on catalytic pyrolysis products of pine wood by Py-GC/MS and principal component analysis. *J. Anal. Appl. Pyrolysis* **2019**, *138*, 145–153. [CrossRef]

43. Kim, J.-Y.; Hwang, H.; Park, J.; Oh, S.; Choi, J.W. Predicting structural change of lignin macromolecules before and after heat treatment using the pyrolysis-GC/MS technique. *J. Anal. Appl. Pyrolysis* **2014**, *110*, 305–312. [CrossRef]

44. Chen, W.-H.; Wang, C.-W.; Kumar, G.; Rousset, P.; Hsieh, T.-H. Effect of torrefaction pretreatment on the pyrolysis of rubber wood sawdust analyzed by Py-GC/MS. *Bioresour. Technol.* **2018**, *259*, 469–473. [CrossRef]

45. Cai, W.; Fivga, A.; Kaario, O.; Liu, R. Effects of Torrefaction on the Physicochemical Characteristics of Sawdust and Rice Husk and Their Pyrolysis Behavior by Thermogravimetric Analysis and Pyrolysis–Gas Chromatography/Mass Spectrometry. *Energy Fuels* **2017**, *31*, 1544–1554. [CrossRef]

46. Apicella, B.; Tregrossi, A.; Popa, C.; Mennella, V.; Ciajolo, A.; Russo, C. Study on the separation and thin film deposition of tarry aromatics mixtures (soot extract and naphthalene pitch) by high-vacuum heating. *Fuel* **2017**, *209*, 795–801. [CrossRef]

47. Horike, S.; Ayano, M.; Tsuno, M.; Fukushima, T.; Koshiba, Y.; Misaki, M.; Ishida, K. Thermodynamics of ionic liquid evaporation under vacuum. *Phys. Chem. Chem. Phys.* **2018**, *20*, 21262–21268. [CrossRef]

48. Yao, M.; Woo, Y.C.; Tijing, L.D.; Choi, J.-S.; Shon, H.K. Effects of volatile organic compounds on water recovery from produced water via vacuum membrane distillation. *Desalination* **2018**, *440*, 146–155. [CrossRef]

49. Rahman, S.; Helleur, R.; MacQuarrie, S.; Papari, S.; Hawboldt, K. Upgrading and isolation of low molecular weight compounds from bark and softwood bio-oils through vacuum distillation. *Sep. Purif. Technol.* **2018**, *194*, 123–129. [CrossRef]

50. Nsaful, F.; Collard, F.-X.; Görgens, J.F. Lignocellulose thermal pretreatment and its effect on fuel properties and composition of the condensable products (tar precursors) from char devolatilization for coal substitution in gasification application. *Fuel Process. Technol.* **2018**, *179*, 334–343. [CrossRef]

51. Tumuluru, J.S.; Boardman, R.D.; Wright, C.T. Response surface analysis of elemental composition and energy properties of corn stover during torrefaction. *J. Biobased Mater. Bioenergy* **2012**, *6*, 25–35. [CrossRef]
52. Pimchuai, A.; Dutta, A.; Basu, P. Torrefaction of agriculture residue to enhance combustible properties. *Energy Fuels* **2010**, *24*, 4638–4645. [CrossRef]
53. Chew, J.J.; Doshi, V. Recent advances in biomass pretreatment—Torrefaction fundamentals and technology. *Renew. Sustain. Energy Rev.* **2011**, *15*, 4212–4222. [CrossRef]
54. Martín-Lara, M.A.; Ronda, A.; Zamora, M.C.; Calero, M. Torrefaction of olive tree pruning: Effect of operating conditions on solid product properties. *Fuel* **2017**, *202*, 109–117. [CrossRef]
55. Chen, W.-H.; Liu, S.-H.; Juang, T.-T.; Tsai, C.-M.; Zhuang, Y.-Q. Characterization of solid and liquid products from bamboo torrefaction. *Appl. Energy* **2015**, *160*, 829–835. [CrossRef]
56. Adams, P.; Bridgwater, T.; Lea-Langton, A.; Ross, A.; Watson, I. Chapter 8—Biomass conversion technologies. In *Greenhouse Gases Balances of Bioenergy Systems*; Academic Press: Cambridge, MA, USA, 2018; pp. 107–139. ISBN 9780081010365. [CrossRef]
57. Roddy, D.J. Biomass in a petrochemical world. *Interface Focus* **2013**, *3*, 20120038. [CrossRef] [PubMed]
58. Popa, V.I. 1-Biomass for fuels and biomaterials. In *Biomass as Renewable Raw Material to Obtain Bioproducts of High-Tech Value*; Elsevier: Amsterdam, The Netherlands, 2018; pp. 1–37. [CrossRef]
59. Chen, D.; Gao, A.; Cen, K.; Zhang, J.; Cao, X.; Ma, Z. Investigation of biomass torrefaction based on three major components: Hemicellulose, cellulose, and lignin. *Energy Convers. Manag.* **2018**, *169*, 228–237. [CrossRef]

 *energies*

Review

Potential of Rice Industry Biomass as a Renewable Energy Source

M. Mofijur [1,*], T.M.I. Mahlia [1], J. Logeswaran [2], M. Anwar [3], A.S. Silitonga [4,*],
S.M. Ashrafur Rahman [5] and A.H. Shamsuddin [2]

[1] School of Information, Systems and Modelling, Faculty of Engineering and IT, University of Technology Sydney NSW 2007, Australia; tmindra.mahlia@uts.edu.au
[2] Institute of Sustainable Energy, Universiti Tenaga Nasional, Kajang 43000, Malaysia; indra@uniten.edu.au (J.L.); abdhalim@uniten.edu.my (A.H.S.)
[3] School of Engineering and Technology, Central Queensland University, Rockhampton QLD 4701, Australia; m.anwar@cqu.edu.au
[4] Department of Mechanical Engineering, Politeknik Negeri Medan, Medan 20155, Indonesia
[5] Biofuel Engine Research Facility (BERF), Queensland University of Technology, Brisbane QLD 4000, Australia; s2.rahman@qut.edu.au
[*] Correspondence: mdmofijur.rahman@uts.edu.au (M.M.); ardinsu@yahoo.co.id (A.S.S); Tel.: +61-469851901 (M.M.)

Received: 3 September 2019; Accepted: 24 October 2019; Published: 28 October 2019

Abstract: Fossil fuel depletion, along with its ever-increasing price and detrimental impact on the environment, has urged researchers to look for alternative renewable energy. Of all the options available, biomass presents a very reliable source due to its never-ending supply. As research on various biomasses has grown in recent years, waste from these biomasses has also increased, and it is now time to shift the focus to utilizing these wastes for energy. The current waste management system mainly focuses on open burning and soil incorporation as it is cost-effective; however, these affect the environment. There must be an alternative way, such as to use it for power generation. Rice straw and rice husk are examples of such potential biomass waste. Rice is the main food source for the world, mostly in Asian regions, as most people consume rice daily. This paper reviews factors that impact the implementation of rice-straw-based power plants. Ash content and moisture content are important properties that govern combustion, and these vary with location. Logistical improvements are required to reduce the transport cost of rice husk and rice straw, which is higher than the transportation cost of coal.

Keywords: rice straw; rice husk; power generation; gasification; alternative fuel

---

1. Introduction

The usage of fossil fuels has increased rapidly since the 19th century due to the increased population and technological development in many countries; neither the population nor the technological development are able to be controlled, and the development of technology is important for sustaining society and the economy [1]. Many scientists are worried that due to their limited reserves, fossil fuels will become incapable of supporting global energy needs in the coming years [2–4]. The increasing price of fossil fuels is due to the high market demand, excessive use, and phasing out of fossil fuel based technologies which have caused climate change over a period of years [5–8]. It has been reported that global carbon dioxide (CO_2) emissions increased from 30,295 million metric tons in 2008 to 33,234.8 million metric tons in 2018. In 2018, the largest CO_2 emitter globally is the Asia Pacific region (16.27 billion metric tons of CO_2 emissions, and China alone contributed 27% of the world total fossil fuel CO_2 emissions) [9]. Currently, the coal industry dominates the Chinese power generation industry, especially in electricity production [10,11]. Malaysian power generation industries generally rely on fossil fuels. High consumption of fossil fuels for combustion processes in power generation increases the emissions of carbon dioxide (CO_2) [12]. During the year 1990, there was an emission of 3.1 metric tons of CO_2 per capita, whereas in 2011, it was 7.8 metric tons per capita, representing an increase of 155% in emission rate [13]. The major environmental and health issues of fossil fuels and increasing energy requirements have drawn attention from all over the world to the search for renewable fuels as these fuels are assumed to be neutral, as shown in Figure 1. During their development cycle, plants use photosynthesis to retain and change sunlight and carbon dioxide from the atmosphere into nutrients and energy. Therefore, when biomass is combusted as a power generation fuel, it does produce carbon, but the carbon dioxide is again absorbed by plants during the next crop cycle [14].

It is crucial to maintain energy usage and find a more sustainable source of energy, such as renewable energy, to fulfil global needs [15–17]. Most countries have come up with policies which offer benefits and funding to companies that seek to develop the renewable energy technology and eliminate fossil-fuel-based technologies [8]. For example, the European Union has set the target for its energy distribution to have a 20% contribution from renewable energies by the year 2020 [18]. The government in Spain has pushed the deployment of renewable energy to lower the impacts of pollution on the environment [19]. In Denmark, Spain, and Germany, innovations and inventions related to renewable energy and the production of energy at lower cost have been highly welcomed and appreciated. In Taiwan, many energy conservation patents have been established to lead the country towards the use of renewable energy sources [20]. In Indonesia, renewable energy has been widely used for biofuel and power generation. Renewable energy is going to be the main alternative energy source in Africa to overcome hydra-headed problems such as climate change and the lack of energy access in certain secluded areas [21]. There is a clear understanding that energy is the need and pivot upon which society turns.

There are various types of renewable energy source, such as biomass, hydro, wind, solar photovoltaic, and solar thermal, which have been implemented in many countries around the world [22–26]. However, some of renewable energy, like wind and solar, is not continuously available, and energy storage devices are therefore required [27]. To date, batteries are the only energy storage equipment available commercially. Hence, many researchers have attempted to find a material that stores energy in significant amounts to be used when necessary [28–30]. For this reason, researchers and the private sector still interested in deriving biofuel from biomass [31–35]. Biomass is classified into two types: waste biomass and energy crops [36]. Harvesting energy from biomass also depends on its fuel capacity and availability. It is only wise for a country to invest in biomass if the supply is abundant and able to be sustained continuously in future years [36]. In recent decades, the application of agricultural residues such as rice husk and rice straw has gained much more attention than other energy crops, as they are cheaper in cost and do not affect the food price, and circumvent the food versus fuel controversy. In addition, both rice husk and rice straw are the main residues from the most essential foods crop: rice. Both the abundant availability of agricultural residues and the technological development of biomass conversion techniques have enabled the biomasses from the rice industry to be converted into an important source of renewable power.

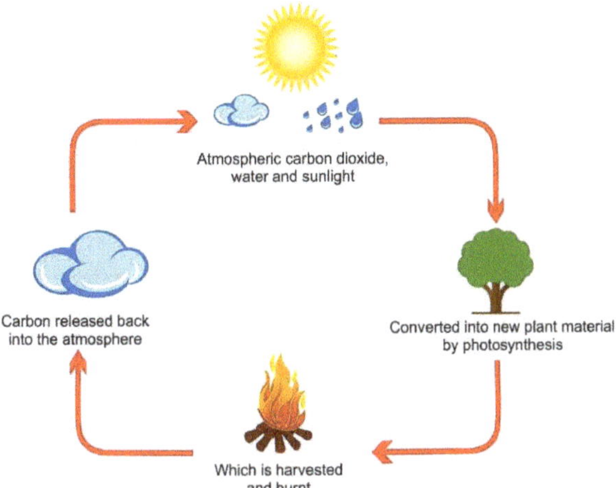

Figure 1. Illustration of the carbon-neutral cycle [37].

It is predicted that the consumption of rice will continue in coming decades due to the increasing population and economic development in Africa and Asia especially. It has been predicted that the total global rice consumption will be 450 million tons by 2020; the rice industry will be sustained long terms and the availability of its agricultural waste will remain high [38]. The development of rice husk and straw utilization for energy purposes has been seriously researched and studied worldwide for years, but few researchers have reviewed and analyzed these works. This paper provides a comprehensive review of the technology available for the production of renewable energy from rice crops and their residues.

2. Advantages of Biomass

There are many agricultural waste products classified under energy crops, such as rice straw, rice husk, wheat, potatoes, and residues from processing fruits [39]. The Association of Southeast Asian Nations (ASEAN) countries have an abundant supply of biomass, but it is unevenly distributed. Malaysia, as one of the top palm oil producers, could be a major contributor to renewable energy from biomass [40,41]. Other countries rich in biomass are Cambodia, Myanmar, and Laos. Due to a lack of financial means and slow progression of technology, agricultural residues are normally disposed of in these countries by open burning, where their energy potential is not being utilized [42].

Biomass offers some benefits compared to other sources of power generation. The advantages of biomass are [43–46] as follows:

(i)   Renewable energy source
      This source of biomass will never run out as crops, manure, and garbage are continuously produced by human activities.
(ii) Carbon-neutral
      Utilizing biomass in the power generation industry basically follows the principle of the carbon-neutral cycle. Therefore, there is no contribution to greenhouse gas emissions when biomass-based fuel is burnt in power plants.
(iii) Cost-effective
      Biomass energy is cheaper compared to other forms of renewable energy generation. It has been reported that utilization of 70% of rice husk residues could contribute 1328 GWh electricity production annually, and the cost of per unit electricity generated using rice husk is 47.36 cents/kWh, compared to 55.22 cents/kWh of electricity generated by coal [47].

(iv) Ability to have small scale power production
  By using the gasification method, power production can be done on a small scale, especially in rural areas.

(v) The large variety of feedstock
  Biomass power is capable of incorporating a variety of feedstock, such as rice straw, rice husk, wood pellets, bagasse, etc.

(vi) Reduces methane gas
  The decomposition process of organic matter indirectly releases methane gas; combustion of the biomass to produce energy could control the release of methane gas.

3. Agricultural Residues from Rice Crops

In general, agricultural wastes are biomass residues that can be divided into two categories, namely the crop residues and the agro-industrial residues. Crop residues can be further divided into different sub-groups, such as rice straw and rice husk. Rice husk is the outer layer of a rice seed. Rice husk is removed from the rice seed as a byproduct during the milling process. Rice straw is the stalk of the rice plant, which is left in the field as a waste product upon harvesting of the rice grain (i.e., the seeds of rice). Rice straw is produced when the paddy plant is cut at grain during harvest [48]. Rice straw actually makes up almost 50% of the clean weight of rice plants, and this may vary in the range of 40–60% depending on the method used during cultivation. Therefore, it is clear that almost half of the weight of the plant is contributed by the straw. Figure 2a,b shows the potential application of rice husk and rice straw for power generation.

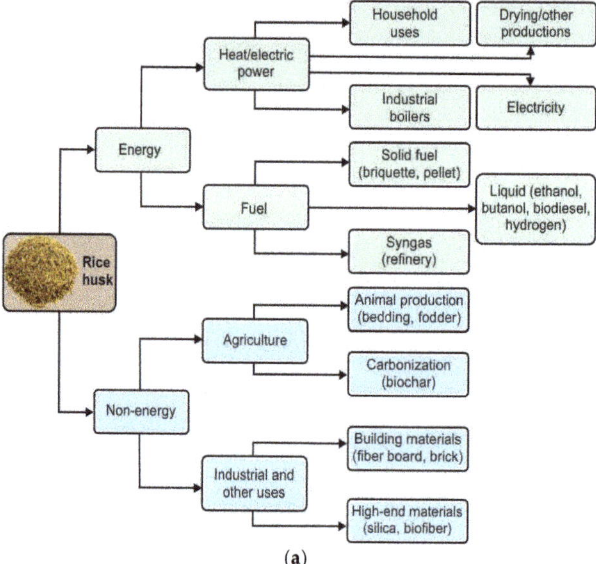

(**a**)

Figure 2. *Cont.*

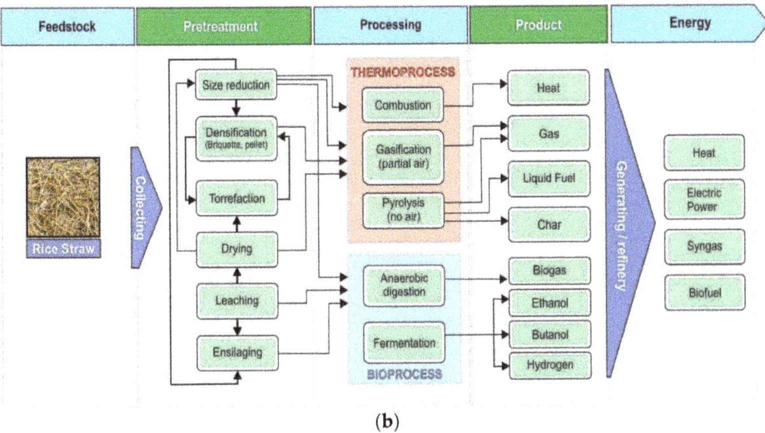

(b)

Figure 2. The potential application of (**a**) rice husk and (**b**) rice straw for fuel and power generation.

Kadam et al. [49] stated that for every ton of rice grain obtained, there is at least 1.35 tons of rice straw residue remaining in the fields. Therefore, looking at the paddy production reported in previous studies, it becomes clear that the amount of agricultural rice straw residue produced is extremely high. According to Zhiqiang Liu [50], rice straw waste represents about 62% of the rice production in China, and it is not utilized properly for energy generation. In Taiwan, during 2007, based on the paddy field area of 9375 ha, around 0.0563 million tons of wet and fresh rice straw remained at farms following the rice harvest. The total planted area was 38,862 ha, producing up to 0.233 million tons of rice straw [51]. Different researchers have reported different residue ratios of rice straw and rice husk, varying from 1.0 to 3.96 and 0.2 to 0.33, respectively. Table 1 shows the rice crop, rice straw, and rice husk production of the top 20 rice-producing countries. Globally, 769.75 million tons of rice straw and 153.95 million tons of rice husk was produced in 2017, which could have been used to produce 638.03 PJ of energy [38].

Table 1. Rice crop, rice straw, and rice husk production in the top 20 rice-producing countries in 2017 [52].

Countries	Rice Crop (million tons)	Predicted Rice Husk [a]	Predicted Rice Straw [b]	Energy Potential (PJ)
China, mainland	212.68	42.54	212.68	638.03
India	168.50	33.70	168.50	505.50
Indonesia	81.38	16.28	81.38	244.15
Bangladesh	48.98	9.80	48.98	146.94
Vietnam	42.76	8.55	42.76	128.29
Thailand	33.38	6.68	33.38	100.15
Myanmar	25.62	5.12	25.62	76.87
Philippines	19.28	3.86	19.28	57.83
Brazil	12.47	2.49	12.47	37.41
Pakistan	11.17	2.23	11.17	33.52
Cambodia	10.35	2.07	10.35	31.05
Nigeria	9.86	1.97	9.86	29.59
Japan	9.78	1.96	9.78	29.34
USA	8.08	1.62	8.08	24.25
Egypt	6.38	1.28	6.38	19.14
Republic of Korea	5.28	1.06	5.28	15.85
Nepal	5.23	1.05	5.23	15.69
Lao People's Democratic Republic	4.04	0.81	4.04	12.12
Madagascar	3.10	0.62	3.10	9.30
World total	**769.75**	**153.95**	**769.75**	**638.03**

[a] Predicted residue ratio of 0.2 and [b] predicted residue ratio of 1.0.

4. Characteristics of Rice Crop Residues

The size of rice husks is normally uniform, similar to the size of grain, and they are normally very dry with a very low moisture content. Husks are normally collected at the factory level where grain is processed. Rice husk has good market access and can be traded, because there are already established rice husk power plants (e.g., the 2.5 MW rice-husk-based cogeneration plant at Hanuman Agro Industries Limited), and it can also be used in the co-firing process. Husks do not require preprocessing due to their low moisture content and high ash content, and so can be directly used for heat generation or energy generation in power plants [18,53,54]. Rice straw is usually very bulky and it is normally dry, although under circumstances of the rainy season, it is normally wet. It is a field-based resource that most farmers do not collect from the field after the harvesting process. However, both rice straw and rice husk are composed of hemicellulose (35.7% and 28.6%), cellulose (32% and 28.6%), lignin (22.3% and 24.3%), and extractive matter (10% and 18.4%), respectively.

In addition, rice crop residues have several properties which can be determined through proximity analysis, ultimate analysis, and elemental analysis. Both residues of rice crops also have high heating values of 15.84 MJ/kg and 15.09 MJ/kg, respectively, which indicates their energy content and potential as a source of power generation. Table 2 lists the important properties of rice crop residues [38,55,56]. The properties of crop residues affect combustion performance during power generation. For example, extractive content characteristics play a role in higher heating values (HHV) and lower ash contents. The presence of Na, K, and P also lowers the melting point of rice husk and rice straw, which may cause fouling and corrosion.

Table 2. Properties of rice crop residues.

Analysis	Properties	Rice Husk	Rice Straw	References
Constant volume	HHV MJ/kg	15.84	15.09	[38]
Proximate analysis (% dry fuel)	FC	16.22	15.86	[55]
	VM	63.52	65.47	
	AC	20.26	18.67	
Ultimate analysis (% dry fuel)	C	38.83	38.24	[55]
	H	4.75	5.20	
	O_2	35.47	36.26	
	N	0.52	0.87	
	S	0.05	0.18	
	Cl	0.12	0.58	
	AC	20.26	18.67	
Elemental analysis of ash (%)	SiO_2	91.42	74.67	[55,56]
	Al_2O3	0.78	1.04	
	TiO_2	0.02	0.09	
	Fe_2O_3	0.14	0.85	
	CaO	3.21	3.01	
	MgO	<0.01	1.75	
	Na_2O	0.21	0.96	
	K_2O	3.71	12.30	
	SO_3	0.72	1.24	
	P_2O_5	0.43	1.41	

5. The Conversion Process of Rice Crop Residues into Power

Rice straw has three key elements: silica, high cellulose, and a long decomposition period. Rice straw is widely used, including as animal feed and to produce non-wood fibers for newsprint production and corrugated mediums. Rice straw can be converted into bioethanol, a clean-burning fuel [57]; however, this requires a costly chemical conversion process. Rice straw can also be used as a fuel source in a combustor or utility boiler. It is necessary to avoid slagging and fouling during the boiling and combustion processes. There are two processes for harvesting energy from rice straw [58]: thermochemical and biochemical processes. The choice of conversion process depends upon the type and quantity of biomass feedstock; the desired form of the energy, i.e., end-user requirements; environmental standards; economic conditions; and project-specific factors. Figure 3 shows a typical rice husk power plant. The details of both thermochemical and biochemical processes are discussed in the following sections.

5.1. Thermochemical Processes

The thermochemical conversion processes can be classified into two categories: direct conversion of biomass to energy products, and conversion of biomass into another form which can be used later to produce energy. The thermochemical processes include direct combustion, gasification, and pyrolysis.

5.1.1. Direct Combustion

In this process, the chemical energy stored into the biomass is converted into heat, electricity, or mechanical power by burning the biomass in the presence of air. For feasible energy production, the biomass should have a moisture content of less than 50%. This process has some disadvantages. In most cases, pretreatment of biomass is important before burning. Some pretreatment processes include drying, chopping, grinding, etc. Pretreatment processes increase financial costs and energy expenditure [59,60]. Combustion is the most widely used thermochemical process, especially in developing countries, accounting for some 97% of the bioenergy obtained worldwide [61,62]. Indirect combustion processes refer to the use of biomass as a source of fuel in a boiler, contributing to the production of steam in the presence of oxygen in the boiler. The steam is used to generate heat and electricity concurrently using a turbine.

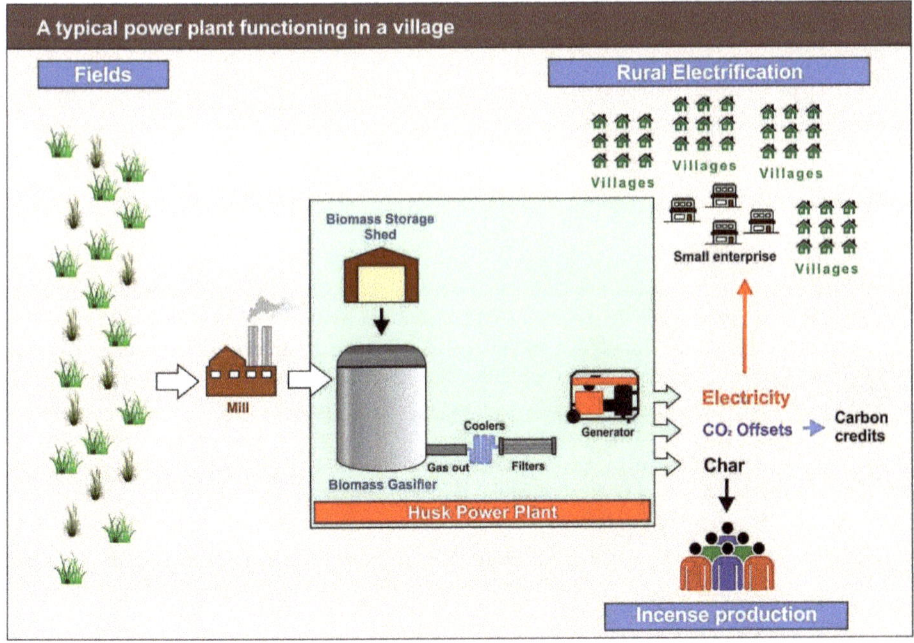

Figure 3. A typical rice husk power plant.

5.1.2. Gasification

The biomass is heated at high temperatures (800–900 °C) with an insufficient supply of air (partial oxidation); as a result, a combustible gas mixture is produced. Rice husk gasification power generation has become very popular in many Asian countries like China, Indonesia, India, Thailand, Cambodia, and the Philippines. The advantages of gasification are that the gas produced can be used in gas engines, gas turbines, and fuel cells for electricity generation at higher efficiency. Pode et al. [63] mentioned that one ton of rice husk can produce 800 kWh of electric power and can save about 1 ton of CO_2 emissions compared to current uses. Prasara et al. [64] mentioned that an ideal gasification process yields only non-condensable gas and ash residue. Susastriawan et al. [65] investigated the compatibility of a downdraft gasifier using different feedstocks such as rice husk, sawdust, and their mixture, and found that the optimum equivalence ratios of the producer gas were 3.13, 2.69, and 0.35 MJ/Nm^3, respectively. The potential of rural electricity generation using biomass gasification system was discussed by Abe et al. [66]. However, tree farming is required to provide a long-term biomass supply for gasification.

5.1.3. Pyrolysis

Pyrolysis occurs at high temperatures and in the absence of air with biomass decomposition. The nominal operating temperature range is 350–550 °C. The ratio of the products from pyrolysis, gas-, liquid-, and carbon-rich residues, depends on the operating conditions of the process. Bridgwater et al. [67] indicated that a fast pyrolysis process is very important and consists of the principal reaction systems and processes along with the resulting liquid products. Depending on the temperature, heating rate, and residence time, there are three main types of pyrolysis process found: namely, slow, fast, and flash. The main product of slow pyrolysis process is char, with a small amount of oil and gas. The fast pyrolysis process produces mainly pyrolysis oil, whereas the main product of flash pyrolysis is gas. Fukuda et al. [68] found that a maximum 50% wt. pyrolysis oil yield could be achieved through a fast pyrolysis process using rice husk. Another study showed that about 75% wt. pyrolysis oil can be obtained on a dry feed basis using a fast pyrolysis process [69].

5.2. Biochemical Processes

Different useful products can be obtained through the conversion of biomass using the following bio-chemical processes.

5.2.1. Anaerobic Digestion

An anaerobic digestion process refers to the conversion of biomass into biogas by microorganisms (the combination of carbon dioxide and methane) in the absence of oxygen. This biogas is an excellent source of fuel for the generation of heat and energy. Anaerobic digestion of rice straw is not a new concept; however, its renewable energy potential has barely been analyzed. Mussoline et al. [70] mentioned that optimum digestion conditions of pH (6.5–8.0), temperature (35–40 °C), and nutrients (C:N ratio of 25–35) could produce methane yields of 92–280 l/kg of volatile solids. Matin et al. [71] investigated the development of biogas production from rice husk by solid-state anaerobic digestion (SSAD) and found that lignin content was very difficult to degrade using microbes. They also found that a better biogas yield was produced by using the SSAD method compared with liquid anaerobic digestion (LAD). Haryanto et al. [72] investigated the effect of urea addition on the biogas yield from co-digestion of rice straw and cow dung using a semi-continuous anaerobic digester, and concluded that urea addition positively influenced the biogas yield and its quality.

5.2.2. Fermentation

Bioethanol produced from lignocellulose biomass generally involves three main steps: (i) pretreatment, (ii) enzymatic hydrolysis, and (iii) fermentation. The first step involves sieving and pelletizing which is very important for reduction of transportation expenses and handling fees. The second step involves the transformation of cellulose and hemicellulose biomass into glucose, pentoses, and hexoses [73]. In the third step, glucose is fermented into ethanol by a chosen microorganism. However, the conversion of rice straw and rice husk is possible through the different simultaneous processes, including saccharification and fermentation (SF) and separate enzymatic hydrolysis and fermentation processes. An overview of the fermentation process of biomass to produce ethanol was provided by Binod et al. [74]. They mentioned that the high ash and silica content in rice straw made it an inferior feedstock for bioethanol production. Swain et al. [75] used rice straw and wheat straw for bioethanol production and found some challenges, such as the lignin, ash, and silica content of rice straw, which required an appropriate pretreatment process. However, other researchers found that ethanol production from rice straw could be achieved up to 83.1% [76]. A summary of rice crop residue research in recent years is presented in Table 3.

Table 3. Summary of recent research in agricultural residues.

Technology	Details	Year	References
Gasification	Experimental investigation on the role of operating conditions on gas and tar composition and product distribution of rice husk gasification in the existence of dolomite.	2019	[77]
	Described the repossession of hydrogen gas from rice husk aided by nanoparticles.	2019	[78]
	Presented an investigational reference for biomass gasifier design and operation.	2016	[79]
	Investigated the feasibility of enhanced energy yield in gasification of rice straw by using a prepared iron-based catalyst.	2016	[80]
	Investigated the characteristics of rice husk from varieties of rice and the benefits of power generation using rice husk compared to a diesel generator.	2016	[81]
	Investigated the experimental statistics of air-staged cyclone gasification of rice crops.	2009	[82]
	Studied the role of different parameters on the gasification performance using rice husk biomass in power generation plant.	2009	[83]
	Studied the role of equivalence ratio on the performance of two-stage gasifier using rice straw.	2009	[84]
Pyrolysis	The physicochemical and toxicological characteristics of rice husk (RH) and rice husk ash (RHA) pyrolysis were investigated.	2019	[85]
	Bio-oil made from rice husk by a commercial-scale biomass fast pyrolysis plant was utilized to investigate the effects of long-period storage (two years) under three different conditions.	2019	[86]
	Investigated the potential tertiary treatment of wastewater by adsorption using rice husk biochar (RHC) obtained from microwave pyrolysis of rice husk.	2019	[87]
	Focused on liquid fuel production through co-pyrolysis of polythene waste and rice straw in varying compositions and characterized the liquid products.	2019	[88]
	Evaluated the effects of different demineralization processes on the pyrolysis behaviour and pyrolysis product properties of raw/terrified rice straw.	2018	[89]
	Investigated the pyrolysis process in a fluidized-bed reactor using rice husk biomass under different operating conditions.	2010	[90]
	Studied the effect of pyrolysis of rice straw biomass in a microwave-induced reactor for the production of hydrogen gas.	2010	[91]
	Produced bio-oil through fast pyrolysis using rice husk biomass in a fluidized-bed reactor.	2011	[92]

Table 3. *Cont.*

Technology	Details	Year	References
Anaerobic digestion	Evaluated the effect of pretreated rice straw on high-solid anaerobic co-digestion with swine manure, focusing on biogas production and kinetics.	2019	[93]
	Investigated rice straw physicochemical characteristics and anaerobic digestion (AD) performance via ammonia pretreatment.	2019	[94]
	Analyzed anaerobic digestion of heavy-metal-contaminated rice straw inoculated with waste-activated sludge.	2018	[95]
	Provided useful parameters to evaluate biogas production via anaerobic digestion of rice straw	2018	[96]
	Investigated the effect of trace element (TE) addition and NaOH pretreatment on the anaerobic digestion of rice straw.	2018	[97]
	Studied the role of sodium hydroxide composition and extractives on the improvement of biogas yield through anaerobic digestion.	2009	[98]
	Studied the influence of different operating condition and method on the production of biogas using rice straw through anaerobic digestion.	2009	[99]
	Studied the use of rice straw and swine feces for anaerobic co-digestion in a fed-batch single-phase reactor.	2009	[100]
	Investigated the co-digestion process for biogas production using cow dung and rice husk.	2009	[101]
	Studied anaerobic digestion performance using rice straw with acclimated sludge and phosphate at room temperature.	2010	[102]
	Reported on the variation of efficiency of anaerobic digestion using rice straw with the variation of solid concentration at different temperatures.	2010	[103]
Fermentation	Studied the ethanol production process through the synchronized saccharification and co-fermentation of rice straw biomass with *Candida tropicalis*.	2010	[104]
Direct combustion	Reported on the performance of rice husk combustion with bituminous coal in a cyclonic fluidized-bed reactor.	2009	[105]
	Investigated the co-firing performance of rice husk with coal in a fluidized-bed reactor with a small combustion chamber.	2009	[106]

6. Current Rice Crop Residue Management

Rice straw is the waste product of paddy production. It comprises up to 50–60% of the paddy itself, and is produced during every harvesting process. There must be a way to manage all the resulting waste. The current methods used by most farmers include open burning, soil incorporation, animal feed, and removal from the field. There is no productive method used due to the cost issues and lack of development in rural areas. Each method has different impacts on the environment and on the nutrient balance and long-term soil fertility due to continuous plantation activities [107].

The straw removal method is used widely in India, Bangladesh, and Nepal. The repeated process of straw removal has resulted in field soils with low potassium (K) and silicon (S), which will present a major problem in the coming years. Apart from that, the removed straw also serves different purposes, such as fuel for cooking in rural areas, ruminant fodder, and stable bedding. It could also be used in the papermaking industry. Most rice straw removal involves loose straw racking, baling it in small bales, and road-siding the bales. Processed rice straw is normally baled and hauled. The baling format, e.g., round, square, large, or small, mostly depends on the farmer and what method is preferable for disposal [108]. If the paddy field is small, straw removal will not be sufficient due to the lack of cost-effectiveness in using vehicle to remove straw from the field. Compared to other management methods, straw removal is the best choice for reducing pollutants; the only setback is the emissions caused by the use of the vehicles transporting the rice straw [109].

The most common method of rice straw disposal is burning. It is the easiest method, involves no additional management cost, and is one of the fastest ways to dispose of all rice straw during harvesting time. This method might be helpful for farmers and paddy field operators but it has a major effect on the environment. According to Dobermann, the burning disposal method ignites atmospheric pollution which also results in nutrient loss, but is the most cost-effective method and has the capability to reduce pests and diseases [107].

Normally, during combustion, very high amounts of CO_2 are released along with carbon monoxide (CO), methane, nitrogen oxides, sulfur dioxide, etc. Some of these gases are classified as toxic and may be human carcinogens. There have been medical issues faced by communities in Japan, India, and California due to continuous exposure to open burning, mainly asthma and pulmonary morbidity. This practice has been banned in several paddy-producing countries, including California. It is still widespread in Asian countries due to the lack of rules regulating open burning enforced by the government. Not only does open burning cause air pollution, but there are also high nutrient and energy losses. Most importantly, it is a clear waste of potential energy where straw energy could be dissipated into heat [110–112].

An alternative to open burning is soil incorporation, which is widely practiced among farmers to re-fertilize the soil. The nutrients can be recycled for the next crop cycle and the waste is turned to good use and helps to maintain soil quality. The incorporation method mostly uses rice straw and stubble by ploughing (wet soil during land preparation or dry soil during fallow periods). This particular method has caused a serious increase in methane gas emissions [108,110]. Table 4 shows the greenhouse gas (GHG) emission factor for different rice straw management.

Table 4. GHG emission factor (EF) for different rice straw management practices [113].

Rice Straw Management Practice	Name of Pollutant	Emission Factor
Open burning	CH_4 (methane) N_2O (nitrogen dioxide) Combustion factor	1.2 g/kg (dry fuel) 0.07 g/kg (dry fuel) 0.8
Scattering and incorporation of rice stubble and straw in the soil (wet condition)	CH_4 (wet soil) CH_4 (dry soil) Baseline EF for continuously flooded fields without organic amendment Conversion factor for rice straw amendment Scaling factor to account for differences in water regime during the cultivation period Scaling factor to account for differences in water regime in the pre-season before the cultivation period	129.77 kg/ton yield 36.99 kg/ton yield 1.3 1.0 for straw incorporated <30 days before cultivation 0.39 for straw incorporated >30 days before cultivation 0.78 for irrigated 1 for irrigated (<180 days); 1.22 for rain-fed
Composting and incorporation	CH_4 (wet soil) CH_4 (dry soil)	13.37 kg/ton yield 2.1 kg/ton yield
Rice straw used as animal feed	CH_4	10,000–20,000 gCH_4/ton dry weight
Rice straw for mushroom production	CH_4	7.27 gCH_4/ton dry weight

7. Challenges in Rice-Crop-Residue-Based Power Generation

The concept rice-crop-residue-based power generation represents an effective method of converting waste into electricity. The first ever rice-crop-based power plant was built in Jai Kheri village in the Patiala district of Punjab in 2006, with a capacity of 10 MW. In China, the first power plant based on rice straw was built, also in 2006, and since then, straw-based power generation has been developing in a remarkable way. The demand for straw in China is roughly 2.13 million tons/year, with 10 direct-fired power plants, one gasification plant, and other mixed-fired plants [114]. However, some challenges still exist in the establishment of rice-crop-based power plants. There are still high chances that these plants could face major problems due to ongoing changes in the state's agricultural and industrial activities. If there is any diversion in agricultural activity, such as from rice paddy to higher-value crops, then the amount of rice straw will drop, causing the plant to fail [115–117]. Zhiqiang [50] explained the challenges for the combustion of rice straw in China; the main challenges are harvesting issues, process and system considerations, technical improvement, and policy support. The details of the challenges are as follows.

7.1. High Ash Content

The main challenge is managing the high ash content of in rice straw, and also the alkali metals such as sodium and potassium that are present. During the combustion process, these chemicals could cause slagging, fouling, and corrosion in superheaters. Slagging refers to deposits of slag material and fouling is accumulation of unwanted particles on solid surfaces. In addition, there is the deterioration of catalysts for nitrogen oxide reduction. Ash from coal has properties useful for cement production, whereas ash from straw cannot be used for any other industries; therefore, it becomes an unmanaged waste [118].

7.2. Harvesting Issues

The collection of rice straw brings up several concerns. First, pests and disease infecting the rice straw could reduce the quality of it. The timeliness of operation has to be short to avoid high expenditures on labour cost; therefore, a system has to be developed to make the collection time just sufficient. Continuous removal of rice straw from paddy fields could cause loss of soil nutrients in the field, and extra fertilizer will be required to restore the minerals lost. However, the addition of fertilizer will have some negative impacts on the soil and may result in increased emissions. The usability of the machines in the field is also an important consideration. The machines should be capable of operating all conditions including muddy fields, and the field should be able to withstand the weight of the machinery used during the harvesting process. Grower attitudes mainly concern farmers' understanding of the collection system, for which the best way forward is to offer incentives to the farmers that provide rice straw.

7.3. Process and System Consideration

There must be a system established for each process, including collection, processing, and transportation of rice straw to the power plant. These systems could ensure that the cost of using machinery for harvesting and the quantity of rice straw could be determined. Drying is definitely required to reduce the moisture content in wet rice straw from 60–70%. It is necessary to wait for the moisture content to drop to 25% for complete combustion to be achieved. Apart from that, if the straw retains a high moisture content, the chance of fermentation increases, which would lessen the quality of the straw. It is more costly to transport wet rice straw compared to dry rice straw; this represents another reason that the drying process is necessary. Besides drying, rice straw should go through densification to increase the bulk density, which will reduce the logistical costs. Moisture content plays an important role in power generation; baled straw has a moisture content of 10–18%, and when the moisture content exceeds 13%, the power output is reduced by 2% for the same feedstock amount [119].

8. Logistical Analysis

Logistics practically describe a distributed flow of things from one point to another. Logistical analysis is very important due to the effect of the cost of transportation, which is relatively high due to the bulk density of rice straw [120]. In this case, we are referring to rice straw transportation from the paddy field to the power plant. Before being transported to the power plant, there are several pit stops it has to go through for processing and collection purposes. The four main processes are harvesting, collection, storage, and transportation. The main concern in harvesting would be increased speed and efficiency of harvesting, instead of focusing on increasing the straw yield. The current rice straw storage facilities in California are suitable, but there are several drawbacks in maintaining the quality of the rice straw.

Many end-users prefer good quality straws with low moisture content and low contamination. These end-users mostly use the straw for power generation purpose. When harvesting, collection, and storage activities increase, the transportation of rice straw to the corresponding end-user will eventually increase as well. Therefore, suitable transportation infrastructure should be established to make sure that sufficient quantities of rice straw can be distributed in an acceptable time [121].

Due to the bulk density discussed earlier, the use of rice straw for power generation has issues related to transportation cost. It is very expensive to transport rice straw compared to coal [122–129]. Besides rice straw, there are many other logistical impacts related to transportation, as discussed in Table 5. The cost of operating a rice straw power plant depends solely on the harvesting, processing, and transporting cost. The use of machinery in forming bales of rice straw is normally powered by diesel, where the cost of bales varies in different countries according to their operating cost. Forming rice straw into bales is very important due to the resulting size reduction and increase in bulk density. Much space is saved by compacting the rice straw into bale form before transporting.

The power requirement for a tractor is about 30 kW, which increases if a higher number of bales need to be produced in a short period of time. The authors mainly focused on the work rate, power requirement, size, quality and mass of bale on a small area of 50 ha of paddy production to determine the techno-economic analysis [130–133].

Table 5. Main parameters impacting logistical costs [134].

Parameter	Impact on Cost	Type of Biomass	Country
Distribution of efficient biomass management	Minimizing transportation cost	Any type	Spain
Effective and efficient planning of logistical operations	Minimizing transportation cost	Agricultural crop	Canada
Increased transport vehicle capacities	The minimum cost of transportation	Cotton plant stalks	Greece and Europe
Maximized truck utilization factor	Minimizing transportation cost	Herbaceous biomass	United States
Increased size of power plants	Logistical constraints on economic performance became less restrictive	Agricultural crop, agro-industrial, and wood waste	Italy
Biomass storage	Has a significant role in biomass logistics	Cotton stalk and almond tree pruning	Greece
Site productivity	A high productivity plantation would reduce the transportation cost	Eucalyptus	United States
Increased bulk density	The minimum cost of transportation	Agricultural and woody biomass	Canada
Development of more efficient collection and transport systems	The minimum cost of transportation	Corn stover	United States

On top of that, the most crucial criterion in baling is that the moisture content of the particular rice straw has to be below 20%. The best method by which to achieve this is to carry out the baling process 2–3 days after harvesting. There are two main baling techniques: small rectangular bales and big rectangular bales. Small rectangular bales are already being implemented in Thailand, whereas study has focused more on big rectangular bales for their higher density. Bales have to be dense to ease the transportation cost [135]. Figure 4 shows an overview of the logistical model. It explains how rice straw is transported to a power plant [134].

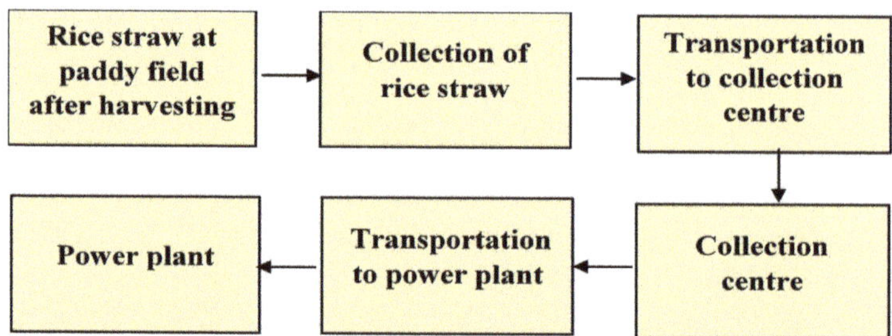

Figure 4. Overview of the logistical model [134].

The first process involved is rice straw being left to sun dry after harvesting; it is then baled in a rectangular form to ease transportation. Secondly, all rice straw in a particular field is collected and transported to a collection center, where all other straw bale from other paddies are also collected. Once the capacity of the collection center is almost reached, the rice straw is checked for quality and moisture content before being transported to a power plant for energy generation purposes. The quality and moisture content are checked to avoid any drop in power plant efficiency. Therefore, the collection center has to have equipment able to check the quality of rice straw, and also equipment to dry bales of rice straw if their moisture content exceeds the required limit.

9. Conclusions

The paper reviewed the possibility of power generation using rice husk and rice straw, and the various factors that impact their production. Based on the literature, the following can be concluded:

- Rice straw has very high potential for power generation; however, sustainable energy production depends on the availability of rice straw.
- Properties of rice straw vary between countries due to relative humidity and the use of different types of fertilizer.
- Though rice straw power plants are eco-friendly, the major problems preventing wide implementation are logistical issues, which contribute to about 35–50% of the total operation cost. Therefore, a detailed cost–benefit study must be conducted to avoid losses and make sure the power plants are sustainable in the future.

There are few things to be considered in the establishment of rice-crop-based power plants:

- The amount of rice straw available in an area, because it is the main supply of feedstock for the plant. A simple method could be used to determine the straw amount in an area, namely the straw to grain ratio. In most studies a ratio of 0.75 is used.
- Ash content and alkali metal contents affect the combustion process.
- The supply of rice straw should have a low moisture content to enable complete combustion and increase efficiency. If the moisture content exceeds 13%, the power output is reduced by 2%.

Before rice straw is fed into a power plant, it has to undergo preprocessing to reduce the moisture content, and the easiest method for this is open sun drying before collection. The next step is to have a decent rice straw supply chain to the power plant. Therefore, the power plant should be located in a strategic location. Finally, it can be concluded that rice husk and rice straw could be a potential source of renewable energy if various factors such as straw properties and logistics can be improved.

Author Contributions: Original draft preparation, M.M., J.L. and M.A.; Supervision, T.M.I.M., A.S.S. and A.H.S.; Review & Editing, S.M.A.R.

Funding: This research received no external funding.

Acknowledgments: This work was supported by Centre for Advanced Modeling and Geospatial Information Systems (CAMGIS) [Grant no. 321740.2232397]; School of Information, Systems and Modelling, University of Technology Sydney, Australia; Direktorat Jenderal Penguatan Riset dan Pengembangan Kementerian Riset, Teknologi dan Pendidikan Tinggi Republik Indonesia, [Grant no. 147/SP2H/LT/DRPM/2019] and Politeknik Negeri Medan, Medan, Indonesia.

Conflicts of Interest: The authors declare no conflict of interest.

References

1. Malecki, E.J. *Technology and Economic Development: The Dynamics of Local, Regional and National Change*; University of Illinois at Urbana-Champaign's Academy for Entrepreneurial Leadership Historical Research Reference in Entrepreneurship: Champaign, IL, USA, 1997.
2. Mofijur, M.; Masjuki, H.H.; Kalam, M.A.; Atabani, A.E. Evaluation of biodiesel blending, engine performance and emissions characteristics of Jatropha curcas methyl ester: Malaysian perspective. *Energy* **2013**, *55*, 879–887. [CrossRef]
3. Mofijur, M.; Hasan, M.M.; Mahlia, T.M.I.; Rahman, S.M.A.; Silitonga, A.S.; Ong, H.C. Performance and Emission Parameters of Homogeneous Charge Compression Ignition (HCCI) Engine: A Review. *Energies* **2019**, *12*, 3557. [CrossRef]
4. Milano, J.; Ong, H.C.; Masjuki, H.H.; Silitonga, A.S.; Chen, W.-H.; Kusumo, F.; Dharma, S.; Sebayang, A.H. Optimization of biodiesel production by microwave irradiation-assisted transesterification for waste cooking oil-Calophyllum inophyllum oil via response surface methodology. *Energy Convers. Manag.* **2018**, *158*, 400–415. [CrossRef]
5. Mofijur, M.; Atabani, A.E.; Masjuki, H.H.; Kalam, M.A.; Masum, B.M. A study on the effects of promising edible and non-edible biodiesel feedstocks on engine performance and emissions production: A comparative evaluation. *Renew. Sustain. Energy Rev.* **2013**, *23*, 391–404. [CrossRef]
6. Mofijur, M.; Masjuki, H.H.; Kalam, M.A.; Atabani, A.E.; Fattah, I.M.R.; Mobarak, H.M. Comparative evaluation of performance and emission characteristics of Moringa oleifera and Palm oil based biodiesel in a diesel engine. *Ind. Crops Prod.* **2014**, *53*, 78–84. [CrossRef]
7. Mahlia, T.M.I.; Syaheed, H.; Abas, A.E.P.; Kusumo, F.; Shamsuddin, A.H.; Ong, H.C.; Bilad, M.R. Organic Rankine Cycle (ORC) System Applications for Solar Energy: Recent Technological Advances. *Energies* **2019**, *12*, 2930. [CrossRef]
8. Singh, R.; Tevatia, R.; White, D.; Demirel, Y.; Blum, P. Comparative kinetic modeling of growth and molecular hydrogen overproduction by engineered strains of Thermotoga maritima. *Int. J. Hydrogen Energy* **2019**, *44*, 7125–7136. [CrossRef]
9. Statistica. World Carbon Dioxide Emissions from 2008 to 2018, by Region (in Million Metric Tons of Carbon Dioxide). Available online: http://www.statista.com/statistics/271748/the-largest-emitters-of-co2-in-the-world/ (accessed on 26 August 2018).
10. Hast, A.; Alimohammadisagvand, B.; Syri, S. Consumer attitudes towards renewable energy in China—The case of Shanghai. *Sustain. Cities Soc.* **2015**, *17*, 69–79. [CrossRef]
11. Zhang, W.; Liu, S.; Li, N.; Xie, H.; Li, X. Development forecast and technology roadmap analysis of renewable energy in buildings in China. *Renew. Sustain. Energy Rev.* **2015**, *49*, 395–402. [CrossRef]
12. Norhasyima, R.S.; Mahlia, T.M.I. Advances in CO2 utilization technology: A patent landscape review. *J. CO2 Util.* **2018**, *26*, 323–335. [CrossRef]
13. Renew Energy World. Punjab Gets World's First Rice Straw Power Station. Available online: http://www.renewableenergyworld.com/articles/2006/12/punjab-gets-worlds-first-rice-straw-power-station-46979.html (accessed on 5 August 2019).
14. Uddin, M.N.; Techato, K.; Taweekun, J.; Rahman, M.M.; Rasul, M.G.; Mahlia, T.M.I.; Ashrafur, S.M. An Overview of Recent Developments in Biomass Pyrolysis Technologies. *Energies* **2018**, *11*, 3115. [CrossRef]
15. Anwar, M.; Rasul, M.G.; Ashwath, N.; Rahman, M.M. Optimisation of Second-Generation Biodiesel Production from Australian Native Stone Fruit Oil Using Response Surface Method. *Energies* **2018**, *11*, 2566. [CrossRef]
16. Silitonga, A.S.; Mahlia, T.M.I.; Ong, H.C.; Riayatsyah, T.M.I.; Kusumo, F.; Ibrahim, H.; Dharma, S.; Gumilang, D. A comparative study of biodiesel production methods for Reutealis trisperma biodiesel. *Energy Sour. Part A Recover. Util. Environ. Eff.* **2017**, *39*, 2006–2014. [CrossRef]

17. Hossain, N.; Zaini, J.; Mahlia, T.; Azad, A.K. Elemental, morphological and thermal analysis of mixed microalgae species from drain water. *Renew. Energy* **2019**, *131*, 617–624. [CrossRef]

18. Hasan, M.H.; Muzammil, W.K.; Mahlia, T.M.I.; Jannifar, A.; Hasanuddin, I. A review on the pattern of electricity generation and emission in Indonesia from 1987 to 2009. *Renew. Sustain. Energy Rev.* **2012**, *16*, 3206–3219. [CrossRef]

19. Gallego-Castillo, C.; Victoria, M. Cost-free feed-in tariffs for renewable energy deployment in Spain. *Renew. Energy* **2015**, *81*, 411–420. [CrossRef]

20. Rexhäuser, S.; Löschel, A. Invention in energy technologies: Comparing energy efficiency and renewable energy inventions at the firm level. *Energy Policy* **2015**, *83*, 206–217. [CrossRef]

21. Ackah, I.; Kizys, R. Green growth in oil producing African countries: A panel data analysis of renewable energy demand. *Renew. Sustain. Energy Rev.* **2015**, *50*, 1157–1166. [CrossRef]

22. Coh, B.H.H.; Ong, H.C.; Cheah, M.Y.; Chen, W.H.; Yu, K.L.; Mahlia, T.M.I. Sustainability of direct biodiesel synthesis from microalgae biomass: A critical review. *Renew. Sustain. Energy Rev.* **2019**, *107*, 59–74. [CrossRef]

23. Ismail, M.S.; Moghavvemi, M.; Mahlia, T.M.I. Techno-economic analysis of an optimized photovoltaic and diesel generator hybrid power system for remote houses in a tropical climate. *Energy Convers. Manag.* **2013**, *69*, 163–173. [CrossRef]

24. Ismail, M.S.; Moghavvemi, M.; Mahlia, T.M.I. Characterization of PV panel and global optimization of its model parameters using genetic algorithm. *Energy Convers. Manag.* **2013**, *73*, 10–25. [CrossRef]

25. Silitonga, A.S.; Masjuki, H.H.; Ong, H.C.; Sebayang, A.H.; Dharma, S.; Kusumo, F.; Siswantoro, J.; Milano, J.; Daud, K.; Mahlia, T.M.I.; et al. Evaluation of the engine performance and exhaust emissions of biodiesel-bioethanol-diesel blends using kernel-based extreme learning machine. *Energy* **2018**, *159*, 1075–1087. [CrossRef]

26. Silitonga, A.S.; Masjuki, H.H.; Mahlia, T.M.I.; Ong, H.C.; Chong, W.T. Experimental study on performance and exhaust emissions of a diesel engine fuelled with Ceiba pentandra biodiesel blends. *Energy Convers. Manag.* **2013**, *76*, 828–836. [CrossRef]

27. Mofijur, M.; Mahlia, T.M.I.; Silitonga, A.S.; Ong, H.C.; Silakhori, M.; Hasan, M.H.; Putra, N.; Rahman, S.M.A. Phase Change Materials (PCM) for Solar Energy Usages and Storage: An Overview. *Energies* **2019**, *12*, 3167. [CrossRef]

28. Amin, M.; Putra, N.; Kosasih, E.A.; Prawiro, E.; Luanto, R.A.; Mahlia, T.M.I. Thermal properties of beeswax/graphene phase change material as energy storage for building applications. *Appl. Therm. Eng.* **2017**, *112*, 273–280. [CrossRef]

29. Latibari, S.T.; Mehrali, M.; Mehrali, M.; Mahlia, T.M.I.; Metselaar, H.S.C. Synthesis, characterization and thermal properties of nanoencapsulated phase change materials via sol-gel method. *Energy* **2013**, *61*, 664–672. [CrossRef]

30. Mehrali, M.; Latibari, S.T.; Mehrali, M.; Mahlia, T.M.I.; Metselaar, H.S.C.; Naghavi, M.S.; Sadeghinezhad, E.; Akhiani, A.R. Preparation and characterization of palmitic acid/graphene nanoplatelets composite with remarkable thermal conductivity as a novel shape-stabilized phase change material. *Appl. Therm. Eng.* **2013**, *61*, 633–640. [CrossRef]

31. Ong, H.C.; Masjuki, H.H.; Mahlia, T.M.I.; Silitonga, A.S.; Chong, W.T.; Leong, K.Y. Optimization of biodiesel production and engine performance from high free fatty acid Calophyllum inophyllum oil in CI diesel engine. *Energy Convers. Manag.* **2014**, *81*, 30–40. [CrossRef]

32. Silitonga, A.S.; Mahlia, T.M.I.; Kusumo, F.; Dharma, S.; Sebayang, A.H.; Sembiring, R.W.; Shamsuddin, A.H. Intensification of Reutealis trisperma biodiesel production using infrared radiation: Simulation, optimisation and validation. *Renew. Energy* **2019**, *133*, 520–527. [CrossRef]

33. Kusumo, F.; Silitonga, A.S.; Ong, H.C.; Masjuki, H.H.; Mahlia, T.M.I. A comparative study of ultrasound and infrared transesterification of Sterculia foetida oil for biodiesel production. *Energy Sources Part. A Recovery Util. Environ. Eff.* **2017**, *39*, 1339–1346. [CrossRef]

34. Ong, H.C.; Milano, J.; Silitonga, A.S.; Hassan, M.H.; Shamsuddin, A.H.; Wang, C.T.; Mahlia, T.M.I.; Siswantoro, J.; Kusumo, F.; Sutrisno, J. Biodiesel production from Calophyllum inophyllum-Ceiba pentandra oil mixture: Optimization and characterization. *J. Clean. Prod.* **2019**, *219*, 183–198. [CrossRef]

35. Silitonga, A.; Shamsuddin, A.; Mahlia, T.; Milano, J.; Kusumo, F.; Siswantoro, J.; Dharma, S.; Sebayang, A.; Masjuki, H.; Ong, H.C. Biodiesel synthesis from Ceiba pentandra oil by microwave irradiation-assisted transesterification: ELM modeling and optimization. *Renew. Energy* **2020**, *146*, 1278–1291. [CrossRef]

36. Biomass-Using Anaerobic Digestion. Biomass Types. Available online: http://www.esru.strath.ac.uk/EandE/Web_sites/03-04/biomass/background%20info4.html (accessed on 10 July 2019).

37. RM Wheildon Limited. Biomass Log Burning Boilers. Available online: http://www.wheildons.co.uk/biomass-log-burning-boilers/ (accessed on 4 August 2019).

38. Lim, J.S.; Abdul Manan, Z.; Wan Alwi, S.R.; Hashim, H. A review on utilisation of biomass from rice industry as a source of renewable energy. *Renew. Sustain. Energy Rev.* **2012**, *16*, 3084–3094. [CrossRef]

39. Aditiya, H.B.; Chong, W.T.; Mahlia, T.M.I.; Sebayang, A.H.; Berawi, M.A.; Hadi, N. Second generation bioethanol potential from selected Malaysia's biodiversity biomasses: A review. *Waste Manag.* **2015**, *47*, 46–61. [CrossRef]

40. Tan, S.X.; Lim, S.; Ong, H.C.; Pang, Y.L. State of the art review on development of ultrasound-assisted catalytic transesterification process for biodiesel production. *Fuel* **2019**, *235*, 886–907. [CrossRef]

41. Ong, H.C.; Masjuki, H.H.; Mahlia, T.M.I.; Silitonga, A.S.; Chong, W.T.; Yusaf, T. Engine performance and emissions using Jatropha curcas, Ceiba pentandra and Calophyllum inophyllum biodiesel in a CI diesel engine. *Energy* **2014**, *69*, 427–445. [CrossRef]

42. Chang, Y.; Li, Y. Renewable energy and policy options in an integrated ASEAN electricity market: Quantitative assessments and policy implications. *Energy Policy* **2015**, *85*, 39–49. [CrossRef]

43. Energy Informative. Biomass Energy Pros and Cons. Available online: http://energyinformative.org/biomass-energy-pros-and-cons/ (accessed on 5 August 2019).

44. Energy Alternatives India. Benefits of Biomass power. Available online: http://www.eai.in/ref/ae/bio/ben/benefits_biomass_power.html (accessed on 5 August 2019).

45. CFF Conserve Energy Future. Biomass Energy. Available online: http://www.conserve-energy-future.com/Advantages_Disadvantages_BiomassEnergy.php (accessed on 8 August 2019).

46. Bioenergy Technology LTD. The Advantages and Disadvantages of Biomass. Available online: http://www.bioenergy.org/information/advantages-disadvantages-biomass/ (accessed on 5 August 2019).

47. Mohiuddin, O.; Mohiuddin, A.; Obaidullah, M.; Ahmed, H.; Asumadu-Sarkodie, S. Electricity production potential and social benefits from rice husk, a case study in Pakistan. *Cogent Eng.* **2016**, *3*, 1177156. [CrossRef]

48. Feedipedia. Rice Straw. Available online: http://www.feedipedia.org/node/557 (accessed on 5 August 2019).

49. Kiran, L.K.; Loyd, H.F.; Alan Jacobson, W. Rice straw as a lignocellulosic resource: Collection, processing, transportation, and environmental aspects. *Biomass Bioenergy* **2000**, *18*, 369–389.

50. Liu, Z.; Xu, A.; Long, B. Energy from Combustion of Rice Straw: Status and Challenges to China. *Energy Power Eng.* **2011**, *03*, 325–331. [CrossRef]

51. Shie, J.L.; Lee, C.H.; Chen, C.S.; Lin, K.L.; Chang, C.Y. Scenario comparisons of gasification technology using energy life cycle assessment for bioenergy recovery from rice straw in Taiwan. *Energy Convers. Manag.* **2014**, *87*, 156–163. [CrossRef]

52. Food and Agriculture Organization of the United Nations Statistical Database. 2017. Available online: http://www.fao.org/statistics/en/ (accessed on 30th July 2019).

53. Werner Siemers. Technical and Economic Prospects of Rice Residues for Energy in Asia. In Proceedings of the Sustainable Bioenergy Symposium, Bangkok, Thailand, 2 June 2012.

54. Shafie, S.M.; Mahlia, T.M.I.; Masjuki, H.H.; Rismanchi, B. Life cycle assessment (LCA) of electricity generation from rice husk in Malaysia. *Energy Procedia* **2012**, *14*, 499–504. [CrossRef]

55. Jenkins, B.M.; Baxter, L.L.; Miles, T.R., Jr.; Miles, T.R. Combustion properties of biomass. *Fuel Proc. Technol.* **1998**, *54*, 17–46. [CrossRef]

56. Beidaghy Dizaji, H.; Zeng, T.; Hartmann, I.; Enke, D.; Schliermann, T.; Lenz, V.; Bidabadi, M. Generation of High Quality Biogenic Silica by Combustion of Rice Husk and Rice Straw Combined with Pre-and Post-Treatment Strategies—A Review. *Appl. Sci.* **2019**, *9*, 1083. [CrossRef]

57. Zheng, J.; Zhu, X.; Guo, Q.; Zhu, Q. Thermal conversion of rice husks and sawdust to liquid fuel. *Waste Manag.* **2006**, *26*, 1430–1435. [CrossRef]

58. Matsumura, Y.; Minowa, T.; Yamamoto, H. Amount, availability, and potential use of rice straw (agricultural residue) biomass as an energy resource in Japan. *Biomass Bioenergy* **2005**, *29*, 347–354. [CrossRef]

59. McKendry, P. Energy production from biomass (part 1): Overview of biomass. *Bioresour. Technol.* **2002**, *83*, 37–46. [CrossRef]

60. Goyal, H.B.; Seal, D.; Saxena, R.C. Bio-fuels from thermochemical conversion of renewable resources: A review. *Renew. Sustain. Energy Rev.* **2008**, *12*, 504–517. [CrossRef]

61. Bridgwater, A.V. Renewable fuels and chemicals by thermal processing of biomass. *Chem. Eng. J.* **2003**, *91*, 87–102. [CrossRef]

62. Zhang, L.; Xu, C.; Champagne, P. Overview of recent advances in thermo-chemical conversion of biomass. *Energy Convers. Manag.* **2010**, *51*, 969–982. [CrossRef]

63. Pode, R. Potential applications of rice husk ash waste from rice husk biomass power plant. *Renew. Sustain. Energy Rev.* **2016**, *53*, 1468–1485. [CrossRef]

64. Prasara-A, J.; Gheewala, S.H. Sustainable utilization of rice husk ash from power plants: A review. *J. Clean. Prod.* **2017**, *167*, 1020–1028. [CrossRef]

65. Susastriawan, A.A.P.; Saptoadi, H. Comparison of the gasification performance in the downdraft fixed-bed gasifier fed by different feedstocks: Rice husk, sawdust, and their mixture. *Sustain. Energy Technol. Assess.* **2019**, *34*, 27–34. [CrossRef]

66. Abe, H.; Katayama, A.; Sah, B.P.; Toriu, T.; Samy, S.; Pheach, P.; Adams, M.A.; Grierson, P.F. Potential for rural electrification based on biomass gasification in Cambodia. *Biomass Bioenergy* **2007**, *31*, 656–664. [CrossRef]

67. Bridgwater, A.V.; Peacocke, G.V.C. Fast pyrolysis processes for biomass. *Renew. Sustain. Energy Rev.* **2000**, *4*, 1–73. [CrossRef]

68. Fukuda, S. Pyrolysis Investigation For Bio-Oil Production From Various Biomass Feedstocks In Thailand. *Int. J. Green Energy* **2015**, *12*, 215–224. [CrossRef]

69. Bridgwater, A.V.; Toft, A.J.; Brammer, J.G. A techno-economic comparison of power production by biomass fast pyrolysis with gasification and combustion. *Renew. Sustain. Energy Rev.* **2002**, *6*, 181–246. [CrossRef]

70. Mussoline, W.; Esposito, G.; Giordano, A.; Lens, P. The Anaerobic Digestion of Rice Straw: A Review. *Critic. Rev. Environ. Sci. Technol.* **2013**, *43*, 895–915. [CrossRef]

71. Hawali Abdul Matin, H. Biogas Production from Rice Husk Waste by using Solid State Anaerobic Digestion (SSAD) Method. *E3S Web Conf.* **2018**, *31*, 02007. [CrossRef]

72. Haryanto, A.; Sugara, B.P.; Telaumbanua, M.; Rosadi, R.A.B. Anaerobic Co-digestion of Cow Dung and Rice Straw to Produce Biogas using Semi-Continuous Flow Digester: Effect of Urea Addition. *IOP Conf. Ser. Earth Environ. Sci.* **2018**, *147*, 012032. [CrossRef]

73. Taherzadeh, M.J.; Niklasson, C. Ethanol from Lignocellulosic Materials: Pretreatment, Acid and Enzymatic Hydrolyses, and Fermentation. In *Lignocellulose Biodegradation*; American Chemical Society: Washington, DC, USA, 2004; Volume 889, pp. 49–68.

74. Binod, P.; Sindhu, R.; Singhania, R.R.; Vikram, S.; Devi, L.; Nagalakshmi, S.; Kurien, N.; Sukumaran, R.K.; Pandey, A. Bioethanol production from rice straw: An overview. *Bioresour. Technol.* **2010**, *101*, 4767–4774. [CrossRef] [PubMed]

75. Swain, M.R.; Singh, A.; Sharma, A.K.; Tuli, D.K. Chapter 11—Bioethanol Production From Rice- and Wheat Straw: An Overview. In *Bioethanol Production from Food Crops*; Ray, R.C., Ramachandran, S., Eds.; Academic Press: Cambridge, MA, USA, 2019; pp. 213–231. [CrossRef]

76. Ko, J.K.; Bak, J.S.; Jung, M.W.; Lee, H.J.; Choi, I.-G.; Kim, T.H.; Kim, K.H. Ethanol production from rice straw using optimized aqueous-ammonia soaking pretreatment and simultaneous saccharification and fermentation processes. *Bioresour. Technol.* **2009**, *100*, 4374–4380. [CrossRef] [PubMed]

77. Zhang, G.; Liu, H.; Wang, J.; Wu, B. Catalytic gasification characteristics of rice husk with calcined dolomite. *Energy* **2018**, *165*, 1173–1177. [CrossRef]

78. Saravana Sathiya Prabhahar, R.; Nagaraj, P.; Jeyasubramanian, K. Enhanced recovery of H2 gas from rice husk and its char enabled with nano catalytic pyrolysis/gasification. *Microchem. J.* **2019**, *146*, 922–930. [CrossRef]

79. Zhai, M.; Xu, Y.; Guo, L.; Zhang, Y.; Dong, P.; Huang, Y. Characteristics of pore structure of rice husk char during high-temperature steam gasification. *Fuel* **2016**, *185*, 622–629. [CrossRef]

80. Chiang, K.-Y.; Liao, C.-K.; Lu, C.-H. The effects of prepared iron-based catalyst on the energy yield in gasification of rice straw. *Int. J. Hydrogen Energy* **2016**, *41*, 21747–21754. [CrossRef]

81. Olupot, P.W.; Candia, A.; Menya, E.; Walozi, R. Characterization of rice husk varieties in Uganda for biofuels and their techno-economic feasibility in gasification. *Chem. Eng. Res. Des.* **2016**, *107*, 63–72. [CrossRef]

82. Sun, S.; Zhao, Y.; Ling, F.; Su, F. Experimental research on air staged cyclone gasification of rice husk. *Fuel Proc. Technol.* **2009**, *90*, 465–471. [CrossRef]

83. Chen, L.; Su, Y.; Chen, Y.; Luo, Y.-H.; Lu, F.; Wu, W.-G. Effect of equivalence ratio on gasification characteristics in a rice straw two-stage gasifier. *Proc. CSEE* **2009**, *29*, 102–107.

84. Wu, C.Z.; Yin, X.L.; Ma, L.L.; Zhou, Z.Q.; Chen, H.P. Operational characteristics of a 1.2-MW biomass gasification and power generation plant. *Biotechnol. Adv.* **2009**, *27*, 588–592. [CrossRef]

85. Almeida, S.R.; Elicker, C.; Vieira, B.M.; Cabral, T.H.; Silva, A.F.; Sanches Filho, P.J.; Raubach, C.W.; Hartwig, C.A.; Mesko, M.F.; Moreira, M.L.; et al. Black SiO2 nanoparticles obtained by pyrolysis of rice husk. *Dyes Pigments* **2019**, *164*, 272–278. [CrossRef]

86. Cai, W.; Kang, N.; Jang, M.K.; Sun, C.; Liu, R.; Luo, Z. Long term storage stability of bio-oil from rice husk fast pyrolysis. *Energy* **2019**. [CrossRef]

87. Shukla, N.; Sahoo, D.; Remya, N. Biochar from microwave pyrolysis of rice husk for tertiary wastewater treatment and soil nourishment. *J. Clean. Prod.* **2019**, *235*, 1073–1079. [CrossRef]

88. Hossain, M.S.; Ferdous, J.; Islam, M.S.; Islam, M.R.; Mustafi, N.N.; Haniu, H. Production of liquid fuel from co-pyrolysis of polythene waste and rice straw. *Energy Procedia* **2019**, *160*, 116–122. [CrossRef]

89. Dong, Q.; Zhang, S.; Ding, K.; Zhu, S.; Zhang, H.; Liu, X. Pyrolysis behavior of raw/torrefied rice straw after different demineralization processes. *Biomass Bioenergy* **2018**, *119*, 229–236. [CrossRef]

90. Heo, H.S.; Park, H.J.; Dong, J.I.; Park, S.H.; Kim, S.; Suh, D.J.; Suh, Y.W.; Kim, S.S.; Park, Y.K. Fast pyrolysis of rice husk under different reaction conditions. *J. Ind. Eng. Chem.* **2010**, *16*, 27–31. [CrossRef]

91. Huang, Y.F.; Kuan, W.H.; Lo, S.L.; Lin, C.F. Hydrogen-rich fuel gas from rice straw via microwave-induced pyrolysis. *Bioresour. Technol.* **2010**, *101*, 1968–1973. [CrossRef]

92. Chen, T.; Wu, C.; Liu, R.; Fei, W.; Liu, S. Effect of hot vapor filtration on the characterization of bio-oil from rice husks with fast pyrolysis in a fluidized-bed reactor. *Bioresour. Technol.* **2011**, *102*, 6178–6185. [CrossRef]

93. Qian, X.; Shen, G.; Wang, Z.; Zhang, X.; Chen, X.; Tang, Z.; Lei, Z.; Zhang, Z. Enhancement of high solid anaerobic co-digestion of swine manure with rice straw pretreated by microwave and alkaline. *Bioresour. Technol. Rep.* **2019**, *7*, 100208. [CrossRef]

94. Yuan, H.; Guan, R.; Wachemo, A.C.; Zhang, Y.; Zuo, X.; Li, X. Improving physicochemical characteristics and anaerobic digestion performance of rice straw via ammonia pretreatment at varying concentrations and moisture levels. *Chin. J. Chem. Eng.* **2019**. [CrossRef]

95. Xin, L.; Guo, Z.; Xiao, X.; Xu, W.; Geng, R.; Wang, W. Feasibility of anaerobic digestion for contaminated rice straw inoculated with waste activated sludge. *Bioresour. Technol.* **2018**, *266*, 45–50. [CrossRef] [PubMed]

96. Candia-García, C.; Delgadillo-Mirquez, L.; Hernandez, M. Biodegradation of rice straw under anaerobic digestion. *Environ. Technol. Innov.* **2018**, *10*, 215–222. [CrossRef]

97. Mancini, G.; Papirio, S.; Riccardelli, G.; Lens, P.N.L.; Esposito, G. Trace elements dosing and alkaline pretreatment in the anaerobic digestion of rice straw. *Bioresour. Technol.* **2018**, *247*, 897–903. [CrossRef] [PubMed]

98. He, Y.; Pang, Y.; Li, X.; Liu, Y.; Li, R.; Zheng, M. Investigation on the changes of main compositions and extractives of rice straw pretreated with sodium hydroxide for biogas production. *Energy Fuels* **2009**, *23*, 2220–2224. [CrossRef]

99. Fu, Y.; Luo, T.; Mei, Z.; Li, J.; Qiu, K.; Ge, Y. Dry Anaerobic Digestion Technologies for Agricultural Straw and Acceptability in China. *Sustainability* **2018**, *10*, 4588. [CrossRef]

100. Chen, G.; Zheng, Z.; Zou, X. Anaerobic co-digestion of rice straw and swine feces. *J. Agro-Environ. Sci.* **2009**, *28*, 185–188.

101. Iyagba, E.T.; Mangibo, I.A.; Mohammad, Y.S. The study of cow dung as co-substrate with rice husk in biogas production. *Sci. Res. Essays* **2009**, *4*, 861–866.

102. Lei, Z.; Chen, J.; Zhang, Z.; Sugiura, N. Methane production from rice straw with acclimated anaerobic sludge: Effect of phosphate supplementation. *Bioresour. Technol.* **2010**, *101*, 4343–4348. [CrossRef]

103. Li, L.; Li, D.; Sun, Y.; Ma, L.; Yun, Z.; Kong, X. Effect of temperature and solid concentration on anaerobic digestion of rice straw in South China. *Int. J. Hydrogen Energy* **2010**, *35*, 7261–7266. [CrossRef]

104. Oberoi, H.S.; Vadlani, P.V.; Brijwani, K.; Bhargav, V.K.; Patil, R.T. Enhanced ethanol production via fermentation of rice straw with hydrolysate-adapted Candida tropicalis ATCC 13803. *Process. Biochem.* **2010**, *45*, 1299–1306. [CrossRef]

105. Madhiyanon, T.; Sathitruangsak, P.; Soponronnarit, S. Co-combustion of rice husk with coal in a cyclonic fluidized-bed combustor (ψ-FBC). *Fuel* **2009**, *88*, 132–138. [CrossRef]

106. Sathitruangsak, P.; Madhiyanon, T.; Soponronnarit, S. Rice husk co-firing with coal in a short-combustion-chamber fluidized-bed combustor (SFBC). *Fuel* **2009**, *88*, 1394–1402. [CrossRef]

107. Dobermann, A.; Fairhust, T.H. Rice Straw Management. *Better Crops Int.* **2002**, *16*, 7–11.

108. Steven, C.B.; Karen, M.J.; Carl, M.W.; Williams, J.F. With a ban on burning, incorporating rice straw into soil may become disposal option for growers. *Calif. Agricult.* **1993**, *47*, 8–12.

109. Monteleone, M.; Cammerino, A.R.B.; Garofalo, P.; Delivand, M.K. Straw-to-soil or straw-to-energy? An optimal trade off in a long term sustainability perspective. *Appl. Energy* **2015**, *154*, 891–899. [CrossRef]

110. Laura, D.-E.; Porcar, M. Rice straw management: The Big Waste. *Biofuels Bioprod. Bioref.* **2010**, *4*, 154–159.

111. Arai, H.; Hosen, Y.; Hong Van, N.P.; Nga, T.T.; Chiem, N.H.; Inubushi, K. Greenhouse gas emissions from rice straw burning and straw-mushroom cultivation in a triple rice cropping system in the Mekong Delta. *Soil Sci. Plant. Nutr.* **2015**, 1–17. [CrossRef]

112. Kumar, P.; Kumar, S.; Joshi, L. *Valuation of the Health Effects. Socioeconomic and Environmental Implications of Agricultural Residue Burning: A Case Study of Punjab, India*; Springer: New Delhi, India, 2015; pp. 35–67.

113. Cheryll, C.L.; Constancio, A.A.; Rowena, G.M.; Evelyn, F. *Javier. Economic Analysis of Rice Straw Management Alternatives and Understanding Farmers' Choices*; Philippine Rice Research Institute: Nueva Ecija, Philippines, 2013.

114. Zhang, Q.; Zhou, D.; Zhou, P.; Ding, H. Cost Analysis of straw-based power generation in Jiangsu Province, China. *Appl. Energy* **2013**, *102*, 785–793. [CrossRef]

115. Koshy Cherail. Straw Can Generate Power, If Available. Available online: http://www.downtoearth.org.in/news/straw-can-generate-power-if-available-29753 (accessed on 5 August 2019).

116. Singh, J. Overview of electric power potential of surplus agricultural biomass from economic, social, environmental and technical perspective—A case study of Punjab. *Renew. Sustain. Energy Rev.* **2015**, *42*, 286–297. [CrossRef]

117. Renew Energy World. Punjab Biomass Plans 96 Megawatts of Renewable Projects in India. Available online: http://www.renewableenergyworld.com/news/2012/10/punjab-biomass-plans-96-megawatts-of-renewable-projects-in-india.html (accessed on 15 August 2019).

118. Wirseniues, S. The Biomass Metabolism of the Food System: A Model-Based Survey of the Global and Regional Turnover of Food Biomass. *J. Ind. Ecol.* **2003**, *7*, 47. [CrossRef]

119. Suramaythangkoor, T.; Gheewala, S.H. Potential alternatives of heat and power technology application using rice straw in Thailand. *Appl. Energy* **2010**, *87*, 128–133. [CrossRef]

120. Pulkit, A.G.; Vedant, S.; Mohan Krishna, G. Rice Husk Power Systems: Exploring Alternate Source of Energy. In *Promoting Socio-Economic Development through Business Integration*; IGI Global: Hershey, PA, USA, 2015; pp. 112–123.

121. Michael, P.K. *Recommendations for Rice Straw Supply*; California Energy Commission: Sacramento, CA, USA, 2001.

122. Arifa, S.; Amit, K. Optimal configuration and combination of multiple lignocellulosic biomass feedstocks delivery to a biorefinery. *Bioresour. Technol.* **2011**, *102*, 9947–9956.

123. Gonzaleza, R.; Treasurea, T.; Wrighta, J.; Salonia, D.; Phillipsa, R.; Abtb, R.; Jameela, H. Exploring the potential of Eucalyptus for energy production in the Southern United States: Financial analysis of delivered biomass. Part, I. *Biomass Bioenergy* **2011**, *35*, 755–766. [CrossRef]

124. Athanasios, A.R.; Athanasios, J.T.; Ilias, P. Tatsiopoulos. Logistics issues of biomass: The storage problem and the multi-biomass supply chain. *Renew. Sustain. Energy Rev.* **2009**, *13*, 887–894.

125. Antonio, C.C.; Palumbo, M.; Pacifico, M.P.; Scacchia, F. Economics of biomass energy utilization in combustion and gasification plants: Effects of logistic variables. *Biomass Bioenergy* **2005**, *28*, 35–51.

126. Sokhansanja, S.; Manib, S.; Tagorec, S.; Turhollowa, A.F. Techno-economic analysis of using corn stover to supply heat and power to a corn ethanol plant—Part 1: Cost of feedstock supply logistics. *Biomass Bioenergy* **2010**, *34*, 75–81. [CrossRef]

127. Ravula, P.P.; Grisso, R.D.; Cundiff, J.S. Comparison between two policy strategies for scheduling trucks in a biomass logistic system. *Bioresour. Technol.* **2008**, *99*, 5710–5721. [CrossRef]

128. Mahmood, E.; Taraneh, S.; Shahab, S.; Mark, S.; Lawrence, T.-S. A new simulation model for multi-agricultural biomass logistics system in bioenergy production. *Biosyst. Eng.* **2011**, *110*, 280–290. [CrossRef]

129. Perpina, C.; Alfonso, D.; Perez-Navarro, A.; Penalvo, E.; Vargas, C.; Cardenas, R. Methodology based on Geographic Information Systems for biomass logistics and transport optimisation. *Renew. Energy* **2009**, *34*, 555–565. [CrossRef]

130. Mangaraj, S.; Kulkarni, S.D. Field Straw Management—A Techno Economic Perspectives. *J. Inst. Eng.* **2010**, *8*, 153–159. [CrossRef]

131. Yiljep, Y.D.; Bilanski, W.K.; Mittal, G.S. Porosity in Large Round Bales of Alfalfa Herbage. *Am. Soc. Agricult. Eng.* **1993**, *36*. [CrossRef]
132. Gupta, P.D.; Goyal, R.K.; Chattopadhyay, P.S. High density baling machine for economic transport and storage of grasses and crop residues-A step towards fodders bank. *Indian Farming* **1994**, 10–13.
133. Indian Agricultural Statistical Research Institute. *Agricultural Research Data Book*; Indian Agricultural Statistical Research Institute: New Dehli, India, 2010.
134. Shafie, S.M.; Mahlia, T.M.I.; Masjuki, H.H.; Chong, W.T. Logistic Cost Analysis of Rice Straw to Optimize Power Plant in Malaysia. *J. Technol. Innov. Renew. Energy* **2013**, *2*, 67–75. [CrossRef]
135. Kami Delivand, M.; Barz, M.; Shabbir, H.G. Logistics cost analysis of rice straw for biomass power generation in Thailand. *Energy* **2011**, *36*, 1435–1441. [CrossRef]

Article

Optimization, Transesterification and Analytical Study of *Rhus typhina* Non-Edible Seed Oil as Biodiesel Production [†]

Inam Ullah Khan, Zhenhua Yan * and Jun Chen *

Key Laboratory of Advanced Energy Materials Chemistry (Ministry of Education), Renewable Energy Conversion and Storage Centre, College of Chemistry, Nankai University, Tianjin 300071, China; 1120176003@mail.nankai.edu.cn
* Correspondence: yzh@nankai.edu.cn (Z.Y.); chenabc@nankai.edu.cn (J.C.)
† Dedicated to the 100th Anniversary of Nankai University.

Received: 9 October 2019; Accepted: 7 November 2019; Published: 11 November 2019

Abstract: Production of biodiesel from non-edible oils is one of the effective methods to reduce production costs and alleviate the obstacle of traditional raw material supply. *Rhus typhina L.* (RT) is a promising non-edible plant because it grows fast and has abundant seeds. But previously reported oil content of RT was only 9.7% and 12%. Further research into improving the biodiesel production of RT seed oil is urgently needed. Here we obtained the biodiesel production of RT with a maximum oil content of 22% with a low free fatty acid content of 1.0%. The fatty acid methyl ester (FAMEs) of the RT seed oil was produced by a standard optimized protocol use KOH as a catalyst with the highest yield of 93.4% (*w/w*). The quality and purity of RT FAMEs, as well as the physio-chemical characterizations of the biodiesel products, were investigated and compared with the international standard of ASTM D6751 and EN 14214. The values of fuel properties are comparable with mineral diesel and environmentally friendly. Overall, the proposed RT seed oil could be a potential source of raw materials for producing high-quality biodiesel after the optimization and transesterification.

Keywords: *Rhus typhina* biodiesel; non-edible oil; base-catalyzed transesterification; Physico-chemical properties

---

1. Introduction

The speedy reduction of traditional diesel oil resources, as well as the environmental problems epitomized by global warming, has led to a worldwide demand for renewable energy [1]. In response to the crisis, alternative fuels such as biodiesel, bioethanol, biomass, biogas, and synthetic fuels are being developed worldwide [2]. Among them, biodiesel is paid much attention because it is renewable, with reduced emissions of CO_x, SO_x, and particulate matter into the atmosphere compared to diesel, with a higher flash point, ensuring greater safety in transportation and storage [3]. It is estimated that biodiesel/bioethanol can replace about 10% of diesel consumption in Europe and 5% of total fuel demand in Southeast Asia [3,4]. Generally, biodiesel, also named fatty acid methyl ester (FAME), was derived from triglycerides, which are produced from vegetable oils and animal fats with a variety of lipid parameters [5]. Direct use of triglycerides (crude oil) is severely restricted due to its high viscosity and poor ignition quality [4,6]. Therefore, a variety of treatment methods are developed to overcome the viscosity and combustibility of vegetable oils [4]. The strategies including dilution, micro emulsification, pyrolysis, and transesterification for biodiesel fuel production. Among them, the base-catalyzed transesterification reaction can converts triglycerides to methyl ester in shorter reaction times with high efficiency [4,6,7]. Therefore, this method has been widely used in the production of biodiesel fuel.

Currently, 95% of commercial biodiesel is made through different edible vegetable oil sources such as palm, sunflower, rapeseed, and soybean oil [7,8]. But the industrial expense, especially raw feedstock, is the major obstacle in the moneymaking of edible oil biodiesel [9]. Due to the competition with food raw materials and fuel dispute, the continuous and large-scale production of biodiesel from edible oil is severely challenged [10]. A promising alternative is to shift the focus to the use of non-edible oil resources and/or waste cooking oil to produce biodiesel [11–13]. Many non-edible oil crops have been used for biodiesel production, evaluation, and commercialization such as Jatropha curcas and Karanja [12,14,15], waste tallow [16], *Silybum marianum L.* [17], wild *Brassica Juncea L.* [18], and prominently, algae [19]. To further promote it, there is an urgent need to shift the focus to planting non-edible oil plants on poor land, which are often unsuitable for human crops, cost-effective, and have no impact on biodiesel production in the food market [8].

R. typhina (sumac staghorn) is one of the non-edible plant's sources, is a member of family Anacardiaceae. Sumac trees are distributed in the subtropical and temperate regions of the world, especially in Africa and North America. It is native to America that can reach a height of 30–35 feet deciduous shrub to a small tree. The *R. typhina* seed has a good economic potential as feedstock for biodiesel. The reasons are as follows: RT plant growth quickly and produces a huge amount of seeds. Three to four hundreds of *R. typhina* trees can be planted per hectare. Approximately one-hectare area will produce 78,000 kg seeds, the productivity of oil is about 17,160 kg per hectare. Some studies have been carried out on the extraction of *R. typhina* (RT) seed oil. Ruan et al., 2012 [20] reported that based on acetone/water extract 9:1 *v/v*, the yield of RT seed oil was 9% *w/w*, and Zhang et al., 2018 [21] improved the seed oil content to 12%. However, the studies warrant further research into improving the biodiesel production of RT seed oil.

Here we produced biodiesel from the RT seed oil (RTSO) and optimized their potential for energy production, such as methanol to oil molar ratio, catalyst concentration, reaction temperature, reaction time, and stirring intensity. The techniques including FTIR (Fourier Transform Infra-Red Spectroscopy), NMR (^{1}H and ^{13}C) (Nuclear Magnetic Resonance Spectroscopy), GC-MS (Gas Chromatography and Mass Spectroscopy) were used to confirm the ester conversion and fatty acid composition. The physic-chemical properties of the FAME were investigated and compared with petrodiesel and standards of ASTM D6751 and EN14214 i.e., density, kinematic viscosity, cloud point, pour point, flash point, fire point, Cetane number, oxidation stability, saponification value, iodine value, acid value, specific gravity, ash content, and cold filter plugging point. Elements analysis by inductively coupled plasma atomic emission spectroscopy (ICP-OES) and elemental analyzer (EA) show that biodiesel is environmentally friendly. Overall, a prospective inedible raw material for biodiesel making was discovered and scrutinized.

2. Materials and Methods

2.1. Source Collection and Preparation of Seeds for Oil Extraction

The *R. typhina* seeds were collected from the wild/wasteland in Tianjin Binhai New Area, China. The seeds were separated by hand, washed with distilled water, and then dried in the sunlight first for 48 h and then at 60 °C oven-dried (to remove the moisture) until the seeds reached an unchanged weight. A grinder (XIANTAOPAI XTP-10000A, Zhejiang, China) was used to grind the seeds. The ground seeds were oven-dried at 60 °C for 1 h and 30 min to finish the moisture and then extracted to obtain the oil. Photographs of plants and seeds processing are shown in Figures S1 and S2 (Supplementary Materials).

2.2. Oil Extraction

R. typhina seeds were shelled, dried, and pulverized with a grinder (Xiantaopai XTP-10000A, Zhejiang, China). The oil was extracted using soxhlet [22] and mechanical oil extractor (Fangtai Shibayoufang FL-S2017 China and Fangtai Shibayoufang J508, Guangdong, China) (Table S1,

Supplementary Materials). The oil extraction was occurring in a soxhlet instrument at 90 °C for 7 h by various kinds of solvent were used, including petroleum ether, acetone, dichloromethane, and ethyl acetate. The particles and impurities were removed by filter paper (pore size 30–50 µm), and the crude RT oil was recovered at 80 °C under reduced pressure by using a rotary evaporator (Tokyo Rikakikai Co. Ltd. N-1210B, Tokyo Japan). The obtained RT oil was stored and dried over anhydrous sodium sulfate before use.

2.3. Biodiesel Production Procedure

By applying 50 g of RT crude oil, 10 mL of methanol at molar ratio 5:1, and 2.3 *w/w%* of KOH catalyst, the RT seed oil biodiesel was prepared. The transesterification reaction is shown in Figure 1. At 65 °C for 1 h under reflux, the reaction was carried out, and 600 rpm was the agitation rate [23]. By using 99.99% excess of methanol [24], the process was performed, such as methanol to oil ratio was 5:1, KOH ratio was 2.9 *w/w%*. The reflux condenser was furnished with a reactor to cool down the methanol as it comes out from the reaction process mixture. The reaction mixture was then put in a separating funnel and keep for a one-night stay that the biodiesel, soap, and glycerol can become separate from each other by layers making, the bottom phase is glycerol, and the upper phase is biodiesel. After reaction completion, crude glycerine can separate through gravity, and KOH can be detached with 3–4 time washing with hot distilled water. The complete removal of the catalyst can be check through a phenolphthalein indicator. The leftover unreacted methanol and moisture were finished through vacuum distillation and continue the process till when it confirms that there is no unreacted methanol and moisture in the final product, and the FAMEs weight loss is constant. The crude FAMEs were further washed 3–4 times with heated deionized water, centrifuging and dry with a vacuum dryer to confirm its purity. The phase separation was usually starting very quickly and can be observed in start 10 min, but the biodiesel phase was cloudy, it shows incomplete detachment. The result shows that the cloudy biodiesel layers can become clean and clear, giving them enough time for settling, and complete impartiality needs up to 20 h. In fact, during the resolution process, the transesterification process was still in progress. Hence, to give them a long time for separation is more beneficial, the separation and conversion yields. All analytical reagent grade chemicals were used.

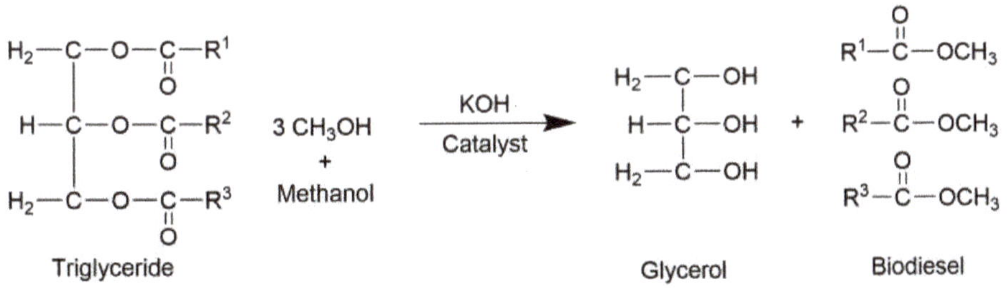

Figure 1. Transesterification reaction.

2.4. FTIR Study

The Fourier Transform Infrared (FT-IR) bands spectroscopy were verified by a (Bruker Vertex 70 FT-IR spectrometer, Germany) with a resolution of 1 cm^{-1}, scanning 15 times, and using Nujol mull as a dispersive medium in the range of 400–4000 cm^{-1}, to originate the obtained biodiesel describe through various functional groups characterization.

2.5. NMR Study

The RT FAMEs NMR spectrum was obtained by (Bruker Avance III400 NMR Spectrometer, Karlsruhe, Germany) at 400 MHz (^{1}H-NMR) or 100 MHz (^{13}C-NMR). Deuterated chloroform was used as a solvent, and tetramethylsilane was used as the internal standard. The RT biodiesel ^{1}H NMR

(300 MHz) spectrum was documented with a cycle delay of 1.0 s, and several scans of 8 times, with a pulse duration of 30°. A carbon ^{13}C NMR (75 MHz) spectrum was recorded with a pulse duration of 30° and a cycle delay of 1.89 s, and a scan of 160 times [25,26].

2.6. GCMS Study

The obtained RT FAMEs result was checked and tested by GCMS (QP2010SE, Shimadzu, Tokyo, Japan). GC-MS conditions are listed in Table 1. Detail procedure is present in Figure S2 (Supplementary Materials).

Table 1. Gas chromatograph conditions.

Parameter.	Descriptions
Column	QP2010SE, Shimadzu PEG-20M Length: 30 m Internal diameter: 0.32 mm Film thickness: 1 um
Injector temperature	220 °C
Detector temperature (EI 250)	210 °C
Carrier gas	Helium, flow rate = 1.2 mL min^{-1}
Injection	V = 1 uL
Split	Flow rate = 40:1
Temperature program	Initial temperature = 100 °C Rate of progression = 10 °C min^{-1}. Final temperature = 210 °C, 20 min.

2.7. ICP-OES and EA Study of RT Biodiesel for Elemental Analysis

The presence of metals in the RTSO FAMEs was studied using Inductively Coupled Plasma Spectrometer (Spectro-blue, Kleve, Germany) and Elemental Analyser (Vario EL CUBE, Hanau, Germany). The procedure is presented in Figure S3 (Supplementary Materials).

3. Results and Discussion

Rhus typhina plants and their pre-treated seeds are shown in Figure 2. Several varieties of solvent were used to extract oil from the RT seed powder, including petroleum ether, acetone, dichloromethane, and ethyl acetate. Seed collection at the right time, proper pre-treatment of seeds (put in sunlight first for 48 h and then at 60 °C oven-dried for 1.5 h to remove the moisture), use of petroleum ether as the extraction solvent and environmental conditions have a significant role in obtaining the high oil contents. The results are summarised in Table 2. Since the lower percentage of FFAs in the oil is the significant point to process direct transesterification of RT seed oil, this data indicates that petroleum ether is the best solvent for extracting oil from RT seeds because it provides high yields of oil and less FFAs. Besides, petroleum ether is less expensive than other solvents and can be recycled from the process. The oil and FFAs content were related to the polarity of the solvent extracted, in the order of petroleum ether < ethyl acetate < acetone < dichloromethane. Less polar solvents can extract larger amounts of non-polar oils (triglycerides) and less amount highly polar FAA. By soxhlet extraction [22], petroleum ether gives the highest extracted oil content of 22 wt. % (FFAs = 1.0 wt. %), which is significantly higher than the reported seed oil contents (Ruan et al. [20] reported that based on acetone/water extract 9:1 *v/v*, the yield of RT seed oil was 9.7% *w/w*, ZHANG et al. [21] reported that based on AOAC method the seed oil content was 12 *w/w*%). Ethyl acetate gives low oil content of 17.8 wt. % (FFAs = 1.3 wt. %), while more polar solvents like acetone and dichloromethane give somewhat lower oil content of 16–14.2 wt. % (FFAs = 1.5–1.8 wt. %). Based on these extraction results using petroleum ether and the ability to produce RT seeds oil, the RT seed could be an efficient source for biodiesel production as an alternative energy.

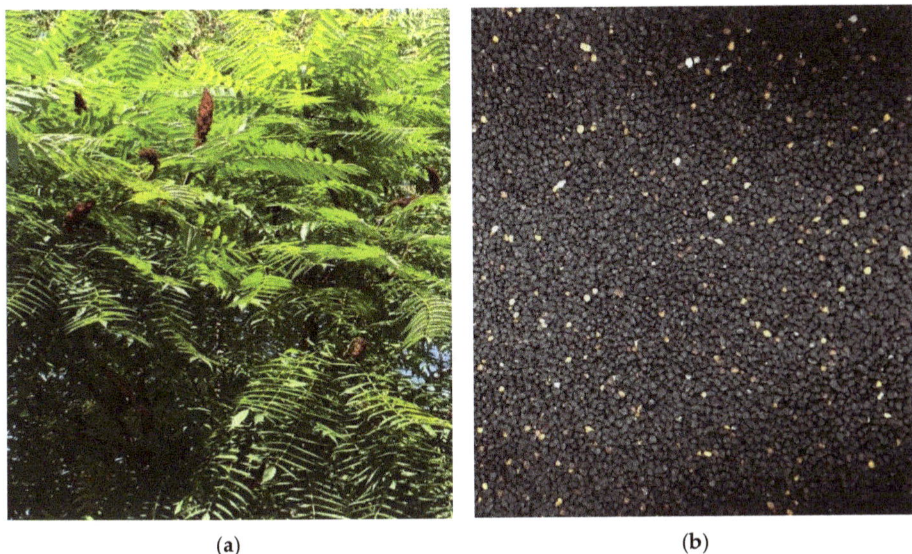

(a)   (b)

Figure 2. (**a**) Optical photographs of RT mature plant and (**b**) pre-treated seeds.

Table 2. FFAs and oil contents of extracted RT seed oil.

Solvent	FFAs Content (wt. %)	Oil Content (wt. %)
Petroleum ether	1.0	22
Acetone	1.5	16
Dichloromethane	1.8	14.2
Ethyl acetate	1.3	17.8

3.1. Optimizations of Reaction Variables on Conversion Yield

We have studied five different parameters through which we have applied different conditions, and each parameter was tested in three conditions to check that what is the most proper and suitable condition for obtaining the optimum result. The five parameters are as follow and discuss one by one, including methanol/oil ratio, catalyst concentration, temperature, stirring intensity, and time. The detail experimental results were presented in Table S2.

3.1.1. Effect of Methanol to Oil Molar Ratio on Yield

Methanol to oil molar ratio is one of the main factors affecting the ester conversion during the transesterification process. The transesterification of RT oil was evaluated and investigated at different methanol to oil range. The added amount of methanol in the reaction was varied from 4:1, 5:1, 6:1, and 7:1 (as shown in Figure 3). While temperature (65 °C), stirring intensity (700 rpm), and reaction time (60 min) was kept constant. The best proportion of methanol and oil was 6:1, in which the conversion yield was 91.66%. But at proportion 7:1, the conversion percentage was as low as to 70%. It was observed that when the proportion of methanol to oil rises from 4:1–6:1, the product yield was increased. Whereas in an additional increasing proportion of methanol to oil, a reduction in biodiesel production was observed. Excessive methanol can make it difficult to separate glycerol from the biodiesel phase. The higher methanol ratio restricts the separation of glycerol due to high solubility. Thus, in the presence of residual glycerol in the biodiesel layer allows the equilibrium reaction to start the backward reaction and again met with the methyl ester to form a monoglyceride [27]. The results show that excessive use of methanol has no substantial effect on biodiesel production; in fact, the separation of ester and glycerol is complicated. Therefore, a molar ratio of 6:1 is the optimum ratio of RT methyl ester yield.

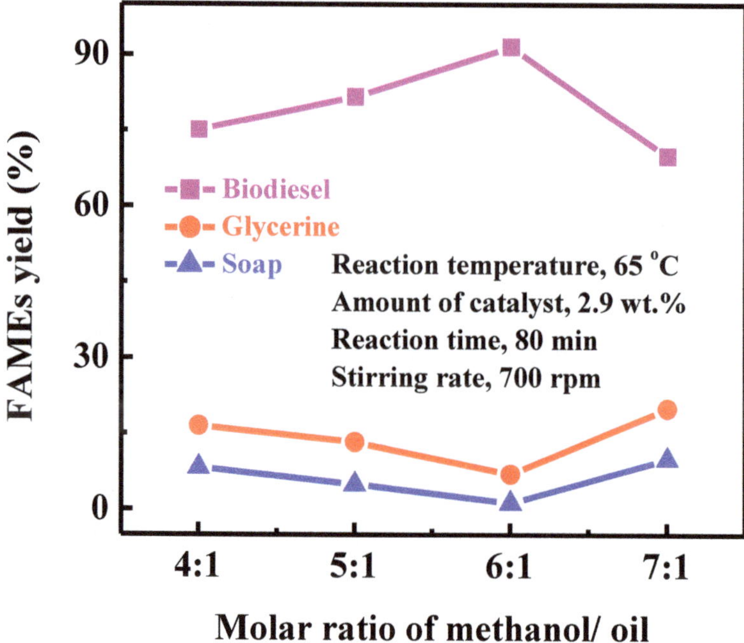

Figure 3. The effect of methanol to oil molar ratio on RTOB yield.

3.1.2. Effect of Reaction Temperature on FAMEs Yield

Temperature is a very important factor in the optimization process and has a great impact on product yield. In this work, constant parameters such as KOH concentration (2.9 wt. %), methanol to oil proportion (6:1) reaction time (60 min), and stirring speed (700 rpm) were used. A graph of biodiesel conversion (%) versus temperatures such as 60, 65, and 70 °C is shown in Figure 4. The product yield was increased from 60 °C to 65 °C; it is clear from the reaction process that the temperature had a progressive effect on the transesterification of the RT oil to biodiesel. Though, when the reaction temperature raised from 65 °C to 70 °C, we noticed a slight decrease in the reaction yield, which possibly due to the accelerated saponification reaction at a high temperature. When the transesterification reaction temperature is higher than the boiling point of the alcohol, the methanol cannot evaporate, and a huge number of the bubble will produce, which make fast the reaction and increase the saponification of glycerides [28]. Therefore, 65 °C is the optimum temperature for the reaction.

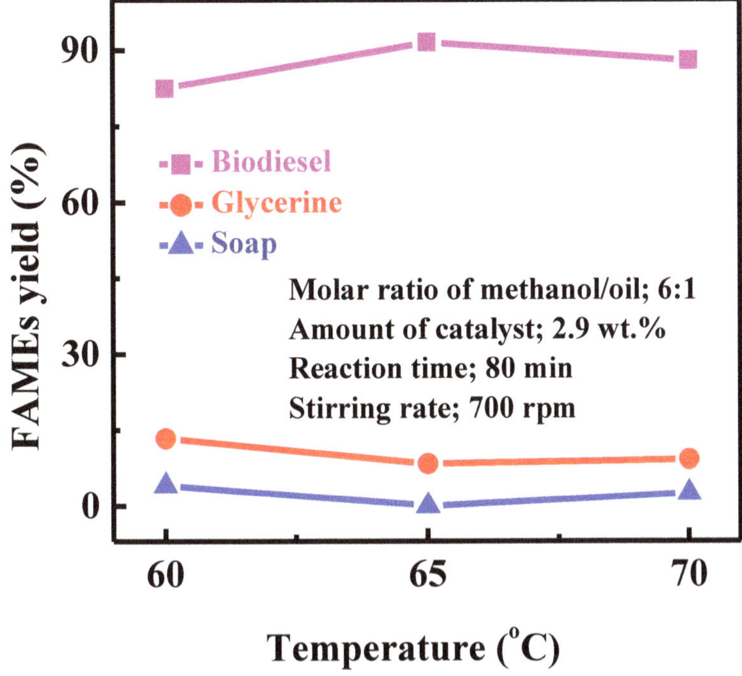

Figure 4. Effect of temperature on RTOB yield.

3.1.3. Effect of the Catalyst Concentration on FAMEs Yield

The process of variables that can affect transesterification, in addition to saponification and hydrolysis are catalyst concentration. In this reaction, methanol to oil molar ratio (6:1), temperature (65 °C), reaction time (60 min), stirring intensity (700 rpm) was kept constant, respectively. The catalyst (KOH) range of 2.0, 2.4, 2.9, and 3.8 wt. % has applied for the methanol decomposition in the RT seed oil. Figure 5 demonstrates the production of FAMEs at various KOH range. A small amount of catalyst 2.0 and 2.4 wt. % was not enough for the completion of the reaction. The highest conversion percentage rate was achieved (91.66%) at 2.9 wt. % of the catalyst. An additional increase in the catalyst concentration will negatively affect the conversion ratio. When the KOH amount was increased from 2.9 to 3.8 *w/w*%, the production of the methyl ester decreases, resulting in an increase in the viscosity of the reactants, and a decrease in the yield, a large amount of soap was observed. This observation has been explained by an increase in saponification due to the excess of the basic catalyst rather than the esterification of the triglyceride [29]. By increasing the saponification process reaction, the extra amount of catalyst will improve and help in soap formation so that the FAMEs solubility chances will increase in the glycerol, FAMEs, and glycerol mixture will appear [30].

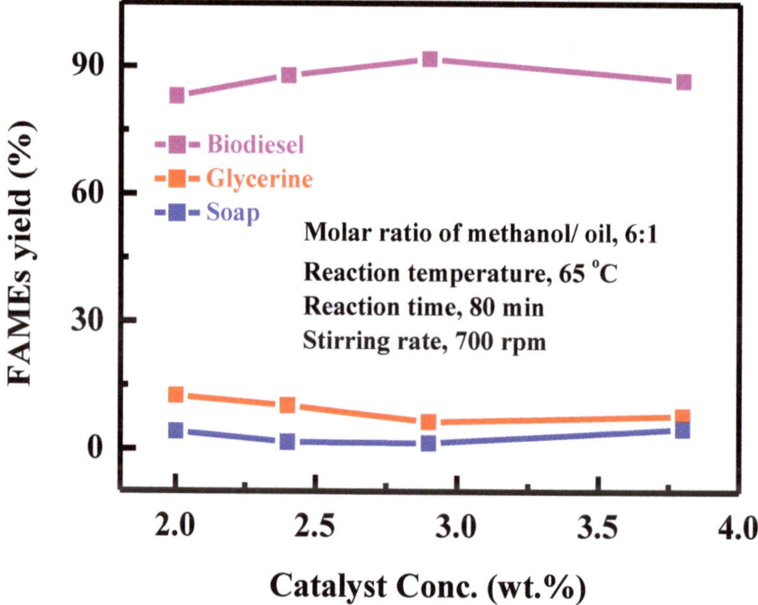

Figure 5. The effect of catalyst concentration on RTOB yield.

3.1.4. Effect of Agitation Speed on FAMEs Yield

Stirring is a key factor in the optimizations because it affects biodiesel production in both ways. In the KOH and methanol mixture, the fat and oil are not soluble, so in the transesterification process, the mixing is very important. The decomposition of methanol was carried out at various agitation speeds, i.e., 500, 600, 700, and 800 rpm. While the methanol to oil molar ratio (6:1), KOH concentration (2.9 wt.%), reaction temperature (65 °C), and reaction time (60 min) was kept constant. The FAMEs production at diverse mixing speed is shown in Figure 6. The reaction was witnessed to be incomplete at 500 rpm, and the mixing rate was not significant for methanol decomposition. The yields of the methyl esters at 600 and 700 rpm were 90% and 92.5%, respectively. As clarified by the results, a direct correlation was elucidated between the agitation rate and the RTOME yield, i.e., as the agitation rate increased, an increase in yield was observed. Therefore, a mixing rate of 700 rpm provides the best conversion rate (92.5%) of RT oil to RTOME. This is consistent with the previous studies [31–33], which concluded that increasing the agitation rate promotes the homogenization of the reactants, resulting in higher methyl ester yields. Once the two phases are mixed, and the reaction is started, stirring is no longer needed. It is clear from the results that a further stirring rate (800 rpm) has no significant effect on RTOMEs yield, but the result was the same as 700 rpm (95.5%).

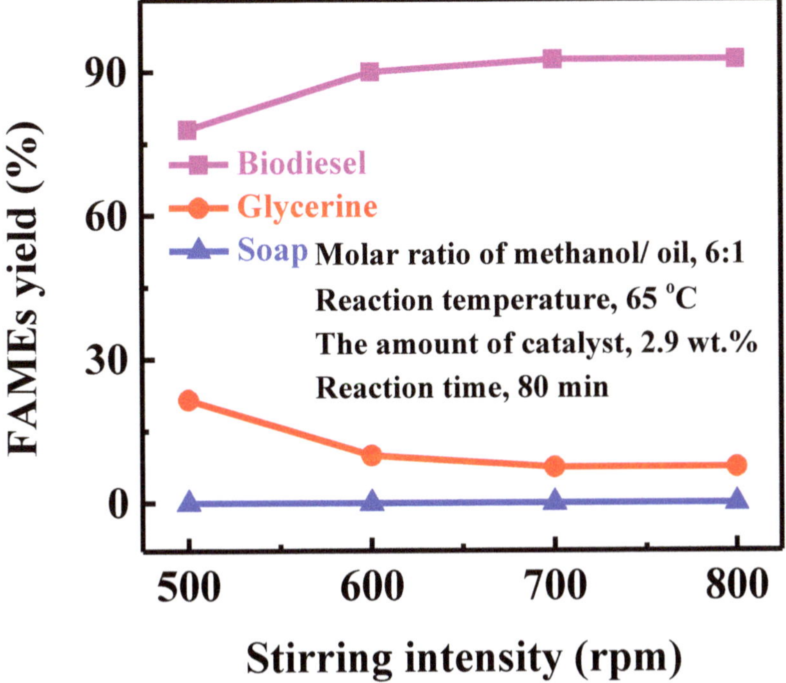

Figure 6. The effect of stirring intensity on RTOB yield.

3.1.5. Influence of Reaction Time on FAMEs Yield

The methyl ester transformation percentage rises with the reaction time. Also, the influence of reaction time on methyl ester alteration is shown in Figure 7. The influence of biodiesel yield was studied at different time rates of 40, 60, 80, and 100 min. Though methanol to oil molar ratio (6:1), reaction temperature (65 °C), catalyst concentration (2.9 wt.%), and stirring intensity (700 rpm) was kept constant. The current results presented that biodiesel conversion reached at the optimum rate in 80 min. Experimental results show that biodiesel production increase as the reaction time increases. The reaction time has no significant effect after 80 min, and the methyl ester production starts decreasing, it is indicating that the FAMEs reached the distribution equilibrium. As the duration increases, further, the reaction process will begin to react backward, and the ester yield will certainly begin to decrease by 88% at 100 min. This is consistent with the literature data showing that longer reaction times will result in lower yields due to the reverse reaction of transesterification (hydrolysis), which tends to produce more fatty acid to forming soaps [34,35]. According to the experimental results, the reaction time has an important role in the increase and decrease of biodiesel conversion. The minimum yield was obtained at 40 min, and the optimum yield occurred at 80 min.

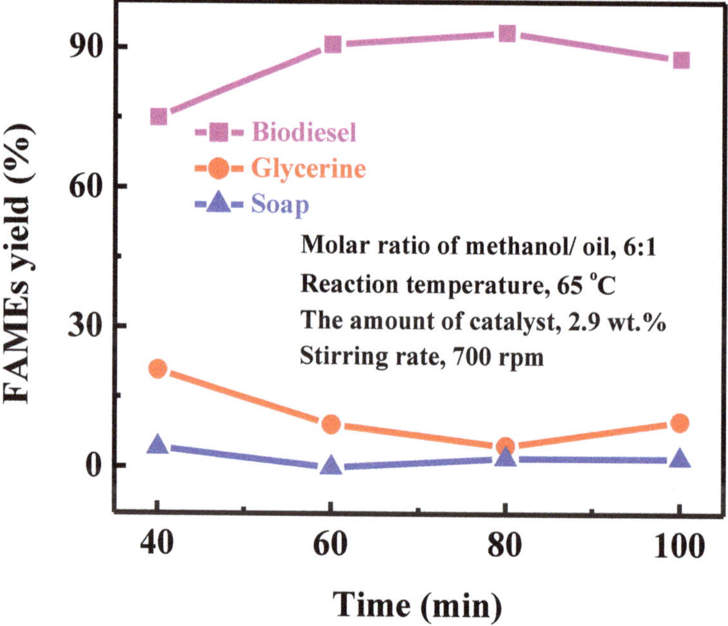

Figure 7. The effect of reaction time on RTOB yield.

3.2. Physio-Chemical Characterization of R. typhina Methyl Ester

The physicochemical properties of the *Rhus typhina* seed oil biodiesel (RTSOBD) were evaluated and examined according to ASTM and EN standards and matched well except kinematic viscosity. The kinematic viscosity (6.3 mm^2/s) is slightly higher than ASTM D6751 (1.9–6.0 mm^2/s). Kinematic viscosity needs further modification to improve its value under the limits. Accordingly, the kinematic viscosity of the BD produced from the RT seed oil is slightly high than those of conventional biodiesels, and thus must be blended with other less viscosities diesel fuels (biodiesels or petrodiesel fuel) to lessen its properties to the acceptable limits. Also, the results obtained were matched with reported data from the same source of biodiesel mentioned by other researchers and compared with petrodiesel. The results are shown in Table 3.

Table 3. Physiochemical characterizations of *R. typhina* seed oil biodiesel.

Studied Parameters	EN 14214	ASTM D6751	Petro-Diesel	RT Experimental Result	ZHANG et al., 2018 [21]	Ruan et al., 2012 [20]
Oil content%	-	-	-	20–22%	12%	9.7%
Density @ 15 °C (g/cm^3)	0.86–0.90	0.86–0.90	0.81–0.87	0.879	0.879	-
Kinematic viscosity @ 40 °C (mm^2/S)	3.5–5.0	1.9–6.0	1.3–4.1	6.3	6.87	-
Flash point, (°C)	Min 120.0	Min 130	≥52	168	165	-
Free fatty acid (%)	Max 0.50	<1	-	1.0	-	-
Saponification value (mg KOH/g)	-	-	-	175.6	-	-
Iodine value (g I$_2$/100 mg)	Max. 120	Max. 120	-	85	-	-
Cloud point (°C)	-	-	−15–5	7	-	-
Pour point (°C)	-	-	−2.0	−11	-	-
Fire point (°C)	-	-	-	198	-	-
Oxidation stability (110 °C, h)	Min 6	Min. 3	25.8	18.3	-	-
Ash content (g/100 g)	-	-	-	0.3	-	-
Specific gravity	-	-	-	0.855	-	-
Cold filter plugging point (CFPP, °C)	Max.19	Max.19	−16	14	-	-
Sulphur content (*w/w*%)	<0.01	<0.01	-	0.01	-	-
Phosphorous content (mg/kg)	<10	<10	-	4	-	-
Carbon residue (*w/w*%)	-	-	-	0.19	-	-

3.3. NMR Spectroscopy

3.3.1. ^{1}H NMR Analysis

The biodiesel yield during the transesterification process was determined by using ^{1}H NMR techniques, and the progression of its characteristic spectra was depicted. When ^{1}H NMR was used, the produce methyl ester was used to monitor the product by the protons of the methylene group adjacent to the ester moiety in the TAG and the protons of the alcohol moiety. Though by ^{1}H NMR spectroscopy technique, the RT seed oil biodiesel was characterized, the spectra of which are shown in Figure 8 and Table S3 (Supplementary Materials). The distinguishing proton methyl ester peaks were detected to be a singlet at 3.663 and triplet of the α-CH$_2$ proton at 2.308 ppm correspondingly. The existence of methyl ester in biodiesel was approved by these two distinctive peaks. At 0.895 ppm, the peaks were detected to be terminal methyl proton; at 1.319 ppm, the associated methylene protons of the carbon chain were detected, which shows a strong confirmation. A strong signal was detected for the β-carbonyl methylene proton at 1.623 ppm and olefin hydrogen at 5.342 ppm, respectively. According to the literature review, several studies specified that this diverse standardization curve established on ^{1}H NMR spectroscopy was used to enumerate the reaction yield throughout biodiesel production by transesterification of a mixture of methanol and fatty acids [36–39]. In this regard, the ^{1}H NMR spectrum of RT oil biodiesel can be used to quantify the conversion of triglycerides to methyl esters during transesterification, as evidenced by previous studies [40,41]. In this study of ^{1}H NMR spectra, the divers and distinctive analyses were existing in the *Rhus typhina* seed oil biodiesel methyl ester spectrum. The appropriate indications selected for incorporation are the methoxy group of the methyl ester at 3.66 ppm and the methyl ester group of the α-carbonyl methylene proton at 2.308 ppm. Thus, ^{1}H NMR spectroscopy showed a strong conversion of TAG to *Rhus typhina* methyl ester.

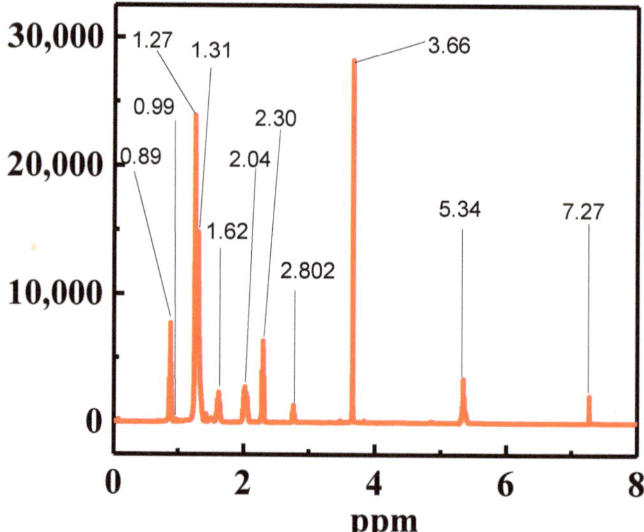

Figure 8. ^{1}H NMR spectrum shows RT FAMEs.

3.3.2. ^{13}C NMR Analysis

Ester carbonyl important peaks (-COO-) 174.24 and (C-O) 51.32 ppm associated with RT biodiesel were depicted by using ^{13}C NMR spectroscopy. The Rhus typhina biodiesel ^{13}C NMR spectra are shown in Figure 9, Table S4 (Supplementary Materials). The unsaturated methyl ester peaks were detected at 130.21 and 127.89 ppm, respectively. The other associated peaks at 14.1 ppm and 22.5–34.1 ppm, correspondingly represent the terminal carbon of the methyl group and the methylene carbon of the long-chain fatty acid ester. The present experimental work spectra were closely matched with the literature work [42] and thus approve our verdicts.

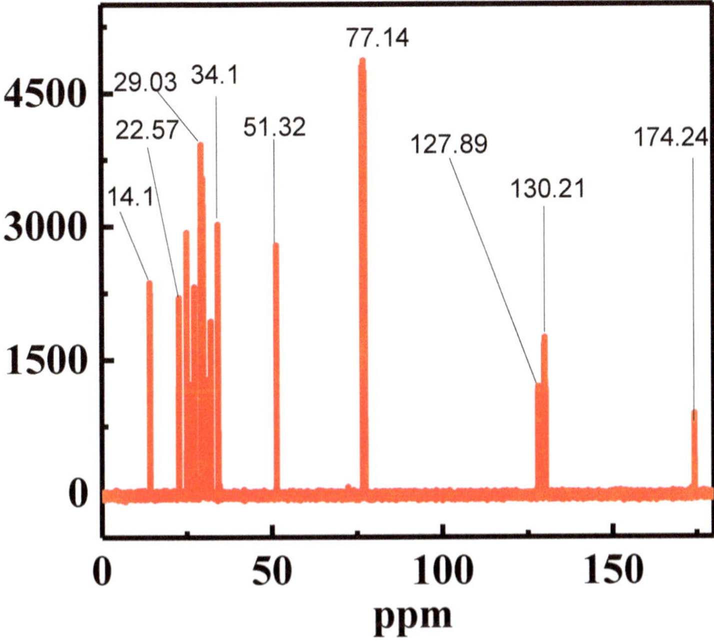

Figure 9. ^{13}C NMR spectrum of RT FAMEs.

3.4. FT–IR Spectroscopy Analysis

To identify the various stretching and bending vibration corresponding functional groups and bands of the pure RT seed oil biodiesel sample, the FTIR spectra mid-infrared regions were used. The FT-IR carbonyl group position is highly sensitive to both molecular structure and substituent effects [43]. The two strong characteristic ester absorption bands which are derived from a carbonyl group are νC=O at 1750–1730 cm^{-1} and C-O at 1300–1000 cm^{-1}. At 2980–2950, 2950–2850, and 3050–3000 cm^{-1}, respectively, the stretching vibrations of CH$_3$, CH$_2$, and CH appear, and at 1475–1350, 1350–1150, and 723 (ρCH$_2$) cm^{-1} respectively the bending vibrations of these groups appear [44]. As presented in Figure 10 and Table S5 (Supplementary Materials), at 1743 cm^{-1}, the Rhus typhina seed oil biodiesel methoxy ester carbonyl group has appeared. The peak 3464 cm^{-1} indicated an overtone of the ester function [45]. The C-O stretching vibration exhibited two asymmetric coupling vibrations at 1170 cm due to O-C-C and 1016 cm^{-1} due to νC-C(=O)-O. The stretching bands of methylene appeared at 2854 cm^{-1}, and methane appeared at 3007, 2925 correspondingly. The bending vibration appeared at 1435, and 1361 cm^{-1} belong to the methyl group, and the bending vibration appeared at 723 cm^{-1} belongs to the methylene group. The peak at 1641 cm^{-1} shows that C=C is not saturated in RT seed oil biodiesel. It is worth noting that the chromatography and spectroscopy techniques can be used to assess biodiesel, making observing, and its superiority. GC-MS, IR, ^{1}H NMR, and ^{13}C NMR was in the FAME-determined features, and different functional group structure is considered to be effective, fast, and easy to use techniques. Therefore, they are used for precise facts elucidation in biodiesel is sensible and dependable.

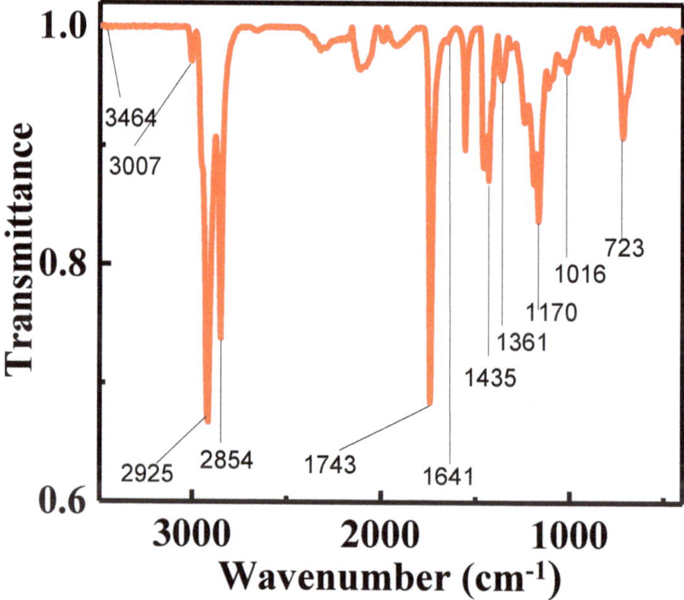

Figure 10. FTIR spectrum of RTME (FAMEs) functional groups.

3.5. Profiling of R. typhina Oil Fatty Acid Methyl Esters Using GC–MS Analysis

For the analysis of biodiesel, the most widely used method is gas chromatography. Generally, it is the perfect method to enumerate the trace components configuration accurately. In the present research work, to evaluate the particular methyl ester conversion of prepared RT seed oil to biodiesel, we have applied the gas chromatography and mass spectroscopy. The data on the GC-MS spectrum was shown in Figure 11. We used the GC spectrum and reported seven major peaks, which is consistent with the literature [46]. By using library matching software (NO. NIST 14), the peak was identified, and every peak matched with a fatty acid methyl ester. The retention time of the identified FAMEs and its position are presented in Table 4. Though by retention time data, the properties of fatty acid methyl ester were prepared and verified by mass spectrometry, and from EI ion source, the mass spectra were obtained [47]. The experimental results were compared and matched with the reported results of the RT seed oil biodiesel.

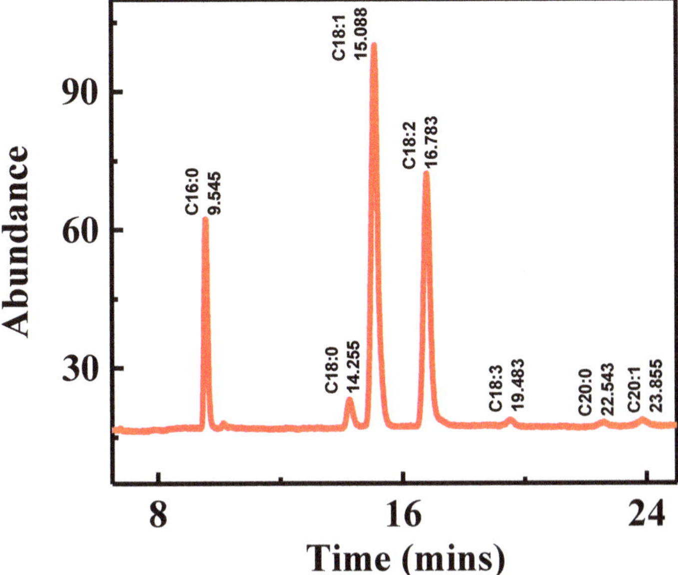

Figure 11. GC–MS analysis of fatty acid compositions of RT seed oil Biodiesel.

Table 4. The detail fatty acid composition of RT seed oil Biodiesel.

S/No	Fatty Acids/Exp. Results	Retention Time	Number of Carbons and Double Bonds	Chemical Name	Chemical Structure	Weight Percentage (%)	Molecular Weight	ZHANG et al., 2018 [21]	Ruan et al., 2012 [20]
1	Palmitic acid	9.545	C16:0	Hexadecanoic acid, methyl ester		14.0	270	Methyl palmitate C16:0	Palmitic acid C16:0
2	Stearic acid	14.248	C18:0	methyl stearate		3.2	298	Methyl stearate C18:0	Stearic acid C18:0
3	Oleic acid	15.042	C18:1	9-Octadecenoic acid (Z)-, methyl ester		47.2	296	Methyl oleate C18:1	Oleic acid C18:1
4	Linoleic acid	16.797	C18:2	9, 12-Octadecadienoic acid (Z, Z)-, methyl ester		32.2	294	Methyl linoleate C18:2	Linoleic acid C18:2
5	α-Linolenic acid	19.565	C18:3	α-Linolenic acid		1.1	292	-	-
6	Arachidic acid	22.587	C20:0	Eicosanoic acid, methyl ester		0.8	326	-	Arachidic acid C20:0
7	Gondoic acid	23.922	C20:1	CiS- 11- Eicosenoic acid, methyl ester		0.5	324	-	Arachidonic acid C20:1

3.6. ICP-OES and EA Analysis

The occurrence of elements within the biodiesel is offensive, as these elements cause many complications such as stimulating biodiesel deprivation, deterioration of engine, operability complications [48]. The essentials whose capacity in biodiesel requires to be controlled are sodium (Na) and potassium (K), which come from the mechanism in biodiesel manufacture. The amount of metal phosphorus (P) is also significant that originate from the raw materials. The elemental concentrations in RT oil biodiesel (RTOBD) were matched with Petro-diesel. It is confirmed from the results that the concentrations of elements were reasonably lower in RT seed oil biodiesel than high-speed diesel (HSD).

Prerequisites (Elements) such as potassium (K), sodium (Na), magnesium (Mg), and calcium (Ca) within biodiesel guide to the injector, stimulate drain, piston and circle wear, locomotive deposit and pass through a filter plug [49]. The recorded concentrations of C, H, N, and O were mentioned in Table 5 and Figure S3 (Supplementary Materials). The high oxygen content results in a low calorific value in the pyrolysis liquid product. The presence of oxygen reduces the ignition delay time, improves the combustibility and burning degree, thereby reducing CO, PM, and other exhaust emissions. Many researchers [50–52] found that high oxygen content in biodiesel can effectively reduce PM emissions from diesel engines. Table 5 shows that the oxygen content of BD100 is 11.12%. The most common biodiesel has an oxygen content of about 10% [53–55]. The high hydrogen contents (13.02%) of biodiesel are attractive for its utilization as fuel [56]. High H to C ratio means higher hydrogen molecules in the fuel. As hydrogen has the highest burning velocity between all fuels, whether gases or liquids. Increasing hydrogen ration in the fuel combination means that the fuel-burning velocity will be better and cleaner. The heating value generally increases as the proportion of H to carbon atoms increases due to the higher heating value of hydrogen than carbon. The HHV is an important property to define the energy content and efficiency of fuels. The HHV of biodiesel (23.73 MJ/kg) was lower than that of diesel (49.65 MJ/kg). The structural oxygen content of fuel improves its combustion efficiency due to an increase in the homogeneity of oxygen with the fuel during combustion. We compared with some reported results (Table 5), our experimental results matched them well. The nitrogen content is moderate. Though K, Na, Mg, and Ca Values Table S6 (Supplementary Materials) RTOBD (3.219, 70.29, 32.74, 7.360 µg/g) were low as compared to HSD (213.3, 868.3, 35.6, and 21.4 µg/g). The highest acceptable concentration of Na and K in biodiesel was 5 mg·kg^{-1}, while P is 10 mg·kg^{-1} [57], Figure 12. The concentration of Na and K was lower in RTOBD, but in HSD, its amount is very high; that's why RT biodiesel will be more capable of utilizing and environmental friendly.

Table 5. RTBD EA (Elemental Analyser) analysis in comparison with other plant seed and shells biodiesel.

Ultimate Analysis	RT-BD	Pistachio Shell [58]	Peach Stones [59]	Apricot Kernel Shells [60]	Cherry Stones [61]	Mahua Seed [62]
C%	74.89	42.41	45.92	47.33	52.48	61.24
H%	13.02	5.64	6.09	6.37	7.58	8.40
N%	1.97	0.070	0.580	0.370	4.54	4.12
O%	10.12	51.87	47.38	45.93	35.30	25.50
HHV (MJ/kg)	23.73	22.21	24.07	24.29	24.11	25.30

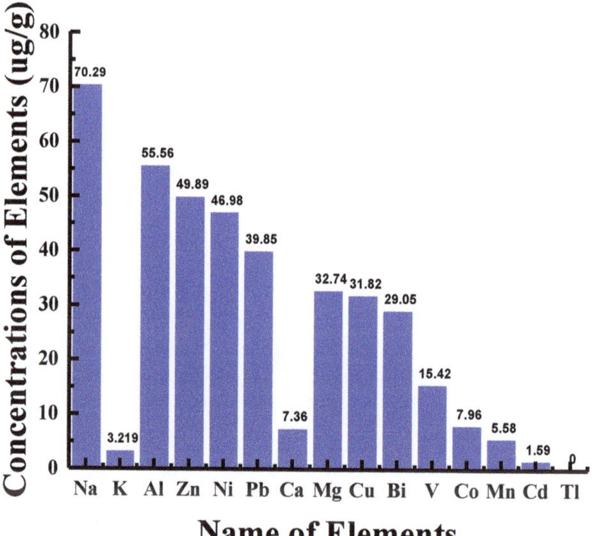

Figure 12. Elemental analysis of RTBD ICP-OES.

4. Conclusions

In conclusion, the crude RT seed oil transesterification was done to produce biodiesel with a highest yield of 93.33% by using a base catalyst. We obtained the optimal transesterification conditions and achieved a maximum oil content of 22%. Seven fatty acids were detected in the RT FAMEs production. ICP-OES and element analyzer indicate it could be environment-friendly energy. RT FAMEs were matched and follow well the recognized biodiesel standards of ASTM D6751 and EN 14214. The RT could be a better choice for renewable energy. This crop can be grown on barren wasteland on a large scale for the mass manufacture of biodiesel to fulfil the energy demand and reduce the energy crises.

Supplementary Materials: The following are available online at http://www.mdpi.com/1996-1073/12/22/4290/s1, Figure S1: R. typhina plant with fruit, Figure S2: (a) RT crude oil filtration (b) optimization (transesterification) (c) BD washing (d) RT biodiesel. Figure S3: RTBD EA (Elemental analysis) for C, H, O, and N, Table S1: Source collection, oil extraction and transesterification of non-edible R. typhina (Sumac fruit) seed oil as biofuel, Table S2: RT FAMEs detail process of optimization, Table S3: 1H NMR spectroscopic data showing the chemical composition of various methyl esters (Methoxy proton) in RT, biodiesel (FAMEs) sample, Table S4: 13C NMR spectroscopic data showing the chemical shift values corresponding to various structural features in RT (Methoxy carbon) FAMEs, Table S5: FTIR data presenting various functional groups in RT FAMEs, Table S6 Comparison ICP-OES elements concentration (ug/g) of RT-BD with petro-diesel.

Author Contributions: J.C. and Z.Y. supervised the research and revised the manuscript, I.U.K., done the experiment and write the manuscript.

Funding: This work was supported by the National Programs for Nano-Key Project (2018YFB1502100, 2017YFA0206700), the National Natural Science Foundation of China (21835004), 111 Project from the Ministry of Education of China (B12015) and the Fundamental Research Funds for the Central Universities, Nankai University (63191711 and 63191416).

Conflicts of Interest: The authors declare no conflicts of interest.

References

1. Hoseini, S.S.; Najafi, G.; Ghobadian, B.; Mamat, R.; Ebadi, M.T.; Yusaf, T. Ailanthus altissima (tree of heaven) seed oil: Characterisation and optimisation of ultrasonication-assisted biodiesel production. *Fuel* **2018**, *220*, 621–630. [CrossRef]

2. Ahmad, M.; Zafar, M.; Sadia, H.; Sultana, S.; Arshad, M.; Irfan, M. Physico-chemical characterization of sunflower oil biodiesel by using base-catalyzed transesterification. *Int. J. Green Energy* **2013**. [CrossRef]

3. Živković, S.B.; Veljković, M.V.; Banković-Ilić, I.B.; Krstić, I.M.; Konstantinović, S.S.; Ilić, S.B. Technological, technical, economic, environmental, social, human health risk, toxicological, and policy considerations of biodiesel production and use. *Renew. Sustain. Energy Rev.* **2017**, *79*, 222–247. [CrossRef]

4. Bueno, A.V. Performance and emissions characteristics of castor oil biodiesel fuel blends. *Appl. Therm. Eng.* **2017**, *125*, 559–566. [CrossRef]

5. Mucak, A. Performance and emission characteristics of a diesel engine fuelled with emulsified biodiesel-diesel fuel blends. *Int. J. Autom. Eng. Technol.* **2016**, *5*, 176–185. [CrossRef]

6. Vahid, B.R.; Haghighi, M. Urea-nitrate combustion synthesis of MgO/MgAl$_2$O$_4$ nanocatalyst used in biodiesel production from sunflower oil: Influence of fuel ratio on catalytic properties and performance. *Energy Convers. Manag.* **2016**, *126*, 362–372. [CrossRef]

7. Gülsen, E.; Olivetti, E.; Freire, F.; Dias, L.; Kirchain, R. Impact of feedstock diversification on the cost-effectiveness of biodiesel. *Appl. Energy* **2014**, *126*, 281–296. [CrossRef]

8. Ashraful, A.M. Production and comparison of fuel properties, engine performance, and emission characteristics of biodiesel from various non-edible vegetable oils: A review. *Energy Convers. Manag.* **2014**, *80*, 202–228. [CrossRef]

9. Sharma, Y.C.; Singh, B. Development of biodiesel: Current scenario. Renew. Sustain. *Energy Rev.* **2009**, *13*, 1646–1651.

10. Kansedo, J.; Lee, K.T. Process optimization and kinetic study for biodiesel production from non-edible sea mango (Cerbera odollam) oil using response surface methodology. *Chem. Eng. J.* **2013**, *214*, 157–164. [CrossRef]

11. Mustafa, B. Potential alternatives to edible oils for biodiesel production-a review of current work. *Energy Convers. Manag.* **2011**, *52*, 1479–1492.

12. Rashed, M.M.; Kalam, M.A.; Masjuki, H.H.; Mofijur, M.; Rasul, M.G. Performance and emission characteristics of a diesel engine fueled with palm, jatropha, and moringa oil methyl ester. *Ind. Crops Prod.* **2016**, *79*, 70–76. [CrossRef]

13. Fadhil, A.B. Evaluation of apricot (Prunus armeniaca L.) seed kernel as a potential feedstock for the production of liquid biofuels and activated carbons. *Energy Convers. Manag.* **2017**, *133*, 307–317. [CrossRef]

14. Sinha, P.; Islam, M.A.; Negi, M.S.; Tripathi, S.B. Changes in oil content and fatty acid composition in Jatropha curcas during seed development. *Ind. Crops Prod.* **2015**, *77*, 508–510. [CrossRef]

15. Foidl, N.; Foidl, G.; Sanchez, M.; Mittelbach, M.; Hackel, S. Jatropha Curcas L. as a source for the production of biofuel in Nicaragua. *Bioresour. Technol.* **1996**, *58*, 77–82. [CrossRef]

16. Bhatti, H.N.; Hanif, M.A.; Qasim, M.; Ata-ur-Rehman. Biodiesel production from waste tallow. *Fuel* **2008**, *87*, 2961–2966. [CrossRef]

17. Fadhil, A.B.; Aziz, A.M.; Altamer, M.H. Biodiesel production from Silybum marianum L. seed oil with high FFA content using sulfonated carbon catalyst for esterification and base catalyst for transesterification. *Energy Convers. Manag.* **2016**, *108*, 255–265. [CrossRef]

18. Aldobouni, I.A.; Fadhil, A.B.; Saied, I.K. Optimized alkali - catalyzed transesterification of wild mustard (Brassica juncea L.) seed oil. *Energy Sources Part A* **2016**, *38*, 2319–2325. [CrossRef]

19. Chisti, Y. Biodiesel from microalgae. *Biotechnol. Adv.* **2007**, *25*, 294–306. [CrossRef] [PubMed]

20. Ruan, C.J.; Xing, W.H.; Jaime, A.; da Silva, T. The potential of five plants growing on unproductive agriculture land as a biodiesel source. *Renew. Energy* **2012**, *41*, 191–199. [CrossRef]

21. Zhang, F.; Ren, H.; Tong, G. Production of biodiesel from the extract of the sumac fruit cluster. *Cellul. Chem. Technol.* **2018**, *52*, 1275–1279.

22. Kpikpi, W.M. Jatropha curcus as a vegetable source of renewable energy. *ANSTI Sub Netw. Meet. Renew. Energy* **2002**, *13*, 18–22.

23. Ahmad, M.; Khan, M.A.; Zafar, M.; Sultana, S. *Practical Handbook on Biodiesel Production and Properties*; Taylor and Francis: London, UK, 2012; pp. 1–157.

24. Antolin, G.; Tinaut, F.V.; Briceno, Y.; Castano, V.; Perez, C.; Ramrez, A.I. Optimization of biodiesel production by sunflower oil transesterification. *Bioresour. Technol.* **2002**, *83*, 111–114. [CrossRef]

25. Bhandari, D.C.; Chandel, K.P.S. Status of rocket germplasm in India: Research accomplishments and priorities. *Rocket Mediterr. Crop World* **1996**, *67*, 13–14.

26. Christie, W.W. *Lipid Analysis*, 3rd ed.; Oily Press: Bridgwater, UK, 2003.

27. Encinar, J.M.; Gonzalez, J.F.; Rodriguez, R.A. Biodiesel from used frying oil. Variables affecting the yields and characteristics of the biodiesel. *Ind. Eng. Chem. Res.* **2005**, *44*, 5491–5499. [CrossRef]

28. Dorado, M.P.; Ballesteros, E.; Lopez, F.J.; Mittelbach, M. Optimization of alkali catalyzed transesterification of Brassica Carinata oil for biodiesel production. *Energy Fuels* **2004**, *18*, 77–83. [CrossRef]

29. Leung, D.; Guo, Y. Transesterification of neat and used frying oil: Optimization for biodiesel production. *Fuel Process. Technol.* **2006**, *87*, 883–890. [CrossRef]

30. Kafuku, G.; Mbarawa, M. Alkaline catalyzed biodiesel production from Moringa oleifera oil with optimized production parameters. *Appl. Energy* **2010**, *87*, 2561–2565. [CrossRef]

31. Rashid, U.; Anwar, F. Production of biodiesel through optimized alkaline-catalyzed transesterification of rapeseed oil. *Fuel* **2008**, *87*, 265–273. [CrossRef]

32. Ma, F.; Clements, L.D.; Hanna, M.A. The effect of mixing on transesterification of beef tallow. *Bioresour. Technol.* **1999**, *69*, 289–293. [CrossRef]

33. Peterson, C.L.; Reece, D.L.; Cruz, R.; Thompson, J. A comparison of ethyl and methyl esters of vegetable oil as diesel fuel substitutes. In Proceedings of the an Alternative Energy Conference, Nashville, TN, USA, 12–15 December 1992; pp. 99–110.

34. Tomasevic, A.V.; Siler-Marinkovic, S.S. Methanolysis of Used Frying Oil. *Fuel Process. Technol.* **2003**, *82*, 1–6. [CrossRef]

35. Eevera, T.; Rajendran, K.; Saradha, S. Biodiesel production process optimization and characterization to assess the suitability of the product for varied environmental conditions. *Renew. Energy* **2009**, *34*, 762–765. [CrossRef]

36. Shu, Q.; Gao, J.; Nawaz, Z.; Liao, Y.; Wang, D.; Wang, J. Synthesis of biodiesel from waste vegetable oil with large amounts of free fatty acids using a carbon-based solid acid catalyst. *Appl. Energy* **2010**, *87*, 2589–2596. [CrossRef]

37. Mello, V.M.; Oliveira, C.C.; Fraga, G.W.; Nascimento, C.J.D.; Suarez, P.A.Z. Determination of the content of fatty acid methyl esters (FAMEs) in biodiesel samples using 1H NMR spectroscopy. *Magn. Reson. Chem.* **2008**, *46*, 1051–1054. [CrossRef] [PubMed]

38. Monterio, M.R.; Ambrozin, A.R.P.; Liao, L.M.; Ferreira, A.G. Determination of biodiesel blends levels in different diesel samples by 1H NMR. *Fuel* **2009**, *88*, 691–696. [CrossRef]

39. Samios, D.; Pedrotti, F.; Nicolau, A.; Martini, D.D.; Dalcen, F.M. A transesterification double process-TDSP for biodiesel preparation from fatty acid tryglycerides. *Fuel Process. Technol.* **2009**, *90*, 599–605. [CrossRef]

40. Liu, C.; Lv, P.; Yuan, Z.; Yan, F.; Luo, W. Biodiesel from different oil using fixed-bed and flow reactors. *Renew. Energy* **2010**, *35*, 1531–1536. [CrossRef]

41. Gelbard, G.; Bres, O.; Vargas, R.M.; Vielfaure, F.; Schuchardt, U.F. 1H nuclear magnetic resonance determination of the yield of the transesterification of rapeseed oil with methanol. *Am. Oil Chem. Soc.* **1995**, *72*, 1239–1241. [CrossRef]

42. Knothe, G. Monitoring a progressing transesterification reaction by fiber optic NIR spectroscopy with correlation to 1H NMR spectroscopy. *Am. Oil Chem. Soc.* **2000**, *77*, 489–493. [CrossRef]

43. Pasto, D.; Johnson, C.; Miller, M. *Experiments and Techniques in Organic Chemistry*, 1st ed.; Prentice-Hall: Upper Saddle River, NJ, USA, 1992.

44. Guillen, M.D.; Cabo, N. Infra-red spectroscopy in the study of edible oils and fats. *Sci. Food Agric.* **1997**, *75*, 1–11. [CrossRef]

45. Safar, M.; Bertrand, D.; Robert, P.; Devaux, M.D.; Genut, C. Characterization of edible oil, butter, and margarine by Fourier Transfer Infra-Red spectroscopy with attenuated total reflectance. *Am. Chem. Soc.* **1994**, *71*, 371–377. [CrossRef]

46. Ahmad, M.; Ullah, K.; Khan, M.A.; Zafar, M.; Tariq, M.; Ali, S. Physico-chemical analysis of hemp oil biodiesel: A promising non-edible new source for bioenergy. *Energy Sources Part A* **2011**, *33*, 1365–1374. [CrossRef]

47. Wang, L.; Yu, H. Biodiesel from Siberian apricot (Prunus sibirica L.) seed kernel oil. *Bioresour. Technol.* **2012**, *112*, 355–358. [CrossRef] [PubMed]

48. Schober, S.; Mittelbach, M. Influence of diesel particulate filter additives on biodiesel Quality. *Eur. J. Lipid Sci. Technol.* **2005**, *107*, 268–271. [CrossRef]

49. McCormick, R.L.; Alleman, T.L.; Ratcliff, M.; Moens, L.; Lawrence, R. *Survey of the Quality and Stability of Biodiesel and Biodiesel Blends in the United States in 2004*; National Renewable Energy Laboratory: Golden, CO, USA, 2005.

50. Song, H.; Quinton, K.S.; Peng, Z.; Zhao, H.; Ladommatos, N. Effects of the oxygen content of Fuels on combustion and emissions of diesel engines. *Energies* **2016**, *9*, 28. [CrossRef]

51. Nakano, M.; Okawa, K. Study of oxygen-containing hydrocarbons in exhaust emission from a spark ignition combustion engine. *Int. J. Eng. Res.* **2014**, *15*, 572–580. [CrossRef]

52. Mwang, J.K.; Lee, W.J.; Chang, Y.C.; Chen, C.Y.; Wang, L.C. An overview: Energy saving and pollution reduction by using green fuel blends in diesel engines. *Appl. Energy* **2015**, *159*, 214–236. [CrossRef]

53. Lin, B.F.; Huang, J.H.; Huang, D.Y. Effects of Biodiesel from Palm Kernel oil on the engine performance, Exhaust emissions and combustion characteristics of a direct injection diesel engine. *Energy Fuels* **2008**, *22*, 4229–4234. [CrossRef]

54. Singh, D.; Subramanian, K.A.; Juneja, M.; Singh, K.; Singh, S. Investigating the effect of fuel cetane number, oxygen content, fuel density, and engine operating variables on NOx emissions of a heavy-duty diesel engine. *Environ. Prog. Sustain. Energy* **2017**, *36*, 214–221. [CrossRef]

55. Demirbas, A. Combustion efficiency impacts of biofuels. *Energy Sources Part A* **2009**, *31*, 602–609. [CrossRef]

56. Huber, G.W.; Iborra, S.; Corma, A. Synthesis of transportation fuels from biomass: Chemistry, catalysts, and engineering. *Chem. Rev.* **2006**, *106*, 4044–4098. [CrossRef] [PubMed]

57. Korn, M.G.A.; Santos, D.S.S.; Welz, B.; Vale, M.G.R.; Teixeira, A.P.; Lima, D.D.C.; Ferreira, S.L.C. Atomic spectrometric methods for the determination of metals and metalloids in automotive fuels. A Review. *Talanta* **2007**, *73*, 1–11. [CrossRef] [PubMed]

58. Acıkalın, K.; Karaca, F.; Bolat, E. Pyrolysis of pistachio shell: Effects of pyrolysis conditions and analysis of products. *Fuel* **2012**, *95*, 169–177. [CrossRef]

59. Uysal, T.; Duman, G.; Onal, Y.; Yasa, I.; Yanik, J. Production of activated carbon and fungicidal oil from peach stone by a two-stage process. *J. Anal. Appl. Pyrolysis* **2014**, *108*, 47–55. [CrossRef]

60. Demiral, I.; Kul, S.C. Pyrolysis of apricot kernel shell in a fixed-bed reactor: Characterization of bio-oil and char. *J. Anal. Appl. Pyrolysis* **2014**, *107*, 17–24. [CrossRef]

61. Duman, G.; Okutucu, C.; Ucar, S.; Stah, R.; Yanik, J. The slow and fast pyrolysis of cherry seed. *Bioresour. Technol.* **2011**, *102*, 1869–1878. [CrossRef] [PubMed]

62. Pradhan, D.; Singh, R.K.; Bendu, H.; Mund, R. Pyrolysis of Mahua seed (Madhuca indica) production of biofuel and its characterization. *Energy Convers. Manag.* **2016**, *108*, 529–538. [CrossRef]

Review

Factors Affecting the Performance of Membrane Osmotic Processes for Bioenergy Development

Wen Yi Chia [1], Kuan Shiong Khoo [1], Shir Reen Chia [1], Kit Wayne Chew [2], Guo Yong Yew [1], Yeek-Chia Ho [3,4,*], Pau Loke Show [1,*] and Wei-Hsin Chen [5,6,7,8,*]

[1] Department of Chemical and Environmental Engineering, Faculty of Science and Engineering, University of Nottingham Malaysia, Jalan Broga Semenyih 43500, Malaysia; wenyichia@gmail.com (W.Y.C.); kuanshiong.khoo@hotmail.com (K.S.K.); shireen.chia127@gmail.com (S.R.C.); yewguoyong@gmail.com (G.Y.Y.)
[2] School of Mathematical Sciences, Faculty of Science and Engineering, University of Nottingham Malaysia, Jalan Broga Semenyih 43500, Malaysia; KitWayne.Chew@nottingham.edu.my
[3] Civil and Environmental Engineering Department, Universiti Teknologi PETRONAS, Seri Iskandar 32610, Malaysia
[4] Institute of Self-Sustainable Building, Centre for Urban Resource Sustainability, Universiti Teknologi PETRONAS, Seri Iskandar 32610, Malaysia
[5] Department of Aeronautics and Astronautics, National Cheng Kung University, Tainan 701, Taiwan
[6] Department of Chemical and Materials Engineering, College of Engineering, Tunghai University, Taichung 407, Taiwan
[7] Department of Mechanical Engineering, National Chin-Yi University of Technology, Taichung 411, Taiwan
[8] Research Center for Energy Technology and Strategy, National Cheng Kung University, Tainan 701, Taiwan
[*] Correspondence: yeekchia.ho@utp.edu.my (Y.-C.H.); PauLoke.Show@nottingham.edu.my (P.L.S.); weihsinchen@gmail.com or chenwh@mail.ncku.edu.tw (W.-H.C.)

Received: 4 December 2019; Accepted: 14 January 2020; Published: 19 January 2020

Abstract: Forward osmosis (FO) and pressure-retarded osmosis (PRO) have gained attention recently as potential processes to solve water and energy scarcity problems with advantages over pressure-driven membrane processes. These processes can be designed to produce bioenergy and clean water at the same time (i.e., wastewater treatment with power generation). Despite having significant technological advancement, these bioenergy processes are yet to be implemented in full scale and commercialized due to its relatively low performance. Hence, massive and extensive research has been carried out to evaluate the variables in FO and PRO processes such as osmotic membrane, feed solutions, draw solutions, and operating conditions in order to maximize the outcomes, which include water flux and power density. However, these research findings have not been summarized and properly reviewed. The key parts of this review are to discuss the factors influencing the performance of FO and PRO with respective resulting effects and to determine the research gaps in their optimization with the aim of further improving these bioenergy processes and commercializing them in various industrial applications.

Keywords: concentration polarization; draw solution; feed solution; forward osmosis; pressure-retarded osmosis; operating conditions; membrane fouling; osmotic membrane

---

1. Introduction

The demand for water and energy has been raising continuously and their shortage has resulted in the development of alternative renewable energies and water treatment systems which require less energy. Two of the osmotically driven membrane processes (ODMPs), namely forward osmosis (FO) and pressure retarded osmosis (PRO), have attracted the most attention in the past years as the emerging technologies for numerous applications such as wastewater treatment and biomass processing for bioenergy [1].

Osmosis refers to the movement of solvent molecules mostly water from the feed solution (FS) which has lower solute concentration into draw solution (DS) which has higher solute concentration across a semipermeable membrane. The osmotic pressure difference ($\Delta\pi$) between the two solutions is the driving force for ODMPs and the same quantity of hydraulic pressure is required to terminate the osmosis process. In contrast, pressure-driven membrane processes (e.g., nanofiltration, RO, etc.) rely on externally applied hydraulic pressure [2]. The key benefits of ODMPs compared to the latter are that they consume less energy and have lower fouling tendency, which results in high water flux, high rejection of salt, and less membrane cleaning [3].

FO processes occur without any applied hydraulic pressure ($\Delta P = 0$) [4]. Once the solvent starts moving from FS to DS, the hydrostatic pressure on the DS side gradually becomes greater and the water flow will stop eventually; mathematically, the water flux (Jw) stops when ΔP equals to $\Delta\pi$ (where $\Delta\pi - \Delta P = 0$; hence Jw = 0) [5]. On the other hand, reverse osmosis (RO) employs hydraulic pressure differential as its driving force to move water from the saltwater to the freshwater, resulting in a negative flux across the membrane. This occurrence is relatively due to the large pressure ($\Delta P > \Delta\pi$) exerted to the concentrated side.

PRO processes are intermediate between FO and RO. Similar to RO, hydraulic pressure is applied to the DS side against the osmotic pressure gradient, but it is smaller than the osmotic pressure difference ($\Delta P < \Delta\pi$). Thus, the net water flux, J_w is still towards concentrated DS, which is in a similar direction as FO [4]. To generate the power steadily, a constant concentration and pressure must be set for the DS so that a constant flow of FS across the membrane can occur. Subsequently, there is more volume flow on the concentrated region and this additional flow can be employed for power generation regarding bioenergy production [5].

There are several differences between FO and PRO in terms of the fundamentals and applications, even though both processes utilize a similar driving force, which is salinity gradient difference [6]. For instance, FO processes are employed to produce clean water, dilute draw solution, or concentrate feed solution in applications including brine treatment and desalination [2]. On the other hand, PRO processes are employed to generate electricity or power using osmotic pressure from the DS. The application difference is caused by the fundamental difference in their operations as FO processes involve direct use of osmotic pressure to transport less concentrated solution across semipermeable membrane barrier while PRO processes utilize hydraulic pressure to pressurize the DS to help in converting the osmotic pressure to hydraulic pressure [7].

Besides PRO processes with power generation as the main application, FO processes can be integrated with other systems for bioenergy production. For instance, bioenergy from organic matter found in wastewater can be converted into electricity via forward osmosis-bioelectrochemical hybrid systems (FO-BESs) which use electrochemically active bacteria [8]. One example of FO-BESs is osmotic microbial fuel cell (OsMFC) which separates anode and cathode with an FO membrane. It produces more electricity compared to the conventional microbial fuel cell with a cation exchange membrane [9]. Moreover, energy recovery (e.g., biogas production) can be carried out through an anaerobic osmotic membrane bioreactor (An-OMBR) which integrates FO with anaerobic treatment by submerging an FO membrane inside the bioreactor. FO processes can be used to concentrate nutrients and harvest algae which is a bioenergy source, where biodiesel can be produced from its lipid extract and bioethanol can be fermented from its carbohydrates [10,11].

Although there are several ODMPs applications for bioenergy production, these processes are so far applied in lab-scale studies only and up-scaling to full scale is essential. To achieve at that stage, optimization of process parameters has to be investigated. Therefore, this paper comprehensively reviews the factors influencing the performance of the ODMPs, which is generally measured by water flux, reverse solute flux, and power density. Various aspects of an osmotic membrane such as concentration polarization, membrane materials, membrane configurations, and fouling are discussed.

The next sections focus on the feed solutions and draw solutions which provide the osmotic pressure difference to drive the ODMPs. Operating conditions including cross-flow rate and temperature are also studied for their effects on the performance of FO and PRO. This review provides insights into the concerned researches in practically designing an optimum osmotic membrane system for bioenergy production.

2. Osmotic Membrane

During the past 40 years, membrane development has been focused on pressure driven processes but the attention in ODMPs, like FO and PRO, has increased in the last decade [12] since they demonstrate significant promise to leverage the global water-energy nexus [13]. It is crucial to formulate customized membranes for these technologies since membrane properties such as structural parameters, water permeability and solute permeability affect the overall performance of FO and PRO processes. For instance, asymmetric FO membrane is an essential component to determine the performance of FO-BES where its properties such as mass transfer resistance and proton transfer ability affect the bioenergy production [8]. Generally, there are two methods to prepare membrane including (1) direct phase inversion, which involves phase inversion of a polymer dope in a non-solvent followed by formation of integrally skinned membranes, and (2) interfacial polymerization (IP), which is utilized to create thin-film composite (TFC) membranes [14]. Modifications on membrane including its microporous substrate and selective layer have been the main research area to improve the performance of FO and PRO processes. For instance, specific functional compounds or groups have been introduced on the surface of the selective layer to tackle the problems of fouling and low productivity [15]. In short, a desired osmotic membrane should be thin, highly selective, and antifouling while compatible with the chosen draw solution, produce high water fluxes, decline dissolved solutes, and tolerate mechanical pressures caused by operation conditions [15,16].

2.1. Concentration Polarization

An osmotic membrane consists of an active layer and a support layer. The former selectively allows movement of solvent molecules but declines at least some dissolved ions whereas the latter with a side bond to the former comprises a phase-inversion sub-layer and an electrospun-fiber sub-layer. Concentration polarization (CP) happens when the salt concentration difference across the active layer is different from the concentration difference of the bulk solutions itself [17]. Since DS is diluted by the incoming water flux, there is an immediate reduction in the osmotic pressure at the surface of membrane facing the concentrated DS, causing the solute concentration on the other membrane side facing the FS to increase, since the diffusion of water is towards the DS [18].

Depending on the membrane orientation, these two CP phenomena can happen either on the active layer or on the support layer since the asymmetric membrane is used in the FO process. Although any of the membrane orientations can be used to operate ODMPs, for clarity, the process is said to be in FO mode (also known as AL-FS orientation) given that the active layer faces FS and support layer faces DS. Generally, this mode can be used for any membrane separation process. In contrast, the process is termed as PRO mode (also known as AL-DS orientation) when the active layer faces the DS and the support layer faces the FS. PRO mode is normally used for osmotic power generation [19].

On the other hand, there are two kinds of CP where the formation of a concentration layer at the membrane surface and in the porous structure of asymmetric membranes is defined as external concentration polarization (ECP) and internal concentration polarization (ICP), respectively [20]. ECP is usually associated with the concentrations of DS and FS whereas ICP is mainly affected by the support layer structure and thick dense membrane [4]. The presence of CP can significantly affect the effective osmotic pressure difference ($\Delta\pi$) across the membrane leading to a sharp decline in water flux and operational performance of FO and PRO processes [21,22].

In PRO mode operation, water from FS moves into the porous support layer and permeates through the active layer. The salt from FS also enters the support layer however, it cannot pass through the active layer, resulting in an increase of concentration within the support layer, which is called concentrative ICP. In FO mode operation, dilutive ICP takes place when water permeate dilutes the DS within the porous support layer, causing slow solute mass transfer [23]. ICP could become acute in ODMPs and it cannot be erased by manipulating hydraulic conditions like increasing turbulence or shear force since it happens within the support layer, thus membranes specially designed for ODMPs are desired for applications [2]. Optimization on porosity, thickness, and tortuosity of the substrate structure is essential for minimizing the severe effects of ICP. For example, an FO membrane that has a highly porous substrate with low tortuosity and a detect-free thin selective layer is the most effective approach to alleviate ICP [15].

2.2. Membrane Materials

To develop a system with optimal flux, the choice of material to fabricate a membrane is significant because membrane fouling can directly affect the membrane flux. The material selection also depends on the purpose of the membrane since the operational parameters of FO and PRO are quite different. They require different selectivity degrees for the membrane skin layer where FO needs highly selective membranes whereas PRO which targets high power density requires just sufficient salt decline to maintain the driving force and govern concentration polarization [24]. Furthermore, most conventional FO membranes, which are usually tailored to be porous and thin to lower ICP and structural parameters, will either be damaged or deformed under high pressure of PRO [14]. This may cause serious leakage of DS and failure towards the entire process [25]. This indicates that PRO membranes must be robust enough to tolerate the operational hydraulic pressure. Micro-void type support layers that tend to reduce CP can be beneficial for the FO process; however, their applications in PRO are restricted to lower pressure processes since the micro-void structure can easily be collapsed under the effects of high hydraulic pressure [26].

Early attempts to fabricate FO and PRO specialized membranes were restricted to cellulose triacetate (CTA) and TFC membranes [25]. Cellulose is the most abundant existing natural polymer and derivatives of cellulose such as cellulose acetate (CA) and CTA are appropriate materials to fabricate membranes due to their good separation, moderate flux, and non-toxicity [27]. In order to produce CTA membranes (as demonstrated in Figure 1A), a dense active layer is formed by casting CTA with an embedded polyester fabric where the thickness of the active layer is minimized to improve the water permeability while maintaining the membrane integrity or contaminant rejection [15]. However, CTA membranes have restrictions in their application for desalination due to their comparatively low water permeability and solute rejection. Severe CP has been induced in CTA membranes by a sponge-like structure which was initially designed to improve flux [15]. In addition, cellulose-based materials not only have low stability to pH and temperature but also undergo biodegradation and hydrolysis [27].

Therefore, enhanced TFC membranes have been developed to keep the high water flux obtained by CTA membranes and to improve the mechanical robustness of the membranes [25]. This is because highly selective and permeable TFC membranes have high flexibility in structural design with the ability to individually optimize the two membrane layers (selective and substrate layer) for specific needs [15,28]. Modification on the structure of the support layer significantly reduces ICP while having the dense thin film layer allows great water permeability and salt rejection [7]. A TFC membrane (as shown in Figure 1B) generally consists of (1) a very thin polyamide active layer, (2) a polymeric support layer, and (3) a fabric layer for mechanical support [29]. Some examples of the artificial polymers are polysulfone (PSF), polyether sulfone (PES), polyvinylidene fluoride (PVDF) and polyacrylonitrile (PAN) [27]. The standard preparation method of TFC membrane layer-by-layer assembly using the microporous substrate obtained through phase inversion and the polyamide layer formed through IP between chloride and amine monomers such as trimesoyl chloride and m-phenylenediamine [15].

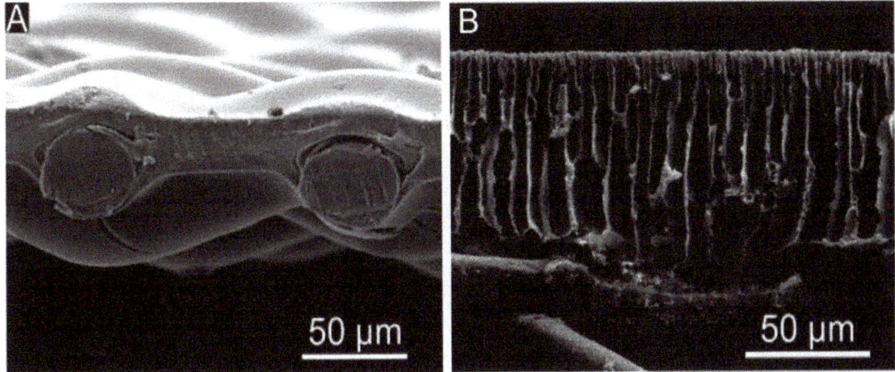

Figure 1. Cross-section SEM micrographs of (**A**) commercial CTA-HTI and (**B**) TFC membrane. Reproduced with permission from Yip, Tiraferri, Phillip, Schiffman and Elimelech [16], Copyright (1969) American Chemical Society.

Wang, et al. [30] had compared the characteristics of CTA and TFC FO membranes provided by Hydration Technologies Innovation (HTI, Albany, NY, USA) and the results are shown in Table 1. The smaller contact angles indicate a higher hydrophilicity degree, which means permeation of water is easier and higher water flux is expected [31]. Indeed, higher water flux was observed in TFC membrane compared to CTA membrane under the same operating conditions [30]. It is noteworthy that in a more recent study by Li, et al. [32], a self-made TFC membrane in comparison with a commercial CTA-HTI membrane also exhibited higher flux loss due to membrane fouling [32]. Flux loss is one of the important criteria to be considered for overall flux performance when designing an osmotic membrane system. Nevertheless, higher reverse salt flux was observed with TFC membranes in both studies [30,32].

Table 1. Comparison of commercial CTA and TFC membrane produced by Hydration Technologies Innovation (HTI, Albany, NY, USA). Adapted with permission from Wang, Tang, Zhu, Dong, Wang and Wu [30], Copyright (2014), Elsevier B.V.

Parameters	Cellulose Triacetate	Thin-Film Composite
Thickness of active layer (μm)	6.1 ± 2.0	4.9 ± 1.1
Thickness of support layer (μm)	51.4 ± 6.7	47.8 ± 2.5
Pore size of SL (μm)	5.3 ± 1.0	3.9 ± 2.0
Contact angle of active layer ($°$)	86.0 ± 4.5	79.2 ± 6.3
Contact angle of support layer ($°$)	72.8 ± 1.9	73.8 ± 6.0
Water permeability (A) (L/(m^2 h bar))	0.70 ± 0.07	1.24 ± 0.04
Salt permeability (B) (L/(m^2 h))	0.53 ± 0.03	0.37 ± 0.08
Salt rejection rate (%)	94.7 ± 0.1	97.7 ± 0.5
Water flux	Lower	Higher
Reverse solute flux	Lower	Higher

In another study [33], which compared the PRO performance of five different membranes including CTA-HTI, CTA-HTI spiral, CTA-FTS, TFC-FTS, and TFC-Aquaporin (Fluid Technology Solutions (FTS) and Aquaporin A/S are commercial membrane companies), it was found that although TFC membranes have higher flux, CTA membranes with higher pressure stability show higher power densities because the latter with a thicker rejection layer is less likely to be damaged at high pressure while the former is more prone to compaction. To summarize, it is obvious that each material has better performance in different aspects and more comparison studies should be carried out to provide evidence that support the hypothesis of one being better than another.

The development of the membrane materials and synthesis methods has been enhanced by the research and commercial attention on FO and PRO, resulting in the use of other materials [25]. For instance, a variety of nanomaterials, as shown in Table 2, has been used to alter the membrane substrates, either only form the active layer, or fabricate the whole thin film nanocomposite (TFN) membranes [34]. Incorporation of hydrophilic nanomaterials such as carbon-based nanomaterials functionalized with hydrophilic moieties can alleviate ICP by forming higher hydrophilicity, bigger porosity, and lower tortuosity [35]. Since the synthesis processes for TFC membranes are well established, any discovery on new membrane materials can be scaled up effortlessly for commercialization [36]. For example, aquaporins, which are water channel membrane proteins, have been incorporated into membrane substrate and this biomimetic Aquaporin membrane has been large scale commercialized for academic and industrial applications [37]. Nonetheless, it must be noted that there is a limited number of pilot-scale research carried out to apply TFC membranes in commercial-scale FO and PRO processes. More extensive researches at a bigger scale are needed for osmotic membranes to find out their feasibility for a sustainable operation to the point of commercialization despite the promising findings in laboratories.

Table 2. Recent nanostructured osmotic membrane and their experimental performance.

Type of Nanomaterials	Nanoparticles Incorporated	Effects	Water Flux (J_w, L/m^2 h)		Ref.
			Unmodified	Incorporated	
Carbon nanotubes	400 ppm sulfonated carbon nanotubes	• Better separation performance • Better permselectivity • Enhanced hydrophilicity • Decreased surface roughness	21.3 ± 2.1	29.9 ± 1.6	Li, et al. [38]
Zeolites	0.4 wt % modified clinoptilolite	• Increased membrane porosity • Higher eccentricity • Reduced structural parameter (S) • Minimized ICP	16.3	24.61	Salehi, et al. [39]
Zwitterions	Poly [3-(N-2-methacry loylxyethyl-N,N-dimethyl)-ammonatopropan esulfonate] (PMAPS)	• Lower water contact angle • Enhanced hydrophilicity • Remarkable anti-fouling properties	12.54	15.79	Lee, et al. [40]
Graphene oxide	0.1% graphene oxide nanosheets	• Increase in the hydrophilicity • Increase in a surface roughness value • Better water permeability	7.9	14.5	Shokrgozar Eslah, et al. [41]
Carbon quantum dots	Na$^+$-functionalized carbon quantum dots	• Higher roughness • Larger effective surface area • Decreased membrane thickness • Increased interstitial space among the polyamide chains	24.25 ± 2.8	34.86 ± 1.41	Gai, et al. [42]

Table 2. *Cont.*

Type of Nanomaterials	Nanoparticles Incorporated	Effects	Water Flux (J_w, L/m^2 h)		Ref.
			Unmodified	Incorporated	
Metal and metal oxide nanoparticles	0.5 wt % molybdenum oxide NPs (MoO$_3$)	• Increased wettability • Enhanced surface roughness, hydrophilicity, and water permeability	~21	67	Amini, et al. [43]
Polyelectrolytes	Layer-by-layer polyvinylidenefluoride (PVDF)	• Enhanced hydrophilicity and porosity • High pure water permeability • Low structural parameter	5.4	24.1	Gonzales, et al. [44]
Metal–organic frameworks	0.12 w/v % copper 1,4-benzenedicarboxylate nanosheets, CuBDC-NS	• Reduced contact angles • Increased surface hydrophilicity • Much lower reverse solute flux	18	28	Dai, et al. [45]

2.3. Membrane Configurations

In addition to membrane materials, membranes have been fabricated into two membrane configurations based on geometry that is flat sheet and hollow fiber configurations to improve performance such as achieving high water fluxes and superior rejection properties [46]. Generally, as compared to the hollow fiber membrane, it is easier to engineer a thinner support layer in a flat sheet membrane, thus it has higher water flux and power density [7]. However, due to the thin support layer and lower packing density, flat sheet membranes are less mechanically stable and more likely to experience structural destructions under pressure, even with a macro-void free support layer [14]. Compaction due to pressure always negatively affects the properties of the flat sheet membranes, such as increasing the resistance of porous substrates, decreasing the permeability of the selective layers, and enhancing the salt leakage by spacer [47].

Therefore, recent approaches in developing flat sheet membranes focus on enhancing their mechanical robustness, which is crucial for the PRO applications since it decides the maximum applicable hydraulic pressure. For example, the selective layer of flat sheet membranes is normally post-treated with chemical and physical modification for better mechanical strength [48]. Another approach to improve the pressure tolerance of flat sheet membranes is by utilizing customized spacers. Membrane with customized tricot fabric spacers was able to tolerate a stable 48 bar hydrostatic pressure for exceeding 10 h and achieve 14.1 W/m^2 at 20.7 bar [14]. The restrictions of pressure observed in numerous previous PRO research are primarily because of the unsuitable selection of spacer material.

Hollow fiber membranes (as shown in Figure 2) are tubular-shaped membranes and this configuration provides key benefits including a self-supporting structure and high packing density in membrane modules [49]. Hollow fiber membranes commonly feature a higher mechanical strength due to their geometrical (circular) structure. The fabric support-free hollow fiber configurations could also give rise to better control of the structural parameters [49]. Furthermore, the hollow fiber membrane module demonstrates higher efficiency (flux) of mass transfer compared to the flat-sheet (spiral-wound) module [50]. Wang, et al. [51] discovered that contact angles of FO hollow fiber membranes were much smaller than the commercial flat sheet membranes suggesting that they have higher hydrophilicity and lower fouling tendency. Besides, Chou, Wang and Fane [47] reported much lower specific reverse salt flux in the hollow fiber membrane as its self-supporting structure can eliminate the deformation-enhanced reverse solute flux which is common for flat-sheet membranes.

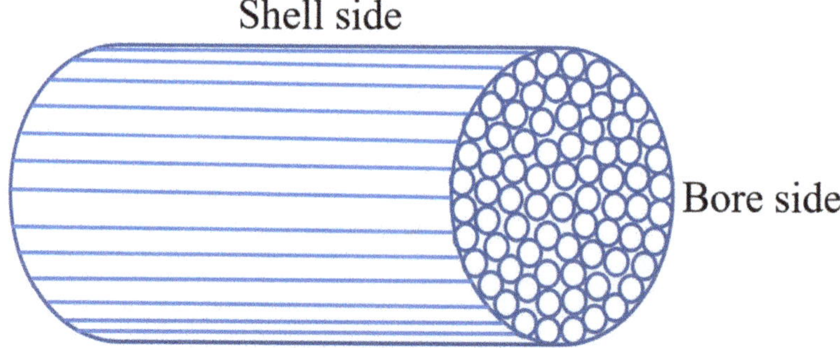

Figure 2. Hollow fiber membranes.

Moreover, hollow fiber membranes do not require spacers to effectively support the flowing of liquids on both sides of the membranes. Without spacers, deformation of the membrane caused by the spacer under high pressures could be minimized and more membranes could be packed into the modules. Also, with a self-supporting structure, higher pressures can be achieved by keeping the pressurized DS on the shell side rather than on the bore side of the fiber [52]. Since a spacer is not required in hollow fiber membranes with DS on the shell side, power density can be improved without the potential energy loss from the interface between membrane and spacer [53]. The Japanese Mega-ton project had tried a prototype by modifying and upscaling CTA-based RO hollow fiber membrane into PRO membrane module and achieved a power density of 7.7 W/m^2 at 25 bar [54]. However, they usually have smaller initial water flux compared to the flat sheet membranes (before deformation takes place), and they are likely to face pressure drop problems within each hollow fiber [6]. In addition, increasing pressure causes a reduction in the dimensions of fiber, leading to an increase in feed pressure loss on the bore side [52]. These are research gaps that researchers can look into so that there will be an enhancement in hollow fiber membranes and the performance of ODMPs.

Power density is a significant parameter in selecting the membrane for the PRO process. Although increasing the membrane area can scale up the energy generation, membranes with high power density which produce greater power per unit area reduce the membrane size required and the capital cost. As demonstrated in Table 3, various enhanced hollow fiber TFC membranes have been developed recently and applied in PRO processes where the highest power density in the literature so far is 38 W/m^2. In contrast, there is limited literature on the application of flat sheet membranes in osmotic power generation recently, after Han et al. [14] had summarized PRO performance of TFC flat sheet membranes up to 2015. One of the recent studies is research by Sharma, et al. [55] which discovered that CA flat sheet membrane modified with PEG 6000 achieved power density of 3.1 W/m^2. In brief, flat sheet membranes are more appropriate for FO applications due to their highly engineered support layer while hollow fiber membranes that have simple fabrication, high-pressure resistant structures, and high membrane surface area per module are comparatively suitable to harvest energy through PRO process [25].

Table 3. PRO performance of recent hollow fiber membranes.

Membrane	Feed Solution	Draw Solution	Hydraulic Pressure (Bar)	Power Density (W/m^2)	Ref.
TFC	DI water	1.2 M NaCl	30	38	Wan, et al. [56]
TFC	DI water	1 M NaCl	16.5	14.6	Park, et al. [57]
TFC	DI water	0.81 M NaCl	20	18.8	Zhang, et al. [58]
TFC	DI water	1.0 M NaCl	20	12.1	Gonzales, et al. [59]
TFC	DI water	1 M NaCl	21	16.7	Lim, et al. [60]
TFC	DI water	1 M NaCl	20	20	Wan, et al. [61]
TFC	DI water	1 M NaCl	20	27	Wan and Chung [62]

2.4. Membrane Fouling

Membrane fouling, which happens due to the deposition of particles and/or solutes at the membrane surface or inside membrane pores, is one of the main factors influencing membrane performance in ODMPs [12]. There are four types of fouling and their respective model foulants are shown in Table 4. The chemical and hydrodynamic interactions between the foulants and the membrane surface either temporarily or permanently decrease membrane water flux [63], leading to lowered permeate quality, reduced water recovery, shortened membrane life, and increased operating cost [13]. This is because foulants not only react with the membrane surface physically but also deteriorate membrane material chemically [63]. In addition, the accumulation of foulant influences power density used to generate energy in PRO. A layered model by Nagy, et al. [64] demonstrated a 50% reduction of power production due to a 500 μm-thick cake layer.

Table 4. Fouling types and model foulants.

Type of Fouling	Model Foulants
Organic	Alginate, humic acid (HA), and bovine serum albumin (BSA)
Inorganic	Calcium salts and silica
Colloidal	Colloidal silica particles
Biological	*Escherichia coli* bacteria suspensions

A lot of academic studies, industrial researches and development efforts have been carried out since the early 1960s to mitigate membrane fouling [63], which is influenced by various factors such as operation conditions, membrane properties, level of pretreatment and solution chemistry [21]. The properties and characteristics of membranes that influence fouling formation are their morphology (i.e., surface roughness, surface pattern, or pore size) and chemical structure (i.e., charge, hydrophilicity, and functional groups). In general, low surface roughness, high hydrophilicity, and negative surface charge are desirable [65].

In ODMPs, the accumulation of foulants takes place on different surfaces of membrane depending on the membrane orientation, hence fouling can be categorized into internal and external fouling [66]. In FO mode, foulants are accumulated on the relatively smooth active layer and a cake-type layer is subsequently formed. The deposition of foulants is influenced by both shear force and permeation drag, resulting from bulk cross-flow and permeate flux, respectively [67]. This manner of fouling is referred to as external fouling.

The mechanism of fouling is more complex in PRO mode. The foulants which are relatively small compared to the pores of the rough support layer can enter the membrane porous support layer and be adsorbed on the walls, or eventually be retained by the dense active layer and deposited at the back of the active layer. Subsequently, foulants which penetrate the support layer attached to the foulants which have been accumulated on the active layer, resulting in clogging of the membrane pores. This is the most serious case of fouling and is very difficult to clean up [68]. At the beginning of fouling in PRO mode, the effect of hydrodynamic shear forces is absent since cross-flow velocity disappears within the support layer [67]. Fouling in this manner is called internal fouling. Under severe conditions of fouling, more foulants continue to accumulate outside of the support layer, resulting in both external and internal fouling [66].

External fouling, in comparison with internal fouling, happens on the surface of the membrane so it is more able to be removed and reversed via optimizing feed water hydrodynamic conditions or chemical cleaning [69]. Therefore, AL-FS orientation is recommended by most researchers for FO application to avoid unwanted internal fouling [70], although there is relatively higher water flux in PRO mode since crossflow shear on the membrane surface can mitigate dilutive ECP [18]. On the other hand, PRO membranes are usually in operation of PRO mode due to higher mechanical stability (to avoid rejection layer damage) and higher power density (in the absence of fouling) [71].

Even though the osmotic backwash method has been developed to clean foulants within the support layer [72], it is essential to develop internal fouling control which is more effective for PRO because the requirement of pretreatment of feed water and membrane cleaning increases energy consumption and incurs additional costs [73]. It is also urgent to develop an appropriate PRO membrane to mitigate support layer fouling where material and roughness of the surface of the support layer should be taken into account [12]. Nevertheless, it should be clarified that both internal and external fouling can be irreversible depending on feed water constituents and their interactions with the membrane [74].

3. Feed Solutions

A variety of FSs, such as sewage [75,76], grey water [77], and microalgae [76,78], have been examined for their feasibility in different FO applications. Study on these commercially available FSs and other unexplored FSs is very important to understand their applicability, sustainability and economic feasibility in FO processes. Since FO is driven by the osmotic pressure difference, increasing feed concentration leads to lower initial flux due to the higher osmotic pressure of FS followed by a thinner foulant layer due to lowered convection of foulants to the membrane [64]. Furthermore, the presence of salt in the FS increases ICP in the membrane and reduces the achievable power density. Madsen et al. [33] reported a 55% reduction in power density (8 W/m^2 to 3.6 W/m^2) when FS was changed from MilliQ ultrapure water to a 2 wt % NaCl solution. The reduction in power density was not linear and the biggest difference occurred when changing MilliQ water to 0.01 wt %, which is the typical salinity of the river water model. This indicates that the suitability of river water as a FS should be investigated and DS with higher salinity should be studied to generate higher power. Moreover, the presence of other solutes (either dissolved or suspended) in FS could directly influence the FO performance [18]. This is because an increase in total dissolved solids of FS results in a reduction of net osmotic pressure, thus reducing water flux [18]. Composition and chemistry of FS—including the availability and concentration of divalent ions, ionic strength, and pH—strongly affect the fouling behavior in FO since characteristics of the foulants (e.g., membrane surface charge) and the foulant–foulant and foulant–membrane interactions are influenced [79]. For instance, comparatively high retention of ammonia was observed at a pH of 5 and the retention gradually decreased when the pH increased. Meanwhile, there was an enhancement in the retentions of nitrite, nitrate, and phosphate when pH was increased from 5–8. This is because the electrostatic repulsion between the membrane and the negatively charged ions was strengthened by the increasing negative charge on the surface of the membrane [76].

Feed waters with different qualities have been researched to determine the major foulants since water flux and osmotic power generation are reduced by membrane fouling, making the overall cost as high as other membrane processes [80]. In addition, for certain quick fouling, high cleaning frequency and long downtime processing are required making the conventional backwash approaches impractical. Also, adding a large amount of chemicals that are expensive to alleviate fouling in PRO processes is not viable economically [81].

Pretreatments

In order to reduce membrane fouling and to improve the performances, various pretreatment approaches—including microfiltration [82], vacuum filtering [83], low-pressure nanofiltration [84], low-pressure reverse osmosis [81], and coagulation [85]—have been investigated. Water flux increases when pretreatment is carried out since the suspended solids in feed solution are removed. However, Yang, Gao, Jang, Shon and Yue [83] reported that the increment associated with sewage pre-filtration decreased with time. Seker, et al. [86] presented similar findings where initial water flux with pretreatment (8.5 L/m^2 h) was lower but water flux of pretreated whey decreased slowly and final flux (4 L/m^2 h) was the same as the flux of non-pretreatment. Therefore, it should be noted that pretreatment only delays, but does not mitigate, the fouling formation.

In addition, the power density increases after the pretreatment. Results shown by Yang, Wan, Xiong and Chung [81] demonstrated that the PRO process without pre-treatment at 15 bar generated a power density of 2.92 W/m^2 (only 40% of its initial power density) while the PRO process using filtrates from nanofiltration could produce a stable power density of 7.3 W/m^2 with only a 10.7% drop. Nonetheless, Kim and Kim [87] suggested omitting pretreatments in designing and operating PRO because pretreatments need more energy (0.1–0.4 kWh/m^3) than the extractable net energy (0.1 kWh/m^3) and net energy extraction from seawater and river water could be done using PRO even with severe fouling. Therefore, an economical and less energy-intensive approach should be developed for pretreatments, otherwise pretreatments might not be that helpful in both FO and PRO processes.

4. Draw Solutions

Identifying an appropriate draw solute/solution is one of the major challenges to the FO and PRO processes. Since the mid-1960s, a variety of DS has been proposed and examined to maximize the efficiency of these processes but none of them have been successful for commercialization [88]. There are several important criteria to be fulfilled by an ideal DS, which are (i) high water flux (high osmotic pressure), (ii) simple and cheap recovery from the diluted DS, and (iii) low reverse solute flux (RSF) [89]. However, the first two requirements are in fact a conflict between one another since good osmotic potential requires a strong association between water molecules and draw solutes, which subsequently complicates the recovery process [90]. The recovery of DS remains one of the main challenges even though various approaches have been proposed including RO, heating/distillation, ultrafiltration, nanofiltration, magnetic separation, physical triggers, precipitation, and membrane distillation [91].

There are various additional criteria for an ideal draw solute, such as having no damage, scaling or fouling on the membrane surface even after prolonged use, intrinsic properties (e.g., at or near neutral pH), low viscosity even at high concentrations, low toxicity without adverse effect on environment and human health, a high diffusion coefficient to reduce ICP, being chemically stable for repeated use as well as cost-effective [91]. However, some contradictions between or deviations to some of these criteria may exist. For instance, solutes with small size like NaCl are more likely to achieve a high diffusion coefficient which reduces ICP, but there is also relatively high RSF due to their size [92].

Considering the diversity of feed solutions, it may not be possible that a single universal DS can treat all kinds of feed solutions [93]. Hence, before applying practically to the real-world industries, a draw solute should be extensively evaluated regarding their suitability through examining its various aspects and the priority order of assessment measure is entirely based on each application. For example, the primary property of a draw solute for power generation via PRO should be water induction ability (high osmotic pressure) since the relationship between power density and water flux is direct proportional [89].

As demonstrated in Figure 3, DS is widely classified into two groups which are synthetic compounds and commercially available compounds that can be further categorized into four main kinds in accordance with their physicochemical properties: volatile compounds, nutrient compounds, inorganic salts, and organic salts but overlapping may occur as this grouping is not strict [94].

Commercially available draw solutes were dominant before 2007 and there have been efforts in exploring them but the outcomes are still far from satisfactory, therefore various synthetic DSs that exhibit superiority with lower RSF and less energy consumption in recycling have been proposed recently [95].

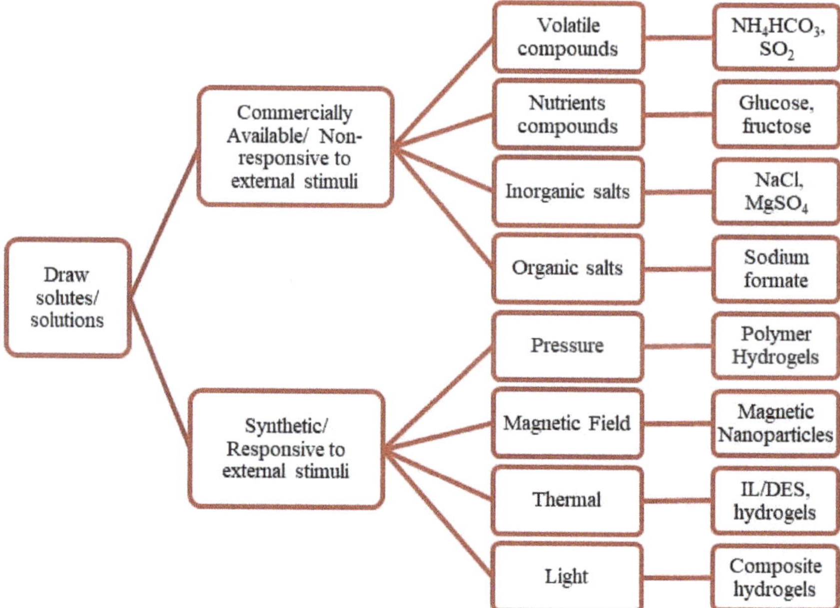

Figure 3. Classification of draw solutes/solution.

There is another classification of DSs based on the recovery methods of draw solute which are either responsive or non-responsive to some external stimuli [93]. The non-responsive draw solutes which include the commercially available compounds refer to those without significant variation in their water affinity upon exposure to stimuli like a magnetic field, thermal, pressure, or light whereas the responsive draw solutes are mostly the synthetic compounds which respond to stimuli by undergoing substantial changes in water affinity. Phase transitions between two states with different water affinities usually happen. Draw solutes can then be easily regenerated or separated from the diluted draw solutions while maintaining drawing ability which is sufficiently high [90].

When the concentration of DS increases, the osmotic pressure difference also increases, resulting in greater water flux. However, this is temporary because there is an enhanced dilutive ICP inside the membrane support layer caused by the increased incoming water flux, reducing the available driving force drastically [18,96]. In addition, increased concentration of DS results in a greater draw solutes loss because of RSF, despite being less significant in comparison to the increased Jw. It is noteworthy that if DS concentration is further increased beyond a certain level, the pumping energy will be increased due to the increased viscosity. The use of highly concentrated DS could also result in other implications; for example, chemical exposure to increased ionic concentrations could affect membrane stability [18]. Hence, it is essentially important to develop draw solutes which can produce high water flux at low concentration as this could lead to momentous energy and cost savings.

5. Operating Conditions

Furthermore, operating conditions are another factor that highly affects the treatment efficiency of FO. Examples of operating conditions are draw solution concentration, membrane orientation, cross-flow velocity and temperature, which are shown in Figure 4. This is also applicable to PRO where the extractable power per unit membrane area (i.e., the specific power) is affected by local conditions such as concentration and temperature of seawater [97]. Moreover, the combined effects of operating conditions should be investigated extensively and the relationship between each variable should be understood so that the performance of ODMPs can be maximized. For example, Hawari, et al. [98] reported a 93.3% increase in the membrane flux when both the DS temperature and the flow rate are increased from 20 to 26 °C and from 1.2 to 3.2 L/min, respectively.

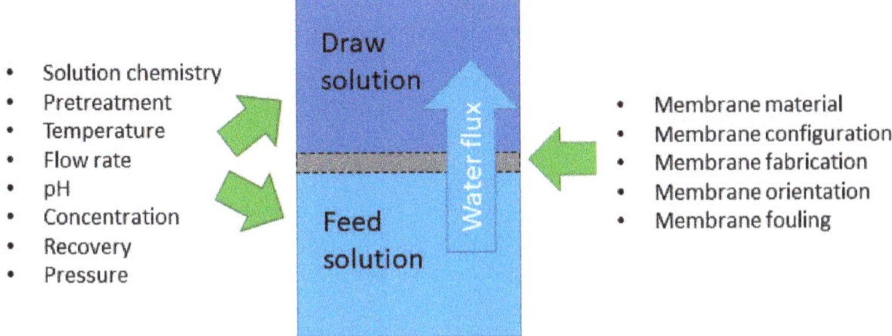

Figure 4. Factors affecting osmotic membrane processes.

Nonetheless, it is a challenge to define a general set of best-operating parameters since they are highly dependent on a particular system where each system has its own membrane parameters, equipment specifications, element geometry, site conditions, and system losses [99]. Maximum power point tracking (MPPT) has been proposed recently to have real-time control of PRO systems by either controlling of only one parameter which is the pressure exerted [100–102] or simultaneously controlling of few parameters including feed flow rates, draw flow rates and pressure exerted [103]. Maximum net power and maximum specific energy can then be extracted from the PRO system at both the design stage and in the field with varying conditions [103]. Nevertheless, laboratory-scale studies are still in great need even though numerous simulation and model-based optimization studies [104–106] have been carried out.

5.1. Cross-Flow Rate

According to film theory, the thickness of the mass transfer boundary layer at the membrane surface is changed by varying the flowrate of solutions [107]. When the flow rates are higher, there is an increase in permeate flux because the boundary layer is thinner, resulting in a higher rate of mass transfer, thus reducing ECP and fouling; however, in some cases, it was found that at higher flow rates, the residence time along the membrane will be shorter which will result in lower permeation across the membrane [23,108]. Increasing cross-flow velocity induces mixing in the porous support layer making the ICP less severe, especially for those membranes with thin and highly porous support layers [109]. In a cross-flow system, two main forces are applied to the feed particles. Hydrodynamic drag which is perpendicular to the surface of the membrane forces the foulant molecules to move towards the membrane while the shear rate which is tangential to the surface of the membrane causes the molecules to move back towards the bulk solution. Higher tangential shear force at high cross-flow velocities reduces the accumulation of foulants [110]. The removal rate of foulant is also higher [64]. Furthermore, the power density can be influenced by changing the flow rate ratio due to the higher water fluxes obtained from the membrane with a higher flow rate ratio. By understanding the influence of the input flow rate ratio, the potential power output from a PRO system can be increased [111].

A relatively low DS flowrate compared to the FS flowrate will cause dilution effect affecting the permeate flux, which is also more serious with a higher concentration of DS. There should be a sufficiently high cross-flow rate of the DS (dilution factor <2) to restrict the adverse dilution effect [96]. In fact, by altering the flow rates of both the FS and the DS could achieve a greater increment in water flux compared to altering the flow rate of either one. An increase in both flow rates causes a simultaneous reduction of CP effect at both sides of the membrane [98]. It shall be noted that high ratios of flow rate can induce great force to the membrane surface which has the potential to damage the membrane [111]. Hence, research works should be conducted to determine the optimal ratio of flow rates to be implemented.

The study conducted by Phuntsho, Sahebi, Majeed, Lotfi, Kim and Shon [18] has shown higher FO water flux at higher crossflow rates with optimum crossflow rate between 400 and 800 mL/min, although the increase is logarithmic and the effect of crossflow is insignificant beyond this level. Seker, et al. [112] supported this wherein their study, there was no variation in water flux while increasing flow rate from 2500–7500 mL/min as the flow rates used were higher than the optimum. There was a reduction in water fluxes when the flow rate was further increased to 10,000 mL/min. Also, Gulied, et al. [113] found out that a small increase in water flux (~1.5%) and %W recovery (1.2%) were resulted from increasing the flow rate from 1600–2800 mL/min. However, when the circulation flow rate increased, the ECP effect could become smaller reducing the water flux where the WF and %W recovery decreased gradually with increasing flow rates from 2800–3200 mL/min.

In addition, there is a requirement of extra pumping energy for high cross-flow rates yet the enhancement on water flux may be limited. For instance, the study reported by Gulied, Al Momani, Khraisheh, Bhosale and AlNouss [113] observed there was no significant effect on FO performance by increasing circulation flow rate (from 1.6 mL/min to 3.2 mL/min) but energy consumption increased from 36.1 W/m^3 to 144.2 W/m^3. This indicates that energy consumption increased by ~4-fold when doubling the flow rates. The efficiency of the PRO process is reduced as a result of increased energy need for pumping even though increasing cross-flow rates leads to increased power densities [114]. In certain cases and up to certain limits, improvement in osmotic performance and power generation can justify the energy invested for additional pumping [103]. At the optimal cross-flow, a balance of pumping energy and osmotic performance shall be accomplished [96].

5.2. Temperature

Operating temperature is a crucial factor which affects the osmotic performance. This is because it has a direct influence on the thermodynamic properties of both solutions, (e.g., osmotic pressure, diffusion, and viscosity) [115]. In fact, according to the van 't Hoff equation ($\pi = \beta CRT$), there is a direct proportional relationship between osmotic pressure and the temperature. Although the osmotic pressure is not proportional to the concentration if the solutions are highly concentrated, there is still an assumption of the proportionality between the osmotic pressure and the temperature [116]. Variation in the thermodynamic properties not only affects the water flux but also influences membrane fouling and solute rejection/diffusion [117]. Heating both sides (FS and DS) simultaneously is more effective since the difference among the water fluxes with increased temperature of either one solution is not significant [115].

An increase in the temperature improves the water flux due to reduced water viscosity in solutions (and/or solubility), and increased water solubility and diffusivity within the membrane [117]. Besides variations in properties of solvent, solute, and solution, there is higher water flux because the polymeric membrane expands upon heating producing larger effective pore sizes, which also unfortunately gives rise to more serious solute diffusion [118]. RSF increases with the increase of temperature as the membrane has less resistance to draw solute transmembrane diffusion [115]. Although the increased solute diffusion alleviates the concentrative ICP in the support layer and increases the water flux, dilutive ECP can be enhanced because such an improved water flux carries more solutes from the feed bulk phase to the surface of the support layer, thereby reducing the available driving force [117].

Furthermore, an increase in water flux leads to a reduction of operation time; for example, Yang, Gao, Jang, Shon and Yue [83] demonstrated reducing sewage to one-quarter volume using FO filtration took 46 h and 15 h at 15 °C and 35 °C, respectively. In addition to increasing water flux, higher temperature generally results in higher power density. Simulated model by Anastasio, et al. [119] predicted that power density is greater at elevated temperature since membrane hydraulic permeance, draw osmotic pressure and solute diffusivity all raise with temperature. Table 5 demonstrates how an increase in temperature affects water flux and/or specific power in previous studies. Most of the studies reported the improvement with increasing temperature, but higher temperatures can be tested for the resulting performance together with their effects on membrane and energy efficiency.

Table 5. Increase in water flux and/or specific power due to increase in temperature.

Membrane	FS	DS	Temperature (°C)	Increase in Water Flux Per °C (%)	Increase in Specific Power Per °C (%)	Ref.
CTA-HTI	Brackish water	1.5 M Na$_2$SO$_4$	25–35 35–45	3.1 1.2	-	Zhao and Zou [120]
CTA-HTI	10 mM NaCl	1 M NaCl	25–35	4.1	3.4	She, Jin and Tang [71]
CTA-HTI	Brackish water	0.5 M KCl	25–35 25–45	1.7 3.2	-	Phuntsho, et al. [121]
CTA-HTI	0.5 M NaCl	1 M NaCl 1.5 M NaCl 2 M NaCl	20–30	7.1 3.9 5.0	6.5 3.3 4.7	Kim and Elimelech [122]
CA-HTI	DI water	1.5 M NaCl	20–40	-	5.2	Anastasio, Arena, Cole and McCutcheon [119]
CTA-HTI	DI water	2 M NaCl	5–20 20–45	2.9 3.3	-	Heo, et al. [123]
TFC-TCK	0.01 M NaCl	0.6 M NaCl 1.2 M NaCl	25–50	2.9 3.0	3.2 3.1	Wang, et al. [124]
TFC	Municipal wastewater	Synthetic seawater concentrate	15–25 25–35	4.0 1.9	-	Yang, Gao, Jang, Shon and Yue [83]

6. Conclusions

There is an extensive potential in ODMPs, specifically FO and PRO, to act as a sustainable bioenergy solution on water and energy scarcity. However, technological developments and large-scale investigations are required to improve their feasibility and performance. This review provides a comprehensive summary of factors affecting their performance so that these efficient and green bioenergy processes can be further improved and implemented in the industrial applications in the near future. Membrane FS and DS, which are the major components of ODMPs, have high impacts on the performance of FO and PRO. These components which are commercially available should be modified for better performance. For example, research should be carried out to improve the pressure stability of TFC membranes for higher power density. Efforts must also be carried out on exploring new materials to fabricate membrane as well as to synthesize FS and DS so that higher water flux and greater power density can be achieved, thereby improving the bioenergy production. Draw solution with high salinity should be explored and examined to enhance the power generation to compete with other renewable energies. Also, optimization of operating parameters and understanding of the relationship between these parameters are essential to maximizing the bioenergy output. Although the performance of osmotic membrane processes (i.e., water flux and power density) is not affected by their economics, the cost aspect—including membrane cost and pretreatment cost—is an important criterion for consideration when designing an osmotic system, especially for bioenergy applications. Therefore, a techno-economic assessment of these osmotic membrane processes is essential to examine their feasibility even though there is a lack of full-scale FO and PRO applications to validate the cost assumption. In short, this paper has summarized the factors affecting FO and PRO processes which help in bioenergy production while outlining research needs and future directions on each factor.

Author Contributions: Writing—original draft, W.Y.C.; Writing—review & editing, K.S.K. and S.R.C.; Conceptualization, K.W.C. and G.Y.Y.; Supervision & Funding acquisition, Y.-C.H., P.L.S. and W.-H.C. All authors have read and agreed to the published version of the manuscript.

Funding: This work was supported by the Fundamental Research Grant Scheme, Malaysia (FRGS/1/2019/STG05/UNIM/02/2) and funded by PETRONAS under the YUTP grant (015LC0-169). The authors acknowledge the financial support of the Ministry of Science and Technology, Taiwan, R.O.C., under the contracts MOST 106-2923-E-006-002-MY3 and MOST 108-3116-F-006-007-CC1 for this research. This research was also supported in part by Higher Education Sprout Project, Ministry of Education to the Headquarters of University Advancement at NCKU.

Conflicts of Interest: The authors declare no conflict of interest.

References

1. Chung, T.-S.; Luo, L.; Wan, C.F.; Cui, Y.; Amy, G. What is next for forward osmosis (FO) and pressure retarded osmosis (PRO). *Sep. Purif. Technol.* **2015**, *156*, 856–860. [CrossRef]
2. Zhang, B.; Gao, H.; Tong, X.; Liu, S.; Gan, L.; Chen, Y. Chapter 6—Pressure Retarded Osmosis and Reverse Electrodialysis as Power Generation Membrane Systems. In *Current Trends and Future Developments on (Bio-) Membranes*; Basile, A., Cassano, A., Figoli, A., Eds.; Elsevier: Amsterdam, The Netherlands, 2019; pp. 133–152.
3. Zhao, S.; Zou, L.; Tang, C.Y.; Mulcahy, D. Recent developments in forward osmosis: Opportunities and challenges. *J. Membr. Sci.* **2012**, *396*, 1–21. [CrossRef]
4. Cath, T.Y.; Childress, A.E.; Elimelech, M. Forward osmosis: Principles, applications, and recent developments. *J. Membr. Sci.* **2006**, *281*, 70–87. [CrossRef]
5. Helfer, F.; Lemckert, C.; Anissimov, Y.G. Osmotic power with Pressure Retarded Osmosis: Theory, performance and trends—A review. *J. Membr. Sci.* **2014**, *453*, 337–358. [CrossRef]
6. Sarp, S. Chapter 11—Fundamentals of Pressure Retarded Osmosis. In *Current Trends and Future Developments on (Bio-) Membranes*; Basile, A., Curcio, E., Inamuddin, Eds.; Elsevier: Amsterdam, The Netherlands, 2019; pp. 271–283.
7. Sarp, S.; Li, Z.; Saththasivam, J. Pressure Retarded Osmosis (PRO): Past experiences, current developments, and future prospects. *Desalination* **2016**, *389*, 2–14. [CrossRef]
8. Yang, E.; Kim, K.-Y.; Chae, K.-J.; Lee, M.-Y.; Kim, I.S. Evaluation of energy and water recovery in forward osmosis–bioelectrochemical hybrid system with cellulose triacetate and polyamide asymmetric membrane in different orientations. *Desalin. Water. Treat.* **2016**, *57*, 7406–7413. [CrossRef]
9. Zhang, F.; Brastad, K.S.; He, Z. Integrating Forward Osmosis into Microbial Fuel Cells for Wastewater Treatment, Water Extraction and Bioelectricity Generation. *Environ. Sci. Technol.* **2011**, *45*, 6690–6696. [CrossRef] [PubMed]
10. Chia, S.R.; Chew, K.W.; Show, P.L.; Yap, Y.J.; Ong, H.C.; Ling, T.C.; Chang, J.-S. Analysis of Economic and Environmental Aspects of Microalgae Biorefinery for Biofuels Production: A Review. *Biotechnol. J.* **2018**, *13*, 1700618. [CrossRef] [PubMed]
11. Phwan, C.K.; Chew, K.W.; Sebayang, A.H.; Ong, H.C.; Ling, T.C.; Malek, M.A.; Ho, Y.-C.; Show, P.L. Effects of acids pre-treatment on the microbial fermentation process for bioethanol production from microalgae. *Biotechnol. Biofuels* **2019**, *12*, 191. [CrossRef]
12. Alsvik, L.I.; Hägg, M.-B. Pressure Retarded Osmosis and Forward Osmosis Membranes: Materials and Methods. *Polymers* **2013**, *5*, 303–327. [CrossRef]
13. She, Q.; Wang, R.; Fane, A.G.; Tang, C.Y. Membrane fouling in osmotically driven membrane processes: A review. *J. Membr. Sci.* **2016**, *499*, 201–233. [CrossRef]
14. Han, G.; Zhang, S.; Li, X.; Chung, T.-S. Progress in pressure retarded osmosis (PRO) membranes for osmotic power generation. *Prog. Polym. Sci.* **2015**, *51*, 1–27. [CrossRef]

15. Goh, P.S.; Ismail, A.F.; Ng, B.C.; Abdullah, M.S. Recent Progresses of Forward Osmosis Membranes Formulation and Design for Wastewater Treatment. *Water* **2019**, *11*, 2043. [CrossRef]

16. Yip, N.Y.; Tiraferri, A.; Phillip, W.A.; Schiffman, J.D.; Elimelech, M. High performance thin-film composite forward osmosis membrane. *Environ. Sci. Technol.* **2010**, *44*, 3812–3818. [CrossRef]

17. McCutcheon, J.R.; Elimelech, M. Influence of concentrative and dilutive internal concentration polarization on flux behavior in forward osmosis. *J. Membr. Sci.* **2006**, *284*, 237–247. [CrossRef]

18. Phuntsho, S.; Sahebi, S.; Majeed, T.; Lotfi, F.; Kim, J.E.; Shon, H.K. Assessing the major factors affecting the performances of forward osmosis and its implications on the desalination process. *Chem. Eng. J.* **2013**, *231*, 484–496. [CrossRef]

19. McCutcheon, J.R.; Elimelech, M. Influence of membrane support layer hydrophobicity on water flux in osmotically driven membrane processes. *J. Membr. Sci.* **2008**, *318*, 458–466. [CrossRef]

20. Population Distribution and Water Scarcity. *Recent Developments in Forward Osmosis Processes*; Linares, R.V., Li, Z., Elimelech, M., Amy, G., Vrouwenvelder, H., Eds.; IWA Publishing: London, UK, 2017.

21. Klaysom, C.; Cath, T.Y.; Depuydt, T.; Vankelecom, I.F. Forward and pressure retarded osmosis: Potential solutions for global challenges in energy and water supply. *Chem. Soc. Rev.* **2013**, *42*, 6959–6989. [CrossRef]

22. Majeed, T.; Lotfi, F.; Phuntsho, S.; Yoon, J.K.; Kim, K.; Shon, H.K. Performances of PA hollow fiber membrane with the CTA flat sheet membrane for forward osmosis process. *Desalin. Water Treat.* **2015**, *53*, 1744–1754. [CrossRef]

23. Wong, M.C.Y.; Martinez, K.; Ramon, G.Z.; Hoek, E.M.V. Impacts of operating conditions and solution chemistry on osmotic membrane structure and performance. *Desalination* **2012**, *287*, 340–349. [CrossRef]

24. Lee, K.L.; Baker, R.W.; Lonsdale, H.K. Membranes for power generation by pressure-retarded osmosis. *J. Membr. Sci.* **1981**, *8*, 141–171. [CrossRef]

25. Sarp, S.; Hilal, N. Chapter 8—Membrane Modules for Large-Scale Salinity Gradient Process Applications. In *Membrane-Based Salinity Gradient Processes for Water Treatment and Power Generation*; Sarp, S., Hilal, N., Eds.; Elsevier: Amsterdam, The Netherlands, 2018; pp. 223–242.

26. Li, X.; Zhang, S.; Fu, F.; Chung, T.-S. Deformation and reinforcement of thin-film composite (TFC) polyamide-imide (PAI) membranes for osmotic power generation. *J. Membr. Sci.* **2013**, *434*, 204–217. [CrossRef]

27. Qasim, M.; Darwish, N.A.; Sarp, S.; Hilal, N. Water desalination by forward (direct) osmosis phenomenon: A comprehensive review. *Desalination* **2015**, *374*, 47–69. [CrossRef]

28. Wei, J.; Qiu, C.; Tang, C.Y.; Wang, R.; Fane, A.G. Synthesis and characterization of flat-sheet thin film composite forward osmosis membranes. *J. Membr. Sci.* **2011**, *372*, 292–302. [CrossRef]

29. Lim, S.W.; Mah, S.-K.; Lee, Z.H.; Chai, S.-P. A study of water permeation using glycerol as the draw solution with thin film composite membranes in forward osmosis and pressure retarded osmosis configurations. *AIP Conf. Proc.* **2018**, *2031*, 020020.

30. Wang, Z.; Tang, J.; Zhu, C.; Dong, Y.; Wang, Q.; Wu, Z. Chemical cleaning protocols for thin film composite (TFC) polyamide forward osmosis membranes used for municipal wastewater treatment. *J. Membr. Sci.* **2015**, *475*, 184–192. [CrossRef]

31. Fam, W.; Phuntsho, S.; Lee, J.H.; Shon, H.K. Performance comparison of thin-film composite forward osmosis membranes. *Desalin. Water Treat.* **2013**, *51*, 6274–6280. [CrossRef]

32. Li, J.-Y.; Ni, Z.-Y.; Zhou, Z.-Y.; Hu, Y.-X.; Xu, X.-H.; Cheng, L.-H. Membrane fouling of forward osmosis in dewatering of soluble algal products: Comparison of TFC and CTA membranes. *J. Membr. Sci.* **2018**, *552*, 213–221. [CrossRef]

33. Madsen, H.T.; Nissen, S.S.; Muff, J.; Søgaard, E.G. Pressure retarded osmosis from hypersaline solutions: Investigating commercial FO membranes at high pressures. *Desalination* **2017**, *420*, 183–190. [CrossRef]

34. Wan, C.F.; Cui, Y.; Gai, W.X.; Cheng, Z.L.; Chung, T.-S. Chapter 14—Nanostructured Membranes for Enhanced Forward Osmosis and Pressure-Retarded Osmosis. In *Sustainable Nanoscale Engineering*; Szekely, G., Livingston, A., Eds.; Elsevier: Amsterdam, The Netherlands, 2020; pp. 373–394.

35. Sakai, M.; Seshimo, M.; Matsukata, M. Hydrophilic ZSM-5 membrane for forward osmosis operation. *J. Water Process Eng.* **2019**, *32*, 100864. [CrossRef]

36. Chung, T.-S.; Zhao, D.; Gao, J.; Lu, K.; Wan, C.; Weber, M.; Maletzko, C. Emerging R&D on membranes and systems for water reuse and desalination. *Chin. J. Chem. Eng.* **2019**, *27*, 1578–1585.

37. Sahebi, S.; Sheikhi, M.; Ramavandi, B. A new biomimetic aquaporin thin film composite membrane for forward osmosis: Characterization and performance assessment. *Desalin. Water Treat.* **2019**, *148*, 42–50. [CrossRef]

38. Li, Y.; Zhao, Y.; Tian, E.; Ren, Y. Preparation and characterization of novel forward osmosis membrane incorporated with sulfonated carbon nanotubes. *RSC Adv.* **2018**, *8*, 41032–41039. [CrossRef]

39. Salehi, T.M.; Peyravi, M.; Jahanshahi, M.; Lau, W.-J.; Rad, A.S. Impacts of zeolite nanoparticles on substrate properties of thin film nanocomposite membranes for engineered osmosis. *J. Nanoparticle Res.* **2018**, *20*, 113. [CrossRef]

40. Lee, W.J.; Goh, P.S.; Lau, W.J.; Ong, C.S.; Ismail, A.F. Antifouling zwitterion embedded forward osmosis thin film composite membrane for highly concentrated oily wastewater treatment. *Sep. Purif. Technol.* **2019**, *214*, 40–50. [CrossRef]

41. Shokrgozar Eslah, S.; Shokrollahzadeh, S.; Moini Jazani, O.; Samimi, A. Forward osmosis water desalination: Fabrication of graphene oxide-polyamide/polysulfone thin-film nanocomposite membrane with high water flux and low reverse salt diffusion. *Sep. Sci. Technol.* **2018**, *53*, 573–583. [CrossRef]

42. Gai, W.; Zhao, D.L.; Chung, T.-S. Thin film nanocomposite hollow fiber membranes comprising Na+-functionalized carbon quantum dots for brackish water desalination. *Water Res.* **2019**, *154*, 54–61. [CrossRef]

43. Amini, M.; Shekari, Z.; Hosseinifard, M.; Seidi, F. Preparation and Characterization of Thin-Film Nanocomposite Membrane Incorporated with MoO3 Nanoparticles with High Flux Performance for Forward Osmosis. *ChemistrySelect* **2019**, *4*, 7832–7837. [CrossRef]

44. Gonzales, R.R.; Park, M.J.; Tijing, L.; Han, D.S.; Phuntsho, S.; Shon, H.K. Modification of Nanofiber Support Layer for Thin Film Composite Forward Osmosis Membranes via Layer-by-Layer Polyelectrolyte Deposition. *Membranes* **2018**, *8*, 70. [CrossRef]

45. Dai, R.; Zhang, X.; Liu, M.; Wu, Z.; Wang, Z. Porous metal organic framework CuBDC nanosheet incorporated thin-film nanocomposite membrane for high-performance forward osmosis. *J. Membr. Sci.* **2019**, *573*, 46–54. [CrossRef]

46. Subramani, A.; Jacangelo, J.G. Emerging desalination technologies for water treatment: A critical review. *Water Res.* **2015**, *75*, 164–187. [CrossRef] [PubMed]

47. Chou, S.; Wang, R.; Fane, A.G. Robust and High performance hollow fiber membranes for energy harvesting from salinity gradients by pressure retarded osmosis. *J. Membr. Sci.* **2013**, *448*, 44–54. [CrossRef]

48. Goh, P.S.; Ismail, A.F. Chapter 5—Flat-Sheet Membrane for Power Generation and Desalination Based on Salinity Gradient. In *Membrane-Based Salinity Gradient Processes for Water Treatment and Power Generation*; Sarp, S., Hilal, N., Eds.; Elsevier: Amsterdam, The Netherlands, 2018; pp. 155–174.

49. Wang, Y.-N.; Goh, K.; Li, X.; Setiawan, L.; Wang, R. Membranes and processes for forward osmosis-based desalination: Recent advances and future prospects. *Desalination* **2018**, *434*, 81–99. [CrossRef]

50. Crowder, M.L.; Gooding, C.H. Spiral wound, hollow fiber membrane modules: A new approach to higher mass transfer efficiency. *J. Membr. Sci.* **1997**, *137*, 17–29. [CrossRef]

51. Wang, R.; Shi, L.; Tang, C.Y.; Chou, S.; Qiu, C.; Fane, A.G. Characterization of novel forward osmosis hollow fiber membranes. *J. Membr. Sci.* **2010**, *355*, 158–167. [CrossRef]

52. Bajraktari, N.; Hélix-Nielsen, C.; Madsen, H.T. Pressure retarded osmosis from hypersaline sources—A review. *Desalination* **2017**, *413*, 65–85. [CrossRef]

53. Lau, W.-J.; Lai, G.-S.; Li, J.; Gray, S.; Hu, Y.; Misdan, N.; Goh, P.-S.; Matsuura, T.; Azelee, I.W.; Ismail, A.F. Development of microporous substrates of polyamide thin film composite membranes for pressure-driven and osmotically-driven membrane processes: A review. *J. Ind. Eng. Chem.* **2019**, *77*, 25–59. [CrossRef]

54. Saito, K.; Irie, M.; Zaitsu, S.; Sakai, H.; Hayashi, H.; Tanioka, A. Power generation with salinity gradient by pressure retarded osmosis using concentrated brine from SWRO system and treated sewage as pure water. *Desalin. Water Treat.* **2012**, *41*, 114–121. [CrossRef]

55. Sharma, M.; Mondal, P.; Chakraborty, A.; Kuttippurath, J.; Purkait, M. Effect of different molecular weight polyethylene glycol on flat sheet cellulose acetate membranes for evaluating power density performance in pressure retarded osmosis study. *J. Water Process Eng.* **2019**, *30*, 100632. [CrossRef]

56. Wan, C.F.; Yang, T.; Gai, W.; Lee, Y.D.; Chung, T.-S. Thin-film composite hollow fiber membrane with inorganic salt additives for high mechanical strength and high power density for pressure-retarded osmosis. *J. Membr. Sci.* **2018**, *555*, 388–397. [CrossRef]

57. Park, M.J.; Lim, S.; Gonzales, R.R.; Phuntsho, S.; Han, D.S.; Abdel-Wahab, A.; Adham, S.; Shon, H.K. Thin-film composite hollow fiber membranes incorporated with graphene oxide in polyethersulfone support layers for enhanced osmotic power density. *Desalination* **2019**, *464*, 63–75. [CrossRef]

58. Zhang, Y.; Li, J.L.; Cai, T.; Cheng, Z.L.; Li, X.; Chung, T.-S. Sulfonated hyperbranched polyglycerol grafted membranes with antifouling properties for sustainable osmotic power generation using municipal wastewater. *J. Membr. Sci.* **2018**, *563*, 521–530. [CrossRef]

59. Gonzales, R.R.; Park, M.J.; Bae, T.-H.; Yang, Y.; Abdel-Wahab, A.; Phuntsho, S.; Shon, H.K. Melamine-based covalent organic framework-incorporated thin film nanocomposite membrane for enhanced osmotic power generation. *Desalination* **2019**, *459*, 10–19. [CrossRef]

60. Lim, S.; Park, M.J.; Phuntsho, S.; Mai-Prochnow, A.; Murphy, A.B.; Seo, D.; Shon, H. Dual-layered nanocomposite membrane incorporating graphene oxide and halloysite nanotube for high osmotic power density and fouling resistance. *J. Membr. Sci.* **2018**, *564*, 382–393. [CrossRef]

61. Wan, C.F.; Li, B.; Yang, T.; Chung, T.-S. Design and fabrication of inner-selective thin-film composite (TFC) hollow fiber modules for pressure retarded osmosis (PRO). *Sep. Purif. Technol.* **2017**, *172*, 32–42. [CrossRef]

62. Wan, C.F.; Chung, T.-S. Osmotic power generation by pressure retarded osmosis using seawater brine as the draw solution and wastewater retentate as the feed. *J. Membr. Sci.* **2015**, *479*, 148–158. [CrossRef]

63. Rana, D.; Matsuura, T. Surface Modifications for Antifouling Membranes. *Chem. Rev.* **2010**, *110*, 2448–2471. [CrossRef]

64. Nagy, E.; Hegedüs, I.; Tow, E.W.; Lienhard, V.J.H. Effect of fouling on performance of pressure retarded osmosis (PRO) and forward osmosis (FO). *J. Membr. Sci.* **2018**, *565*, 450–462. [CrossRef]

65. Le, N.L.; Nunes, S.P. Materials and membrane technologies for water and energy sustainability. *Sustain. Mater. Technol.* **2016**, *7*, 1–28. [CrossRef]

66. Ibrar, I.; Naji, O.; Sharif, A.; Malekizadeh, A.; Alhawari, A.; Alanezi, A.A.; Altaee, A. A Review of Fouling Mechanisms, Control Strategies and Real-Time Fouling Monitoring Techniques in Forward Osmosis. *Water* **2019**, *11*, 695. [CrossRef]

67. Mi, B.; Elimelech, M. Chemical and physical aspects of organic fouling of forward osmosis membranes. *J. Membr. Sci.* **2008**, *320*, 292–302. [CrossRef]

68. Zhao, P.; Gao, B.; Yue, Q.; Liu, P.; Shon, H.K. Fatty acid fouling of forward osmosis membrane: Effects of pH, calcium, membrane orientation, initial permeate flux and foulant composition. *J. Environ. Sci.* **2016**, *46*, 55–62. [CrossRef] [PubMed]

69. Abbasi-Garravand, E.; Mulligan, C.N.; Laflamme, C.B.; Clairet, G. Investigation of the fouling effect on a commercial semi-permeable membrane in the pressure retarded osmosis (PRO) process. *Sep. Purif. Technol.* **2018**, *193*, 81–90. [CrossRef]

70. Zhao, S.; Zou, L.; Mulcahy, D. Effects of membrane orientation on process performance in forward osmosis applications. *J. Membr. Sci.* **2011**, *382*, 308–315. [CrossRef]

71. She, Q.; Jin, X.; Tang, C.Y. Osmotic power production from salinity gradient resource by pressure retarded osmosis: Effects of operating conditions and reverse solute diffusion. *J. Membr. Sci.* **2012**, *401–402*, 262–273. [CrossRef]

72. Yip, N.Y.; Elimelech, M. Influence of natural organic matter fouling and osmotic backwash on pressure retarded osmosis energy production from natural salinity gradients. *Environ. Sci. Technol.* **2013**, *47*, 12607–12616. [CrossRef]

73. Chun, Y.; Mulcahy, D.; Zou, L.; Kim, I.S. A Short Review of Membrane Fouling in Forward Osmosis Processes. *Membranes* **2017**, *7*, 30. [CrossRef]

74. Jiang, S.; Li, Y.; Ladewig, B.P. A review of reverse osmosis membrane fouling and control strategies. *Sci. Total Environ.* **2017**, *595*, 567–583. [CrossRef]

75. Singh, N.; Dhiman, S.; Basu, S.; Balakrishnan, M.; Petrinic, I.; Helix-Nielsen, C. Dewatering of sewage for nutrients and water recovery by Forward Osmosis (FO) using divalent draw solution. *J. Water Process Eng.* **2019**, *31*, 100853. [CrossRef]

76. Xue, W.; Tobino, T.; Nakajima, F.; Yamamoto, K. Seawater-driven forward osmosis for enriching nitrogen and phosphorous in treated municipal wastewater: Effect of membrane properties and feed solution chemistry. *Water Res.* **2015**, *69*, 120–130. [CrossRef]

77. Wang, J.; Xiao, T.; Bao, R.; Li, T.; Wang, Y.; Li, D.; Li, X.; He, T. Zwitterionic surface modification of forward osmosis membranes using N-aminoethyl piperazine propane sulfonate for grey water treatment. *Process Saf. Environ. Prot.* **2018**, *116*, 632–639. [CrossRef]

78. Son, J.; Sung, M.; Ryu, H.; Oh, Y.-K.; Han, J.-I. Microalgae dewatering based on forward osmosis employing proton exchange membrane. *Bioresour. Technol.* **2017**, *244*, 57–62. [CrossRef]

79. Yun, T.; Kim, Y.-J.; Lee, S.; Hong, S.; Kim, G.I. Flux behavior and membrane fouling in pressure-assisted forward osmosis. *Desalin. Water Treat.* **2014**, *52*, 564–569. [CrossRef]

80. Nave, F.; Kommalapati, R.; Thompson, A. Introductory Chapter: Osmotically Driven Membrane Processes. In *Osmotically Driven Membrane Processes—Approach, Development and Current Status*; IntechOpen: London, UK, 2018.

81. Yang, T.; Wan, C.F.; Xiong, J.Y.; Chung, T.-S. Pre-treatment of wastewater retentate to mitigate fouling on the pressure retarded osmosis (PRO) process. *Sep. Purif. Technol.* **2019**, *215*, 390–397. [CrossRef]

82. Choi, Y.; Vigneswaran, S.; Lee, S. Evaluation of fouling potential and power density in pressure retarded osmosis (PRO) by fouling index. *Desalination* **2016**, *389*, 215–223. [CrossRef]

83. Yang, S.; Gao, B.; Jang, A.; Shon, H.K.; Yue, Q. Municipal wastewater treatment by forward osmosis using seawater concentrate as draw solution. *Chemosphere* **2019**, *237*, 124485. [CrossRef] [PubMed]

84. Chen, Y.; Liu, C.; Setiawan, L.; Wang, Y.-N.; Hu, X.; Wang, R. Enhancing pressure retarded osmosis performance with low-pressure nanofiltration pretreatment: Membrane fouling analysis and mitigation. *J. Membr. Sci.* **2017**, *543*, 114–122. [CrossRef]

85. Wan, C.F.; Jin, S.; Chung, T.-S. Mitigation of inorganic fouling on pressure retarded osmosis (PRO) membranes by coagulation pretreatment of the wastewater concentrate feed. *J. Membr. Sci.* **2019**, *572*, 658–667. [CrossRef]

86. Seker, M.; Buyuksari, E.; Topcu, S.; Babaoglu, D.S.; Celebi, D.; Keskinler, B.; Aydiner, C. Effect of pretreatment and membrane orientation on fluxes for concentration of whey with high foulants by using NH3/CO2 in forward osmosis. *Bioresour. Technol.* **2017**, *243*, 237–246. [CrossRef]

87. Kim, M.; Kim, S. Practical limit of energy production from seawater by full-scale pressure retarded osmosis. *Energy* **2018**, *158*, 373–382. [CrossRef]

88. Chekli, L.; Phuntsho, S.; Shon, H.K.; Vigneswaran, S.; Kandasamy, J.; Chanan, A. A review of draw solutes in forward osmosis process and their use in modern applications. *Desalin. Water Treat.* **2012**, *43*, 167–184. [CrossRef]

89. Gwak, G.; Hong, S. Chapter 3—Draw Solute Selection. In *Membrane-Based Salinity Gradient Processes for Water Treatment and Power Generation*; Sarp, S., Hilal, N., Eds.; Elsevier: Amsterdam, The Netherlands, 2018; pp. 87–122.

90. Cai, Y.; Hu, X.M. A critical review on draw solutes development for forward osmosis. *Desalination* **2016**, *391*, 16–29. [CrossRef]

91. Alejo, T.; Arruebo, M.; Carcelen, V.; Monsalvo, V.M.; Sebastian, V. Advances in draw solutes for forward osmosis: Hybrid organic-inorganic nanoparticles and conventional solutes. *Chem. Eng. J.* **2017**, *309*, 738–752. [CrossRef]

92. Johnson, D.J.; Suwaileh, W.A.; Mohammed, A.W.; Hilal, N. Osmotic's potential: An overview of draw solutes for forward osmosis. *Desalination* **2018**, *434*, 100–120. [CrossRef]

93. Dutta, S.; Nath, K. Prospect of ionic liquids and deep eutectic solvents as new generation draw solution in forward osmosis process. *J. Water Process Eng.* **2018**, *21*, 163–176. [CrossRef]

94. Ge, Q.; Ling, M.; Chung, T.-S. Draw solutions for forward osmosis processes: Developments, challenges, and prospects for the future. *J. Membr. Sci.* **2013**, *442*, 225–237. [CrossRef]

95. Chen, Q.; Xu, W.; Ge, Q. Synthetic draw solutes for forward osmosis: Status and future. *Rev. Chem. Eng.* **2018**, *34*, 767. [CrossRef]

96. Xu, Y.; Peng, X.; Tang, C.Y.; Fu, Q.S.; Nie, S. Effect of draw solution concentration and operating conditions on forward osmosis and pressure retarded osmosis performance in a spiral wound module. *J. Membr. Sci.* **2010**, *348*, 298–309. [CrossRef]

97. Sivertsen, E.; Holt, T.; Thelin, W.R. Concentration and Temperature Effects on Water and Salt Permeabilities in Osmosis and Implications in Pressure-Retarded Osmosis. *Membranes* **2018**, *8*, 39. [CrossRef]

98. Hawari, A.H.; Kamal, N.; Altaee, A. Combined influence of temperature and flow rate of feeds on the performance of forward osmosis. *Desalination* **2016**, *398*, 98–105. [CrossRef]

99. Maisonneuve, J.; Pillay, P.; Laflamme, C.B. Pressure-retarded osmotic power system model considering non-ideal effects. *Renew. Energy* **2015**, *75*, 416–424. [CrossRef]

100. He, W.; Luo, X.; Kiselychnyk, O.; Wang, J.; Shaheed, M.H. Maximum power point tracking (MPPT) control of pressure retarded osmosis (PRO) salinity power plant: Development and comparison of different techniques. *Desalination* **2016**, *389*, 187–196. [CrossRef]

101. He, W.; Wang, Y.; Shaheed, M.H. Maximum power point tracking (MPPT) of a scale-up pressure retarded osmosis (PRO) osmotic power plant. *Appl. Energy* **2015**, *158*, 584–596. [CrossRef]

102. Chen, Y.; Vepa, R.; Shaheed, M.H. Enhanced and speedy energy extraction from a scaled-up pressure retarded osmosis process with a whale optimization based maximum power point tracking. *Energy* **2018**, *153*, 618–627. [CrossRef]

103. Maisonneuve, J.; Chintalacheruvu, S. Increasing osmotic power and energy with maximum power point tracking. *Appl. Energy* **2019**, *238*, 683–695. [CrossRef]

104. Manzoor, H.; Selam, M.A.; Abdur Rahman, F.B.; Adham, S.; Castier, M.; Abdel-Wahab, A. A tool for assessing the scalability of pressure-retarded osmosis (PRO) membranes. *Renew. Energy* **2019**. [CrossRef]

105. Ettouney, H.; Al-Hajri, K. Modeling and performance analysis of forward and pressure-retarded osmosis. *Desalin. Water Treat.* **2019**, *154*, 1–13. [CrossRef]

106. Chen, Y.; Alanezi, A.A.; Zhou, J.; Altaee, A.; Shaheed, M.H. Optimization of module pressure retarded osmosis membrane for maximum energy extraction. *J. Water Process Eng.* **2019**, *32*, 100935. [CrossRef]

107. Touati, K.; Tadeo, F. Study of the Reverse Salt Diffusion in pressure retarded osmosis: Influence on concentration polarization and effect of the operating conditions. *Desalination* **2016**, *389*, 171–186. [CrossRef]

108. Jung, D.H.; Lee, J.; Kim, D.Y.; Lee, Y.G.; Park, M.; Lee, S.; Yang, D.R.; Kim, J.H. Simulation of forward osmosis membrane process: Effect of membrane orientation and flow direction of feed and draw solutions. *Desalination* **2011**, *277*, 83–91. [CrossRef]

109. Bui, N.-N.; Arena, J.T.; McCutcheon, J.R. Proper accounting of mass transfer resistances in forward osmosis: Improving the accuracy of model predictions of structural parameter. *J. Membr. Sci.* **2015**, *492*, 289–302. [CrossRef]

110. Tang, C.Y.; Chong, T.H.; Fane, A.G. Colloidal interactions and fouling of NF and RO membranes: A review. *Adv. Colloid Interface Sci.* **2011**, *164*, 126–143. [CrossRef] [PubMed]

111. Tran, T.T.D.; Park, K.; Smith, A.D. Performance Analysis for Pressure Retarded Osmosis: Experimentation With High Pressure Difference and Varying Flow Rate, Considering Exposed Membrane Area. In Proceedings of the ASME 2016 International Mechanical Engineering Congress and Exposition IMECE, Phoenix, AZ, USA, 11–17 November 2016.

112. Seker, M.; Buyuksari, E.; Topcu, S.; Sesli, D.; Celebi, D.; Keskinler, B.; Aydiner, C. Effect of process parameters on flux for whey concentration with NH3/CO2 in forward osmosis. *Food Bioprod. Process.* **2017**, *105*, 64–76. [CrossRef]

113. Gulied, M.; Al Momani, F.; Khraisheh, M.; Bhosale, R.; AlNouss, A. Influence of draw solution type and properties on the performance of forward osmosis process: Energy consumption and sustainable water reuse. *Chemosphere* **2019**, *233*, 234–244. [CrossRef] [PubMed]

114. Hickenbottom, K.L.; Vanneste, J.; Elimelech, M.; Cath, T.Y. Assessing the current state of commercially available membranes and spacers for energy production with pressure retarded osmosis. *Desalination* **2016**, *389*, 108–118. [CrossRef]

115. Wang, C.; Li, Y.; Wang, Y. Treatment of greywater by forward osmosis technology: Role of the operating temperature. *Environ. Technol.* **2019**, *40*, 3434–3443. [CrossRef] [PubMed]

116. Touati, K.; Tadeo, F.; Hänel, C.; Schiestel, T. Effect of the operating temperature on hydrodynamics and membrane parameters in pressure retarded osmosis. *Desalin. Water Treat.* **2015**, *57*, 10477–10489. [CrossRef]

117. Shin, S.; Kim, A.S. Temperature Effect on Forward Osmosis. In *Osmotically Driven Membrane Processes—Approach, Development and Current Status*; IntechOpen: London, UK, 2018.

118. Touati, K.; Tadeo, F. Chapter Three—Effects of the Temperatures on PRO. In *Pressure Retarded Osmosis*; Touati, K., Tadeo, F., Chae, S.H., Kim, J.H., Alvarez-Silva, O., Eds.; Academic Press: Cambridge, MA, USA, 2017; pp. 97–128.

119. Anastasio, D.D.; Arena, J.T.; Cole, E.A.; McCutcheon, J.R. Impact of temperature on power density in closed-loop pressure retarded osmosis for grid storage. *J. Membr. Sci.* **2015**, *479*, 240–245. [CrossRef]

120. Zhao, S.; Zou, L. Effects of working temperature on separation performance, membrane scaling and cleaning in forward osmosis desalination. *Desalination* **2011**, *278*, 157–164. [CrossRef]
121. Phuntsho, S.; Vigneswaran, S.; Kandasamy, J.; Hong, S.; Lee, S.; Shon, H.K. Influence of temperature and temperature difference in the performance of forward osmosis desalination process. *J. Membr. Sci.* **2012**, *415–416*, 734–744. [CrossRef]
122. Kim, Y.C.; Elimelech, M. Potential of osmotic power generation by pressure retarded osmosis using seawater as feed solution: Analysis and experiments. *J. Membr. Sci.* **2013**, *429*, 330–337. [CrossRef]
123. Heo, J.; Chu, K.H.; Her, N.; Im, J.; Park, Y.-G.; Cho, J.; Sarp, S.; Jang, A.; Jang, M.; Yoon, Y. Organic fouling and reverse solute selectivity in forward osmosis: Role of working temperature and inorganic draw solutions. *Desalination* **2016**, *389*, 162–170. [CrossRef]
124. Wang, Q.; Zhou, Z.; Li, J.; Tang, Q.; Hu, Y. Investigation of the reduced specific energy consumption of the RO-PRO hybrid system based on temperature-enhanced pressure retarded osmosis. *J. Membr. Sci.* **2019**, *581*, 439–452. [CrossRef]

Review

Nanomaterials Utilization in Biomass for Biofuel and Bioenergy Production

Kuan Shiong Khoo [1], Wen Yi Chia [1], Doris Ying Ying Tang [1], Pau Loke Show [1], Kit Wayne Chew [2] and Wei-Hsin Chen [3,4,5,6,*]

[1] Department of Chemical and Environmental Engineering, Faculty of Science and Engineering, University of Nottingham Malaysia, Jalan Broga, Semenyih 43500, Malaysia; kuanshiong.khoo@hotmail.com (K.S.K.); wenyichia@gmail.com (W.Y.C.); doristangyingying@gmail.com (D.Y.Y.T.); PauLoke.Show@nottingham.edu.my (P.L.S.)

[2] School of Mathematical Sciences, Faculty of Science and Engineering, University of Nottingham Malaysia, Jalan Broga, Semenyih 43500, Malaysia; KitWayne.Chew@nottingham.edu.my (K.W.C.)

[3] Department of Aeronautics and Astronautics, National Cheng Kung University, Tainan 701, Taiwan

[4] Department of Chemical and Materials Engineering, College of Engineering, Tunghai University, Taichung 407, Taiwan

[5] Department of Mechanical Engineering, National Chin-Yi University of Technology, Taichung 411, Taiwan

[6] Research Center for Energy Technology and Strategy, National Cheng Kung University, Tainan 701, Taiwan

* Correspondence: weihsinchen@gmail.com or chenwh@mail.ncku.edu.tw

Received: 6 January 2020; Accepted: 11 February 2020; Published: 17 February 2020

Abstract: The world energy production trumped by the exhaustive utilization of fossil fuels has highlighted the importance of searching for an alternative energy source that exhibits great potential. Ongoing efforts are being implemented to resolve the challenges regarding the preliminary processes before conversion to bioenergy such as pretreatment, enzymatic hydrolysis and cultivation of biomass. Nanotechnology has the ability to overcome the challenges associated with these biomass sources through their distinctive active sites for various reactions and processes. In this review, the potential of nanotechnology incorporated into these biomasses as an aid or addictive to enhance the efficiency of bioenergy generation has been reviewed. The fundamentals of nanomaterials along with their various bioenergy applications were discussed in-depth. Moreover, the optimization and enhancement of bioenergy production from lignocellulose, microalgae and wastewater using nanomaterials are comprehensively evaluated. The distinctive features of these nanomaterials contributing to better performance of biofuels, biodiesel, enzymes and microbial fuel cells are also critically reviewed. Subsequently, future trends and research needs are highlighted based on the current literature.

Keywords: bioenergy; biofuel; nanotechnology; nano-catalysts; nano-additives

1. Introduction

The current primary energy consumption is dominated by conventional fossil fuels including coal, oil and gas [1], leading to sustainability problems such as a declining amount of fossil fuels, environmental impacts and huge price fluctuations [2]. Greenhouse gas emissions, global climate change as well as intense energy demand have driven a number of professionals to develop novel solutions to replace fossil fuels. Among the alternative energy sources, biomass accounts for around 80% of the energy produced by global renewable energy carriers [3]. It can be stored and employed to generate heating, fuels and electricity when required. These are called bioenergy which is defined as solid, liquid or gaseous fuels produced from biological origin.

Bioalcohol derived from corn, wheat, sugar beet and sugarcane as well as biodiesel produced by transesterification of oils extracted from rapeseed, palm, soybean and sunflower are examples of bioenergy generated from first-generation feedstock. Non-food feedstocks like lignocellulosic and microalgae biomass are second- and third-generation feedstocks used to produce bioenergy, respectively. The biomass can be processed using thermal conversion technologies such as combustion, gasification and pyrolysis which converts biomass to bio-oil (a liquid fuel) and biochar (a solid residue), while syngas, which can be further processed to biofuels and electricity, is produced by the gasification of biomass [4]. Furthermore, anaerobic digestion has been employed commercially to produce biogas, which is also a type of bioenergy. The biogas produced has been utilized to generate heat and electricity. It plays an important role to provide the necessary energy to rural areas for cooking and lighting [4]. In addition, bioenergy can be produced by microbial fuel cells (MFCs) which uses naturally occurring microorganisms with biological electricity-generation ability.

Despite having numerous scientific breakthroughs, there are still various technical barriers to tackle for bioenergy production so that it can compete with fossil fuels. For instance, for microalgal biofuels production, cultivation of algae on a large scale effectively and efficiently, maintenance of the desired culture with alien species, harvesting cost of algae and energy efficiency and the best conversion method to biofuels remain uncertain [5]. Moreover, pretreatment methods are required to extract fermentable sugars from lignocellulosic biomass before proceeding to biofuel generation processes [6]. Other than technological barriers in production, challenges like insufficient existing infrastructure for the production process and high production cost compared to first-generation biofuels are apparent [7]. These drawbacks necessitate the development of production and optimization strategies to achieve high quality and high yield of bioenergy. For example, processes regarding pre-treatment, enzymes and fermentation can be looked into so that bioenergy production can be made to more energy-efficient and cost-effective [8].

Nanomaterials are the fundamental principle of nanoscience and nanotechnology application. The application of nanostructure science and technology covers a wide interdisciplinary area of research and development activity which has grown explosively worldwide in the past research discipline. Nanoscale materials are defined as a set of substances, where at least one dimension is less than approximately 100 nm. This extremely tiny size gives a large ratio of surface area to volume and increases the number of active sites for various reactions and processes. These nanoparticles (NP) also have the ability to exhibit different morphologies that have broadened their applications in different fields [9]. In addition, nanostructured materials, in comparison with large particles, have a faster reaction rate with other molecules [10]. The existence of nanomaterials has already influenced significant commercial impact, and the awareness of nanomaterials will raise due to its unique optical scale properties which are impactful for various fields such as bioenergy, electronics, mechatronics, medicine, pharmaceutical, ionic liquids, polymer and many more.

A number of direct and indirect applications of nanomaterials in bioenergy production have been reported. Nanomaterials are exceptional candidates in numerous biofuel systems due to their large surface areas and special characteristics including high catalytic activity, crystallinity, durability, efficient storage, stability as well as adsorption capacity [11]. The effects on metabolic reactions of bioprocesses producing biofuel are enhanced with nanoparticles such as nanofibers, metallic nanoparticles and nanotubes [9]. Nanoparticles, which are usually used as catalytic agents, take part in enhancing the activity of anaerobic consortia, reducing inhibitory compounds and transferring electrons in order to improve the process yields. Nanomaterials such as nano-crystals, nano-droplets and nano-magnets are also used as nano-additives in order to enhance the blending efficiency of biofuel with petrol and diesel [12].

This paper comprehensively reviews the recent approaches and applications related to nanotechnology incorporated processes for bioenergy production. The fundamentals of nanotechnology are introduced with the uses and benefits of various nanoparticles. Then the next section covers applications of nanotechnology on biomass including microalgal biomass and lignocellulosic biomass. Then, recent advances in the nanotechnology-based biofuel industry including nano-catalysts for higher biofuel yields and nano-additives for better fuel blends performance are presented. The conversion of chemical energy to electrical energy via MFCs with the aid of nanoparticles is also discussed. Moreover, future works and challenges are highlighted to provide insights for the future development of bioenergy production using nanomaterials. This study with in-depth comparison analysis will contribute to the bioenergy field where it provides a clear outline to the concerned researchers in how nanotechnology can improve bioenergy production.

2. Fundamental of Nanomaterials

Redesigning a material at the molecular level state is also known as engineered nanomaterials in which modification is made toward their small size and novel properties which are generally not visualized in their conventional and bulk counterparts. The distinct properties of these materials at the nanoscale are their relatively large surface area which triggers the novel theory of quantum effects. Nanomaterials provide a much greater surface area to volume ratio compared to their conventional forms, which is beneficial as this can provide greater chemical reactivity created by their specialty [13]. Considering the reaction on a nanoscale level, the properties and characteristics of materials including novel optical, electrical and magnetic behaviors can be more vital due to the quantum effects [14].

The most common terms for nanostructured materials are classified as zero-dimensional (0-D), one-dimensional (1-D), two-dimensional (2-D) and three-dimensional (3-D) nanostructures [15]. These dimensionalities of nanomaterials are characterized using an ultrafine grain size less than 50 nm or limited to 50 nm. Various modulation dimensionalities can be formed such as 0-D (e.g., atomic clusters, filaments and cluster assemblies), 1-D (e.g., multilayers), 2-D (e.g., ultrafine-grained overlayers or buried layers) and 3-D (e.g., nanophase materials composed of equiaxed nanometer-sized grains). Common types of nanomaterials include nanotubes, dendrimers, quantum dots and fullerenes. Nanomaterials have applications in the field of nanotechnology where they display different physical and chemical characteristics from normal-sized chemicals.

The most fundamental component in a nanostructure fabrication is nanoparticles. NPs with diverse size and morphology can be fabricated via several synthetic routes which offer superior quality of NPs, but the fabrication procedures such as biosynthesis are still under development for further improvement [16]. Organic nanoparticles have been widely investigated up-to-date, with liposomes, polymersomes, polymer constructs and micelles, all being employed for imaging or drug and gene delivery techniques [17]. Meanwhile, inorganic nanoparticles have also attracted attention in recent years attributed to their unique material- and size-dependent physicochemical properties, which are incomparable with traditional lipid- or polymer-based nanoparticles. A common reason to what makes inorganic nanoparticles attractive is their physical properties (e.g., optical and magnetic), in addition to their chemical properties such as inertness, stability and ease of functionalization [18]. Thus, inorganic nanoparticles such as magnetic, gold, quantum dots and carbon nanotubes have vast potential in various modern applications. For example, carbon nanotubes, metal-oxide and magnetic nanoparticles (MNPs) are employed for bioenergy production. Table 1 summarizes the advantages and disadvantages of magnetic nanoparticles.

Table 1. Advantages and disadvantages of magnetic nanoparticles [19].

Advantages	Disadvantages
• Excellent biodegradability • Readily to be customized • Ease of separation • Low cytotoxicity to biomass cell • Ease of synthesis • Ability to bind multiple targeted compounds • Large surface-to-volume ratio • Maintain stability after mechanical, physical and chemical modification	• Poor dispersion abilities • High cost of synthesis material • Limitation in scale up production processes • Mobility dependent on environment compatibilities

MNPs which comprise of a magnetic core (e.g., magnetite (Fe_3O_4) or maghemite (g-Fe_2O_3)) are some of the most profound inorganic nanomaterials [20]. The MNPs are most frequently used over all the nanoparticles examined for bioenergy production since their magnetic properties give them easy recoverability. Enzymes used in biodiesel or bioethanol generation can be immobilized with MNPs as a carrier. High coercivity and great paramagnetic property of MNPs during the methanogenesis process also make them useful for biogas production [21]. However, metals such as cobalt and nickel which are incorporated in the synthesis exhibit toxic and susceptible compounds when subjected to the oxidation process, thereby more research studies are required to overcome these problems [22].

3. Biomass

The world energy production of CO_2 has been tremendously rising due to the exhaustion of fossil fuels. On top of that, the concern relating to energy safety and environmental pollution has been an appointment in searching for alternative sources for bioenergy production. Lignocellulosic biomass involves plants and agricultural residues composed of cellulose, hemicellulose, lignin as well as other components (i.e., proteins, pectins and extractives). It was estimated that only 3% out of 13 billion t/y of plant residues were fabricated into manufacturing goods and the remaining were left for decomposition [23]. Thereby, these lignocellulosic residues should be properly managed by converting them into bioenergy where researchers have proven that the composition of lignocellulose has the capability to transform their monomers or building block into biofuels (e.g., bioethanol and biodiesel). On the other hand, ongoing studies have also shown the ability of microscopic filamentous photosynthetic microorganism, or well-known as "microalgae", regarding the conversion of algal lipids into biofuels. These microalgae-based biofuels have similar chemical properties compared to those from fossil fuels which are deemed to be a promising natural source for bioenergy production. Thus, incorporating nanotechnology into these alternative biomasses could greatly contribute to bioenergy production by acting as an aid to improve efficiency in various applications such as manufacturing, energy resources, transportation, mechatronics, health care and pharmaceutical technologies.

3.1. Lignocellulose for Conversion of Cellulose to Biofuel

The conversion of cellulose to biofuels faces some difficulties such as the recalcitrant structure of cellulose and the rigidity of the cell wall from lignocellulose biomass. The preliminary step involves depolymerization of cellulose polymer into its monomers, delignification of cellulose into cellulolytic enzymes, hydrolysis of cellulolytic enzymes into carbohydrates and fermentation of hydrolyzed sugars into biofuel production [24]. In some cases, enzymatic hydrolysis and fermentation (e.g., simultaneous saccharification (SS), simultaneous saccharification and co-fermentation (SSCF) and consolidated bioprocessing (CB)) were combined to reduce the major steps involved in biofuel production from lignocellulose biomass.

A study has shown the capabilities of nanotechnology where acid-functionalized magnetic nanoparticles (MNPs) are used as catalysts to hydrolyze the cellobiose (β-glucose) from lignocellulose biomass [25]. However, the disadvantage of the dispersion from some nanoparticles is the difficulty to disperse in the aqueous solution where the hydronium ions were ineffective in the solution due to the wettability of the sample. The results showed that the acid-functionalized MNPs with 6% of sulfur content achieved cellobiose conversion up to 96.0% more than the conventional conversion (32.8%) without catalyst [25]. The presence of these acid-functionalized MNPs could enhance the hydrolysis reaction by their nanobiocatalyst properties for the immobilization of different enzymes. Aside from that, the high surface-to-volume ratio of these MNPs facilitates the rate of hydrolysis compared to the chemical pretreatment approach. Likewise, the separation of these magnetic nanoparticles can be recycled for the subsequent hydrolysis process which is more preferable in minimizing the process cost as they can be separated from the reaction medium magnetically. This was supported by Lai et al. [26] and Erdem et al. [27] who demonstrated the sulfonate-supported silica MNPs for the hydrolysis of lignocellulose biomass which was deemed as a promising hydrolysis catalyst.

The presence of the silica-coated on the MNPs accelerates the mass transport in the acidic reaction due to its relatively porous structure and higher stability of the magnetic core. On the other hand, propyl-sulfonic acid-functionalized MNPs were subject to pretreatment and hydrolysis of wheat straw lignocellulose biomass, however the efficiency was not as promising compared to the above studies [25]. Recent technologies using microwaves-, ultrasonication- and electricity-assisted approaches are also recommended to be explored in the field of nanotechnology. A study by Su et al. [28] has demonstrated the potential of incorporating microwave-assisted technology with carbonaceous acid MNPs for the pretreatment and hydrolysis of sugarcane bagasse, *Jatropha* hulls and *Plukenetia* hulls. The hydrolysis performance obtained for sugarcane bagasse, *Jatropha* hulls and *Plukenetia* hulls was 58.3%, 35.6% and 35.8%, respectively. Table 2 summarizes the efficiencies of the MNPs for the hydrolysis of lignocellulose biomass.

Table 2. Comparison studies on the efficiencies of magnetic nanoparticles.

Magnetic Nanoparticles (MNPs)	Biomass Strain	Operating Condition	Yield (%)	References
Sulfonate-supported silica MNPs, FE_3O_4-SBA-SO_3H	Amorphous cellulose	1.0 g, 15 mL H_2O at 150 °C for 3 h	50	[26]
Sulfonate-supported silica MNPs, FE_3O_4-SBA-SO_3H	Cellulose	1.0 g, 15 mL H_2O at 150 °C for 3 h	26	[26]
Sulfonate-supported silica MNPs, FE_3O_4-SBA-SO_3H	Starch	1.0 g, 15 mL H_2O at 150 °C for 3 h	95	[26]
Sulfonate-supported silica MNPs, FE_3O_4-SBA-SO_3H	Corn cob	1.5 g, 15 mL H_2O at 150 °C for 3 h	45	[26]
Perfluoroalkylsulfonic MNPs, PFS-MNPs	Wheat straw	2.5% (*w/w*) biomass, 160 °C for 24 h	66.3 ± 0.9	[25]
Alklysulfonic MNPs, AS-MNPs	Wheat straw	2.5% (*w/w*) biomass, 160 °C for 24 h	61.0 ± 1.2	[25]
Carbonaceous acid MNPs, C-SO_3H-Fe_3O_4-MNPs	Sugarcane bagasse	0.027 g, 15 mL H_2O, 160–200 °C (0.5–2.2 MPa) for 3 min	58.3	[28]
Carbonaceous acid MNPs, C-SO_3H-Fe_3O_4-MNPs	*Jatropha* hulls	0.027 g, 15 mL H_2O, 160–200 °C (0.5–2.2 MPa) for 3 min	35.6	[28]
Carbonaceous acid MNPs, C-SO_3H-Fe_3O_4-MNPs	*Plukenetia* hulls	0.027 g, 15 mL H_2O, 160–200 °C (0.5–2.2 MPa) for 3 min	35.8	[28]

Yet, the limitation of MNPs has evolved nanobiocatalysts through using silica-based NPs (Si-NPs), nickel-based NPs and carbon nanotubes. Si-NPs are usually coated on the surface of the nanoparticles which functions to immobilize a lignocellulolytic enzyme such as cellulase. It has been reported that Si-NPs improved the catalytic activity in the simultaneous saccharification reaction for bioethanol production from *Trichoderma viride* cellulase [29]. Factors such as particle size, pore size and surface area are also crucial points as stated by Chang et al. [30], who evaluated two mesoporous silica NPs (MS-NPs) on commercial cellulose. These MS-NPs have the chemical binding ability to immobilized cellulase on their porous size surface for cellulose-to-glucose conversion up to 80%. Alternatively, nickel-based NPs (Ni-NPs) are also commonly used for the hydrogenation process for the conversion of glucose to a sorbitol molecule [31]. Gasification of biomass for the production of synthesis gas, also known as syngas ($CO + H_2$), can be useful as their intermediates can be further converted into biofuels. Subjecting these unstable enzymes into a high operating temperature and pressure catalytic processes would result in a lower productivity yield of biofuels. Another study has shown the use of nickel-cobaltite NPs on the stability of *Aspergillus fumigatus* cellulases at different concentrations of synthesized NPs. The results showed that the addition of 1 mM of nickel-cobaltite NPs increased enzyme activity of endoglucanase, β-glucosidase and xylanase by 49%, 53% and 19.8%, respectively [32]. Contrast to these, carbon nanotubes (CNTs) are widely known for their attractive features in electricity, thermal properties and mechanical strength [33]. Most studies have reported that multi-walled carbon nanotubes (MWCNTs) performed more effectively than single-walled carbon nanotubes (SWCNTs) as the immobilization of enzymes is compatible with their structural arrangement which enhanced catalytic activities of immobilized enzymes [33,34]. These MWCNTs outperformed the hydrolysis of cellulose from *Aspergillus niger* within 85%–97% efficiency and retained its recyclable activity at 52%–75% after six cycles of hydrolysis process [33,34].

3.2. Nanotechnology for Bioenergy Production from Microalgal Biomass

Microalgae are widely researched as the third-generation biofuels feedstock with a diversity of photosynthetic species [35,36]. The ability of microalgae to grow in harsh environments, high carbon dioxide uptake and rapid productivity have made them an alternative biofuel feedstock. In addition, the composition of microalgae which is rich in proteins, carbohydrates, lipids and carotenoids makes them an excellent choice compared to lignocellulosic biomass [37]. However, these microalgae-based biofuels face some challenges such as being difficult to manage in industrial-scale production, in addition to the high cost for biomass production and harvesting, which requires efficient technologies for biofuel conversion.

Previous studies have shown the feasibility of using MNPs for the hydrolysis of the microalgae cell wall by immobilizing cellulase on MNPs followed by lipid extraction [38]. Subjecting the immobilized cellulase to MNPs allows the microalgae cell wall composed of polysaccharide cellulose to be hydrolyzed for the release of lipid composition. Under optimal conditions, the maximum yield (93.56%) of biodiesel was achieved. A similar study also utilized MNPs replaced with metal-oxide MgO as an aid linked with the cellulase enzyme to improve the hydrolysis of cellulose from *Chlorella* sp. CYB2. The results showed that the glucose yield obtained was 91% performed by the mechanism of metal-oxides. Meanwhile, Nematian et al. [39] reported the use of superparamagnetic nano-biocatalysts for the conversion of bio-oil extraction from *Chlorella vulgaris* microalgae to biodiesel production. The results showed that the transesterification reaction using 3-aminopropyl triethylenesilane-glutaraldehyde (MNPs-AP-GA) was 69.8 wt %. The study also claimed that the covalent bonding of lipase showed a reliable method for improving enzyme loading and productivity. Apart from that, microwave-assisted with MNPs for the enhancement of biogas and biohydrogen production from microalgae were also studied. The biogas and hydrogen productions were 328 mL and 51.5%, respectively [40]. However, there is still a lack of study regarding the feasibility and economic analysis of these MNPs in large-scale production which is a gap to-be-filled for researchers dealing with nanomaterials.

4. Impacts of Nanomaterial for Enhancement of Biofuels Production

The development of nanomaterial has expanded by modification with different functionalized groups (e.g., amino-based, nickel-based, hydrophobic-based, gold-based) for the enhancement of biofuel production. Nanomaterials are also capable of improving these enzymes' activities by introducing it into the cultivation medium [41]. The properties of these nanomaterials have proven to generate stress during cultivation conditions such as high metal concentration (Fe), which affects lipid accumulation in *Chlorella vulgaris* microalgae [42]. Introducing these nanoparticles with silica and iron oxide composition in the cultivation medium would result in a strong sheer between the nanoparticles and the cell as these nanoparticles act as a competitor for nutrients uptake. In terms of extraction and recovery of lipids, these nanoparticles have also demonstrated excellent extraction ability to replace these conventional solvents (e.g., chloroform, methanol and hexane) in the extraction. The benefits of these nanoparticles are to prevent algae from dying and bring up the re-cultivation process from these extracted microalgae [43].

4.1. Nanomaterial Incorporation as Nanocatalyst in Microalgae Processing

These unique nanomaterials stimulate the photosynthesis growth of microalgae by inducing a mild stress condition for the accumulation of lipid without harming the cells. Several studies have implemented nanoparticles as a nutrient in the culture medium (e.g., iron and magnesium) [42]. The iron nanoparticles also generate various reactive oxygen species (ROS) via Fenton-type reaction which causes oxidative stress to the microalgae [44]. However, a study by Kang et al. [45] reported that a high concentration of TiO_2 nanoparticles in the presence of light would induce the viability of the cell. On the other hand, Mg-aminoclay nanoparticles have also been tested positively for the growth of *Chlorella* sp. KR-1 and *Chlorella vulgaris* as the amino clay nanoparticles are composed of metal cations such as Ca^{2+}, Mg^{2+} and Fe^{3+} covalent bonded on the center of the nanoparticles [46]. The supplement of $MgSO_4$-NPs has shown its enhancement activity in both photosynthesis and reduced glycerol consumption in the mixotrophic cultivation of *C. vulgaris* [47]. The implementation of $MgSO_4$-NPs induces the flocculation of microalgae by reducing the penetration of light, resulting in an increase of chlorophyll content.

The impact of these nanoparticles has also been applied as an enzyme immobilizer via covalent bonding for biodiesel production. The previous study has reported that *Porcine pancreas* lipase, *Candidarugosa* lipase and *Pseudomonas cepacia* lipase were subjected to amino-functionalized MNPs for enzymatic transesterification reaction and achieved a high conversion of biodiesel up to 67% [48,49]. A similar study also modified the amino-functionalized MNPs with a glutaraldehyde crosslinker which aided a higher immobilized lipase on the surface of the MNPs [50]. Results showed that the presence of this crosslinker coated on the MNPs was efficient to achieve a biodiesel conversion of 90% where the superior properties of glutaraldehyde activated the surface availability for enzyme immobilization. Aside from that, hydrophobic MNPs were also investigated for the transesterification of immobilized lipase to biodiesel. These hydrophobic MNPs have the ability to adsorb lipases or immobilize lipase on their hydrophobic interfaces by their lids and protein chains [51]. The conversion of extracted oil to fatty acid methyl esters was 70% along with a biodiesel production rate of 43.5 g/L/h under optimized conditions [52]. The enhancement and separation of C-phycocyanin from *Spirulina platensis* microalgae using a fabricated chitosan-modified nanofiber membrane has also shown a purification factor of 3.3-fold and 66% recovery, respectively [53]. The function of this fabricated chitosan-modified nanofiber membrane enables the coordination binding of contaminated proteins via electrostatic interaction by separating and purifying the targeted C-phycocyanin molecules during the separation process. A recent study by Cheah et al. [54] also utilized a fabricated chitosan-modified nanofiber on the antibacterial activity with *Escherichia coli* which exerted antibacterial activity up to 99.5% effectively. The presence of a polycationic charge from the fabricated chitosan-modified nanofiber membrane forms an electrostatic bond with the negatively charged site on a bacterium cell wall. This deformed the permeability of the cell wall due to the stress condition, hence leading to cell lysis and death. Other fabricated nanomaterials such as carbon nanotubes, mesoporous, nanofibers, electrospun nanofiber, ferric-silica and gold-based support were also incorporated as engineered nanoparticles to enhance the immobilization of enzymes for higher biofuel production [51,55,56]. Yet, there is still a lack of insight and studies regarding its optimized condition for an ideal immobilized enzyme for biodiesel conversion.

The controversy in utilizing nanomaterials especially carbon nanotubes (e.g., Al_2O_3, CuO, ZnO and TiO_2) faces challenges due to its toxicity to microalgae which covers oxidative stress, agglomeration and the inconsistent supply of nutrients and synthesis cost of these nanomaterials. In addition, on a molecular chemistry level, the internalization mechanisms of these functionalized NPs are still not clearly understood. This calls for researchers to further evaluate these problems where to date, the economic analysis, environmental safety and life cycle analysis (LCA) are subjects of interest as standardized processes are much more preferable for an appropriate assessment of these lacking issues.

4.2. Nano-Additives Blended Biodiesel in Diesel Engines

Nanomaterials have also been tested as nano-additives on fuel properties due to their distinctive properties of nanofluid which enhance various properties such as viscosity, flash point density, cetane number and many more. An experiment has evaluated the effect of physicochemical properties of biodiesel using metal-oxide NPs as a fuel additive [57]. The presence of these metal-oxide NPs acts as an oxygen buffer resulting in a simultaneous oxidation process of hydrocarbons by reducing the emission of oxides from nitrogen. Metal-oxide NPs exhibit a high surface-to-volume ratio that improves the fuel efficiencies of this biodiesel compared to those of a conventional powder form. The results showed that a higher dosage of these cerium oxide NPs increases the fluid layer resistance and viscosity where lower fuel viscosity is incapable of lubricating the fuel injection pump which will cause leakage and easily wear off, reducing the fuel delivery performances. Clearly, this showed that metal-oxides NPs are thermally stable to promote the oxidation of hydrocarbon and the reduction of nitrogen oxide.

The addition of nano-additives blended with fuel improves the cetane number and calorific value resulting in better performance of combustion. Studies have shown that aluminum- and silicon-based NPs improve the combustion quality of biodiesel engines [58]. This was also supported by another study using zinc oxide-based NPs diesel–pomoplion stearin wax biodiesel blends where the improvement of calorific value and cetane index were observed significantly [59]. Carbon nanotubes were also used by Singh and Bharj [60] who reported that the cetane index improves when the concentration of carbon nanotubes increases. Other nanoparticles such as iron oxide-based NPs were evaluated showing the benefits from these NPs in enhancing both cetane number and calorific value for an ideal combustion quality as well as reducing the emission release from diesel engines [61,62]. Nanofluids as additives are also promising for the improvement of brake thermal efficiency of diesel engines, as these additives promote complete combustion due to the higher evaporation rates, reduced ignition delay, high flame temperatures and lengthy flame sustenance [58,63]. Other effects such as carbon monoxide emission, hydrocarbon emission, NO_x emission, combustion and evaporation can be resolved by adding nano-additive blends into biodiesel fuel.

Despite its benefits, the major issues of NPs remain at the production cost that hindered the commercialization of nanofluids. Studies regarding NPs as fuel additives are still limited to be implemented at this point in time. Problems regarding nanoparticle aggregation, settling and erosion are yet to be resolved and this requires better characterization of nanofluids to boost its effective usage. On top of that, insufficient experimental results and poor understanding of the theoretical mechanism of heat transfer are the main points to be tackled before commercializing these nano-additives in diesel engines.

5. Bioelectrochemical System (BES)

Bioelectrochemical system (BES) is defined as the combination of biological and electrochemical processes, involving the use of electrochemically-active bacteria to degrade organic matters in various sources, such as industrial wastewater and biomass wastes [64,65]. The end products obtained are electricity, hydrogen or other valuable compounds such as ethanol, hydrogen peroxide (H_2O_2) and formic acid (CH_2O_2). BES is widely applied for wastewater treatment, and at the same time, the production of bioenergy. Therefore, BES is a promising technology for managing water pollution and global energy crisis [65]. Basic microbial fuel cells (MFC), photosynthetic MFC, plant MFC and biophotovoltaics [66] are examples of several forms of BES. Due to the simple operation and mild conditions, MFC nowadays attracts great interest from researchers worldwide as a new source of renewable bioenergy of the future.

5.1. What Is MFC

In 1911, Potter first came up with the idea of utilizing the microbes to produce electricity [67] followed by the development of first microbial half fuel cells by Cohen in 1931. MFC involves the electrochemical interactions between the microorganisms or electrogenic microbes and organic matters in which the electrons are transferred from the substrate to the anode electrode. This process is known as extracellular electron transport (EET) [68]. The electrogenic microbes are the microorganisms that serve as the main biocatalysts by transferring the electron produced from the metabolism of organic compounds to the electrode through a series of chemical reactions, for instance, c-type cytochrome or nanowires of the bacteria [69]. The examples of electrogenic microbes are bacteria (e.g., *Geobacter sulfurreducens*, *Shewanella putrefaciens*, *Clostridium cellulolyticum*, *Enterobacter cloacae*, *Rhodoferax ferrireducens*, *Clostridium butyricum*) and fungi (e.g., *Aspergillus awamori*, *Phanerochaete chrysosporium*) [70–73]. The microbes on the anode are responsible for generating electrons and protons by utilizing the organic substrates. The protons (H$^+$) produced will pass through the membrane and the electrons then flow through the electric circuit to the cathode at which oxygen reduction reaction (ORR) occurs [74,75]. As a result, bioelectricity is produced.

A typical MFC (Figure 1) is made up of two electrodes (anode and cathode) and a semi-permeable membrane known as the proton exchange membrane. Different types of materials such as metals have been used in the fabrication of electrodes. The most common materials for electrodes are carbon and graphite [64]. For the production of non-carbon based electrodes, the metals used are stainless steel, cobalt, copper, silver, nickel, titanium and gold [76]. There are some factors that affect the generation of bioelectricity by MFC including the surface area, stability, porosity, durability of the anode, cathode and membrane. The ideal electrodes should have the following characteristics [77]:

- Good electrical conductivity;
- Good thermal stability;
- Low resistance;
- Good biocompatibility with the system;
- Strong stability and anti-corrosion toward the chemical used in MFC;
- Large surface area;
- Good mechanical strength;
- Low cost.

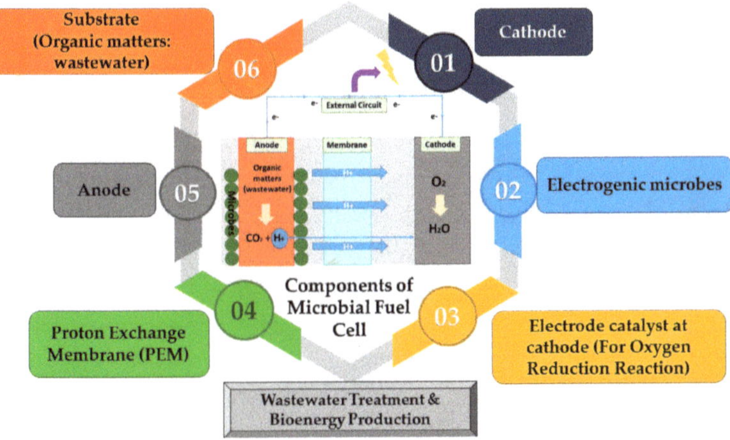

Figure 1. The component of microbial fuel cells (MFC).

The anodic chamber of MFC is made up of an anode, microbes (bacteria) and a substrate (wastewater) [66]. Being the most significant component of MFC, anode with the microbes attached to it allows the electrons flow via the electrochemical reactions of the microbes through the degradation of substrates. An essential aspect of the anode is the ability of the microbes to facilitate the formation of biofilms and increase the probability of EET to occur [66]. The most common materials used in the fabrication of anode are graphite or carbon that comes in various sizes or geometrics, for instance, carbon nanotubes, rods, felt, cloth, paper and plates [66,77–81]. On the other hand, the cathode is the electrode where the oxygen reduction reaction (ORR) will happen [75]. Overall, the ORR occurred at the cathode is the limiting factor of MFC and will affect the maximum power density, efficiency and performance of the entire MFC [66]. The catalysts have been integrated to improve the cathodic ORR [82]. The preferred cathode catalyst is platinum (Pt) due to high surface areas but the production cost is high [66]. Hence, graphite, which is cheaper than platinum, and possesses a large surface area is utilized as the cathode material to increase the efficiency of MFC [83,84]. Another component of MFC is the proton exchange membrane (PEM), a physical membrane that separates anode and cathode. There are various types of membranes that exist; for instance, cation or anion exchange membrane, nylon fibers, ultrafiltration membrane, microfiltration membrane, glass fibers and porous fabrics, but most of the membranes are not cost-effective [66]. This urges researchers to explore the low-cost material for use in the production of membrane and simultaneously increase its efficiency as the barrier between anode and cathode and proton transfer rate.

5.2. Modification of MFC Components with Nanomaterials

There are a few challenges that need to be addressed to produce bioenergy in pilot-scale by MFC. First, low power density that is insufficient to support a large population is the main bottleneck faced by MCF [85–88]. The maximum power density achieved by the conventional electrodes is about 26 mWm^{-2} for 3D graphite rods [89] or 611.5 ± 6 mWm^{-2} for 2D carbon cloth [90]. The performance of MFC is also affected by temperature as microbes cannot grow and carry out their activities at extremely low or high temperatures [91,92]. Thus, MFC needs to be conducted at an optimal temperature that is suitable for microbes. Besides, the complex, toxicity and high-cost process of manufacturing of components of MFC could hinder the practical applications and economical usage of MFC [93]. Therefore, researchers have been searching for a replacement or new materials in the production of the main components of MFC, such as anode, cathode and separator in order to improve performance and enhance the conductivity of electrons. Figure 2 demonstrates the advantages and disadvantages of MFC.

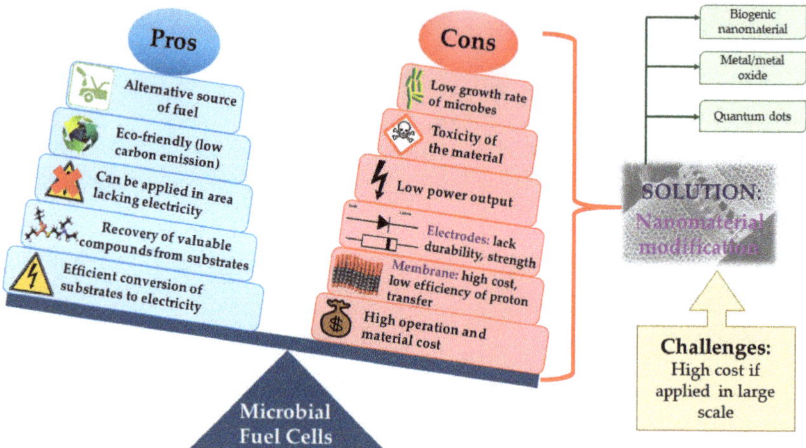

Figure 2. Advantages and disadvantages of microbial fuel cells (MFC).

Recently, nanotechnology or use of nanomaterial has revolutionized the fabrication of components of MFC to improve the performance and efficacy of traditional MFC in terms of electron conductivity, power density, cost, thermal stability, ORR rate and anti-corrosion [94,95], particularly the modification of electrodes (anode and cathode) by nanomaterials. This is because the materials used in the production of electrodes that are the major constituent of MFC are essential to determine the overall performance of MFC [96]. Thus, the production of bioenergy through MFC will be enhanced. The examples of nanomaterials for electrodes that will improve the function of MFC are metal nanoparticles (i.e., copper, gold, platinum, palladium and silver), quantum dots (i.e., Cds, CdSe, ZnS), metal-oxides (i.e., CeO$_2$, TiO$_2$, ZnO, SiO$_2$, Al$_2$O$_3$ and MnO$_2$) [93,97], graphene (2D-nanomaterials) [76], carbon nanotubes and nanocomposites (multiphase materials). However, the use of nanomaterials in the modification of components of MFC in the pilot-scale is still in progress of development due to the high production cost. Table 3 illustrates the examples of nanomaterials used in the modification of components of MFC.

Table 3. The use of nanomaterials in the modification of components of MFC.

Nanomaterial	Modified Part of MFC	Advantages	Description	Reference
Fabrication of bio-palladium nanoparticles using pure strain *Shewanella oneidensis* on carbon cloth	Anode	• Less chemicals required • Good biocompatibility • High catalytic activity • Simple and easy to be produced • Gentle reaction condition	The maximum power output and coulombic efficiency were improved by 14% and 31% as compared to unmodified anode.	[69]
Spinel type Ni-ferrite ($NiFe_2O_4$) modified composite anode	Anode	• High conductivity • Good reaction activity	As compared to control, the maximum power density achieved was increased by 26% and the internal resistance was lowered by 39%.	[98]
TiO_2 nanotubes (TN) on the surface of titanium anode	Anode	• Good biocompatibility • Stable and low cost • Resistance to corrosion • Can be synthesized in situ	The maximum current density achieved was 12.7 Am^{-2}, which was up to 190-fold as compared with bare titanium anode electrode.	[99]
Bimetallic core-shell gold-palladium nanoparticles as cathode catalyst	Cathode	• High durability • Low bulk resistance	High stability (can stable over 150 days), high durability and the power output produced was 15.98 Wm^{-3}, twice the power obtained with hollow structures-based platinum (Pt) cathodes (7.1 Wm^{-3}).	[100]
Fe_3O_4 nanoparticles/polyethersulfone (PES) nanocomposite membrane	Proton exchange membrane	• Eco-friendly • Low cost	High thermal stability and mechanical properties as well as the maximum power output produced was 9.59 mWm^{-2}, higher than commercial membrane.	[101]
Graphene oxide (rGO)/manganese oxide (MnO_2) composite on carbon felt surface	Anode	• Large surface area • High electric conductivity • Good electrocatalytic activity	The internal resistance was lowered, and maximum power density achieved was 2065 mWm^{-2}, 154% higher as compared with carbon felt anode.	[102]
Nickel oxide (NiO)/graphene nanocomposite with the addition of pectin into NiO	Anode	• Good conductivity • Appropriate pore size for movement of bacteria • Good ions accessible surface	The maximum power density achieved was 3.632 mWm^{-2}, higher than NiO anode and Pt/C anode.	[103]
Fabrication of two graphite based composite electrodes using graphene paste modified with TiO_2 (GP-TiO_2) or hybrid graphene (GP-HG)	Cathode	• Good electrocatalytic activity	The power density was increased above 80 mWm^{-2} for GP-TiO_2 and above 220 mWm^{-2} for GP-HG as compared to the control, graphite paste bare electrode (30 mWm^{-2}).	[104]

Table 3. *Cont.*

Nanomaterial	Modified Part of MFC	Advantages	Description	Reference
Polyaniline/graphene modified carbon cloth	Anode	• High electrical conductivity • Eco-friendly • Simple and easy	The maximum power density achieved was 884 ± 96 mWm^{-2}, 1.9 times higher than the unmodified carbon cloth anode (454 ± 47 mWm^{-2}). The voltage achieved was 573 ± 37 mV, higher than CC anode, 454 ± 34 mV.	[105]
Fabrication of polyaniline hybridized disorderly large mesoporous carbon nanocomposite with aid of nanoparticles, $CaCO_3$	Anode	• Low cost • High electric conductivity • Good electrochemical activity	The maximum power density obtained was 1280 mWm^{-2}, 1.5-fold and 10-fold higher than LMC anode (878 mWm^{-2}) plain carbon cloth anode (127 mWm^{-2}).	[94]
Nitrogen-doped carbon nanotube/reduced graphene oxide (rGO) composite with polyaniline as nitrogen source	Anode	• Good electric conductivity • Biocompatible • Stable	The power density achieved was 1137 mWm^{-2}, 8.9 times higher than carbon cloth anode.	[106]
3D graphene macroporous scaffold anode	Anode	• Large surface area • Good electrical conductivity	The MFC system was able to accommodate a high population of microbes and the power density achieved was 5.61 Wm^{-2}/11220 Wm^{-3}, 3-fold higher than planar 2D control counterparts. The highest power achieved was 3320 Wm^{-3}.	[107]
Graphene material (RGO$_{HI-AcOH}$) and graphene nanoparticles composite (RGO/Ni) as cathode catalyst	Cathode	• Large surface area • Eco-friendly	The power generated was 1683 mWm^{-2} and had good stability (can operate for 30 days and around 27 cycles) as compared to non-metal cathode MFCs.	[93]
Integration of carbon nanotube-gold-platinum nanomaterial with osmium redox polymer and *Gluconabacter oxydans* DSM 2343 in carbon felt electrode	Anode	• Stable • Can be reused for long time	The maximum power density and current density achieved were 32.1 mWm^{-2} and 1032 mAm^{-2}, respectively, showed the improvement of the system.	[108]

6. Future Works

To develop a cost-effective system of bioenergy production, a techno-economic assessment must be carried out with consideration on the cost of nanoparticle synthesis which can influence the overall production process. This also emphasizes the development of economically viable nanoparticles to make the whole process economically feasible for commercialization. Pilot-scale research is necessary to examine the viability of incorporating nanoparticles on a large-scale bioenergy production basis. Furthermore, future researches are not limited to sources and production of bioenergy where nanotechnology could address the technical limitations in science and engineering by contributing to areas of transformation, transportation, energy efficiency and storage, as well as the use of bioenergy end-product [11]. Apart from that, there are still limited studies on using NPs as fuel additives up-to-date where approaches to solving nanoparticle aggregation, erosion and settling are still required. There is a lack of practical results and an understanding of heat transfer mechanisms to commercialize these nano-additives in diesel engines. In addition, safety assessment must be carried out because nanoparticles have demonstrated obvious exposure effect in terms of human and environment with the increasing use in biofuel applications. The toxicity of nanoparticles has been examined using several approaches wherein in vitro investigation of nanotoxicity is mostly involved [109]. However, in vivo interaction should be studied extensively focusing on the nanoparticles particularly used to produce bioenergy as well as biofuel [21]. This also applies to microorganisms as nanoparticles that are safe, non-toxic and compatible with enzymes and microbes should be synthesized. For instance, NPs are toxic to microalgae as they result in agglomeration, oxidative stress and inconsistent nutrient supply. Therefore, screening studies of nanoparticles are required to investigate their broad range of concentrations with an influence on microbial and enzymatic activity. Research at the molecular level should be conducted to study the mechanism involving nanoparticles and proteins in the production process. Subsequently, the optimum process conditions of bioenergy production can be established.

7. Conclusions

The depletion of fossil fuels and intensive energy demand has motivated researchers to develop alternative energy sources. Among the renewable energy technologies, bioenergy from biomass has its unique advantages. To meet future energy requirements while overcoming the technological barriers of bioenergy production, incorporation of nanomaterials in bioenergy production has been investigated since it can improve both quality and quantity of bioenergy produced by biomass, biofuel and microbial fuel cells. The bioenergy production process can be enhanced by NPs in different approaches. Firstly, acid-functionalized MNPs could be used to improve the hydrolysis reaction of biomass using different immobilized enzymes. Furthermore, metal-oxide NPs have been tested as nano-additives to enhance the performance of combustion as well as the blending performance of biofuel and conventional diesel. BES, or more specifically MFC, which is widely employed for wastewater treatment and bioenergy production, has also been modified by fabricating its components with nanomaterials in order to promote better performance and efficacy. However, there are still technical gaps in the world of nanotechnology-based bioenergy, whereby there are limited studies on the application of NPs as fuel-additives, in vivo toxicity of NPs and molecular-scale mechanism of NPs-protein. Last but not least, economic analysis, safety assessment and life cycle analysis (LCA) on the incorporation of nanomaterials in bioenergy production are essential for providing insights and outlines for future research.

Author Contributions: Writing—original draft, review and editing, K.S.K., W.Y.C. and D.Y.Y.T.; conceptualization, K.W.C.; supervision and funding acquisition, P.L.S. and W.-H.C. All authors have read and agreed to the published version of the manuscript.

Funding: This work was supported by the Fundamental Research Grant Scheme, Malaysia (FRGS/1/2019/STG05/UNIM/02/2). The authors acknowledge the financial support of the Ministry of Science and Technology, Taiwan, under the contracts MOST 106-2923-E-006-002-MY3 and MOST 108-3116-F-006-007-CC1 for this research. This research is also supported in part by Higher Education Sprout Project, Ministry of Education to the Headquarters of University Advancement at NCKU.

Conflicts of Interest: The authors declare no conflicts of interest.

Abbreviations

SS	Simultaneous saccharification
BES	Bioelectrochemical system
SSCF	Simultaneous saccharification and co-fermentation
CB	Consolidated bioprocessing
Si-NPs	Silica-based NPs
MS-NPs	Mesoporous silica NPs
EET	Extracellular electron transport
H+	Proton
Ni-NPs	Nickel-based NPs
MFC	Microbial fuel cells
NPs	Nanoparticles
MNPs	Magnetic nanoparticles
ORR	Oxygen reduction reaction
CNTs	Carbon nanotubes
MWCNTs	Multi-walled carbon nanotubes
SWCNTs	Single-walled carbon nanotubes
ROS	Reactive oxygen species
LCA	Life cycle analysis

References

1. Senjyu, T.; Howlader, A.M. Chapter 3—Operational aspects of distribution systems with massive DER penetrations. In *Integration of Distributed Energy Resources in Power Systems*; Funabashi, T., Ed.; Academic Press: Cambridge, MA, USA, 2016; pp. 51–76.
2. Lijó, L.; González-García, S.; Lovarelli, D.; Moreira, M.T.; Feijoo, G.; Bacenetti, J. Life Cycle Assessment of Renewable Energy Production from Biomass. In *Life Cycle Assessment of Energy Systems and Sustainable Energy Technologies: The Italian Experience*; Basosi, R., Cellura, M., Longo, S., Parisi, M.L., Eds.; Springer International Publishing: Cham, Switzerland, 2019; pp. 81–98.
3. Strzalka, R.; Schneider, D.; Eicker, U. Current status of bioenergy technologies in Germany. *Renew. Sustain. Energy Rev.* **2017**, *72*, 801–820. [CrossRef]
4. Li, Y.; Khanal, S.K. *Bioenergy: Principles and Applications*; Wiley: Hoboken, NJ, USA, 2016.
5. Hallenbeck, P.C.; Grogger, M.; Mraz, M.; Veverka, D. Solar biofuels production with microalgae. *Appl. Energy* **2016**, *179*, 136–145. [CrossRef]
6. Zheng, Y.; Zhao, J.; Xu, F.; Li, Y. Pretreatment of lignocellulosic biomass for enhanced biogas production. *Prog. Energy Combust. Sci.* **2014**, *42*, 35–53. [CrossRef]
7. Rai, M.; dos Santos Júlio, C.; Soler Matheus, F.; Franco Marcelino Paulo, R.; Brumano Larissa, P.; Ingle Avinash, P.; Gaikwad, S.; Gade, A.; da Silva Silvio, S. Strategic role of nanotechnology for production of bioethanol and biodiesel. *Nanotechnol. Rev.* **2016**, *5*, 231–250. [CrossRef]
8. Patumsawad, S. 2nd Generation biofuels: Technical challenge and R and D opportunity in Thailand. *J. Sustain. Energy Environ. Spec. Issue* **2011**, 47–50.
9. Sekoai, P.T.; Ouma, C.N.M.; du Preez, S.P.; Modisha, P.; Engelbrecht, N.; Bessarabov, D.G.; Ghimire, A. Application of nanoparticles in biofuels: An overview. *Fuel* **2019**, *237*, 380–397. [CrossRef]
10. Contreras, J.E.; Rodriguez, E.A.; Taha-Tijerina, J. Nanotechnology applications for electrical transformers—A review. *Electr. Power Syst. Res.* **2017**, *143*, 573–584. [CrossRef]
11. Nizami, A.-S.; Rehan, M. Towards nanotechnology-based biofuel industry. *Biofuel Res. J.* **2018**, *5*, 798–799. [CrossRef]
12. Palaniappan, K. An overview of applications of nanotechnology in biofuel production. *World Appl. Sci. J.* **2017**, *35*, 1305–1311.

13. Eustis, S.; El-Sayed, M.A. Why gold nanoparticles are more precious than pretty gold: Noble metal surface plasmon resonance and its enhancement of the radiative and nonradiative properties of nanocrystals of different shapes. *Chem. Soc. Rev.* **2006**, *35*, 209–217. [CrossRef]

14. Bogani, L.; Wernsdorfer, W. Molecular spintronics using single-molecule magnets. *Nat. Mater.* **2008**, *7*, 179–186. [CrossRef]

15. Tiwari, J.N.; Tiwari, R.N.; Kim, K.S. Zero-dimensional, one-dimensional, two-dimensional and three-dimensional nanostructured materials for advanced electrochemical energy devices. *Prog. Mater. Sci.* **2012**, *57*, 724–803. [CrossRef]

16. Siddiqi, K.S.; Husen, A. Fabrication of Metal and Metal Oxide Nanoparticles by Algae and their Toxic Effects. *Nanoscale Res. Lett.* **2016**, *11*, 363. [CrossRef]

17. Qiu, L.Y.; Bae, Y.H. Polymer architecture and drug delivery. *Pharm. Res.* **2006**, *23*, 1–30. [CrossRef]

18. Lohse, S.E.; Murphy, C.J. Applications of Colloidal Inorganic Nanoparticles: From Medicine to Energy. *J. Am. Chem. Soc.* **2012**, *134*, 15607–15620. [CrossRef]

19. Lamberti, M.; Zappavigna, S.; Sannolo, N.; Porto, S.; Caraglia, M. Advantages and risks of nanotechnologies in cancer patients and occupationally exposed workers. *Expert Opin. Drug Deliv.* **2014**, *11*, 1087–1101. [CrossRef]

20. Pantic, I. Magnetic nanoparticles in cancer diagnosis and treatment: Novel approaches. *Rev. Adv. Mater. Sci.* **2010**, *26*, 67–73.

21. Antunes, F.A.F.; Gaikwad, S.; Ingle, A.P.; Pandit, R.; dos Santos, J.C.; Rai, M.; da Silva, S.S. Bioenergy and Biofuels: Nanotechnological Solutions for Sustainable Production. In *Nanotechnology for Bioenergy and Biofuel Production*; Rai, M., da Silva, S.S., Eds.; Springer International Publishing: Cham, Switzerland, 2017; pp. 3–18.

22. Valko, M.; Morris, H.; Cronin, M.T. Metals, toxicity and oxidative stress. *Curr. Med. Chem.* **2005**, *12*, 1161–1208. [CrossRef]

23. Morganti, P. Saving the environment by nanotechnology and waste raw materials: Use of chitin nanofibrils by EU research projects. *J. Appl. Cosmetol.* **2013**, *31*, 89–96.

24. Rahim, A.H.A.; Khoo, K.S.; Yunus, N.M.; Hamzah, W.S.W. Ether-functionalized ionic liquids as solvent for Gigantochloa scortechini dissolution. *AIP Conf. Proc.* **2019**, *2157*, 020025. [CrossRef]

25. Peña, L.; Hohn, K.; Li, J.; Sun, X.; Wang, D. Synthesis of propyl-sulfonic acid-functionalized nanoparticles as catalysts for cellobiose hydrolysis. *J. Biomater. Nanobiotechnol.* **2014**, *5*, 241. [CrossRef]

26. Lai, D.M.; Deng, L.; Guo, Q.X.; Fu, Y. Hydrolysis of biomass by magnetic solid acid. *Energy Environ. Sci.* **2011**, *4*, 3552–3557. [CrossRef]

27. Erdem, S.; Erdem, B.; Öksüzoğlu, R.M. Magnetic Nano-Sized Solid Acid Catalyst Bearing Sulfonic Acid Groups for Biodiesel Synthesis. *Open Chem.* **2018**, *16*, 923–929. [CrossRef]

28. Su, T.-C.; Fang, Z.; Zhang, F.; Luo, J.; Li, X.-K. Hydrolysis of selected tropical plant wastes catalyzed by a magnetic carbonaceous acid with microwave. *Sci. Rep.* **2015**, *5*, 17538. [CrossRef]

29. Papadopoulou, A.; Zarafeta, D.; Galanopoulou, A.P.; Stamatis, H. Enhanced Catalytic Performance of Trichoderma reesei Cellulase Immobilized on Magnetic Hierarchical Porous Carbon Nanoparticles. *Protein J.* **2019**, *38*, 640–648. [CrossRef]

30. Chang, R.H.-Y.; Jang, J.; Wu, K.C.W. Cellulase immobilized mesoporous silica nanocatalysts for efficient cellulose-to-glucose conversion. *Green Chem.* **2011**, *13*, 2844–2850. [CrossRef]

31. Kobayashi, H.; Hosaka, Y.; Hara, K.; Feng, B.; Hirosaki, Y.; Fukuoka, A. Control of selectivity, activity and durability of simple supported nickel catalysts for hydrolytic hydrogenation of cellulose. *Green Chem.* **2014**, *16*, 637–644. [CrossRef]

32. Srivastava, N.; Rawat, R.; Sharma, R.; Oberoi, H.S.; Srivastava, M.; Singh, J. Effect of nickel–cobaltite nanoparticles on production and thermostability of cellulases from newly isolated thermotolerant Aspergillus fumigatus NS (Class: Eurotiomycetes). *Appl. Biochem. Biotechnol.* **2014**, *174*, 1092–1103. [CrossRef]

33. Ahmad, R.; Khare, S.K. Immobilization of Aspergillus niger cellulase on multiwall carbon nanotubes for cellulose hydrolysis. *Bioresour. Technol.* **2018**, *252*, 72–75. [CrossRef]

34. Mubarak, N.; Wong, J.; Tan, K.; Sahu, J.; Abdullah, E.; Jayakumar, N.; Ganesan, P. Immobilization of cellulase enzyme on functionalized multiwall carbon nanotubes. *J. Mol. Catal. B Enzym.* **2014**, *107*, 124–131. [CrossRef]

35. Khoo, K.S.; Chew, K.W.; Ooi, C.W.; Ong, H.C.; Ling, T.C.; Show, P.L. Extraction of natural astaxanthin from Haematococcus pluvialis using liquid biphasic flotation system. *Bioresour. Technol.* **2019**, *290*, 121794. [CrossRef]

36. Chew, K.W.; Yap, J.Y.; Show, P.L.; Suan, N.H.; Juan, J.C.; Ling, T.C.; Lee, D.-J.; Chang, J.-S. Microalgae biorefinery: High value products perspectives. *Bioresour. Technol.* **2017**, *229*, 53–62. [CrossRef]

37. Khoo, K.S.; Lee, S.Y.; Ooi, C.W.; Fu, X.; Miao, X.; Ling, T.C.; Show, P.L. Recent Advances in Biorefinery of Astaxanthin from Haematococcus pluvialis. *Bioresour. Technol.* **2019**, *288*, 121606. [CrossRef]

38. Duraiarasan, S.; Razack, S.A.; Manickam, A.; Munusamy, A.; Syed, M.B.; Ali, M.Y.; Ahmed, G.M.; Mohiuddin, M.S. Direct conversion of lipids from marine microalga C. salina to biodiesel with immobilised enzymes using magnetic nanoparticle. *J. Environ. Chem. Eng.* **2016**, *4*, 1393–1398. [CrossRef]

39. Nematian, T.; Salehi, Z.; Shakeri, A. Conversion of bio-oil extracted from Chlorella vulgaris micro algae to biodiesel via modified superparamagnetic nano-biocatalyst. *Renew. Energy* **2020**, *146*, 1796–1804. [CrossRef]

40. Zaidi, A.; Feng, R.; Malik, A.; Khan, S.; Shi, Y.; Bhutta, A.; Shah, A. Combining microwave pretreatment with iron oxide nanoparticles enhanced biogas and hydrogen yield from green algae. *Processes* **2019**, *7*, 24. [CrossRef]

41. Windt, W.D.; Aelterman, P.; Verstraete, W. Bioreductive deposition of palladium (0) nanoparticles on Shewanella oneidensis with catalytic activity towards reductive dechlorination of polychlorinated biphenyls. *Environ. Microbiol.* **2005**, *7*, 314–325. [CrossRef]

42. Liu, Z.-Y.; Wang, G.-C.; Zhou, B.-C. Effect of iron on growth and lipid accumulation in Chlorella vulgaris. *Bioresour. Technol.* **2008**, *99*, 4717–4722. [CrossRef] [PubMed]

43. Lin, V.; Mahoney, P.; Gibson, K. Nanofarming technology extracts biofuel oil without harming algae. News released from Office of Public Affairs.

44. Lee, Y.-C.; Huh, Y.S.; Farooq, W.; Han, J.-I.; Oh, Y.-K.; Park, J.-Y. Oil extraction by aminoparticle-based H_2O_2 activation via wet microalgae harvesting. *RSC Adv.* **2013**, *3*, 12802–12809. [CrossRef]

45. Kang, N.K.; Lee, B.; Choi, G.-G.; Moon, M.; Park, M.S.; Lim, J.; Yang, J.-W. Enhancing lipid productivity of Chlorella vulgaris using oxidative stress by TiO_2 nanoparticles. *Korean J. Chem. Eng.* **2014**, *31*, 861–867. [CrossRef]

46. Lee, Y.-C.; Lee, K.; Oh, Y.-K. Recent nanoparticle engineering advances in microalgal cultivation and harvesting processes of biodiesel production: A review. *Bioresour. Technol.* **2015**, *184*, 63–72. [CrossRef] [PubMed]

47. Sarma, S.J.; Das, R.K.; Brar, S.K.; Le Bihan, Y.; Buelna, G.; Verma, M.; Soccol, C.R. Application of magnesium sulfate and its nanoparticles for enhanced lipid production by mixotrophic cultivation of algae using biodiesel waste. *Energy* **2014**, *78*, 16–22. [CrossRef]

48. Wang, X.; Dou, P.; Zhao, P.; Zhao, C.; Ding, Y.; Xu, P. Immobilization of lipases onto magnetic Fe_3O_4 nanoparticles for application in biodiesel production. *ChemSusChem Chem. Sustain. Energy Mater.* **2009**, *2*, 947–950.

49. Kaieda, M.; Samukawa, T.; Kondo, A.; Fukuda, H. Effect of methanol and water contents on production of biodiesel fuel from plant oil catalyzed by various lipases in a solvent-free system. *J. Biosci. Bioeng.* **2001**, *91*, 12–15. [CrossRef]

50. Xie, W.; Ma, N. Immobilized lipase on Fe_3O_4 nanoparticles as biocatalyst for biodiesel production. *Energy Fuels* **2009**, *23*, 1347–1353. [CrossRef]

51. Wang, Z.-G.; Wang, J.-Q.; Xu, Z.-K. Immobilization of lipase from Candida rugosa on electrospun polysulfone nanofibrous membranes by adsorption. *J. Mol. Catal. B Enzym.* **2006**, *42*, 45–51. [CrossRef]

52. Liu, C.-H.; Huang, C.-C.; Wang, Y.-W.; Lee, D.-J.; Chang, J.-S. Biodiesel production by enzymatic transesterification catalyzed by Burkholderia lipase immobilized on hydrophobic magnetic particles. *Appl. Energy* **2012**, *100*, 41–46. [CrossRef]

53. Ng, I.-S.; Tang, M.S.; Show, P.L.; Chiou, Z.-M.; Tsai, J.-C.; Chang, Y.-K. Enhancement of C-phycocyanin purity using negative chromatography with chitosan-modified nanofiber membrane. *Int. J. Biol. Macromol.* **2019**, *132*, 615–628. [CrossRef]

54. Cheah, W.Y.; Show, P.-L.; Ng, I.-S.; Lin, G.-Y.; Chiu, C.-Y.; Chang, Y.-K. Antibacterial activity of quaternized chitosan modified nanofiber membrane. *Int. J. Biol. Macromol.* **2019**, *126*, 569–577. [CrossRef]

55. Sakai, S.; Liu, Y.; Yamaguchi, T.; Watanabe, R.; Kawabe, M.; Kawakami, K. Production of butyl-biodiesel using lipase physically-adsorbed onto electrospun polyacrylonitrile fibers. *Bioresour. Technol.* **2010**, *101*, 7344–7349. [CrossRef]

56. Tran, D.-T.; Chen, C.-L.; Chang, J.-S. Immobilization of Burkholderia sp. lipase on a ferric silica nanocomposite for biodiesel production. *J. Biotechnol.* **2012**, *158*, 112–119. [CrossRef] [PubMed]

57. Sajith, V.; Sobhan, C.; Peterson, G. Experimental investigations on the effects of cerium oxide nanoparticle fuel additives on biodiesel. *Adv. Mech. Eng.* **2010**, *2*, 581407. [CrossRef]

58. Mehta, R.N.; Chakraborty, M.; Parikh, P.A. Impact of hydrogen generated by splitting water with nano-silicon and nano-aluminum on diesel engine performance. *Int. J. Hydrog. Energy* **2014**, *39*, 8098–8105. [CrossRef]

59. Karthikeyan, S.; Elango, A.; Prathima, A. Performance and Emission Study on Zinc Oxide Nano Particles Addition with Pomolion Stearin Wax Biodiesel of CI Engine. *J. Sci. Ind. Res.* **2014**, *73*, 187–190.

60. Singh, N.; Bharj, R. Effect of CNT-emulsified fuel on performance emission and combustion characteristics of four stroke diesel engine. *Int. J. Curr. Eng. Technol.* **2015**, *5*, 477–485.

61. Aalam, S.; Saravanan, C.; PremAnand, B. Reduction of emissions from CRDI diesel engine using metal oxide nanoparticles blended diesel fuel. *Int. J. Appl. Eng. Res.* **2015**, *10*, 3865–3869.

62. Aalam, C.S.; Saravanan, C.; Premanand, B. Influence of Iron (II, III) oxide nanoparticles fuel additive on exhaust emissions and combustion characteristics of CRDI system assisted diesel engine. *Int. J. Adv. Eng. Res. Sci.* **2015**, *2*, 23–28.

63. Mehta, R.N.; Chakraborty, M.; Parikh, P.A. Nanofuels: Combustion, engine performance and emissions. *Fuel* **2014**, *120*, 91–97. [CrossRef]

64. Santoro, C.; Arbizzani, C.; Erable, B.; Ieropoulos, I. Microbial fuel cells: From fundamentals to applications. A review. *J. Power Sources* **2017**, *356*, 225–244. [CrossRef]

65. Li, Z.; Fu, Q.; Kobayashi, H.; Xiao, S. Biofuel Production from Bioelectrochemical Systems. In *Bioreactors for Microbial Biomass and Energy Conversion*; Springer: Berlin/Heidelberg, Germany, 2018; pp. 435–461.

66. Gul, M.M.; Ahmad, K.S. Bioelectrochemical systems: Sustainable bio-energy powerhouses. *Biosens. Bioelectron.* **2019**, *142*, 111576. [CrossRef]

67. Potter, M.C. Electrical effects accompanying the decomposition of organic compounds. *Proc. R. Soc. Lond. Ser. B Contain. Pap. Biol. Character* **1911**, *84*, 260–276.

68. Kalathil, S.; Pant, D. Nanotechnology to rescue bacterial bidirectional extracellular electron transfer in bioelectrochemical systems. *RSC Adv.* **2016**, *6*, 30582–30597. [CrossRef]

69. Quan, X.; Sun, B.; Xu, H. Anode decoration with biogenic Pd nanoparticles improved power generation in microbial fuel cells. *Electrochim. Acta* **2015**, *182*, 815–820. [CrossRef]

70. Ray, S.G.; Ghangrekar, M. Enhancing organic matter removal, biopolymer recovery and electricity generation from distillery wastewater by combining fungal fermentation and microbial fuel cell. *Bioresour. Technol.* **2015**, *176*, 8–14.

71. Kalathil, S.; Lee, J.; Cho, M.H. Granular activated carbon based microbial fuel cell for simultaneous decolorization of real dye wastewater and electricity generation. *New Biotechnol.* **2011**, *29*, 32–37. [CrossRef]

72. Rezaei, F.; Xing, D.; Wagner, R.; Regan, J.M.; Richard, T.L.; Logan, B.E. Simultaneous cellulose degradation and electricity production by Enterobacter cloacae in a microbial fuel cell. *Appl. Environ. Microbiol.* **2009**, *75*, 3673–3678. [CrossRef]

73. Ren, Z.; Steinberg, L.; Regan, J. Electricity production and microbial biofilm characterization in cellulose-fed microbial fuel cells. *Water Sci. Technol.* **2008**, *58*, 617–622. [CrossRef]

74. Hernández-Fernández, F.; De Los Ríos, A.P.; Salar-García, M.; Ortiz-Martínez, V.; Lozano-Blanco, L.; Godínez, C.; Tomás-Alonso, F.; Quesada-Medina, J. Recent progress and perspectives in microbial fuel cells for bioenergy generation and wastewater treatment. *Fuel Process. Technol.* **2015**, *138*, 284–297. [CrossRef]

75. Rahimnejad, M.; Adhami, A.; Darvari, S.; Zirepour, A.; Oh, S.-E. Microbial fuel cell as new technology for bioelectricity generation: A review. *Alex. Eng. J.* **2015**, *54*, 745–756. [CrossRef]

76. Slate, A.J.; Whitehead, K.A.; Brownson, D.A.; Banks, C.E. Microbial fuel cells: An overview of current technology. *Renew. Sustain. Energy Rev.* **2019**, *101*, 60–81. [CrossRef]

77. Logan, B.E.; Hamelers, B.; Rozendal, R.; Schröder, U.; Keller, J.; Freguia, S.; Aelterman, P.; Verstraete, W.; Rabaey, K. Microbial fuel cells: Methodology and technology. *Environ. Sci. Technol.* **2006**, *40*, 5181–5192. [CrossRef]

78. Majid, S.; Ahmad, K.S. Analysis of dopant concentration effect on optical and morphological properties of PVD coated Cu-doped Ni_3S_2 thin films. *Optik* **2019**, *187*, 152–163. [CrossRef]

79. Majid, S.; Ahmad, K.S. Optical and morphological properties of environmentally benign Cu-Tin sulphide thin films grown by physical vapor deposition technique. *Mater. Res. Express* **2018**, *6*, 036406. [CrossRef]

80. Ahmad, K.S.; Hussain, Z.; Majid, S. Synthesis, characterization and PVD assisted thin film fabrication of the nano-structured bimetallic Ni_3S_2/MnS_2 composite. *Surf. Interfaces* **2018**, *12*, 190–195.

81. Erbay, C.; Pu, X.; Choi, W.; Choi, M.-J.; Ryu, Y.; Hou, H.; Lin, F.; de Figueiredo, P.; Yu, C.; Han, A. Control of geometrical properties of carbon nanotube electrodes towards high-performance microbial fuel cells. *J. Power Sources* **2015**, *280*, 347–354. [CrossRef]

82. Aryal, N.; Ammam, F.; Patil, S.A.; Pant, D. An overview of cathode materials for microbial electrosynthesis of chemicals from carbon dioxide. *Green Chem.* **2017**, *19*, 5748–5760. [CrossRef]

83. Zhang, Y.; Sun, J.; Hu, Y.; Li, S.; Xu, Q. Bio-cathode materials evaluation in microbial fuel cells: A comparison of graphite felt, carbon paper and stainless steel mesh materials. *Int. J. Hydrog. Energy* **2012**, *37*, 16935–16942. [CrossRef]

84. Zhang, G.; Zhao, Q.; Jiao, Y.; Wang, K.; Lee, D.-J.; Ren, N. Efficient electricity generation from sewage sludge usingbiocathode microbial fuel cell. *Water Res.* **2012**, *46*, 43–52. [CrossRef] [PubMed]

85. Salar-García, M.; Ortiz-Martínez, V. Nanotechnology for Wastewater Treatment and Bioenergy Generation in Microbial Fuel Cells. In *Advanced Research in Nanosciences for Water Technology*; Springer: Berlin/Heidelberg, Germany, 2019; pp. 341–362.

86. Kaur, R.; Marwah, A.; Chhabra, V.A.; Kim, K.-H.; Tripathi, S. Recent developments on functional nanomaterial-based electrodes for microbial fuel cells. *Renew. Sustain. Energy Rev.* **2019**, *119*, 109551. [CrossRef]

87. Mohammadifar, M.; Choi, S. A solid phase bacteria-powered biobattery for low-power, low-cost, internet of Disposable Things. *J. Power Sources* **2019**, *429*, 105–110. [CrossRef]

88. Prasad, J.; Tripathi, R.K. Energy harvesting from sediment microbial fuel cell to supply uninterruptible regulated power for small devices. *Int. J. Energy Res.* **2019**, *43*, 2821–2831. [CrossRef]

89. Liu, H.; Ramnarayanan, R.; Logan, B.E. Production of electricity during wastewater treatment using a single chamber microbial fuel cell. *Environ. Sci. Technol.* **2004**, *38*, 2281–2285. [CrossRef] [PubMed]

90. Liu, W.; Cheng, S.; Guo, J. Anode modification with formic acid: A simple and effective method to improve the power generation of microbial fuel cells. *Appl. Surf. Sci.* **2014**, *320*, 281–286. [CrossRef]

91. Shantaram, A.; Beyenal, H.; Veluchamy RR, A.; Lewandowski, Z. Wireless sensors powered by microbial fuel cells. *Environ. Sci. Technol.* **2005**, *39*, 5037–5042. [CrossRef] [PubMed]

92. Zhang, D.; Li, Z.; Zhang, C.; Zhou, X.; Xiao, Z.; Awata, T.; Katayama, A. Phenol-degrading anode biofilm with high coulombic efficiency in graphite electrodes microbial fuel cell. *J. Biosci. Bioeng.* **2017**, *123*, 364–369. [CrossRef] [PubMed]

93. Valipour, A.; Ayyaru, S.; Ahn, Y. Application of graphene-based nanomaterials as novel cathode catalysts for improving power generation in single chamber microbial fuel cells. *J. Power Sources* **2016**, *327*, 548–556. [CrossRef]

94. Zou, L.; Qiao, Y.; Zhong, C.; Li, C.M. Enabling fast electron transfer through both bacterial outer-membrane redox centers and endogenous electron mediators by polyaniline hybridized large-mesoporous carbon anode for high-performance microbial fuel cells. *Electrochim. Acta* **2017**, *229*, 31–38. [CrossRef]

95. Zou, L.; Lu, Z.; Huang, Y.; Long, Z.E.; Qiao, Y. Nanoporous Mo_2C functionalized 3D carbon architecture anode for boosting flavins mediated interfacial bioelectrocatalysis in microbial fuel cells. *J. Power Sources* **2017**, *359*, 549–555. [CrossRef]

96. Liu, Z.; Zhou, L.; Chen, Q.; Zhou, W.; Liu, Y. Advances in graphene/graphene composite based microbial fuel/electrolysis cells. *Electroanalysis* **2017**, *29*, 652–661. [CrossRef]

97. Lead, J.R.; Valsami-Jones, E. *Nanoscience and the Environment*; Elsevier: Amsterdam, The Netherlands, 2014; Volume 7.

98. Peng, X.; Chu, X.; Wang, S.; Shan, K.; Song, D.; Zhou, Y. Bio-power performance enhancement in microbial fuel cell using Ni–ferrite decorated anode. *RSC Adv.* **2017**, *7*, 16027–16032. [CrossRef]

99. Feng, H.; Liang, Y.; Guo, K.; Chen, W.; Shen, D.; Huang, L.; Zhou, Y.; Wang, M.; Long, Y. TiO_2 nanotube arrays modified titanium: A stable, scalable, and cost-effective bioanode for microbial fuel cells. *Environ. Sci. Technol. Lett.* **2016**, *3*, 420–424. [CrossRef]

100. Yang, G.; Chen, D.; Lv, P.; Kong, X.; Sun, Y.; Wang, Z.; Yuan, Z.; Liu, H.; Yang, J. Core-shell Au-Pd nanoparticles as cathode catalysts for microbial fuel cell applications. *Sci. Rep.* **2016**, *6*, 35252. [CrossRef] [PubMed]

101. Di Palma, L.; Bavasso, I.; Sarasini, F.; Tirillò, J.; Puglia, D.; Dominici, F.; Torre, L. Synthesis, characterization and performance evaluation of Fe_3O_4/PES nano composite membranes for microbial fuel cell. *Eur. Polym. J.* **2018**, *99*, 222–229. [CrossRef]

102. Zhang, C.; Liang, P.; Yang, X.; Jiang, Y.; Bian, Y.; Chen, C.; Zhang, X.; Huang, X. Binder-free graphene and manganese oxide coated carbon felt anode for high-performance microbial fuel cell. *Biosens. Bioelectron.* **2016**, *81*, 32–38. [CrossRef]
103. Wu, X.; Shi, Z.; Zou, L.; Li, C.M.; Qiao, Y. Pectin assisted one-pot synthesis of three dimensional porous NiO/graphene composite for enhanced bioelectrocatalysis in microbial fuel cells. *J. Power Sources* **2018**, *378*, 119–124. [CrossRef]
104. Mashkour, M.; Rahimnejad, M.; Pourali, S.; Ezoji, H.; ElMekawy, A.; Pant, D. Catalytic performance of nano-hybrid graphene and titanium dioxide modified cathodes fabricated with facile and green technique in microbial fuel cell. *Prog. Nat. Sci. Mater. Int.* **2017**, *27*, 647–651. [CrossRef]
105. Huang, L.; Li, X.; Ren, Y.; Wang, X. In-situ modified carbon cloth with polyaniline/graphene as anode to enhance performance of microbial fuel cell. *Int. J. Hydrog. Energy* **2016**, *41*, 11369–11379. [CrossRef]
106. Wu, X.; Qiao, Y.; Shi, Z.; Tang, W.; Li, C.M. Hierarchically porous N-doped carbon nanotubes/reduced graphene oxide composite for promoting flavin-based interfacial electron transfer in microbial fuel cells. *ACS Appl. Mater. Interfaces* **2018**, *10*, 11671–11677. [CrossRef]
107. Ren, H.; Tian, H.; Gardner, C.L.; Ren, T.-L.; Chae, J. A miniaturized microbial fuel cell with three-dimensional graphene macroporous scaffold anode demonstrating a record power density of over 10,000 W m^{-3}. *Nanoscale* **2016**, *8*, 3539–3547. [CrossRef]
108. Aslan, S.; Ó Conghaile, P.; Leech, D.; Gorton, L.; Timur, S.; Anik, U. Development of a Bioanode for Microbial Fuel Cells Based on the Combination of a MWCNT-Au-Pt Hybrid Nanomaterial, an Osmium Redox Polymer and Gluconobacter oxydans DSM 2343 Cells. *ChemistrySelect* **2017**, *2*, 12034–12040. [CrossRef]
109. Malorni, L.; Guida, V.; Sirignano, M.; Genovese, G.; Petrarca, C.; Pedata, P. Exposure to sub-10nm particles emitted from a biodiesel-fueled diesel engine: In vitro toxicity and inflammatory potential. *Toxicol. Lett.* **2017**, *270*, 51–61. [CrossRef] [PubMed]

Review

Organic Carbonate Production Utilizing Crude GlycerolDerived as By-Product of Biodiesel Production: A Review

Saifuddin Nomanbhay [1], Mei Yin Ong [1], Kit Wayne Chew [2], Pau-Loke Show [3], Man Kee Lam [4] and Wei-Hsin Chen [5,6,7,8,*]

[1] Institute of Sustainable Energy, Universiti Tenaga Nasional, Jalan IKRAM-UNITEN, Kajang 43000, Selangor, Malaysia; saifuddin@uniten.edu.my (S.N.); me089475@hotmail.com (M.Y.O.)

[2] School of Mathematical Sciences, Faculty of Science and Engineering, University of Nottingham Malaysia, Jalan Broga, Semenyih 43500, Selangor, Malaysia; KitWayne.Chew@nottingham.edu.my

[3] Department of Chemical and Environment Engineering, Faculty of Science and Engineering, University of Nottingham Malaysia, Jalan Broga, Semenyih 43500, Selangor, Malaysia; PauLoke.Show@nottingham.edu.my

[4] Chemical Engineering Department, HICoE-Centre for Biofuel and Biochemical Research, Institute of Self-Sustainable Building, Universiti Teknologi PETRONAS, Seri Iskandar 32610, Perak, Malaysia; lam.mankee@utp.edu.my

[5] Department of Aeronautics and Astronautics, National Cheng Kung University, Tainan 701, Taiwan

[6] Department of Chemical and Materials Engineering, College of Engineering, Tunghai University, Taichung 407, Taiwan

[7] Department of Mechanical Engineering, National Chin-Yi University of Technology, Taichung 411, Taiwan

[8] Research Center for Energy Technology and Strategy, National Cheng Kung University, Tainan 701, Taiwan

[*] Correspondence: weihsinchen@gmail.com or chenwh@mail.ncku.edu.tw

Received: 5 February 2020; Accepted: 20 March 2020; Published: 21 March 2020

Abstract: As a promising alternative renewable liquid fuel, biodiesel production has increased and eventually led to an increase in the production of its by-product, crude glycerol. The vast generation of glycerol has surpassed the market demand. Hence, the crude glycerol produced should be utilized effectively to increase the viability of biodiesel production. One of them is through crude glycerol upgrading, which is not economical. A good deal of attention has been dedicated to research for alternative material and chemicals derived from sustainable biomass resources. It will be more valuable if the crude glycerol is converted into glycerol derivatives, and so, increase the economic possibility of the biodiesel production. Studies showed that glycerol carbonate plays an important role, as a building block, in synthesizing the glycerol oligomers at milder conditions under microwave irradiation. This review presents a brief outline of the physio-chemical, thermodynamic, toxicological, production methods, reactivity, and application of organic carbonates derived from glycerol with a major focus on glycerol carbonate and dimethyl carbonate (DMC), as a green chemical, for application in the chemical and biotechnical field. Research gaps and further improvements have also been discussed.

Keywords: crude glycerol; glycerol carbonate; dimethyl carbonate; microwave irradiation; reaction kinetics

---

1. Introduction

Throughout history, the survival of human beings has been fully dependent on the resources found on the Earth. In recent times, the world has been threatened with increasing environmental problems, particularly global warming. Statistics show that the carbon dioxide (CO_2) levels in the atmosphere have been increasing every year and touched a record of 411.93 ppm in February 2020 [1]. Moreover, it is projected that conventional petroleum in Malaysia will run out by 2050 with the current consumption rate [2]. Hence, sustainable development by replacing conventional fossil fuels with alternative renewable and sustainable resources is a must to overcome the fossil fuel depletion issue and protect the global environment. Governments have also shown their initiatives by adopting new laws and regulations to ensure that the emission standards are being constricted for off-road, marine and stationary engines. Hence, alternative fuels have been explored and fostered, especially for transportation, construction, cultivation and electricity generation. Among the available renewable energy resources, biomass plays a significant role as feedstock in most biofuel production [3]. In the chemical industry, raw material origins are usually not the main issue. In contrast, cost and process effectiveness are the major driven of innovation. Nevertheless, the growth of sustainable consciousness, reinforced by both general society attitude and government initiatives, has pushed the researcher/chemist to take the naturally available raw material into consideration. As a result, the interest of researchers in exploring the alternative material and chemicals derived from sustainable biomass resources has been increased. The main aim of this article is to review the recent publications on the process of glycerol upgrading into glycerol carbonate. The catalytic activities, such as hydrotalcites, zeolites, heteropoly acids, and oxides, are emphasized. Further, the effect of different reaction conditions on the catalytic activities and selectivity of glycerol carbonate are also discussed in this article. Furthermore, issues that require further investigation have also been highlighted.

2. Biodiesel and Renewable Diesel

Petroleum diesel, i.e., conventional fossil fuel, is produced from petroleum-based crude oil. Crude oil is the combination of hydrocarbons, organic compounds, and small amounts of metal, which formation takes millions of hundreds of years. Crude oil is removed from the ground and transferred into a refinery. The crude oil is then passed through a heat-and-pressure based process within the refinery. To replace the conventional non-renewable petroleum diesel, biodiesel and renewable diesel are introduced (Figure 1). The major difference between biodiesel and renewable fuel will be explained shortly. As biodiesel and renewable diesel are in the form of liquid, unlike other renewable energy (i.e., hydro energy and solar energy), they can be directly used to operate current engines, especially transportation vehicles and industrial machines that are mostly operated by liquid fuel, without modification to their design.

Unlike petroleum diesel, biodiesel does not depend on fossil fuel. Biodiesel is usually produced from vegetable oils and/or animal fats through the transesterification process. In 2019, petrol stations in Malaysia switched from B7 to B10, to increase the sustainability of energy sources [4]. The next renewable alternative fuel is called renewable diesel. Renewable diesel, almost similar to biodiesel, is formed from waste agricultural products [5], mainly waste vegetable oils and animal fats. The main difference between renewable diesel and biodiesel is the production method. The production process of renewable diesel and petroleum diesel is the same, and hence, it is chemically similar to each other. However, the greenhouse gas emissions of renewable diesel (also known as hydro-treated vegetable oil or HVO) and traditional biodiesel are both smaller than those from fossil diesel. From an environmental perspective, the usage of biodiesel and renewable diesel fuel can significantly reduce the emissions of hydrocarbon, carbon monoxide, and particulate matter [6]. However, it is well-known that the combustion of biodiesel in diesel engine emits 10% more of NO_x in comparison to petroleum diesel. On the other hand, this increase in the NO_x emission issue can be solved by using renewable fuel combustion as it was reported that the use of renewable fuel leads to lower NO_x emission when compared to petroleum diesel [7,8]. Furthermore, as the chemical properties of renewable diesel and petroleum diesel are similar, that would mean a few things:

- The hydrogenation process makes the renewable diesel devoid of oxygen. Hence, the problem related to freezing temperature and storage, which are faced by biodiesel, will be avoided.
- Due to hydrogenation, the combustion of renewable diesel is cleaner than biodiesel.

- The engines that are designed for conventional diesel fuel are also compatible with renewable diesel, with no blending necessary since they are chemically the same.
- Along with the above, since the process is not a transesterification reaction, there will be no production of waste glycerol. Hence, the crude glycerol issue, which is the major concern of biodiesel production, will be eliminated.

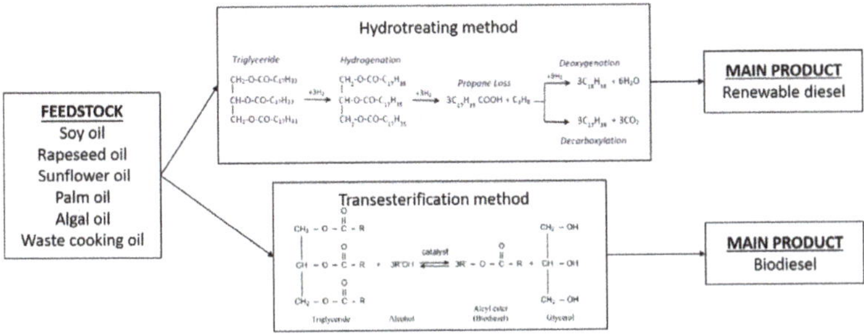

Figure 1. Processing for renewable diesel and biodiesel production.

The demand for biodiesel, as a renewable liquid biofuel, has been increasing. Global biodiesel production is expected to reach 23.57 billion litres by 2025 [9]. Biodiesel is produced from vegetable oils or animal fats along with methanol via transesterification reaction [10,11]. The transesterification process is usually catalysed by sodium methoxide (CH_3NaO) for a single-feedstock biodiesel plant, and potassium hydroxide (KOH) or sodium hydroxide (NaOH) for a multiple-feedstock biodiesel plant. As an important source of sustainable energy fuel, the annual biodiesel production increased year by year. Worldwide, biodiesel production uses primarily vegetable oil as its raw feedstock. In the European Union, rapeseed oil is used as main feedstock, while in the United States it is soybean oil, and in Asia it is mainly palm oil. Moreover, algae has also been identified as a potential feedstock for biodiesel production [12]. Crude glycerol is the major by-product formed from biodiesel production via the transesterification reaction. The simplified biodiesel production process is shown in the flowchart (Figure 2).

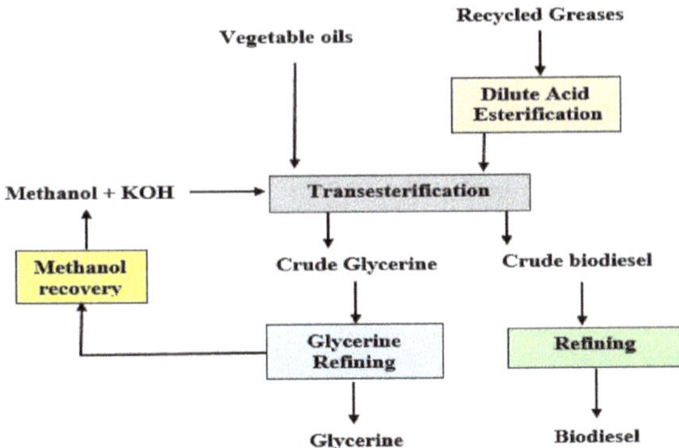

Figure 2. An overview of a standard biodiesel production process - Biodiesel and glycerol are the two main products.

3. Crude Glycerol – By-product of Biodiesel Production

Although the growth of the biodiesel industry has increased dramatically, especially between 2005 and 2015 (from 10% of total biofuel output to 25% in 2015), the biodiesel production still highly depend on the government policy and economic subsidies due to the high production costs of biodiesel [13].

Moreover, another major issue of biodiesel production is the generation of a by-product, crude glycerol. It is reported that 10 g of glycerol is generated for every 100 g of biodiesel produced [14–16]. Hence, the development of sustainable methods for exploiting this low cost organic raw material is imperative. Nevertheless, as there are impurities present in the crude glycerol, making the direct applications of crude glycerol restricted. The world markets for biodiesel are going into a period of precipitous, transitional growth, creating both uncertainty and opportunity [17]. Hence, an urgent solution is required to mitigate the problem of oversupply of this crude glycerol. Over the long term, biodiesel producers that are best able to evolve and adapt to advancement in technology, markets, feedstock, and government policies are most likely to succeed. One of the obvious ways to significantly improve the economic aspect of overall biodiesel production is the utilization and conversion of the glycerol by-product into value-added products [18]. With the estimate of the world market for biodiesel reaching 150 billion litres per year by 2016, this potentially means a stockpile of over 15 billion litres of crude glycerol available for refining and converting into value-added products per year [19]. The crude glycerol stream from the biodiesel process typically is about 50% glycerol or less and also contains an unused catalyst, alcohol, soap, water and salts. As mentioned previously, the presence of impurities lowers the value of the crude glycerol. Its value varies according to the level of impurities, which in turn is directly dependent on the type of feedstock used and the biodiesel synthesis method. The untreated crude glycerol can be used as animal feed, and for the co-digestion/co-gasification process but this can only be done on a short-term basis. The by-product crude glycerol will require a stringent purification process before it can be utilised in food, pharmaceutical or cosmetics fields. There are three main procedures in purifying crude glycerol. The first stage is the removal of free fatty acids, soap, and carried over catalyst via neutralization. The second step involves the evaporation of the excess methanol present in the crude. The third procedure, however, includes selective purification with the purpose to achieve the final product purity, through several methods, such as distillation (vacuum or azeotropic type), cation exchangers, which remove positively charged ions, or anion exchangers, which remove negatively charged ions, membrane separation with vaporization (pervaporation), and absorption [20]. In a recent review, a more detailed description of the purification of crude glycerol was reported [21]. The typical steps are summarised in Table 1 [22].

Table 1. Typical steps in crude glycerol purification [22].

Steps	Methods	Description
1	Neutralization	• Remove catalyst and soap by using acids. • If using sulphuric or hydrochloric acid, the final product of this stage will be in two phases. • If phosphorous acid is used, three phases of the final product will be obtained. √ Upper fraction: free fatty acids; middle fraction: glycerol and methanol; lower fraction: catalyst. √ Advantage: less harmful to the environment and tri-potassium phosphate, that is broadly used as fertilizer, is obtained as a by-product.
2	Methanol removal	• It is a must to remove methanol from glycerol fraction as it is toxic. • There is a huge amount of methanol in crude glycerol as a result of the excess addition of methanol during transesterification. • To remove the methanol, the evaporation process is conducted at 50–90 °C for at least 2 hours under vacuum.
3	Vacuum distillation	• Basically, glycerol can be polymerized into polyglycerol at a temperature above 200 °C. Hence, vacuum distillation is essential so that the purification process can be conducted without the polymerization of polyglycerol. • This method is optimal for small and medium-sized companies due to its high energy consumption as glycerol has high heat capacity.
	Ion exchange resins	• Removal of inorganic salts, free fatty acids, and free ions from crude glycerol.

One of the major drawbacks of the methods mentioned above is the high energy consumption. Therefore, it has triggered the researchers to find different ways for crude glycerol purification, such as membranes separation. The membrane separation process is driven by the concentration or electric potential difference between the two mediums [23,24]. As membrane separation is an eco-friendly technology, it has become more well-known recently in terms of biodiesel's waste glycerol purification. Despite all the processes mentioned, they are economically not viable for the smaller biodiesel producers. It is vital to develop a more eco-friendly, more efficient and economical process for crude glycerol purification, which could improve the cost-effectiveness of the overall biodiesel production [25,26].

Innovative methods for transforming glycerol into high value-added platform chemicals are being developed. Glycerol is one of the most versatile known chemicals with a wide variety of uses and applications such as in food, pharmaceutical, cosmetic, coating, and other industries. In the literature, several protocols for the synthesis of value-added products from glycerol have been reported and a significant number of platform molecules have been synthesized. Some of them have high potential in replacing petroleum-based products. As stated by Behr et al., all chemical products derived from glycerol are a result of one of many processes [27], as shown in Figure 3. These processes include the synthesis of glycerol esters, ethers, acetals and ketals, propanediols, epoxides, the oxidation and dehydration products of glycerol, and the production of synthesis gas. Several important chemicals can be obtained, among the more common ones are propane-diols, glycerol carbonate, epichlorohydrin, acrolein, esters of glycerol, fuel additives and glyceric acid [18,28–30]. Among these products, glycerol carbonate is one of the most attractive derivatives of glycerol reported in the last couple of years because of its high reactivity with amines, alcohols, carboxylic acids, ketones, and isocyanates. According to the market source, the cost for glycerol carbonate in 2017 ranged from US$2.4 to US$3 per kilogram, which means US$2400 to US$3000 per metric tonne [31].

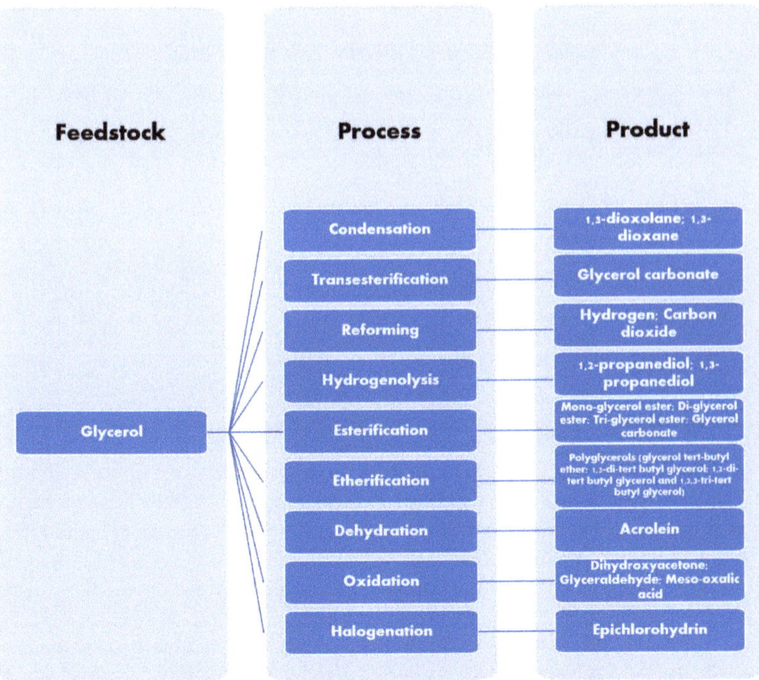

Figure 3. Different Chemical pathway for glycerol conversion to chemical derivatives (Adapted from [32]).

4. Glycerol Carbonate

Among the valuable raw chemicals derived from glycerol is glycerol carbonate (Gly-C) (i.e., 4-(hydroxymethyl)-1,3-dioxolan-2-one). Glycerol carbonate is a cyclic carbonate and its molecular structure presents a structural duality. It bears a hydroxyl group and 2-oxo-1,3-dioxolane group that gives Gly-C wide reactivity [33]. The Gly-C has excellent properties, such as good water-solubility, high boiling point, low toxicity, low flammability, good biodegradability, and a high flash point [34]. The glycerol carbonate is advocated as a useful and green building block in the field of organic chemistry as it is bio-based and has wide reactivity. Gly-C has a molar mass of 118.09 g/mol, and is widely used as a polar high boiling solvent, a surfactant component, a membrane component for gas separation and a component for industrial of coating, detergent, polymers, ink, paint, lubricant and electrolyte [28,35,36]. Furthermore, glycerol carbonate can also be used as chemical intermediates for the synthesis of other chemical compounds such as glycidol, which is employed in textile, plastics, pharmaceutical, and cosmetics industries [28,37].

Figure 4 shows the structural formula of glycerol carbonate. The hydroxyl group of glycerol carbonate consists of H-bonding and hence, glycerol carbonate has higher density properties. Besides, it has the potential to substitute petrochemically derived propylene carbonate. The general properties of glycerol carbonate and pure glycerol are shown in Table 2.

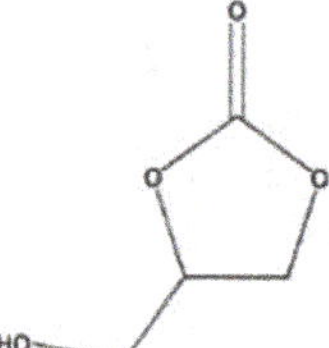

Figure 4. The glycerol carbonate molecular structure.

Table 2. The comparison of properties between glycerol carbonate and pure glycerol.

Properties	Unit	Glycerol Carbonate	Pure Glycerol
Molecular weight	g/mol	118.09	92.094
Density at 25 °C	g/ml	1.4	1.261
Boiling point	°C	354	290
Melting point	°C	−69	17.8
Vapour pressure at 177 °C	bar	0.008	0.003
Flash point	°C	190	177
Dielectric constant at 20 °C	ESU	111.5	42.5
Hansen solubility parameter delta D at 25 °C	MPa$^{1/2}$	17.9	17.4
Hansen solubility parameter delta P at 25 °C	MPa$^{1/2}$	25.5	12.1
Hansen solubility parameter delta H at 25 °C	MPa$^{1/2}$	17.4	29.3
Hildebrand solubility parameter at 25 °C	MPa$^{1/2}$	34.1	36.1
Viscosity at 25 °C	cP	85.4	1500

Glycerol carbonate (Gly-C) has several distinctive reactive locations: (a) the dioxolane ring with three carbon atoms, and (b) the suspended hydroxyl moiety. Hence, Gly-C has the ability to react not only as a nucleophile via its hydroxyl group, but also as an electrophile through its ring carbon atoms. These reactive sites provide potentials for utilizing glycerol carbonate as a precursor molecule for conversion to other intermediate chemicals, which ultimately are used in numerous direct and indirect applications. It has vast industrial applications and some of the potential industrial uses of Gly-C are presented in Table 3.

A sustainable route is to react urea with glycerol to produce glycerol carbonate, which has a vast perspective to be used as a replacement for fossil-fuel derived compounds. More studies were carried out to find environmentally friendly alternative paths in upgrading the waste glycerol into glycerol carbonate. Among the other promising routes for glycerol carbonate production, include reactions based on easily available raw materials, which are inexpensive and bio-based. Using enzymes as catalytic systems could generate more environmentally friendly alternatives. Based on an in-depth analysis of the recent publications on the subject, the present review will focus on glycerol carbonate syntheses routes, it's broad reactivity, and current applications.

Table 3. The potential application of glycerol carbonate in various industries.

Chemical	Industry Application	References
	• Polymers and plastics • Polyesters • Polycarbonates • Polyamides • Polyurethane plastic coatings • Hyperbranched polyethers • Solvent for plastics and resins	[38]
	• Cosmetics and personal care • Emollient and solvent in nail polish remover, lipsticks, anti-perspirant sticks • Wetting agent for cosmetic clays	[33]
Glycerol Carbonate	• Chemicals • Chemical intermediate • Glycidol • Biolubricants • Biobased polar solvents • Liquid membrane in gas separation • Surfactants and detergents • Blowing agent	[39]
	• Pharmaceutical • Solvent for medicinally active species • Carrier in pharmaceutical preparations	[36]
	• Semiconductor • Electrolytes in lithium and lithium-ion batteries	[40]
	• Agricultural • Plant-activating agent	[41]
	• Building and Construction • Curing agent in cement and concrete	[42]

5. Existing Production Routes for Synthesis of Glycerol Carbonate

As mentioned previously, the conventional glycerol market is limited and any large increase in biodiesel production will cause a sharp decrease in its market price. In the last few years, there have been numerous reports in the literature concerning the synthesis of glycerol carbonate from glycerol. While some production methods are being applied at an industrial scale, several new promising synthesis methods were reported recently which are more sustainable. The greener pathway for synthesis of glycerol carbonate involves chemicals such as CO/H_2, organic carbonate (e.g., ethylene carbonate, dimethyl carbonate, and diethyl carbonate), and also carbon dioxide [36,37,43]. These pathways, overall, results in the chemical fixation of CO_2. Among these pathways, the transesterification between glycerol and dimethyl carbonate is one convenient method that could be performed under mild conditions (at 50–100 °C; under atmospheric pressure) in the presence of a catalyst, as shown in Figure 5 [37,43,44]. The catalysts used to convert glycerol to glycerol carbonate can be homogeneous catalysts such as K_2CO_3, NaOH, and H_2SO_4 [44,45], enzymatic catalysts [44], or heterogeneous catalysts such as CaO, NaOH/γ-Al_2O_3, and Mg/Al/Zr mixed oxide [28,35,45]. Table 4 summarizes the reaction conditions for various types of synthesis pathway for glycerol carbonate production. Meanwhile, Table 5 presents the advantages and limitations of the different routes for the synthesis of glycerol carbonate.

Table 4. Different routes for the synthesis of glycerol carbonate (Gly-C) and their experimental conditions.

Reference	Reactants Mole Ratio	Solvent	Temperature (°C)	Pressure (MPa)	Reaction Time (h)	GC Yield (%)	Catalyst
		Glycerol + CO					
[46]	5 : 1 : 5 (K_2CO_3 toSe to glycerol)	Dimethylformamide	20	0.1	6	83	None
[46]	5 : 1.5 : 3 : 1 (Et_3N to CuBr to S to glycerol)	Dimethylformamide or Dimethyl sulfoxide	80	1	21	92	None
[41]	Excess CO and O_2	Dimethylformamide	140	3	2	85	0.25 mol% $PdCl_2$(phen) + 2.5 mol% KI
		Glycerol + CO_2					
[47]	1.5 : 1 : 1 ($N(CH_2CH_3)_3$ to HCl to glycerol)	Free	100	2.5	1	90	-
[48]	1 : 3 : 3 (K_2CO_3 to HCl to glycerol)	Free	80	0.1	30	80	KOH then HCl
[49]	Excess CO_2	Methanol	140	5.0	59	0.24	$RhCl_3$ + PPh_3 + KI
		Glycerol + EC					
[50]	2 : 1 (EC to glycerol)	Free	80	0.1	1.5	92	RNX-MCM41
[51]	2 : 1 (EC to glycerol)	Free	50	0.1	5	82	7 wt% Al/MgO hydrotalcite
[51]	2 : 1 (EC to glycerol)	Free	50	0.1	5	78	7 wt% MgO
[51]	2 : 1 (EC to glycerol)	Free	50	0.1	5	68	7 wt% Al/Mg hydrotalcite
		Glycerol + DMC					
[52]	39 : 1 (DMC to corn oil)	Supercritical DMC	380	15-25	0.5	By-product	None
[53,54]	42 : 1 (DMC to rapeseed Oil)	Supercritical DMC	350	20	0.2	By-product	None
[55]	6 : 1 (DMC to soybean oil)	tert-Butanol	60	0.1	48	92	10 wt% lipase (Novozyme 435)
[56]	10 : 1 (DMC to corn oil)	Free	60	0.1	15	62	10 wt% lipase (Novozyme 435)
[34]	10 :1 (DMC to glycerol)	Free	60	0.1	4	59	12 wt% lipase (Aspergillus niger)
[57]	10 : 1 (DMC to glycerol)	Free	70	0.1	48	90	Lipase (Novozyme 435)
[58]	5 : 1 (DMC to glycerol)	Dimethylformamide	100	0.1	0.5	79	Calcined hydrotalcite-hydromagnesite
[28]	5 : 1 (DMC to glycerol)	Free	75	0.1	1.5	95	Mg/Al/Zr
[59]	2 : 1 (DMC to glycerol)	Free	78	0.1	1	99	3 wt% KF/hydroxyapatite
[60]	5 : 1 (DMC to glycerol)	Dimethylformamide	100	0.1	1	75	Uncalcined Mg–Al hydrotalcite
[61]	2.5 : 1 (DMC to glycerol)	Benzene	60	0.1	2	95	4 mol% CaO

Table 4. *Cont.*

Reference	Reactants Mole Ratio	Solvent	Temperature (°C)	Pressure (MPa)	Reaction Time (h)	GC Yield (%)	Catalyst
	Glycerol + DEC						
[62]	3 : 1 (DEC to camellia oil)	Free	50	0.1	24	95	Lipases (Lipozyme TL IM and Novozym 435)
[63]	21 : 1 (DEC to glycerol)	Dimethyl sulfoxide	-	0.1	8	84	Hydrotalcites supported on Al_2O_3
[64]	17 : 1 (DEC to glycerol)	Free	130	0.1	60	97	Mg/Al hydrotalcite-like
	Glycerol + Urea						
[65]	1 : 1 (urea to glycerol)	Free	140	1.4×10^{-2}	6	46	Ionic liquids immobilized onto a structurally modified Merrifield peptide resin
[66]	1 : 3 (urea to glycerol)	Free	140	3.0×10^{-3}	1	91	0.5 wt% calcined La_2O_3
[67]	1.5 : 1 (urea to glycerol)	Free	150	0.1	4	55	Gold, gallium, and zinc supported on oxides and zeolite ZSM-5
[68]	1 : 1 (urea to glycerol)	Free	145	0.1	4	69	Co_3O_4/ZnO nanodispersion
[51]	1 : 1 (urea to glycerol)	Free	145	3.9×10^{-3}	5	72	Calcined Zn hydrotalcite

Table 5. Advantages and disadvantages of different routes of glycerol carbonate production.

Synthesis Methods	Merit	Demerit	References
Direct Carbonation • Carbon monoxide+ Crude glycerol	• High selectivity enhancing the chances of commercialisation.	• Large amount of by-products produced. • Carbon monoxide is toxic so it needs to be handled extra safely in the laboratory and industrial stages. Hence, this has restricted its usage. • Catalyst poisoning due to the use of carbon monoxide.	[46,69,70]
• Carbon dioxide+ crude glycerol	• Direct sequestration of CO_2.	• This reaction is thermodynamically restricted. • Carbon dioxide is too stable and leads to a very low conversion rate. So, this method is currently not feasible. • Require highly reactive catalysts and high energy consumption for the conversion of CO_2.	[69,71,72]

Table 5. *Cont.*

Synthesis Methods	Merit	Demerit	References
Transcarbonation • Phosgene + crude glycerol	• Simple and effective oldest method for the production of organic carbonates. • Glycerol carbonate can be obtained by this method with a good yield.	• Require highly toxic and corrosive gas and hence, need extra precaution during handling it. Therefore, it makes this route very restricted. • Process is characterized by a low atom economy and is recognized as unsafe and highly polluting.	[33,51]
• Urea + crude glycerol	• Urea is a cheap and readily available raw material and no azeotrope formation is required.	• High quantity of ammonia produced as a by-product, which leads to poor selectivity, so limiting its industrial implementation. • The relatively high reaction temperatures and the utilization of homogenous and/or uneasily recoverable catalysts- have negative economical impacts on this method.	[29,67,73,74]
• Ethylene carbonate + crude glycerol	• Production of desired higher valued carbonates. • Reaction temperature used for this method is relatively low.	• Ethylene carbonate is expensive • Facing difficulty during the purification process of glycerol carbonate as ethylene carbonate used in this process have a high boiling point (261°C).	[33,75]
• Dimethyl carbonate (DMC) + crude glycerol, or diethyl carbonate (DEC) + crude glycerol	• DMC/DEC is a versatile chemical and eco-friendly. DMC/DEC has high chemical reactivity and superior physical properties. It can be used in a one-step-one-pot reaction system. • With DMC/DEC, rigorous separation and high energy consumption are not required. Besides, this method is less time-consuming in comparison with the reaction involve phosgene and alkylene carbonates. • DMC has low boiling point (90°C) and so, easier the distillation separation process.	• Industrial grade DMC is approximately three times more expensive than methanol (MeOH).	[76–78]

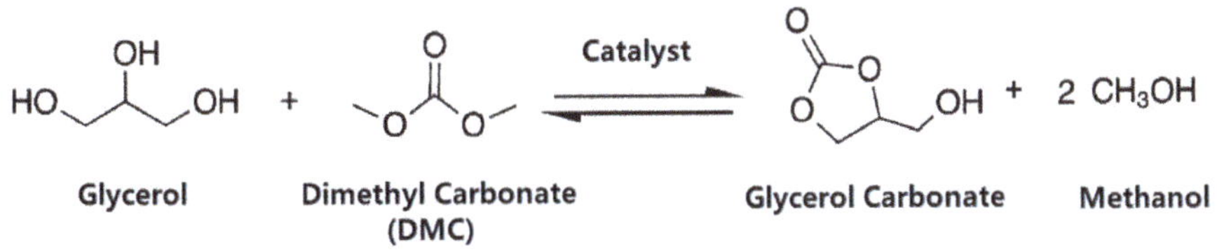

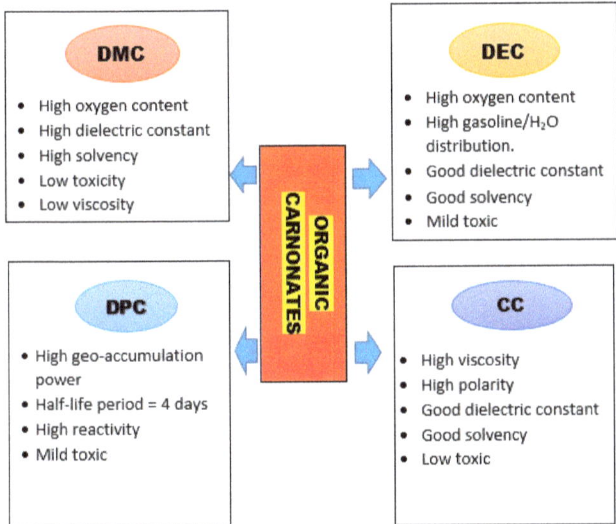

Figure 5. Reaction mechanism between glycerol and dimethyl carbonate for glycerol carbonate production [37].

6. Microwave Assisted Conversion of Glycerol/Glycerol Derivatives to Organic Carbonates

Organic carbonates are very important compounds having a simple structure, high polarity, low viscosity, low toxicity and degradability [79]. These compounds are able to replace hazardous reagents in some organic processes. Two main classes of organic carbonates are cyclic and linear carbonates. The total market production of both cyclic and linear organic carbonates is about 18 million tons per year and mainly produced from phosgene [80]. Figure 6 shows some of the more commercially important organic carbonates (dimethyl carbonate (DMC), diethyl carbonate (DEC), cyclic carbonates (CC), diphenyl carbonate (DPC)) and their properties.

Figure 6. The properties of various organic carbonates (Adapted from [80]).

Different methods have been deployed for the typical synthesis of glycerol carbonate from glycerol as depicted in Table 4 (such as the reaction of glycerol with urea, direct carbonation of glycerol with phosgene, or carbon monoxide and oxygen reaction). However, there are technical and environmental drawbacks accompanying the reaction of glycerol with urea, such as high reaction temperature (150 °C) required. The carbonation reaction is known as an environmentally hazardous process due to the toxicity of phosgene and CO [70]. Given the toxicity of phosgene, alternative routes for the synthesis from carbon dioxide are becoming more and more relevant. A greener method for the production of glycerol carbonate is the carbonation of glycerol with carbon dioxide (catalysed by Sn). This method, however, has low yield (below 35%) due to the thermodynamic limitation [81]. Most of the other methods also have disadvantages such as low Gly-C yield, vacuum operation condition to separate ammonia continuously, or difficulty in the separation of products [82,83]. Several new non-phosgene methods including alcoholysis of urea, carbonylation of alcohols using CO_2, oxy-carbonylation of alcohols, and transesterification of alcohols and carbonates have been developed for synthesizing organic carbonates. If applied, some of these new routes would lower the emission of CO_2 into the atmosphere about seven times (0.92 tons per ton of product, compared to 6.62 tons at present) [28,35,36]. Carbon dioxide (the most significant and long-lived greenhouse gas) is produced by most significantly by combustion of organic materials and fossil fuels. Instituting chemical processes and industries based on carbon as a building block would be paramount to reduce CO_2 levels in the atmosphere. As stated earlier, a specific problem in this context is the issue of CO_2 reactivity (the carbon atom in CO_2 is in its most oxidized state, - hence, makes the molecule highly stabile with low reactivity). The suitable catalyst system can overcome the high reaction barrier, but in general, these reactions still require rather harsh conditions. Researchers are discovering new, environmentally benign, green processes for the synthesis of value-added chemicals and platform molecules.

Microwave irradiation has been used for the synthesis of many organic compounds. The uptake of this technique was sluggish for about 12–14 years after the original publications on the benefits of conducting organic reactions in a microwave in the mid-1980s. However, since the year 2000, there has been an upsurge in publications describing the utilization of microwave synthesis. This trend is most probably correlated to the availability of instruments designed specifically for organic synthesis, hence, allowing microwave reactions to be conducted in a safe and reproducible manner. Among the obvious advantages of microwave heating include more uniform and rapid heating resulting in reduced reaction time for the synthesis of chemicals. Since glycerol is a very good absorber of microwave radiations, microwaves could play a greater role in the synthesis of Gly-C. In microwave irradiation, efficient internal heating is produced by the direct coupling of microwave energy with the solvent, reagents or catalysts in the reaction mixture. The radiation passes through the walls of the vessel directly into the whole reaction mixture, since the reaction vessels are made out of microwave-transparent materials. In typical microwave reactors, the magnetrons (microwave generators) produce microwave radiation of wavelength 12.25 cm, which corresponds to a frequency of 2.45 GHz. Two mechanisms are responsible for the microwave heating, dipolar polarisation and ionic conduction. By using closed vessels, higher pressures can be attained and the superheating effects are greatly magnified, hence, it is possible to maintain solutions at temperatures above their conventional reflux temperature. The higher purity of products often observed in microwave-assisted reactions is largely attributed to the homogeneous and smooth in situ heating. The heating process is also easily controllable because the energy input stops immediately when the power is turned off. It is important to point out that microwave energy cannot break chemical bonds. Compared to classical heating, microwave irradiation provides the advantage of allowing high-temperature reaction. Another advantage of microwave heating is the non-thermal effects due to specific heating of polar intermediates produced during the reaction. These polar intermediates will lead to modified selectivity, enabling polymerization which otherwise could not be performed with thermal heating. For a more detailed account of microwave-assisted reactions, some excellent reviews can be referred [84–87].

Despite all the studies on the synthesis of Gly-C, only a few studies mentioned the microwave-assisted synthesis of Gly-C. In a microwave-assisted reaction, the rate of a reaction involving glycerol as a reactant can be greatly improved due to the high dielectric properties of glycerol [88]. Along with similar scope, polyglycerols has been evaluated over several years as a way of utilizing excess crude glycerol. Polyglycerol have received a lot of attention as a hydrophilic component for neutral surfactants and emulsifiers for food, cosmetic and pharmaceutical. They comprise mainly of glycerol oligomers of low molecular mass. The conventional synthesis of polyglycerols is from glycerol. It involves drastic conditions (high temperature and alkaline conditions), resulting in a complex mixture of oligomers with no well-defined chemical composition [89]. Recently, the production of polyglycerols derivatives and glycerol carbonate using microwave irradiations has been demonstrated [88]. By using microwave irradiation, the synthesis of glycerol oligomers, can be performed under relatively mild conditions using glycerol carbonate as a valuable starting material. In this case, the reaction did not require any solvent or reactant and was completed in a relatively short time. Hence, microwave-assisted polymerization of glycerol carbonate presents the advantages of using less-hazardous conditions and better energy efficiency. On top of that, the experimental conditions allow no usage of any solvent, reactant, or purification steps [90].

In a recent report the transesterification of crude glycerol (GLY) at 70% and 86% purity with dimethyl carbonate (DMC) using calcium oxide (CaO) as catalyst was performed with both the conventional heating and microwave-assisted process [91]. In both the processes, it was found that 70% purity of crude GLY gave a higher yield of Gly-C, with the microwave technique showing better energy efficiency. In their work the highest yield of Gly-C (93.4%) was obtained under the microwave system (crude GLY purity of 70%, 1 wt% of CaO as catalyst, 2:1 M ratio of DMC: GLY at 65 °C, 5 minutes reaction). While the impurities and leftover catalyst residue in the crude GLY are generally undesirable for the conventional transesterification process, it is interesting to note that some impurities (methanol, soap, salt, and fatty acids) in the crude glycerol had demonstrated positive effect [92]. According to the authors, crude GLY performs better than pure GLY in microwave heating because of greater energy efficiency. Thus, direct utilization of crude GLY from the biodiesel plant to produce GLY-C via microwave irradiation transesterification is a viable and economical option [91].

More microwave efficiency in synthesis was demonstrated in a reaction for the production of 2,3-dihydroxypropyldecanoate using glycerol derivatives. In this process the microwave-assisted synthesis (solvent-free) was performed by esterification of decanoic acid in the presence of two distinct glycerol derivatives, glycidol, and glycerol carbonate. The process uses microwave irradiation with an output power of 200–400 W, involving decanoic acid and glycerol derivatives (stoichiometric proportions), with an organo-catalyst. The microwave-assisted synthesis notably enhances the selectivity in 2,3-dihydroxypropyl decanoate (at 300 W, 91%), reinforcing the efficiency and selectivity of the microwave-assisted method [93].

7. Dimethyl Carbonate

Dimethyl carbonate (DMC) is a valuable and green platform chemical and /or solvent which continues to attract a lot of attention. Dimethyl carbonate (DMC) is an environmentally benign chemical and is widely used as a carbonylation agent. DMC has been classified as one of the greenest solvents in terms of safety, health and environmental criteria [94,95]. It is a nonpolar aprotic solvent having good miscibility with water. It is non-toxic and biodegrades readily in the atmosphere. In terms of its usage, DMC can be a potential replacement for methyl ethyl ketone, ethyl acetate, methyl isobutyl ketone, and most other ketones. Moreover, DMC is a green substitute for highly toxic and hazardous compounds such as (i) dimethyl sulfate and halohydrocarbons in methylation reactions and (ii) phosgene ($COCl_2$) in carboxymethylation (methoxycarbonylation) reactions [96]. Among the reaction routes for the production of DMC, the transesterification route and direct synthesis from methanol and CO_2 is the most attractive one due to the inexpensive raw material and the avoidance of corrosive reagents, such as phosgene and dimethyl sulphate [97]. Various catalysts have been studied extensively. Different dehydrating agents and additives have been used to minimize the effect of water produced during the reaction and improving the catalytic performance of the catalyst. Currently, the most established commercial pathway for the production of DMC is through oxidative carbonylation of methanol using O_2. In addition, new alternative processes for DMC from CO_2 are being developed. Direct synthesis of dimethyl carbonate (DMC) from CO_2 and methanol is a very attractive reaction. This is because CO_2 as a greenhouse compound can be consumed in this process [96]. Patents have already been registered for use of DMC and DEC as a fuel additive in gasoline due to their excellent blending properties, high octane number, and oxygen content. Diesel engines are much more efficient than gasoline engines, however, they suffer from NOx and particulate emissions. Many studies have already been reported on the reduction of hydrocarbons, CO, NOx, and particulate emission from diesel engines because of the use of organic carbonates as oxygenate in the fuel [98–100].

The conventional way for the direct synthesis of DMC requires the use of an autoclave along with solid base K_2CO_3 and CH_3I additive, high pressure and long reaction time of over 2 h. In this reaction, the dimethyl ether (DME) appeared as a by-product and its proportion reached 7%–31% of the total products [101]. More studies, including the mechanism of reaction are directed towards the conversion of CO_2 and methanol to DMC. The synthesis of dimethyl carbonate (DMC) from CO_2 and methanol under milder reaction conditions using reduced cerium oxide catalysts and reduced copper-promoted Ce oxide catalysts were reported recently [102]. The report stated that the conversion of methanol was low (0.005%–0.11%) when the milder condition was used. The mild reaction conditions for DMC synthesized were as follows: the reaction time was 2 h, low temperature of 353 K, and low total pressure of 1.3 MPa using reduced $Cu–CeO_2$ as catalyst (0.5 wt% of Cu). New development in organic synthesis is the increasing interest in applying microwave irradiation. With high heat efficiency, microwave irradiation can accelerate the reaction and decrease the run time resulting in fewer by-products than by conventional heating, but there was very few literatures on the synthesis of DMC concerning microwave energy. Conventional heating methods are not an effective method to transfer energy to the reactants of a reaction mixture because the thermal conductivity of the various materials affects the energy transfer process. On the other hand, during microwave irradiation, molecules are efficiently heated internally by the direct coupling of energy from microwaves. Furthermore, current microwave instruments can precisely control the temperature, even allowing reaction mixtures to be superheated in a closed vessel. A study reported that the synthesis of dimethyl carbonate (DMC) from CO_2 and methanol required about 2 min for maximum yield of DMC when a power of 800 W was used. In the case of decreasing microwave power to 450 W, this reaction could be completed in 10 min and the yield of DMC was at least the same as in non-microwave reaction using conventional heating (autoclave) [101]. The reaction temperature under microwave irradiation was obviously lower than that with conventional heating, but the corresponding DMC yield was higher. It is very likely that microwave energy could be efficiently absorbed by the reactants, especially by methanol leading to a rate acceleration effect. The authors also claimed that in the reaction conditions used, about 2.5% of DME formed as a by-product in the conventional heating (autoclave). In contrast, only 0.4% of DME (by-product) was produced when the reaction was irradiated by microwave [101]. In the microwave irradiation, due to the interaction of the microwave energy with the molecules at a very fast rate, the real reaction temperature is higher than the average temperature of the medium. Hence, both energy consumption and reaction time are reduced by means of microwave irradiation. On the down-side, the microwave-assisted reaction has some disadvantages such as difficulty in the scale-up from laboratory to industrial scale due to the penetration depth of microwave radiation into the material is only a few centimeters. There are safety issues concerning the industrial vessel that need to be addressed, especially for uncontrolled heating.

Recently, there were reports on the usage of dimethyl carbonate instead of using alcohol as an alternative method to produce biodiesel. In this reaction, the product of triglyceride and dimethyl carbonate reaction is fatty acid methyl ester (FAME) and fatty acid glycerol carbonate (FAGC). Subsequently, FAGC then reacts with another molecule of dimethyl carbonate to generate another FAME molecule and glycerol dicarbonate. Accordingly, the overall reaction involves one molecule of triglyceride and two molecules of dimethyl carbonate to generate three molecules of FAME and one molecule of glycerol dicarbonate. This would enable biodiesel production without the production of waste glycerol as a by-product. Ilham and Saka firstly investigated biodiesel production from rapeseed oil in supercritical dimethyl carbonate using a batch-type reactor [103]. The reaction steps for biodiesel production under supercritical dimethyl carbonate conditions are shown in Figure 7.

A study about biodiesel production under supercritical dimethyl carbonate conditions is still limited. A complete conversion to biodiesel about 94% (w/w) was made at 350 °C, 20 MPa, oil-to-dimethyl carbonate molar ratio of 1:42 after 12 min [103]. In addition, there is no previous report about reaction kinetics of oil conversion to biodiesel in supercritical dimethyl carbonate. Optimization studies on biodiesel production using supercritical dimethyl carbonate method have also been carried out in recent years. Recently, Kwon et al. performed non-catalytic biodiesel production from coconut oil using dimethyl carbonate under ambient pressure [76]. They conducted an experiment via a continuous flow mode using a tubular reactor. They reported that a complete conversion to biodiesel was achieved in a short reaction time of 1–2 min at 365-450 °C under ambient pressure, obtaining 98%. By using DMC to produce biodiesel, it is envisaged that the glut of crude glycerol in the market will be reduced. The glycerol carbonate obtained from the reaction is a much valuable product. It has higher economic value and it is used in many reactions as green chemical or solvent. The properties and benefits of glycerol carbonate have been mentioned earlier.

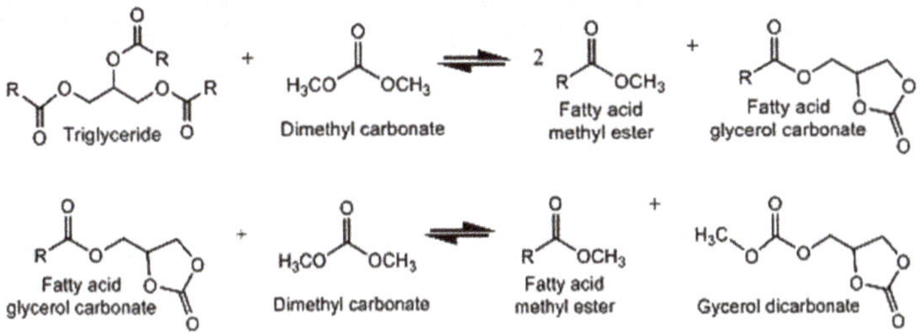

Figure 7. Reaction scheme for biodiesel production under supercritical dimethyl carbonate condition.

8. Kinetic Model of Transesterification of Glycerol with Dimethyl Carbonate

Transesterification of glycerol is also known as transcarbonation. As its name implies, it involves the carbonate exchange reaction between the alcohols and carbonate sources [33]. One of the carbonate sources is dimethyl carbonate, DMC. DMC is an environmentally friendly chemical and has been widely studied in producing the value-added glycerol carbonate with different catalysts such as, potassium methoxide (CH_3OK) [104], trisodium phosphate [78], calcium oxide (CaO) [91], Mg-Al hydrotalcite [105], guanidine ionic liquids [106], lipase [107], K-zeolite [108], and others. Notice that the transcarbonation of glycerol with DMC is a reversible reaction as shown in Figure 5. Thus, an excess amount of DMC should be used in order to shift the chemical equilibrium towards the formation of glycerol carbonate. Moreover, Jiabo Li and Tao Wang concluded that the chemical equilibrium constant of the reaction of glycerol and DMC increased as the reaction temperature increased. In order words, the formation of glycerol carbonate is more favourable by increasing the reaction temperature [61].

To gain more understanding about the transesterification of glycerol, its kinetic study has been conducted and reported by several researchers [104,109,110]. The kinetic study is usually performed to obtain the kinetic parameters (reaction rate) for the ease of comparison with other works. Typically, the kinetic analysis includes the investigation of the effect of different factors on reaction speed, which is essential for the system design, and the optimization of chemical reaction [111]. In addition, more information regarding the characteristics of a reaction mechanism can be obtained [112]. Moreover, it allows the construction of a mathematical model that can represent a specific chemical reaction. In determining the unknown parameters for a mathematical model, a series of experiments with high accuracy should be conducted. However, recent times different types of modelling technique has been introduced and eventually reduce the number of experiments [113].

Based on previous work there are four suitable kinetic models in describing the reaction of transesterification of glycerol with DMC, as summarized in Table 6 [104]. Each model involves two equations for the reaction before and after a certain conversion value, the critical conversion, X_{crit} to account for the effect of different phase regime on the catalytic behaviour. The first equation describes the reaction under the biphasic regime in which the catalysts are soluble only in the glycerol-rich stage with certain concentration, C_{cat}'. Moreover, the first order was assumed with respect to the concentration of glycerol, C_{gly}. The DMC, however, reacts at a constant concentration, C_{DMCsol}. The second equation of the proposed models describes the reaction under the single-phase stage (at X larger than X_{crit}) where the catalysts are assumed to be dissolved in the entire reaction medium at C_{cat}. In addition, unlike the previous equation, the first-order effect was estimated for both concentrations of glycerol, C_{gly} and DMC, C_{DMC}. As mentioned previously, the transesterification of glycerol with DMC is a reversible reaction. With the use of excess DMC, the reaction tends to shift to the production of glycerol carbonate. Nevertheless, the effect of reversible should be taken into consideration in proposing the kinetic model, and thus, kinetic models 2 and 4 are suggested.

In addition, more complex kinetic models (kinetic models 3 and 4) were contemplated by accounting for the nature of the catalyst and its potential deactivation. This is because it had already been investigated that some of the catalysts suffered from deactivation in the transesterification process of glycerol with DMC [104,108,114,115]. Catalyst deactivation is defined as the loss of catalytic activity over time. It is a continuing concern in the catalytic reaction, especially in the practice of industrial catalytic process. Generally, the causes of catalyst deactivation can be divided into three groups, which are chemical, mechanical and thermal. More specifically, the catalyst deactivation process can be explained by six basic mechanisms, including poisoning, fouling, thermal degradation and sintering, vapor formation, vapor-solid and solid-solid reactions, and crushing. Despite the best effort to avoid catalyst deactivation, it is inevitable that all catalyst will decay. To restore the catalytic activity, regeneration of the catalyst is usually the first choice, and disposal of the catalyst is the last resort. However, the ability to regenerate the catalyst highly relies on the reversibility of the deactivation process mentioned previously. As an example, sintering is basically irreversible. On the other hand, some poisons can be removed through chemical washing, oxidation and/or mechanical and heat treatment. Further details can be obtained from [116].

Li and Wang investigated the deactivation effect of the alkali solid catalyst (calcium oxide, calcium hydroxide and calcium methoxide) in the transesterification process of glycerol with DMC [114]. They found that the alkali solid catalysts reacted with glycerol and glycerol carbonate, transformed into basic calcium carbonate and eventually, deactivated the catalysts as the basic calcium carbonate consists of less catalytic strength. In the transesterification of glycerol with DMC that utilized K-zeolite as the catalysts, the deactivation phenomenon observed might be attributed to the reduction of the available active spaces for the conversion of glycerol to glycerol carbonate [108]. Moreover, it is reported that the use of crude glycerol led to rapid catalyst deactivation, which was not observed for runs with pure glycerol [117]. This is because the impurities present in the crude glycerol has limited the catalytic activity, resulting in significant deactivation of the catalyst. However, pure glycerol is far more expensive than crude glycerol. Hence, understanding the effect of impurities in crude glycerol on catalyst performance is crucial to optimize the overall production cost of glycerol carbonate.

Table 6. Kinetic models for the transesterification of glycerol with dimethyl carbonate.

Model Number	Reaction Type	Potential Catalyst Deactivation	If $X \leq X_{crit}$	If $X > X_{crit}$
1	Irreversible	Excluded	$r_1 = k_1 \cdot C_{cat'} \cdot C_{Gly} \cdot C_{DMCsol}$	$r_1 = k_1 \cdot C_{cat} \cdot C_{Gly} \cdot C_{DMC}$
2	Reversible			$r_1 = k_1 \cdot C_{cat} \cdot C_{Gly} \cdot C_{DMC}$ $r_2 = k_2 \cdot C_{cat} \cdot C_{GC} \cdot C_{MeOH}$
3	Irreversible	Included	$r_1 =$ $k_1 \cdot C_{cat'} \cdot \left[(1-\beta) \cdot e^{-k_d \cdot t} + \beta \right] \cdot C_{Gly} \cdot C_{DMCsol}$	$r_1 = k_1 \cdot C_{cat} \cdot \beta \cdot C_{Gly} \cdot C_{DMC}$
4	Reversible			$r_1 = k_1 \cdot C_{cat} \cdot \beta \cdot C_{Gly} \cdot C_{DMC}$ $r_2 = k_2 \cdot C_{cat} \cdot \beta \cdot C_{GC} \cdot C_{MeOH}$

In conjunction with the kinetic models stated in Table 6, the Arrhenius equation is applied to consider the temperature effect, T as in equation (1):

$$\ln k = \ln A - \frac{E_a}{R} \cdot \frac{1}{T} \tag{1}$$

where k indicates the kinetic constants of the reaction, A is the pre-exponential factor, E_a symbolizes the activation energy (kJ/mol), and lastly, R is the ideal gas constant (8.314 J/mol.K).

9. Summary and Future Perspectives

The worldwide inclination to embrace a sustainable and bio-compatible process has led to the development of systems focused on the use of CO_2 and CO_2-based compounds as feedstocks, promoters, and reaction media. In the present article, the synthesis of various carbonates has been reviewed with the main emphasis on the usage of crude glycerol as the raw material. Production of surplus glycerol in the biodiesel industry is a pressing issue and the industry is looking for ways to utilize and transform crude glycerol into valuable products, such as glycerol carbonate, with green solvent, the dimethyl carbonate (DMC) especially. DMC is one of the simplest carbonates and is termed as the green compound of the 21st century due to its low toxicity, high biodegradability and peculiar reactivity. Although studies on organic carbonates have been done extensively, there remain some shortcomings that must be resolved. In this regard, different ways to improve the thermodynamically limited equilibrium conversion to shift towards DMC formation is crucial (DMC yield is only around 1%, even at thermodynamically favourable, high-pressure conditions). Therefore, the development of high activity and high stability catalysts are crucial. Determining the mechanistic kinetics in the synthesis of organic carbonates has the potential to inspire improved catalyst design, which will result in mitigating the problems of long reaction times and high reaction temperatures. Furthermore, microwave-assisted organic synthesis is well known due to its shorter reaction time, lower operating temperature, volumetric heating mechanism, specific microwave effect, and rapid heating. The specific microwave effect is the non-thermal effect of microwave irradiation which causes the specific heating of polar intermediates and leads to a modified selective reaction, that normally cannot be performed using conventional heating. However, non-thermal microwave effects should be further investigated as their presence is still a controversial issue, especially in the chemical synthesis area. Moreover, mechanisms and process improvements using microwave irradiation are very limited and further work is needed in order to develop an understanding of these synthesis routes.

Author Contributions: Conceptualization, S.N. and M.Y.O.; writing—original draft preparation, M.Y.O.; writing—review and editing, K.W.C. and W.-H.C.; project administration, S.N., P.-L.S., M.K.L.; funding acquisition, S.N., P.-L.S. and W.-H.C. All authors have read and agreed to the published version of the manuscript.

Funding: This research was funded by AAIBE Chair of Renewable Energy (Grant code: 201801 KETTHA). This work was also supported by the Fundamental Research Grant Scheme, Malaysia [FRGS/1/2019/STG05/UNIM/02/2]. The authors acknowledge the financial support of the Ministry of Science and Technology, Taiwan, R.O.C., under the contracts MOST 106-2923-E-006-002-MY3 and MOST 109-3116-F-006-016-CC1 for this research.

Acknowledgments: The authors would like to acknowledge UNITEN for the research facilities. M.Y.O would also like to thank UNITEN for the UNITEN Postgraduate Excellence Scholarship 2019. This research is also supported in part by Higher Education Sprout Project, Ministry of Education to the Headquarters of University Advancement at NCKU.

Conflicts of Interest: The authors declare no conflict of interest.

References

1. US Department of Commerce, N.E.S.R.L. *Recent Global CO$_2$ Trend*; US Department of Commerce: Washington, DC, USA, 2020.
2. Office of YB Wong Chen. *Oil and Gas and Then?* Pusat Khidmat P104 Kelana Jaya: Subang Jaya, Malaysia, 2017.
3. Perea-Moreno, M.A.; Samerón-Manzano, E.; Perea-Moreno, A.J. Biomass as renewable energy: Worldwide research trends. *Sustainability* **2019**, *11*, 863. [CrossRef]
4. Federation of Malaysian Manufacturers. *Implementation of Biodiesel B10 & B7 and Procedure for Exemption*; Federation of Malaysian Manufacturers: Kuala Lumpur, Malaysia, 2019.
5. Heo, S.; Choi, J. Potential and environmental impacts of liquid biofuel from agricultural residues in Thailand. *Sustainability* **2019**, *11*, 1502. [CrossRef]
6. Silitonga, A.S.; Masjuki, H.H.; Ong, H.C.; Sebayang, A.H.; Dharma, S.; Kusumo, F.; Siswantoro, J.; Milano, J.; Daud, K.; Mahlia, T.M.I.; et al. Evaluation of the engine performance and exhaust emissions of biodiesel-bioethanol-diesel blends using kernel-based extreme learning machine. *Energy* **2018**, *159*, 1075–1087. [CrossRef]

7. Dimitriadis, A.; Natsios, I.; Dimaratos, A.; Katsaounis, D.; Samaras, Z.; Bezergianni, S.; Lehto, K. Evaluation of a Hydrotreated Vegetable Oil (HVO) and effects on emissions of a passenger car diesel engine. *Front. Mech. Eng.* **2018**, *4*, 4. [CrossRef]

8. Heikkilä, J.; Happonen, M.; Murtonen, T.; Lehto, K.; Sarjovaara, T.; Larmi, M.; Keskinen, J.; Virtanen, A. Study of miller timing on exhaust emissions of a Hydrotreated Vegetable Oil (HVO)-fueled diesel engine. *J. Air Waste Manag. Assoc.* **2012**, *62*, 1305–1312. [CrossRef] [PubMed]

9. OECD/FAO. *OECD-FAO Agricultural Outlook, Edition 2019*; OECD: Paris, France, 2020.

10. Demirbas, A.; Karslioglu, S. Biodiesel production facilities from vegetable oils and animal fats. *Energy Sources, Part A Recover. Util. Environ. Eff.* **2007**, *29*, 133–141. [CrossRef]

11. Amini, Z.; Ilham, Z.; Ong, H.C.; Mazaheri, H.; Chen, W.H. State of the art and prospective of lipase-catalyzed transesterification reaction for biodiesel production. *Energy Convers. Manag.* **2017**, *141*, 339–353. [CrossRef]

12. Goh, B.H.H.; Ong, H.C.; Cheah, M.Y.; Chen, W.H.; Yu, K.L.; Mahlia, T.M.I. Sustainability of direct biodiesel synthesis from microalgae biomass: A critical review. *Renew. Sustain. Energy Rev.* **2019**, *107*, 59–74. [CrossRef]

13. Douvartzides, S.L.; Charisiou, N.D.; Papageridis, K.N.; Goula, M.A. Green diesel: Biomass feedstocks, production technologies, catalytic research, fuel properties and performance in compression ignition internal combustion engines. *Energies* **2019**, *12*, 809. [CrossRef]

14. Ilham, Z.; Saka, S. Esterification of glycerol from biodiesel production to glycerol carbonate in non-catalytic supercritical dimethyl carbonate. *Springerplus* **2016**, *5*, 923. [CrossRef]

15. Garlapati, V.K.; Shankar, U.; Budhiraja, A. Bioconversion technologies of crude glycerol to value added industrial products. *Biotechnol. Rep.* **2016**, *9*, 9–14. [CrossRef] [PubMed]

16. Anand, P.; Saxena, R.K. A comparative study of solvent-assisted pretreatment of biodiesel derived crude glycerol on growth and 1,3-propanediol production from Citrobacter freundii. *N. Biotechnol.* **2012**, *29*, 199–205. [CrossRef] [PubMed]

17. Sorda, G.; Banse, M.; Kemfert, C. An overview of biofuel policies across the world. *Energy Policy* **2010**, *38*, 6977–6988. [CrossRef]

18. Roschat, W.; Kacha, M.; Yoosuk, B.; Sudyoadsuk, T.; Promarak, V. Biodiesel production based on heterogeneous process catalyzed by solid waste coral fragment. *Fuel* **2012**, *98*, 194–202. [CrossRef]

19. Rastegari, H.; Ghaziaskar, H.S. From glycerol as the by-product of biodiesel production to value-added monoacetin by continuous and selective esterification in acetic acid. *J. Ind. Eng. Chem.* **2015**, *21*, 856–861. [CrossRef]

20. Abdul Raman, A.A.; Tan, H.W.; Buthiyappan, A. Two-step purification of glycerol as a value added by product from the biodiesel production process. *Front. Chem.* **2019**, *7*, 774. [CrossRef]

21. Tan, H.W.; Abdul Aziz, A.R.; Aroua, M.K. Glycerol production and its applications as a raw material: A review. *Renew. Sustain. Energy Rev.* **2013**, *27*, 118–127. [CrossRef]

22. Rodrigues, A.; Bordado, J.C.; Dos Santos, R.G. Upgrading the glycerol from biodiesel production as a source of energy carriers and chemicals-A technological review for three chemical pathways. *Energies* **2017**, *10*, 1817. [CrossRef]

23. Pal, P.; Chaurasia, S.P.; Upadhyaya, S.; Agarwal, M.; Sridhar, S. Glycerol purification using membrane technology. In *Membrane Processes*; John Wiley & Sons, Inc.: Hoboken, NJ, USA, 2018; pp. 431–463.

24. Gomes, M.C.S.; Pereira, N.C.; de Barros, S.T.D. Separation of biodiesel and glycerol using ceramic membranes. *J. Memb. Sci.* **2010**, *352*, 271–276. [CrossRef]

25. Saifuddin, N.; Refal, H.; Kumaran, P. Rapid purification of glycerol by-product from biodiesel production through combined process of microwave assisted acidification and adsorption via Chitosan immobilized with yeast. *Res. J. Appl. Sci. Eng. Technol.* **2014**, *7*, 593–602. [CrossRef]

26. Shen, Z.; Jin, F.; Zhang, Y.; Wu, B.; Kishita, A.; Tohji, K.; Kishida, H. Effect of alkaline catalysts on hydrothermal conversion of glycerin into lactic acid. *Ind. Eng. Chem. Res.* **2009**, *48*, 8920–8925. [CrossRef]

27. Behr, A.; Eilting, J.; Irawadi, K.; Leschinski, J.; Lindner, F. Improved utilisation of renewable resources: New important derivatives of glycerol. *Green Chem.* **2008**, *10*, 13–30. [CrossRef]

28. Malyaadri, M.; Jagadeeswaraiah, K.; Sai Prasad, P.S.; Lingaiah, N. Synthesis of glycerol carbonate by transesterification of glycerol with dimethyl carbonate over Mg/Al/Zr catalysts. *Appl. Catal. A Gen.* **2011**, *401*, 153–157. [CrossRef]

29. Zhou, C.H.; Zhao, H.; Tong, D.S.; Wu, L.M.; Yu, W.H. Recent advances in catalytic conversion of glycerol. *Catal. Rev.* **2013**, *55*, 369–453. [CrossRef]

30. Bagheri, S.; Julkapli, N.M.; Yehye, W.A. Catalytic conversion of biodiesel derived raw glycerol to value added products. *Renew. Sustain. Energy Rev.* **2015**, *41*, 113–127. [CrossRef]

31. Nguyen, N.; Demirel, Y. Economic analysis of biodiesel and glycerol carbonate production plant by glycerolysis. *J. Sustain. Bioenergy Syst.* **2013**, *3*, 209–216. [CrossRef]

32. Pagliaro, M.; Ciriminna, R.; Kimura, H.; Rossi, M.; Della Pina, C. From glycerol to value-added products. *Angew. Chemie Int. Ed.* **2007**, *46*, 4434–4440. [CrossRef]

33. Sonnati, M.O.; Amigoni, S.; Taffin de Givenchy, E.P.; Darmanin, T.; Choulet, O.; Guittard, F. Glycerol carbonate as a versatile building block for tomorrow: Synthesis, reactivity, properties and applications. *Green Chem.* **2013**, *15*, 283–306. [CrossRef]

34. Tudorache, M.; Protesescu, L.; Coman, S.; Parvulescu, V.I. Efficient bio-conversion of glycerol to glycerol carbonate catalyzed by lipase extracted from Aspergillus niger. *Green Chem.* **2012**, *14*, 478. [CrossRef]

35. Jagadeeswaraiah, K.; Kumar, C.R.; Prasad, P.S.S.; Loridant, S.; Lingaiah, N. Synthesis of glycerol carbonate from glycerol and urea over tin-tungsten mixed oxide catalysts. *Appl. Catal. A Gen.* **2014**, *469*, 165–172. [CrossRef]

36. Teng, W.K.; Ngoh, G.C.; Yusoff, R.; Aroua, M.K. A review on the performance of glycerol carbonate production via catalytic transesterification: Effects of influencing parameters. *Energy Convers. Manag.* **2014**, *88*, 484–497. [CrossRef]

37. Gade, S.M.; Munshi, M.K.; Chherawalla, B.M.; Rane, V.H.; Kelkar, A.A. Synthesis of glycidol from glycerol and dimethyl carbonate using ionic liquid as a catalyst. *Catal. Commun.* **2012**, *27*, 184–188. [CrossRef]

38. Ang, G.T.; Tan, K.T.; Lee, K.T. Recent development and economic analysis of glycerol-free processes via supercritical fluid transesterification for biodiesel production. *Renew. Sustain. Energy Rev.* **2014**, *31*, 61–70. [CrossRef]

39. Ochoa-Gómez, J.R.; Gómez-Jiménez-Aberasturi, O.; Maestro-Madurga, B.; Pesquera-Rodríguez, A.; Ramírez-López, C.; Lorenzo-Ibarreta, L.; Torrecilla-Soria, J.; Villarán-Velasco, M.C. Synthesis of glycerol carbonate from glycerol and dimethyl carbonate by transesterification: Catalyst screening and reaction optimization. *Appl. Catal. A Gen.* **2009**, *366*, 315–324. [CrossRef]

40. Wang, J.; Yong, T.; Yang, J.; Ouyang, C.; Zhang, L. Organosilicon functionalized glycerol carbonates as electrolytes for lithium-ion batteries. *RSC Adv.* **2015**, *5*, 17660–17666. [CrossRef]

41. Hu, J.; Li, J.; Gu, Y.; Guan, Z.; Mo, W.; Ni, Y.; Li, T.; Li, G. Oxidative carbonylation of glycerol to glycerol carbonate catalyzed by PdCl2(phen)/KI. *Appl. Catal. A Gen.* **2010**, *386*, 188–193. [CrossRef]

42. Stark, J. Recent advances in the field of cement hydration and microstructure analysis. *Cem. Concr. Res.* **2011**, *41*, 666–678. [CrossRef]

43. Pan, S.; Zheng, L.; Nie, R.; Xia, S.; Chen, P.; Hou, Z. Transesterification of glycerol with dimethyl carbonate to glycerol carbonate over Na–based Zeolites. *Chinese J. Catal.* **2012**, *33*, 1772–1777. [CrossRef]

44. Bai, R.; Wang, Y.; Wang, S.; Mei, F.; Li, T.; Li, G. Synthesis of glycerol carbonate from glycerol and dimethyl carbonate catalyzed by NaOH/γ-Al2O3. *Fuel Process. Technol.* **2013**, *106*, 209–214. [CrossRef]

45. Arzamendi, G.; Campo, I.; Arguiñarena, E.; Sánchez, M.; Montes, M.; Gandía, L.M. Synthesis of biodiesel with heterogeneous NaOH/alumina catalysts: Comparison with homogeneous NaOH. *Chem. Eng. J.* **2007**, *134*, 123–130. [CrossRef]

46. Mizuno, T.; Nakai, T.; Mihara, M. New synthesis of glycerol carbonate from glycerol using sulfur-assisted carbonylation with carbon monoxide. *Heteroat. Chem.* **2010**, *21*, 99–102. [CrossRef]

47. Ochoa-Gómez, J.R.; Gómez-Jiménez-Aberasturi, O.; Ramírez-López, C.A.; Nieto-Mestre, J.; Maestro-Madurga, B.; Belsué, M. Synthesis of glycerol carbonate from 3-chloro-1,2-propanediol and carbon dioxide using triethylamine as both solvent and CO2 fixation–activation agent. *Chem. Eng. J.* **2011**, *175*, 505–511. [CrossRef]

48. Gómez-Jiménez-Aberasturi, O.; Ochoa-Gómez, J.R.; Pesquera-Rodríguez, A.; Ramírez-López, C.; Alonso-Vicario, A.; Torrecilla-Soria, J. Solvent-free synthesis of glycerol carbonate and glycidol from 3-chloro-1,2-propanediol and potassium (hydrogen) carbonate. *J. Chem. Technol. Biotechnol.* **2010**, *85*, 1663–1670. [CrossRef]

49. Ezhova, N.N.; Korosteleva, I.G.; Kolesnichenko, N.V.; Kuz'min, A.E.; Khadzhiev, S.N.; Vasil'eva, M.A.; Voronina, Z.D. Glycerol carboxylation to glycerol carbonate in the presence of rhodium complexes with phosphine ligands. *Pet. Chem.* **2012**, *52*, 91–96. [CrossRef]
50. Cho, H.-J.; Kwon, H.-M.; Tharun, J.; Park, D.-W. Synthesis of glycerol carbonate from ethylene carbonate and glycerol using immobilized ionic liquid catalysts. *J. Ind. Eng. Chem.* **2010**, *16*, 679–683. [CrossRef]
51. Climent, M.J.; Corma, A.; De Frutos, P.; Iborra, S.; Noy, M.; Velty, A.; Concepción, P. Chemicals from biomass: Synthesis of glycerol carbonate by transesterification and carbonylation with urea with hydrotalcite catalysts. The role of acid–base pairs. *J. Catal.* **2010**, *269*, 140–149. [CrossRef]
52. Tan, K.T.; Lee, K.T.; Mohamed, A.R. Optimization of supercritical dimethyl carbonate (SCDMC) technology for the production of biodiesel and value-added glycerol carbonate. *Fuel* **2010**, *89*, 3833–3839. [CrossRef]
53. Ilham, Z.; Saka, S. Production of biodiesel with glycerol carbonate by non-catalytic supercritical dimethyl carbonate. *Lipid Technol.* **2011**, *23*, 10–13. [CrossRef]
54. Ilham, Z.; Saka, S. Optimization of supercritical dimethyl carbonate method for biodiesel production. *Fuel* **2012**, *97*, 670–677. [CrossRef]
55. Seong, P.-J.; Jeon, B.W.; Lee, M.; Cho, D.H.; Kim, D.-K.; Jung, K.S.; Kim, S.W.; Han, S.O.; Kim, Y.H.; Park, C. Enzymatic coproduction of biodiesel and glycerol carbonate from soybean oil and dimethyl carbonate. *Enzyme Microb. Technol.* **2011**, *48*, 505–509. [CrossRef]
56. Min, J.Y.; Lee, E.Y. Lipase-catalyzed simultaneous biosynthesis of biodiesel and glycerol carbonate from corn oil in dimethyl carbonate. *Biotechnol. Lett.* **2011**, *33*, 1789–1796. [CrossRef] [PubMed]
57. Lee, K.H.; Park, C.-H.; Lee, E.Y. Biosynthesis of glycerol carbonate from glycerol by lipase in dimethyl carbonate as the solvent. *Bioprocess Biosyst. Eng.* **2010**, *33*, 1059–1065. [CrossRef] [PubMed]
58. Kumar, A.; Iwatani, K.; Nishimura, S.; Takagaki, A.; Ebitani, K. Promotion effect of coexistent hydromagnesite in a highly active solid base hydrotalcite catalyst for transesterifications of glycols into cyclic carbonates. *Catal. Today* **2012**, *185*, 241–246. [CrossRef]
59. Bai, R.; Wang, S.; Mei, F.; Li, T.; Li, G. Synthesis of glycerol carbonate from glycerol and dimethyl carbonate catalyzed by KF modified hydroxyapatite. *J. Ind. Eng. Chem.* **2011**, *17*, 777–781. [CrossRef]
60. Takagaki, A.; Iwatani, K.; Nishimura, S.; Ebitani, K. Synthesis of glycerol carbonate from glycerol and dialkyl carbonates using hydrotalcite as a reusable heterogeneous base catalyst. *Green Chem.* **2010**, *12*, 578. [CrossRef]
61. Li, J.; Wang, T. Chemical equilibrium of glycerol carbonate synthesis from glycerol. *J. Chem. Thermodyn.* **2011**, *43*, 731–736. [CrossRef]
62. Wang, Y.; Cao, X. Enzymatic synthesis of fatty acid ethyl esters by utilizing camellia oil soapstocks and diethyl carbonate. *Bioresour. Technol.* **2011**, *102*, 10173–10179. [CrossRef]
63. Álvarez, M.G.; Plíšková, M.; Segarra, A.M.; Medina, F.; Figueras, F. Synthesis of glycerol carbonates by transesterification of glycerol in a continuous system using supported hydrotalcites as catalysts. *Appl. Catal. B Environ.* **2012**, *113–114*, 212–220. [CrossRef]
64. Alvarez, M.G.; Segarra, A.M.; Contreras, S.; Sueiras, J.E.; Medina, F.; Figueras, F. Enhanced use of renewable resources: Transesterification of glycerol catalyzed by hydrotalcite-like compounds. *Chem. Eng. J.* **2010**, *161*, 340–345. [CrossRef]
65. Kim, D.-W.; Park, M.-S.; Selvaraj, M.; Park, G.-A.; Lee, S.-D.; Park, D.-W. Catalytic performance of polymer-supported ionic liquids in the synthesis of glycerol carbonate from glycerol and urea. *Res. Chem. Intermed.* **2011**, *37*, 1305–1312. [CrossRef]
66. Wang, L.; Ma, Y.; Wang, Y.; Liu, S.; Deng, Y. Efficient synthesis of glycerol carbonate from glycerol and urea with lanthanum oxide as a solid base catalyst. *Catal. Commun.* **2011**, *12*, 1458–1462. [CrossRef]
67. Hammond, C.; Lopez-Sanchez, J.A.; Hasbi Ab Rahim, M.; Dimitratos, N.; Jenkins, R.L.; Carley, A.F.; He, Q.; Kiely, C.J.; Knight, D.W.; Hutchings, G.J. Synthesis of glycerol carbonate from glycerol and urea with gold-based catalysts. *Dalt. Trans.* **2011**, *40*, 3927. [CrossRef] [PubMed]
68. Rubio-Marcos, F.; Calvino-Casilda, V.; Bañares, M.A.; Fernandez, J.F. Novel hierarchical Co3O4/ZnO mixtures by dry nanodispersion and their catalytic application in the carbonylation of glycerol. *J. Catal.* **2010**, *275*, 288–293. [CrossRef]

69. Shukla, K.; Srivastava, V.C. Synthesis of organic carbonates from alcoholysis of urea: A review. *Catal. Rev.* **2017**, *59*, 1–43. [CrossRef]

70. Li, J.; Wang, T. Coupling reaction and azeotropic distillation for the synthesis of glycerol carbonate from glycerol and dimethyl carbonate. *Chem. Eng. Process. Process Intensif.* **2010**, *49*, 530–535. [CrossRef]

71. Su, X.; Lin, W.; Cheng, H.; Zhang, C.; Wang, Y.; Yu, X.; Wu, Z.; Zhao, F. Metal-free catalytic conversion of CO_2 and glycerol to glycerol carbonate. *Green Chem.* **2017**, *19*, 1775–1781. [CrossRef]

72. Ozorio, L.P.; Mota, C.J.A. Direct carbonation of glycerol with CO_2 catalyzed by metal oxides. *ChemPhysChem* **2017**, *18*, 3260–3265. [CrossRef]

73. Lertlukkanasuk, N.; Phiyanalinmat, S.; Kiatkittipong, W.; Arpornwichanop, A.; Aiouache, F.; Assabumrungrat, S. Reactive distillation for synthesis of glycerol carbonate via glycerolysis of urea. *Chem. Eng. Process. Process Intensif.* **2013**, *70*, 103–109. [CrossRef]

74. Kondawar, S.E.; Mane, R.B.; Vasishta, A.; More, S.B.; Dhengale, S.D.; Rode, C.V. Carbonylation of glycerol with urea to glycerol carbonate over supported Zn catalysts. *Appl. Petrochemical Res.* **2017**, *7*, 41–53. [CrossRef]

75. Lanjekar, K.; Rathod, V.K. Utilization of glycerol for the production of glycerol carbonate through greener route. *J. Environ. Chem. Eng.* **2013**, *1*, 1231–1236. [CrossRef]

76. Kwon, E.E.; Yi, H.; Jeon, Y.J. Boosting the value of biodiesel byproduct by the non-catalytic transesterification of dimethyl carbonate via a continuous flow system under ambient pressure. *Chemosphere* **2014**, *113*, 87–92. [CrossRef] [PubMed]

77. Syamsuddin, Y.; Hameed, B.H. Synthesis of glycerol free-fatty acid methyl esters from Jatropha oil over Ca–La mixed-oxide catalyst. *J. Taiwan Inst. Chem. Eng.* **2016**, *58*, 181–188. [CrossRef]

78. Okoye, P.U.; Abdullah, A.Z.; Hameed, B.H. Glycerol carbonate synthesis from glycerol and dimethyl carbonate using trisodium phosphate. *J. Taiwan Inst. Chem. Eng.* **2016**, *68*, 51–58. [CrossRef]

79. Aresta, M.; Dibenedetto, A. Carbon dioxide as building block for the synthesis of organic carbonates: Behavior of homogeneous and heterogeneous catalysts in the oxidative carboxylation of olefins. *J. Mol. Catal. A Chem.* **2002**, *182–183*, 399–409. [CrossRef]

80. Tamboli, A.H.; Chaugule, A.A.; Kim, H. Catalytic developments in the direct dimethyl carbonate synthesis from carbon dioxide and methanol. *Chem. Eng. J.* **2017**, *323*, 530–544. [CrossRef]

81. George, J.; Patel, Y.; Pillai, S.M.; Munshi, P. Methanol assisted selective formation of 1,2-glycerol carbonate from glycerol and carbon dioxide using nBu2SnO as a catalyst. *J. Mol. Catal. A Chem.* **2009**, *304*, 1–7. [CrossRef]

82. Aresta, M.; Dibenedetto, A.; Nocito, F.; Pastore, C. A study on the carboxylation of glycerol to glycerol carbonate with carbon dioxide: The role of the catalyst, solvent and reaction conditions. *J. Mol. Catal. A Chem.* **2006**, *257*, 149–153. [CrossRef]

83. Aresta, M.; Dibenedetto, A.; Nocito, F.; Ferragina, C. Valorization of bio-glycerol: New catalytic materials for the synthesis of glycerol carbonate via glycerolysis of urea. *J. Catal.* **2009**, *268*, 106–114. [CrossRef]

84. De La Hoz, A.; Díaz-Ortiz, A.; Moreno, A. Review on non-thermal effects of microwave irradiation in organic synthesis. *J. Microw. Power Electromagn. Energy* **2006**, *41*, 45–66. [CrossRef]

85. de la Hoz, A.; Loupy, A. *Microwaves in Organic Synthesis*; Wiley-VCH: Hoboken, NJ, USA, 2012; ISBN 3527331166.

86. Sun, J.; Wang, W.; Yue, Q. Review on microwave-matter interaction fundamentals and efficient microwave-associated heating strategies. *Materials* **2016**, *9*, 231. [CrossRef]

87. Nomanbhay, S.; Ong, M.Y. A Review of microwave-assisted reactions for biodiesel production. *Bioengineering* **2017**, *4*, 57. [CrossRef] [PubMed]

88. Bookong, P.; Ruchirawat, S.; Boonyarattanakalin, S. Optimization of microwave-assisted etherification of glycerol to polyglycerols by sodium carbonate as catalyst. *Chem. Eng. J.* **2015**, *275*, 253–261. [CrossRef]

89. Ruppert, A.M.; Meeldijk, J.D.; Kuipers, B.W.M.; Ernø, B.H.; Weckhuysen, B.M. Glycerol etherification over highly active CaO-based materials: New mechanistic aspects and related colloidal particle formation. *Chem. Eur. J* **2008**, *14*, 2016–2024. [CrossRef] [PubMed]

90. Iaych, K.; Dumarçay, S.; Fredon, E.; Gérardin, C.; Lemor, A.; Gérardin, P. Microwave-assisted synthesis of polyglycerol from glycerol carbonate. *J. Appl. Polym. Sci.* **2011**, *120*, 2354–2360. [CrossRef]

91. Teng, W.K.; Ngoh, G.C.; Yusoff, R.; Aroua, M.K. Microwave-assisted transesterification of industrial grade crude glycerol for the production of glycerol carbonate. *Chem. Eng. J.* **2016**, *284*, 469–477. [CrossRef]

92. Hu, S.; Li, Y. Polyols and polyurethane foams from base-catalyzed liquefaction of lignocellulosic biomass by crude glycerol: Effects of crude glycerol impurities. *Ind. Crops Prod.* **2014**, *57*, 188–194. [CrossRef]

93. Mhanna, A.; Chupin, L.; Brachais, C.-H.; Chaumont, D.; Boni, G.; Brachais, L.; Couvercelle, J.-P.; Lecamp, L.; Plasseraud, L. Efficient microwave-assisted synthesis of glycerol monodecanoate. *Eur. J. Lipid Sci. Technol.* **2018**, *120*, 1700133. [CrossRef]

94. Prat, D.; Wells, A.; Hayler, J.; Sneddon, H.; McElroy, C.R.; Abou-Shehada, S.; Dunn, P.J. CHEM21 selection guide of classical- and less classical-solvents. *Green Chem.* **2016**, *18*, 288–296. [CrossRef]

95. Jessop, P.G. Searching for green solvents. *Green Chem.* **2011**, *13*, 1391. [CrossRef]

96. Jin, S.; Hunt, A.J.; Clark, J.H.; McElroy, C.R. Acid-catalysed carboxymethylation, methylation and dehydration of alcohols and phenols with dimethyl carbonate under mild conditions. *Green Chem.* **2016**, *18*, 5839–5844. [CrossRef]

97. Bian, J.; Xiao, M.; Wang, S.; Lu, Y.; Meng, Y. Direct synthesis of DMC from CH_3OH and CO_2 over V-doped Cu–Ni/AC catalysts. *Catal. Commun.* **2009**, *10*, 1142–1145. [CrossRef]

98. Bruno, T.J.; Wolk, A.; Naydich, A. Composition-explicit distillation curves for mixtures of gasoline and diesel fuel with γ-Valerolactone. *Energy Fuels* **2010**, *24*, 2758–2767. [CrossRef]

99. Zhu, R.; Cheung, C.S.; Huang, Z. Particulate emission characteristics of a compression ignition engine fueled with diesel–DMC blends. *Aerosol Sci. Technol.* **2011**, *45*, 137–147. [CrossRef]

100. Kim, P.S.; Cho, B.K.; Nam, I.-S.; Choung, J.W. Bifunctional Ag-based catalyst for NO_x reduction with E-diesel fuel. *ChemCatChem* **2014**, *6*, 1570–1574. [CrossRef]

101. Kiadó, A.; Vol, B. *Microwave-Assisted Synthesis of Dimethyl Carbonate*; Kluwer Academic Publishers: London, UK, 2001; Volume 74.

102. Wada, S.; Oka, K.; Watanabe, K.; Izumi, Y. Catalytic conversion of carbon dioxide into dimethyl carbonate using reduced copper-cerium oxide catalysts as low as 353 K and 1.3 MPa and the reaction mechanism. *Front. Chem.* **2013**, *1*, 8. [CrossRef]

103. Ilham, Z.; Saka, S. Dimethyl carbonate as potential reactant in non-catalytic biodiesel production by supercritical method. *Bioresour. Technol.* **2009**, *100*, 1793–1796. [CrossRef]

104. Esteban, J.; Domínguez, E.; Ladero, M.; Garcia-Ochoa, F. Kinetics of the production of glycerol carbonate by transesterification of glycerol with dimethyl and ethylene carbonate using potassium methoxide, a highly active catalyst. *Fuel Process. Technol.* **2015**, *138*, 243–251. [CrossRef]

105. Zheng, L.; Xia, S.; Hou, Z.; Zhang, M.; Hou, Z. Transesterification of glycerol with dimethyl carbonate over Mg-Al hydrotalcites. *Chinese J. Catal.* **2014**, *35*, 310–318. [CrossRef]

106. Wang, X.; Zhang, P.; Cui, P.; Cheng, W.; Zhang, S. Glycerol carbonate synthesis from glycerol and dimethyl carbonate using guanidine ionic liquids. *Chin. J. Chem. Eng.* **2017**, *25*, 1182–1186. [CrossRef]

107. Waghmare, G.V.; Vetal, M.D.; Rathod, V.K. Rathod Ultrasound assisted enzyme catalyzed synthesis of glycerol carbonate from glycerol and dimethyl carbonate. *Ultrason. Sonochem.* **2015**, *22*, 311–316. [CrossRef]

108. Algoufi, Y.T.; Hameeda, B.H. Synthesis of glycerol carbonate by transesterification of glycerol with dimethyl carbonate over K-zeolite derived from coal fly ash. *Fuel Process. Technol.* **2014**, *126*, 5–11. [CrossRef]

109. Esteban, J.; Fuente, E.; Blanco, A.; Ladero, M.; Garcia-Ochoa, F. Phenomenological kinetic model of the synthesis of glycerol carbonate assisted by focused beam reflectance measurements. *Chem. Eng. J.* **2015**, *260*, 434–443. [CrossRef]

110. Herseczki, Z.; Marton, G.; Varga, T. Enhanced use of renewable resources: Transesterification of glycerol, the byproduct of biodiesel production. *Hungarian J. Ind. Chem.* **2011**, *39*, 183–187.

111. Almquist, J.; Cvijovic, M.; Hatzimanikatis, V.; Nielsen, J.; Jirstrand, M. Kinetic models in industrial biotechnology–Improving cell factory performance. *Metab. Eng.* **2014**, *24*, 38–60. [CrossRef] [PubMed]

112. Sabbe, M.K.; Reyniers, M.-F.; Reuter, K. First-principles kinetic modeling in heterogeneous catalysis: An industrial perspective on best-practice, gaps and needs. *Catal. Sci. Technol.* **2012**, *2*, 2010. [CrossRef]

113. Smadbeck, P.; Kaznessis, Y. Stochastic model reduction using a modified Hill-type kinetic rate law. *J. Chem. Phys.* **2012**, *137*, 234109. [CrossRef]

114. Li, J.; Wang, T. On the deactivation of alkali solid catalysts for the synthesis of glycerol carbonate from glycerol and dimethyl carbonate. *React. Kinet. Mech. Catal.* **2011**, *102*, 113–126. [CrossRef]

115. Liu, Z.; Wang, J.; Kang, M.; Yin, N.; Wang, X.; Tan, Y.; Zhu, Y. Synthesis of glycerol carbonate by transesterification of glycerol and dimethyl carbonate over KF/γ-Al$_2$O$_3$ catalyst. *J. Braz. Chem. Soc.* **2014**, *25*, 152–160.
116. Argyle, M.D.; Bartholomew, C.H. Heterogeneous catalyst deactivation and regeneration: A review. *Catalysts* **2015**, *5*, 145–269. [CrossRef]
117. Boga, D.A.; Liu, F.; Bruijnincx, P.C.A.; Weckhuysen, B.M. Aqueous-phase reforming of crude glycerol: Effect of impurities on hydrogen production. *Catal. Sci. Technol.* **2016**, *6*, 134–143. [CrossRef]

MDPI

St. Alban-Anlage 66

4052 Basel

Switzerland

Tel. +41 61 683 77 34

Fax +41 61 302 89 18

www.mdpi.com

Energies Editorial Office

E-mail: energies@mdpi.com

www.mdpi.com/journal/energies